INSECT AND MITE NUTRITION

INSECT AND MITE NUTRITION

SIGNIFICANCE AND IMPLICATIONS IN ECOLOGY AND PEST MANAGEMENT

edited by J. G. RODRIGUEZ

THIS BOOK RESULTED FROM TOPICS OR PAPERS DISCUSSED AT THE INTERNATIONAL CONFERENCE ON THE
SIGNIFICANCE OF INSECT AND MITE NUTRITION
APRIL 25–28, 1972
UNIVERSITY OF KENTUCKY
LEXINGTON, KENTUCKY
USA

1972

NORTH-HOLLAND PUBLISHING COMPANY – AMSTERDAM • LONDON

Library of Congress Catalog Card Number: 72-96711

North-Holland ISBN: 0 7204 4126 9
American Elsevier ISBN: 0 444 10437 2

PUBLISHERS:
NORTH-HOLLAND PUBLISHING COMPANY – AMSTERDAM
NORTH-HOLLAND PUBLISHING COMPANY, LTD. – LONDON

SOLE DISTRIBUTORS FOR THE U.S.A. AND CANADA:
AMERICAN ELSEVIER PUBLISHING COMPANY, INC.
52 VANDERBILT AVENUE, NEW YORK, N.Y. 10017

PRINTED IN THE NETHERLANDS

CONTENTS

PREFACE

This volume comes as a result of the Conference on Significance of Insect and Mite Nutrition held at the University of Kentucky Carnahan House Conference Center in April 1972. This conference developed after colleagues had been queried and had responded relative to its justification and relevance. Response was enthusiastic and convincing that insect nutritionists would support a conference and many problem areas were suggested for discussion. Our beliefs were confirmed, for example, that advances in the field have been rapid, even dramatic in recent years, that many scientists are working in specialties that impinge on the general area of nutritional physiology via a wide array of pathways and that many specialists are searching for a greater integration of the results of their investigations into other areas or disciplines.

An Organizing Committee was then named and this group developed the structure of the conference. International in scope and the first of its kind, the meeting sought to bring about much needed discourse on problems confronting scientists working in arthropod nutrition. The stated aim was to evaluate the role of insect and mite nutrition in the solution of biological problems and its unique, significant contribution to other biological disciplines and to environmental questions. The areas germane to the conference objective were identified and developed with the overall view to not only delineate and evaluate what has been accomplished but also to suggest where future research should lead. Each member of the Organizing Committee was given the responsibility of developing a particular area and section; this colleague subsequently served as moderator of the section and as section editor.

The papers that comprise this volume were presented at the conference in an informal manner by the discussants. Later the discussants, having had the benefit of reappraising their work, submitted a manuscript for publication to their section editor. As general editor it was my duty to arrange and focus material more effectively. Any editorial errors that may appear in this volume occurred through oversight in my part.

Grateful acknowledgement is made of the cheerful assistance and splendid cooperation rendered by numerous colleagues at the University of Kentucky. To my fellow members of the Organizing Committee, named separately in these pages, who were most cooperative and unstinting of their energies and enthusiastic over bringing the whole endeavor into fruition, I cannot express enough gratitude. Other colleagues gave valuable support and assistance to this undertaking but especially I would like to mention H. L. House and Calvin A. Lang. I would also be remiss if I did not acknowledge the spirited response of all conference participants and especially I would like to thank the discussants who submitted papers. My task in editing was lightened with the generous assistance of my daughters, Carmen and Teresa, and my wife, Lorraine, who gave of their time to provide a photo-ready manuscript. Many thanks are due Mrs. Alice Kidd who typed long hours and accomplished a professional job. Lastly, I would like to acknowledge the sponsoring agencies whose support made the international conference possible. These were:

National Science Foundation
Cooperative State Research Service, USDA
Environmental Protection Agency
University of Kentucky:Research Fund and Biomed Fund

October, 1972

J. G. Rodriguez
Lexington

ORGANIZING COMMITTEE

Stanley D. Beck
Marion A. Brooks
R. H. Dadd
Ernest Hodgson
G. P. Waldbauer
B. S. Wostmann
J. G. Rodriguez, Chairman

DEDICATION

The following resolution was approved by the Conference:

"This conference, on the Significance of Insect and Mite Nutrition meeting at the Carnahan House, Lexington, Kentucky, April 25-28, 1972 dedicates the book resulting from the Conference to our colleague, Gottfried Fraenkel, as a token of appreciation of his pioneering contributions to the field."

NUTRITION, ADAPTATION AND ENVIRONMENT

Stanley D. Beck
Department of Entomology
University of Wisconsin,
Madison, Wisconsin 53706

The planning and organization of a conference entitled "Significance of Insect and Mite Nutrition" necessarily involves adoption of the premise that the subject does indeed have significance. Thus, we are not gathered here to test the validity of that premise, or to convince each other of its validity. Rather, we have come together in order to exchange information and to assess the current status of our varied research interests in diverse aspects of arthropod nutrition. Out of these three and a half days, we may individually gain some concept of the most significant problems and the most productive avenues that call for greater research emphasis, even though we may not reach some collective consensus. Such a collective consensus is not particularly important in any case. What is important is that there be a spirit of informal communication; that there be open and frank mutual exchange of insights and experimental information. This is the real significance of our conference, and the degree to which this is accomplished will be the measure of our success.

The conference program shows a wide variety of subjects, from metabolic specificities to aspects of pest management, all of which have been forced under one umbrella--the general rubric of nutrition. Obviously, the conference organizers have taken a very broad view of how the term 'nutrition' may be defined. This was done deliberately, of course, in order to bring into focus not only the central subject of specific biochemical nutritional requirements, but also all of the biological implications and peripheral aspects that contribute so much to the real significance of the broad subject of food habits, behavior, and nutritional adaptions that contribute to the impressive biological success of the insects and related arthropods.

Early 20th century workers, in viewing the enormous range of insect food habits, tended to postulate that both sensory and nutritional factors were operative in delimiting feeding specificity. The very great range of food habits was thought to reflect, in part, differences in the food value of the various substrates and corresponding differences in the specific chemical nutrients required by the insects subsisting successfully on those substrates. This hypothesis was reflected, at least to a degree, in Uvarov's 1928 review of insect nutrition and metabolism. Some of the early biologists took a surprisingly modern view of the problem of insect food habit specificity, including especially Brues (1920) and Folsom & Wardle (1934). In the 1920's and 1930's, however, nutrition as a field of biochemical endeavor was in its infancy, and very little work on insect nutritional requirements had been carried out. That insects might differ in some very significant ways from higher animals in respect to their nutritional requirements was shown by Hobson's 1935 demonstration that blowfly larvae required a dietary source of cholesterol.

Concurrently with the isolation and identification of the water-soluble vitamins, our knowledge of insect nutrition was magnificently advanced by the work of Gottfried Fraenkel, beginning in the early 1940's. In a classical series of papers published between 1941 and 1947, Fraenkel and his associates elucidated the principal specific biochemical requirements of several stored products insects (for a summary, see Fraenkel, 1959a). This research, more than any other, really established the significance of the field of insect nutrition. And to a large extent it also set the pattern of experimental design and fixed the criteria of evaluation according to which most subsequent nutritional research has been pursued.

It is not my intention to review step-by-step the history of research on insect nutrition during the thirty years that have elapsed since the early 1940's. But a few important points merit emphasis, as they have direct bearing on the concepts that guide us in this conference in 1972. The work of those past years has firmly established the concept that insects do not differ markedly in their fundamental qualitative requirements for biochemical nutrients that must be obtained by ingestion and assimilation. Nor do they show fundamental differences from other animal forms. To be sure, some differences and specializations occur, as for example, in respect to sterol identity, ascorbic acid requirement, and requirements for choline, carnitine, and some of the lesser B-vitamins. But these specialized differences do not seem to be of a magnitude any greater than the differences found among species of mammals or species of birds. It would seem, then, that the great range of insect food habits cannot be dealt with meaningfully by any experimental program that investigates only the insect's nutritional requirements in the strict sense. The general similarities in insect nutritional requirements were so impressive that Fraenkel (1959b) was led to postulate that the basic nutritional requirements of all plant-feeding insects are identical, and that any phytophagous insect could thrive on the tissues of any green plant, if the insect could be induced to eat enough of it. Food habits and host plant specificity were postulated to be determined by the effects of secondary plant chemicals that attracted or repelled the insects and influenced the insect's locomotor, ovipositional, and feeding behavior patterns.

Although simplistic in its original form, Fraenkel's theory served well to encourage a shift of emphasis from the purely biochemical determination of minimum requirements for various amino acids, vitamins, etc., to a broader consideration of what we might call "insect dietetics." As reflected by the program of this conference, we are dealing here with insect dietetics rather than with the more narrowly defined nutrition in the strict sense. The simplistic nature of Fraenkel's original theory was apparent when it was shown that even if insects were induced to feed on a variety of nonhost plant tissues, they grew and survived much better on some plant species than on others (Waldbauer, 1962, 1964; House, 1961, 1969; Bongers, 1970). Such results showed that insect dietetics involve more than a more-or-less standard array of nutrient factors plus appropriate behavioral releasers. This had also been demonstrated earlier when we (Beck, 1957; Beck and Stauffer, 1957; Beck and Smissman, 1960) had shown that growth inhibiting and toxic substances were present in corn foliage and a number of other plants, and that these substances exerted adverse effects on the survival and growth of European corn borer, *Ostrinia nubilalis.* Much of the advances in our understanding of insect dietetics over a ten year span was reflected in Fraenkel's 1969 much-needed re-evaluation of the role of secondary plant substances in insect-host plant relationships.

We have now reached a point where we are beginning to appreciate realistically that the effects of an insect's dietary substrate are not simply nutritional in the strict sense. We must also deal with the influence of factors affecting digestion, utilization, and conversion as well as factors affecting metabolism, form determination, reproduction, longevity, and general behavior.

The feeding insect must not only ingest a dietary substrate, but the material ingested must also be suitable for conversion into the energy and structural substances required for development and other biological functions. Digestibility of the diet is an important factor determining utilizability, and has been shown to be of particular significance in insect-host plant relationships. Protease inhibitors occur in some plant tissues as part of the plant's defense against herbivores. Such inhibitors have long been known to occur in legumes and grains, and have been recently reported in solanaceous plants. Green and Ryan (1972) found that feeding activity of larvae and adults of the Colorado potato beetle, *Leptinotarsa decemlineata,* on potato and tomato plants induced the formation of a protease inhibitor that was a powerful antagonist of the major intestinal proteases of insects and mammals. Applebaum (1964) studied the effects of plant-borne and antiproteases on the host specificity of Bruchid beetles, and has postulated that some of the specificity of these insects may be dependent on their having overcome the barrier of such a plant defense mechanism.

Assimilation and conversion into insect tissue must follow digestion, and plant tissues differ greatly in the degree to which they meet this requirement. Waldbauer (1964, 1968) studied assimilation and conversion in the tobacco hornworm, *Manduca sexta,* on a wide variety of solanaceous and nonsolanaceous plants, and demonstrated that these factors are of real importance in the dietetics of the hornworm. Even relatively polyphagous species such as grasshoppers and armyworms do not utilize all hosts equally efficiently. Among the plants fed on by larvae of the southern armyworm, *Prodenia eridania*, digestibility ranged from a highly satisfactory 76% down to only 36% in the poorest host plant; conversion into insect tissue showed a range of efficiency of from 56 to 16% (Soo Hoo and Fraenkel, 1966).

Although insect feeding habits are not determined primarily by the insect's specific biochemical nutritional requirements, it must certainly be recognized that not all dietary substrates are equally nutritious. Plant parts and plant products vary in their content of nutrients required by insects, with such variation being dependent on developmental stage, physiological condition, and plant genotype. These variations have been shown to influence both the behavior and developmental success of plant-feeding insects. In studies of the pea aphid, *Acyrthosiphon pisum*, Auclair *et al.* (1957) found that the amino acid content of an aphid-resistant variety of peas was quantitatively different from that of an aphid-susceptible variety; they postulated that resistance was caused by a relatively low content of free amino acids. Similarly, Colorado potato beetle larvae were shown to grow faster with highest survival rates on young potato foliage than on older foliage, presumably because of a more favorable amino acid content in the younger tissue (Cibula *et al.*, 1967). Grison (1958) observed that adult Colorado potato beetles showed egg production rates that were positively correlated with the phospholipid content of the host plant foliage; senescent foliage was deficient in the phospholipids required for egg production. A number of other investigators have also reported differences in nutrient content of plant structures of different ages, and their apparent effects on growth, survival, and reproduction of insects. Whether in a host plant, synthetic diet, or other nutritional substrate, the importance of the dietary proportions of required nutrients may be of greater importance than their absolute quantities (House, 1969, 1971). Such nutrient ratios may also influence feeding behavior, and they are almost certain to be important factors in the efficiency of the conversion of ingested food into energy and insect tissues.

This very short survey of the different facets of insect nutrition has dealt with but a few of the ways in which the subject has significance. Because of the reviewer's particular bias, emphasis has been on insect-plant interactions, and equally important aspects--such as parasitic, medically important, storage product, and household insects--have been largely ignored. The practical importance of scientific knowledge concerning insect nutrition is apparent in all of these areas of concern. The economic importance of insect nutrition studies has come most sharply into focus in the development and use of mass-rearing programs that underlie the conception and execution of pest management systems involving sterile-male releases, pheromonal manipulation, and lethal gene introductions (Smith, 1966).

REFERENCES

APPLEBAUM, S. W. (1964). Physiological aspects of host specificity in the Bruchidae. I. General considerations of developmental compatibility. *J. Insect Physiol.* **10**:783-788.

AUCLAIR, J. L., MALTAIS, J. B. and CARTIER, J. J. (1957). Factors in resistance of pease to the pea aphid, *Acyrthosiphon pisum* (Harr.) (Homoptera: Aphididae). II. Amino acids. *Canad. Ent.* **89**:457-464.

BECK, S. D. (1957). The European corn borer, *Pyrausta nubilalis* (Hubn.), and its principal host plant. VI. Host plant resistance to larval establishment. *J. Insect Physiol.* **1**:158-177.

BECK, S. D. and SMISSMAN, E. E. (1960). The European corn borer, *Pyrausta nubilalis*, and its principal host plant. VIII. Laboratory evaluation of host resistance to larval growth and survival. *Ann. Ent. Soc. Amer.* **53**:755-762.

BECK, S. D. and STAUFFER, J. F. (1957). The European corn borer, *Pyrausta nubilalis* (Hubn.), and its principal host plant. III. Toxic factors influencing larval establishment. *Ann. Ent. Soc. Amer.* **50**:166-170.

BONGERS, W. (1970). Aspects of host-plant relationship of the Colorado beetle. *Meded. Landbouwh. Wageningen, Nederl.* **70-10**:1-77.

BRUES, C. T. (1920). The selection of food plants by insects, with special reference to lepidopterous larvae. *Amer. Nat.* **54**:313-332.

CIBULA, A. B., DAVIDSON, R. H., FISK, F. W. and LAPIDUS, J. B. (1967). Relationship of free amino acids of some solanaceous plants to growth and development of *Leptinotarsa decemlineata* (Coleoptera: Chrysomelidae). *Ann. Ent. Soc. Amer.* **60**:626-631.

FOLSOM, J. W. and WARDLE, R. A. (1934). "Entomology with Special Reference to its Ecological Aspects." Blakistons & Son. Philadelphia.

FRAENKEL, G. (1959a). A historical and comparative survey of the dietary requirements of insects. *N. Y. Acad. Sci.* **77**:267-274.

FRAENKEL, G. (1959b). The Raison d'etre of secondary plant substances. *Science* **129**:1466-1470.

FRAENKEL, G. (1969). Evaluation of our thoughts on secondary plant substances. *Ent. Exp. & Appl.* **12**:473-486.

GREEN, T. R. and RYAN, C. A. (1972). Wound-induced proteinase inhibitor in plant leaves: A possible defense mechanism against insects. *Science* **175**:776-777.

GRISON, P. (1958). L'influence de la plant-hote sur le fecondite de insectes phytophage. *Ent. Exp. & Appl.* **1**:73-93.

HOBSON, R. P. (1935). On a fat-soluble growth factor required by blowfly larvae. II. Identity of the growth factor with cholesterol. *Biochem. J.* **29**:2023-2026.

HOUSE, H. L. (1961). Insect Nutrition. *Ann. Rev. Ent.* **6**:13-26.

HOUSE, H. L. (1969). Effects of different proportions of nutrients on insects. *Ent. Exp. & Appl.* **12**:651-669.

HOUSE, H. L. (1971). Relations between dietary proportions of nutrients, growth rate, and choice of food in the fly larva *Agria affinis*. *J. Insect Physiol.* **17**:1225-1238.

SMITH, C. N. (ed.) (1966). "Insect Colonization and Mass Production." Academic Press, New York.

SOO HOO, C. F. and FRAENKEL, G. (1966). The consumption, digestion, and utilization of food plants by a polyphagous insect, *Prodenia eridania* (Cramer). *J. Insect Physiol.* **12**:711-730.

UVAROV, B. P. (1928). Insect nutrition and metabolism. *Trans. Ent. Soc. London* **76**:255-343.

WALDBAUER, G. P. (1962). The growth and reproduction of maxillectomized tobacco hornworms feeding on normally rejected non-solanaceous plants. *Ent. Exp. & Appl.* **5**:147-158.

WALDBAUER, G. P. (1964). The consumption, digestion, and utilization of solanaceous and non-solanaceous plants by larvae of the tobacco hornworm, *Protoparce sexta* (Johan.). *Ent. Exp. & Appl.* **7**:253-269.

WALDBAUER, G. P. (1968). The consumption and utilization of food by insects. *Adv. Insect Physiol.* **5**:229-288.

ASPECTS OF INSECT AND ANIMAL NUTRITION

INTRODUCTION
by

Bernard S. Wostmann, Section Editor
Lobund Laboratory, Department of Microbiology
University of Notre Dame
Notre Dame, Indiana 46556

Besides providing data on nutrient requirements pertaining to a specific insect species, the study of insect nutrition appears to offer two distinct potentialities. On the one hand, because of the often quoted similarity in nutritional requirements among the various insects, and even between insects and mammals, nutritional data obtained with insects may have a much wider applicability. On the other hand, differences in requirements have become apparent, especially when mutant strains are considered, which may present a specific insect as an almost tailor-made model for the study of a certain nutritional problem.

Considering nutrition in general, even while assuming much more similarity than dissimilarity in metabolic systems throughout the more complex representatives of the animal kingdom, our final emphasis will often be on mammalian nutrition. However, in our studies, any small size animal with a relatively short life cycle may prove to be a tool of importance, provided that a reasonable generalization of experimental data appears warranted. Years ago, protozoans of the genus Tetrahymena were used extensively to assess the nutritional quality of proteins. Davis' paper, "Application of Insect Nutrition in Solving General Nutrition Problems," follows similar principles, describing the use of growth data from *Tenebrio molitor* larvae as a criterion for nutritional adequacy of various vegetable proteins.

But nutrition, nowadays, cannot be studied only in terms of what, or how much. Nutrition studies must eventually lead to an acceptable answer of the question why. As such, nutrition is only part of a larger picture, that of the totality of a coordinated metabolism. Here, insects, with their vast potential to produce specific, recognizable mutants appear to offer a welcome alternative to *in vitro* studies with mammalian tissues or tissue homogenates, or even cell constituents on the one hand, and the use of unicellular organisms on the other hand. As pointed out by Sang in his study, "The Use of Mutants in Nutritional Research," judicious use of mutants may often be advantageous, especially since, with the use of an integrated multicellular organism, the influence of metabolic control mechanisms is retained. Some of the efforts in the field of genetics that are involved in obtaining 'nutritionally dependent' mutants are described by Falk and Nash in their paper, "The Search for Auxotrophic Mutants in *Drosophila melanogaster.*"

In all living systems involving multicellular organization, the role of involuntary, uncontrolled associates must be considered. Work with germfree and gnotobiotic mammals has time and again pointed to the vast influence of the mammalian intestinal flora. Years ago, Howe already stated that, in insect nutrition studies, associated microorganisms should be eliminated "to avoid the possible intervention of symbiotic forms." In line with this thinking, and in an effort to achieve maximal experimental control, many of the insect nutrition studies were done axenically, using chemically defined diets.

However, comparative studies of function and metabolism of germfree and conventional rodents thus far have indicated differences of such importance that a quantitative extrapolation of nutritional and metabolic data obtained with axenic rodents at this time appears dubious at best. It would seem that in insect nutrition, as in the mammalian field, a comparison between axenic and conventional insects may be needed to obtain insight into the possible role of an intestinal microflora, and to obtain data that not only provide a theoretical insight into the metabolic needs of the animal, but which reflect actual nutritional requirements under normally existing conditions.

THE USE OF MUTANTS IN NUTRITIONAL RESEARCH

J. H. Sang
Sussex University
Brighton, England

INTRODUCTION

I want to advance a particular thesis in this paper, so I must start by explaining that I have an ulterior motive in doing so. The thesis is that: mutant strains of organisms could frequently be employed with advantage in nutritional research. Too often, well established stocks are used (sometimes without regard to their excessive inbreeding and atypical constitution), and a mutant strain is discarded because it is phenotypically abnormal in one character although it may be perfectly sound in its remaining 9,999 or so genes! It is saluatory to remember the wild populations of *Drosophila* (and of man) are polymorphic at about one third of all genetic loci (Harris, 1969): we cannot escape from genetic variability. I shall argue that we may as well use it. And my reason for suggesting this is not only because we have no choice (in one sense) but that if we exercise choice (in a different sense) we may contribute to the solution of another puzzle: the problem of how genes act in higher organisms. At present we have some theories about the regulation of gene action (e.g. Britten and Davidson, 1969), but remarkably little information to allow a choice among them, or even for the critical formulation of experiments. In short, I would suggest that a judicious use of mutants for nutritional studies will generally not be disadvantageous, it may sometimes be advantageous, and occasionally it will give two results for the price of one! My hope also is that the nutritional-genetic results will open up new areas of understanding of nutrition. I am convinced that we are about to move forward from what one might call, in shorthand, the "metabolic pathway" view of nutrition, to a finer understanding of the regulation of nutrition; the control of nutrilite usage during development, maturity and age. There are good grounds for thinking this regulation must depend on the phased activity of gene arrays in time.

It is now easy to make mutations using the well established nitrogen mustards and the like. These can be fed to males (Lewis and Bacher, 1968) and, even lacking the trick stocks of the Drosophilist, it is not too difficult to procure a great number of mutants as has been done with *Musca domestica*, or with *Tribolium confusum.* Some of these mutants will be auxotrophs, unable to synthesise a particular nutrient (e.g. purines and pyrimidines, as shown by Vyse and Sang (1971). Such mutants have obvious advantages for metabolic studies, as the bacterial and fungal biochemists have known for a long time. It is somewhat surprising that this class of

Insect and mite nutrition – North-Holland – Amsterdam, (1972)

mutation has been so long ignored in our studies. One reason is that insects synthesise so little for themselves, and another is that metabolism can sometimes be followed more readily using isotopically labelled materials. However, there is no good reason for treating this as an either/or situation. Since it has now been shown that some *Diptera* synthesise most of their folic acid requirement (Venters, 1971), it should be instructive to combine labelling techniques with mutants, not only those affecting folic synthesis but also, say, mutations altering the pteridine pigments, in order to disentangle this very important area of insect metabolism. Obviously, there are many other instances where auxotrophic mutations should now be exploited.

If one surveys the post-DNA era of genetic research (Watson, 1965), it will be seen that another class of mutation has also been exploited. This is mutations affecting, for instance, plaque type in the bacteriophage-bacterium relationship. Here, the metabolic step is not known, but the phenotypic difference between mutant and wild type is used as an indicator. With insects, we have an advantage in that this class of mutation is sometimes found to give a graded series of responses of the kinds considered below, simply because they are multicellular. Of course, all-or-none response systems can be used in higher organisms, at least in principle, but I should like to suggest that continuous, or graded, response systems may have some advantages that have been ignored. But before we look at two examples, it is important to emphasise again that the metabolic lesion caused by the gene mutation need not be known in either case. The thesis is that such mutations, which cause measurably altered phenotypic characteristics of unknown origin, can be exploited to further our understanding of nutritional metabolism.

Of course, it is not easy to find examples to support a proposal of this kind since its essence is that the work has not yet been done. In what follows, I shall consider two cases where some data exist, but since they were collected with other purposes in mind the information is necessarily incomplete. It will be sufficient, therefore, if I merely raise problems rather than answer them. In fact, in so far as it is possible to ask new questions my case will be made, for you can take it for granted that the common nutritional questions (assessed by growth rates and survival - which I shall ignore) have already been comprehended (see Sang (1956) and Burnet and Sang (1968)). But before looking at this situation involving genes of variable penetrance and expressivity (i.e. modifiable by nutrition), it is worth noting that auxotrophic mutants have been isolated in *Drosophila*, (Nørby, 1970 and Vyse and Sang, 1971) and that they can be exploited in nutritional research.

METHODS

The culture medium used was essentially that described by Sang (1956), as modified by Bryant and Sang (1969).

The argument for using mutants is that we are dealing with a defective enzyme system which results in some failure of normal metabolism (often localised) which causes the malformation which we identify as an abnormal phenotype. It is this failure of normal metabolism which concerns us since this "natural operation" can be used to explore the normal. A nutritional deficiency which exaggerates the mutant effect may be assumed (at a first approximation) to interfere with the same area of metabolism as the mutant gene. Deficiencies which "cure" the defect may do so in one of three ways: (a) by generally delaying development so that the defective process has time to catch up, (b) by short circuiting the defect in some way, or (c) by causing the accumulation of a precursor and thus loading the defective enzyme so that it functions more normally. And there may be other processes which we have still to understand.

The data in Table 1 summarise the information from two examples where we have some comprehensive information. The first is for the single gene mutation *eyeless* which, unlike its name, causes a reduction of eye size, frequently unequally on the two sides of the head. Since eye size can be measured either as area, or by counting facets, it is easy to gauge treatment effects. Two points should be made: this is an instance of failure to form structures (i.e. of cell multiplication) which is localised to the developing eye disc, and the system is sensitive to deficiencies at two stages in development, about 15 and 40 hours of larval development. We shall ignore this latter fact. The second example is of a melanotic tumor which is undoubtedly a genetically complex example since a number of non-allelic genes can influence its appearance (Lindsley and Grell, 1967). The melanoma arises from an aggregation of haemocytes, around 71 hours of larval development (the sensitive period) and these aggregates melanise just prior to pupation. Populations can be scored for the proportion of tumors they carry, and individuals likewise, since under extreme treatments they may have up to a dozen aggregates. Both characters are thus easy to quantify.

As we should expect for such different kinds of character, there is little or no relationship between their reactions to the same (or similar) treatments, perhaps less than one might have expected by chance (Table 1). And this makes my first obvious point: some mutants will be useful for some problems, but not for others. The second interesting point is that an array of deficiencies has no influence on eye size (implying a specificity among the effective treatments), whereas nearly all treatments affect tumors, one way or the other. The two mutants behave very differently under nutritional stress. Notice, too, that lecithin and biotin have doubtful effects on eye size.

This raises an interesting point, also illustrated by Hunt and Burnet's (1969) data. These workers tested a number of alleles of *eyeless* in the same wild-type genetic background and one allele in a different background, and both situations showed response variability such that in one background a treatment might be significant whereas it would be insignificant in another. This illustrates the point made earlier: we take our putative wild strains too much for granted. There is a great diversity among wild genotypes which is exposed only under stress.

Table 1. Responses of the mutant character to nutrient deficiencies in *eyeless* and *melanotic tumor* strains of *Drosophila melanogaster.*

Deficient nutrient	Eye reduction	Tumor penetrance
Casein	0	+
Cholesterol	+	+
Sucrose	0	-
Lecithin	+?	-
RNA	-	-
Thiamine	-	+
Riboflavine	0	-
Niacin	0	-
Pyridoxine	0	+
Pantothenic acid	0	-
Folic acid	0	-
Biotin	+?	-
Excess nutrient		
Casein	0	-
Cholesterol	0	-
Sucrose	0	0
Lecithin	0	0
RNA	0	+

An increased mutant effect is shown by +, a decreased (i.e. normalising) effect by -, and 0 indicates no significant alteration. All treatments slowed growth and were detrimental to survival.

Detailed data are given in Hunt & Burnet (1969) and in Sang & Burnet (1967).

Of the treatments which affect *eyeless*, only one (cholesterol) enhances the mutant expression. Unfortunately, it has not been studied at all. Two treatments normalise development: low RNA and low thiamine. Hunt (1970) has confirmed Sang's & Burnet's (1963) finding that the RNA effect is dependent on the purine half of the RNA molecule (low pyrimidine having the opposite effect), although the situation is complex. The problem, then, is why should a deficient supply of adenylic acid (guanylic is scarcely used (Burnet and Sang, 1963),) permit improved growth of the eye disc when it has the opposite effect on growth generally? Similarly, it is fairly obvious that thiamine deficiency is not operating by way of its effects on energy metabolism, since riboflavine and niacin deficiencies might then also be expected to behave likewise, but they do not. What particular role of thiamine might be involved here? And why should it also improve eye disc growth while slowing general development?

The situation is quite different if we turn to tumor responses, but again raises problems of significance. In the first place, all the vitamin deficiencies affecting energy supplies reduce tumor frequency, except thiamine. This may be explained if one assumes that the formation of the haemocyte aggregate is an energy-requiring process; certainly it involves a cellular transformation (Rizki, 1960). The interesting effects are thus those which increase tumor penetrance: thiamine, pyridoxine, casein and cholesterol, a curious grouping. This last is dealt with elsewhere (this volume, page 493) and we shall return to thiamine in the context of both systems. Possibly, low pyridoxine means low protein metabolism, and is equivalent to low casein.

The effects of protein provision were examined by supplementary media with an excess of each amino acid individually (Table 2). Surprisingly, only the 1-aspartic acid addition improved one mutant: there were no other significant improvements, although they might have been expected. Equally surprisingly, three amino acids had effects which exaggerated the genetic phenotype in both examples, 1-lysine, 1-phenylalanine and 1-tryptophane. This is again a curious combination, but only the latter two have been studied in any detail.

Although tyrosine is ineffective in both systems it seemed possible that the products of phenylalanine and tryptophane (Fig. 1) might prove to be involved in the regulation of the gene expression, and some of these products were tested (Table 3). The derivatives of phenylalanine (Dopa, dopamine and noradrenalin) enhanced the effects of both genes, i.e. increased tumor number and decreased eye facet number. Contrariwise, the products of tryptophane (serotonin and tryptamine) normalised both phenotypes. Thus these biogenic amines behave in an antagonistic manner, just as they do in nerve function, and it is tempting to include the thiamine effects, as Hunt (1971) has done, as also involving nerve action. However, as we shall see, there is no good proof that this is so, at least with respect to tumors.

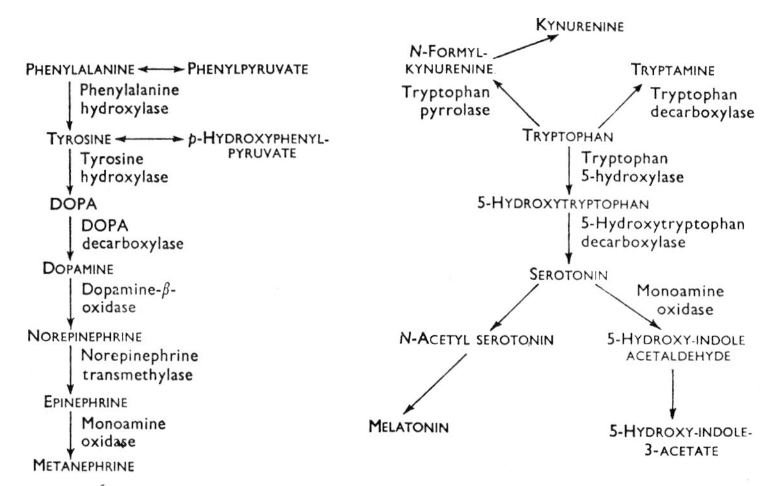

Figure 1. Pathways of phenylalanine and tryptophan metabolism.

Table 2. Effects on mutant expression of adding particular amino acids to the normal diet of *eyeless* and *melanotic tumor* strains of *Drosophila melanogaster*.

Amino acid addition	Eye reduction	Tumor penetrance
Glycine	+	0
1-Aspartic acid	–	+
1-Histidine	+	0
1-Lysine	+	+
1-Methionine	+	0
1-Phenylalanine	+	+
1-Threonine	+	0
1-Tryptophan	+	+

All amino acids were tested as 40 mM additions (except lysine which was used at 25 mM on *eyeless*). Only those additions having effects significant at the 1 per cent level are listed. Data on eye reduction from Hunt (1971) and on tumors from Burnet and Sang (1968). Symbols as in Table 1.

Table 3. Responses of mutant expression to phenylalanine and tryptophan derivatives in *eyeless* and *melanotic tumor* strains of *D. melanogaster.*

Additive	Eye facet number	Tumors per cent
Dopa	– 76.3	+ 47
Dopamine	– 42.0	+ 20
Noradrenaline	– 46.7	– 8 n.s.
Tranylcypromine	—	– 61
+ serotonin	+ 66.3	– 20*
Tryptamine	+ 65.4	– 26

*Difference from control containing trancylcypromine.
All figures are given as differences from controls and only the difference marked n.s. is not significant at the 5 per cent level, or less.

If we look at the interaction between thiamine deficiency and excess supply of the two aromatic amino acids (Table 4) it is clear that their modes of action must be different. Low thiamine and high tryptophan behave more or less additively, whereas the interaction with phenylalanine is synergistic. This implies that the first response is the sum of two separate pathways and that there is a heterogeneity of cause; whereas the second may share the same step in one pathway. In fact, I have been unable to find any reference to an intimate interrelationship between phenylalanine metabolism and thiamine.

Table 4. Interactions between thiamine, and phenylalanine and tyrosine as they affect tumor expression in the *melanotic tumor* strain of *D. melanogaster.*

Additive	Normal thiamine	Low thiamine	Difference
Nil	2.10	3.96	1.86
1-phenylalanine	2.61	6.77	4.16*
1-tryptophan	2.52	4.77	2.25

The figures are for mean tumors per fly. The amino acids were added in 40 mM quantities to the media. * indicates a significantly greater difference than the nil addition control set.

CONCLUSION

As I stated in the beginning, there are insufficient available data to permit a comprehensive analysis of the two examples considered; but that was not the point, anymore than I am here concerned with the exploration of the metabolism of biogenic amines in *Drosophila*. My objective was simply to show that mutants can be used to add an extra dimension to nutritional research. It would also be tedious to recite the many problems which these results pose, but it is perhaps worth emphasising that many problems have been ignored, such as the eye reducing role of methionine and the functions of pyridoxine in tumor development, the contradiction between the effects of tryptophan and its product serotonin, and so on. Indeed, I hope it is clear without further detailed elaboration that there are such obvious advantages in departing from the traditional use of "standard laboratory strains" of insects and mites in our studies that the genetical journals will be flooded with new kinds of material. It would certainly be to their advantage. But I believe nutritional research will also gain from such a development, which would help to bring the subject into a more central position in biology. The fact that the two examples quoted above appear to involve psychoactive substances is both relevant and surprising. At least one of these, serotonin, has recently been shown to have morphogenetic effects during sea urchin embryogenesis (Gustafson and Toneby, 1971), and it may be only our present ignorance that prevents us from seeing the general significance of these very separate groups of observations.

REFERENCES

BRITTEN, R. J. and DAVIDSON, E. H. 1969. Gene regulation for higher cells: a theory. *Science* **165**:349-358.

BRYANT, P. J. and SANG, J. H. 1969. Physiological genetics of melanotic tumors in *Drosophila melanogaster*. VI. The tumorigenic effects of juvenile hormone-like substances. *Genetics* **62**:321-336.

BURNET, B. and SANG, J. H. 1963. Dietary utilisation of DNA and its derivatives by *Drosophila melanogaster* (Meig.) *J. Insect Physiol.* **9**:553-562.

BURNET, B. and SANG, J. H. 1968. Physiological genetics of melanotic tumors in *Drosophila melanogaster*. V. Amino acid metabolism and tumor formation in the *tu bw; st su-tu* strain. *Genetics* **59**:211-235.

GUSTAFSON, T. and TONEBY, M. I. 1971. How genes control morphogenesis. *American Scientist* **59**:452-462.

HARRIS, H. 1969. Genes and isozymes. *Proc. Roy. Soc. Lond. B.* **174**:1-31.

HUNT, D. M. 1971. The physiological control of gene action in the *eyeless* and *eyegone* mutants of *Drosophila melanogaster*. *Genet. Res. Camb.* **17**:195-208.

HUNT, D. M. and BURNET, B. 1969. Gene-environment interactions of the *eyeless* mutant in *Drosophila melanogaster*. *Genet. Res. Camb.* **13**:251-265.

LEWIS, E. B. and BACHER, F. 1968. Method of feeding ethyl methan sulfonate (EMS) to *Drosophila* males. *Drosophila Information Service* 43:193.

LINDSLEY, D. L. and GRELL, E. H. 1967. "Genetic variations of *Drosophila melanogaster.*" *Carnegie Institution of Washington, U.S.A.*

NØRBY, S. 1970. A specific nutritional requirement for pyrimidines in the rudimentary mutant of *Drosophila Melanogaster. Hereditas* 66:205-214.

RIZKI, M. T. M. 1960. Melanotic tumor formation in *Drosophila. J. Morphol.* 106:147-158.

SANG, J. H. 1956. The quantitative nutritional requirements of *Drosophila melanogaster. J. Exp. Biol.* 33:45-72.

SANG, J. H. and BURNET, B. 1963. Environmental modification of the eyeless phenotype in *Drosophila melanogaster. Genetics* 48:1683-1699.

SANG, J. H. and BURNET B. 1967. Physiological genetics of melanotic tumors in *Drosophila melanogaster.* IV. Gene environment interactions of *tu-bw* with different third chromosome backgrounds. *Genetics* 56:743-754.

VENTERS, D. 1971. Folate synthesis in *Ae. Aegypti* and *Drosophila melanogaster larvae. Trans. Roy. Soc. Trop. Med. and Hyg.* 65:687-688.

VYSE, E. R. and SANG, J. H. 1971. A purine and pyrimidine requiring mutant of *Drosophila melanogaster. Genet. Res., Camb.* 18:117-121.

WATSON, J. D. 1965. "Molecular biology of the gene." W. A. Benjamin, Inc. New York.

THE SEARCH FOR AUXOTROPHIC MUTANTS IN *DROSOPHILA MELANOGASTER*

D. R. Falk and David Nash
Department of Genetics, University of Alberta, Edmonton, Alberta, Canada

INTRODUCTION

Ever since Beadle and Tatum (1941) first demonstrated and explained the auxotrophic phenotype in *Neurospora*, auxotrophs have played a primary role in biological investigations. They have provided sophisticated tools for analysis of inheritance, exposed the nature of many metabolic pathways, aided in understanding the genetic code and broadened our knowledge of regulation of gene action.

Operationally, an auxotrophic strain requires an extra nutritional component compared with the "wild-type" organism. Functionally, such strains usually bear a mutation in a gene responsible for some step in an essential anabolic pathway, although the operational definition can also encompass other situations. Bacteria and fungi can be cultured axenically, have simple nutritional requirements and can usually be analysed rapidly from the genetic standpoint. It is thus not surprising that most of the known auxotrophs are found in microorganisms, where most of the advances described above have also been made. However, there remain several major problem areas which could be attacked with auxotrophs in more complex organisms. In this laboratory we have spent some years in producing strains of *Drosophila melanogaster* with abnormal nutritional requirements for an approach to gene activity during development.

Production of auxotrophs in higher organisms (other than fungi) is not easy. Even in unicellular photosynthetic plants such as *Chlamydomonas* very few are known (Li, Redei and Gowans, 1967). In multicellular plants a number of well-defined mutants involving thiamine synthesis (Li, Redei and Gowans, 1967) are found in angiosperms and, although Carlson (1969) has described a number of nutritionally dependent strains in the fern species *Todea barbara* and *Osmunda cinnamomea*, no further work has been reported on them.

In animals, a narrower range of probable auxotrophies would be expected since their normal nutritional requirements are more complex than those of plants. Some auxotrophic strains of vertebrate tissue culture lines are known (Kao and Puck, 1968) but their genetic analysis is difficult. Very few animals which are easily manipulable genetically can be studied in the axenic, defined culture conditions necessary for auxotroph isolation. *Drosophila melanogaster* fulfills all these requirements.

Sporadic attempts at isolation of fruitfly auxotrophs have been made during the last twenty years. The results have been encouraging but not spectacular. Early workers screened existing strains for nutritional dependencies, and discovered extra requirements in two strains. Both requirements were pH dependent (Ellis, 1959) and due to complex genetic factors. Norby (1970) has described a pyrimidine dependency in a known mutant strain, *rudimentary*, and has demonstrated the auxotrophy to be a pleiotropic effect of the rudimentary wing mutant.

Vyse and Nash (1969) made the first successful attempt at auxotroph production by mutation breeding. Using a nutritional selective regime similar to that described below, they obtained three strains with nutritional conditional lethality. One of the three mutants, 1308 has since been shown to require double supplementation with adenosine and a pyrimidine nucleoside (Vyse and Sang, 1971).

PRODUCTION OF MUTANTS

Our mutation and breeding experiments are carried out under continuous axenic culture conditions; initial sterilizations are made using calcium hypochlorite dechorionation of eggs and sterility is maintained by transfer and handling of flies in sterile chambers. Our media contain antibiotics and/or the fungicide propionic acid (see Table 1).

The breeding protocol for our experiments is shown in Figure 1. Adult male flies are fed the mutagen ethyl methane-sulfonate. (6.4 mM in 1% sterile sucrose solution) according to the method of Lewis and Bacher (1968) and mated to females bearing an "attached-X" chromosome (two similar X-chromosomes arms attached to the same centromere) and a Y chromosome, as well as the normal autosomal complement. Individual males from this cross, which receive an X-chromosome from their father, can be mated to further attached-X females to establish strains of flies which can subsequently be tested for nutritional abnormalities. Since these strains may contain males each with an identical mutant X-chromosome and females without mutagenized X-chromosomes, the system is particularly convenient for studying sex-linked mutants, with the non-mutants (females) providing an excellent internal control for quantitative measurements of the behavior of mutants.

To minimise problems associated with possible genetic heterogeneity, the males in the test are from a highly inbred line from Amherst College, Amherst, Massachusetts (see Drosophila Information Service, 1968, stock 1).

Table 1. Media used in screening and testing nutritional conditional mutants.

Defined medium (modified slightly from Bryant and Sang, 1969)			
Agar (Oxoid No. 3)	3.00 g.	Biotin	.016 mg.
Casein (Vitamin Free)	5.50 g.	Folic acid	.3 mg.
Sucrose	750 mg.	$NaHCO_3$ (anhydrous)	140 mg.
Cholesterol	30 mg.	KH_2PO_4 (anhydrous)	183 mg.
Lecithin	400 mg.	K_2HPO_4 (anhydrous)	189 mg.
Thiamine	.2 mg.	$MgSo_4$ (anhydrous)	62 mg.
Riboflavin	.1 mg.	Streptomycin	17 mg.
Nicotinic acid	1.2 mg.	Penicillin[a]	25,000 I.U.
Ca pantothenate	1.6 mg.	Water	to 100 ml.
Pyridoxine	.25 mg.	RNA (when added)	400 mg.
Dead yeast-sucrose medium (modified from Nash and Bell, 1968)			
Brewers yeast	12.5 g.	Penicillin[a]	25,000 I.U.
Sucrose	10.0 g.	Proprionic acid[a]	1.0 ml.
Granulated Agar	2.0 g.	Water	90 ml.
Streptomycin	25 mg.		

[a] - added after autoclaving.

The female autosomal complement was derived from the same inbred line by five successive backcross generations and, although this technique of chromosome substitution is not fool-proof, we have several reasons to believe that it was successful in this case. Thus the strains derived from our mutagenesis experiment are probably uniform, except for differences resulting from the mutagenic treatment.

To this point, all cultures have been grown on a sterile, enriched medium containing dead yeast and sucrose (Table 1). After sufficient expansion of the culture, usually in one generation, a sample of flies is placed on modified Bryant and Sang's (1969) defined medium (see Table 1). Cultures carrying auxotrophic mutations on their sex-chromosomes should lack males, but will yield females, all of which bear the unmutated attached-X chromosome. We have considered it worthwhile expanding our mutant screen to include strains with low (less than 10%), rather than zero, male viability or slow male development, on the ground that *Drosophila*, being more complex than bacteria, may exhibit compensatory genetic, cellular or tissue interactions which will, to some extent, mask clear auxo-

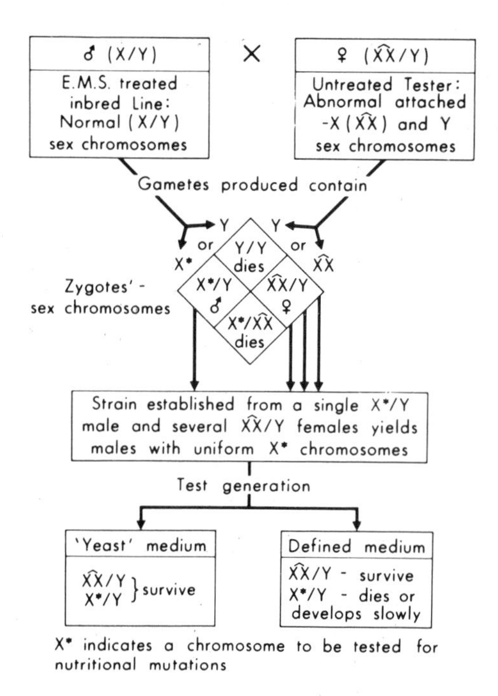

Figure 1. Protocol for the production and isolation of sex-linked nutritional conditional lethal mutations using an attached-X mutagenesis system.

trophy and replace it with these 'leaky' phenotypes.

When a strain rechecked as mutant on defined medium and as reasonably normal on the dead yeast/sucrose medium, it was set aside for later detailed scrutiny. The results presented below represent some of our initial findings in this latter process.

To summarise: We have isolated strains of *Drosophila* in which only males carry potentially mutant X-chromosomes and have screened these cultures for abnormal growth of males on defined medium as compared with a dead yeast/sucrose medium. We are thus analysing sex-linked nutritional conditional mutants, as did Vyse and Nash (1969), but our genetic techniques are simpler, yet perhaps more sophisticated.

NUTRITIONAL RESPONSES OF THE MUTANT STRAINS

The first stage in characterizing our mutants, after the initial screening, is a comparison of their eclosion patterns with that of a control strains, on both defined and yeast/sucrose media. In all experiments our control strain is similar to the mutant strains except in not having been subjected to mutagenesis. Thus it, too, segregates inbred males and XX/Y females with autosomes derived from the inbred line. In order to characterize a given strain many replicate cultures were made; for each culture approximately 20 females were allowed to oviposit for 24 hours or less in a culture vial containing about 5cc of the appropriate medium. The cultures were maintained at 25°C and the number and sex of flies emerging recorded daily until eclosion ceased.

The data were used to produce the diagrammatic "eclosion curves" for males shown in figures 2-4. The curves show the temporal patterns of eclosion for mutant males relative to their non-mutant sisters and have been normalised to compensate for variation between experiments by use of control (wild-type) data. For a full explanation of the calculations see the appendix. Two aspects of the mutants are evident in the curves; their viability and their development rate, both relative to the non-mutant females of the same strain.

On the basis of their viability, we have divided the mutants into five classes. Table 2 summarizes this classification and the mutants so far described in each class. (In the Table and elsewhere each strain has been denoted by the letters FNC and a number which indicates the order of its isolation. The letters "FNC" indicate "First (X-) chromosome Nutritional Conditional mutant).

Not all strains initially isolated subsequently show mutant characteristics as extreme as might be expected from the original screening criteria; some (about 25%) now appear nutritionally unresponsive. We do not know whether the strains have changed or were insufficiently characterized at first. However, our more detailed characterization has already taken several months and our strains now appear stable in repeated tests. Table 3 and Figure 2 show characteristic behavior of a selected group of mutants from each class. The class I mutants show little or no eclosion on defined medium, hence no eclosion curves have been constructed. Of the three mutants FNC8, FNC22 and FNC26 (Table 3), FNC22 differs from the others in being strongly supplementable by addition of RNA to the medium, insofar as they show different relative male viability on defined medium. All the mutants shown have essentially normal viability and development rate on yeast-sucrose medium. It can be seen that, in each of the strains shown in Figure 2, except FNC32, those males which do survive are delayed in development on defined medium. The length of this delay varies from strain to strain.

One other aspect of nutritional behavior does differentiate some of these mutants; whilst all of them except FNC8 respond somewhat to the addition of RNA to defined medium, in some it is a relatively low response (FNC26, Table 3; FNC32, Figure 3) whereas in others the response is almost comparable to their behavior on yeast medium (FNC22, Figure 3; FNC30,

Table 2. Auxotrophic and nutritional conditional mutations classified according to viability on defined medium.

CLASS I (<5% Viability)		CLASS II (5-20% Viability)		CLASS III (20-40% Viability)		CLASS IV (40-60% Viability)		CLASS V (>60% Viability)	
Mutant designation	P[a]	Mutant designation	P	Mutant designation	P	Mutant designation	P	Mutant designation	P
[b] 1308	?	FNC 32	0.11	FNC 9	0.20	FNC 20	0.46	FNC 23	0.69
[b] 11523	?	FNC 30	0.125	FNC 27	0.225	FNC 10	0.53	FNC 28	0.70
[b] r	?	FNC 31	0.19	FNC 6	0.32	FNC 7	0.54	FNC 15	0.78
FNC 8	0			FNC 29	0.35	FNC 5	0.56	FNC 25	0.78
FNC 22	.003			FNC 4	0.38				
FNC 26	0								

[a] The proportion of males relative to females in a given strain.

[b] These mutants have been described by other authors (Vyse and Nash, 1969; Norby, 1970; Vyse and Sang, 1971). They have not been tested under our conditions, but are probably in class I.

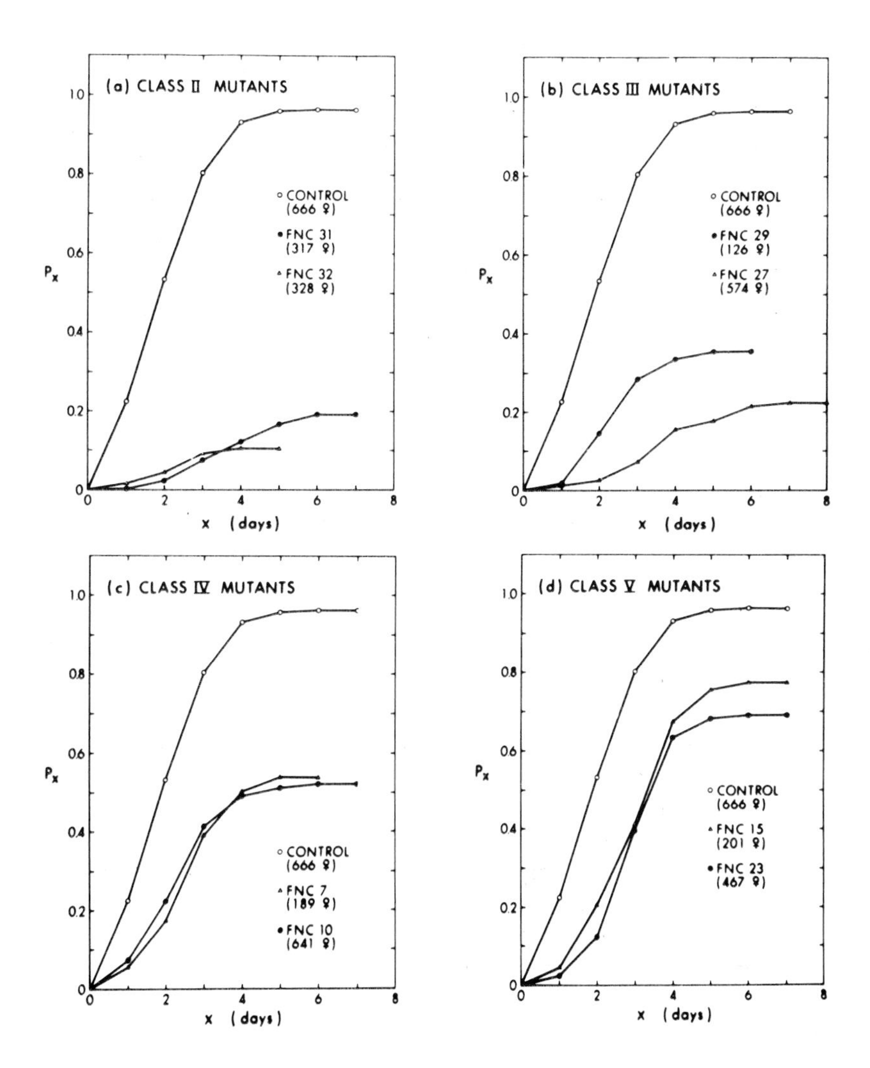

Figure 2. Eclosion curves for males of control and of selected mutant strains of classes II-V grown on defined medium without RNA. The number of females (T_f) upon which each curve is based is shown in parentheses.

Px, the cumulative proportion of males eclosing up to day x relative to the overall production of females, T_f, is defined in the appendix. All curves are normalised using control female data to allow direct comparison between strains.

The first day of eclosion in individual cultures has been defined as day 1 (x = 1).

Table 3. Comparitive productivity of males and females in three class I mutants and a control, on defined medium with and without RNA.

	- RNA		+ RNA[a]	
	Male	Female	Male	Female
FNC 8	0	> 500	0	> 100
FNC 22	1	249	334	304
FNC 26	0	109	8	60
Control	643	666	252	243

[a]4mg/ml of medium.

Figure 4). For the present our tests with RNA have used only Sang's (1956) optimal concentration of RNA (4mg/ml of medium). It is possible that other concentrations will prove optimal for mutants.

In cases where RNA supplements, but incompletely, it is a moot point whether the more interesting aspect of nutritional lethality will prove to be that which is remediable with RNA or that which is not. In passing, inspection of the data of Vyse and Nash (1969) make it seem likely that the mutant 1308 is only partially supplementable with RNA and yet, it is that supplementation which has been analyzed with respect to requirements for a purine and a pyrimidine (Vyse and Sang, 1971).

Not all strains are completely normal on yeast-sucrose medium. For example FNC32 males probably develop faster than controls and FNC30 males (Figure 4) have low viability.

In conclusion, all that can be said at the present is that we have a variety of strains, with different levels of nutritional dependency and different development rates. We are presently embarking upon a more detailed examination of those strains with large responses to dietary RNA.

NON-NUTRITIONAL CHARACTERISTICS OF THE MUTANT STRAINS

Several other aspects of the behavior of the strains have become evident. Four strains, FNC 4, 7, 8 and 15 carry female sterility mutations on their X-chromosomes. FNC10 is a female semi-sterile. Two other strains, FNC12 (semi-sterile) and FNC18 (sterile), were also isolated by our screening technique but are not now nutritionally responsive. These seven female sterile mutants which it should be remembered were selected on the basis of male growth patterns, were found amongst one batch of 23 strains (FNC4 -26). This rate of female sterile production is at least a five-fold increase over

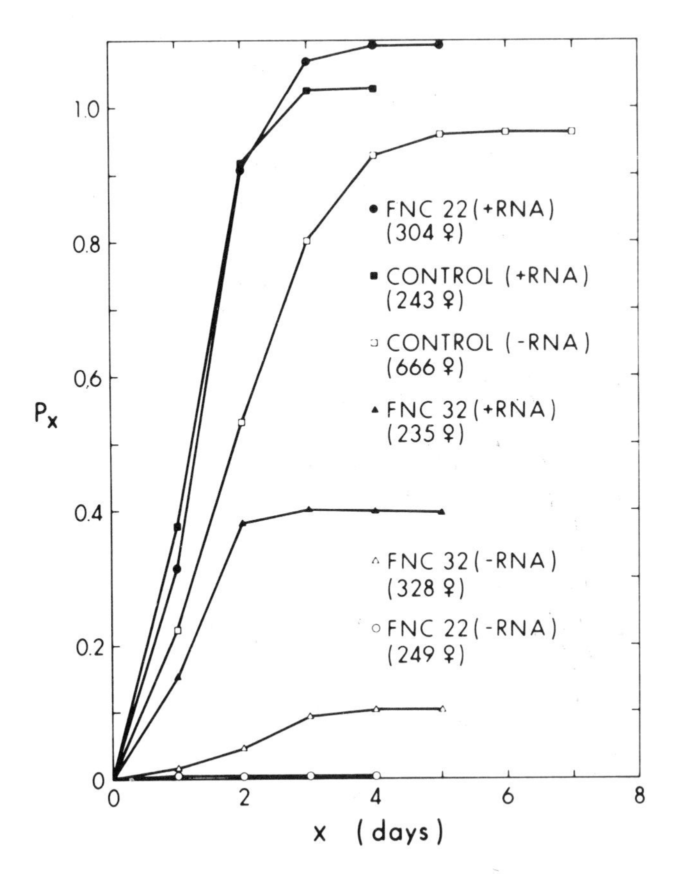

Figure 3. The effect of addition of 4 mg/ml RNA to defined medium on the eclosion curves of two mutant strains and the control. FNC 22 (class I) shows a strong response; FNC32 (Class II) shows a less strong response. Computation of the value of x from the first day of eclosion removes a consistent 1-2 day decrease in development time due to addition of RNA. See figure 2 for additional notes.

the basic rate of female sterile production using identical or more severe (up to 12 mM EMS) mutagenic treatments (Romans, unpublished), without nutritional screening.

We are presently mapping both nutritional and sterility effects genetically. These tests could well indicate that both phenotypes are commonly pleiotropic effects of the same mutations. If so, it might be suggested that some metabolic functions which are dispensible (given correct nutrition) during larval development, are indispensible to the production of viable eggs. All seven female sterile strains lay eggs. A somewhat similar explanation of

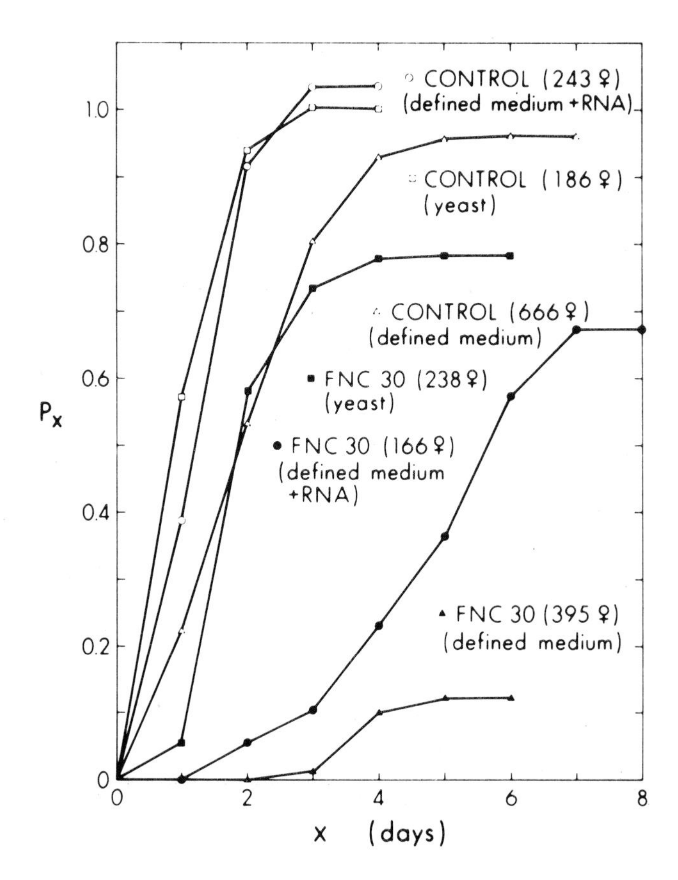

Figure 4. Eclosion curves for the control and for FNC 30 on defined medium, defined medium plus 4mg/ml RNA and yeast-sucrose medium. FNC30 is exceptional insofar as it does not behave like the control on yeast-sucrose medium. Notice also that RNA increases survival of FNC30 males without increasing their development rate relative to females. Computation of the value of x from the first day of eclosion removes a consistent 1-2 day decrease in development time due to addition of RNA and a 3 day decrease due to growth on yeast-sucrose medium rather than RNA-free defined medium. See figure 2 for additional notes.

the female sterility of the *rudimentary* auxotroph has been advanced by Norby (1970).

A variety of morphological and temperature-sensitive phenotypes have also been observed in our strains and these are also being studied to determine their possible relationship to the nutritional defects.

CONCLUSION

Differential responsiveness to nutritional conditions is now well established as an inherited phenotype. Earlier work has shown that such responses are at times attributable to classical auxotrophy (Norby 1970; Vyse and Sang 1971). Our present aim is to expand the spectrum of nutritional conditional mutants, both by more extensive work and by using a wider mutant screen. In this we have succeeded.

To date, the work is not complete enough to answer many crucial questions. However, it does seem possible to hazard a calculation of the number of loci present in the genome which, in mutant form, will yield nutritional conditional lethals of the kinds we have been studying.

In one experiment we studied some eight-hundred chromosomes which lack recessive lethal effects. Among these, we found three sex-linked, class I mutants. Unpublished data suggest that the rate of sex-linked recessive lethal mutations in males under the conditions used by us is approximately 35%, so that the ratio of lethal to conditional lethal mutants is of the order of 100:1. Recent findings of Hochman (1971) and Judd, Shen, and Kaufman (1972) suggest that the upper limit to the number of potentially lethal loci is given by the number of "bands" found in polytene chromosomes and that the lower limit is probably no less than half that number. Using Thomas' (1972) estimate of the number of bands in the genome, 5149 we can conclude that 25-50 class I loci may be found. We would also predict, perhaps, a greater than five-fold increase in this number if we include class II-V mutants. These estimates are, of course, subject to gross errors, but are probably correct well within an order of magnitude.

In summary, this work shows that a large scale isolation of nutritionally supplementable, low viability strains is possible. Whether or not the abnormalities are a result of single mutations (as one would expect) is currently being tested. The real value of this work depends upon successful biochemical analysis of the mutants and this work is just beginning.

However judging from the success of analysis of nutritional mutations in lower organisms, we would predict that such analysis will be both possible and profitable.

APPENDIX

The procedure for production of eclosion curves

For a given mutant strain the cumulative proportion of males (P_x) on a given day (x) is derived from the absolute number of males (n_m) which emerged on or before day x by the following transformation:

$$P_x = \frac{n_m}{T_f} \times \frac{n_{cf}/T_{cf}}{n_f / T_f},$$

where T_f = the total number of females produced *overall* by the mutant strain, n_f = the absolute number of females produced on or before day x, T_{cf} = the total number of females produced overall by the control XX/Y line under similar culture conditions and n_{cf} = the absolute number of females produced by the control line on or before day x.

The values of x indicate the time since the first day of eclosion for each culture individually, which is day 1. The absolute time from egg laying to first eclosion differs according to culture conditions; for example, it is usually 14 days on defined medium, but on yeast it is usually 11 days. There is also variation between cultures on a given medium and this is the reason why we have chosen to calculate eclosion patterns from the first day of eclosion, rather than in absolute terms. From inspection of the data it is evident that whatever the causes of the variation, it usually affects both sexes similarly.

Acknowledgements.—This work was supported by NRC grant A-3269. D. R. Falk was supported by an NRC Postgraduate Scholarship.

REFERENCES

BEADLE, G. W. and TATUM, E. L. (1941). Genetic control of biochemical reactions in *Neurospora*. *Proc. Natl. Acad. Sci., U. S. A.* 27:499-506.

BRYANT, P. J. and SANG, J. H. (1969). Physiological genetics of melanotic tumors in *Drosophila melanogaster*. VI. The tumorigenic effects of juvenile hormone-like substances. *Genetics* 62:321-336.

CARLSON, P. S. (1969). Production of auxotrophic mutants in ferns. *Genet. Res., Camb.* 14:337-339.

ELLIS, J. F. (1959). Reversal of an adenine and a cytidine requirement in axenic *Drosophila* culture. *Physiological Zoology* 32:29-39.

HOCHMAN, B. (1971). Analysis of chromosome 4 in *Drosophila melanogaster*. II. Ethyl methane-sulphonate induced lethals. *Genetics* 67:235-252.

KAO, F. and PUCK, T. T. M. (1968). Genetics of somatic cells. VII. Induction and isolation of nutritional mutants in Chinese hamster cells. *Proc. Natl. Acad. Sci., U. S. A.* 60:1275-1282.

JUDD, B. H., SHEN, M. W. and KAUFMAN, T. C. (1972). The anatomy and function of a segment of the X-chromosome of *Drosophila melanogaster*. *Genetics*, in press.

LEWIS, E. B. and BACHER, F. (1968). Method of feeding ethyl methane-sulphonate (EMS) to *Drosophila* males. Drosophila Information Service 43, 193.

LI, S. L., REDEI, G. P. and GOWANS, C. S. (1967). A phylogenetic comparison of mutation spectra. *Molec. Gen. Genetics.* **100**:77-83.

NASH, D. and BELL, J. (1968). Larval age and the pattern of DNA synthesis in polytene chromosomes. *Canadian J. of Genet. and Cytol.* **10**:82-90.

NORBY, S. (1970). A specific nutritional requirement for pyrimidines in rudimentary mutants of *Drosophila melanogaster. Hereditas* **66**:205-214.

SANG, J. H. (1956). The quantitative requirements of *Drosophila melanogaster. J. Exptl. Biol.* 33:45-72.

THOMAS, C. A. JR. (1971). The genetic organization of chromosomes. *Ann. Rev. Gen.* 5:237-256.

VYSE, E. R. and NASH, D. (1969). Nutritional conditional mutants of *Drosophila melanogaster. Genet. Res., Camb.* 13:281-287.

VYSE, E. R. and SANG, J. H. (1971). A purine and pyrimidine requiring mutant of *Drosophila melanogaster. Genet. Res., Camb.* **18**:117-121.

APPLICATION OF INSECT NUTRITION IN SOLVING GENERAL NUTRITION PROBLEMS

G. R. F. Davis
Entomology Section, Research Station,
Research Branch, Canada Agriculture
University Campus, Saskatoon, Sask., S7N OX2, Canada

Although organisms at either extreme of the animal kingdom are used routinely in nutrition investigations, organisms intermediate to these have generally been neglected as test animals for this purpose. Fundamental research with insects over the past 25 years has indicated that these organisms are similar to the others in many of their dietary requirements. The evolutionary process, however, has imposed a major difference on this group: that of a dietary sterol requirement. In the area of protein nutrition, insects require the same 10 essential amino acids as other organisms, although quantitative requirements may vary to some degree. Despite a wealth of knowledge of comparative insect nutrition, insects have been overlooked as test animals in nutrition investigations. The time seems ripe to bridge the gap between comparative insect nutrition and applied general nutrition.

House & Graham (1967) indicated that *Tribolium confusum* grew differently on 20 different dried foodstuffs. Although these authors were interested in susceptibility of food products to insect attack, their results indicated that this insect grew better on some products than on others. The implication is that the nutritional quality of these products influenced the growth of this insect. That this situation actually occurs was later shown in an assessment of the nutritive value of barley varieties with the same insect (Loschiavo *et al.* 1969).

Earlier than these investigations, the use of the yellow mealworm, *Tenebrio molitor*, as a test animal for nutritional evaluation of proteins, had been suggested (Leclercq & De Bast 1965). The reasons for such a suggestion were as follows:

1. larvae of *Tenebrio molitor* will feed on almost any powdered substrate, even of manifestly poor nutritional quality;
2. these larvae do not require a moisture content of greater than 10% in the food;
3. the larvae are small and can be used with small quantities of dietary material;
4. during active growth, the larvae are large enough to be handled

Insect and mite nutrition – North-Holland – Amsterdam, (1972)

without risk of damaging them and to be weighed very accurately; and

5. once larvae have attained a fresh weight of 10 mg, they can quadruple or quintuple this weight in a 4-week period on an optimal diet.

Preliminary investigations (Leclercq 1965) showed that larvae of *T. molitor* grew differentially on a selection of proteins, presumably in relation to the nutritional value of these proteins. In a 4-week period, larvae of *T. molitor* gained 0.5 times their initial weight on the poorest protein of 4.8 times their initial weight on the best. Other proteins were responsible for a continuous range of weight gains between these extremes.

Subsequent studies (Leclercq *et al.* 1967), performed to classify the nutritional value of plant leaves using larvae of *T. molitor*, indicated that none were as efficient as casein. Fed leaves of stinging nettle, *Urtica dioica*, these larvae gained 4.7 times their initial weight during the experimental period, compared with 5.3 times when fed a casein diet. Gains in weight by these larvae on 15 other sources of protein ranged from 0.5 to 2.6 times the initial weights. Although a classification according to nutritional values of the leaves was obtained, the investigators suggested that it might be biased because of the lack of availability to the insect of bound leaf protein.

Neither investigation provided comparisons of the nutritional values of these products with those determined with other organisms fed similar diets.

Recently (Davis & Sosulski 1972), tests have been performed with larvae of *T. molitor*, Gembloux strain, race F, using defatted oilseed meals prepared for feeding trials with weanling mice, Carworth Farms, No. 1 Strain (Sarwar 1971). This study has provided some comparative data regarding nutritional values as measured by larvae of *T. molitor* and by weanling mice.

Defatted meals were prepared from seeds of sunflower, soybean, rape (*Brassica napus*), turnip rape (*B. campestris*), safflower and flax. Seeds of sunflower, soybean and safflower were dehulled before grinding; the others were not. Seeds of the two rape species were treated to inactivate myrosinase, which may hydrolyze seed glucosinolates into toxic compounds. These meals were analyzed for proximate principles and for amino acid composition. Protein score, using vitamin-free casein (Nutritional Biochemicals Corp., Cleveland, Ohio) supplemented with methionine as the reference protein, and essential amino acid index were calculated for each meal.

Diets containing 3% or 10% of dietary protein were prepared for larvae of *T. molitor*, using both unautoclaved and autoclaved meals. The dietary formulation was that published by Leclercq & De Bast (1965) and indicated in Table 1. To these components, a mixture of B-vitamins and zinc chloride was added in the concentration indicated in Table 2. A control diet consisting of 90 parts of ground wheat and 10 parts of Brewer's yeast was also prepared.

Table 1. Composition of experimental diets fed to larvae of *Tenebrio molitor* L.

Dietary Component	% of Diet
Protein (Oilseed Meals)	3 or 10
Carbohydrate (Oilseed Meals + Glucose)	94.09 or 87.09
McCollum-Davis Salt Mixture 185	1.94
Cholesterol	0.97

Larvae, ranging in weight from 8.4 to 12.5 mg, were selected from stock rearing in the ground wheat - Brewer's yeast medium. They were starved for 48 hr to clear the alimentary canal of this diet and to insure readiness to feed on the experimental diets. After starvation, they were weighed again and were selected within the same weight range as indicated above. Thirty larvae, having an initial average weight of approximately 10 mg, were reared individually in plastic vials on each diet at 27 $\pm$ 2°C and 65.0 $\pm$ 0.5% relative humidity. The vials were closed with minutely-perforated snap-on plastic caps, which permitted free exchange of air between the incubator and the vials. After 4 wk, the larvae were weighed to the nearest 0.1 mg and change in fresh weight and mortality were recorded at this time.

Table 2. Composition of vitamin solution added to diets for larvae of *Tenebrio molitor* L. per gram of dry diet

Component	Amount (μg)
Choline chloride	$\pm$500.0
Niacin	50.0
Thiamine HCl	25.0
Ca Pantothenate	25.0
Pyridoxine HCl	12.5
Riboflavin	12.5
DL-Carnitine	6.0
Folic acid	2.5
Biotin	0.25
$ZnCl_2$	40.0

For mice, only autoclaved oilseed meals were incorporated into isocaloric protein-free based rations, to produce diets of 10% protein content. Weanling mice were fed the experimental diets or a casein control diet for a 2-wk period. Protein efficiency ratio (PER) was calculated from feed consumption and the weight gains on each diet. Dry matter intake and protein digestibility were determined using a chromic oxide technique. Net protein utilization (NPU) was calculated from feed intake, protein digestibility and body composition of the mice. Energy gains of mice on experimental diets were obtained by bomb calorimetry, comparing mice of the same initial weights and ages with those on various diets at the end of the experiment.

Data obtained with larvae of *T. molitor* on all unautoclaved meals were pooled and compared with the pooled data obtained on all autoclaved meals. Larvae of *T. molitor* gained more weight ($P<0.05$) on diets containing autoclaved meals than on those containing unautoclaved meals (Table 3). Similarly, larvae reared on diets containing meals of 4% residual fat gained more weight ($P<0.05$) than those reared on diets containing meals of 1% residual fat, when the meals were autoclaved or unautoclaved. Again, data obtained with diets containing 3% protein, both from unautoclaved and autoclaved meals, were compared with those obtained with diets containing 10% protein. Growth response of larvae of *T. molitor* were greater with 10% dietary protein ($P<0.05$) than with 3%.

Although heat treatment and residual fat content of the meals had significant effects, the principal controlling factor of weight gains was the source of protein (Table 4). Weight gains with soybean, sunflower and turnip rape approached 70% of that with the control diet. Larval growth on safflower and rape diets was less. On flax meal diets, it was less than 20% of that obtained with the control diet.

When average gains in weight of larvae of *T. molitor* were compared with average gains of mice on the same autoclaved meals, the correlation coefficient was not significant (Table 5). However, the utilization of proteins of oilseed meals by mice (PER) was nearly of the same order as the gain in fresh weight by *T. molitor* on the basis of gains in fresh weight, so that a positive, significant correlation was obtained for these parameters. The values obtained for net protein utilization (NPU) by mice ranked the oilseed meals in much the same way as the gains in fresh weight by *T. molitor*. However, the correlation coefficient for these factors was not significant.

This preliminary study also indicates that as for vertebrate animals, autoclaving has a favorable effect on protein digestibility and on the destruction of specific anti-nutritive factors, such as trypsin inhibitor. It strongly suggests that when techniques have been refined to measure food ingestion and utilization in *T. molitor*, this insect can be used as a reliable test organism to determine nutritive values of novel proteins.

Table 3. Average initial weights and gains in weight and survival of larvae of *Tenebrio molitor* L. on oilseed meal diets

Treatment of Oilseed meal	Initial weight (mg)	Avg gain in weight[a] (mg/4 wk)	Avg gain Avg wt gain[b] Initial weight (% of control)	Survival (No./30)
Unautoclaved	10.2	25.1 b	46.2	29
Autoclaved	10.2	34.6 a	63.2	28
1% residual fat	10.0	27.4 b	51.2	28
4% residual fat	10.3	32.3 a	58.1	29
3% protein	10.1	26.2 b	48.0	28
10% protein	10.2	33.5 a	61.3	29
White-Yeast[c]	10.4	55.4	100.0	27

[a]For each treatment, the values followed by different letters are significantly different at P = 0.05.

[b]Average weight gain divided by initial weight as percent of control.

[c]An untreated control diet of ground wheat + Brewer's yeast was included in each replicate, but was not included in the analysis of variance.

Table 4. Average gains in weight and survival of larvae of *Tenebrio molitor* L. fed oilseed meal diets

Treatment of oilseed meal	Initial weight (mg)	Avg gain in weight[a] (mg/4 wk)	Avg gain Initial weight (% of control)	Survival (No./30)
Wheat-Yeast[b]	10.4	55.4	100.0	27
Soybean	10.5	39.2 a	69.0	28
Sunflower	10.0	37.5 a	69.5	28
Turnip rape	10.0	36.2 a b	68.0	29
Safflower	10.1	30.2 b c	56.4	28
Rape	10.3	26.4 c	47.8	30
Flax	10.1	9.7 d	18.1	29

[a]For each treatment, the values followed by different letters are significantly different at P = 0.05.

[b]An untreated control diet of ground wheat + Brewer's yeast was included in each replicate but was not included in the analysis of variance.

Table 5. Relationship of gains in fresh weight of larvae of *Tenebrio molitor* L. to gains in weight and to protein utilization of autoclaved oilseed meals by mice

Oilseed meal	*T. molitor* Weight gain in 4 wk (mg)	Mice gains in 2 wk Weight gain (g)	Energy gain (Kcal)	Protein utilization by mice P.E.R.	N.P.U.
Soybean	55.9	10.1	22.2	2.67	39.3
Sunflower	40.2	9.1	17.3	2.53	38.4
Turnip rape	35.1	7.5	14.6	2.41	39.4
Safflower	29.2	8.2	17.4	2.29	33.4
Rape	34.1	7.4	13.5	2.25	36.1
Flax	13.1	10.2	19.7	2.22	36.2
Correlation with avg gains in weight of *Tenebrio molitor* larvae on autoclaved oilseed meals					
r =		+0.03	+0.24	+0.90*	+0.58

*Significant at P = 0.05.

However, its importance is not limited to simply testing diets before feeding them to other animals. Because of the small quantities of material required (3 g of protein in 30 g of diet for 30 larvae) for testing, use of *T. molitor* could be of invaluable assistance in plant-breeding programs. Its value in this regard would likely be even greater than indicated, because even smaller amounts of protein than those indicated are likely to be required for testing.

This insect is also very sensitive to glucosinolates, which act as feeding deterrents for it (Dr. N. S. Church, Canada Agriculture Research Station, Saskatoon, personal communication). A specific assay might well be developed using *T. molitor*, therefore, to indicate the presence of such substances in rape seed or in other plant products.

Possibly this insect or others could be used for rapid, accurate quantitative assays of particular amino acids, vitamins, alkaloids or other chemical components of natural products. The only impediment arises from lack of knowledge regarding how these substances affect the physiology of the insect. Application of such determinations with insects to general physiology is also complicated by the lack of knowledge of how these substances affect the physiology of other animals.

Use of insects for determining nutritional information does not preclude experimentation with other animals. Such investigations are essential in defining the nutritional physiology of those animals. However, insects appear capable of giving rapid, accurate answers, especially where minute quantities of material are initially available for testing. They also require a minimum of time, space, and facilities as compared with other animals, for extensive and well-replicated investigations.

Acknowledgements.—This work is contribution No. 480, Canada Agriculture Research Station, Saskatoon, Saskatchewan, Canada.

REFERENCES

DAVIS, G. R. F. and SOSULSKI, F. W. (1972). Use of larvae of *Tenebrio molitor* L. to determine nutritional value of proteins in six defatted oilseed meals. *Arch. Internat. Physiol. Biochem.* **80**: *(in press).*

HOUSE, H. L. and GRAHAM, A. R. (1967) Nutritional pest control: the "self-protection" of foodstuffs against *Tribolium confusum* (Coleoptera: Tenebrionidae) often presumably through nutritional factors. *Can. Ent.* **99**:1082-1087.

LECLERCQ, J. (1965) Premiers essais d'utilisation des larves de *Tenebrio molitor* pour comparer la valeur nutritive des proteines. *Ann. Nutr. Alim.* **19**:47-58.

LECLERCQ, J. and DE BAST, D. (1965) Projet d'utilisation des larves de *Tenebrio molitor* pour comparer la valeur nutritive des proteines. *Ann. Nutr. Alim.* **19**:*19-25.*

LECLERCQ, J., DUYCKAERTS, C. and GASPAR, C. (1967) Utilisation des larves de *Tenebrio molitor* pour comparer la valeur nutritive des proteines de feuilles de vegetaux. *Ann. Nutr. Alim.* **21**:189-197.

LOSCHIAVO, S. R., McGINNIS, A. J. and METCALFE, D. R. (1969) Nutritive value of barley varieties assessed with the confused flour beetle. *Nature Lond.* **224**:288.

SARWAR, G. (1971) Nutritive value of oilseed meals and protein isolates. M. Sc. Thesis, University of Saskatchewan, Saskatoon.

THE GNOTOBIOTIC APPROACH TO NUTRITION: EXPERIENCE WITH GERMFREE RODENTS

Bernard S. Wostmann
Lobund Laboratory, Department of Microbiology,
University of Notre Dame
Notre Dame, Indiana 46556

INTRODUCTION

In his review of insect nutrition House (1965) points out that some of the apparent differences in nutritional requirements between insects species may be due to the action of intestinal microorganisms, and cites studies by Baines (1956), Pant and Fraenkel (1954), Clayton (1960) and Henry (1962) to support this view. He then states that "one should know the chemical composition of the diet and be able to vary the chemical composition precisely so that the effects on the insect can be related accordingly. Microorganisms must be eliminated to avoid the possible intervention of symbiotic forms." In the light of the above House concludes that "it is reassuring that much of our understanding of insect nutrition is founded on work done axenically on chemically defined diets" (1965). Apparently mammalian nutritionists and entomologists not only work with animal systems of great nutritional similarity, as Sang (1956) and others have repeatedly indicated, but also face similar difficulties and envision solutions along similar lines (Rodriguez, 1966; Wostmann *et al.*, 1967).

The work with germfree mammals was the direct result of a statement by Pasteur (1885) who voiced his interest in raising experimental animals from birth on "pure nutritive substances", and in complete absence of interfering microbes. Already in 1895 this lead to the production of the first germfree guinea pigs by Nuttal and Thierfelder (1895, 1896). Nutritional developments had to wait, however, until the late 1950s, after the pioneering work of Rose (1957) and others in the field of human and animal nutrition. At the Lobund Laboratory of the University of Notre Dame, Pasteur's original concept eventually lead to the formulation of chemically defined, low molecular weight, water soluble diets for germfree rodents that are sterilized by filtration. These diets have since maintained germfree rats from birth into old age, and colonies of germfree Swiss Webster and C3H mice for 5 or more generations (Pleasants, 1970, 1972).

The direct incentive to our work on the development of defined, low molecular weight diets originated from the fact the caesarian derived germfree baby rat experiences difficulties in digesting any handfed formula containing more than 5 to 6% protein (Wostmann, 1959). At that time we thought that a water soluble diet based on glucose and free amino acids would provide the answer to our predicament. However, we recognized, beyond our direct problem, the enormous potential of the combination of germfree techniques with the feeding of low molecular weight, chemically defined diets. Such an approach would not only make total control possible in nutritional studies (Pleasants *et al.*, 1970), but would also eliminate the diet as a major source of uncontrolled antigenicity. As a result immunological studies could be carried out with full control of antigenic and synergistic stimulation (Wostmann *et al.*, 1971).

In all of our nutritional studies with gnotobiotes we have found it necessary to ask 2 basic questions:

1. What are the basic metabolic requirements of the organism in the absence of an associated microflora?
2. How are these requirements modified, directly or indirectly, by various possible microfloras?

We consider both questions of equal importance because eventually our research should not be aimed at solving problems in a highly quixotic ecosystem, but should yield information necessary to approach situations as they exist in the world in which most of us must live.

I profess to almost total ignorance about the role that extracellular microorganisms might play in function and metabolism of insects and mites. I am, however, very much aware of the vast influence associated microfloras may exert on the metabolic performance of the rodent. Many of the studies underlying this awareness have been carried out by the Lobund Laboratory of the University of Notre Dame, and at the Department of Pharmacology of the University of Kentucky Medical Center. The investigations at these institutions entail both the purely basic approach aimed at the understanding of the influence of associated microfloras (or the absence thereof) on function and metabolism, and more applied studies aimed at specific biomedical problems such as heart disease, malnutrition, immune potential as it relates to infectious disease and to cancer, etc. The examples cited hereafter will illustrate how in rodents nutritional requirements and the underlying intermediary metabolism may deviate from conventional patterns in the absence of an intestinal microflora. In all instances comparisons were made between gnotobiotic and conventional animals which, by virtue of especially designed husbandry procedures, were genetically closely related.

MICROFLORA EFFECTS ON CHOLESTEROL AND BILE ACID METABOLISM (IN RELATION TO HEART DISEASE)

A number of years ago we were able to show that germfree rats catabolize cholesterol much more slowly than their conventional counterparts (Wostmann *et al.*, 1966). This is a result of reduced bile acid elimination in the absence of a deconjugating and dehydroxylating microflora (Gustafsson *et al.*, 1957; Kellogg and Wostmann, 1969a). Consequently a build-up of bile acids occurs in the enterohepatic circulation which in turn appears to shift hepatic production away from taurocholic acid towards tauro-β-puricholic acid. This is illustrated in the concentration and composition of biliary bile acids which show not only a much higher total concentration in germfree rats, but especially an increase in-β-muricholic acid accounting for approximately half of the total bile acid (Fig. 1). The bile acid pattern and concentration in the small intestine of the germfree rat follow this change in similar fashion (Table 1). The result is that the germfree rat will absorb up to

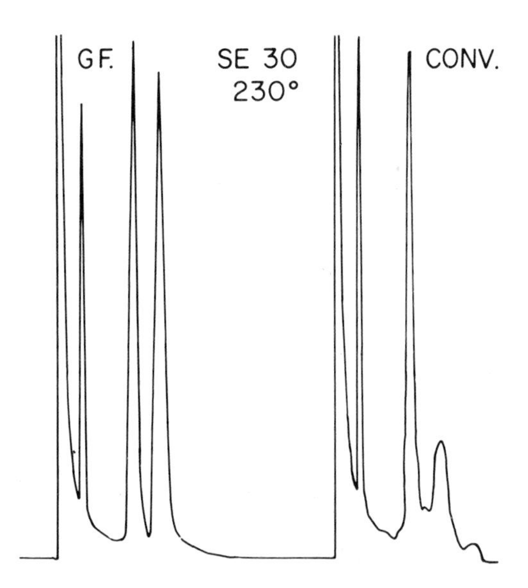

Figure 1. Gas-liquid chromatographic pattern of biliary bile acids of germfree and conventional male Wistar rats indicating equal amounts of cholic acid and β-muricholic acid in germfree bile, and much less total bile acid consisting mostly of cholic acid in conventional bile (see text).

50% more cholesterol from its diet than its conventional counterpart. Thus in the absence of a normal microflora the germfree rat will absorb more cholesterol, and excrete less of its catabolic products. This causes a build-up of cholesterol pools which in the rat becomes especially obvious in the liver (Table 1; also Kellogg and Wostmann, 1969b).

Table 1. Main bile acid components and total bile acid content of germfree and conventional rat bile and small intestinal content in relation to the absorption of dietary cholesterol and the cholesterol content of the liver (Mean ± S.E. of 6 or more observations.)

	Germfree	Conventional
Bile Acids		
Bile (mg/ml)		
Cholic acid	5.00 ± 0.15	3.32 ± 0.39
β-Muricholic acid	4.98 ± 0.37	0.63 ± 0.18
Total	10.11 ± 0.49	4.16 ± 0.50
Intestine (mg total)		
Cholic acid	22.5 ± 2.8	16.6 ± 1.8
β-Muricholic acid	23.8 ± 3.6	4.5 ± 1.6
Total	47.9 ± 5.6	21.8 ± 2.3
Absorption dietary cholesterol[1] (percent)	56.6 ± 3.5	44.0 ± 1.0
Liver cholesterol (mg total/100 g body wt)	23.3 ± 1.0	13.0 ± 0.5

[1] 0.5% cholesterol added to a semi-synthetic diet based on rice starch, casein, vitamins and minerals.

MICROFLORA EFFECTS ON AMINO ACID REQUIREMENTS (AS RELATED TO MARGINAL PROTEIN NUTRITION)

It has been known for a long time that corn is a poor source of protein for the mammal because of its low content of tryptophane and lysine (Longenecker, 1963). In a classical experiment, Dubos and Schaedler (1960) have shown that ex-germfree mice populated with a controlled non-pathogenic microflora could survive and even grow on corn, while the conventional mouse harboring a "normal" flora would lose weight and eventually die. Even when the diet contained protein of more adequate quality but in marginal quantity (e.g. 15% casein), growth of the "clean" mice was always much better than of the conventional "dirty" mice (Table 2). In a refinement of these studies, Stoewsand *et al.* (1968) fed defined diets containing increasing amounts of lysine to germfree mice, ex-germfree mice associated with a defined microflora (Schaedler *et al.*, 1965), and conventional mice. Both the germfree and the selectively associated mice grew maximally at 2/3 of the lysine level needed by the conventional mice for optimal growth.

Thus in mammalian nutrition the intestinal microflora may be a major influence in determining amino acid requirements. Since at least 2/3 of the world population subsists on diets marginal in protein, this effect should be of prime importance in the consideration of human nutrition.

Table 2. Three week cumulative weight gain of 28 days old Swiss mice freed from common mouse pathogens.[a]

	NCS mice[b]	SS mice[c]
Diet (*ad lib*)	g per mouse	
Mouse pellets	+ 8.2	+ 7.8
Corn	+ 1.0	– 3.4
15% gluten[d]	+ 7.9	+ 1.5
15% casein[d]	+ 9.0	+ 4.8

[a]Compiled from data taken from Dubos and Schaedler (1960).
[b]Mice from colony originally derived by caesarian section, and subsequently protected from contact with common mouse pathogens.
[c]Standard Swiss colony.
[d]Semi-synthetic diet containing 15% gluten or 15% casein as sole source of protein.

REQUIREMENTS FOR UNSATURATED FATTY ACIDS

Eyssen and coworkers have demonstrated that the intestinal microflora of the rat has the capacity to hydrogenate 18:1 and 18:2 fatty acids (Evrard, *et al.*, 1964; De Somer *et al.*, 1972). As a result the feces of the conventional rat fed a corn oil containing diet may contain less than 10% of the linoleic acid (18:2) found in the feces of a comparable germfree rat. These investigators found that a synergism between at least 2 microbial species usually occurring in the gut was needed to reduce linoleic (18:2) to stearic acid (18:0). At present it is not clear, however, at which level of the intestinal tract this loss of essential fatty acids takes place. Further studies will be required to establish whether this phenomenon could influence the requirements for essential fatty acids in a significant way.

INFLUENCE OF THE MICROFLORA ON THE METABOLIC RATE OF THE RAT

Far more basic than the above phenomena were a series of findings which were triggered by the observation that in germfree rats we consistently found smaller hearts than in conventional rats (Gordon *et al.*, 1966). Further study then revealed that both cardiac output and blood volume were approximately 3/4 of normal value (Gordon *et al.*, 1963). This aroused suspicion about the metabolic rate of the germfree rat, especially since earlier data had indicated a low thiamine pyrophosphate content in germfree rat liver, suggestive of a reduced level of high-energy phosphate production (Wostmann *et al.*, 1963). Subsequently 3 laboratories reported independently that the metabolic rate of the adult germfree rat is reduced by approximately 25% (Desplaces *et al.*, 1963; Wostmann *et al.*, 1968; Levenson, 1971), although food intake and the resulting growth were found to be quite comparable (Combe, 1972).

Since this appeared to be a very fundamental deviation from the normal pattern, we subsequently investigated its possible relationship with that other conspicuous abnormality of the germfree rodent, the enlarged cecum. In the absence of an intestinal microflora the cecum of the germfree rat starts to enlarge rather abruptly during the suckling period, and eventually reaches a size 4 to 7 times that found in the conventional control (Wostmann and Bruckner-Kardoss, 1959). To relieve the animal of this abnormal abdominal mass, the enlarged cecum can be removed surgically, leaving the passage from small to large intestine intact. This operation results in an apparently quite normal and healthy germfree rat (Bruckner-Kardoss and Wostmann, 1967). However, this rat now shows normal cardiac output values and an almost normal metabolic rate (Wostmann *et al.*, 1968). For reasons still under investigation, the cecal enlargement typical for the intact germfree rodent produces a major shift in metabolic patterns, resulting from, or maybe resulting in a strongly reduced cardiovascular capacity.

Some of the changes in intermediary metabolism of the rat that relate to the germfree state are illustrated in Table 3. A comprehensive report will be published shortly (Reddy *et al.*, 1972). They suggest that as a result of reduced tricarbonic acid cycle activity additional citrate is shunted from the mitochondrion to the cytoplasm. Both the citrate lyase and the fatty acid synthetase complex activity in the cytoplasm appear to be increased accordingly. However, germfree rats show no increase in fat deposit (Levenson, 1971) indicative of an enhanced conversion of citrate to fatty acids, presumably because of a paucity of NADPH needed for chain elongation. Instead citric acid appears in the urine in abnormally large amounts (Gustafsson and Norman, 1962). Thus in the absence of an intestinal microflora the germfree rat, with a food intake comparable to that of its conventional counterpart, appears to utilize food calories less effectively for the production of chemical energy. Using less O_2 than normal, it excretes excess utilizable calories like those represented by citric acid in the urine.

CAUSES AND POSSIBLE CONSEQUENCES OF CECAL ENLARGEMENT IN RODENTS

The cecal enlargement which appears to be at the basis of above syndrome results in first instance from the accumulation of high molecular

Table 3. Metabolic data in adult male germfree and conventional Wistar rats[a] (means ± SE).

	GF	Conv	GF/Conv
Cardiac output[b] (ml/min/kg)	137 ± 7	203 ± 11	(.80)
O_2 consumption[b] (ml/min/kg)	11.5 ± 0.4	15.2 ± 0.4	(.76)
Liver			
Regional arterial blood flow[c] (ml/min/kg)	0.27 ± 0.04	0.53 ± 0.04	(.51)
Thiamine[c] (μg/g)	6.8 ± 0.1	9.1 ± 0.3	(.75)
Glucose 6-P dehydrogenase[d] (units/g protein)	17.8 ± 0.7	29.6 ± 2.6	(.60)
6-P gluconate dehydrogenase[d] (units/g protein)	36.4 ± 1.4	45.3 ± 2.7	(.80)
Succinic dehydrogenase[d] (units/mg prot.)	179 ± 2	264 ± 10	(.68)
Malic enzyme[d] (units/g prot.)	19.4 ± 1.1	17.0 ± 1.7	(1.14)
Cytochrome oxidase[d] (units/mg prot.)	758 ± 37	628 ± 35	(1.21)
ATP-citrate lyase[d] (units/g prot.)	12.1 ± 1.0	7.2 ± 0.5	(1.68)
FA-synthetase complex[d] (units/g prot.)	21.4 ± 1.0	14.4 ± 1.7	(1.49)

[a]Age: 3-4 months, 6 or more animals per observation.
[b]Wostmann *et al.* (1968).
[c]Wostmann *et al.* (1963).
[d]Reddy *et al.* (1972).

weight mucopolysaccharides in the cecum of the germfree rodent (Gordon and Wostmann, 1972). In the absence of degrading enzymes of microbial origin these negative polyions will attract water into the lumen of the cecum, while at the same time sequestering Na^+ ions which normally would transport water out of the cecum via solute coupled water transport. The cecum of the germfree rodent thus enlarges to a size determined by the accumulation of water attracting negatively charged macromolecules, the ensuing limitation set upon water efflux, and a reduced muscular tone resulting from the absence of stimulation by certain unspecified microorganisms (Gordon *et al.*, 1966). In this enlarged cecum, in the absence of microbial degradation, high concentrations of the various proteolytic enzymes tend to persist

(Reddy *et al.*, 1969). Thus an ideal environment exists for the digestion of those protein materials that reach the lower gut, and which to a large extent consist of desquamated intestinal tissue.

Gordon and coworkers (Gordon, 1967; Gordon and Kokas, 1968) have reported that the contents of the germfree rodent cecum contain substances that affect smooth muscle in a number of preparations. Cecal contents from comparable conventional animals displayed such activity only on a greatly reduced scale. They regard a refractoriness to catecholamines as a major sequela to the germfree state that relates to the presence of these substance(s) (Gordon and Kokas, 1968; Baez and Gordon, 1971). In further studies this group reported the chromatographic isolation from germfree cecal contents of α-pigment, a presumably ferritin related material which causes refractoriness to catecholamines in both the dog villus preparation. (Kokas *et al.*, 1972) and the mesenteric microvessel preparation (Baez *et al.*, 1972).

Further purification of α-pigment via Sephadex G 50 resulted in an Fe containing peptide of a molecular weight of approximately 4800 which, according to our latest tests, seems to be the active principle as far as the norepinephrine refractoriness is concerned. At this time we consider α-pigment to be a degradation product of desquamated ferritin. Incubation of ferritin *in vitro* with trypsin and chymotrypsin and subsequent gel filtration via Sephadex G 50 have produced distinct fractions of similar molecular weight, of which thus far one has proved to be epinephrine antagonistic on the dog villus preparation (Wostmann *et al.*, 1972). At present the physiological role of α-pigment is studied in relation to its potential effect on the cardiovascular system and to possibly related functional and/or metabolic characteristic of the germfree rodent mentioned earlier.

CONCLUSION

Studies in the field of mammalian nutrition appear to unfold along similar lines as in the field of entomology. In order to gain utmost control of nutritional and immunological parameters Dr. Pleasants, Dr. Reddy and I have developed defined, low molecular weight diet from which, by additional Diaflo filtration, we have eliminated even traces of any material with a molecular weight over 10,000 (Pleasants, 1970, 1972).

This diet has shown quite dramatically the strong antigenicity of the usual experimental diets. Only this most extensive nutritional control made it possible to produce an adult germfree mouse in which perceptible amounts of immune globulins were absent (Wostmann *et al.*, 1971). Consequently we now have, for the first time, a system in which we can control not only nutritional, but also antigenic and non-specific synergistic input.

On the other hand, these and related studies have demonstrated the major influence of the intestinal microflora in determining the homeostasis of the animal. The germfree rodent has often been termed an almost different species; and even at this moment we are only vaguely aware of the microbial associates necessary for "normal life" (Schaedler *et al.*, 1965, Levenson, 1971). Insect nutrition and the metabolism of insects "should be regarded as a key both to the successful control on injurious insects, and to the progress of the industries dependent upon the products of useful insects" (Uvarov, 1928). For reasons stated above it therefore may not be *that*, "reassuring that much of our understanding of insect nutrition is founded on work done axenically on chemically defined diets" (House, 1965) and, strictly as an uninitiated outsider, I would like to make a strong plea for further consideration of the effects of the intestinal microflora on insect nutrition and metabolism.

Acknowledgements.—The studies at the University of Notre Dame were supported by U.S.P.H.S. HD 00855 and HE 08351, the Indiana Heart Foundation and the Fannie E. Rippel Foundation.

REFERENCES

BAEZ, S. and GORDON, H. A. (1971). Tone and reactivity of vascular smooth muscle in germfree rat mesentery. *J. Exp. Med.* **134**:846.

BAEZ, S., BRUCKNER, G. and GORDON, H. A. Responsiveness of jejunal-ileal mesentery microvessels in unoperated and cecectomized germfree rats to some smooth muscle antagonist. Proc. IV Internat. Symposium on Germfree Life Research, New Orleans, 1972 (Ed. J. B. Heneghan), Academic Press, New York. In press.

BAINES, S. (1956). The role of symbiotic bacteria in the nutrition of *Rhodnius prolixus*. *J. Exptl. Biol.* **33**:533.

BRUCKNER-KARDOSS-KARDOSS, E. and WOSTMANN, B. S. (1967). Cecectomy of germfree rats. *Lab. Animal Care.* **17**:542.

CLAYTON, R. B. (1960). The role of intestinal symbionts in the sterol metabolism of *Blattella germanica*. *J. Biol. Chem.* **235**:3421-3425.

COMBE, E. Nutritional studies using germfree animals. Proc. IV Internat. Symposium on Germfree Life Research, New Orleans, 1972 (Ed. J. B. Heneghan), Academic Press, New York. In press.

DE SOMER, G., DE PAUW and EYSSEN, H. Biohydrogenation of long-chain fatty acids of intestinal microorganisms. Proc. IV Internat. Symposium on Germfree Life Research, New Orleans, 1972 (Ed. J. B. Heneghan), Academic Press, New York. In Press.

DESPLACES, A., ZAGURY, D. and SAQUET, E. (1963). Etude de la fonction thyroidienne du rat prive de bacteries. *C. R. Acad. Sci. Paris.* **257**:756.

DUBOS, R. J. and SCHAEDLER, R. W. (1960). The effect of intestinal flora on the growth rate of mice and on their susceptibility to experimental infection. *J. Exp. Med.* **111**:407.

EVRARD, E., HOLT, P. P., EYSSEN, H., CHARLIER, H. and SACQUET, E. (1964). Faecal lipids in germfree and conventional rats. *Brit. J. Exp. Pathol.* **45**:409-414.

GORDON, H. A., WOSTMANN, B. S. and BRUCKNER-KARDOSS, E. (1963). Effects of microbial flora on cardiac output and other elements of blood circulation. *Proc. Soc. Exp. Biol. Med.* **114**:301.

GORDON, H. A., BRUCKNER-KARDOSS, E., STALEY, T. E., WAGNER, M. and WOSTMANN, B. S. (1966). Characteristics of the germfree rat. *Acta. Anat.* **64**:367.

GORDON, H. A. (1967). A substance acting on smooth muscle in intestinal contents of germfree animals. *Ann. N. Y. Acad. Sci.* **147**:83-106.

GORDON, H. A. and KOKAS, E. (1968). A bioactive pigment ("Alpha pigment") in cecal contents of germfree animals. *Biochem. Pharmac.* **17**:2333.

GORDON, H. A. and WOSTMANN, B. S. Chronic mild diarrhea in germfree rodents: a model portraying host-flora synergism. Proc. IV Internat. Symposium on Germfree Life Research, New Orleans, 1972 (Ed. J. B. Heneghan), Academic Press, New York. In press.

GUSTAFSSON, B. E., BERGSTÖM, S., LINDSTEDT, S. and NORMAN, A. (1957). Turnover and nature of fecal bile acids in germfree and infected rats fed cholic acid-24-C^{14}. *Proc. Soc. Exp. Biol. Med.* **94**:467-471.

GUSTAFSSON, B. E. and NORMAN, A. (1962). Urinary calculi in germfree rats. *J. Exp. Med.* **116**:273.

HENRY, S. M. (1962). The significance of microorganisms in the nutrition of insects: Trans. N. Y. *Acad. Sci.* **24**:676-683.

HOUSE, H. L. (1965). In: the Physiology of Insecta, Vol. 2. Academic Press, New York. pp. 769-813.

KELLOGG, T. F. and WOSTMANN, B. S. (1969a). Fecal neutral steroids and bile acids from germfree rats. *J. Lipid Res.* **10**:495-503.

KELLOGG, T. F. and WOSTMANN, B. S. (1969b). The response of germfree rats to dietary cholesterol. *Av. Exp. Biol. Med.* **3**:293-305.

KOKAS, E., WOSTMANN, B. S. and GORDON, H. A. Effects of germfree rodent cecal contents on spontaneous villus movement. Proc. IV Internat. Symposium on Germfree Life Research, New Orleans, 1972 (Ed. J. B. Heneghan), Academic Press, New York. In press.

LEVENSON, S. M. (1971). Some effects of *E. coli* and other microbes in mammalian metabolism and nutrition. *Ann. N. Y. Acad. Sci.* **176**:273-283.

LONGENECKER, J. B. (1963). Utilization of dietary protei. In: Newer Methods of Nutritional Biochemistry (Ed. A. A. Albanese, Academic Press, New York). pp. 113-141.

NUTTAL, G. H. F. and THIERFELDER, H. (1895-1896). Thierisches Leben ohne Bakterien im Verdauungs-kanal. *Hoppe Seyler's Z. Physiol. Chem.* **21**:109-121.

NUTTAL, G. H. F. and THIERFELDER, H. (1896-1897). Thierisches Leben ohne Bakterien im Verdauungs-kanal II. Mitt. *Hoppe Seyler's Z. Physiol. Chem.* **22**:62-73.

PANT, N. C. and FRAENKEL, G. (1954). Studies on the symbiotic yeasts of two insect species *Lasioderma sericorne* F. and *Stegobium paniceum* L. *Biol. Bull.* **107**: 420-452.

PASTEUR, L. (1885). Observations relatives a la Note precedente de M. Declaux. *C. R. Acad. Sci. Paris* **100**:68.

PLEASANTS, J. R., REDDY, B. S. and WOSTMANN, B. S. (1970). Qualitative adequacy of a chemically defined liquid diet for reproducing germfree mice. *J. Nutr.* **100**:498-508.

PLEASANTS, J. R., WOSTMANN, B. S. and REDDY, B. S. Improved lactation in germfree mice following changes in amino acid and fat components of a chemically defined diet. Proc. IV Internat. Symposium on Germfree Life Research, New

Orleans, 1972 (Ed. J. B. Heneghan), Academic Press, New York. In press.
REDDY, B. S., PLEASANTS, J. R. and WOSTMANN, B. S. (1969). Pancreatic enzymes in germfree and conventional rats fed chemically defined, water-soluble diet free from natural substrates. *J. Nutr.* **97**:327-334.
REDDY, B. S., PLEASANTS, J. R. and WOSTMANN, B. S. (1972). Submitted for publication.
RODRIGUEZ, J. G. (1966). Axenic arthropoda: Current status of research and future possibilities. *Am. N. Y. Acad. Sci.* **139**:53-64.
ROSE, W. C. (1957). The amino acid requirements of adult man. *Nutrition Abstr. & Rev.* **27**:631-647.
SANG, J. H. (1956). The quantitative nutritional requirement of *Drosophila melanogaster*. *J. Exptl. Biol.* **33**:45.
SCHAEDLER, R. W., DUBOS, R. J. and COSTELLO, R. (1965). Association of germfree mice with bacteria isolated from normal mice. *J. Exp. Med.* **122**:77-82.
STOEWSAND, G. S., DYMSZA, H. A., AMENT, D. and TREXLER, P. C. (1968). Lysine requirement of the growing gnotobiotic mouse. *Life Sciences* **7**:689-694.
UVAROV, B. P. (1928). Insect nutrition and metabolism. *Trans. Ent. Sc. of London.* **76**:255-343.
WOSTMANN, B. S. (1959). Nutrition of the germfree mammal. *Ann. N. Y. Acad. Sci.* **78**:175-182.
WOSTMANN, B. S. and BRUCKNER-KARDOSS, E. (1959). Development of cecal distention in germfree baby rats. *Am. J. Physiology* **197**:1345-1346.
WOSTMANN, B. S., KNIGHT, P. L., KEELEY, L. L. and KAN, D. F. (1963). Metabolism and function of thiamine and naphthoquinones in germfree and conventional rats. *Fed. Proc.* **22**:120-124.
WOSTMANN, B. S., WIECH, N. L. and KUNG, E. (1966). Catabolism and elimination of cholesterol in germfree rats. *J. Lipid Res.* **7**:77-82.
WOSTMANN, B. S., PLEASANTS, J. R. and REDDY, B. S. (1967). Water-soluble, non-antigenic diets. *Proc. Int. Congress of Lab. Animals,* Dublin, Ireland. pp. 187-195.
WOSTMANN, B. S., BRUCKNER-KARDOSS, E. and KNIGHT, P. L. (1968). Cecal enlargement, cardiac output and oxygen consumption in germfree rats. *Proc. Soc. Exp. Biol. Med.* **128**:137-141.
WOSTMANN, B. S., PLEASANTS, J. R. and BEALMEAR, P. (1971). Dietary stimulation of immune mechanisms. *Fed. Proc.* **30**:1779-1784.
WOSTMANN, B. S., REDDY, B. S., BRUCKNER-KARDOSS, E., GORDON, H. A. and SINGH, B. Causes and possible consequences of cecal enlargement in germfree rats. Proc. IV Internat. Symposium on Germfree Life Research, New Orleans, 1972 (Ed. J. B. Heneghan), Academic Press, New York. In press.

FOOD UTILIZATION

INTRODUCTION
by

G. P. Waldbauer, Section Editor
Department of Entomology
University of Illinois
Urbana, Illinois 61801

Efforts to elucidate the qualitative nutritional requirements of insects have been remarkably successful, and for some time it has been apparent that insects are relatively uniform with respect to these requirements. Many researchers have recognized that there is not likely to be a fundamental change in our understanding of qualitative nutritional requirements and have, therefore, suggested a shift in emphasis to quantitative studies such as the determination of absolute requirements and the efficiency of utilization.

Quantitative studies have appeared, but a widespread effort has yet to emerge. This is at least partly due to technical difficulties in food utilization measurement. Quantitative studies inevitably come down to the measurement of intake and assimilation, which in turn necessitates food weighing and the quantitative collection and weighing of feces—tedious, time consuming activities fraught with opportunities for error. We will probably have to live with this tedious business, but the use of chemical markers may offer hope for avoiding some of it occasionally. The chromic acid technique, adapted for use with insects by McGinnis and Kasting (1964a and b), is practical under some circumstances but has not often been used. It remains to be seen whether or not the uric acid method for determining the proportions of food and feces in an inseparable mixture (Bhattacharya and Waldbauer, 1969, 1970) will be widely useful.

Terms used to express nutritional efficiencies (see Gordon and McGinnis and Kasting in this volume and Waldbauer, 1968) need to be standardized, but we must recognize that the application to insects of certain terms used in mammalian nutrition would involve serious ambiguities. The difficulty arises because the feces of mammals consist largely of unassimilated food while insects pass a practically inseparable mixture of unassimilated food and urine. Thus, calculations involving the dry weight, caloric value or other measures of the feces yield values which represent different biological and biochemical parameters for insects and mammals. For example, the difference between the dry weight of the food ingested and the dry weight of the feces is a close approximation of the *food assimilated* by mammals, but with insects represents the *food assimilated less the waste products resulting from the metabolism of that food.* By assimilated food is understood the total quantity available to be incorporated as body substance or to be metabolized for energy. It is obvious that indices of digestibility (A.D.) and of the conversion of digested (assimilated) food to body substance (E.C.D.) (Waldbauer, 1968) by insects measure different rates than do indices similarly calculated for mammals and that the two sets of indices are only approximately comparable. Prudence is advisable in the matter of terminology since ambiguity will eventually lead to misunderstanding.

The publication of Stepien and Rodriguez's paper in this volume affords an opportunity to comment on some areas of overlap between the fields of insect nutrition and ecological energetics. Insect nutritionists usually express their data in terms of dry weights while ecologists, focusing their attention on energy flow in communities, express their data in terms of calories. Dry weights are easily converted to calories if the caloric value of the substance in question is known; however, dry weight and caloric data are not comparable since the caloric values of food, feces and insect bodies are not likely to be the same. Some of the indices used in ecological energetics are the precise caloric equivalents of dry weight indices commonly used by insect nutritionists (Waldbauer, 1968). Stepien and Rodriguez's U^{-1} is equivalent to A.D., K_1 to E.C.I. and K_2 to E.C.D. Ecologists usually calculate the caloric value of R (respiration) from the oxygen consumption of the organism in question. R is considered to include the energy expended for metabolism and activity and also the energy lost in the urine (Kozlovsky, 1968). It does not seem to be generally recognized that R can also be arrived at gravimetrically. R is precisely equivalent to the caloric content of the ingested food less the caloric contents of the feces, the gain in mass of the organism and any products produced by the organism. Gordon's discussion makes this point. His O is the dry weight equivalent to R. Note that an estimate of R based on oxygen consumption is subject to sampling error unless oxygen consumption is measured during the entire period of the experiment. R determined gravimetrically is not subject to sampling error since it is calculated from the total food intake, fecal production and gain in mass during the experiment. It would be interesting to compare R's arrived at by the two methods since they serve as checks on each other.

REFERENCES

BHATTACHARYA, A. K. and WALDBAUER, G. P. (1969). Faecal uric acid as an indicator in the determination of food utilization. *J. Insect Physiol.* 15:1129-1135.

BHATTACHARYA, A. K. and WALDBAUER, G. P. (1970). Use of the faecal uric acid method in measuring the utilization of food by *Tribolium confusum. J. Insect*

Physiol. **16**:1983-1990.

KOZLOVSKY, D. G. (1968). A critical evaluation of the trophic level concept. I. Ecological efficiencies. *Ecology* **49**:48-60.

MCGINNIS, A. J. and KASTING, R. (1964a). Chromic oxide indicator method for measuring food utilization in a plant-feeding insect. *Science*, N.Y. **144**:1464-1465.

MCGINNIS, A. J. and KASTING, R. (1964b). Comparison of gravimetric and chromic oxide methods for measuring percentage utilization and consumption of food by phytophagous insects. *J. Insect Physiol.* **10**:989-995.

WALDBAUER, G. P. (1968). The consumption and utilization of food by insects. *Adv. Insect Physiol.* **5**:229-288.

QUANTITATIVE NUTRITION AND EVALUATION OF PROTEIN IN FOODS OF PHYTOPHAGOUS INSECTS

A. J. McGinnis
Canada Agriculture Research Station
Box 185, Vineland Station, Ontario, Canada

R. Kasting
Canada Agriculture Research Station
Lethbridge, Alberta, Canada

The French chemist, Lavoisier, regarded by many as the father of the science of nutrition was the first to use the balance and the thermometer in nutritional studies. From his work he concluded that, "La vie est une fonction chemique" (Maynard, 1937). Indeed since his time chemistry has been an important discipline in studies of nutrition and tremendous advances have been made in our knowledge of nutritional requirements for both man and his domestic animals.

When we consider deficiency diseases such as scurvy, beri-beri, and pellagra, for example, it is clear that the science of nutrition has contributed much to the health and welfare of mankind. Furthermore, through nutritional research with domestic animals we can now produce meat and other animal products more efficiently than ever before. The broiler industry provides an outstanding example of what nutritional work has contributed to efficiency of production. Twenty years ago 4.5 pounds of feed were needed to produce one pound of gain; today it is common to obtain one pound of gain from 2.3 pounds of feed and lower ratios have been achieved (Hodgson, 1972). Clearly man has also benefited from nutritional research where the objective has been economic production.

What are the objectives of nutritional studies on insects? Of course there is the academic interest relative to general nutrition and metabolic processes of insects and the comparative aspects with other organisms. There are three further reasons of more immediate practical significance. First, such knowledge can ultimately result in increased productivity of insects such as bees and silkworms. Second, nutritional knowledge is essential for economical mass production of parasites and predators and those pest species that can be controlled by autocidal techniques. Third, detailed knowledge of the nutritional needs of pest species may be used in devising new control procedures. This latter possibility has been the major objective of our work with cereal insects over the years (Kasting & McGinnis, 1963; McGinnis & Kasting, 1967).

Insect nutrition is not a new field of study as evidenced by the bibliography compiled by Uvarov (1928). The early work was concerned primarily with insect rearing; the relationship between nutritional adequacy and specific chemical components came later. During the first half of the twentieth century, however, it was experimentally demonstrated that certain vitamins and some of the amino acids were essential in the diets of higher animals. The concept of dietary essentiality of particular chemicals resulted from work primarily with small laboratory animals: rats, mice, and guinea pigs, and microorganisms. From these studies, the insect nutritionist knew that certain components were probably essential in insect diets. Moreover, he had the advantage that supplies of these chemicals were available commercially and because of the small size of insects, it was economically feasible to prepare chemically defined diets and test their nutritional adequacy. The simplicity of the deletion procedure and the availability of suitable nutrient chemicals may well have led many insect nutritionists away from the quantitative approach to insect nutrition.

Early studies with farm animals were quantitative in nature. Nutritionists measured food intake, percentage digestibility, nitrogen balance, weight gain, milk production, and so forth. Such studies emphasized differences in efficiency of rations for production of the animal product of concern. With farm livestock the importance of quality of ration was recognized but quantitative factors were of major concern.

Quantitative nutritional data are limited for insects because of technical difficulties in measuring quantities of food consumed and the quantities of excreta produced. Notable quantitative studies were conducted with the silkworm by Peligot, (1867) about 100 years ago, and Hiratsuka (1920) about 50 years ago. The number of quantitative studies has increased recently but because such studies are time consuming and subject to error they seem not to be favored. As more suitable procedures for obtaining quantitative data become available (McGinnis & Kasting, 1964; Bhattacharya & Waldbauer, 1969) it is certain that more quantitative results will be produced. With small insects, those with piercing and sucking mouth parts, and those that frequent an aquatic habitat, however, major technical difficulties remain to be overcome. Before nutritional science can exert its full effect on practical problems, whether they be insect rearing or manipulating dietary components for insect control, more quantitative data must be available. Moreover, to maximize our returns on such quantitative studies it is imperative that, whenever possible, they be related to the concepts, nomenclature, and techniques established with large animals. Waldbauer (1968) recognized the confusion in quantitative nutritional terms that exists in the entomological literature and attempted to simplify and unify the terminology. In further attempts to unify terms and concepts the excellent summaries on nomenclature with large animals provided by Harris (1966) and Harris and co-workers (1968) and Crampton and Harris, (1969) should be consulted. These reports offer an excellent basis for comparing large animals with insects and, in addition, related quantitative nutrition to energy factors, a relationship that to date has been little considered for insects.

Despite obvious deficiencies we should not be unduly despondent about the current state of the art with insects. It was only 30 years ago that Mitchell (1942) pleaded for more detailed quantitative information for large animals. Proximate analyses for feedstuffs were available but the relationships between these values and the productive capacity of the feedstuff for any particular species of farm animal were not known. Quantitative experimental values that relate composition and energy content of a feedstuff to its productive capacity for different animal species are now available (Harris, 1966). In addition, the energy flow within the animal system relative to food intake is becoming clearer. This quantitative approach to nutrition has made possible the formulation of rations that yield the narrow feed:weight-gain ratios mentioned earlier. Moreover, feed manufacturers today consider all types of feedstuffs as merely alternate sources of nutrients needed in a ration. A computer facility coupled with knowledge of prices per pound and composition of feedstuffs permits formulation of the least expensive standard ration at that point in time. Perhaps, in the years ahead, this technique will be used in formulating diets upon which insects will be grown. There is also the possibility that the systematic tabulation of data on insect requirements and composition of their host plants will reveal interrelationships not presently apparent that can be exploited in control procedures.

Insect nutritionists need not despair because all the answers for the various species of insect are not yet available. One needs only to remember that legions of scientists in numerous laboratories have worked for decades on the nutrition of the white rat, and even there the story is still incomplete. In contrast, an insect nutritionist may be working independently with one, two, or more species of insects. We do have the advantage, however, that work with higher animals is being consolidated which should point the way toward effective accumulation and interpretation of nutritional information with insects.

We feel that much can be gained in insect nutrition by following the concepts and terminology commonly used in large animal studies. With this thought in mind, therefore, we will consider some of the terminology and techniques in protein nutrition that may be applicable to insects. Again we appeal for the use of consistent and well defined terms in all reports on evaluation of proteins and protein-feeds for insects. The terms and concepts used in studies on domestic animals and humans have been thoroughly reviewed (Harris, 1966; NAS-NRC, 1963) and are well standardized. In so far as possible this glossary of terms (NAS-NRC, 1963) should be applied to any work with insects as rigorous use of these terms should greatly improve communications among insect nutritionists, whose backgrounds are so diverse. Published reports, including some of our own, indicate the lack of consistency in terminology between laboratories and even within them

(Kasting & McGinnis, 1961; McGinnis & Kasting, 1966). More serious, possibly because of a lack of awareness, new terminologies are developing (Waldbauer, 1968) that do not consider the earlier standardization done with higher animals.

An extensive literature has been developed over the last 25-30 years on the nutritional evaluation of proteins or protein-containing feeds for mammals (Albanese, 1959; Mitchell & Block, 1946; Munroe & Allison, 1964) and we have drawn heavily upon this. Studies with livestock (Crampton & Harris, 1970) have contributed to the development and standardization of techniques for evaluating proteins in foods of man (FAO Nutritional Studies, 1970) and animals but there is little or no information on use of these newer methods (Eggum, 1970b) for evaluating proteins in the foods of insects.

Indeed, Waldbauer's review of consumption and utilization of food by insects (1968) referred to about a dozen examples of the nitrogen-balance method applied to insects. Basically this method is used to determine the nutritional value of proteins by the "biological value" procedure (NAS-NRC, 1963). For plant-feeding insects therefore, a beginning has been made but further refinements and developments are required. Other methods (Eggum, 1970a, 1970b) not yet tested with insects, but used successfully to evaluate proteins for domestic and experimental animals, should be tried. Such information may help to explain observed differences in the growth and development of insects that are fed on different foods (House, 1969). Perhaps the lack of techniques sufficiently sensitive for use with insects (Bhattacharya & Waldbauer, 1969; McGinnis & Kasting, 1964) is one reason that insect diets have rarely been evaluated. Moreover, as Waldbauer (1968) has pointed out it is difficult to measure true digestibility with insects because urinary wastes cannot be separated from fecal wastes. Of course, this same problem exists with poultry unless the birds are surgically modified to permit urine and feces to be collected separately. Bragg *et al.* (1969) compared results obtained with normal and surgically modified birds and concluded that normal birds more truly assessed availability of dietary amino acids. In studies with insects, therefore, perhaps this source of error is really less significant than had been anticipated. In any case the concepts and techniques for determining nutritional values of proteins are well known and are used by the livestock nutritionist (Crampton & Harris, 1969) but seem to have been neglected by nutritionists working with insects.

With all animals the nutritional value of proteins resides in: (1) their amino acid composition, (2) the quantities of each amino acid that are liberated during digestion of the protein and (3) the quantities absorbed through the gut wall. After absorption, the amino acids from a feed protein are recombined into new proteins which at that moment are needed to support growth and development of the animal. To synthesize proteins within the animal tissues, all those amino acids required in the synthesis must be available.

The crude protein content in the feeds of animals, usually determined by measuring nitrogen by the Kjeldahl method and multiplying by the appropriate factor (FAO/WHO, 1965; NAS-NRC, 1963), provides an estimate of the total quantity of amino acids that are available to a feeding animal. However, because proteins differ in patterns and quantities of amino acids, the estimate of crude protein does not reflect the amino acid make-up of the proteins. Thus to determine the nutritional value of a protein its amino acid composition with particular reference to essential or indispensable amino acids (Meister, 1957) must be known.

There are numerous methods for evaluating proteins for nutritional value. Some of the methods less commonly used for higher animals, including repletion, enzymatic (Mauron, 1970), pepsin digest residue, and plasma amino acid ratio (PAA) and other methods (Eggum, 1970a, 1970b) appear to have little promise for insect studies. Others which have their own particular advantages and limitations may be useful for studies with insects. As defined by the FAO/WHO Expert Committee on Proteins (1965) we would suggest that the following procedures can be useful:

Biological value
Digestibility
Protein Efficiency Ratio (PER)
Net Protein Utilization (NPU)
Protein Score

We have chosen to consider in detail only the "protein score" method. It has been used widely for assessing the nutritional value of many human foods (FAO, 1970) and in work with livestock (Crampton & Harris, 1969). Because the "protein scores" for many feeds are highly correlated with results of other methods for protein evaluation (Block & Mitchell, 1946, Crampton & Harris, 1969) it appears to have value for insect studies. Prior to this conference we had no knowledge of this procedure having been applied to insects. At this conference, however, Davis (1972) reported having used the "protein score" procedure with *Tenebrio molitor* and Rock (1972) reported on use of this general approach with *Argyrotaenia velutinana*.

The "protein score" is based on the concept that the most limiting indispensable amino acid in a diet limits the overall efficiency with which the protein can be used for tissue synthesis. If, for example, tryptophan is the most limiting amino acid then tissue synthesis can only occur to the extent of the tryptophan available. The other essential amino acids present in concentrations greater than those needed for tissue synthesis at the level dictated by the tryptophan concentration must necessarily be deaminated and oxidized, or be excreted. The extra protein is wasted! Unfortunately the treatment to which a commodity is subjected and the "associative effects" of other dietary components also affect the nutritional value of the protein (Mitchell & Block, 1946; Sibbald *et al.*, 1961a,b, Vanschovbroek, 1971). This is believed to be the result of altering the availability of certain indispensable amino acids. In any case it confronts the nutritionist with added difficulties as he attempts to rate proteins according to nutritional value.

A general definition of "protein score" (NAS-NRC, 1963) follows: "The content of each of the essential amino acids of a protein is expressed as a percentage of a standard; the lowest percentage is taken as the score. The value will depend upon the method of analysis for amino acids and the standard chosen." Either casein or egg protein has been used most frequently as the standard in studies with experimental animals and humans (Crampton & Harris, 1969). This protein would appear to have little relevance for studies with the phytophagous insects we have worked with and indeed phytophagous insects generally. Until further information is available it must be assumed that each insect species has an unique protein requirement that is optimum for its growth, development, and reproduction. A limitation of the "protein score" method is that of finding a fully satisfactory protein standard. As a first approximation, that diet either natural or artificial, which permits the insect to grow best, can be used as the standard. It is possible that the amino acid composition of the whole carcass (Munroe & Allison, 1964) would provide an even better basis for "protein score" calculation.

To apply the "protein score" method to any organism, those amino acids indispensable in its diet must be known. In addition, the concentration of each of these amino acids in the diet being evaluated must also be known. For illustrative purposes we will use data obtained in our studies with the pale western cutworm, *Agrotis orthogonia* Morr. The indispensable amino acids for this insect were determined (Table 1) by using the indirect radiometric procedure (Kasting & McGinnis, 1966) with glucose U-C^{14} as substrate (Kasting & McGinnis, 1962). Threonine and tryptophan are included

Table 1. Indispensable amino acids for larvae of the pale western cutworm.

Arginine	Sulfur amino acids
Histidine	Threonine
Isoleucine	Tryptophan
Leucine	Tyrosine
Lysine	Valine
Phenylalanine	

although their essentiality for the cutworm has not been irrevocably established. The concentrations of the amino acids in the foods were measured with a Technicon amino acid analyser after hydrolysis under reflux with 6M hydrochloric acid. Tryptophan was measured by a colorimetric method (Opienska-Blauthetal, 1963) after alkali hydrolysis *in vacuo*.

It has been well documented that the pale western cutworm grows well on fresh Thatcher wheat leaves and also after the leaves have been lyophilized, ground, and reconstituted with water (McGinnis & Kasting, 1960). In contrast, under similar conditions pith from the solid stems of Rescue wheat fails to support good larval growth (Kasting & McGinnis, 1961; McGinnis & Kasting, 1962). Moreover, supplementation of the pith with either casein or a mixture of amino acids sufficient to increase the nitrogen content as much as eightfold resulted in very little improvement in growth and development (McGinnis & Kasting, 1962). An artificial diet (Table 2) has been developed

Table 2. Composition of "Diet F."[a]

Components	Percentage
Casein	2.7
Leaf	1.6
Nitrogen sources	0.6
Sucrose	2.7
Salt	1.2
Agar	1.9
Cellulose	7.2
Water	74.0
Vitamins, solution[b]	8.1

[a]Includes wheat germ oil, choline chloride
[b]Kasting & McGinnis, 1967

that permits growth and development of cutworm larvae but it is also less satisfactory that Thatcher leaves (Kasting & McGinnis, 1967). Weights of larvae fed on these diets are related to protein contents (Table 3). The best diet, Thatcher wheat leaves, was used as the standard and "protein scores" are presented for the other two.

"Protein scores" were calculated according to the procedure recommended by the FAO/WHO expert group on protein requirements (FAO, 1965). The amino acid composition of a food can be calculated in terms of either mg of amino acid per g of nitrogen or mg of amino acid per 100 g of total diet (FAO, 1970). The former method has been used in the present calculations.

Table 4 compares the contents of essential amino acids in the standard diet (Thatcher leaf) and Diet F, for which "protein score" is being

Table 3. Weight gains of larvae of the pale western cutworm fed diets with different concentrations of crude protein

Diet	Crude protein[a] (% dry wt)	Wt 2 days after second molt (mg)
Thatcher leaf	36.6	14
Diet F	17.1	6
Rescue pith	2.5	< 0.5

[a]% Nitrogen X 6.25.

Table 4. Essential amino acid contents of Thatcher leaf and "Diet F" mg/g nitrogen.

Amino acids	Leaf (mg)	Diet F (mg)
Arginine	204	256
Histidine	55	173
Isoleucine	84	301
Leucine	213	609
Lysine	226	423
Phenylalanine	146	357
Sulphur amino acid	96	179
Threonine	159	259
Tryptophan	77	70
Tyrosine	110	302
Valine	166	418
Total	1536	3347

determined. Tryptophan is the only amino acid with a concentration lower in Diet F than in Thatcher leaves. Tryptophan has been used, therefore, to illustrate the calculations necessary in establishing "protein score" for a diet (Table 5). Each of the essential amino acids is assessed in this manner and the lowest value constitutes the "protein score."

The values for Diet F (Table 6) show that tryptophan is, in fact, the most limiting amino acid in the diet. Proportionally it provides only 42 per cent of that present in Thatcher leaves. However, arginine, threonine, lysine, and the sulfur amino acids also yield values less than unity, indicating that they also are present in less than adequate amounts. These data suggest that

Table 5. Calculation of chemical score for tryptophan in "Diet F."[a]

	Leaf (mg)	Diet F (mg)
Total essential amino acid content	1536	3347
Tryptophan content	77	70
Tryptophan as % of total	5.02%	2.09%

[a] $\text{Chemical score} = \frac{\text{Diet F}}{\text{Leaf}} = \frac{2.09}{5.02} = 0.42$

Table 6. Chemical scores for essential amino acids in "Diet F."

Amino acid	Score
Arginine	0.58
Histidine	〉1.00
Isoleucine	〉1.00
Leucine	〉1.00
Lysine	0.86
Phenylalanine	〉1.00
Sulphur amino acid	0.85
Threonine	0.75
Tryptophan	0.42[a]
Tyrosine	〉1.00
Valine	〉1.00

[a] Most limiting amino acid; value = "protein score."

Diet F supports relatively poor growth of larvae because it is severely deficient in tryptophan and arginine and moderately deficient in threonine, lysine and the sulfur amino acids. Of course, the validity of this explanation can only be established by showing that supplementation of the diet with tryptophan and the other four amino acids does enhance larval growth and development. By using a similar approach, however, Rock (1972) identified the limiting amino acids in a diet for *A. velutinana* and succeeded in improving the diet by supplementing with those amino acids so identified.

The "protein score" for the pith of Rescue wheat stems is 0.79 with arginine the most limiting amino acid (Table 7). The graphical method of Oser as described by Crampton and Harris (1969) shows that the pith contained all the essential amino acids, except tyrosine, at concentrations equivalent to those in Diet F (Fig. 1). Hence the poor growth of cutworms

Table 7. Chemical scores for essential amino acids in "Pith Tissue"

Amino acid	Score
Arginine	0.79[a]
Histidine	〉1.00
Isoleucine	〉1.00
Leucine	〉1.00
Lysine	0.88
Phenylalanine	0.98
Sulphur amino acid	〉1.00
Threonine	0.94
Tryptophan	〉1.00
Tyrosine	0.84
Valine	〉1.00

[a]Most limiting amino acid; value = "protein score."

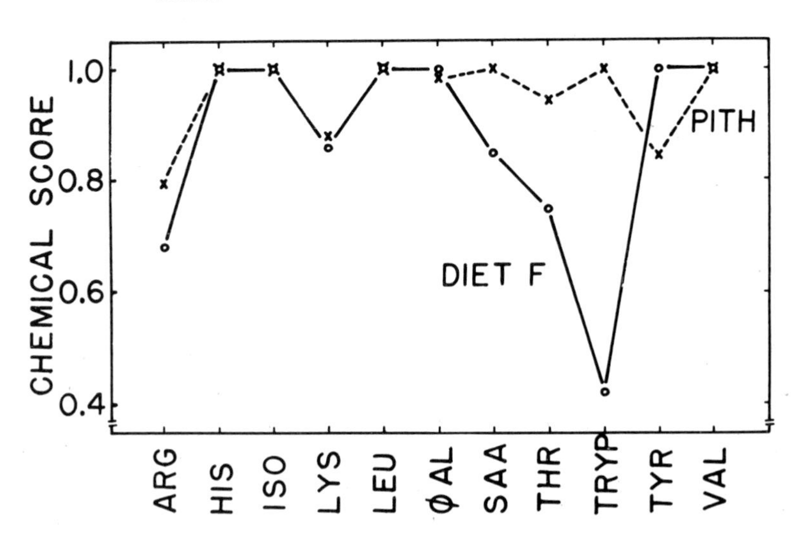

Figure 1. Graphical comparison of protein scores for "Diet F" and pith tissue.

fed the pith diet seems not to result primarily from any major deficiency of the essential amino acids as identified by "protein score." Perhaps the poor growth of cutworms on this diet is the direct result of too little of it being consumed.

This raises an important question relative to protein quality evaluation. Does the amount of food consumed affect availability of amino acids and their efficiency of use? It is known with large animals that the biological efficiency of the dietary protein declines as larger amounts are consumed (Lofgreen *et al.*, 1963) and that the biological value of a particular protein is altered according to the other components present in the ration. What happens with insects? Is it justifiable to compare rates of growth on two diets, pith and leaf, for example, one consumed slightly and the other consumed readily and to conclude that the difference is primarily a reflection of quality? What difference would have resulted had the insect population consumed equal quantities of the two diets? By using the *ad libitum* approach are we overestimating "quality" differences? Can the paired feeding technique, which is cumbersome and time consuming, be used with insects? Certainly if studies of insect nutrition are to approach the sophistication of those with large animals the consumption factor must be eliminated experimentally.

As a means of evaluating protein quality the chemical score procedure offers several advantages over other methods. First, it does not require long, detailed, time-consuming feeding trials with insects that may be difficult to rear. Second, it is possible to evaluate dietary protein against carcass composition which may prove useful. Third, a simple, single value is used that rates the relative nutritional adequacy of different dietary proteins. The major disadvantage with this procedure is the complete lack of any biological evaluation. Presence of an amino acid in a food protein is not necessarily directly related to its availability. This was clearly shown when different sorghum lines were evaluated as poultry feeds (Stephenson *et al.*, 1971). In contrast digestibility and net protein utilization values, for example, do implicate the physiological capacities of the organism with the result that both quantity and availability of the amino acids are measured.

From the foregoing it is apparent that there is no single "best" system for evaluating the adequacy of dietary proteins. We are now in the initial stages of such studies with insects and many different procedures will be tested. For any particular species of insect the method of choice will be that one which best meets the needs imposed by the species and its habitat. We have attempted here to identify some of the procedures that have been found useful with large animals and have provided some detail on the "protein score" method along with experimental results from studies with the pale western cutworm. Perhaps this report will be of assistance as insect nutritionists become more deeply involved in assessing dietary proteins for nutritional adequacy. It seems evident, however, that acceptance of the terms and concepts of large animal nutritionists can expedite progress in studies with insects.

SUMMARY

Nutritional research has contributed greatly to the health and welfare of man and to economic production of animal products. With farm livestock in particular, the success achieved has resulted from the presence of economically definable goals - more meat, milk or eggs for the same feed input. There have also been many scientists working on any one species. Only in recent years with the advent of autocidal techniques for insect control has the insect nutritionist really been concerned with the costs of production.

Much nutritional work with insects only considers quality of the diet. This approach has been feasible because of the relatively small amounts of food needed to support growth and development of experimental insects and because the pure nutritional chemicals have been available commercially.

Today, however, more quantitative nutritional studies are being conducted with insects. Indeed, if insect nutritionists are to make advances comparable to those with man and large animals quantitative data must be obtained. It seems unreasonable to study nutritive quality, as assessed by growth and development of insects, without considering the quantities consumed. Results of some of our studies suggest that quality judgements based on *ad libitum* feeding are not well founded. Ideally, the "paired feeding" technique should be employed but with insects, unfortunately, the technique is cumbersome, time-consuming, and difficult to use.

Nevertheless if insect nutritionists are to achieve success comparable to that obtained by large animal nutritionists we should be aware of what they have done. The procedures for evaluating quality of dietary proteins are considered in this context. The "protein score" technique is considered as a potentially useful means of evaluating proteins in insect diets and is discussed in some detail. "Protein score" is defined as follows: "The content of each of the essential amino acids of a protein is expressed as a percentage of a standard; the lowest percentage is taken as the score. The value will depend upon the method of analysis and the standard chosen." In studies with large animals and humans egg protein and casein have been commonly used as standard proteins. With insects, carcass composition can be readily determined and may be a useful standard against which to evaluate dietary proteins.

For illustrative purposes data from studies with the pale western cutworm, *Agrotis orthogonia* Morr. are used; the best diet, Thatcher wheat leaf, was used as the standard against which the protein quality of two other diets was evaluated. The "protein score" for the artificial diet was 0.42 with tryptophan, the most limiting amino acid. For the wheat stem pith diet, arginine was the most limiting amino acid and the "protein score" was 0.79. This technique is simple to use because single values are obtained based entirely on chemical analyses. It has the major disadvantage, however, of a complete lack of biological evaluation. In contrast, values for "digestibility" and "net protein utilization" are based on feeding trials which are time consuming and often subject to error. They do, however, assess the combination of quantity and availability of amino acids in the diet.

It is apparent that there is no single "best" system for evaluating nutritional adequacy of dietary proteins. For any particular species of insect it will be that procedure that best meets the needs imposed by the nature of the insect and its habitat. The success achieved in nutritional studies with large animals suggests that as much as possible the same approaches, concepts, and terminology should be used with insects. Perhaps this report, in its attempt to relate nutritional work with insects to that with large animals, can assist insect nutritionists as they continue their studies.

REFERENCES

ALBANESE, A. A. (1959). Protein and amino acid nutrition. Academic Press. New York.

BHATTACHARYA, A. K. and WALDBAUER, G. P. (1969). Faecal uric acid in an indicator in the determination of food utilization. *J. Insect Physiol.* **15**:1129-1135.

BLOCK, R. J. and MITCHELL, H. H. (1946). The correlation of the amino acid composition of proteins with their nutritive value. *Nutr. Abs. Revs.* **16**:249-278.

BRAGG, D. B., IVY, C. A. and STEPHENSON, E. L. (1969). Methods for determining amino acid availability of feeds. *Poultry Sci.* **48**:2135-2137.

CRAMPTON, E. W. and HARRIS, L. E. (1969). Applied Animal Nutrition, W. H. Freeman Co., San Francisco, California pp. 57-103.

DAVIS, G. R. F. (1972). Application of Insect Nutrition in Solving General Nutrition Problems. This publication.

EGGUM, B. O. (1970a). Nutritional evaluation of proteins by laboratory animals. In evaluation of novel protein products. Proc. International Biological Program. Sept. 1968, Permagon Press.

EGGUM, B. O. (1970b). Current methods of nutritional protein evaluation. In Proceedings of Symposium on Improving Plant Protein by Nuclear Techniques. IAEA/FAO Vienna. pp. 289-302.

FAO Nutritional Studies, no. 24. (1970). Amino-acid content of foods and biological data on proteins. 285 p.

FAO Nutritional studies, no. 37. (1965). Protein requirements. Report of a Joint FAO/WHO Expert Group. 71 p.

HARRIS, L. E. (1966). Nutrient requirements of domestic animals. Biological energy interrelationships and glossary of energy terms. Publ. 1411 Nat'l Acad. Sci. Nat'l. Res. Council, Washington, D. C.

HARRIS, L. E., ASPLUND, J. M. and CRAMPTON, E. W. (1968). In International feed nomenclature and methods for summarizing and using feed data to calculate diets. Bull. 479, Agriculture Experiment Station, Utah State University, Logan, Utah.

HIRATSUKA, E. (1920). Researches on the nutrition of the silkworm. *Bull. Imper. Exper. Sericult. Station* **1**:257-315.

HODGSON, G. C. (1972). Personal communication. University of Manitoba.

HOUSE, H. L. (1969). Effects of different proportions of nutrients on insects. *Ent. Exp. Appl.* **12**:651-669.

KASTING, R. and MCGINNIS, A. J. (1961). Comparison of tissues from solid- and hollow-stemmed spring wheats during growth. II. Food values determined with the pale western cutworm, *Agrotis orthogonia* Morr. *Can. J. Zool.* **39**:273-280.

KASTING, R. and MCGINNIS, A. J. (1962). Nutrition of the pale western cutworm, *Agrotis orthogonia* Morr. IV. Amino acid requirements determined with glucose-U-C^{14}. *J. Insect Physiol.* **8**:97-103.

KASTING, R. and MCGINNIS, A. J. (1963). Resistance of plants to insects - the role of insect nutrition. *Agric. Institute Rev.* **18**(5):9-11.

KASTING, R. and MCGINNIS, A. J. (1966). Radioisotopes and the determination of nutrient requirements. *Ann. New York Acad. Sci.* **139**: Art. **1**:98-107.

KASTING, R. and MCGINNIS, A. J. (1967). An artificial diet and some growth factor requirements for the pale western cutworm. *Can. J. Zool.* **45**:787-796.

LOFGREEN, G. P., BATH, D. L. and STRONG, H. T. (1963). Net energy of successive increments of feed above maintenance for beef cattle. *J. Animal Sci.* **22**:598-603.

MAURON, J. (1970). Nutritional evaluation of proteins by enzymatic methods. In Proceedings of a Symposium on Improving Plant Protein by Nuclear Techniques. IAEA/FAO, Vienna. pp. 303-318.

MAYNARD, L. A. (1937). McGraw-Hill Inc., N. Y. and London, pp. 8-9.

MCGINNIS, A. J. and KASTING, R. (1960). Nutrition of the pale western cutworm, *Agrotis orthogonia* Morr. III. Lyophilized sprouts and leaves of wheat as a basal diet for larvae and effects of supplementation with L-leucine. *Can. J. Zool.* **38**:585-592.

MCGINNIS, A. J. and KASTING, R. (1962). Comparison of tissues from solid- and hollow-stemmed spring wheats during growth. III. An ether-soluble substance toxic to larvae of the pale western cutworm, *Agrotis orthogonia* Morr. *Ent. Exp. Appl.* **5**:313-321.

MCGINNIS, A. J. and KASTING, R. (1964). Chromic oxide indicator method for measuring food utilization in a plant feeding insect. *Science* **144**:1464-1465.

MCGINNIS, A. J. and KASTING, R. (1966). Comparison of tissues from solid- and hollow-stemmed spring wheats during growth. IV. Apparent dry-matter and nitrogen balance in the two-striped grasshopper, *Melanoplus bivittatus* (Say.). *J. Insect. Physiol.* **12**:671-678.

MCGINNIS, A. J. and KASTING, R. (1967). Wheat production and insects. In Proceedings of the Canadian Centennial Wheat Symposium. Ed. Kenneth F. Nielsen, Modern Press, Saskatoon, Saskatchewan. pp. 291-313.

MEISTER, A. (1957). Biochemistry of the amino acids. Academic Press, New York. pp. 98-103.

MITCHELL, H. H. (1942). The evaluation of feeds on the basis of digestible and metabolizable nutrients. *J. Animal Sci.* **1**:159-173.

MITCHELL, H. H. and BLOCK, R. J. (1946). Some relationships between amino acid contents of proteins and their nutritive values for rats. *J. Biol. Chem.* **163**:599-620.

MUNROE, H. W. and ALLISON, J. B. (1964). Mammalian protein metabolism. Vol. 2. Academic Press, New York.

National Academy of Sciences - National Research Council. (1963). Publ. 1100. Evaluation of protein quality. Food and Nutrition Board. 119 p.

OPIENSKA-BLAUTH., CHARENZINSKI, M. and BERBEC, H. (1963). A new rapid method of determining tryptophan. *Anal. Biochem.* 6:69-76.

PELIGOT, E. (1967). Études chimiques et physiologiques sur les vers á soie. *Ann. Chim. et Phys.* 12:445-463.

ROCK, G. C. (1972). Optimal Proportions of Dietary Amino Acids. This publication.

SIBBALD, I. R., SLINGER, S. J. and ASHTON, G. C. (1961a). Factors affecting the metabolizable energy content of poultry feeds. 2. Variability in the M.E. values attributed to samples of tallow and ungummed soybean oil. *Poultry Sci.* 40:303-308.

SIBBALD, I. R., SLINGER, S. J. and ASHTON, G. C. (1961b). The influence of dietary calorie:protein ratios on the weight gain and feed efficiency of growing chicks. *Poultry Sci.* 40:308-313.

STEPHENSON, E. L., YORK, J. O., BRAGG, D. B. and IVY, C. A. (1971). The amino acid content and availability of different strains of grain sorghum to the chick. *Poultry Sci.* 50:581-584.

UVAROV, B. P. (1928). Insect nutrition and metabolism. A summary of the literature. *Trans. Ent. Soc. London* 74:255-343.

VANSCHOVBROEK, F., VERMEERSCH, G. and DE SCHRIJVER, R. (1971). A comparison of the effects of lard and soybean oil on food and water intake, body weight, food conversion efficiency and mortality of broiler chicks. *Poultry Sci.* 50:495-501.

WALDBAUER, G. P. (1968). The consumption and utilization of food by insects. In *Advances in Insect Physiology* 5:229-288.

INTERPRETATIONS OF INSECT QUANTITATIVE NUTRITION

H. T. Gordon
Division of Entomology and Parasitology
University of California
Berkeley, California 94720

INTRODUCTION

"Ce qui est simple est toujours faux. Ce qui ne l'est pas est inutilisable." Although I would prefer the milder word "insuffisant" to "faux," and the more temperate ending "tres difficilement utilisable," this aphorism of Paul Valery is often valid. In the domain of insect nutrition, I would paraphrase it as: We still know too little to form a true picture of the trophic interaction of insects with the environment. But the day may come when we are unable to make sense out of an inchoate mountain of data. It is already difficult to draw useful information from the mass of details being accumulated by scientists. In this paper I shall add only a little data, and devote much time to certain aspects of nutritional experimentation and interpretation that have captured my attention in recent years, and that are relevant to the question: What are the nature, needs, and uses of insect nutrition? I justify this on the ground that the conference to which this paper contributes was meant to deal with the *significance* of nutrition.

In his review, Dadd (1970) interprets nutrition "in the narrow sense" as the study of feeding, growth, and at least some aspects of metabolism on diets of known and controllable composition. Although the scientific merit of such studies is undeniable, they have distracted many nutritionists' attention from the foods that are the natural dietary of insects. Such foods—complex and often not very stable or reproducible—have been left to naturalists, behaviorists, ecologists (who tend to view foods as packets of stored solar energy), and applied entomologists interested in the phenomena of host specificity or resistance to insect attack.

The narrow interpretation of nutrition has also created—in the minds of many entomologists and elsewhere—an image of irrelevance. I have been asked, by colleagues and students, questions such as: "What uses does insect nutrition have, comparable to the triumphs of mammalian nutrition in animal husbandry and in clinical medicine? What basic scientific ideas can emerge from work on insects that could not as or more easily be developed in the strong fields of microbial and vertebrate nutrition? What, in these tedious but simple compoundings of diets and weighings, is there to challenge human ingenuity and technical skill—so that one can take pride in the difficulty and sophistication of techniques, even if the results are neither useful nor novel? What are the job and grant opportunities for nutritionists?" The answer to the last question is, of course, "not very many." Among the reasons for this (the predominance of insecticide technology, the general decline in funding, etc.) one ought to include the fact that there are no self-evident answers to the other questions. For example, it is to me inconceivable that nutritional methods alone could ever destroy all insect (and sometimes other animal) life over immense acreages, as many insecticides temporarily can.

In the search for answers, I believe it is easier not to start with applications, since these inevitably emerge when a critical mass of principles and techniques has come into existence. Not just any mass, but a critical one in which all elements form a whole. Nutrition research springs from the belief that insects, because of the eons of differential evolution, must differ from mammals and microorganisms in many fundamental ways. The progress of insect nutrition has been aided by the great advances in vertebrate nutrition, but perhaps also hindered from finding its own path, since the viewpoint of conventional nutrition is strongly oriented toward man and his domestic animals (the rat, source of much of our knowledge, is a commensal of man). Unquestionably, the dispersal of sparse resources on so many diverse species of insects has been a major hindrance, dictated by the pest-orientation of entomological research in general. Classical genetics would still be a primitive science if research had not been focused on *Drosophila melanogaster,* and our present understanding of inset endocrinology rests on the study in depth of a few species. Most of insect nutrition is still in the realm of possibility. There may be an insect species as ideal for the study of metazoan nutritional genetics as *Drosophila* was for classical genetics, but not much thought has been given to this problem. Some insect embryos may be tough enough to tolerate yolk removal and allow us to study developmental nutrition, but we tend to ignore all events prior to larval emergence, as outside our field and sphere of interest. In the broadest sense, insect nutrition is not limited to the feeding stages of the life cycle, certainly not to man-made diets, and especially not to one-man projects. Collaboration with other specialists may be essential. But if there is to be a future, the small number of scientists now interested in insect nutrition must work together toward common goals. Common goals imply a community of outlook, techniques, terminology and interpretation that needs to be sustained in many ways: research papers, reviews, personal communications, and perhaps also critical commentaries of the kind I have included in this paper.

The parameters of nutrition

What is the key quantitative parameter of nutrition? I believe that it is the rate of utilization of the metabolite(s) that are most limiting the metabolic activity of the animal. "Most limiting" means that a small increment or decrement in the level of the metabolite causes a large change in activity–

either an increase if the metabolite is deficient or a decrease if it is in excess. Seldom will there be a state of "metabolic balance," in which all or most metabolites are equally limiting. In the continuous molding of thousands of genes by natural selection into a harmonious whole, some reactions will prove more difficult to synchronize and will slow down the whole system. Selection pressure will be most intense on the genes controlling these reactions, since adaptive accelerations or decelerations are most easily managed by minor variations at these loci, while maintaining a general stability elsewhere. When the food sources are also living organisms, there will be selection for biochemical variations in them that directly or indirectly act on these limiting reactions. The identification of these critical areas is one of the important tasks of nutritional analysis.

Metabolite utilization can be subdivided into a few major classes. The first is incorporation. This is most easily studied as post-embryonic growth, since growing animals are utilizing many metabolites at very high rates and must obtain them (or suitable precursors) from an environment that man can control. It is important to clarify the distinction between a nutrient and a metabolite. The former is present in the food supply and can, sometimes after digestion only, be absorbed and advantageously utilized; thus, neither protein nor glucose-6-phosphate would be nutrients, but nutrient precursors, the actual nutrients being amino acids, glucose, and phosphate. A metabolite is any substance utilized (i.e., converted into something else by, or necessarily associated with, any enzyme) in metabolism. A nutrient starts a metabolic chain, but its effects may not become rate-limiting until a metabolite begins to accumulate; interpretation of nutritional effects is possible only at a biochemical level. In a growing animal many nutrients are incorporated, with a considerable increase in mass. The overall rate can be expressed in many ways. *I have previously given reasons for preferring the term* $\underline{G}$, *defined as mg gain in dry weight per g live body weight per day* (Gordon, 1968b). This includes a very large number of incorporation reactions, but can be limited by any one of these; the study of such limitations has been so far carried out by omission of nutrients from artificial diets, thereby determining "nutritional requirements." When a diet is "complete and optimal," the limitation of $\underline{G}$ becomes one of enzyme reaction rates in the critical areas referred to above.

Although incorporation is easier to study during growth, it of course occurs during embryonic development (transfer of nutrients from yolk reserves to the embryo) and in non-growing adults (in which incorporation of a nutrient from the diet is nullified by the loss of previously incorporated substance). Therefore, a rate of "de-incorporation" can be neglected in growing animals, as it is small relative to the rate of incorporation, but will be a part of any complete nutritional study. The growth curve, describing the change in live body weight with time, has so far been the primary data input of all nutritional analysis. Much attention has been given to mathematical equations fitting growth curves (Gray, 1929; Zucker *et al.*, 1941; Brody, 1945; Weiss and Kavanau, 1957; von Bertalanffy, 1960; Laird *et al.*, 1965). All involve a theoretical exponential growth rate, plus a variety of ingenious modifiers to account for the decrease in this rate (eventually to zero) as the organism increases in size. Although such studies are by no means useless, the underlying metabolic significance of the empirical "growth constants" they generate is often unclear. In fact, the information content in a growth curve is too slight to support any elaborate mathematical anslysis.

In a separate compartment of scientific development, using partly respirometric and partly calorimetric methods, a much less analytical understanding of the second major parameter of nutrition gradually emerged. Energy was clearly essential, and was derived from the oxidation of certain nutrients, a concept given its major impulse by the great Otto Warburg. This can be measured in the same units as $\underline{G}$, and can be called $\underline{O}$, the rate of dry body weight loss by oxidation to CO_2 and H_2O.

The third component S_i, is the rate of dry body weight loss by excretion of solids, quantitatively less important than $\underline{G}$ and $\underline{O}$. The subscript indicates that the substances excreted (or their precursors) have been absorbed and may have been metabolized. The overall metabolic rate, expressed in mass-rate units, is therefore

$$\underline{G} + \underline{O} + \underline{S}_i = \underline{I}$$

where $\underline{I}$ must be the intake rate of nutrients required to compensate for the metabolic expenditures. I believe that this expression is the simplest possible description of the metabolic rate, although this term is commonly restricted to $\underline{O}$ and its associated phenomena of oxygen consumption and heat production. The intake rate of nutrients does not fully correspond to the intake rate of food, however, since part of the food intake is excreted without having been absorbed. The complete equation is therefore

$$\underline{G} + \underline{O} + \underline{S}_i + \underline{S}_e = \underline{I} + \underline{S}_e = \underline{F}$$

which links metabolism to nutrition. The directly measurable parameters are $\underline{F}$, $\underline{G}$, and $\underline{S}$ (the sum of $\underline{S}_i$ and $\underline{S}_e$). The problem of the 2 components of $\underline{S}$ is discussed by Waldbauer (1968), but has no exact solution. The equation also omits the quantitatively very important nutrient, atmospheric oxygen, which can add to and be retained in components of both $\underline{G}$ and $\underline{S}$; This problem will be discussed in a later section of this paper. Although $\underline{O}$ can be calculated by subtracting the sum of $\underline{G}$ and $\underline{S}$ from $\underline{F}$, this is not yet a common practice. Its existence is implied in the E. C. D. ratio, calculated as $\underline{G}/(\underline{F} - \underline{S})$ and therefore equal to $\underline{G}/(\underline{G} + \underline{O})$. The $\underline{G}$ and $\underline{O}$ systems must be to some extent independent, since otherwise the E. C. D. would be constant. The other ratio usually calculated is the E. C. I., equal to $\underline{G}/\underline{F}$ (Waldbauer, 1968). Such ratios reflect the almost exclusive interest of nutritional research in growing animals, where $\underline{G}$ is large. The basic equation is not restrictive; in a non-growing adult, $\underline{G}$ becomes zero and $\underline{F}$ flows only into $\underline{O}$ and $\underline{S}$, while in a starved animal $\underline{F}$ is zero and $\underline{G}$ becomes negative.

The significance of this formulation is that it relegates the food intake rate, $\underline{F}$, to its proper place. It is not an independent parameter but the sum of several important mass-flow rates, determined primarily by the genetic constitution of the organims, secondarily by the composition of the food, and tertiarily by the phagostimulant and phagoinhibitant effects of some dietary constituents and by a variety of factors influencing metabolic activity (such as illumination, crowding, the search for food and water, etc.). The many studies on the role of sensory reflexes in the control of $\underline{F}$, principally by Dethier and his students on adult *Phormia regina* fed aqueous solutions of carbohydrates (Gelperin, 1971), assign a dominant role to receptors and the nervous system, instead of the subservient one that I believe they play. These reflexes are thought to regulate—with a precision that must command our admiration—the intake of food at some level above the minimal metabolic needs and below the metabolic capacity. Metabolic utilization (in the *Phormia* work, exclusively the $\underline{O}$ system) somehow adjusts itself to the level ordained by the central computer, which also monitors internal factors (such as hemolymph osmotic pressure) that allow correction of any miscalculations. There can be no doubt that the neural system determines the timing and size of the first meal, since only a sensory evaluation is then possible, but the rate of utilization of this meal then provides truly crucial information about the capacity of the rate-limiting metabolic system to utilize the food. This capacity then takes control of the neural machinery, presumably by some kind of satiety signal, to maintain $\underline{F}$ at that level. Such a mechanism will ensure the most efficient possible utilization of the food supply.

Waldbauer (1968) listed 3 possible ways of expressing the food intake rate (C. I. in his terminology) but did not prefer any one of these. It is desirable to clear up this ambiguity. A C. I. based on fresh (wet) weight of food to fresh weight of insect is undesirable because it sums the water and food intake rates. Even if water is the rate-limiting dietary constituent (a condition that is probably uncommon except in stored-product insects largely dependent on metabolic water) it would surely be better to state the 2 rates independently. The other 2 variants of C. I. are fundamentally similar if the water content of the larva is well-regulated, so that $\underline{d}$ (the ratio of dry to wet body weight) is constant. Waldbauer's "dry-dry C. I." can then be multiplied by $\underline{d}$ to give his "dry-wet C. I." This last expression, multiplied by 1000 to convert to units of mg/g·day, is equivalent to $\underline{F}$ but more precise if mean body weight is calculated by the method of Waldbauer. When larvae have an increasing lipid content, which tends to increase $\underline{d}$, $\underline{F}$ (a more concise term by far than dry-wet C. I.!) is the best parameter since lipid storage makes up a large fraction of the dry weight but contributes little to metabolic activity. It would probably be better still to use the FFWW or "fat-free wet weight," popular among mammalian nutritionists, although its determination involves much more work than the total wet weight on which F is based. Since the purpose of calculating such parameters is to allow comparison between different stages and species on different diets, and since I have come to feel that a measurement of lipid content is an essential element in quantitative nutrition, I am tempted to advocate a redefinition of F on the basis of FFWW. Obviously, all other parameters would be likewise redefined.

The variability and instability of nutritional parameters

In his thorough review of the technical difficulties, errors, and uncertainties in obtaining and analysing quantitative data, Waldbauer (1968) discussed only incidentally the problem of variability, which has 2 aspects. The first is the random variation in a series of measurements, made at the same time in presumably identical conditions, that ought to be (but never are) identical. The second can be called "instability," a broad term that can include phenomena such as abrupt "triggered" discontinuities (e.g., periodic feeding, molting, or egg-laying), rhythmicities (circadian, etc.), maturation and aging, and "transition effects," which are states of disequilibrium during the period following a change in some environmental parameter (temperature, diet, etc.) and will be discussed in a later section.

Biological variance is presumably partly of genetic and partly of environmental origin. The major phenotypic characters show a remarkable stability over a fairly wide range of environmental change. In fact, it is difficult to produce, by changing the temperature or diet composition, a very great change in phenotype in most insect species. The ability of genotypically diverse zygotes to develop into quite similar adults, even when their environments vary, has been termed homeorhesis by Waddington (1957). Although the complex regulatory mechanisms are still a matter for speculation, they may eventually be resolved into enzyme inductions, repressions, and feedback inhibitions of the kind demonstrated in microorganisms (Davis, 1961; Serra, 1966). Variance exists because even within normal environmental limits homeorhesis is imperfect; outside these limits variance increases, many individuals die, and phenotypic abnormalities (sometimes obvious, sometimes not) become more common. Some systems are more strongly stabilised than others (in Waddington's terminology they show "cusped" as opposed to "smooth" canalisation). They may be systems of greatest survival value, that arise earlier in development (e.g., the nervous as opposed to the reproductive system) and that command a high priority when there are quantitative or qualitative restrictions of the food supply (Hammond, 1962).

The variance of a complex parameter such as dry body weight may ultimately prove to be largely that of a few "unstable" systems, such as fat body or female reproductive organs, in which storage of nutrient reserves plays a dominant role. Hirano (1964) showed that a change in the dietary protein/carbohydrate ratio from 4.1 to 0.85 caused an increase in the fat content of *Chilo suppressalis* larvae from 11 to 38%. The nitrogen content declined from 11.5 to 7.5%, but on a fat-free basis it was far more stable, declining from 12.7 to 11.9% (with most of the difference in the free amino acid pool). Something akin to the homeorhesis of embryonic growth therefore persists during larval growth, and the variations in nutritional parameters on foods of different composition result from effort of the genotype to construct the same phenotype from different raw materials. Mammalian nutritionists have of course studied body composition in much more detail, and have developed many indirect methods applicable to human beings (Brozek, 1963), although direct analysis on experimental animals is more useful. In general, most compositional parameters (except body fat) seem to be quite well stabilized and there is reason to believe that large departures from the norm would be pathological. The small departures, due to imperfect homeorhesis, may be of no practical importance; in the words of Weil and Wallace (1963): "A refinement of technique increases, more and more differences of smaller and smaller magnitude will be observed. The question must then be asked whether such differences have biological significance or are purely of interest to the biostatistician." An interesting point emerges from the vertebrate work: some minor body constituents seem to increase with age instead of with body size, and therefore are present in higher concentration in animals that have grown at a slow rate. Except for these, and of course the variable fat content, the composition of $\underline{G}$ is probably relatively constant at any given body size; growth represents a selection from the materials available in $\underline{F}$, with the remainder being disposed of in either $\underline{O}$ or $\underline{S}$. Unfortunately, there is much less information that can be brought to bear on the "composition" of $\underline{O}$ and $\underline{S}$.

The significance of "transition effects"

The genotype of an organism "programs" development of the phenotype through any transitional states (embryonic, larval, etc.) that follow one another in a predictable normal sequence and may include periods of relative stability (exponential growth, diapause, egg production at a steady rate, etc.). Biological parameters are often measured during such periods, or else during fairly well-defined intervals such as an instar. To determine the effect of environmental factors, the experimental animals are taken from their customary environment and abruptly plunged into an environment in which one or more of the parameters (temperature, diet, illumination, etc.) are different. The animals are given little or no time to adjust to the new environment, but it is tacitly assumed that a stable state is being measured, independent of the nature of the previous environment and the duration of the experiment itself. Students of biological rhythms never assume this. Their experiments are never short-term and use rather stable biological material.

Occasionally an environmental transition is immediately followed by a new stable state, in which a long series of successive measurements shows no upward or downward trend, or followed by a fixed unstable state in which the trend is the same regardless of the previous environment. However, an earlier study (Gordon, 1968a) of adult *Blattella germanica* males taken from a colony maintained on a yeast diet plus cubes of pure sucrose showed a different trend. When transferred to a diet of sucrose only, they consumed it at a rate of 20-25 mg/g·day for a long period. When transferred to cellobiose, an F of a few mg/g·day was maintained during the 4-5 days of survival. Transfer to trehalose, however, caused a "transition effect"; F was only a few mg/g·day during the first 2 days but then increased to a high level, fluctuating between 18 and 33 mg/g·day for a long time.

A "transition effect" is a transient state that, while interesting, might be considered aberrant and surprising. Its occurrence depends on the previous state of the insects adaptation, on the existence of unsuspected temporary barriers or drives that delay the onset of a stable state or may even make its attainment impossible. That is, a direct transition from state A to state C may fail, while a transition from A to B and then to C may succeed; this is the well-known phenomenon of acclimatization. Short-term experiments involving abrupt transitions may therefore lead to erroneous conclusions, either implicit or overt, especially if they seem plausible or fit generally-accepted preconceptions.

There is no reason why the duration of a "transition effect" should be limited to a few days. Several instars or even generations may be required to attain a terminal state (either death or a new dynamic equilibrium). Not until several generations have been reared on a new diet—with biological parameters showing no drift—can one assess its nutritive value (Dadd, 1970). If there is mortality or differential fecundity, an even longer time may be required to establish a stable genetic equilibrium. This is the "transient polymorphism" of Ford (1965). It may then be argued that comparisons are between different environments and additionally, between genetically different populations. This is true only if one thinks of a population as a particular set of gene frequencies in equilibrium with a given control environment, but not if one prefers to think of it as a gene pool. The first approach seems excessively rigid, an extension at the population level of the idea that an individual is incapable of adaptation, and can be used to measure an environment as a thermometer does. In fact, the surprise element in a "transition effect" results from the violation of an experimenter's expectancy of stability, the expectancy that when the environment is stable the animal should also be stable.

Expectancy lies at the root of biological adaptation. Hasty response to most sudden environmental changes is undesirable, as these are usually temporary departures from the norm. As the length of the duration of the change increases, so does the probability that it has become the norm, and similarly the desirability of adjusting to it. It is interesting to consider the problem of a newly-hatched larva of a monophagous species. It has a very high expectancy of finding itself on a suitable host and of being able to exploit it effectively. If it is (by accident or by human intervention) on an unsuitable host, the wisest course may be to wander off in search of the expected one. I have watched many larvae of the olive fruit fly, *Dacus oleae,* take this course when I set them on artificial diets; others, after some delay, accepted the unusual food, thereby giving to the population as a whole access to a variety of strategies and making possible the novel phenomenon of rearing monophages on man-made diets. In nature, such behavior may be useful in the "tracking" of variations in the host species so that it cannot escape, or (very rarely) in the extension of the host range. The problem of host specificity of stenophages (strictly speaking there may be no truly monophagous insects) is interestingly reviewed by Zwolfer and Harris (1971). They seem to be confident that the nutritional barriers between plant species will prove insurmountable, even after the acceptability barriers are broken. This confidence is justifiable only if these trophic barriers are strong and the stenophages have lost most of their biochemical plasticity. We do not yet understand the differences at the nutritional level between monophagous and polyphagous species. Waldbauer (1968) noted that the oligophage *Manduca [Protoparce] sexta* digested its preferred host plants less efficiently but converted the digestion products into body weight more efficiently than the polyphage *Prodenia eridanea.* Since there is no obvious advantage in inefficient digestion (unless the soluble and easily digestible components of plants are more nutritious), it can be imagined that the inefficiency of *Manduca* is caused by its very high F (ca 300-400 mg/g·day on both host and non-host plants) that would allow less time for digestion, while *Prodenia* (with F in the range of 100-200) has more time. This is not necessarily true. The physical determinants of F are the volume of the digestive tract and the transit time. If F is doubled by halving the transit time, then the time available for digestion and absorption is also halved. If the volume of the digestive tract is doubled, however, F will be doubled with no decrease in the processing time; nor need there be a sacrifice in absorptive surface area, if the volume increase involves not only a larger diameter but a greater length. There is some added cost in the higher proportion of body weight devoted to the digestive system. Possibly the digestive machinery of *Manduca* has been adjusted (by natural selection) to the F range where I (the product of F x the % digestion) is a max. Simultaneously, selection must have acted to improve the efficiency of conversion of this specific (solanaceous) composition of I into G. The proof that *Manduca* cannot metabolize the I from non-host plants as efficiently is the first evidence of the specialization of oligophages at the trophic level. Regrettably Waldbauer's researches, the only work in which nutritional phenomena have been freed from the interference of chemosensory reflexes, have not been pursued. Valuable knowledge of the difference in composition of I from host and non-host plants, could be easily obtained by comparing the composition of F and of S.

Nevertheless, *Manduca* does rather well on a non-host such as *Taraxacum*, at least for one instar, and one wonders whether much greater differences might not show up in the next instar, the pupa, the adult, or the next generation (cf. Morris, 1967). A single instar, lasting only a few days, is likely to be a "transition" effect of uncertain predictive value. This may also be true of some of the data of Soo Hoo and Fraenkel (1966a, b), since they seem to have transferred *Prodenia* larvae from a colony on *Chenopodium album* to leaves of other plant species. The lower F of *Prodenia* is accompanied by lower values of I and G but no great loss of overall efficiency since the inefficient conversion is largely compensated by more efficient digestion. However, a relatively low G implies either a longer larval life or a reduced adult size. In effect, *Prodenia* could produce a larger population of smaller adults from a given food supply in the same time that *Manduca* produces a smaller population of larger adults. An oligophage need not be large, since it could use its high G to produce a large population of small adults in a shorter time, thereby improving its population r in the most effective way (Lewontin, 1965).

Two of the observations in the study of Soo Hoo and Fraenkel (1966b) could be "transition effects." The first is the abnormally high F, I, and O of *Prodenia* fed leaves of *Lycopersicon.* The second is the fact that leaves of *Antirhinum*, (a moderately good host plant) were utilized less efficiently than leaves of *Nicandra* (a very poor host plant); this puzzling observation is not interpreted by Soo Hoo and Fraenkel, perhaps because it runs counter to the general thought trend in their work. Their criterion of host plant quality (Soo Hoo and Fraenkel, 1966a) was the degree of success in rearing from the beginning of the 5th instar (transferred from the rearing host, presumably *Chenopodium*) to the pupal stage, but the quantitative work was restricted to the 5th instar only. In as yet unpublished studies on *Oncopeltus fasciatus* I have noted that satisfactory feeding and growth may be obtained when nymphs in the penultimate instar are transferred to seeds that cannot support continuous growth, i.e., the "transition effect" can be a temporarily good performance instead of a temporarily bad one, as in the trehalose example cited earlier or a (permanently?) bad one, as in the examples of "conditioning" cited by Zwolfer and Harris (1971). The true host plant range of *Prodenia* may be more narrow than it is now assumed to be, and most of its reproductive potential may derive from a number of host species not much larger than that of oligophages, although scattered among several plant families. This in no way lessens the interest aroused by the biochemical versatility of polyphages and the selective forces that have molded them.

When living plant or animal food is used in experiments, everything that has been said in the previous section on variability and instability is applicable, and "transition effects" may also occur. When a leaf is cut it is reasonable to expect compositional changes, and these could well be both important and rapid. Nutritional differences of some kind have been shown to exist between leaves on intact plants and cut leaves (Moreau, 1971) but there is no information on the relative intake rates and utilization efficiency.

Rates, limits, and priorities of major food conversions in animals

Table 1 presents the stoichiometry of the major metabolic conversions of the 3 biochemical types (carbohydrate, fat, and protein) that constitute most of the mass of common foods. I am indebted to a recent textbook (McGilvery, 1970) for its useful quantitative formulation of metabolism; this is based on mammalian biochemistry, but there is every reason to believe that it will be applicable to insects. Studies of insect biochemistry have revealed many similarities and very few differences. Even the metabolic differences of insect flight muscle (Sacktor, 1970) do not seem to involve modification of basic substrates or efficiencies but the addition of auxiliary mechanisms that make possible phenomenal rates of energy production (consumption as high as 3000 μl. O_2/g/min, O_2/g/min., representing a glucose oxidation rate of 5700 mg/g$\cdot$day).

Table 1 lists 4 ways of utilizing dietary glucose (or other carbohydrate that can readily enter the same pathways). The first is complete oxidation. On a glucose-rich diet this is likely to be the primary source of high-energy phosphate, and the rate of glucose utilization will be limited by the rate of energy utilization (in muscular contraction, in the *de novo* biosynthesis required for growth, and in the "turnover" resynthesis of maintenance metabolism). When the rate of glucose inflow exceeds that required for energy production (e.g., after a meal) the concentration of glucose-6-phosphate will rise and activate glycogen synthase. Much glucose will be stored as glycogen and a little oxidized to provide the necessary energy. There is an upper limit to glycogen storage, presumably genetically determined, but the limiting mechanism is unknown.

When the rate of glucose and ATP utilization by glycogen synthesis falls, the ATP/ADP ratio rises, slowing the flow of electrons in the mitochondrial transport chain and causing an accumulation of acetyl CoA and citrate in the tricarboxylic acid cyle. Citrate activates acetyl CoA carboxylase and activates the next priority level, fatty acid synthesis. Since fatty acids must be stored as triglyceride, glycerol is simultaneously drained from triose metabolism. Very large amounts of glucose can be utilized in this

Table 1. Major food conversions in animals.[a]

Food	O_2 cons.	CO_2 prod.	H_2O prod.	Solid prod.	Energy ($\sim$P)	Weight loss	$\frac{QO_2}{O}$
	33.3	33.3	33.3	None	200		
	(1.07)	(1.47)	(0.6)			(1.0)	31
Glucose							
5.55	1.9	1.9	7.1	5.23 Glycogen[b]	0		
(1.0)	(0.06)	(0.08	(0.13)			(0.15)	12
	4.95	13.6	13.6	1.23 Palmitic	11		
	(0.16)	(0.60)	(0.24)	(0.32)		(0.68)	6.8
+ 3.7 NH_3	3.7	3.7	14.8	3.7 Chitin	11		
(0.06)	(0.12)	(0.16)	(0.27)	(0.75)		(0.31)	11
Glycogen[b]				None			
6.15	36.9	36.9	30.8		220		
(1.0)	(1.19)	(1.63)	(0.56)			(1.0)	34
Palmitic[c]				None			
3.9	89.5	62.5	62.5		504		
(1.0)	(2.88)	(2.75)	(1.13)			(1.0)	80
Tripalmitin[c]				Tripalmitin			
1.24	1.2	1.2	1.2	1.24	0		
(1.0)	(0.04)	(0.05)	(0.02)	(1.0)		(0.04)*	31
	45.4	43.5	17.4	12.4 NH_3	250		
	(1.45)	(1.91)	(0.31)	(0.21)		(1.0)	42
Protein[d]							
9.2	45.4	37.4	23.3	6.2 Urea	215		
(1.0)	(1.45)	(1.64)	(0.42)	(0.37)		(1.0)	42
	39.7	27.4	29.3	3.1 Uric acid	160	(0.48)	77[e]
	(1.27)	(1.20)	(0.53)	(0.52)		(1.0)	37
	4.5	4.5	4.5	9.2 Protein	0		
	(0.14)	(0.20)	(0.08)	(1.0)		(0.14)*	31

[a]Calculations largely rely on basic information in McGilvery (1970). The first 2 columns represent input (food and O_2). Each substance has 2 equivalent values: the number of μmoles and (in parentheses) the corresponding wt in mgms. Energy produced is given in μmoles of high-energy phosphate. Whenever a conversion is endergonic (e.g., glucose to glycogen), it is assumed to be coupled to glucose oxidation so as to yield zero energy; if the food being converted is not glucose, the value of the dry body wt loss is marked with an asterisk to indicate that it is glucose loss, and the necessary O_2, CO_2, and H_2O values are included in the Table. All food conversions therefore involve oxidative metabolism and some dry wt loss; the true energy cost is surely larger but not easy to calculate since the energetics of acquisition, digestion, and transport are unknown. The last column gives the factor by which the rate of dry wt loss (O, in mg/g·day) must be multiplied to give the corresponding QO_2 in μl/mg hr; it is included to rectify previous errors in calculation (Gordon, 1968b) and to stress the linkage between quantitative nutrition and respirometry.

[b]Moles expressed as the monomer (glucose less 1 H_2O, mol wt 162). The number of

monomeric subunits in the polymer may be as high as 20,000. Although glycogen storage does not add to the intracellular osmotic pressure the molecule is strongly hydrated (2 g H_2O/g glycogen) and the storage wt is about 13 g/mole of high-energy phosphate, compared to 2.5 g for triglyceride storage (McGilvery, 1970).

[c]Before oxidation, dietary or body reserve triglyceride must be hydrolysed to free fatty acids and glycerol, but inclusion of this in the oxidation would not much alter the numerical values. In the conversion of dietary to reserve triglyceride, complete hydrolysis (assumed in the calculations here) may not be necessary; if so, the energy cost will be less. Since there is continuous "turnover," few of the original ester linkages are likely to remain intact for long periods.

[d]The amino acid composition is that of "beef muscle protein" (p. 396, McGilvery (1970)). For other dietary proteins the major difference is likely to be in N content (11% lower for casein). Although total oxidative degradation of dietary proteins is unusual, the selective destruction of those amino acids present in excess (relative to the most deficient ones) is common. The conversion leading to urea production is given only for comparison since urea is not an important waste product in insects. Insects that normally have an ample water supply often hydrolyse uric acid to allantoin or allantoic acid before excretion (Razet, 1961); in such cases the lower yield of H_2O from the oxidation does not matter. It is important to note that NH_3 is essential for synthesis of chitin and the purine nucleotides that play a central role in metabolism; since in many diets protein is the major precursor of ammonia, some oxidative degradation is necessary whatever the "amino acid balance" in the protein. However, conversion of a large fraction of dietary protein to carbohydrate or fat is necessary only in insects adapted to diets abnormally rich in protein (e.g., keratin, meat and blood feeders like *Tineola, Phormia*, and *Rhodnius* respectively). If uric acid is the end-product, roughly 0.6 g of glucose plus 0.1 g of palmitic acid can be produced from 1 g of protein. The metabolic problem is the increased rate of the many complex pathways that degrade the essential amino acids, relative to the rate of protein synthesis. Many devices must have been evolved to ease these problems, e.g., the poor gut absorption of arginine and histidine by *Glossina*; these 2 amino acids contain about 20% of the total N in many proteins but a minor fraction of the useful energy, so that their conversion to uric acid is a large net loss.

[e]In many of the insect Orders, large amounts of uric acid are retained in the body (so-called "storage excretion"). In his extensive survey of the excretion of end-products of N metabolism by the Insecta, Razet (1961) noted that *Blattella* excretes little or none. The internal accumulation of solid uric acid complicates the mass-balance in many ways. It has less associated H_2O than other body components and may cause an increase in the ratio of dry wt to live body wt. It may bind Ca and mg ions and so increase the ash content and alter its composition. Since its wt is 52% of the wt of the protein precursor, its retention (if not corrected for) can give rise to calculated values of growth and food efficiency larger than the true values. Also, since 2 of the O_2 atoms in uric acid are derived from molecular O_2, there is a significant error in a mass-balance based exclusively on the input and output of dry solids. This error will be a reduction of the true rate of oxidative mass-loss, O. If the terminal products to which atmospheric O_2 has been added are retained, calculations of food efficiency will be too high; if they are excreted, calculated efficiency will be too low. Uric acid is not the only terminal metabolic product to which atmospheric oxygen is added. Many phytophagous insects have extremely high levels of oxidase systems that presumably add O_2 to some organic toxins present in the food (Krieger *et al.*, 1971). In such cases, however, the (%) of added O_2 will be much less than in uric acid (10%), and the major cause of inefficiency may be the loss of amino acids and glucose required for conjugation and excretion of the oxidation products.

*See explanation footnote a.

relatively slow pathway, but no ATP since the overall reaction is slightly exergonic. Since body fat content may be very high, its accumulation rate can be a major (and limiting) component of G on high-carbohydrate diets, and the 67% weight loss in the conversion will be a major component of O.

When the triglyceride store attains its limit, palmitoyl CoA will accumulate and inhibit acetyl CoA carboxylase. Glucose-6-phosphate will then accumulate and inhibit glucose phosphorylase, finally allowing the accumulation of free glucose. However, if the rate of glucose inflow exceeds the rate of utilization by triglyceride synthesis there will be glucose accumulation long before the fat depots are filled, so that mechanisms to prevent this will be activated. One can expect any phenotypic adaptability to act so as to increase the rate of fatty acid synthesis, on a carbohydrate-rich diet, to a level above that on a low-carbohydrate diet. Insects of the same species, maintained on very different diets, will therefore be somewhat—perhaps very—different.

A non-feeding stage in the insect life-cycle (eggs, pupae, or diapausing larvae or adults) must have an ample reserve of triglyceride, since this is not only a compact store of potential energy but also a rich source of potential metabolic water. The preceding stage must therefore accumulate triglycerides. This may involve changes in rates, limits, and priorities and possibly in food sources and feeding behavior.

The direct incorporation of dietary fatty acids into body triglycerides is not only much more economical but also probably much faster than fatty acid synthesis. One can therefore expect more rapid and efficient utilization if part of the carbohydrate in a carbohydrate-rich diet is replaced by 1/3 of its weight of triglyceride. Replacement of all carbohydrate by triglyceride, on the other hand, produces a carbohydrate deficiency since fatty acid cannot be reconverted to carbohydrate; only the glycerol component re-enters carbohydrate pathways and this is usually too little to satisfy the requirement.

Among the other pathways of glucose utilization, it is more difficult to evaluate quantities and priorities. Trehalose seems to be a higher-priority but much smaller reserve than glycogen (Sacktor, 1970). Synthesis of pentoses (for nucleosides and some polysaccharides) is associated with protein synthesis but the quantitative aspects are poorly known. Chitin synthesis is, of course, quantitatively very important in insects. If body volume is trebled at a molt, surface area and chitin content are doubled. At an imaginal molt a larger chitin increase is necessitated by the addition of large wing areas and the need for thicker cuticle. Since the N-acetylglucosamine precursor is synthesized during the non-feeding pre-molt period, the synthesis draws on reserves of glucose (presumably as glycogen), ammonia (presumably as glutamine), and acetyl groups (possibly from fatty acid).

The utilization of protein involves many complex pathways, and only an oversimplified presentation is possible here. In a growing or reproducing animal, the incorporation of dietary amino acids into body or egg proteins is the highest-priority major component of G. When there is an ample inflow of "balanced" amino acids, protein synthesis will proceed at some finite rate to an upper limit. If inflow exceeds this utilization level, the excess amino acids will be converted by a battery of enzymes to keto acids and ammonia. Ammonia can be utilized in glutamine synthesis, chitin synthesis and purine synthesis, the last being also the disposal pathway for excess ammonia by excretion of uric acid. Keto acids from the so-called "non-essential amino acids" can be rapidly oxidized or reversibly converted into glucose. Those from the essential amino acids, however, enter into irreversible pathways (that can easily become rate-limiting) and ultimately flow into either carbohydrate or fatty acid metabolism. As Table 1 shows, much energy can be drawn from protein oxidation, but this does not include the maintenance cost of the complex machinery involved; if this could be included protein would probably prove to be a much less satisfactory energy source. Many of these enzymes are adaptive, and their level is much reduced on low-protein diets, presumably to reduce the maintenance cost.

When better energy sources are available, it is imbalance—a large difference in composition, especially of the essential amino acids, between dietary and body proteins—that compels insects to maintain the oxidative pathways. There must be great selection pressure for similarity in amino acid composition between insect and food source, especially in essential amino acids, since this would minimize protein wastage. However, this can be effective only in relatively monophagous associations of long duration. Polyphages must be prepared to encounter a variety of amino acid imbalances, making use of an adaptable armament of oxidative enzymes and excreting much uric acid. A striking difference in the level of microsomal oxidases (which are also adaptive enzymes) exists between larvae of monophagous and polyphagous Lepidoptera (Krieger *et al.*, 1971), presumably because the polyphages encounter a much greater variety of toxic substances. One would expect a generally lower efficiency of utilization of food by polyphages, not only because of the high maintenance cost of all this defensive machinery at the phenotypic level, but also because of the chaos created in the gene pool by wildly fluctuating selection pressures. The genetic system of a polyphagous species must be a storehouse of genetic variability protected by balanced polymorphism; one can expect not only greater variability among individuals and populations than in monophages, but also a protean irreproducibility in their quantitative responses.

The effects of dietary amino acid imbalance are predictable. A mixture of only the essential amino acids will be a severe test of the metabolic capacity to oxidize the keto acids and will (if ample glucose is available) to a lesser extent test the biosynthetic pathways for all the non-essentials. Since these and other syntheses will utilize most of the ammonia, very little uric acid need be synthesized; since 1/2 of the amino acid intake must be degraded, most variations in the composition of the mixture will not matter, with the exception of an increase in the rate-limiting acids. The overall energy recovered from the oxidations will vary with composition, but will be less than that expected because much glucose will be utilized in the synthesis of non-essentials. Few insect species can grow on essentials alone (Dadd, 1970). Despite its general toughness, *Blattella* is not one of them. It is unlikely that such an imbalance occurs in nature, since no organism could function well with such unusual proteins.

The common natural imbalances are primarily deficiencies of one or more essential amino acids. A 25%-optimal level of any one requires the elimination of about 75% of all the others and so overloads both the oxidative degradation and uric acid synthesis pathways. The metabolic problem may be aggravated in various ways, e.g., by a deficiency of serine-glycine needed for ammonia disposal by the purine pathway. As Dadd (1970) has emphasized, some non-essentials are synthesized relatively slowly and can be rate-limiting when all other factors are optimal; these are proline and glycine, neither of which seems to be an efficient precursor of the other non-essentials. Proline resembles the essential amino acids in that it is synthesized and degraded by different, irreversible enzyme reactions; mutation and selection can therefore independently alter the enzyme activities and even lead to loss of the biosynthesis, as in *Bombyx*. Since proline is at the tip of a metabolic branch, seldom needing fast inflows or outflows, relatively slow metabolism makes sense. Glycine is a problem of a very different kind. It is often said to be "freely interconvertible" with serine, an amino acid that has often been shown to be a good precursor of the other non-essentials. Such statements seem to be largely based on the very fast exchange of isotopic carbon between the serine and glycine of the metabolic pool. Such exchanges prove nothing about any large-scale net conversion of one of these into the other; in fact there is on the serine side a large surplus of formyl groups that cannot be stored when serine is converted in large amounts of glycine, and so are irreversibly oxidized to CO_2. If glycine is to be reconverted to serine, the formyl groups needed must be derived from glycine itself, with liberation of ammonia and CO_2. Glycine therefore is not a good precursor of serine, and its amino group apparently cannot be readily transferred to pyruvic or other keto acids, so that it is an end-point, rather than a starting-point in amino acid metabolism. It is likely, however, that the conversion of serine to glycine can be fast (unless oxidation of the one-carbon fragments is rate–limiting). A deficiency of serine-glycine would then have no nutritional effect unless the synthesis of serine is slow. The fast utilization of serine as a precursor for other non-essentials probably depends on its easily reversible transamination with pyruvate to form alanine, with slower utilization of the hydroxypyruvate by reversible reactions into carbohydrate metabolism. The absence of such efficient direct transamination may be the major metabolic difficulty of both glycine and proline. The steps in serine synthesis between the fast pathways of carbohydrate metabolism and the final transamination are mere connecting links and there is no reason why they should be unusually fast.

Imbalances need to be more clearly understood at all levels: biochemical, nutritional, and evolutionary, since they are the result of natural selection. They are among the most important factors that have favored the evolution of very specific plant-insect-symbiote associations, and perhaps also of large, slow-growing, polyphagous animals that dispose of ammonia as urea.

To sum up, compilations like Table 1 (and the more elaborate ones that will some day become possible) offer hope that these problems, however, complex, are not beyond human understanding. They show that "construction costs"–that fraction of $\underline{Q}$ necessarily associated with $\underline{G}$–are a minor budget item, and open up the question of other costs. Maintenance costs may be estimated from turnover values, i.e., the frequency of resynthesis, that adds to $\underline{Q}$ without apparent change in $\underline{G}$, but these will not be exorbitant, at least on an optimum diet. Eventually we may reach a full cost-accounting, on marginal natural diets, and comprehend how this determines the success or failure of an insect population in nature.

EXPERIMENTAL

Objectives

The work reported here started from the observation that *Blattella germanica* reared on yeast-sucrose mixtures had a high rate of food consumption and growth at 30° C but excreted surprisingly little solid matter. The tedious collection and weighing of feces, essential when diet utilization is very inefficient, might therefore be dispensed with or allowed for by a minor correction. The earliest experiments were done by Bandal (1964), as part of a MS thesis under my direction, but he did not pursue this line of work. At this time I became aware of the difficulties in the interpretation of such data. Subsequent experiments were done at intervals, over a period of several years, and the data were transferred to IBM cards for later computer analysis. Limited resources discouraged any systematic application of techniques that might have allowed more detailed interpretation of data, although there were occasional determinations of dry body weight, fat-free body weight, and weight of excreta. The intent of the work was exploratory, essentially to see how far nutritional analysis could go by supplementing the usual data on live weight gain with correlated data on food intake.

Materials and methods

The basic component of all diets was dry powdered brewer's yeast, enriched with 40 μmoles/g of choline chloride added as an ethanol solution. This is also the main constituent of our stock colony diet, although this is further enriched with 3% of corn oil and supplemented with cubes of cane sugar. The advantage of yeast is that it is a very homogeneous mixture of all the nutrients required for first-generation growth, reproducible because of the relatively precise control of the brewing process and stable for long periods. Since corn oil is not as stable, it was omitted. The basic composition of yeast is also quite well-known and reproducible from sample to sample. With synthetic diets, ensuring uniform composition is a major problem.

For the 1st series of experiments dilutions of yeast with powdered confectioner's sugar (sucrose with a small amount of starch) were prepared by thorough grinding in a mortar. The 5 diets contained 100, 75, 50, 25 and 12.5% yeast. In the 2nd series various mixtures of yeast with other carbohydrates, sometimes with sucrose as a 3rd component were used. These were ground as finely as possible before mixing and then further ground after mixing. The cellulose used was the extremely fine microcrystalline form, "avicel."

The isolation of 1st-instar nymphs and the basic rearing techniques were as described by Gordon (1959). The initial live weight was always assumed to be 1 mg per nymph, although various samples had mean weights in the range from 1 to 1.2 mg and older nymphs that had fed on sugar and water for 1-2 days tended to be heavier than freshly-hatched ones. The food was weighed into EP-16 cups (polyethylene Caplugs from Protective Closures Co., Buffalo, N. Y.). In the early experiments, 50 mg of food was provided in the first stage and 100 mg in subsequent stages. In later work 25 mg was provided in the first, 50 mg in the second, and 100 mg in subsequent stages. The jars were observed every few days and then every day as the food supply neared exhaustion. On the day of complete disappearance the surviving insects were counted and weighed under CO_2 anesthesia. Since CO_2 is heavier than air and dissolves in the body water of the insects, care is needed to ensure that a min of this gas is being weighed. The elapsed time in days was recorded, and fresh food was weighed into the same cup for the following stage. The procedure was continued until adults appeared. All rearing was at 30° C.

Although Brooks (1965) claimed that CO_2 anesthesia greatly slowed the growth of *Blattella* nymphs, her data show that the effect occurred only in 1 stage and I believe that this erratic result may have been caused by some unidentified factor rather than by CO_2, since all other stages (although also treated with CO_2) were normal.

The advantage of this experimental design is its great convenience and simplicity. The disadvantages are the loss of information about individuals involved in the use of groups (with occasional cannibalism also), inaccuracy in time (which will be consistently overestimated), and the poor correlation between the arbitrary growth stages and the natural instars, which terminate at about 2.5, 5, 11, 24, and 40 mg with a 6th female instar terminating at 70 mg.

Each feeding stage was considered as an independent experiment with 6 measurements: the initial and final number of insects (N_i and N_f), the initial and final total live weights (W_i and W_f mg), the dry weight of food consumed (ΔF, mg) and the time (T days). *From these data, 2 basic rates were calculated by the method of Gordon (1968b): $\underline{F}$, the food intake rate, and $\underline{G}$, the growth rate, both in terms of mg/g/day.* To calculate $\underline{G}$, an estimate of $\underline{d}$, the ratio of dry to wet body weight, was necessary. This was obtained by drying samples of nymphs of various sizes from the stock colony, that gave values ranging from 0.27 to 0.33. Brooks (1965) had found similar values for *Blattella.* The variability in $\underline{d}$ makes the value of $\underline{G}$ less reliable; it may not be wholly random, since a few tests on insects with a very high lipid content (reared on diets very rich in sugar) gave some $\underline{d}$ values as high as 0.4, and there may be sexual dimorphism in the late stages of growth.

A rough estimate of the quantity of solid excretion was possible on 100% yeast diets, on which it represented ca 6% of the yeast intake; since recovery was not quantitative, it is probably nearer 8%. It may be less on mixtures with a low percentage of yeast, but no attempt was made to collect the minute amount of excretion observed in the rearing jars. The assumption was made that the contribution of the yeast component of a diet to the intake rate $\underline{I}$ was 0.92 $\underline{Y} \cdot \underline{F}$, where $\underline{Y}$ is the % yeast in the diet. Since yeast contains about 35% carbohydrate, it was assumed that this would be metabolically equivalent to any additional sucrose, leaving as a new parameter $\underline{I}_y$, the inflow rate of the non-carbohydrate fraction of yeast into metabolism, defined as 0.57 $\underline{Y} \cdot \underline{F}$. The ratio $\underline{G}/\underline{I}_y$ was included in all analyses as a potential indicator of the efficiency of conversion of the non-carbohydrate fraction of yeast into body mass. If this ratio is below 1, some of the yeast intake is being wasted. If it is above 1, some of the body weight gain must be derived from the carbohydrate intake. The parameter $\underline{O}+\underline{S}$ was calculated by subtracting $\underline{G}$ from $\underline{F}$. It is usually mostly $\underline{O}$, the oxidative mass-loss rate; despite the uncertainties involved in its estimation, I believe it is desirable to bring this parameter into the light of day instead of concealing it in a ratio.

All these assumptions can be justly criticized. However, because of the limited support available for this work, any gain in depth and precision would have involved a great loss in breadth and interest. Studies that aim at the greatest possible precision tend to be limited in scope (van Herrewege, 1971) and are so far unable to reduce the high intrinsic biological variation.

Analysis of data for yeast-sucrose mixtures

Table 2 presents data for each feeding stage at 5 concentrations of yeast, as means of coefficients of variation. Because of the difference in design between the earlier studies using 100 and 25% yeast and the later ones at other concentrations, it is not possible to compare the stage 1 values, which include roughly the first 3 instars in the early and the first 2 in the later work. For this reason, an additional analysis of the pooled data for stages 1 and 2 (total food supply 75 mg) is included for 75, 50, and 12.5% yeast. These values are somewhat more comparable to the stage 1 values for 100 and 25% yeast.

Even for experiments of similar design, however, the high variation of the mean values (often $>$ 10-20%) indicate that no conclusions can be based on small differences. The variation seems to be random, since medians (not included in the table) were generally close to the means and the 95% confidence range was near the actual range of values. Although there are many causes of variation, the major one may be inherent in the experimental design. Most of the food intake and weight gain occur during the first half of each instar. If the food supply is exhausted early in the instar, the calculated $\underline{F}$ and $\underline{G}$ will be high, but near the end of the instar they will be at a minimum. I had originally expected that the individual variation in instar duration would smooth out this effect, but with groups of only 10 insects at the start the natural periodicity may persist. Van Herrewege (1971), using groups of 3 to 5 carefully reared and selected adult male *Blattella* and a relatively precise technique, found that the coefficient of variation of the mean for daily measurements of food intake ranged from 30 to 40%. When group means were averaged, the coefficient was reduced to 10%, which is also the variability of body weight at the time of the imaginal molt; it is therefore possible that irreducible biological variability is of this order of magnitude.

Waldbauer (1964) obtained coefficients of variation of only a few percent in experiments with *Manduca* larvae during a single instar, but the application of the same technique to other lepidopteran larvae by Soo Hoo and Fraenkel (1966b) gave coefficients near 10%. McGinnis and Kasting (1959) found that the live body weight of *Agrotis orthogonia* larvae at the beginning of an instar has a coefficient usually above 10%, and it may be that the correspondingly high variability in daily food intake of individual larvae (Kasting and McGinnis, 1959) is correlated with the initial body weight differences so that F values for individual larvae would be less variable. This is not certain, however, since some of the variance in body size may arise from genetic differences affecting $\underline{G}$ and $\underline{O}$, and therefore $\underline{F}$; i.e., small individuals could (in proportion to their body weight) eat as much or even more than large ones but utilize the food less efficiently. There is no reason to assume that identical "physiological age" or even body size will

Table 2. Condensed data analysis for food intake and growth of *Blattella germanica* nymphs reared on yeast-sucrose diets of varying composition.[a]

% Y	Stage	n	ΔF	W_i/N_i	W_f/N_f	G/F	G/I_y	F	G	O + S	% Survival
100	1	9	50	1	9.4(9)	.40(8)	.70	96(9)	39(9)	57(12)	83(11)
	2	9	100	9.4(8)	23(13)	.29(11)	.51	68(11)	20(20)	49(10)	94(9)
	3	9	100	23(11)	38(12)	.29(14)	.51	43(17)	13(27)	30(15)	97(6)
	4	7	100	39(6)	52(6)	.25(10)	.44	44(12)	11(10)	33(14)	100
75	1	8	25	1	5.0(7)	.41(11)	.96	99(12)	41(13)	59(16)	89(9)
	2	8	50	5.1(6)	12(14)	.30(15)	.70	76(19)	23(22)	53(21)	96(8)
	1+2	8	75	1	12(19)	.35(7)	.82	90(15)	31(18)	59(15)	84(16)
	3	11	100	12(17)	26(17)	.30(15)	.70	60(14)	18(15)	42(18)	95(6)
	4	7	100	26(19)	38(11)	.22(13)	.51	46(18)	10(18)	36(20)	91(6)
50	1	10	25	1	4.7(10)	.36(14)	1.26	113(8)	41(13)	72(14)	87(12)
	2	10	50	4.7(10)	12(14)	.33(14)	1.16	71(11)	23(15)	47(14)	94(8)
	1+2	8	75	1	12(19)	.36(7)	1.26	90(9)	32(6)	58(9)	83(13)
	3	10	100	12(12)	25(7)	.27(12)	.95	59(12)	16(22)	43(10)	93(9)
	4	11	100	26(7)	37(8)	.23(19)	.81	51(15)	11(13)	39(19)	99(4)
	5	9	100	38(7)	47(6)	.20(16)	.70	42(20)	8(19)	34(22)	100
25	1	10	50	1	7.4(12)	.30(7)	2.10	104(14)	31(15)	73(14)	82(14)
	2	10	100	7.7(12)	18(9)	.23(9)	1.61	75(6)	17(11)	58(7)	93(9)
	3	6	100	18(14)	30(11)	.17(11)	1.19	53(15)	9(7)	44(17)	85(10)
	4	7	100	29(10)	39(13)	.16(15)	1.12	41(9)	6.7(11)	34(11)	96(7)
12.5	1	7	25	1	3.4(11)	.21(10)	2.95	104(4)	22(7)	82(6)	83(14)
	2	11	50	3.6(14)	7.6(11)	.14(9)	1.96	64(11)	8.9(12)	55(12)	86(14)
	1+2	8	75	1	7.5(11)	.17(10)	2.38	87(7)	14(6)	73(8)	70(19)
	3	6	50	7.8(8)	13(12)	.14(13)	1.96	48(8)	6.8(7)	41(7)	92(12)
	4	10	100	13(14)	20(11)	.14(14)	1.96	52(14)	7.1(20)	45(14)	97(6)
	5	4	100	21(8)	29(8)	.12(18)	1.69	42(18)	4.9(10)	37(21)	96(6)

[a]Table heading abbreviations: % Y, percent yeast in diet. Stage, feeding stage. n, number of experiments. Δ F, mg of diet consumed. W_i/N_i, W_f/N_f, mean live wt (mg) of nymphs at beginning and end of stage. G/F, ratio of dry body wt gain (assumed to be 30% of live wt gain) to food intake. G/I_y, ratio of dry body wt gain to non-carbohydrate yeast absorption (assumed to be 57% of total yeast intake). F, G, food intake and growth rates in mg dry wt/g live body wt/day. O + S, rate of food loss by oxidative metabolism (O) and solid excretion (S), obtained by subtracting G from F; S could be estimated by multiplying F by 0.08 and by (% Y)/100, but its highest value (for stage 1 on 100% yeast) would be less than 8 % Survival, percent of number of nymphs at beginning of stage that survived to end of stage. Values in parentheses are coefficients of variation.

ensure uniformity in any other parameter.

In the very thorough study of Hirano (1964) on growth of *Chilo suppressalis* on stems of rice plants grown under varying conditions, the coefficient of variation of the live body weight of groups of 30-day old larvae was often less than 5%, but tended to rise about 10% in experiments involving food of low nutritive value (indicated by poor growth and survival). A similar increase in variability occurred in his experiments using synthetic diets of varying composition when survival and growth rates were reduced by lowering the protein content. It is noteworthy that mean larval weights in replicate experiments using living rice plants could differ by more than 20%, in spite of variations below 5% in individual experiments. In Waldbauer's (1964) work on *Manduca* fed living leaves of *Lycopersicon*, mean $\underline{G}$ values for presumably identical experiments (separated by a 2-week interval) differed by 25%.

Although it is regrettable that values for nutritional parameters cannot yet be determined with great precision, the means in Table 2 do show certain consistencies. $\underline{F}$ values for the early instars are not far from 100 for all 5 diet compositions and decline to about 40 during the 5th instar. The data of van Herrewege and David (1969) on the intake of a casein-glucose synthetic diet (with 25% alumina added) can be recalculated to give an $\underline{F}$ value (at 30° and for the non-alumina constituents only) of 110 for 1st instar nymphs and 35 for adult males. These workers found that the mean intake (of non-alumina constituents) was 0.3 mg/insect on both days 1 and 2 of the 1st instar, 0.19 on day 3, 0.11 on day 4, and 0.0 on day 5 when molting occurred. Their main objective was the measurement of the "coefficient of digestibility," which was not far from 84% for all instars and both sexes and represents $\underline{S}$ values (corrected to exclude alumina) of 19 for 1st instar nymphs and 5.6 for adult males. These are very high for a theoretically highly-digestible diet, and it is conceivable that the high alumina content was not inert but may have interfered with digestion or absorption.

Unlike the $\underline{F}$ values, the $\underline{G}$ values in Table 2 are affected by diet composition. The $\underline{G}/\underline{F}$ ratio is relatively high in the range of 50-100% yeast but declines by ca 25 to 50% at 25% and 12.5% yeast. One might expect the yeast content to be severely rate-limiting at 12.5% yeast, but this would be true only if the ratio $\underline{G}/\underline{I}_y$ were constant. In fact, this ratio rises as the yeast content falls. If we neglect the 1st feeding stage because of possible "transition effects," we see that nearly half of the intake of non-carbohydrate yeast must be somehow excreted on the 100% yeast diet, while on the 12.5% yeast diet nearly half of the dry body weight gain must be derived from the carbohydrate intake. This implies a drastic change in the composition of $\underline{G}$.

There have been no systematic studies of the body composition of *Blattella*. The analyses of Lipke *et al.* (1965a, b) on *Periplaneta americana* suggest a maximum content of glucosamine (mostly in the cuticle) of 8% of the dry body weight, and not more than 2% of glycogen plus trehalose. It seems unlikely that much more than 10% of $\underline{G}$ consists of substances with the unaltered hexose carbon skeleton. The well-known exchange of glucose carbon with that of non-essential amino acids cannot involve any large *net* incorporation of carbohydrate into protein. Most of the incorporation must be into lipids.

The lipid content of yeast is only about 1.5%, so that even on a 100% yeast diet the retention of all the lipid intake would contribute less than 4% of $\underline{G}$ while on 12.5% yeast it would be negligible. The early work of Melampy and Maynard (1937) reported lipid contents from 1.7% of the live body weight in adult males to 5.7% in large nymphs on their standard diet, although values ranged from 7.8% to 21% when the diet contained a high proportion of fat. This would correspond to from 6 to more than 40% of the dry body weight. Rough analyses in our laboratory (based on weight loss after chloroform extraction) indicate lipid contents of 10% of the dry body weight in adult males and 13% in large nymphs reared on 100% yeast. For nymphs reared on 25% yeast-75% sucrose the lipid content was 33% of the dry body weight, and an additional 4% could be extracted with 2:1 chloroform:methanol. The major lipid constituent on thin-layer chromatograms was triglyceride, as in *Periplaneta* (Kinsella and Smyth, 1966).

If the lipid content of nymphs reared on 25% yeast is in fact in the range of 35-40%, one might expect $\underline{G}/\underline{I}_y$ to be about 1.5, but in the later stages it is less than 1.2. There are 2 possible explanations for this discrepancy. Either a large fraction of the non-carbohydrate yeast intake is being excreted—not likely when the yeast content is low—or the true value of $\underline{G}$ is 25-30% higher than the calculated one. This would be possible if the assumed value of 0.3 for $\underline{d}$ is incorrect; a change to 0.4 would increase $\underline{G}$ by 33%. Probably both phenomena occur, since 100% efficiency of yeast utilization is unlikely. If the dry body weight (and also its lipid content and N content) were precisely known, some useful conclusions could be derived by calculation. In order to illustrate the principle, a hypothetical analysis will be outlined. In stage 3 on 25% yeast, the true value of $\underline{G}$ is greater than the listed value of 9 but not higher than 12 (based on a d of 0.4). $\underline{I}_y$ is 7.5, to which we can add about 10% direct incorporation of the carbohydrate intake (as glucosamine, glycogen, trehalose), or 0.8 as $\underline{G}_c$, for a total fat-free $\underline{G}$ of 8.3. The contribution to $\underline{G}$ from fat synthesis then lies between 0.7 and 3.7; this $\underline{G}_f$ is the rate of fatty acid synthesis, which must be limiting the total $\underline{G}$. If it is 3.7, the fat content will be 31%. Using the same reasoning we can compare stage 5 on 12.5% yeast, which has about the same body weight range of 20-30 mg. Here $\underline{G}$ lies between 4.9 and 6.5, while the 3.0 $\underline{I}_y$ and 0.3 $\underline{G}_c$ add up to 3.3, so that $\underline{G}_f$ lies between 1.6 and 3.2. The fat content would be ca 50%, but the rate of synthesis would be about the same (especially if calculated on the fat-free wet weight instead of the total live weight). Protein synthesis would be $<$ 50% as fast as at 25% yeast. The importance of this approach is that the true rate-limiting reaction, lipid synthesis, is identified, although the carbohydrate precursor is not the rate-limiting nutrient by conventional standards. The increase in $\underline{G}$ on 25% yeast relative to 12.5% yeast involves an increase in $\underline{I}_y$ of 4.5 mg/g·day (from 3.0 to 7.5) and an increase in $\underline{I}_c$ of 7 (from 39 to 46). The effect is an increase in $\underline{G}_y$, that portion of growth metabolism requiring the components provided by the non-carbohydrate fraction of yeast, that was operating far below capacity on the 12.5% yeast diet, but probably no change in $\underline{G}_f$ and only an equal increase in $\underline{G}_c$. It is because there are 2 important kinds of growth–in active tissues and in lipid reserves–that *Blattella*, like many other animals, can consume and utilize at a high rate diets of very different composition.

For the sake of completeness we can also try to extend this analysis to the $\underline{O}$ component of the examples given above. Each will have an $\underline{O}_f$ of about 7, the carbohydrate weight loss involved in the $\underline{G}_f$ synthesis of about 3.5. Since the body weight range is the same, a maintenance $\underline{O}$ of about 20 can be assumed. This leaves 17 $\underline{O}$ unaccounted for at 25% yeast and 10 at 12.5% yeast, which may be the energy cost of the corresponding $\underline{G}$ values of 12 and 6.5.

The above would be considered by mammalian nutritionists a rather crude 2-compartment analysis of growth. Although studies of insect body composition were initiated many years ago (Teissier, 1929), they have lagged far behind the richly-subsidized mammalian work, which has even subdivided the fat-free body mass into compartments, e.g., fat-free adipose tissue, muscle, etc. (Pitts, 1963; Anderson, 1963).

During growth on any one diet, $\underline{F}$ tends to decline to about 50% of the stage 1 value when the nymphs approach maturity and $\underline{G}$ to about 25%, so that $\underline{G}/\underline{F}$ must decline to about 50%. Therefore $\underline{O}$ declines more slowly, probably because a considerable fraction is a relatively constant maintenance metabolism. If we apply the analysis used above for the later stages of growth to the earlier stages, however, this is roughly correct; e.g., for stage 2 on 12.5% yeast $\underline{G}$ lies between 8.9 and 12, with an $\underline{I}_y$ 4.5 and $\underline{G}_c$ of 0.5, giving a $\underline{G}_f$ between 3.9 and 7. Although it is not impossible that the rate of fat synthesis is higher in young *Blattella*, a fat content above 50% of the dry weight seems impossible, so a $\underline{G}_f$ of 5 or less is more reasonable. The corresponding $\underline{O}_f$ of 10 and estimated maintenance of 20 leaves a residual $\underline{O}$ of ca 25. This could represent a somewhat higher cost of growth than that of larger nymphs, but it is also possible that the maintenance cost could be as high as 30. In either case, the maintenance $\underline{O}$ would be the most stable of all the nutritional parameters. There has been only marginal interest in this area of research because of the isolation between work in insect nutrition and insect biochemistry.

It is clear that merely adding data on food intake to growth data does not allow more than a speculative analysis of the phenomena, and I believe that this is equally true of work that also measures solid excretion and respiratory exchanges or energy content by calorimetry. Information on food and body composition is essential. This presents a hazard, since modern analytical technology can all too easily generate a bewildering mass of compositional details, of which only a few may be of exceptional interpretive significance, and which may still be insufficient or incomplete. Although the work of Hirano on the utilization of rice stems by *Chilo* is perhaps the most serious attempt to obtain a balance sheet so far made in insect nutrition, he acknowledged (personal communiction, 1965) that "over 50% of utilized food remains unaccounted for."

The effect of adding various carbohydrates to yeast or yeast-sucrose mixtures on food intake and growth

This study was a blind search for unusual phenomena, using carbohydrates as tools to induce them. Most of the large number of experiments were not replicated more than once, so that only fairly large changes from the basic patterns of Table 2 are meaningful and only a few selected experiments were of sufficient interest to present in Table 3. The experiment numbers, from the original protocols, are included only for identification, while the order of tabulation is intended to facilitate the discussion.

Dilution with an equal weight of cellulose (320 and 326) caused a roughly 2-fold increase in $\underline{F}$ and a concomitant increase in $\underline{S}$, the rate of solid excretion. All the ingested cellulose was recovered (at the end of an experiment) as hard, dry fecal pellets. The "compensatory" increase in $\underline{F}$ is in general insufficient to provide the same nutrient intake attained on undiluted diet, but there is no clear discriminatory effect on $\underline{G}$ and $\underline{O}$. At 75% cellulose, all nymphs died during the first feeding stage when the nutrient was yeast or 1 : 1 yeast : sucrose, but they survived and grew with a 1 : 3 yeast : sucrose mixture (336). The reason may be that $\underline{F}$ cannot much exceed 200 mg/g·day that is attainable only on the 1 : 3 : 12 yeast : sucrose : cellulose mixture. The effect of reduced nutrient intake is clear in stages 2 to 4 of exp. 336 (stage 1 is not interpretable because of the "transition effect"). When compared to nymphs in the same weight range on an undiluted diet, $\underline{G}$ is less than 1/2 and $\underline{O}$ about 2/3 normal. My earlier interpretation (Gordon, 1968b) that $\underline{G}$ has a "higher priority" than $\underline{O}$ relied on first-stage values and now seems nonsensical. At the time I was not fully aware of the universality of "transition effects," and I naively believed that roaches had some ill-defined control over $\underline{O}$ that would allow them to "burn away" substrates within wide limits (or to refrain from doing so). The abnormally low rates of stages 2 to 4 (336) are also a "transition effect," since mortality in stage 5 was so high that the experiment was terminated. The calculated values of $\underline{O}$ in this experiment are unreliable and probably too

Table 3. Effect of adding various carbohydrates to yeast-sucrose diets.[a]

Expt. No. and component % Yeast	% Sucrose	% Carbohydrate	W_f/N_f	G/I_y	F	G	O
50	0	50	8	0.7	201	41	60
	(320, cellulose)		17	1.0	118	21	38
			25	0.6	91	10	36
			29	0.06	51	0.5	26
12.5	37.5	50	6	2.3	163	26	56
	(326, cellulose)		12	1.8	132	17	49
			16	0.4	73	2	35
			26	2.7	84	16	26
6.25	18.75	75	5	3.0	302	32	43
	(336, cellulose)		7	0.8	178	5	39
			10	1.2	176	8	36
			13	1.0	200	7	43
50	0	50	5	1.9	105	56	49
	(863, D-glucose)		12	1.2	63	21	42
			24	1.1	70	22	48
			38	0.9	76	20	56
25	0	75	3	2.4	170	57	(113)
	(992, inulin)		6	0.2	124	3	(121)
			11	0.8	119	13	(106)
			18	0.9	88	12	(76)
			23	0.8	136	16	(120)
			28	0.8	60	6	(54)
			31	0.5	79	5	(74)
			36	0.7	52	5	(47)
50	0	50	5	1.6	97	45	52
	(842, D-ribitol)		12	0.8	75	18	57
			25	0.9	45	12	33
			39	1.0	55	16	39
50	0	50	4	1.1	75	24	51
	(942, L-rhamnose)		10	0.7	48	9	39
			22	0.6	40	7	33
			32	0.6	32	5	27
50	0	50	5	0.8	56	12	44
	(782, D-xylose)		24	0.9	43	11	32
50	0	50	5	0.5	64	8	56
	(802, L-xylose)						
75	0	25	4	0.8	84	28	56
	(944, L-rhamnose)		10	0.5	70	15	55
			26	0.7	59	18	41
			45	0.7	55	15	40
75	0	25	3	0.7	91	26	65
	(805, L-xylose)		9	0.4	75	13	62
			22	0.7	68	19	49
			36	0.7	54	15	39
			51	0.7	54	17	37
75	0	25	3	0.6	72	17	55
	(745, D-ribonolactone)		10	0.6	59	16	43
			24	0.5	51	11	40

Table 3. (cont'd.)

Expt. No. and component % Yeast % Sucrose % Carbohydrate			W_f/N_f	G/I_y	F	G	O
25	25	50	4	2.7	65	25	(40)
(966, D-cellobiose)			9	1.7	80	20	(60)
			21	1.9	37	10	(27)
			29	1.3	60	11	(49)
			37	1.2	41	7	(34)
12.5	37.5	50	4	2.6	53	12	(41)
(968, D-cellobiose)			11	1.9	41	7	(34)
			20	0.9	38	2	(36)
			34	0.8	29	2	(27)

[a]Column headings are the percentages of yeast, sucrose, and of the other carbohydrate (named in parentheses together with the experiment number) in the diet; other symbols are defined in Table 2. The O values for the 3 cellulose dilutions have been corrected by subtracting the S component, assuming complete excretion of the cellulose intake. This correction has not been done (for reasons explained in the text) for inulin or cellobiose; therefore the O values are in parentheses to indicate the inclusion of an unknown level of S. In the cellulose experiments 100 mg of food was provided at each stage; in all the others, 25 mg was provided in the first stage, 50 mg in the second, and 100 mg in all succeeding stages.

high, since some of the nutrients may have been swept out of the gut, trapped in the large mass of fast-flowing cellulose. The low values of G/I_y correspond to a yeast : sucrose ration in Table 2 of 1 : 1 instead of 1 : 3, and it is possible that some sucrose has been excreted. Sucrose diluted with cellulose was much less satisfactory than pure sucrose for maintenance of adult male *Blattella*, despite a compensatory increase in F (Gordon, 1968a). Without a careful analysis of the nutrients present in the excreta, no definite conclusions are possible; this is of course one of the major difficulties in all experiments where S is large and only its mass or energy content is known. It should be noted that dilution of well-defined synthetic diets with either 25% cellulose or 25% alumina allows normal growth and reproduction for several generations (Gordon, 1959; van Herrewege and David, 1969), but in neither case have there been comparisons with undiluted diets.

The fructose polymer, inulin, gave quite different results (exp. 992). F was abnormally high and there was much solid excretion (as soft, rather moist pellets) but survival was good, G was sustained, and G/I_y was similar to that on a 3 : 1 yeast : sucrose diet (Table 2). In the body weight range from 28 to 36 mg., G was apparently limited by the yeast intake rate. The slow growth may reflect the difficulty of passing the gelatinous hydrated inulin through the digestive tract; when the experiment was terminated (because the original food sample was used up) the nymphs were in good health. Similar experiments with 75% starch are not shown because the results were very similar to 75% sucrose, as one might expect from the complete digestibility of starch and the very low $\underline{S}$.

The next group of experiments involved dilution of yeast with an equal weight of various soluble sugars. D-glucose (exp. 863) was comparable to sucrose, as were many other sugars (melibiose, raffinose) known to be well utilized by *Blattella.* D-ribitol (exp. 842), L-arabinose, D-galactose, D-galactitol, and L-rhamnose (exp. 942) also allowed growth, although it is probable that multi-replicated experiments would reveal statistically significant inferiority to sucrose. The difference was most striking with L-rhamnose, for which $\underline{G}$ values were only about half of those for the 1 : 1 yeast : sucrose in Table 2. The low $\underline{G}/\underline{I}_y$ values suggest that rhamnose was mostly wasted; rhamnose proved useless for survival of adult males and depressed $\underline{F}$ in a 1 : 1 mixture with sucrose (Gordon, 1968a). In the earlier study of Gordon (1959) glucose, L-arabinose and D-ribitol could completely replace sucrose as sole carbohydrate source in a synthetic diet for *Blattella*; in the present work it seems that the last 3 sugars have at least some sparing action when added to the carbohydrate present in yeast, and so must enter into some useful metabolic pathways. With D-xylose (exp. 782) and L-xylose (exp. 802), which caused reduction of $\underline{G}$ to less than 1/3 and of $\underline{O}$ to less than 2/3 of the sucrose control values, death soon ended the experiments. Many other sugars (D-arabinose, D-arabonolactone, D-ribonolactone, D-galactonolactone, and D-glucuronolactone) were so deleterious that even the first feeding stage could not be completed; however, these effects were much less severe at a 25% level of these sugars. With 25% L-rhamnose (exp. 944) $\underline{G}$ was lowered only during the early period of growth (to ca 10 mg) and subsequently, was in the normal range. The effect of 25% L-xylose (exp. 805) was similar, suggesting that some marginal utilization may be possible (largely because $\underline{G}/\underline{I}_y$ is consistently above the values for pure yeast diet). However, 25% D-ribonolactone (exp. 745) seems to exert a purely inhibitory effect, although its inflow rate is at a level within the capacity of the enzyme systems. One can roughly calculate inflow rates (during the first feeding stage) for the 50 and 25% dietary levels: 37 and 21 mg/g·day for L-rhamnose and 32 and 33 mg/g·day for L-xylose, while for 25% D-ribonolactone it is only 18 mg/g·day. These are presumably the maximal metabolic rates of these exotic sugars, that limit the value of $\underline{F}$. How they are metabolized is still unknown, but some unpublished studies with D-arabinose-U-^{14}C have shown that only part of the radioactivity is recoverable as CO_2; the remainder is incorporated in many long-lived body constituents, including some of high molecular weight. With glucose-U-^{14}C one recovers even less CO_2 and finds even more widespread incorporation. Such studies, however, prove only that there is intermingling of metabolic pathways and exchange of carbon, not that there is any *net* incorporation of the sugar into $\underline{G}$, or any *net* gain in high-energy phosphate. It is easy to misinterpret the results of radiotracer experiments, e.g., the fact that the carbon atoms of acetate flow freely into body carbohydrates lent credence to the erroneous belief that a *net synthesis* of glucose from acetate was possible.

Cellobiose had very unusual effects. When an equal weight was added to a 1 : 1 yeast : sucrose diet (exp. 966), $\underline{F}$ was depressed to nearly 1/2 the normal but $\underline{G}/\underline{I}_y$ was abnormally high during the first stage. We must conclude that much of the ingested cellobiose was retained. Since there is evidence that it is not hydrolysed or metabolised, the retention must be as unchanged cellobiose, probably in the gut with associated water. Unfortunately, the excreta produced were not easy to collect and measure, but they contained cellobiose. The effect was similar when the nutrient component was a 1 : 3 yeast : sucrose (exp. 968), although growth was exceptionally slow, since the yeast inflow was so greatly reduced. Interpretation of the phagodepressant effect of cellobiose can only be speculative. It may slow the rate of crop-emptying, the rate of water extraction from the rectum, and other activities of the digestive tract. If some enters the hemolymph it will tend to accumulate there (since the rate of clearance of cellobiose injected into the hemolymph is extremely slow) and the rise in osmotic pressure may reduce hunger in some way. In any case, cellobiose cannot be disposed of at a rate much above 30 mg/g·day.

It is of interest that the most unfavorable modifications in Table 3 – 75% cellulose, 50% D-xylose, 50% cellobiose - all seem to be approaching, in very different ways, some lower limit of nutrient intake below which the insects cannot long survive. This limit is of the order of 30 mg/g·day, a large fraction of which must be a carbohydrate that can yield high-energy phosphate. Despite many uncertainties, there seems to be a roughly 3-fold range between this minimal metabolic rate and the maximal rates given in Table 2. It is even possible to discern, in vague outline, its organization. At the core is a fairly stable energy requirement involving a carbohydrate $\underline{O}$ of ca 20 mg/g·day. Once this is satisfied, the $\underline{G}$ mechanism can utilize a well-balanced complex of nutrients (amino acids, salts, vitamins, etc.) in a wide range of rates, from 10 to 40 mg/g·day in the early instars of 2 to 10 mg/g·day in the late instars; somehow associated with this $\underline{G}$ is an $\underline{O}$ of the same order of magnitude (and so much higher than the simple cost of biosynthesis of Table 1). When there is a dietary surplus of carbohydrate, a new component of $\underline{G}$ becomes important; this is triglyceride synthesis, associated with an $\underline{O}$ about twice as large, and perhaps representing a "tolerance" rather than a requirement like the basic $\underline{G}$ and $\underline{O}$ mechanisms. Additional tolerances are the ability to ingest and excrete cellulose (and presumably some other indigestible polymers) at a rate of 100 mg/g·day and to metabolize many exotic sugars at a rate of ca 20 mg/g·day. Much of the future development of nutrition, at least for polyphagous insects, will require definition of tolerances rather than mere nutritional requirements; one can reasonably expect that areas of tolerance will not show the depressing monotony of conventional nutrition.

SUMMARY

Nutritional data (except H_2O and O_2 inflow and H_2O and CO_2 outflow) are elements of the dry-solid mass-flow rates in the equation $\underline{G} + \underline{O} + \underline{S}_i + \underline{S}_e = F$. The size of $\underline{F}$ and its partition among the 4 other rates are determined by the effort of the genotype to optimize the size and composition of $\underline{G}$. Part of $\underline{G}$ is strongly stabilized in composition, the major component being protein synthesis, that can utilize only well-balanced inflows of amino acids. The part of $\underline{G}$ that comprises nutrient reserves is more variable, the major component being triglyceride synthesis, into which unbalanced surpluses from amino acid or carbohydrate metabolism can irreversibly flow, in addition to deitary fatty acids. Each of the many subsystems of $\underline{G}$ and $\underline{O}$ has priorities and upper limits that can, on various diet compositions, control the overall rate. Some have lower limits below which prolonged survival is not possible. Measurements of the nutritional parameters have high coefficients of variation, and are also complicated by the instability of parameters during normal development and by "transition effects," transient adaptive states determined by the dietary and other environmental conditions antecedent to the experiment.

Experiments on *Blattella germanica* nymphs reared from the 1st instar on mixtures of yeast and sucrose yielded F values near 100 mg/g·day that declined to about 40 mg/g·day in the last instar, in the entire range from 12.5 to 100% yeast. $\underline{G}$ values declined from 40 to 10 mg/g·day during growth on 50 to 100% yeast, from 30 to 7 on 25% yeast, and from 22 to 5 on 12.5% yeast. Only tentative interpretations were possible, since data on the composition of $\underline{G}$ were lacking. One can estimate that, at yeast levels of 25% or less, fat synthesis from carbohydrate contributed ca 3 mg/g·day to $\underline{G}$ and utilized dietary carbohydrate at $<$ 10 mg/g·day. In $\underline{O}$ the presence of 2 components that oxidize carbohydrate for energy can be inferred: a maintenance component in the range of 20 - 30 mg/g·day and a growth component with a rate of 1.5 - 2 times the total fat-free $\underline{G}$.

Experiments in which other carbohydrates were included in the diet indicated that $\underline{F}$ can be limited by $\underline{S}_e$ that, for the indigestible cellulose, cannot much exceed 200 mg/g·day. With various unusual monosaccharides, $\underline{F}$ seems to be limited by the capacity of the enzyme systems to oxidize (or otherwise metabolize) the intake. Some, such as D-ribitol or L-rhamnose, can be utilized at rates of ca 40 mg/g·day and apparently contribute to $\underline{G}$; others, such as D-ribonolactone, can be utilized at rates $<$ 20 mg/g·day and seem to be useless. Cellobiose had a phagodepressant effect that cannot yet be interpreted. Many experiments indicate that the minimal intake of useful nutrients must be ca 30 mg/g·day, most of which flows into the maintenance $\underline{Q}$.

Acknowledgements.—The experimental work was done at Berkeley, but the writing was continued at the Democritus Nuclear Research Center in Athens, Greece, and completed at Laboratoire d'Entomologie et d'Ecophysiologie Universite Paris-Sud, 91 Centre d'Orsay, France. Among the many individuals who have contributed to the development of this paper, I wish to cite especially R. F. Smith, D. D. Jensen, R. H. Dadd, G. Fraenkel, G. P. Waldbauer, A. P. Economopoulos, J. R. Le Berre, and R. W. McGilvery.

REFERENCES

ANDERSON, E. C. (1963). Three-component body composition analysis based on potassium and water determinations. *Ann. N. Y. Acad. Sci.* **110**:189-210.

BANDAL, S. K. (1964). Effects of various chenicals on growth and food intake in the German roach. M. S. Thesis, University of California, Berkeley.

BERTALANFFY, L. VON (1960). Chap. 2, pp. 137-259, in "Fundamental Aspects of Normal and Malignant Growth," edited by W. W. Nowinski, Elsevier Pub. Co., Amsterdam.

BRODY, S. (1945). Bioenergetics and growth: with special reference to the efficiency complex in domestic animals. 1023 pp. Reinhold, N. Y.

BROOKS, M. A. (1965). The effects of repeated anesthesia on the biology of *Blattella germanica* (Linnaeus). *Ent. Exp. & Appl.* **8**:39-48.

BROZEK, J. (1963). Roots and goals of the symposium. *Ann. N. Y. Acad. Sci.* **110**:7-8.

DADD, R. H. (1970). Arthropod Nutrition. Chap. 2 in Chemical Zoology, vol. 5. Academic Press, Inc., N. Y.

DAVIS, B. D. (1961). The teleonomic significance of biosynthetic control mechanisms. *Cold Spr. Harbor Symp. Quant. Biol.* **26**:1-10.

FORD, E. B. (1965). Ecological Genetics. 335 pp. Methuen & Co. Ltd., London.

GELPERIN, A. (1971). Regulation of Feeding. *Ann. Rev. Entomol.* **16**:365-378.

GORDON, H. T. (1959). Minimal nutritional requirements of the German roach, *Blattella germanica* L. *Ann. N. Y. Acad. Sci.* **77**:290-351.

GORDON, H. T. (1968a). Intake rates of serious solid carbohydrates by male German cockroaches. *J. Insect Physiol.* **14**:41-52.

GORDON, H. T. (1968b). Quantitative aspects of insect nutrition. *Am. Zoologist* **8**:131-138.

GRAY, J. (1929). The kinetics of growth. *Brit. J. Exp. Biol.* **6**:248-274.

HAMMOND, J. (1962). The physiology of growth. Pp. 18-25 in "Nutrition of Pigs and Poultry," edited by J. T. Morgan and D. Lewis. Butterworths, London.

HIRANO, C. (1964). Studies on the nutritional relationships between larvae of *Chilo suppressalis* Walker and the rice plant, with special reference to role of nitrogen in nutrition of larvae. *Bull. Nat. Inst. Agric. Sci.* (Japan), ser. C., **17**:103-180.

KASTING, R. and MCGINNIS, A. J. (1959). Nutrition of the pale western cutworm, *Agrotis orthogonia* Morr. (Lepidoptera: Noctuidae). II. Dry matter and nitrogen economy of larvae fed on sprouts of a hard red spring and a durum wheat. *Can. J. Zool.* **37**:713-720.

KINSELLA, J. E. and SMYTH, T. (1966). Lipid metabolism of *Periplaneta americana* during embryogenesis. *Comp. Biochem. Physiol.* 17:237-244.

KRIEGER, R. I., FEENY, P. P. and WILKINSON, C. F. (1971). Detoxication enzymes in the guts of caterpillars: an evolutionary answer to plant defenses? *Science* 172:579-581.

LAIRD, A. K., TYLER, S. A. and BARTON, A. D. (1965). Dynamics of normal growth. *Growth* 29:233-248.

LEWONTIN, R. C. (1965). Selection for colonizing ability. Pp. 77-94 in "The Genetics of Colonizing Species," edited by H. G. Baker and G. L. Stebbins. Academic Press, N. Y. 588 pp.

LIPKE, H., GRAINGER, M. M. and SAIKOTOS, A. N. (1965a). Polysaccharide and glycoprotein formation in the cockroach. I. Identity and titer of bound monosaccharides. *J. Biol. Chem.* 240:594-600.

LIPKE, H., GRAVES, B. and LETO, S. (1965b). Polysaccharide and glycoprotein formation in the cockroach. II. Incorporation of D-glucose-C^{14} into bound carbohydrate. *J. Biol. Chem.* 240:601-608.

MCGILVERY, R. W. (1970). Biochemistry, a functional approach. 769 pp. W. B. Saunders Company, Philadelphia.

MCGINNIS, A. J. and KASTING, R. (1959). Note on a method of comparing diets for the pale western cutworm, *Agrotis orthogonia* Morr. (Lepidoptera: Noctuidae). *Canadian Entomol.* 91:742-743.

MELAMPY, R. M. and MAYNARD, L. A. (1937). Nutrition studies with the cockroach (*Blattella germanica). Physiol. Zool.* 10:36-44.

MOREAU, J. P. (1971). Das Verhalten des Kartoffelkäfers (*Leptinotarsa decemlineata* Say) gegenüber drei Kartoffelsorten. *Acta Phytopathologica Academiae Scientiarum Hungaricae.* 6:165-168.

MORRIS, R. F. (1967). Influence of parental food quality on the survival of *Hyphantria cunea. Canad. Entomol.* 99:24-33.

PITTS, G. C. (1963). Studies of gross body composition by direct dissection. *Ann. N. Y. Acad. Sci.* 110:11-22.

RAZET, P. (1961). Recherches sur l'Uricolyse chez les Insectes. *Bull. Soc. Scient. de Bretagne* 36:1-206.

SACKTOR, B. (1970). Regulation of intermediary metabolism, with special reference to the control mechanisms in insect flight muscle. Pp. 267-347 in vol. 4 of "Advances in Insect Physiology," edited by J. W. L. Beament, J. E. Treherne and V. B. Wigglesworth.

SERRA, J. A. (1966). Modern Genetics. Vol. 2, Chap. 12, Principles of gene action and phenogenesis. Academic Press, N. Y.

SOO HOO, C. F. and FRAENKEL, G. (1966a). The selection of food plants in a polyphagous insect, *Prodenia eridanea* (Cramer). *J. Insect Physiol.* 12:693-709.

SOO HOO, C. F. and FRAENKEL, G. (1966b). The consumption, digestion, and utilization of food plants by a polyphagous insect, *Prodenia eridanea* (Cramer). *J. Insect Physiol.* 12:711-730.

TEISSIER, G. (1929). Dysharmonies chimiques dans la croissance larvaire de *Tenebrio molitor* L. *C. R. Soc. Biol.* 100:1171-1173.

VAN HERREWEGE, C. (1971). Consommation alimentaire chez les males adultes de *Blattella germanica* (L.). Influence de l'age, de la nourriture larvaire et du jeune. *Arch. Sci. Physiol.* 25:401-413.

VAN HERREWEGE, C. and DAVID, J. (1969). Une nouvelle methode pour mesurer la prise de nourriture et l'utilisation digestive des aliments chez *Blattella germanica*

(L.). *Ann. Nutr. Alim.* **23**:253-268.

WADDINGTON, C. H. (1957). The Strategy of the Genes. 262 pp. Allen & Unwin, London.

WALDBAUER, G. P. (1964). The consumption, digestion, and utilization of normally rejected plants by larvae of the tobacco hornworm, *Protoparce* sexta (Lepidoptera: Sphingidae). *Ent. Exp. & Appl.* **7**:253-269.

WALDBAUER, G. P. (1968). The consumption and utilization of food by insects. Pp. 229-288 in "Advances in Insect Physiology, vol. 5, edited by J. W. L. Beament, J. E. Treherne, and V. B. Wigglesworth.

WEIL, W. B. and WALLACE, W. M. (1963). The effect of variable food intakes on growth and body composition. *Ann. N. Y. Acad. Sci.* **110**:358-373.

WEISS, P. and KAVANAU, J. L. (1957). A model of growth and growth control in mathematical terms. *J. Gen. Physiol.* **41**:1-47.

ZUCKER, L., HALL, L., YOUNG, M. and ZUCKER, T. F. (1941). Animal growth and nutrition, with special reference to the rat. *Growth* **5**:399-413.

ZWOLFER, H. and HARRIS, P. (1971). Host specificity determination of insects for biological control of weeds. *Ann. Rev. Entomol.* **16**:159-178.

INTAKE AND UTILIZATION OF NATURAL DIETS BY THE MEXICAN BEAN BEETLE, *EPILACHNA VARIVESTIS*---A MULTIVARIATE ANALYSIS

Marcos Kogan
Section of Economic Entomology
Illinois Natural History Survey and
Illinois Agricultural Experiment Station, Urbana, Illinois 61801

The Mexican bean beetle (MBB), *Epilachna varivestis* Mulsant, is normally associated with certain species of Leguminosae of the genera *Phaseolus* and *Desmodium*, and has developed considerable affinity for soybeans, *Glycine max* (L.), in certain areas of the United States. In a previous study, adult females of the MBB exhibited a wide range of feeding acceptance for a group of 22 plant types belonging to six species of Leguminosae (Kogan 1972). They also oviposited on the same 22 plants in the laboratory, but no quantitative data were recorded. Among the 22 plants, 17 were lines and varieties of soybeans, 8 of which differed only by one or a few genes that control leaf pubescence and another 9 which had been selected for MBB resistance (Van Duyn, Turnipseed and Maxwell 1971; J. A. Schillinger personal communication). The five other species of cultivated legumes were included because they represented the upper and lower limits of adult food acceptance when compared to a standard soybean variety. The same plant array was used for this larval study in order to correlate adult feeding preferences with the nutritional quality of the food for larval development.

The work discussed here is part of a long-range program to study the nutrition and host-selection behavior of insects associated with soybeans. The research should also assist in evaluating and defining the nature of soybean resistance to insect defoliators. With the larvae on an exclusive diet of each of the 22 plant types previously designated, food intake was measured and a series of nutritional and other biological parameters was computed. One can infer that at comparable rates of food intake, differences in rates of growth and development necessarily derive from nutritional deficiencies in the diet.

Oligophagous insect species display maximum preference for a single plant species or a limited group of related plant species, but at least under experimental conditions, they frequently accept some plant outside their normal host range. A question that has practical importance in the fields of biological weed control (Zwolfer and Harris 1971) and host plant resistance (Saxena 1969, Pathak 1969, Maxwell *et al.* 1969) is how much and under what conditions will the insect depart from this set of preferred hosts and still evince some degree of acceptance. Furthermore, after a plant is accepted for feeding, how will it affect the insect's normal growth and development?

Insect and mite nutrition – North-Holland – Amsterdam, (1972)

It is in the context of this latter proposition that this paper is developed.

MATERIALS AND METHODS

Insect and plant cultures

The stocks for the MBB culture originated from a population breeding on soybeans in south-central Indiana, U.S.A. The insects were maintained on leaves of soybeans of the variety Harosoy--occasionally leaves of snap beans, *Phaseolus vulgaris* L., were also added to the diet. The rearing method was described in Kogan (1972). Larvae used in the experiments were from the 6th–7th generations of this colony.

All plants were grown in a greenhouse, where a 16 hr photophase was maintained throughout the vegetative growth period. Only plants 40 days old or older were used and attempts were made to use completely developed trifoliolates. The species, varieties and lines of the 22 plants included in the experiments are described in Table 1.

Experimental design

A standard growth curve was obtained for 24 MBB larvae reared on leaf cuttings of normal and glabrous Harosoy soybeans. Weights of larvae were recorded daily.

The basis for the experimental design used in the comparison of the nutritional value of the 22 plant types followed the observation of McGinnis and Kasting (1959) that a simplified growth curve could be obtained by measuring the weight of the larvae in each instar on that day when the variability of weights within the population was least. This was found to be 1 day after molting for the MBB during the first 3 stages and 2 days after the molt into the 4th stage and onset of prepupation.

Twenty larvae in 5 groups of 4 were offered each test plant. The day of molting was considered the day that half the larvae in the group completed the molting process. Food was given *ad libitum* but adjusted so that minimum amounts were left over to reduce the relative error in the computations of food consumption (see Waldbauer 1968). Food consumed was calculated from the difference between the corrected weight of food introduced and the weight of food left after a feeding period. Dry feces that occasionally accumulated on the leaves were removed prior to weighing, thus true consumption was measured. Aliquots for the calculation of the dry weight of food consumed were obtained following standard procedures in gravimetric techniques (Waldbauer op. cit). No data on weights of feces were recorded, therefore parameters of digestibility were not computed.

Experiments were conducted at 27 + 2°C, R.H. close to saturation, and in constant darkness.

Table 1. List of plants tested with reference to their nutritional value for the Mexican bean beetle.

Species	Variety	Line	Pubes-cence type	Maturity group	Source	Origin
Glycine max	Harosoy	L-2	Normal	2	USDA Soybean Lab. Urbana, Illinois	a
	"	L62-561	Glabrous	2	"	a
	"	L63-1097	Curly	2	"	a
	"	L62-801	Dense	2	"	a
	Clark	L-1	Normal	4	"	a
	"	L62-1385	Glabrous	4	"	a
	"	L63-2435	Curly	4	"	a
	"	L62-1686	Dense	4	"	a
		PI171,451	Dense	7	USDA-DELTA Res. Sta. Stoneville, Miss.	Japan
		PI227,687	Normal	8	"	Ryukys Is.
		PI229,358	Dense	7	"	Japan
		PI80,837	Dense	4	College Park, Mar.	Japan
		PI181,777	Normal	4	"	Japan
		PI89,784	Normal	3+	"	China
		PI103,091	Appressed	4	"	China
		PI157,482	Normal	4	"	Korea
		PI243,519	Normal	4	"	Japan
Phaseolus vulgaris	Burpees stringless	6163	-	-	Commercial variety	
Phaseolus luntatus	Fordhook	6183	-	-	"	
Phaseolus aureus	b	-	-	-		
Phaseolus mungo	b	-	-	-		
Medicago sativa			-	-	Ill. Nat. Hist. Survey	

[a] Improved varieties. Both Clark and Harosoy soybeans include germplasm of Manchurian origin, through breedings of the varieties Mandarin and/or Mandchu. Each variety was crossed with the same 3 strains containing the gene or genes for glabrous, curly or dense pubescence. Continuous selections have produced in each variety a series of near-isogenic lines differing almost exclusively by the type of pubescence. They are referred to here as isolines of Clark and Harosoy (see Bernard and Singh (1969).

[b] Unidentified varieties obtained from local supplier.

Computation of nutritional parameters

All nutritional parameters were based on the fresh weight of the larvae and dry weight of food consumed. All parameters except the mean body weight were computed according to Waldbauer (1968), as follows:

Mean body weight: $A = \sum_{i=1}^{n} (a_i \times t)/T$: in mg.

Consumption index: CI = F/AT: mg food consumed per mg body weight per day.

Growth rate: GR = G/AT: mg body weight gained per mg body weight per day.

Efficiency of conversion of ingested food: ECI = GR/CI or (G/F) x 100: fresh body weight gained per dry weight of food consumed, expressed on a percent basis.

where: a_i = fresh weight of the ith animal at a certain feeding period.
t = duration in days of the feeding period completed when the animal reached the a_{ith} weight.
T = duration of total feeding from larval emergence through prepupation.
F = dry weight of food consumed.
G = fresh weight gain of the animal during one feeding period.

Together with these parameters were recorded: maximum weight gained by the larvae, total weight of food consumed, total development time, percent survival through prepupation and weight of the pupa.

Statistical analyses

The 22 plants used in each series of tests were considered as a population, the components of which were characterized by the effect they had on the insects feeding upon them. Therefore, each plant type represented "OTU's" (operational taxonomic units) defined by the series of nutritional and other biological parameters computed for the MBB. The relative nutritional value of each plant type was consequently expressed as the total effect the plant had on the insect's feeding, growth, development and survival. A classification or ranking of the plants according to these parameters was attempted by means of techniques of numerical taxonomy (cluster analysis, Sokal and Sneath 1963). The computations were performed using University of Illinois Agronomy Statistical Laboratory programs on an IBM 360/75 computer. The original data (an array of 22 "OTU's" (plant types) by 10 characters) was transformed into standard normal deviates and a Q-correlation matrix (product-moment correlation coefficients) was calculated for all pairs of "OTU's." The Q-correlation coefficients were transformed into Fisher's Z values and clustered by the unweighted pair group method with simple averages (Sokal and Sneath op. cit, and for an example of application and details of the method see Rhodes *et al.* 1968).

RESULTS

A detailed growth curve of the MBB feeding on fresh soybean leaves is shown in Fig. 1. The mean total development time through pupation was 18.3 $\pm$ 1.2 days, and the mean total weight gained by the larvae was 29.1 $\pm$ 3.6 mg.

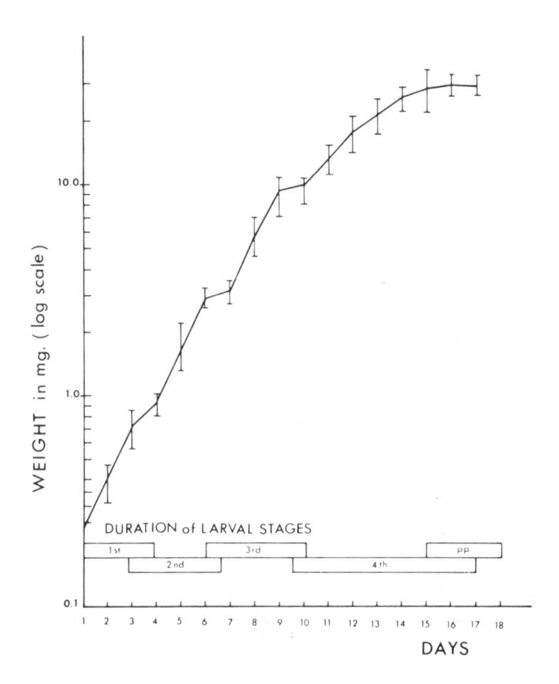

Figure 1. Growth curve and duration of developmental stages of Mexican bean beetle larvae reared on a mixture of normal and glabrous Harosoy soybeans. Vertical bars indicate standard deviations.

The growth curves of the larvae feeding on the 22 test plants are represented in Figs. 2-3, which also include the dry weight of food consumed during each feeding period. The points on the curves represent the weights of the approximate central point in each stage of development, that is, one day after emergence for the 1st stage, one day after molting for the 2nd and 3rd stages, and 2 days after molt to 4th stage and onset of prepupal stage (compare feeding periods to mean duration of larval stages in the detailed growth curve, Fig. 1). Table 2 summarizes the recorded and computed values of the mean fresh body weight (A), total dry weight of food consumed (F), total weight gain (G), development time in days (T), gross efficiency of conversion of ingested food (ECI), gross consumption index (CI), mean growth rate (GR), percent survival through prepupation, mean weight of pupae, and adult preference index (C).

The mean body weight as calculated here differed from the method suggested by Waldbauer (1964) and approached that used by Soo Hoo and Fraenkel (1966). In general the values of A thus calculated were higher than those obtained by Waldbauer's method and therefore reflected on the

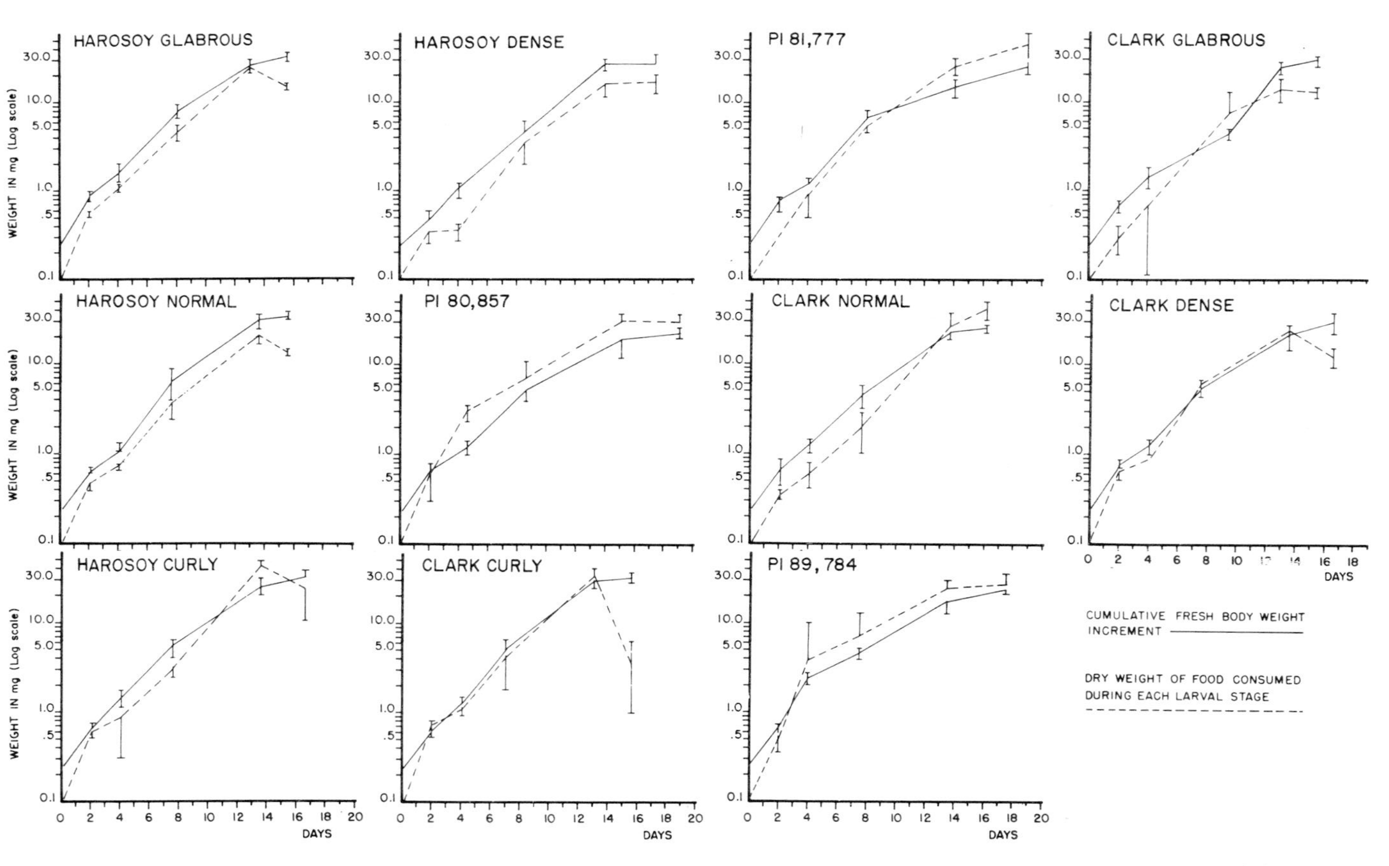

Figure 2. Cumulative fresh body wt increments and dry wt of food consumed during each larval stage. Mexican bean beetle larvae on an exclusive diet of each plant type.

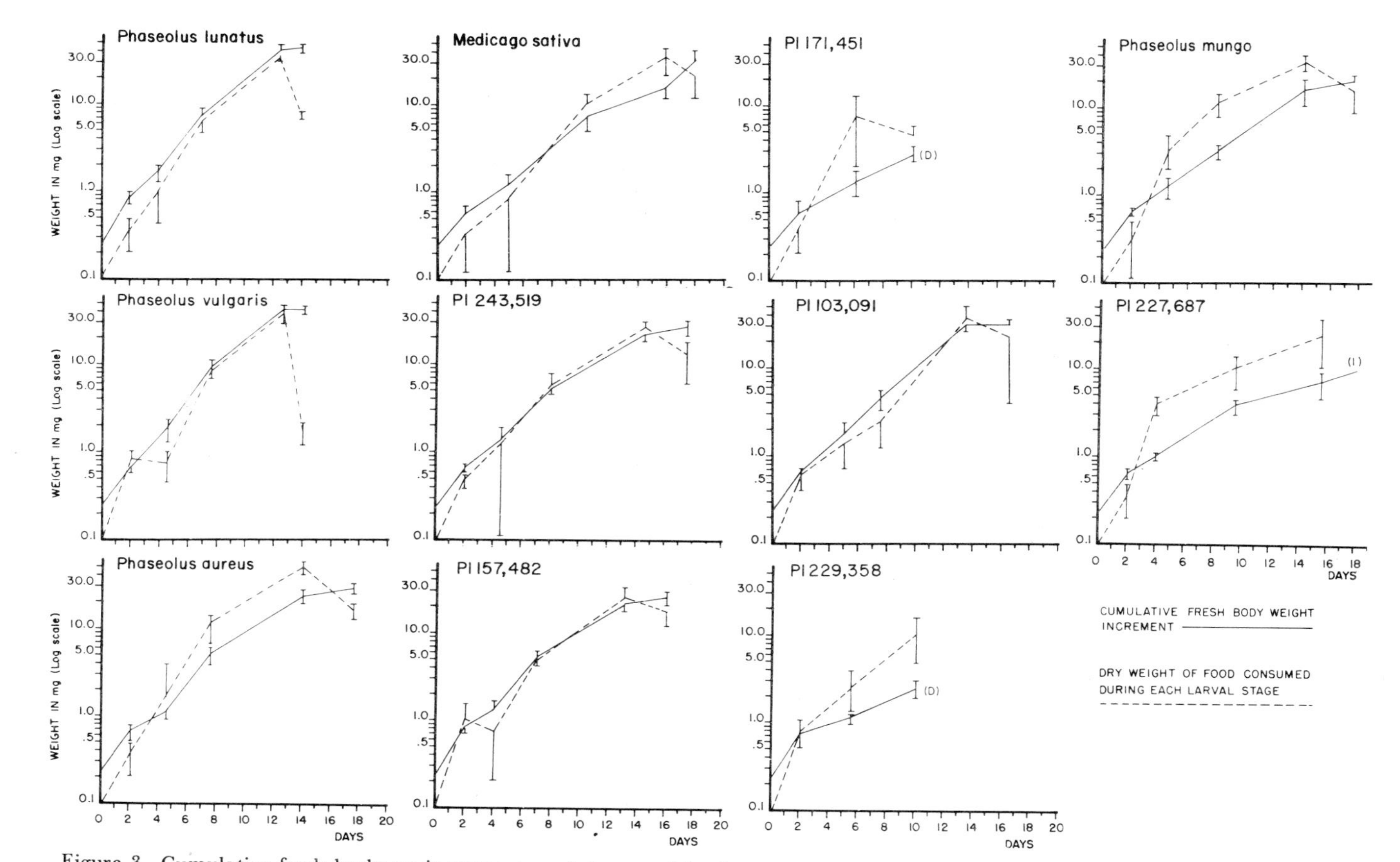

Figure 3. Cumulative fresh body wt increments and dry wt of food consumed during each larval stage. Mexican bean beetle larvae on an exclusive diet of each plant type.

Table 2. Nutritional value and other biological parameters obtained with Mexican bean beetle larvae feeding upon 22 plant types (weights in mg), and preference index for adults.

Food plants[a]	Mean body wt (A)	Total wt gain (G)	Wt food consumed (F)	Develop-ment time (T) days	Efficiency of conversion (ECI)	Consumption index (CI)	Growth rate (GR)	Mean wt of pupae	Sur-vival	Pref. index (C)x10
H. Normal	18.15±2.46	36.27± 2.76	38.51± 8.32	15.76±0.44	92.54±29.84	0.134±0.030	0.128±0.041	33.64±2.42	65.0	10.00
H. Glabrous	16.95±1.29	32.25± 2.33	55.18± 9.49	15.81±0.40	59.98±10.77	0.206±0.033	0.121±0.009	30.81±2.97	70.0	11.22
H. Curly	16.71±3.14	32.38± 2.09	50.50± 8.90	16.23±1.19	67.64±24.06	0.191±0.040	0.121±0.014	29.55±4.69	85.0	11.51
H. Dense	15.58±1.82	30.22± 4.60	38.45± 6.06	17.42±0.79	79.18± 9.94	0.144±0.032	0.112±0.017	27.43±5.45	50.0	9.82
C. Normal	15.13±1.77	27.56± 3.21	56.13± 7.17	16.13±0.64	50.38±11.57	0.232±0.035	0.114±0.017	24.81±1.18	65.0	11.04
C. Glabrous	15.16±2.21	29.54± 2.46	35.84± 8.01	15.41±0.51	86.72±21.00	0.164±0.010	0.135±0.030	29.48±2.96	75.0	11.03
C. Curly	19.59±2.30	34.63± 2.69	44.83±10.09	15.86±0.35	79.75±12.32	0.162±0.053	0.112±0.008	32.66±1.53	75.0	11.44
C. Dense	16.56±1.00	31.41± 4.57	46.82± 1.77	16.67±1.30	67.48±12.32	0.184±0.053	0.126±0.030	28.26±5.15	50.0	11.86
PI 80857	14.06±1.92	25.89± 4.73	70.93± 7.83	18.78±0.43	37.54±11.57	0.276±0.056	0.097±0.004	22.25±1.08	45.0	10.13
PI 81777	12.92±1.35	26.27± 3.51	86.30±15.83	19.00±0.00	31.80± 9.43	0.357±0.084	0.107±0.008	26.10±8.04	75.0	9.31
PI 89784	13.15±1.85	25.33± 3.31	77.13±16.67	17.58±1.54	34.43± 8.42	0.340±0.084	0.110±0.01	23.96±3.16	80.0	9.42
PI 103091	18.27±0.96	31.35± 1.83	54.23±13.58	17.00±0.12	61.16±14.59	0.174±0.041	0.101±0.005	27.96±1.72	50.0	10.07
PI 157482	15.40±2.13	27.77± 2.63	53.07±10.86	16.35±0.49	54.18±11.40	0.154±0.075	0.111±0.007	21.19±3.83	65.0	10.03
PI 245519	14.56±0.47	28.38± 1.56	49.64± 6.45	17.80±0.41	57.76± 5.47	0.192±0.026	0.110±0.009	25.65±1.86	65.0	9.06
PI 171451	2.08±0.62	2.60± 0.35	12.65± 4.43	0[b]	23.03± 7.81[c]	0.661±0.249	0.135±0.020	0[b]	0.0	6.89
PI 227687	4.97±0.75	10.69± 0.0	66.21± 0.0	18.00±0.0	27.95± 8.60[c]	0.516±0.201	0.115±0.014	0[b]	5.0	6.00
PI 229358	1.75±0.73	3.08± 0.0	22.44± 0.0	0[b]	18.15±10.14	0.737±0.211	0.123±0.036	0[b]	0.0	8.63
P. vulgaris	22.95±1.76	43.13± 3.41	51.26± 7.98	14.00±0.0	86.86±19.44	0.160±0.026	0.134±0.009	38.30±2.41	45.0	12.09
P. lunatus	21.47±1.99	42.57± 2.12	53.35± 5.32	14.00±0.0	80.44± 8.12	0.180±0.033	0.142±0.014	40.38±1.92	75.0	11.27
P. mungo	12.31±3.14	23.40± 5.30	74.97±19.31	17.64±1.12	32.37± 7.87	0.351±0.069	0.110±0.017	22.87±3.67	30.0	9.23
P. aureus	16.50±0.97	30.50± 1.52	72.35± 6.03	17.25±1.18	42.28± 2.00	0.258±0.042	0.108±0.074	27.93±1.02	70.0	9.14
M. sativa	12.66±0.91	32.75±14.91	76.95±10.69	18.43±1.40	42.82±18.11	0.338±0.026	0.142±0.055	24.40 –	15.0	9.26

[a]Identified in Table 1.
[b]None completed development on these plants.
[c]ECI computed up to 3rd stage larvae.

computations of GR and CI that were proportionally lowered. The total weight gain was obtained by subtracting the mean weight of newly emerged larvae from the maximum weight recorded. The mean initial weight of the larvae was 0.25 mg.

ECI was computed on the basis of fresh animal body weight and dry weight of food consumed (Waldbauer 1962). This method, of course, yielded very high ECI values, which, for purposes of comparison with values obtained for other insect species can be divided by 4, considering a gross 25% dry matter content in MBB throughout their whole larval period.

For identical reasons the CI was lower than usual and for comparative purposes the values reported here could be multiplied by 4. Since GR is based exclusively on the animal body weight, its values conform with those computed by other authors.

The preference index (C) was determined by comparing all plants to Harosoy normal. A C value of about 1 indicates that test plants were about as acceptable as H. normal. C>1 corresponds to test plants preferred, and C<1 corresponds to test plants partially rejected (or non-preferred) in comparison with H. normal (Kogan 1972).

The mean values of the 10 parameters listed in Table 2 were used to compute a Q-correlation matrix for all 22 plants. The Q-correlation coefficients are presented in Table 3. These correlation coefficients were used in the cluster analysis, the results of which are graphically represented in Fig. 4, in the form of a dendrogram. The levels of affinity, in this case denoting similar effect of the plants on MBB larvae feeding upon them, are defined by the Q-correlation coefficients.

In general the graph in Fig. 4 followed the computed sequence of clustering except in the case of Clark dense. This "OTU" clustered with *P. vulgaris* and *P. lunatus* at a .697 level mainly due to similar percent survival and preference index, but as most nutritional parameters did not follow this trend the .697 cluster was omitted and the position of Clark dense was shifted to the bottom of the cluster formed at the .579 level, including also H. normal, H. dense and Clark glabrous.

The sequence of "OTU's" from top to bottom generally follows the tendencies indicated in the upper left corner of the graph--a decrease in mean body weight, total weight gain, gross ECI, gross GR, weight of pupae, larval survival, and preference index, and an increase in gross CI and developmental time.

For the purpose of the discussion of results, the four large clusters formed above the .40 level and the 9 clusters formed above the .60 level will be considered separately.

The means of parameter values (extracted from Table 3) of all "OTU's" in each cluster formed above the .40 correlation level are presented in Table 4. The means for those clusters formed above the .60 level are shown in Table 5.

Table 3. Q-Correlation matrix of 22 plant types based on their nutritional value for the Mexican bean beetle; decimal points and diagonal of self-comparisons omitted.

Plant types[a]	1.	2.	3.	4.	5.	6.	7.	8.	9.	10.	11.	12.	13.	14.	15.	16.	17.	18.	19.	20.	21.	22.
H. Normal																						
H. Glabrous	44																					
H. Curly	51	90																				
H. Dense	80	36	45																			
C. Normal	-18	72	68	05																		
C. Glabrous	76	40	58	53	-06																	
C. Curly	66	78	81	76	51	50																
C. Dense	52	67	59	46	35	60	57															
PI 80857	-59	-02	-14	-07	52	-73	-00	-28														
PI 81777	-70	-18	-23	-43	32	-79	-34	-67	77													
PI 89784	-68	-13	-11	-47	37	-69	-31	-68	67	97												
PI 103091	23	41	31	63	45	-14	70	15	60	17	09											
PI 157482	20	61	68	45	72	02	65	19	43	29	34	67										
PI 245519	31	21	33	59	24	03	45	-20	36	36	36	64	72									
PI 171451	-32	-66	-57	-61	-62	01	-69	-35	-49	-20	-15	-84	-85	-71								
PI 227687	-62	-92	-85	-49	-50	-57	-87	-68	22	43	38	-41	-43	-15	53							
PI 229358	-50	-61	-54	-68	-41	-12	-65	-33	-28	-08	-05	-73	-79	-76	95	51						
P. vulgaris	72	56	39	49	-01	50	58	77	-43	-76	-79	25	03	-18	-28	-72	-34					
P. lunatus	78	61	51	34	-08	62	48	59	-68	-67	-62	-01	03	-08	-17	-73	-33	84				
P. mungo	-80	-50	-68	-49	-00	-91	-61	-54	74	76	63	07	-14	-03	02	68	14	-58	-74			
P. aureus	-33	07	-02	-09	34	-65	03	-55	72	85	82	53	54	63	-52	10	-46	-42	-33	53		
M. sativa	-28	-50	-69	-45	-54	-36	-77	-11	-11	02	-07	-50	-57	-46	37	58	27	-04	-05	48	-16	

[a]Identified in Table 1.

Table 4. Summary of nutritional parameters that define the suitability for MBB larvae of 22 plant types grouped in 4 clusters formed above the .40 correlation level.[a]

	No. of OTU's	A (mg)	G (mg)	F (mg)	T (days)	ECI	CI	CR	Wt. Pupae	% Surv.	C X10
Cluster I	6	18.31	35.52	44.04	15.54	82.02	.121	.129	32.91	60.0	11.01
Cluster II	7	16.65	30.62	51.94	16.45	61.52	.187	.113	27.51	67.8	10.62
Cluster III	5	13.78	26.28	76.33	18.05	35.68	.316	.106	24.62	60.0	9.44
Cluster IV	4	b	b	b	b	27.98[c]	.563[c]	b	24.40[d]	15.0	7.69

[a] For meaning of symbols see Table 2.

[b] Only 3 larvae completed development through prepupation on 1 plant in this cluster (*M. sativa*). There was no meaning in the avg values for this cluster as they would combine results of animals in different stages of development.

[c] Based on larvae that reached at least the 3rd stage.

[d] Only 1 pupa developed on *M. sativa*, none on the 3 other plants in this cluster.

(Cluster I: *P. vulgaris, P. lunatus*, H. normal, H. dense, Clark glabrous, Clark dense; Cluster II: H. glabrous, H. curly, Clark curly, Clark normal, PI's 157482, 103091, 243519; Cluster III: *P. aureus*, PI's 81777, 89784, 80857, *P. mungo*; Cluster IV: *M. sativa*, PI's 227687, 174451, 229358.)

Table 5. Summary of nutritional parameters that define the suitability for MBB larvae of 22 plant types grouped in 9 clusters formed above the 0.60 correlation level.[a]

	No. of OTU's	A (mg)	G (mg)	F (mg)	T (days)	ECI	CI	GR	Wt. Pupae	% Surv.	C X10
Cluster I	2	22.21	42.85	52.30	14.0	83.65	.170	.138	39.34	60.0	11.68
Cluster II	3	16.29	32.01	37.60	16.19	86.16	.147	.125	30.18	63.3	10.28
Cluster III	1	16.56	31.41	46.82	16.67	67.48	.184	.126	28.26	50.0	11.86
Cluster IV	5	16.75	30.91	51.94	16.07	62.38	.189	.115	27.80	72.0	11.04
Cluster V	2	16.41	29.86	51.93	17.40	59.46	.183	.105	26.80	57.5	9.56
Cluster VI	5	13.78	26.28	76.33	18.05	35.68	.316	.106	24.62	60.0	9.44
Cluster VII	1	12.66	(32.75)[b]	76.95	18.43	(42.82)[b]	.338	(.142)[b]	24.40[c]	15.0	9.26
Cluster VIII	1	4.97[d]	10.69	66.21	18	27.95	.516	.115	e	5.0	6.00
Cluster IX	2	1.91[d]	2.84	17.54	f	20.59	.699	.129	e	0.0	7.76

[a] For meaning of symbols see Table 2.
[b] Only 3 larvae completed development through prepupation, 2 were hypertrophic and did not pupate, hence the high value of G.
[c] Wt of 1 pupa.
[d] Based on larvae that reached at least the 3rd stage.
[e] None reached pupal stage.
[f] None reached prepupal stage.

(Cluster I: *P. vulgaris*, *P. lunatus*; Cluster II: H. normal, H. dense, Clark glabrous; Cluster III: Clark dense; Cluster IV: H. glabrous, H. curly, Clark curly, Clark normal, PI 157482; Cluster V: PI's 103091, 243519; Cluster VI: *P. aureus*, PI's 81777, 89784, 80857, *P. mungo*; Cluster VII: *M. sativa;* Cluster VIII: PI 227682; Cluster IX: PI's 171451, 229358.)

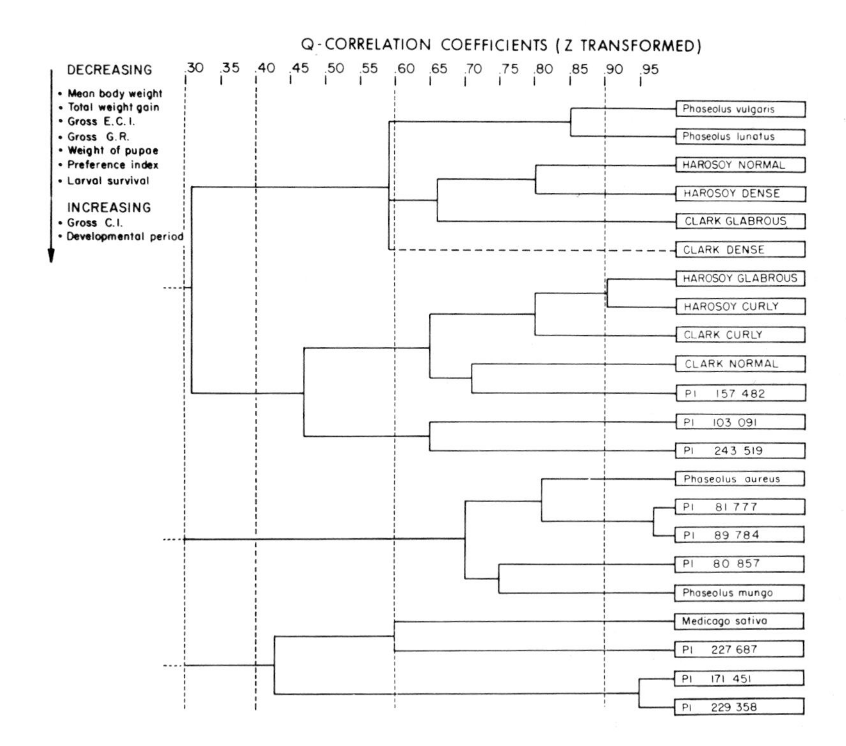

Figure 4. Correlation dendrogram resulting from unweighted pair group method of clustering using simple averages.

DISCUSSION

Proper evaluation of results of food intake and utilization by the MBB must take into consideration the unique feeding mechanisms of this species. Howard (1941) described the MBB larval feeding process as follows: "The larva swings the left mandible to the left, moving the anterior portion of the body slightly to the left also. The mandible scrapes the parenchymatous tissue to the right, or toward an imaginary extension of the median sagittal plane of the insect, and usually leaves the epidermis on the opposite side of the leaf intact. The right mandible moves to the right in a similar manner, usually alternately, and tissue from a distance as far as the mandible can reach is drawn to the left toward the imaginary line mentioned. At intervals both mandibles meet and the tissue is compressed. It is assumed that during this operation the larva ingests the sap and plant juices, for the most part cellulose and chloroplasts are left in a ridge similar to a windrow in a hayfield."

MBB therefore ingests a highly liquefied diet and the bulk of the fiber never enters the animal's digestive tract. Excretion in the first two larval stages is very reduced, and adults and older larvae eliminate a relatively small amount of feces. This feeding process probably results in a kind of ingested food that is intermediate between the liquid ingesta of sucking insects and the highly fibrous ingesta of chewing insects. A consequence of this feeding process is that AD is probably high, and ECI approaches ECD. A second consequence is that the water content of the diet probably plays a very critical role in food acceptance and continuation of feeding. Finally, the analysis of the role of leaf pubescence in the patterns of acceptance of a diet by the larvae has to consider that the cellulose hairs are probably never ingested, and therefore, do not add to the crude fiber content of the diet.

Sokal and Sneath (1963) commented that few Q-type studies conducted to that date have produced negative coefficients of correlation. They explained the generally positive nature of Q-correlation matrices by the unlikelihood of finding "... a pair of OTU's antithetical for an appreciable number of characters." (Sokal and Sneath *op. cit.*) In the current study there are several highly significant negative coefficients of correlation between certain pairs of "OTU's". Negative coefficients probably appear because four of the nutritional parameters used in the analysis, A, ECI, CI, and GR, are ratios calculated on the basis of the original data used to compute three other single parameters, G, F, and T. Thus the differences between diets ("OTU's") are somewhat magnified. Seemingly an inconvenience in taxonomic studies, this was in fact one of the purposes of this investigation, i.e., to expand subtle nutritional differences so as to permit their detection and analysis.

Clusters at the 0.40 correlation level

Cluster I included 6 "OTU's" that were the best food plants for the MBB. The range of maximum larval weight was 43.38 - 29.79 mg, but weights of larvae did not necessarily correlate with survival as the best food plant in the cluster (*P. vulgaris*) produced a high larval mortality (65%). ECI reached a maximum of 92.54% with H. normal and a low of 67.48% with Clark dense, but the majority of the plants in this cluster had ECI's near the 80% level (~20% on a dry weight/dry weight basis), suggesting that despite differences in the growth of larvae, the plants in this group were nutritionally adequate for the MBB.

Cluster 2 included seven "OTU's". Larval weights in this group were more uniform, ranging from 34.88 to 28.83 mg, however, survival varied also rather widely from a high of 85% to as low as 50%. In this group, weights of food consumed were higher than in cluster 1, yielding considerably lower ECI values around the 60% level. H. curly was a discrepant component in this cluster, and by its assessed nutritional value it should rank with plants in cluster 1. This cluster consists of moderately good food plants for the MBB.

Cluster 3 included 5 "OTU's". *P. aureus* was the only nutritionally adequate food plant, but if given a choice the larvae usually prefer H. normal to *P. aureus.* In general pupae produced from larvae feeding on these plants were very small, practically at the limit for survival, although in general survival through prepupation was as variable as in the other groups, suggesting that larval mortality in this experiment was not necessarily correlated to major antibiotic effects in the plants included in these first 3 clusters. Those were plants of low nutritional value for the MBB as reflected by the rather low ECI's of the order of 35%.

Cluster 4 included alfalfa (*M. sativa*) and PI's 227687, 171451, and 229358 that were selected for MBB resistance in South Carolina (Van Duyn, Turnipseed and Maxwell 1971). Only 3 larvae completed pupation on plants of this cluster. The few larvae that completed development on alfalfa were extremely large, as they continued to feed over an unusually long period of time (〉 18 days). None of the larvae passed the third instar on PI's 171451 and 229358; values of ECI, CI, GR were based on larvae surviving through more than two stages. The inadequacy of these plants is reflected in the wide variance in total weight gain and weight of *M. sativa* consumed. These were inadequate food plants for the MBB. They also ranked lowest in the preference scale.

The steep increase of ECI from cluster I to IV reflects the gradual reduction of the nutritional value of the diets. The proportional increase of CI indicates that the insects compensated for the reduced nutritional value, up to a certain degree, by extending development time (T) and increasing daily consumption rates. Compensatory rates of food intake in practically all plants in Clusters I and II suggest that proper feeding excitants were present.

Percent survival on diets of these clusters was essentially the same (ca. 60%), indicating that no antibiotic effects were present to account for the reduced suitability of plants in clusters II and III. It is therefore hypothesized that nutritional inadequacies are probably due to an imbalanced reduction in the level of certain nutrients available to the larvae. The relatively mild effects suggest that the relative proportions of the nutrients must have been within tolerable limits, since according to House (1966, 1969), imbalances in the proportion of nutrients result in a reduction of food intake, weight gain, and efficiency of conversion. Diets in cluster III caused an average of 25% reduction in weight gain, but food consumption increased nearly 70%. Insects which emerged from pupae developed on these diets, however, were frequently malformed and much smaller than normal. The increase in food intake was not accompanied by compensation of ECI, and the signs of nutrient deficiencies expressed by adult malformations, especially of wings, suggest abnormal proportions of nutrients.

One noteworthy aspect of the summarization of the results at the 0.40 correlation level is the almost perfect direct correlation between diet suitability for the larvae and food preferences of the adults, as measured by the preference indices. As the argument on the role of nutrients in host-plant selection seems to be far from settled, results of this experiment hint at the possibility that some discrimination of the nutritional value of the plant does indeed occur, a question that remains open for future investigation.

Clusters at the 0.60 correlation level

At the 0.60 correlation level nine clusters were formed, three of them including one single "OTU". The summary of the average results in each cluster is presented in Table 5.

At this higher level of correlation more complex effects of the diets on the insects become apparent. Again, the general trend toward reduced nutritional value of the diets is demonstrated by the decrease of ECI.

The mean ECI of cluster II, however, was slightly higher than that of cluster I, as reduction in total dry weight of food consumed in the former occurred in the proportion of the decrease in weight gain. House (1965) reported that a series of diets containing a constant relative proportion of nutrients but in increasingly diluted amounts were fed to 5th-instar larvae of *Celerio euphorbiae* (L.). The larvae ate larger amounts of the more diluted diets but weight gain and ECI did not vary significantly. Such observations were later generalized in the rule of nutrient proportionality (House 1966). Mean parameter values in cluster II indicate that both proportions and total levels of nutrients should have been adequate, but that the total amount of food consumed was reduced. As *P. vulgaris* and *P. lunatus* are preferred hosts in nature and in the laboratory (Friend and Turner 1931, Howard and English 1924, Kogan 1972, Thomas 1924), differences in total amounts of food intake can be explained by the lower efficiency of feeding excitants in the soybeans. Lippold (1957) reported that different saponins in *P. vulgaris* and *G. max* elicited positive feeding responses of MBB adults. Nayar and Fraenkel (1963) did not confirm the presence of the active saponins, and reported that the cyanogenetic glycoside, phaseolunatin, was synergistic to glucose, acting as a feeding excitant on leaves of *P. lunatus*.

Augustine *et al.* (1964) stressed the role of sucrose, glucose and fructose as important feeding excitants for MBB. Gas chromatographic determinations of sucrose and reducing sugars (our unpublished data) revealed that indeed the preferred soybean lines were higher in total sugar content, but some of the more resistant lines were higher in sugars than several of the more susceptible ones, leaving the question of the role of these sugars open for further investigation.

Again evidence points to a great deal of interactions of factors, of which sugar concentrations is unquestionably a primary one. It seems, however, that the variety of responses obtained cannot be explained by the levels of this component alone.

In considering results of cluster II the fact must finally be considered that those plants were used in regular maintenance of the cultures and that selection towards improved utilization of these diets may have occurred. Gordon (1959) stated that selection pressure favors genotypes that maximize food utilization.

Clusters II, III and IV include the four isolines each of Harosoy and Clark soybeans, plus PI 157482. It appears that no major differences exist between the Harosoy and Clark basic genotypes, as representative isolines of each variety clustered together in clusters II and IV. Leaf pubescence did not seem to affect food utilization, as Harosoy dense and Clark dense ranked among the best diets in clusters II and III respectively. As dense isolines have over 4 times more hairs per leaf unit area than the normal isolines (Singh, Hadley and Bernard 1971), a typical chewing insect would be expected to reflect the increased cellulose content of the diet by a lower ECI (unpublished data with the corn earworm, *Heliothis zea*, show this effect). The peculiar "scrape and suck" type of feeding of the MBB most likely excludes ingestion of the fibrous materials, which did not seem to interfere with normal intake and utilization. There was, however, a significantly higher than normal mortality on both dense isolines. With Harosoy dense, larval mortality occurred mostly during the first two stages. In this case leaf pubescence may have affected some of the young larvae by blocking their access to the nutritious parenchymatous tissues. Soybean leaf pubescence is a factor in soybean resistance to leafhoppers (Singh, Hadley and Bernard 1971; Broersma, Bernard, and Luckmann 1972). Van Duyn (1971) ascribed considerable importance to leaf pubescence, especially the length of hairs, in the resistance of PI's 171451, 227687 and 229358 to the MBB.

The additional variability in the eight isolines remains unexplained because, at least theoretically, they should be alike in all respects except leaf pubescence. Particularly difficult to explain were results with Clark normal, which was one of the preferred plants for adult feeding, and yet was of reduced nutritional value.

Clusters V and VI included five PI lines selected for MBB resistance in Maryland, and two species of beans--*P. aureus* and *P. mungo*. The two species of *Phaseolus* were reported as almost immune to the MBB (Thomas 1924), and ranked low in a preference scale (Kogan 1972). The low nutritional value of these plants is generally reflected in poor weight gain and low ECI, however, food consumption was high. In this case, even admitting the reduced level or absence of proper feeding excitants, higher food consumption in confinement was probably the result of starvation for some key nutrient that was probably present in disproportionally low amounts. One such nutrient that probably was in short supply was water. Leaves of plants in this cluster were considerably thinner and percent moisture was reduced, thus larger amounts of dry matter may have been ingested to assure water intake in adequate levels. Whereas mean percent dry matter in H. normal was 17.5% and in H. glabrous 15.7%, these PI lines all had over 19.1% dry matter.

Cluster VII included *M. sativa* which is only a marginal food for the MBB. The few larvae that survived beyond the 3rd stage grew exceedingly large and died of bacterial septicemia, indicating a breakdown of defenses resulting from major disruptions of growth regulation. In this case one may suspect the presence of adverse chemical factors in the leaves.

Clusters VIII and IX included the three MBB resistant PI 227687, and PI's 171451 and 229358 respectively. The evident unsuitability of these plants for the MBB seems to be the result of a combination of factors including the absence of proper feeding excitants, presence of physical deterrents (pubescence), and nutritional imbalances.

CONCLUSIONS

A large spectrum of responses was obtained with MBB larvae feeding on these 22 plants. In broad lines (.40 level of analysis) adult preferences closely followed the nutritional suitability of the diets for the larvae. There was evidence that decreasing suitability from *P. vulgaris* to PI 229358 could be accounted for by one or a combination of the following factors: a) improper and/or suboptimal levels of feeding excitants, b) generally reduced levels of nutritional content of the diet, c) imbalanced alteration of nutrient proportionality, d) presence of physical feeding deterrents, and e) presence of chemical antibiotic factors.

Among the resistant lines that were tested, some (PI's 103091, 243519, 81777, 89784, and 80857) were partially resistant and their use as a potential source of resistance should not be neglected. PI's 227687, 171451, and 229358 were highly resistant, with evidence of both non-preference and antibiotic types of resistance.

This investigation was an initial step in the analysis of insect-soybean interactions at the nutritional level. Future progress in this area will largely depend upon the detailed knowledge of the nutritional requirements of insects associated with soybeans.

The cluster analysis was helpful in the interpretation of the results. Available programs put this technique within easy reach of the entomologist and it may certainly become a powerful tool in the analysis of complex data whenever some objective criteria for an orderly classification of elements based on a plurality of characters is desirable.

REFERENCES

AUGUSTINE, M. G., FISK, F. W., DAVISON, R. H., LAPIDUS, J. B., and CLEARY, R. W. (1964). Host-plant selection by the Mexican bean beetle, *Epilachna varivestis*. *Ann. Entomol. Soc. Amer.* 57:127-134.

BERNARD, R. L. and SINGH, B. B. (1969). Inheritance of pubescence type in soybeans: glabrous, curly, dense, sparse, and puberulent. *Crop. Sci.* 9:192-197.

BROERSMA, D. B., BERNARD, R. L., and LUCKMANN, W. H. (1972). Some effects of soybean pubescence on populations of the potato leafhopper. *J. Econ. Entomol.* 65:78-82.

FRIEND, R. B. and TURNER, N. (1931). The Mexican bean beetle in Connecticut. *Conn. Agr. Exp. Sta. Bull.* 332:71-108.

GORDON, H. T. (1959). Minimal nutritional requirements of the German roach, *Blattella germanica*. *Ann. N. Y. Acad. Sci.* 77:290-351.

HOUSE, H. L. (1965). Effects of low levels of the nutrient content of a food and of nutrient imbalance on the feeding and the nutrition of a phytophagous larva, *Celerio euphorbiae* (Linnaeus) (Lepidoptera: Sphingidae). *Can. Entomol.* 97:62-68.

_____(1966). The role of nutritional principles in biological control. *Can. Entomol.* 98:1121-1134.

_____(1969). Effect of different proportions of nutrients on insects. *Ent. Exp. & Appl.* 12:651-669.

HOWARD, N. F. (1941). Feeding of the Mexican bean beetle larva. *Ann. Entomol. Soc. Amer.* 34:359-360.

_____, and ENGLISH, L. L. (1924). Studies of the Mexican bean beetle in the Southeast. *U. S. Dept. Agr. Bull.* 1243:1-50.

KOGAN, M. (1972). Feeding and nutrition of insects associated with soybeans. 2. Soybean resistance and host preferences of the Mexican bean beetle, *Epilachna varivestis*. *Ann. Entomol. Soc. Amer.* 65 (in press).

LIPPOLD, P. C. (1957). The history and physiological basis of host specificity of the Mexican bean beetle, *Epilachna varivestis* Muls. PhD Diss. Univ. Ill., Urbana. 146 pages.

MAXWELL, F. G., JENKINS, J. N., PARROTT, W. L., and BUFORD, W. T. (1969). Factors contributing to resistance and susceptibility of cotton and other hosts to the boll weevil, *Anthonomus grandis*. *Ent. Exp. & Appl.* 12:801-810.

MCGINNIS, A. J. and KASTING, R. (1959). Note on a method of comparing diets for the pale western cutworm, *Argotis orthogonia* Morr. (Lepidoptera: Noctuidae). *Can. Entomol.* 41:742-743.

NAYAR, J. K. and FRAENKEL, G. (1963). The chemical basis of the host selection in the Mexican bean beetle, *Epilachna varivestis* (Coleoptera: Coccinellidae). *Ann. Entomol. Soc. Amer.* 56:174-178.

PATHAK, M. D. (1969). Stem borer and leafhopper-planthopper resistance in rice varieties. *Ent. Exp. & Appl.* 12:789-800.

RHODES, A. M., BEMIS, W. P., WHITAKER, T. W., and CARMER, S. G. (1968). A numerical taxonomic study of Cucurbita. *Brittonia.* 20:251-266.

SAXENA, K. N. (1969). Patterns of insect-plant relationships determining susceptibility or resistance of different plants to an insect. *Ent. Exp. & Appl.* 12:751-766.

SINGH, B. B., HADLEY, H. H., and BERNARD, R. L. (1971). Morphology of pubescence in soybeans and its relationship to plant vigor. *Crop. Sci.* 11:13-16.

SOKAL, R. R., and SNEATH, P. H. A. (1963). Principles of Numerical Taxonomy. W. H. Freeman, S. Francisco. 359 p.

SOO HOO, C. F., and FRAENKEL, G. (1966). The consumption, digestion, and utilization of food plants by a polyphagous insect, *Prodenia eridania* (Cramer). *J. Insect Physiol.* 12:711-730.

THOMAS, F. L. (1924). Life history and control of the Mexican bean beetle. *Ala. Agr. Exp. Sta. Bull.* 221, 99 p.

VAN DUYN, J. W. (1971). Investigations concerning host-plant resistance to the Mexican bean beetle, *Epilachna varivestis* Mulsant, in soybeans, *Glycine max* (L.) Merrill. PhD Diss., Clemson Univ., S. C. 210 p.

VAN DUYN, J. W., TURNIPSEED, S. G. and MAXWELL, J. D. (1971). Resistance in soybeans to the Mexican bean beetle. I. Sources of resistance. *Crop Sci.* 11:572-573.

WALDBAUER, G. P. (1962). The growth and reproduction of maxillectomized tobacco hornworms feeding on normally rejected non-solanaceous plants. *Ent. Exp. & Appl.* 5:147-158.

_____, (1964). The consumption, digestion and utilization of solanaceous plants by larvae of the tobacco hornworm, *Protoparce sexta* (Johan.) (Lepidoptera: Sphingidae). *Ent. Exp. & Appl.* 7:253-269.

_____, (1968). The consumption and utilization of food by insects. *Advan. Insect Physiol.* 5:229-288.

ZWOLFER, H., and HARRIS, P. (1971). Host specificity determination of insects for biological control of weeds. *Annu. Rev. Entomol.* 16:159-178.

FOOD UTILIZATION BY ACARID MITES

Z. A. Stepien and J. G. Rodriguez
Department of Entomology
University of Kentucky
Lexington, Kentucky 40506

INTRODUCTION

Studies of food utilization in acarid mites are scarce. Some information concerning food consumption measurements of acarid mites may be found in Boczek (1957), Jakubowska (1967) and Solomon (1946, 1959). Consequently, little is known about the efficiencies of food utilization and energy transformation in these mites. Klekowski and Stepien (in press), Stepien (1970) and Stepien *et al.* (unpubl.) studied food utilization and energy transformation of two species of acarid mites: *Rhizoglyphus echinopus* (Fumouze and Robin) and *Tyrophagus putrescentiae* (Shrank). These studies indicate that acarid mites are not only voracious feeders but they utilize their food very efficiently.

The objectives of this study were to evaluate food requirements, biomass production, respiration and efficiencies of food and energy utilization of all developmental stages of *Caloglyphus berlesei* (Michael). These objectives could be best met by calculation of energy budgets.

An energy budget is an equation expressed in identical energy units per unit of time, in which the amount of energy inflowing in the form of food to the organism is expressed as a sum of the elements representing further uses of this energy.

The symbols of the energy budget based on Klekowski (1970) are

$$C = P + R + U + F = D + F \qquad (1)$$

$$D = P + R + U \qquad (2)$$

$$A = P + R \qquad (3)$$

where: C represents consumption, food intake; D - digestion, part of the food that is digested and absorbed; P - production (body, exuviae, reproductive products etc.); R-respiration, cost of maintenance; F-unabsorbed part of consumption; U-urinary and other "waste" part of digestion; FU-when difficult to separate U from F, and treated together; A-assimilation, sum of production and respiration, food absorbed less the excreta.

The efficiency of an organism in energy utilization is depicted by the following indices:

$$U^{-1} = \frac{A}{C} \qquad (4)$$ Assimilation efficiency;

$$K_1 = \frac{P}{C} \qquad (5)$$ Index of gross production efficiency or efficiency of utilization of consumed energy for growth;

Insect and mite nutrition – North-Holland – Amsterdam, (1972)

$$K_2 = \frac{P}{P+R} = \frac{P}{A}$$ (6) Index of net production efficiency or efficiency of utilization of assimilated energy for growth.

At this point it is well to call attention to the fact that U^{-1} is precisely the approximate digestibility (A.D.) expressed in calories and that K_1 and K_2 are efficiency of conversion of ingested food to body substance (E.C.I.) and efficiency with which digested food is converted to body substance (E.C.D.) respectively, also expressed in calories. (See Introduction by G. P. Waldbauer, this book).

The energy budget and efficiencies are calculated in two different ways:

1. The elements of the budget and the efficiencies are calculated for every 24 hour period of the life of an animal - such balance is said to be instantaneous;
2. The elements of the budget and the efficiencies are calculated for the entire period from the beginning of life of the animal to any moment of its life - such balance and efficiencies are said to be cumulative.

An instantaneous energy budget and efficiencies illustrate well the physiological effectiveness of an organism especially if they are measured in relation to short periods of an organism's life. A cumulative energy budget and efficiencies characterize well ecological properties of an organism. It describes the position of an organism in energy transformation between its food and the next trophic level. (For details concerning energy budgets and energy transformation see Klekowski, 1970).

METHODS AND PROCEDURES

The measurements of food consumption, respiration and biomass production were used in energy budget calculations. Energy transformation efficiencies were also computed and compared with those for other species of acarid mites.

Food consumption and biomass production

Food consumption and biomass production in all developmental stages of *C. berlesei* were determined by the gravimetric method. An artificial diet (Rodriguez (1972), this publication) was used as the food in this study.

Preliminary tests indicated that this diet dried to constant weight could not be reconditioned to the original moisture and physical characteristics. Hence, small plugs were cut with a 3 mm cork borer from a thin layer of a diet. Pairs of these pellets were adjusted to identical weight and one of them was placed in a rearing cell with mites. The other was used in determining the dry weight of the food by drying at 50°C in a vacuum dessicator with anhydrous $CaCl_2$ for 48 hours, and then weighing.

Table 1. Design of experiments on food consumption and biomass production of *C. berlesei.*

Developmental stage	Number of mites	Reps.	Total no. of mites in expt.
Larva	100	10	1000
Protonymph	80	10	800
Tritonymph	50	12	600
Adult mites (♀ + ♂)	2	40	80

The following experiments were outlined, in order to establish the average amount of food consumed by each developmental stage of the mites. (Table 1). In the experiment dealing with larvae, 100 newly hatched larvae were placed into each rearing cell that contained a known amount of food. Larvae attain growth in about 1 day and go into their quiescent stage. At this point, in the second experiment, 80 quiescent larvae were weighed on a Cahn electro-balance and then placed into each rearing cell with a known amount of food. Similarly 50 quiescent tritonymphs were weighed and then placed in a rearing cell with food.

The adult mites used in the experiment were obtained immediately after they had emerged from the quiescent tritonymphal stage. Larvae, protonymphs and tritonymphs were removed from the experimental cells as soon as they became quiescent, and then stored in separate vials in the freezer until the last specimen from the particular cell completed its development (became quiescent).

All quiescent stages from each replicate were weighed then dried at 50°C in the vacuum dessicator with anhydrous $CaCl_2$ for 48 hrs and weighed again. These weight data determined the average fresh and dry weight of individual mites of each developmental stage. The biomass production of the larval stage was calculated by subtracting the average dry weight of the egg (average from 10 samples of 100 eggs) from the average dry weight of the quiescent larvae in the experiment. The biomass production in the protonymphal and tritonymphal stages was found in similar way, i.e., by subtracting the calculated average dry weight of the mites at the beginning of experiment from the average dry weight of the mites at the end of experiment.

After the last mite had completed its development, the remaining uneaten food was separated microscopically from fecal pellets, and cast skins, then dried in the same way as the check food pellets and mites and weighed. The amount of food eaten was determined by subtracting from the original calculated dry weight of the food pellet, the weight of the uneaten food. In the experiment with adult mites the eggs were counted and removed from the rearing cells daily. Adult mites were weighed on the 3rd and 10th day after they had emerged from a tritonymphal stage, and then every 10 days; the food was changed in each experimental cell concurrently with these weighings.

A group of 150 adult mites of mixed sexes and of similar age as those tested in the food consumption and biomass production experiments were reared under the same conditions as the experimental mites in order to determine the dry matter content of the adult mites of different ages. Each time the experimental mites were weighed, a sample of 10 females and 10 males was weighed alive then dried and weighed again. The percentage of dry matter content found this way was used to calculate the average dry weight of the experimental animals during their life time.

Respirometry

The oxygen consumption of all developmental stages of *C. berlesei* was determined in a microrespirometer (cartesian divers) according to the methods described in Klekowski (1967, 1968). Oxygen consumption was determined for one mite of each developmental stage in one cartesian diver. Each measurement was repeated at least 10 times. Respiration was recorded in $\mu l\ O_2 \times 10^{-3}$ per hr per mite. The relationship between respiration and fresh body weight of mites was calculated as the regression: $QO_2 = a \times W^b$, where: QO_2 - oxygen consumption in $\mu l \times 10^{-3}$ per hr per specimen; W - fresh body weight in μg; a - constant; b - regression coefficient. This formula was then used to calculate daily cost of maintenance of adult mites in energy budget calculations.

It should be noted that R as measured by oxygen consumption should be equal to R as measured gravimetrically; that is, C-FU-P = R.

Caloric equivalents

The caloric values of the artificial diet, adult mites, quiescent tritonymphs and eggs (methods of collecting of different stages of mites are described by Stepien and Rodriguez (in press) were determined with a Parr microbomb calorimeter.

Samples of food, mites and their eggs were dried in a vacuum dessicator with anhydrous $CaCl_2$ at 50°C. After drying they were ground to a very fine powder and then pelletized in a pellet press to form 20-60 mg pellets. These pellets were dried again for 24 hrs in a vacuum at 50°C, weighed and burned in the calorimeter. The results of these measurements were expressed as calories per mg dry weight.

Energy budget calculation procedure

All measured parameters of the energy budget were converted into calories and used in energy budget calculations. The energy budget was calculated in the instantaneous and cumulative form for an average individual, but ignoring sex, as the sex ratio in *C. berlesei* populations is about equal (Rodriguez and Stepien, in press).

Biomass production of preimaginal stages

In the course of its development, this acarid mite has 4 feeding stages which alternate with 3 non-feeding or quiescent stages. Growth takes place in the feeding stages; namely, the larval, protonymphal, tritonymphal and adult stages. Periods of physiological starvation and loss in body weight (energy content) occur in the egg, and in the 3 quiescent stages. During the feeding periods the mite accumulates energy in body tissues and reproductive products (eggs), but during each non-feeding stage, part of the energy accumulated by the previous feeding stage is dissipated in respiration.

The calorific values of eggs and quiescent stages measured in the calorimeter represent the average energy content of each stage for half the time of its development. Assuming that respiration rates of those non-feeding stages are constant it was possible to calculate the initial and final energy value of an average individual of each stage. Thus for those non-feeding stages (egg, quiescent larvae, quiescent protonymph and quiescent tritonymph) the initial caloric value was the average measured caloric value of the specimen plus 1/2 of the total energy used in process of metabolism (respiration) for the entire time of this particular stage. The final caloric value of the average specimen was the initial caloric value of the average specimen less the total energy used for respiration during the entire time of this stage.

Total biomass production of the larval stage was calculated from the difference between the initial caloric value of the quiescent larva and the final caloric value of the egg less the energy contained in the egg shell. The production of the protonymph was calculated from the difference between the initial caloric value of the quiescent protonymphal stage and the final caloric value of the quiescent larva less the caloric value of the larval cast skin. The production of the tritonymphal stage was calculated in a similar manner.

Thus, calculated data on the total biomass production in each developmental stage were then plotted on a graph against the time of development of an average mite as established for this species by Rodriguez and Stepien (in press). The approximate values of daily biomass production were then read from that graph and used in energy budget calculations.

Biomass production of adult mites

As was mentioned previously, adult mites were weighed on the 3rd and the 10th day after they had emerged, and then every 10 days. Those measurements expressed in calories were then plotted on the graph against the time development and the approximate values of daily biomass production were taken from that graph. The total daily biomass production of an average adult specimen was calculated as the sum of the daily body biomass production and the daily egg production.

Assimilation

Assimilation was calculated as the sum of production and respiration (A = P + R).

Consumption

The daily food consumption rates of all actively feeding immature stages and adult mites were calculated from the following relationship:

$$\Delta C = \frac{\Delta C}{U^{-1}} \times 100 \qquad (7)$$

where: ΔC - daily food consumption rate; ΔA - daily assimilation rate; U^{-1} - assimilation efficiency.

Respiration - cost of maintenance

Respiration measured in volumetric units were converted into caloric units by using oxycaloric coefficients based on the respiratory quotients. The value of a respiratory quotient, (RQ=0.9) for the related species *Tyroglyphus farinae* = *Acarus siro* L., as found by Hughes (1943) was used for adult mites, and the active preimaginal stages and eggs.

As far as metabolism in the quiescent stages of mites is concerned, it apparently may be compared with that of the pupal stages of holometabolic insects. Stepien (1970) found considerable decrease in the lipid content in quiescent stages of *R. echinopus*, 9.45% of dry weight, as compared to the lipid content of active nymphal stages, 24.06% of dry weight. This fact suggests that feeding stages accumulate lipids as a source of energy and these are metabolized during quiescence. Klekowski *et al.* (1967) found an RQ = 0.7 for the pupal stage of *Tribolium castaneum* (Herbst) and suggested that metabolism in this stage is dependent on lipids. The same RQ value was used for *C. berlesei*. For RQ = 0.7 the thermal equivalent of oxygen (oxycaloric coefficient) equals 4.686 Kcal/1.0 liter, and for RQ = 0.9 it is 4.924 Kcal/1.0 liter (Ewy, 1964). The rates of oxygen consumption in the immature stages were assumed to be constant for each stage. The daily oxygen consumption of an average adult mite was calculated from the formula: $QO_2 = a \times W^b$, substituting W with the values of fresh body weights of an average specimen, male or female, for each day of its life.

The values of assimilation efficiency (U^{-1}) for each feeding immature stage and for each of the 10 day periods in adult mite life were calculated as:

$$U^{-1} = \frac{A}{C} \times 100 \qquad (8)$$

where: A - total assimilation for given stage or feeding period; C - total food consumption for the same stage or feeding period. The value of assimilation efficiency thus calculated was assumed to be constant for each stage or feeding period in the life of an adult mite.

On the basis of these measurements and calculations the instantaneous (daily) and cumulative energy budgets were compiled for an average individual of *C. berlesei* during its entire life span.

RESULTS

Time of development, longevity

The biology and population dynamics of *C. berlesei* have been studied by Rodriguez and Stepien (in press), and pertinent data from those findings were utilized in this study (Table 2).

Table 2. The average time of development of *C. berlesei* based on observations of 50 specimens (after Rodriguez and Stepien, in press)

Developmental stage	Average time of development (days)	Cumulative time of development (days)	Development in days	
			females	males
Egg	2.42	2.42	2.42	2.42
Larva				
active	1.34	3.76	1.37	1.31
quiescent	0.72	4.48	0.71	0.74
Protonymph				
active	0.67	5.15	0.65	0.68
quiescent	0.52	5.67	0.50	0.53
Tritonymph				
active	1.69	7.36	1.77	1.60
quiescent	0.71	8.07	0.68	0.75
Total	8.07	8.07	8.10	8.03

The energy budgets were calculated till the 38th day of life of an average mite. This time was chosen because oviposition ceased on the 32nd day, but 54 percent of the females were still alive at that time. After the 32nd day there was no egg production and the survival rate of females dropped to 24% at the end of the 38th day.

Caloric values of the food and mites - biomass production

The caloric values of the artificial diet used as a good and different stages of mites are given in Table 3. The caloric value of food was significantly lower (t = 5%) than the caloric values of the biological material. There was no significant difference (5% level) between the energy content of adult mites and quiescent tritonymphs, but the caloric value of eggs differed significantly at the same level from that of adult mites and tritonymphs.

Table 3. Caloric values of the food and different stages of *C. berlesei*.

Material	No. of replicates	Avg caloric value cal/mg (50°C)	± S.E.
Food	8	4.3326	0.0725
Adult mites (♀ + ♂)	8	5.1597	0.3071
Quiescent tritonymphs	7	5.2458	0.1906
Eggs	6	6.0794	0.3101

Stepien (1970) found similar caloric values for adult *R. echinopus* mites and their eggs; 5.1418 cal/mg and 5.9619 cal/mg respectively, but the caloric value of quiescent tritonymphs was higher, namely, 6.0017 cal/mg. Mites in that study were fed rye germ having a 5.0510 cal/mg value.

The production rates of preimaginal stages and adult mites are given in Table 4. The data show that the highest growth rate in the immature stages occurred in the tritonymphal stage. The animal gained 68.8% of its maximum caloric value at the end of that stage.

The teneral adult grows quite rapidly in the first 3 days after emergence from the tritonymphal stage. The caloric value of the body increased more than 300% during this short period, while the total biomass increased by about 700%. This was due to intensive egg production of 47.1% of the total biomass. The highest egg production rate occurred between the 11th and 18th day at which period 83.9% of the total biomass was produced. Between the 19th and 28th day the body biomass production rate decreased while egg production contributed as much as 94.8% of energy to the total biomass production increase.

Determinations were made of the energy dissipated and respiration in each non-feeding stage, of the caloric values of egg shells and cast skins of all immature stages, as well as the changes in caloric values of the average mite during its lifetime (Fig. 1). The energy of the egg shell was assumed to be 2% of the caloric value of freshly deposited egg (2% of 2.55 cal x 10^{-3}). Caloric

Table 4. Biomass production of immature stages and adult mites of *C. berlesei*.

Developmental Stage	Animal age in days	Avg dry wt (μg)	Caloric value of an avg mite cal x 10^{-3}	Biomass production cal x 10^{-3}		
				body	eggs	total
Egg	2.42	0.35	2.13			
Larva						
active	3.76			3.12	------	3.12
quiescent	4.48	1.00	5.25			
Protonymph						
active	5.15			9.02	------	9.02
quiescent	5.67	2.72	14.27			
Tritonymph						
active	7.36			31.47	------	31.47
quiescent	8.07	8.72	45.74			
Adult						
(mixed sexes)	11.0	29.94	154.48	108.74	96.99	205.73
	18.0	43.92	226.61	72.13	375.33	447.46
	28.0	45.72	235.90	9.29	170.36	179.65
	38.0	42.93	221.51	(14.39)[a]	3.97	(10.42)[a]

[a]Negative production.

values of larval, protonymphal and tritonymphal cast skins were taken as for *R. echinopus* (Stepien, 1970); the values were 0.41 cal x 10^{-3}, 0.83 cal x 10^{-3} and 2.93 cal x 10^{-3} respectively. The caloric value of an average mite increased exponentially during each of the feeding periods in the immature stage, and during the first 4 days of its adult life (Fig. 1). Each quiescent stage is characterized by a decrease in caloric value caused by intensive respiratory metabolism and by the lack of feeding and growth at that phase of development. The total respiratory energy expenditure at the end of the quiescent larva was 15.2% of the maximum caloric value at the beginning of this phase. For the quiescent protonymph and tritonymph those values were 14.2% and 7.2% respectively. For *R. echinopus* the respiratory expenditures were as follows: larva, 19.4%; protonymph, 13.9% and tritonymph, 10.8% (Stepien, 1970). A very sharp increase of the caloric value of an average adult mite occurred between days 8 and 12. During that time an average mite gained as much as 60.6% of the maximum caloric value that occurred on the 28th day.

The steady state of the caloric value of an average mite was achieved from the 22nd through the 30th day, after which time the caloric value steadily decreased. On the 38th day this value was 6.1% lower than the

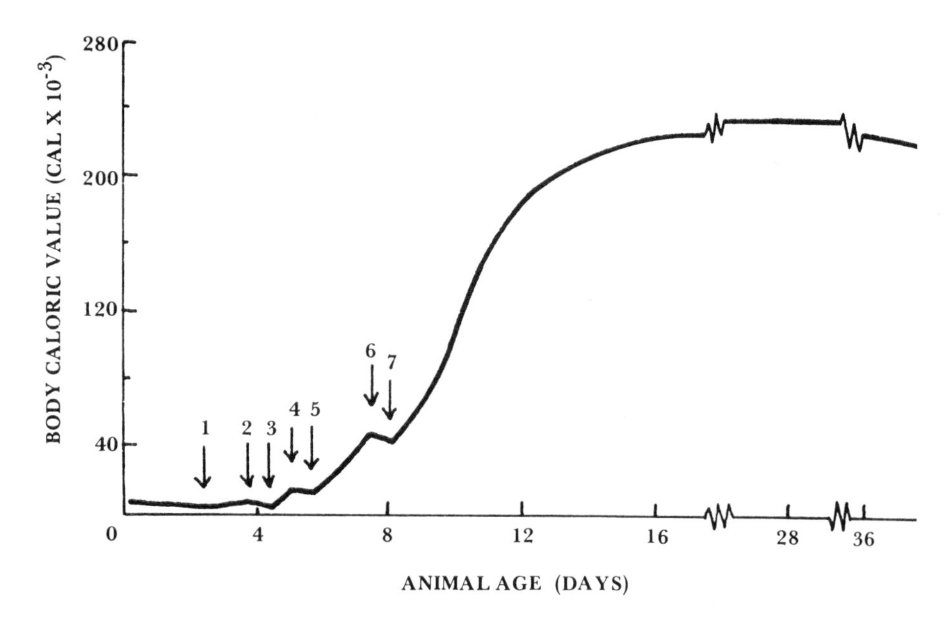

Figure 1. The changes in body caloric value of an average mite, *C. berlesei* during its life cycle. Legend for immature forms: larva active, 1; larva quiescent, 2; protonymph active, 3; protonymph quiescent, 4; tritonymph active, 5; tritonymph quiescent, 6, adult, 7.

maximum caloric value of an average animal on the 28th day. The growth curve of an average mite during each growth period (active larva, protonymph, tritonymph and young adult) has an "S" shape.

Food consumption

The amounts of food consumed by the average mite during each immature stage and during each arbitrarily chosen feeding period in the life of an adult mite are given in Table 5. The total food consumption for the immature period of this species is 111.69 cal x 10^{-3} of which 80.6% of that energy was consumed in the tritonymphal stage, 13.6% in the protonymphal stage and 5.8% in the larval stage. The high consumption rate in the tritonymphal stage was attributed to the intensive body growth that occurred in this stage and this was 68.8% of the total caloric value of an average mite at the end of the immature period. The young adult mite consumed food at the highest rate during the first 10 days of its emergence (2.61 cal per mite). The food consumption rate was lowest during the last feeding period than in the previous period. The average mite consumed a total of 6.14 cal of food energy during its entire life of 38 days.

Table 5. Food consumption rates of an average individual of *C. berlesei.*

Developmental Stage	Mite age in days	Food consumption	
		µg (50°C)	cal x 10^{-3}
Larva	2.42-3.76	1.48	6.42
Protonymph	4.48-5.15	3.51	15.19
Tritonymph	5.67-7.36	20.79	90.08
Total for immature period	0.00-8.07	25.78	111.69
Adult (male and female	8-18	602.47	2610.25
	19-28	408.19	1768.54
	29-38	380.57	1648.87
Total for adult life	8-38	1391.23	6027.66
Total for entire life	0-38	1417.01	6139.35

Respiratory metabolism

The results of oxygen consumption measurements of all developmental stages of *C. berlesei* are given in Table 6. The oxygen consumption, expressed in $\mu l \times 10^{-3} O_2\ hr^{-1}$ per individual, of each successive stage are higher than that for the preceding stage except for the quiescent tritonymph. Females consumed generally more oxygen than males but the body weight of females was 1.6 times that of males. Oxygen consumption rates, expressed in $\mu l \times 10^{-3} O_2 \mu g^{-1}\ hr^{-1}$, of the younger stages, both active and quiescent, were higher than those of the older stages. The metabolic rate of the female was higher than that of the male; this is probably due to the metabolic activity related with egg production.

The correlation between oxygen uptake (QO_2) and body weight (W) for all developmental stages of *C. berlesei* was calculated as the regression: $QO_2 = 2.578 \times W^{0.959}$. This equation was then used in the calculation of daily metabolism (ΔR) of adult mites substituting W with the values of live body weights of an average individual for each successive day of its life.

The values of b fluctuated usually from 0.7, for the dependence of metabolism from the body surface, to 1.0 for the dependence of metabolism of *C. berlesei* is related with body weight. Stepien (1970) and Stepien and Klekowski (in press) studied the oxygen metabolism of *R. echinopus* and *Acarus farris* (Oud.) and found the following equations for those species; $QO_2 = 1.524 \times W^{0.74}$ and $QO_2 = 1.894 \times W^{0.86}$ respectively.

Energy budgets

The cumulative energy budget of an average specimen of *C. berlesei* during its immature life is given in Table 7 and on Figure 2.

Table 6. Oxygen consumption of *C. berlesei* at 25°C.

Developmental Stage	No. of mites	Avg live wt (μg)	$\mu l \times 10^{-3} O_2 hr^{-1}$ per mite (avg)	$\mu l \times 10^{-3} O_2 hr^{-1}$
Egg	10	1.26	2.97	2.36
Larva				
active	20	2.84[a]	7.83	2.76
quiescent	20	4.42	11.15	2.52
Protonymph				
active	20	8.79[a]	21.99	2.50
quiescent	20	13.17	32.44	2.46
Tritonymph				
active	10	27.94[a]	46.03	1.65
quiescent	10	42.71	42.60	1.00
Adult				
female	10	107.85	133.73	1.24
male	10	66.65	61.82	0.93

[a] Values calculated as averages between the body weights of non-active stages.

Production. Cumulative total net production for the immature period consists of the cumulative body and exuviae production (Table 7 and Fig. 2). At the end of the active larval stage the cumulative total net production (P_c) reached the value of 3.18 cal x 10^{-3} and then it decreased by 27.4% as a result of respiratory metabolism of the quiescent phase to 2.31 cal x 10^{-3} at the end of quiescent larval stage.

During the active protonymphal stage, the total cumulative net production value reached the maximum of 13.27 cal x 10^{-3} on the 5.15 day and decreased again to 11.10 cal x 10^{-3}, (16.4%) on the 5.67 day due to utilization of the previously accumulated energy. A very sharp increase of the total cumulative net production occurred in the tritonymphal stage; 75.8% of the total cumulative net production for the entire immature period was achieved in this stage. During the last quiescent stage 7.4% of the energy of the P_c at the end of the active tritonymphal stage was spent in respiration, so the final value of the cumulative total net production at the end of the immature period was 42.75 cal x 10^{-3}.

After day 9 of the life cycle the cumulative total net production of an average mite consisted of the cumulative body and exuviae production and cumulative egg production (Table 7 and Fig. 3). The changes in the rates of cumulative body and exuviae production for the average adult mite are identical with changes in caloric value of the mite body (Fig. 1), as the value

Table 7. Elements of cumulative energy budget of *Caloglyphus berlesei* during its development (in cal x 10^{-3}).

Developmental stage	Time of develop- ment (days)	Cumulative consumption C_c	Cumulative total net production P_c	Cumulative respiration R_c	Cumulative assimilation $A_c = P_c + R_c$	Percent U_c^{-1}	Percent K_{1c}	Percent K_{2c}
Egg	1.00	-	0.15	0.15	0.00	-	-	-
	2.00	-	0.55	0.55	0.00	-	-	-
	2.42	-	0.86	0.86	0.00	-	-	-
Larva,	3.00	1.71	0.02	1.37	1.39	81.3	1.2	1.4
active	3.76	6.42	3.18	2.04	5.22	81.3	49.5	60.9
Larva,	4.00	6.42	2.85	2.37	5.22	81.3	44.4	54.6
quiescent	4.80	6.42	2.31	2.91	5.22	81.3	36.0	44.3
Protonymph,	5.00	17.96	10.61	4.19	14.80	82.4	59.1	71.7
active	5.15	21.61	13.27	4.56	17.83	82.5	61.4	74.4
Protonymph, quiescent	5.67	21.61	11.10	6.73	17.83	82.5	51.4	62.3
Tritonymph,	6.00	31.69	14.24	8.54	22.78	71.9	44.9	62.5
active	7.00	95.93	40.34	13.98	54.32	56.6	42.1	74.3
	7.36	111.69	46.17	15.89	62.06	55.6	41.3	74.4
Tritonymph,	8.00	111.69	43.04	19.02	62.06	55.6	38.5	69.4
quiescent	8.07	111.69	42.75	19.31	62.06	55.6	38.3	68.9
Adult mite,	9.00	214.42	65.82 (0.15[a])	37.23	103.05	48.1	30.7	63.9
(male or female)	18.00	2721.94	700.63 (472.32)	402.92	1103.55	40.5	25.7	63.5
	28.00	4490.48	880.24 (642.68)	905.97	1786.21	39.8	19.6	49.3
	38.00	6139.35	869.80 (646.65)	1442.40	2312.20	37.7	14.2	37.6

[a]Numbers in parentheses present the values of cumulative egg production.

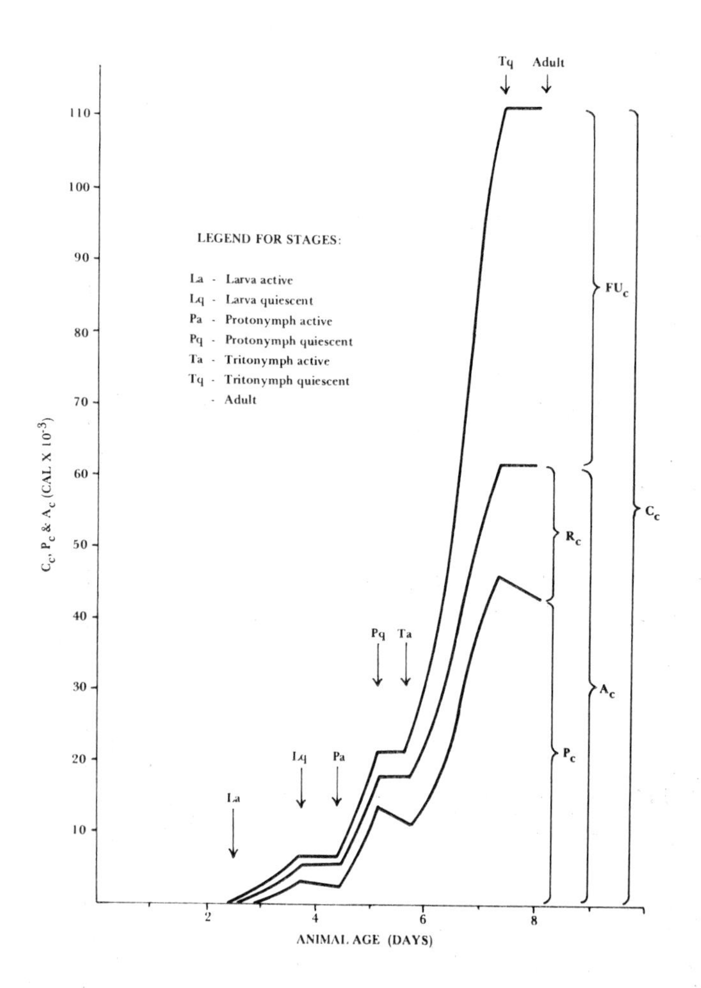

Figure 2. The cumulative budget of an average mite, *C. berlesei*, during its immature life. Legend: A_c, cumulative assimilation; C_c, cumulated food consumption; FU_c, unassimilated food cumulated; P_c, cumulated total net production; R_c, respiration cumulated.

of this component of the cumulative net production includes both, caloric value of body and exuviae.

Egg production began on the 9th day at a very low value of 0.15 cal x 10^{-3} but the rate increased rapidly and reached the peak (75.57 cal x 10^{-3}/day) on the day 13 of the life cycle. From the 15th day the daily egg production rate was practically constant, about 40 cal x 10^{-3}, till the 18th day, after which time the daily egg production rate decreased steadily until it ceased on the 32nd day. The maximal value of the cumulative total net

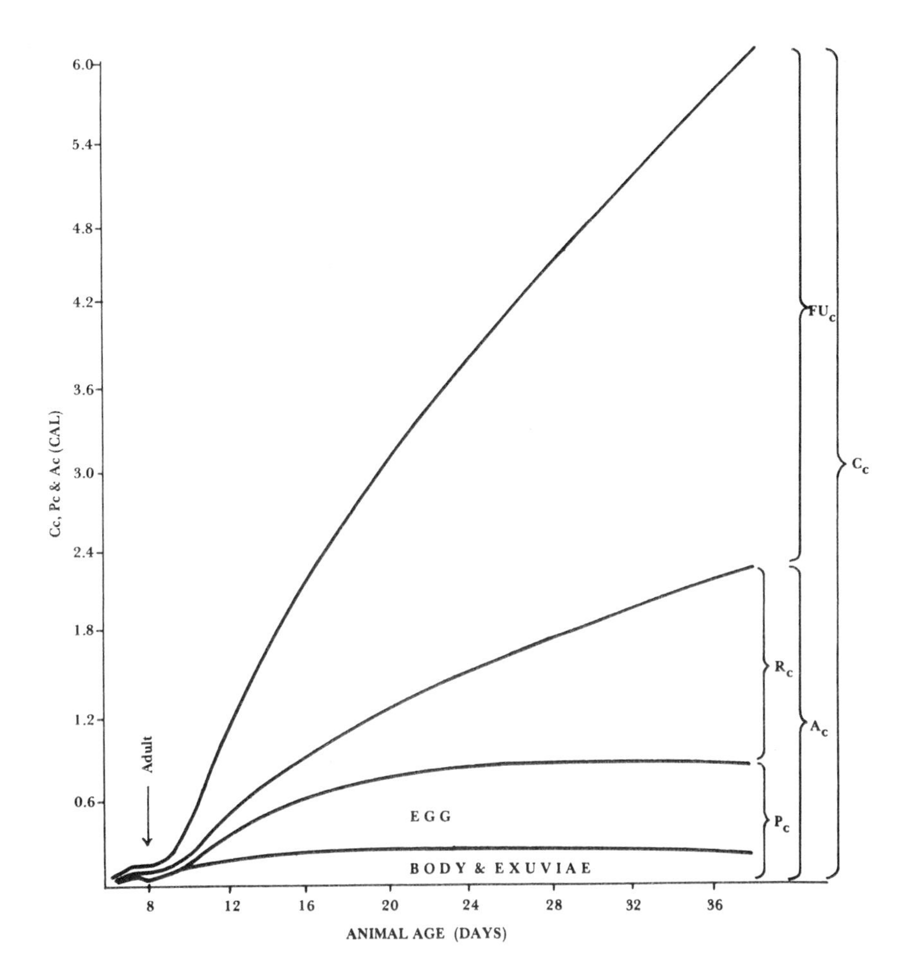

Figure 3. The cumulative energy budget of an average mite, *C. berlesei*, during its adult life. Legend: see description Fig. 2.

production, 83.76 cal x 10^{-3}, occurred on the day 31 of the life cycle and then it decreased to 869.80 cal x 10^{-3} on day 38 due to cessation of egg
production, 83.76 cal x 10^{-3}, occurred on the day 31 of the life cycle and then it decreased to 869.80 cal x 10^{-3} on day 38 due to cessation of egg
production and loss in body caloric value.

Assimilation. The cumulative assimilation calculated as the sum of cumulative total net production (P_c) and cumulative respiration (R_c) is presented in Table 7, and on Fig. 2 and 3). The line representing cumulative assimilation (A_c) during each of the feeding immature stages is parallel to that of cumulative total net production (P_c) as the daily respiration rates respiratory energy expenditures are covered by the loss of body biomass ($R_c = P_c$). The P_c of the active larval stage accounted for 60.9% of the cumulative assimilation; the value for both protonymph and tritonymph was 74.4%.

During the first week of adult life, the cumulative assimilation made up about 68% of the total net production, (P_c); there was intensive body growth and egg production in that period. From that time the P_c in the values of the cumulative assimilation (A_c) decreased from 66.5% on day 16 to 50.7% on day 27, after which cumulative respiration is the biggest factor in cumulative assimilation. At the end of the life cycle of an average mite, the cumulative assimilation value consisted of 62.4% of cumulative respiration and in 37.6% of total cumulative net production.

Consumption. The fastest rate of increase of cumulative food consumption (C_c) during the immature period occurred in the tritonymphal stage; namely, 80.6% of the total food consumed during this period. The highest daily food consumption ($\triangle C$) occurred on the 7th day and concomitantly the daily total net production ($\triangle P$) value was also the highest at this time (Fig. 2). The daily food consumption rates were high during the first 10-day period of adult life (Fig. 3). This was due to intensive daily total net production rates ($\triangle P$) of immatures development, as the food consumption rate is related indirectly with production rate,

$$C = \frac{\triangle A}{U^{-1}} \text{ or } C = \triangle P + \triangle R \times 100$$

and the value of U^{-1} was assumed to be constant during each feeding period. After 20 days the cumulative food consumption generally showed a linear growth (Fig. 3).

Efficiencies of energy utilization

Assimilation efficiency (U^{-1}). It was noted previously that the instantaneous (daily) indices of energy utilization could be calculated only for the periods of "positive" production. The instantaneous assimilation, U^{-1}, for both larval and protonymphal stages was very high; 81.3% and 83.0% of the consumed energy was assimilated in these stages (Fig. 4-A). The instantaneous assimilation value decreased considerably in the tritonymphal stage to 49.1%. The instantaneous assimilation value decreased with aging of the adult from about 40% for the first feeding period through 38.6% for the second to 31.9% for the last 10 days of life cycle of an average mite.

The cumulative assimilation, U_c^{-1}, was the same as the instantaneous for the larval stage; this showed the maximum value of 82.5% during the protonymphal stage and dropped to 55.6% at the end of the immature period. Further decrease in the index value occurred during the first week of adult life. Then for the remainder of the life of an adult mite, the cumulative assimilation remained relatively constand with values between 40.7 and

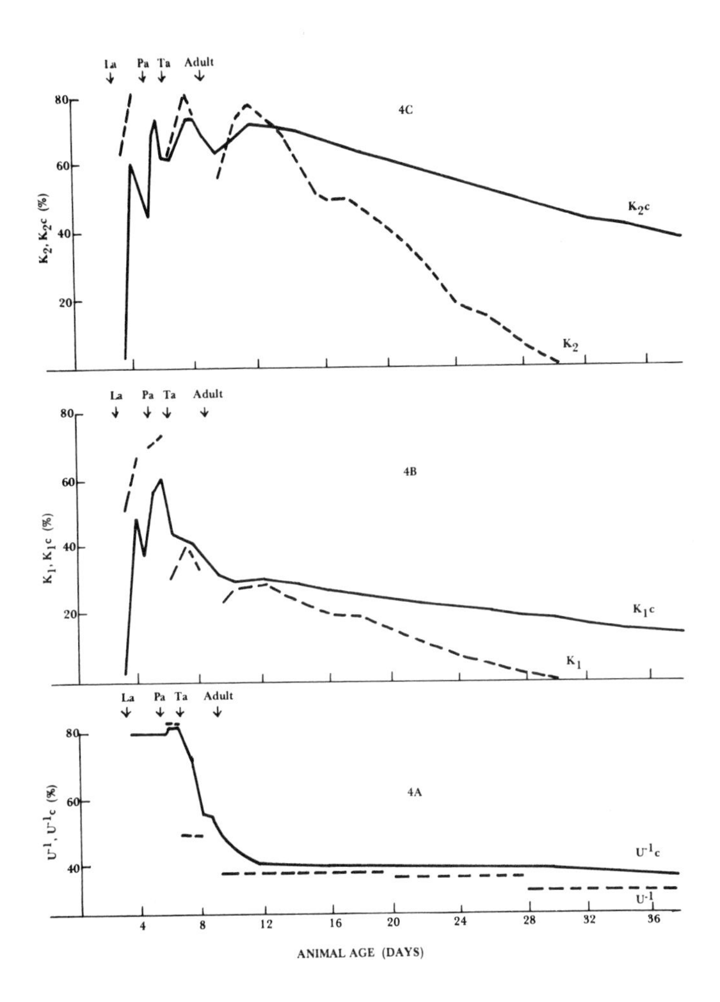

Figure 4. The efficiencies of energy utilization for *C. berlesei* during its life span. 4-A.—Assimilation of efficiency index: U^{-1} and $U^{-1}{}_c$. 4-B.—Index of gross production efficiency: K_1 and K_{1c}. 4-C.—Index of net production efficiency: K_2 and K_{2c}.

37.7%.

Gross production efficiency (K_1). This index representing the utilization of ingested food energy for production, in its instantaneous for, K_1, varied considerably during each feeding stage and quiescent stages (Fig. 4-B). The highest values of this index occurred in protonymphal and larval stages, 72.9 and 67.1% respectively. In the tritonymphal stage this index fluctuated from 31.2 to 40.6%. These high values of instantaneous K_1 indicate that the youngest stages utilized food very efficiently during the feeding periods and served to store energy for the succeeding quiescent (non-feeding) stages.

The first 4 days of adult life, marked a slight increase of the value of instantaneous, K_1, and then it decreased at a relatively constant rate, reaching 0.3% level on day 31 when the daily net production was nil.

Net production efficiency (K_2). The net production efficiency index (K_2, Fig. 4-C) describes the utilization of assimilated energy for production, in its instantaneous form; K_2 was quite high during larval and nymphal development, ranging from 63.3 to 87.8%. The first week of adult life, was a period of intensive body growth and egg production and the index showed maximum values for the adult stage, 56.3 to 78.4%. Then with aging of the animal, this index decreased to zero concomitantly with a decreasing daily total net production.

The cumulative net production efficiency index, K_{2_c}, was also high during preimaginal development when the quiescent phases were relatively short and part of the energy utilized by these stages was small as compared to stored energy reserves. Intensive body growth and egg production during the first week of adult life influenced the cumulative index value of net production efficiency, K_{2_c} (63.9 to 72.8%). With aging the energy used for growth and reproduction decreased steadily, and after the 27th day the cumulative respiration value overcame cumulative energy assimilation and the value of K_{2_c} decreased from 50.7 to 37.6%.

DISCUSSION

In the growth and development of acarid mites, there are 3 immature stages or periods of active feeding and growth: namely, the larva, protonymph and tritonymph. Each of these active forms are followed by their immobile stages of quiescence. The rapid growth of each mobile stage requires a relatively high food intake and stored energy which is then utilized during the quiescent phases. Teneral females grow rapidly and soon begin to produce large number of eggs; these expenditures of energy for biomass production and respiration are balanced by high food intake.

The problem of defining the effectiveness of food and energy utilization by an organism can best be described by indices of efficiency calculated from energy budgets. It is interesting to compare this effectiveness between *C. berlesei* and other species of acarid mites. A complete energy budget in instantaneous and cumulative forms during the entire life cycle of *R. echinopus* is available (Stepien, 1970). The energy budget in both forms for immature *T. putrescentiae* and some information concerning food consumption and biomass production of adult mites of that species is also available (Stepien *et al.*, unpublished).

Hence, the food requirements of the active immature stages and adults of *C. berlesei, R. echinopus* and *T. putrescentiae* were compared in Table 8.

Table 8. Food consumption, biomass production and efficiencies of food utilization for biomass production in 3 species of acarid mites. (Data for *R. echinopus* after Stepien, 1970 and for *T. putrescentiae* after Stepien *et al.*, in press.)

Developmental stage and species	Animal age in days	Consumption cal x 10^{-3}	Body	Egg	Total	$K_1 = \frac{\text{production}}{\text{consumption}}$ (%)
Larva						
C. berlesei	2.42 - 3.76	6.42	3.12	--	3.12	48.6
R. echinopus	5.60 - 11.2	5.35	2.12	--	2.12	39.6
T. putrescentiae	2.7 - 7.2	5.93	2.11	--	2.11	35.6
Protonymph						
C. berlesei	4.48 - 5.15	15.19	9.02	--	9.02	59.4
R. echinopus	12.5 - 15.8	18.53	7.47	--	7.47	40.3
T. putrescentiae	8.2 - 10.3	14.52	7.20	--	7.20	49.6
Tritonymph						
C. berlesei	5.67 - 7.36	90.08	31.47	--	31.47	34.9
R. echinopus	17.1 - 19.8	47.26	16.77	--	16.77	35.5
T. putrescentiae	11.4 - 13.7	24.22	11.88	--	11.88	49.1
Adult (male or female)						
C. berlesei	8 07 - 38.0	6027.66	175.77	646.65	822.42	13.6
R. echinopus	21.2 - 72.0	1539.92	51.33	501.24	552.57	35.9
T. putrescentiae	14.7 - 80.0	528.45	23.91	340.16	364.07	68.9

The data show that the immature stages and adult mites of *C. berlesei* consumed food at higher rates than those of *R. echinopus* and *T. putrescentiae*. The same is true for the biomass production rates. The data also show that protonymphal stages of all 3 mite species utilized food energy very efficiently for body growth.

It should be noted that the values of cumulative indicies of efficiency represent a given moment of life of an average mite and that the instantaneous indices represent only the periods of active feeding and growth. Figures 4-A, 5-A and 6-A present the values of assimilation efficiency indices in

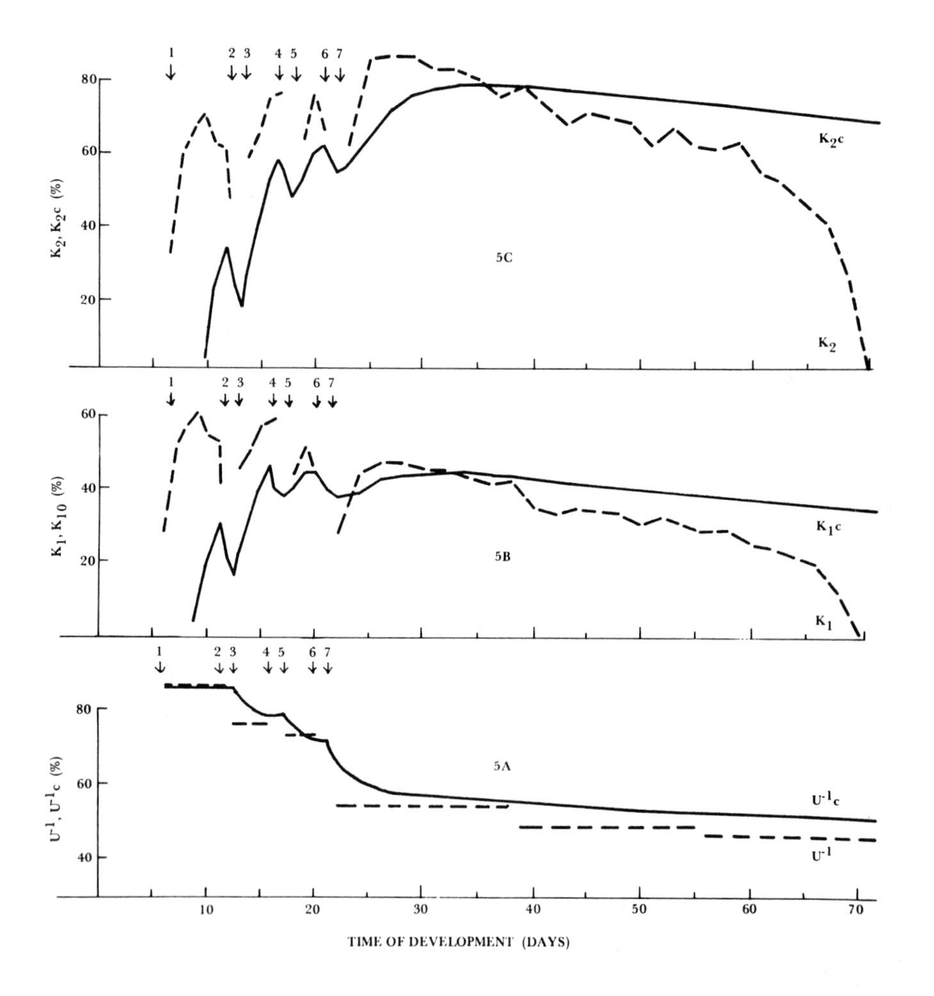

Figure 5. The efficiencies of energy utilization for *R. echinopus* during its life span (after Stepien, 1970). Legend: see description Figs. 1 & 4.

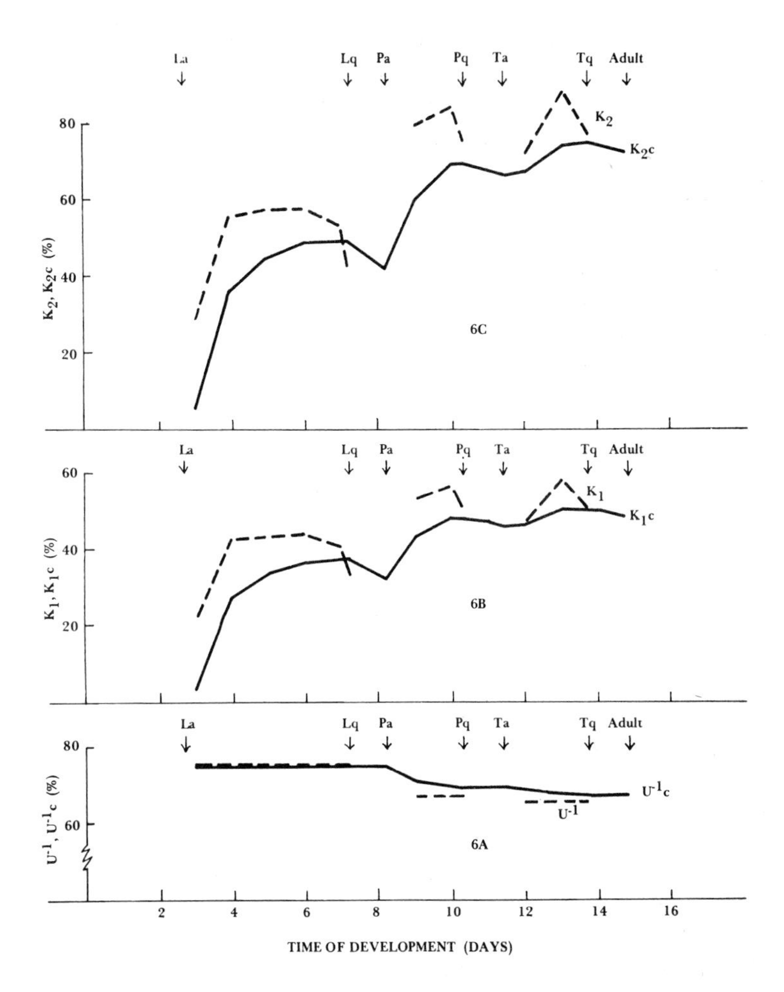

Figure 6. The efficiencies of energy utilization for *T. putrescentiae* during its immature development (after Stepien *et al.*). Legend: see description Figs. 2 & 4.

instantaneous (U^{-1}) and cumulative (U_c^{-1}) forms for *C. berlesei, R. echinopus* and *T. putrescentiae*. These data show that the youngest stages, that is, larvae and protonymphs, assimilated ingested food very efficiently; larval stages of *C. berlesei* and *T. putrescentiae* assimilated 86.3% of ingested food energy. During the subsequent developmental stages, food assimilation decreased, except for the protonymphal stage of *C. berlesei*, which appeared to be the most effective in energy assimilation. There were no large differences in assimilation efficiency among the larval, protonymphal and tritonymphal stages of *R. echinopus* and *T. putrescentiae*, but in the tritonymphal stage of *C. berlesei*, assimilation decreased considerably from about 72% at the beginning to 55.6% at the end of this stage. As aging progressed in the adult mites, the energy assimilation became generally more stable. The adult mites of *C. berlesei* assimilated an average of 43% of the ingested energy. *R. echinopus* utilized its food more effectively; the adult average mite assimilated about 55% of its ingested food energy.

The instantaneous indices of gross production efficiency, K_1, (Fig. 4-B, 5-B and 6-B) varied greatly during the development of the immature stages of these mites, but they were always high, 73% for *C. berlesei*, 61% for *R. echinopus* and 58% for *T. putrescentiae*. Hence, the actively feeding immature stages used their food very efficiently and this served as stored energy for the quiescent periods that followed. Comparison of K_1 values of these 3 species show that *T. putrescentiae* adults were the most efficient in the utilization of ingested energy for biomass production. Egg production is the main component of the total biomass production in adult acarid mites; it contributed 93.4% of this value in *T. putrescentiae*, 90.7% in *R. echinopus*, and 78.6% in *C. berlesei*.

The highest value of cumulative gross production efficiency index, $K_{1\,c}$, was attained in the protonymphal stages of *C. berlesei*, and *R. echinopus*, 61% and 46% respectively, whereas during the immature development of *T. putrescentiae* the highest value of $K_{1\,c}$ (51%) was achieved in the tritonymphal stage. Hence, the protonymphal and tritonymphal stages utilized food energy for growth very efficiently.

The adult mites also utilized food very efficiently during the period of intensive body growth and egg production. With aging, however, the efficiency of utilization of consumed energy for body and egg production decreased slightly. *R. echinopus* utilized food energy for production more effectively than *C. berlesei*. The values of $K_{1\,c}$ for *R. echinopus* fluctuated from 36 to 45%, and for *C. berlesei* from 14 to 30%.

The net production efficiency indices, K_2 and $K_{2\,c}$, (Fig. 4-C, 5-C, and 6-CO that represent the part of assimilated energy used for production, are extremely high during development of the immature stages and also during the period of intensive body growth and egg production of adult mites. The high values of instantaneous net production efficiency indices (K_2) for active immature stages and young and reproducing adult mites indicate how well these animals used assimilated food energy for growth and reproduction during the short periods of feeding and intensive egg production. Instantaneous K_2 reached such high values as 87.8% in the protonymphal stage in *C. berlesei*, as compared to 76.6% in *R. echinopus* and 88.6% in the tritonymphal stage of *T. putrescentiae*. Cumulative K_2 also gave high values during the preimaginal development of these acarid mites because the non-feeding stages are relatively short and the energy expenditures are small in comparison with the stored energy in the bodies of prequiescent stages.

The values of cumulative net production efficiency index, K_{2c}, for adult *R. echinopus* were high and decreased very slowly with age. Values of the same index for *C. berlesei* were about equal with protonymphal and tritonymphal stages, during the first several days, but then they decreased considerably with age. The values of K_{2c} for adult mites were influenced by body growth and egg production. In *C. berlesei* body growth and egg production was concentrated mainly in the first half of adult life, whereas *R. echinopus* grew and produced eggs more steadily during its entire lifetime.

Generally, all coefficients of energy utilization in acarid mites were relatively high; the young developing stages assimilated food energy more efficiently than adult mites. During preimaginal development, the actively feeding stages utilized their food well and energy was stored and later utilized in metamorphosis. Adult mites of *C. berlesei* had a lower efficiency of food energy utilization, in body and egg production as compared to the immature stages of the same species and to adult *R. echinopus.*

The use of assimilated energy for body growth improved with age of the immature form during the development of these mites. Adult mites of *C. berlesei* and *R. ethinopus* have different abilities in their utilization of assimilated energy for body and egg production. *C. berlesei* utilized assimilated energy efficiently in the first period of intensive body growth and egg production, but as aging progressed respiration accounted for the greater part of assimilated food energy. Adult *R. echinopus* mites used from 70 to 80% of assimilated energy for production during its life and only 20 to 30% of it was utilized in respiration.

The main difference between *C. berlesei* and *R. echinopus* was in the distribution of the assimilated energy for body and egg production. Cumulative egg production of *R. echinopus* was about 5 x greater than cumulative body production. In *C. berlesei* cumulative egg production was only 2.7 times that of cumulative body production.

Klekowski *et al.* (1967) studied energy utilization in the red flour beetle, *T. castaneum*. They found that the instantaneous net production efficiency index in the development of the immature stages varied from 10 to 50% and during the first 20 days of its adult life it increased to about 60%. The same index in the cumulative form in preimaginal development reached the maximum value of about 30%, then decreased to 20% during the first 20 days of its adult life and finally it increased to 47% after 200 days. A comparison of the values of these indices show that acarid mites utilized assimilated food energy for biomass production better than *T. castaneum.*

SUMMARY

The bioenergetics of *Caloglyphus berlesei* were studied under controlled laboratory conditions. Mites were fed and artificial diet and food intake and biomass production rates were measured by means of a gravimetric method. Oxygen consumption was determined by means of a cartesian diver micro-respirometer. Caloric values of food, adult mites, quiescent tritonymphs and eggs of this species were evaluated in a micro-bomb calorimeter.

During its immature developmental period, *C. berlesei* consumed a total of 11.69 cal x 10^{-3} of food energy; of which 80.6% was consumed in the tritonymphal, 13.6% in the protonymphal, and 5.8% in the larval stages. Adult mites consumed food at the highest rate during the first 10 day period of their life and the consumption rate decreased with aging. The average mite consumed about 6.14 calories of food energy during its entire life span of 38 days. The highest production rate during preimaginal development occurred in the last nymphal stage; in this tritonymphal stage the mite gained 68.8% of its maximum body caloric value. Adult mites grew very rapidly during the first few days after emerging from the tritonymphal stage. Cumulative egg production was 2.7 times greater than the cumulative body production. Respiratory metabolism (μl x $10^{-3} O_2 hr^{-1}$) varied with the 0.959 power of fresh body weight.

A comparison of the food requirements of the immature stages and adults of *C. berlesei, R. echinopus* and *T. putrescentiae* showed that *C. berlesei* consumed food at higher rates than the other species. The data for assimilation efficiency for these 3 species show that larvae and protonymphs assimilated food very efficiently; larval stages of *C. berlesei* and *T. putrescentiae* assimilated 86.3% of ingested food energy. The latter showed decreased assimilation in the tritonymphal stage, namely from 72 at the beginning to 55.6% at the end of this stage. The adult mites of *C. berlesei* assimilated an average of 43% of ingested energy, compared to 55% for *R. echinopus.*

The gross production efficiency varied greatly during the development of the immature stages of these mites, but these indices were always high, e.g., 73, 61 and 58% for *C. berlesei, R. echinopus* and *T. putrescentiae* respectively. *T. putrescentiae* adults were the most efficient in the utilization of ingested energy for biomass production. Egg production was the main component of total biomass production in the adult mites.

The high values of net production efficiency indices were very high during the development of the immature stages and also during the period of intensive body growth and egg production of adult mites. The instantaneous net production indices reached 87.8% in the *C. berlesei* protonymph compared to 76.6% for *R. echinopus.*

The main difference between *C. berlesei* and *R. ehinopus* was in distribution of the assimilated energy for body and egg production; *R. echinopus* had a cumulative egg production 5x that of the cumulative body production while the comparable value for *C. berlesei* was 2.7.

Energy utilization data with insects are limited but a comparison of acarid mites and *Tribolium castaneum* shows that these acarid mites utilized assimilated food energy for biomass production more efficiently.

Acknowledgements.—The technical assistance of Miss Magdalena Sosnowska and Mrs. Martha Potts in the course of this work is gratefully acknowledged. We would also like to cite Jan Boczek for his contribution to the development of this paper.

REFERENCES

BOCZEK, J. (1957). Rozkruszek maczny *(Tyroglyphus farinae* L.). Morfologia, biologia i ekologia, szkodliwosc oraz proby zwalczania. *Roczn. Nauk Roln.*, **75-A(4)**: 559-644.

EWY, Z. (1964). Zarys fizjologii zwierzat. PWN, Warszawa-Krakow.

HUGHES, T. E. (1943). The respiration of *Tyroglyphus farinae. J. Exp. Biol.* **20**:1-5.

JAKUBOWSKA, J. (1967). Wystepowanie i ekologia rozkruszkow z rodzaju Acarus (Acarina). Ph.D. thesis - Warsaw Agricultural University.

KLEKOWSKI, R. Z. (1967). Cartesian diver technique for microrespirometry. Prepared for Int. Meeting on "Methods of assessment of secondary production in freshwaters" Prague 1967:1-26.

KLEKOWSKI, R. Z. (1968). Cartesian divers microrespirometry for terrestrial animals. In: Gradzinski, W., Klekowski, R. Z./eds/Methods of ecological bioenergetics. 51-66, PAN, Warszawa-Krakow.

KLEKOWSKI, R. Z. (1970). Bioenergetic budgets and their application for estimation of production efficiency. *Pol. Arch. Hydrobiol.* 17:1-2.

KLEKOWSKI, R. Z., PRUS, T., ZYROMSKA-RUDZKA, H. (1967). Elements of energy budget of *Tribolium castaneum* (Hbst) in its developmental cycle. In: K. Petrusewicz (Ed.) Secondary productivity of terrestrial ecosystems. 2:859-879, PWN, Warszawa-Krakow.

KLEKOWSKI, R. Z., STEPIEN, Z. Elements of energy budget of *Rhizoglyphus echinopus.* (in press).

RODRIGUEZ, J. G. (1972). Inhibition of acarid mite development by fatty acids. (this publication).

RODRIGUEZ, J. G., STEPIEN, Z. Biology and population dynamics of *Caloglyphus berlesei.* (in press), *Jour. Kan. Entomol. Soc.).*

SOLOMON, M. E. (1946). Tyroglyphid mites in stored products. Nature and amount of damage to wheat. *Ann. Appl. Biol.* 33:280-289.

SOLOMON, M. E. (1959). Pests in packaging. Weight loss from infestation. *Food.* 28:(329) 43-45,52.

STEPIEN, Z. (1970). Bilans energetyczny rozkruszka korzeniowego *Rhizoglyphus echinopus* (F. et R.) (Acarina: Acaridae) w czasie jego rozwoju. Ph.D. thesis - Warsaw Agricultural University.

STEPIEN, Z., GOSZCZYNSKI, W., and BOCZEK, J. Studies on food and energy utilization of *Tyrophagus putrescentiae* (Schr.) (Acaridae). (Unpublished.)

STEPIEN, Z., KLEKOWSKI, R. Z. Oxygen uptake in the development of *Rhizoglyphus echinopus* (F. et R.) and *Acarus farris* (Oud.) (in press).

STEPIEN, Z., RODRIGUEZ, J. G. Rearing and collecting large quantities of acarid mites. (in press, *Ann. Entomol. Soc. Amer.)*

UTILIZATION OF WATER BY TERRESTRIAL MITES AND INSECTS

G. W. Wharton
Department of Entomology, The Ohio State University
Columbus, Ohio 43210

Larry G. Arlian
Department of Biological Sciences
Wright State University
Dayton, Ohio 45431

Water is unique among major nutrients in a number of ways. It is significantly volatile at temperatures in the biological range and thus terrestrial mites and insects exchange significant amounts of water with the ambient air. The time required for one-half of the water content to be exchanged (the half-life of exchange) has been calculated for a number of forms, (Table 1). During each half-life, at equilibrium conditions, 69% of the water mass passes from the body to the air or is transpired and at the same time an amount equal to this must pass from the air into the body or be sorbed. Thus during each half-life 138% of the water mass of the organism passes across the air-body interface. Actually the concentration of water in mites and insects is almost always higher than the concentration in the air so that mites and insects tend to lose water to the air. The activity of the water, a_w, in the body is about 0.99, (Edney, 1968; Wharton and Devine, 1968). Since air usually has a relative humidity significantly lower than 99%, the terrestrial environment is truly a desiccating one. For air to be in equilibrium with an organism whose body water has an a_w equal to 0.99, the relative humidity of the air would have to be 99%. The activity of water vapor, a_v, is related to the relative humidity and a_w as follows:

$$a_v = rh/100 = a_w \qquad (1)$$

When a net loss of water occurs, it must be replaced by water which has an a_w equal to 0.99 in some manner or another. In many cases it is ingested with the food or imbibed by drinking. Pure water may also be extracted from unsaturated air by some mechanism that is not yet fully understood, (Buxton, 1930; Lees, 1946; Noble-Nesbitt, 1970). It may also be regained by some combination of these.

Water is not only the most concentrated molecular species, it is also the most common in actively metabolizing tissues. In mites and insects more than 99 out of every 100 molecules is a water molecule. Even though water molecules are light compared to the active macromolecules, most terrestrial arthropods are at least 70% water by weight, and the extremes range from 46% to 92%, (Rapoport and Tschapek, 1967). Concentrations of water near a_w = 0.99 must be maintained or metabolism will slow down and stop. When this occurs, death results in most instances, but in some cases a state of

Table 1. Half-life of exchange of water and critical equilibrium activity (CEA) in insects and mites, reported in or calculated from references indicated.

Order & species	CEA	Half-life (hrs)	References
Thysanura			
Ctenolepisma terebrans	0.475		Edney, 1971
Thermobia domestica	0.45		Beament *et al.*, 1964
Corrodentia (Psocoptera)			
Liposcelis bostrychophilus	0.60		Knülle and Spadafora, 1969
Liposcelis knullei	0.70		Knülle and Spadafora, 1969
Liposcelis rufus	0.58		Knülle and Spadafora, 1969
Orthoptera			
Arenivaga sp., nymphs	0.825		Edney, 1966
adult females	< 0.95		Edney, 1966
adult males	> 0.95		Edney, 1966
Blatta nymphs	> 0.95		Edney, 1966
Chortophaga viridifasciata, larvae	0.82		Ludwig, 1937
Periplaneta nymphs & adult females	> 0.95		Edney, 1966
Hemiptera			
Graphosoma lineatum	> 1.00	120[a]	Govaerts and Leclercq, 1946
Coleoptera			
Leptinotarsa decemlineata	< 1.00	216[a]	Govaerts and Leclercq, 1946
Tenebrio molitor, larva	> 0.88		Marcuzzi and Santoro, 1959
		198	Mellanby, 1932
		312[a]	Govaerts and Leclercq, 1946
Tenebrio molitor, adults	> 1.00	216[a]	Govaerts and Leclercq, 1946
Siphonoptera			
Xenopsylla brasiliensis, prepupa	0.50		Edney, 1947
Xenopsylla cheopis, prepupa	0.65		Edney, 1947; Knülle, 1967
Acariformes			
Acarus siro	0.70		Knülle, 1965a
Bryobia praetiosa	> 0.93		Winston and Nelson, 1965
Dermatophagoides farinae	0.70	27.7	Arlian, 1972; Larson, 1969
Tetranychus telarius	> 0.75		McEnroe, 1961
Tyrophagus putrescentiae	ca 0.84	6.9	Cutcher, 1970
Parasitiformes			
Amblyomma americanum, larva	> 0.69		Lancaster & McMillan, 1955
Amblyomma americanum, females	0.85		Sauer and Hair, 1971
Amblyomma americanum, males	0.83		Sauer and Hair, 1971
Amblyomma cajennense, larva	0.80-0.85		Knülle, 1965b
Amblyomma cajennense, females	0.90		Lees, 1946
Amblyomma compressum	ca 0.80		Aeschlimann, 1963
Amblyomma maculatum, females	0.88		Lees, 1946
Dermacentor andersoni, larva	0.80-0.85		Knülle, 1965b
Dermacentor andersoni, females	0.82		Lees, 1946
Dermacentor reticulatus, females	0.86		Lees, 1946
Hyalomma asiatican, females	0.80		Balasgov, 1960

Table 1. (cont'd.)

Order & species	CEA	Half-life (hrs)	References
Parasitiformes (cont'd.)			
Hyalomma dromedarii, larva	0.75		Hafez *et al.*, 1970
Hyalomma dromedarii, nymph	> 0.75		Hafez *et al.*, 1970
Hyalomma dromedarii, adult	> 0.75		Hafez *et al.*, 1970
Ixodes canisuga, females	0.92		Lees, 1946
Ixodes hexagonus, females	0.94		Lees, 1946
Ixodes ricinus, females	0.92		Lees, 1946
Laelaps echidnina	0.90	18.6	Wharton and Devine, 1968; Wharton and Kanungo, 1962
Ornithodoros moubata, nymphs	0.82		Lees, 1946
Ornithodoros porcinus porcinus	0.85		Lees, 1946; Walton, 1964
Ornithodoros savignyi,			
first instar nymph	ca 0.75		Hafez *et al.*, 1970
second instar nymph	> 0.75		Hafez *et al.*, 1970
third instar nymph	0.84		Hafez *et al.*, 1970
female, adult	0.84		Hafez *et al.*, 1970
Rhipicephalus sanguineus, females	0.84		Lees, 1946

[a]Time to reach equilibrium.

cryptobiosis may intervene, (Crowe and Cooper, 1971).

The functions of water are numerous, and as the matrix for metabolism water is responsible for: solvation; the three dimensional configuration of macromolecules; bioelectrical currents associated with the diffusion of ions; movement of nutrients, wastes, and metabolic intermediates by diffusion or bulk flow; lubrication of reproductive, digestive, and other tracts; and in some cases such functions as temperature control through evaporative cooling, (Church, 1959).

Water is a product of catabolic release of energy from nutrients with caloric values; therefore, the possibility exists that water may be replenished by the conversion of O_2 to H_2O through aerobic metabolism. Such water is known as metabolic water and it can be significant in maintaining the a_w of the body fluids, especially in larger organisms, (Schmidt-Nielsen and Schmidt-Nielsen, 1953). In insects and mites it has not been demonstrated to be sufficient to support growth and reproduction, (Edney, 1966). In smaller arthropods such as mites, metabolic water is insignificant in the overall mechanism of water balance, (Kanungo, 1965; Arlian, 1972).

In culturing mites it has been found that ingestion of other nutrients is frequently associated with water relations. In a series of studies on laelapid and dermanyssid mites, it was discovered that feeding took place more readily when blood was offered at high temperatures $>35^\circ$C and low humidities ⟨22%, (Wharton and Cross, 1957; Cross and Wharton, 1964). Loss of water is greater at higher temperatures and when the temperature reaches a certain level, usually between 30-40°C depending on the species, water loss suddenly increases dramatically. The temperature at which this occurs is called the critical temperature, CT, (Edney, 1957). Water balance can be maintained by some species of fasting mites and insects, (Table 1), so long as they remain in an atmosphere of sufficiently high relative humidity. The lowest relative humidity at which a fasting organism can maintain its net water loss at zero has been called the critical equilibrium humidity, CEH, (Knulle and Wharton, 1964). It is more convenient to talk about the activity of the water in the air than it is to talk about the relative humidity. Relative humidity cannot be used to describe the concentration of water in a solution but activity can. Thus the same unit, activity, can be used for vapor and liquid phases. For this reason CEA, critical equilibrium activity, is used to replace CEH, (Wharton and Devine, 1968). In the studies on laelapid and dermanyssid mites it was found that maximum feeding occurred at temperatures above the CT and water vapor activities below the CEA. In these mites it appeared that imbibition of blood occurred most frequently only when water was being lost at a rapid rate or in other words because the mites were thirsty rather than because they were hungry.

Studies on plant feeding mites, *Tetranychus urticae,* suggest a similar motivation for feeding. If mites are fed on plants with a high phosphorus content, they produce more eggs than if fed on plants whose leaves had a low phosphorus content, (Rodriguez, 1954). This suggests that mites were feeding to maintain their water balance and that leaves with higher phosphorus contents contain more nutrients for a given amount of water than plants with a low amount of phosphorus. Tetranychids have also been observed to lay more eggs when exposed to air with a low a_V as opposed to a high a_V, (Boudreaux, 1958). The two- to three-fold increase in eggs per mite observed in tests at different humidities is comparable to the difference observed from more or less phosphorus in the leaves of the food plant. In the first instance it is postulated that the mites that produced the most eggs imbibed more phosphorus with each unit of water while in the second case more water was lost and therefore more water and nutrients were imbibed by mites laying the higher number of eggs. Tetranychids lose water rapidly in the evacuated column of a scanning electron microscope, (Fig. 1). On the other hand, if they have a source of water while in the vacuum, they can temporarily maintain their water balance by imbibing water, (Fig. 2). When imbibing water from leaves, tetranychids also imbibe nutrients. There is, therefore, a positive correlation between the amount of imbibition, amount of nutrients obtained, and the amount of water lost. This relationship between water loss and nutrient uptake can be important in nutritional studies of all types. In most studies where water is not one of the nutrients

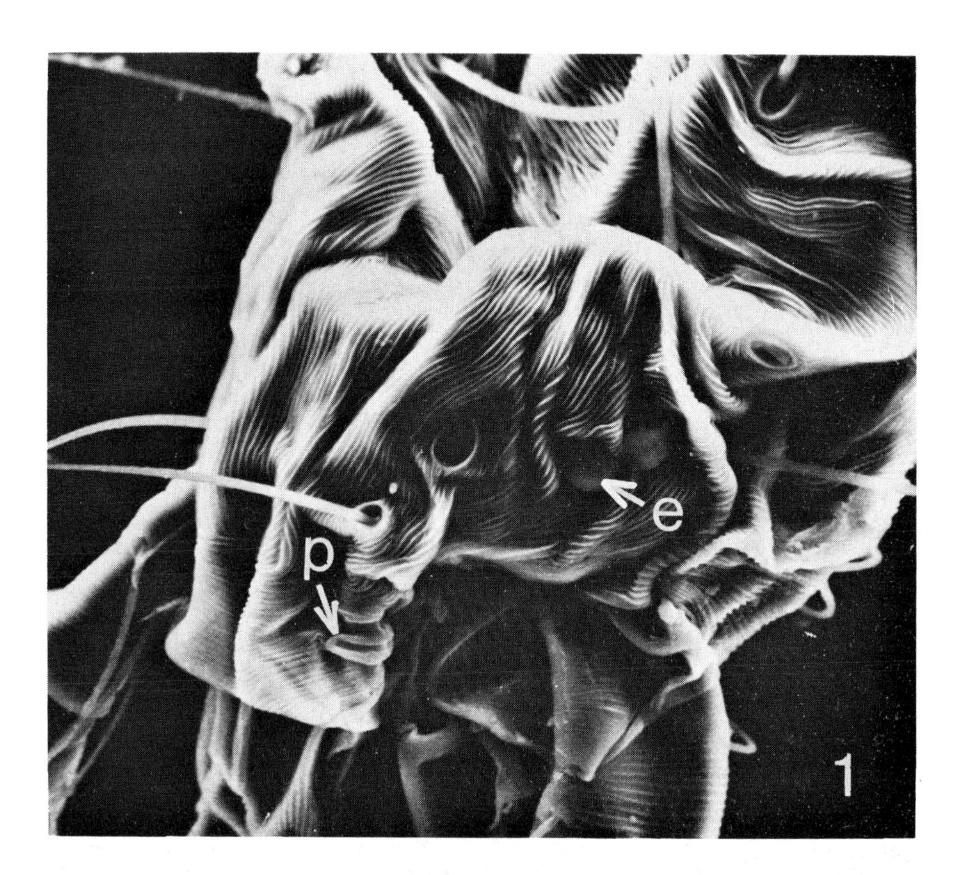

Figure 1. The anterior portion of a desiccating two-spotted spider mite, *Tetranychus urticae.* 670x. e - eye; p - peritreme.

under investigation, its influence can be minimized by providing an atmosphere of high water vapor activity, food with a high water content, and/or pure water for drinking.

In studies on the role of water as a nutrient or as it is more conventionally stated, studies on water balance, the use of fasting animals reduces the complexity of exchanges by eliminating water intake by ingestion and imbibition and water loss by excretion or defecation, (Wharton and Kanungo, 1962). In fasting animals the major exchanges are sorption from the air and transpiration to it. By appropriate use of tritium labelled water, it is possible to measure sorption and transpiration independently of each other. The difference between these two determines the net flow of water, (Wharton and Devine, 1968; Devine, 1969; Arlian, 1972; Devine and Wharton, in press).

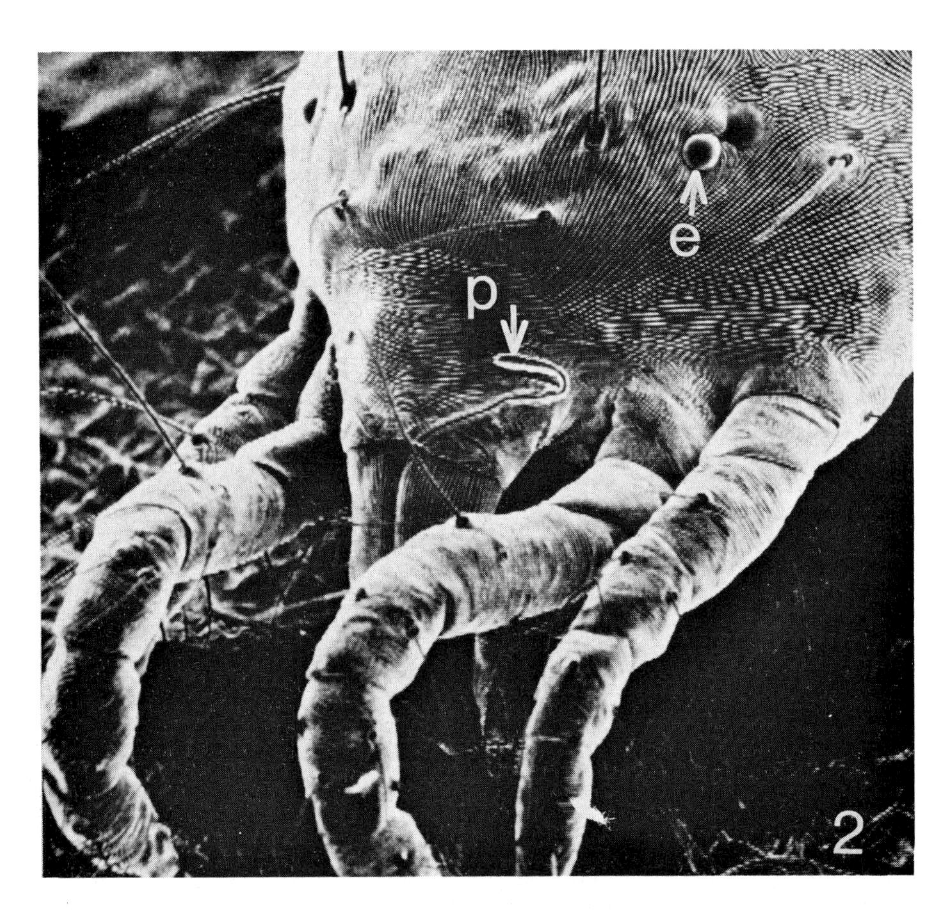

Figure 2. Anterior portion of a two-spotted spider mite feeding on a leaf in the evacuated column of a stereo scan electron microscope. Note that it is maintaining its water balance using the water in the leaf. 600x. e - eye; p - peritreme.

In animals as small as insects and especially mites, the high surface to volume ratio enhances the exchange of water with the surrounding air. The surfaces of these organisms must be extremely impermeable to water but they must, at the same time, be permeable enough to O_2 and CO_2 to provide for respiratory exchange. Since biological membranes are usually more permeable to water than they are to respiratory gases, (Waggoner, 1967), it is obvious that active forms must have surfaces that are permeable to water and, of course, they do, (Table 1). Where the exchange takes place is not known, but it is most improbable that a surface as diverse as that which an insect or mite presents to the ambient air would have a uniform permeability. For this reason it is not very realistic to attempt to express in metric terms the observed flows and permeabilities. Biological units are more practical, and thus the whole organism is the unit; and water masses, transpiration rates, sorption rates, and half-lives are presented on a per mite or per insect basis rather than per unit of surface area.

The mass of water and its a_w in an experimental subject must be known if the exchange of this water with the ambient air is to be understood. Comparison of wet weights and dry weights will give a value for exchangeable water mass. It is also possible to bring a group of organisms into equilibrium with tritium labelled water vapor of a given specific activity. When this is done, the radioactivity and weight of a number of specimens can be determined. If the radioactivity in counts per minute, CPM, per mite is plotted against weight, a straight line is formed that can be extrapolated to CPM = 0. When CPM = 0, the weight is the dry weight, (Wharton and Devine, 1968). The a_w of the water in larger insects can be determined by using conventional osometers. In the case of small forms, it is more difficult. The organism itself can be used as an osometer, (Wharton and Devine, 1968). An indirect method that will give reasonable values is determination of the optimum a_w of culture media for cells of the organism in question. It was by this method that the a_w of the haemolymph of the North American house dust mite was found to be 0.99 in the Acarology Laboratory, (Davidson, 1972).

The CEA can be determined by a number of techniques. A simple one involves exposure of standardized experimental forms to dry air and then to a graded series of water vapor activities. The weight changes experienced in the various activities are determined and change of weight is then plotted against a_v. The a_v when change in weight equals zero is the CEA, (Wharton and Kanungo, 1962). To find the water fluxes at an $a_v > \text{CEA}$, the experimental animal is allowed to pick up tritium labelled water from an atmosphere that contains tritium labelled water vapor. The animal containing the labelled water is then exposed to water vapor in a chamber that contains a relatively large amount of non-tritiated water to serve as a sink for the relatively infinitesimal amount of labelled water in the animal. Under these experimental conditions the amount of tritium will disappear according to the formula:

$$\frac{T_t^*}{T_o^*} = e^{-k_{T^*}t} \tag{2}$$

where T_o^* = CPM at time 0, T_t^* = CPM at time t, e is the base of natural logarithms, k_{T^*} is t^{-1}, and t is time. It is not necessary to know the actual tritium content because the tritium is used only as a tracer for water. The rate constant k_{T^*} for tritium loss will also be the rate constant for water loss or transpiration, (Wharton and Devine, 1968). Therefore, the amount of water transpired at any given time is simply k_{T*} times the water mass at a given time.

Sorption, on the other hand, can be measured by following the decrease in concentration of tritium in time. A measure of the concentration of the tritium is its specific activity, CPM/m, where m is the amount of water in the animal in question. In following the decrease in tritium concentration of an initially tritium labelled animal, it will be found to have the following relationship with time:

$$\frac{S_t}{S_o} = e^{-k_S t} \tag{3}$$

where S_o = CPM/m at time 0, S_t = CPM/m at time t, and k_S = rate constant for sorption, (Wharton and Devine, 1968). Since above the CEA the change in weight is zero, transpiration, T^*, and sorption, S, must be equal. Also, since the change in weight is zero, the amount of water lost or gained will be constant. Under equilibrium conditions then, both transpiration and sorption will proceed at a constant rate and thus both will be zero order functions, even though both tritium loss and decrease in concentration of tritium will be first order functions in time.

Under non-equilibrium conditions, that is at an a_v below the CEA when sorption and transpiration rates are not equal, the water mass of the organism will change in time according to a first order function. The amount of water transpired at any given instant will be the product k_{T*} times the water mass at that instant. Under both equilibrium and non-equilibrium conditions the number of water molecules that move from the water pool of the mite or insect to the ambient air will be a function of the temperature, activity, and amount of water present as well as the characteristics of the exchange surface. The fact that water loss is a first order function means that each molecule in the water pool has the same probability of escape as every other water molecule in the pool. A single water pool is observed only when the exchange between air and the organism is very slow compared to the movement of water from one part of the pool to another. For example, at time t a water molecule in a gut cell will have the same probability of escaping from the surface as a water molecule in the cuticle at time t. While changes in the a_v of the ambient air were found to influence the permeability of mites to water, no evidence that transpiration is interfered with by sorption was encountered.

The sorption of water from the vapor phase is dependent upon the absolute temperature, T; a_v; and pressure, P, of the ambient air. Just as sorption does not interfere with transpiration; transpiration does not interfere with sorption. As long as temperature, water vapor activity, and pressure of the ambient air are constant, the amount of water sorbed in a given time will be constant. Since the amount of water sorbed in most experiments is insignificant compared to the amount of water in the atmosphere, the rate of sorption does not change. Thus sorption is a zero order process and the sorption rate, $\dot{m}_S$, is a constant under both equilibrium and non-equilibrium conditions as long as the properties of the ambient atmosphere are unchanged. When changes in the a_v, T, or P occur, the sorption rate, $\dot{m}_S$, will increase or decrease appropriately, e.g. it will be approximately directly proportional to the a_v, that is:

$$\dot{m}_{Sa_v} = \dot{m}_{Sa_v=1} \cdot a_v \quad (4)$$

where $\dot{m}_{Sa_v}$ is the sorption rate at vapor activity a_v, $\dot{m}_{Sa_v=1}$ is the sorption rate at a_v=1, and a_v is the vapor activity in question.

Except in the cases of zero water content when transpiration is zero and a completely dry atmosphere when sorption equals zero, water will be absorbed. Thus during each half-life 138% of the water mass of the organism reached:

$$\dot{m}_S = m_{\infty} k_{T*} \quad (5)$$

where $\dot{m}_{\infty}$ is the water mass at equilibrium weight or at time infinity and k_{T*} is the rate constant for transpiration, (Devine and Wharton, in press). When equation 5 is satisfied an equilibrium weight is reached and no subsequent change in weight occurs. This is the condition at or above the CEA. Below the CEA, k_{T*} is constant as long as the physiological integrity of the organism is not disrupted by reduced water content. When water loss is excessive, the mite dies and k_{T*} increases dramatically and equation 5 is no longer valid. Nevertheless, it is possible to calculate a virtual rather than a real equilibrium weight or $\dot{m}_{\infty}$ for animals exposed to atmospheres below the CEA. Animals exposed to atmospheres below the CEA will have a constant sorption rate, $\dot{m}_S$, and rate constant for transpiration, k_{T*}, when they are first exposed to desiccating conditions. As long as they maintain their physiological integrity, water content at any time t can be calculated from the equation:

$$\frac{m_t - m_{\infty}}{m_o - m_{\infty}} = e^{-k_{T*}t} \quad (6)$$

where m_t is water mass at time t, m_0 is water mass at time 0, m_{∞} is water mass at time infinity, k_{T*} is the rate constant for transpiration, and t is time, (Devine and Wharton, in press). The water contents at times 0 and t can be determined by weighing, k_{T*} can be determined by using tritiated water as described above, and then m_{∞} can be calculated from equation 6. This is one way in which a virtual as opposed to a real m_{∞} can be obtained. Another method that will work does not require the use of radioactive tracers so long as all of the exchangeable water behaves as a single pool insofar as transpiration is concerned. Using this method m_t and m_0 are determined as before. These values are inserted into equation 6 along with a series of trial values for m_{∞}. Plots of ln $\frac{m_t - m_{\infty}}{m_o - m_{\infty}}$ versus time will give a family of curves. The curves when m_∞ is too large will be convex; when m_∞ is too small they will be concave. The value of m_∞ that results in a straight line when such plots are made is the correct value for the virtual m_∞. This value of m_∞ can be inserted in equation 6 and k_T* calculated. If k_T* calculated in this manner is the same as k_T* determined from using tritiated water as a tracer, then it can be concluded that all of the exchangeable water is in a single pool. If this is not the case, then more than a single pool is involved, (Arlian, 1972).

Energy is required to move water from a low activity to a higher activity. When a mite moves water from an activity of 0.75 to an activity of 0.99 it uses energy as house dust mites do when they are cultured in an atmosphere in equilibrium with a saturated solution of NaCl. The minimum energy required is given by:

$$\triangle G = - nRT \ln a_v/a_w \tag{7}$$

where $\triangle G$ is the change in Gibbs free energy, n is the number of moles of water moved, and R is the gas content. Solving equation 7 for the house dust mite NaCl cultures gives an energy requirement for the maintenance of water balance of 2×10^{-9} calories/mite/hour, (Arlian, 1972). The energy requirements are more significantly influenced by the total amount of water exchanged, n, than they are by the difference in osmotic pressure. The osmotic pressure difference between a_v and a_w in the case of the house dust mite is 400 atmospheres. In the spiny rat mite where the osmotic pressure difference is only about 100 atmospheres, the amount of energy required to maintain water balance is 2×10^{-4} calories/mite/hour, (Kanungo, 1965). The reason that more energy is required by the spiny rat mite is that much more water is moved. In considering the energetics of water as a nutrient in reference to arthropods living in a desiccating terrestrial environment, the amount of water exchanged is critical. Because this is true, studies of the net water flow are not sufficient for the evaluation of the role of water in nutrition. Physicists, chemists, and physiologists usually ignore total fluxes and regularly measure only the net diffusion. They use Fick's diffusion coefficient "D" and measure net flux "J" in terms of space and time. A net flow of 1 ml/hr in a system can result from different amounts of sorption and transpiration. For example, in one instance sorption may amount to 2 ml and transpiration 1 ml or in a second instance sorption may be 100 ml and transpiration 99 ml. Other things being equal, 50 times as much energy must be expended on the second case to maintain the same net flow as was needed in the first instance.

SUMMARY

Water is unique in that it is the only major nutrient that is significantly volatile and thus subject to exchange between the air and the organism. As the matrix for the metabolic machinery water is the most concentrated nutrient; on a mole fraction basis its concentration is 0.99. It serves many functions among which are: solvation, configuration of macromolecules, bioelectrical currents, and circulation.

In some instances the drive to regain lost water may influence the ingestion or imbibition of other nutrients and thus determine feeding, growth, and/or reproductive rates. Exchange of water between the air and mites or insects can be most conveniently studied in fasting animals. If the water mass is in a single pool insofar as transpiration is concerned, the following mathematical model will serve as a basis for understanding the exchanges:

$$\frac{m_t - m_\infty}{m_0 - m_\infty} = e^{-k_{T^*}t}$$

where m_0 is the water mass at time 0, m_t is the water mass at time t, m_∞ is the water mass at equilibrium or time infinity, k_{T^*} is the rate constant for transpiration, and t is time. The energy required to maintain water balance in the desiccating condition of terrestrial existence is primarly determined by the amount of water transpired.

Acknowledgements.—This work was supported by a Training Grant AI-000216 from the National Institute of Allergy and Infectious Disease and a contract, NIH-69-65, with the Division of Biologics Studies, both of NIH.

REFERENCES

AESCHLIMANN, A. (1963). Observations sur la morphologie la biologie et le developpement d'*Amblyomma compressum* (Macalister, 1872) la tique des pangoline d'Afrique occidentale. *Acta Trop.* 20:154-177.

ARLIAN, L. G. (1972). Equilibrium and Non-equilibrium Water Exchange Kinetics in an Atracheate Terrestrial Arthropod, *Dermatophagoides farinae* Hughes. Ph.D. Dissertation, 93+ix, The Ohio State University, Columbus, Ohio.

BALASHOV, YU. S. (1960). Water balance and the behavior of *Hyalomma asiaticum* in desert areas. (In Russian) *Med. Parazitol. Moscow* 29:313-320. (In English: NAMRU-3, T245).

BEAMENT, J. W. L., NOBLE-NESBITT, J. and WATSON, J. A. L. (1964). The waterproofing mechamism of arthropods. III. Cuticular permeability in the firebrat, *Thermobia domestica* (Packard). *J. Exp. Biol.* 41:323-330.

BOUDREAUX, H. B. (1958). The effect of relative humidity on egg-laying, hatching, and survival in various spider mites. *J. Ins. Physiol.* 2:65-72.

BUXTON, P. A. (1930). Evaporation from the mealworm *(Tenebrio*; Coleoptera) and atmospheric humidity. *Proc. R. Soc. (B)* 106:560-577.

CHURCH, S. N. (1959). Heat loss and the body temperature of flying insects. I. Heat loss by evaporation of water from the body. *J. Exp. Biol.* 37:171-185.

CROSS, H. F. and WHARTON, G. W. (1964). A comparison of the number of tropical rat mites and tropical fowl mites that fed at different temperatures. *J. Econ. Ent.*

57:439-443.

CROWE, J. H. and COOPER, A. F. JR. (1971). Cryptobiosis. *Sci. Amer.* **225**:30-36.

CUTCHER, J. J. (1970). The Kinetics of Exchange of Tritiated Water Between the Atmosphere and the Water Pool and the Metabolic Pool of the Mite *Tyrophagus putrescentiae*. Ph.D. Dissertation, 61+vii, The Ohio State University, Columbus, Ohio.

DAVIDSON, E. (1972). Personal communication.

DEVINE, T. (1969). A Systemic Analysis of the Exchange of Water between a Mite *Laelaps echidnina* and the Surrounding Vapor. Ph.D. Thesis, 79+viii, The Ohio State University, Columbus, Ohio.

DEVINE, T. and WHARTON, G. W. In press. The kinetics of water exchange between a mite *Laelaps echidnina* and the surrounding air. *J. Ins. Physiol.*

EDNEY, E. B. (1947). Laboratory studies on the bionomics of the rat fleas, *Zenopsylla brasiliensis* Baker and *X. cheopis* Roths. II. The water relations during the cocoon period. *Bull. Ent. Res.* **38**:263-280.

EDNEY, E. B. (1957). The Water Relations of Terrestrial Arthropods. Cambridge University Press, 109+vi, Cambridge.

EDNEY, E. B. (1966). Absorption of water vapour from unsaturated air by *Arenivaga* sp. (Polyphagidae, Dictyoptera). *Comp. Biochem. Physiol.* **19**:387-408.

EDNEY, E. B. (1968). The effects of water loss on the haemolymph of *Arenivaga* sp. and *Periplaneta americana*. *Comp. Biochem. Physiol.* **25**:149-158.

EDNEY, E. B. (1971). Some aspects of water balance in tenebrionid beetles and a Thysanuran from the Namib Desert of Southern Africa. *Physiol. Zool.* **44**:61-76.

GOVAERTS, J. and LECLERCQ, J. (1946). Water exchange between insects and air moisture. *Nature* **157**:483. London.

HAFEZ, M. S. EL-ZIADY, and HEFNAWY, T. (1970). Biochemical and physiological studies of certain ticks (Ixodoidea). Uptake of water vapor by the different developmental stages of *Hyalomma* (H.) *dromedarii* Koch (Ixodidae) and *Ornithodoros* (O.) *savignyi* (Audouin) (Argasidae). *J. Parasit.* **56**:354-361.

KANUNGO, K. Oxygen uptake in relation to water balance of a mite (*Echinolaelaps echidninus)* in unsaturated air. *J. Ins. Physiol.* **11**:557-568.

KNÜLLE, W. (1965a). Die sorption und transpiration des Wasserdampfes bei der Mehlmilbe (*Acarus siro* L.). *Z. Vergl. Physiol.* **49**:586-604.

KNÜLLE, W. (1965b). Equilibrium humidities and survival of some tick larvae. *J. Med. Ent.* **2**:335-338.

KNÜLLE, W. (1967). Physiological properties and biological implications of the water vapour sorption mechanism in larvae of the Oriental rat flea, *Xenopsylla cheopis* (Roths). *J. Ins. Physiol.* **13**:333-357.

KNÜLLE, W. and SPADAFORA, R. P. (1969). Water vapor sorption and humidity relationship in *Liposcelis* (Insecta: Psocoptera). *J. Stored Prod. Res.* **5**:49-55.

KNÜLLE, W. and WHARTON, G. W. (1962). Equilibrium humidities in arthropods and their ecological significance. Acarologia, fasc. h. S. 1964 (C. R. Ier Congres Int. d'Acarologie, Fort Collins, Col., U.S.A., 1963).

LANCASTER, J. L. and MCMILLAN, H. L. (1955). The effects of relative humidity on the lone star tick. *J. Econ. Ent.* **48**:338-339.

LARSON, D. (1969). The Critical Equilibrium Activity of Adult Females of the House Dust Mite, *Dermatophagoides farinae* Hughes. Ph.D. Dissertation, 35+vi, The Ohio State University, Columbus, Ohio.

LEES, A. D. (1946). The water balance in *Ixodes ricinus* and certain other species of ticks. *Parasitology* **37**:1-20.

LUDWIG, D. (1937). The effect of different relative humidities on respiratory metabolism and survival of the grasshopper *Chortophaga viridifasciata* DeGeer. *Physiol. Zool.* **10**:342-351.

MARCUZZI, G. and SANTORO, V. (1959). Indagini sul ricambio idrico del *Tenebrio molitor* mediante acqua tritiata. *Ric. Sci.* **29**:2576-2581.

MCENROE, W. D. (1961). The control of water loss by the two-spotted spider mite *(Tetranychus telarius). Ann. Ent. Soc. Amer.* **54**:883-887.

MELLANBY, K. (1932). The effect of atmospheric humidity on the metabolism of the fasting mealworm (*Tenebrio molitor* L., Coleoptera). *Proc. R. Soc. B* **III**:376-390.

NOBLE-NESBITT, J. (1970). Water balance in the firebrat, *Thermobia domestica* (Packard). The site of uptake of water from the atmosphere. *J. Exp. Biol.* **52**:193-200.

RAPOPORT, E. H. and TSCHAPEK, M. (1967). Soil water and soil fauna. *Rev. Ecol. Biol. Sol.* **4**:1-58.

RODRIGUEZ, J. G. (1954). Radiophosphorus in metabolism studies in the two-spotted spider mite. *J. Econ. Ent.* **47**:514-517.

SAUER, J. R. and HAIR, J. A. (1971). Water balance in the lone star tick (Acarina: Ixodidae): the effects of relative humidity and temperature on weight changes and total water content. *J. Med. Ent.* **8**:479-485.

SCHMIDT-NIELSEN, K. and SCHMIDT-NIELSEN, B. (1953). The desert rat. *Sci. Amer.* **189**:73-80.

WAGGONER, P. E. (1967). Moisture loss through the boundary layer. IN Biometeorology, *III*. Proc. 4th Biomet. Congr.:41-52.

WALTON, G. A. (1964). The *Ornithodoros "moubata"* group of ticks in Africa. Control problems and implications. *J. Med. Ent.* **1**:53-64.

WHARTON, G. W. and CROSS, H. F. (1957). Studies on the feeding habits of three species of laelaptid mites. *J. Parasit.* **43**:45-50.

WHARTON, G. W. and DEVINE, T. L. (1968). Exchange of water between a mite, *Laelaps echidnina*, and the surrounding air under equilibrium conditions. *J. Ins. Physiol.* **14**:1303-1318.

WHARTON, G. W. and KANUNGO, K. (1962). Some effects of temperature and relative humidity on water balance in females of the spiny rat mite, *Echinolaelaps echidninus* (Acarina: Laelaptidae). *Ann. Ent. Soc. Amer.* **55**:483-492.

WINSTON, P. W. and NELSON, E. (1965). Regulation of transpiration in the clover mite *Bryobia praetiosa* Koch (Acarina: Tetranychidae). *J. Exp. Biol.* **43**:257-269.

NUTRITIONAL REQUIREMENTS

INTRODUCTION
by

R. H. Dadd, Section Editor
Division of Parasitology
University of California
Berkeley, California 94720

Given our present substantial knowledge of insect nutrition, perhaps also some information on the chemical constitution of a particular insect and its natural food, and an awareness that feeding is a complex behavioral sequence involving more than presenting nutritionally appropriate chemicals, how do we set about confecting the first artificial diet for this insect? How do we refine an already workable formulation that supports only suboptimal growth or is insufficiently defined for our eventual purposes? Approaching similar problems from another direction, if our target is large-scale insect rearing for which a satisfactory synthetic diet has been worked out, how may we best develop a cheap, crude diet, or modify an existing mass-culture system when economic vagaries render its ingredients and preparatory adjuncts too expensive?

If our insect is naturally indiscriminate in its dietetics we may perhaps assume that it will chew on anything. We can ignore its possible feeding idiosyncracies and plunge straight in with an initial concoction of a diet culled from the literature of other, apparently similar insects that have already been successfully perverted to unnatural food. If it takes our mixture and grows on it to any extent, our general approach then will doubtless follow the rubric laid down in the discussion of methodologies by G. R. F. Davis: a patient, stepwise testing of modifications, some based on the purest speculation, some more rationally suggested by available chemical analysis of food and insect, but in either case, ultimately subject to the realities of determining empirically what actually works. Davis bases his discussion on two contrasting cases: in one, the empirical approach worked well to result in a good, totally synthetic diet; but in the other, after three decades of nutritional research, the insect still baulks at a totally defined diet. G. C. Rock's discussion of progressive protein and amino acid refinement for some Lepidoptera comprehensively considered an analytical approach that worked according to theory.

If our immediate need is to improve a diet, beneficial modifications will be thankfully embraced, whether or not reasons for their efficacy are clear.

Usually we wish also to comprehend why they worked, and are tempted to offer nutritional explanations of our results. Therein lie possibilities for interpretational errors and ambiguities. This is especially so at two stages of an investigation. Initially, when some growth has been achieved but optimal growth not nearly approached, and when little is yet known of specific nutrient requirements. It is then difficult to distinguish gross dietary shortcomings of a mechanical/physical/behavioral sort that cause underfeeding or gross starvation from truly chemical deficiencies that prevent growth for specific nutritional reasons. My discussion of artificial media for mosquito larvae considers such a situation. These filter-feeders whose food medium is their environment, are sensitive to many dietary factors that are not strictly nutritional. When major nutritional needs have been provided and performances approach optimal, small physical or gustatory effects may become apparent. It is then difficult to interpret exactly why small modifications are beneficial, especially as the variations in test or diet performances tend to become progressively subtler. Such ultimate dietary refinements tend to be expressed by changes in performance late in the insect's development, when diets may have aged in storage or experimental use. At this time subtle physio-chemical interactions between ingredients of the prepared diet may emerge as critical constraints on performance, as adumbrated by T. E. Mittler's dissertation of the pitfalls in interpreting manipulations of dietary ascorbic acid and trace metals.

During this late investigational stage we may also find simple nutritional explanation confounded by the emergence of gustatory effects, even in insects which initially seemed non-fastidious, and indeed, whose apparently indiscriminate feeding was perhaps largely instrumental in allowing us to develop diets to the luxurious point of worrying about optimal, as distinct from just good performances. Insect nutritionists are now sensitively aware how gustatory factors may affect the outcome of nutritional experiments. Their awareness stems from the fact that normal feeding by extremely fastidious feeders, a prerequisite for nutritional study, absolutely necessitates the concommitant provision of suitable phagostimulants, often non-nutrient chemicals. Gustatory stimulants are most extensively characterized for phytophagous insects, discussed especially in relation to optimal diet ingestion by T. H. Hsaio. The more recent appreciation that proper engorgement depends upon specific phagostimulant chemicals in the highly specialized bloodsucking insects is exemplified by a discussion of feeding regulation of *Rhodnius* on artificial diets by W. G. Friend and J. J. B. Smith.

Implicit in my foregoing comments is the assumption that our interest in synthetic diets centers on their use in clarifying insect nutrition *sui generis*. But, many may ask, how is this practically beneficial to the non-nutritionally inclined? R. E. Gingrich's account of the screwworm mass-rearing program makes it clear that rigorous nutritional study is a powerful resource for many endeavors of ultimate utilitarian intent. Mass-rearing of screwworms started as an exercise in common-sense, rule-of-thumb dietetics. In their wisdom, the USDA early commissioned a comprehensive, classical nutritional study of this insect. Information thereby gained has more than justified itself in terms of rationally guiding the changes in mass-culture techniques made mandatory by market fluctuations of even cheap, crude dietary components. I say "more than justified itself" advisedly. This fine example of interlocking pure and applied nutrition has, through the exigencies of the market economics involved, entailed a detailed consideration of the differential requirements of instars within larval development, thereby providing for "pure" nutrition a rare authentication of something we once considered speculatively, perhaps the besetting sin of the overly pure.

REFINING DIETS FOR OPTIMAL PERFORMANCE

G. R. F. Davis
Entomology Section, Research Station
Research Branch, Canada Agriculture
University Campus, Saskatoon, Sask., S7N OX2, Canada

Insects are among the most successful and the most adaptable of organisms. With comparative rapidity, many can modify their metabolism to develop on suboptimal diets or to use toxicants as metabolically useful compounds. Other factors being equal, insects develop normally with adequate nutrition, regardless of the source of dietary components.

Conditioned by the concept of high nutrition quality of milk for man and animals, we readily choose casein or lactalbumin as a protein source in artificial diets for animal and for insect nutritional studies. However, casein is known to be deficient in certain amino acids and has a lower biological value for mammals than some other proteins. Commercial preparations vary from processor to processor (Table 1) and from lot to lot with reference to

Table 1. Average gains in weight of larvae of *Tenebrio molitor,* reared on synthetic diets, containing 3% of vitamin-free casein from different suppliers

	Avg initial wt (mg)	Avg wt after 4 wks (mg)	Percentage wt gained
Hoffman-Laroche[a]	10.6	56.2	430.2
General Biochemicals[b]	10.9	46.4	325.7
Difco[b]	10.7	45.4	324.3
Dajac[b]	10.8	45.5	321.3
Nutritional Biochemicals[b]	10.4	43.7	320.2
Mann[a]	10.5	37.1	253.3

[a]Leclercq, J. (1965).
[b]Davis, G. R. F. & Leclercq, J. (1969).

biological value, suggesting that routine analyses for amino acid composition of caseins used in diets should be performed, especially in view of the ease of current analytical methods. These provisos not withstanding, casein has proved generally a suitable source of protein in artificial diets for insects.

To develop a satisfactory synthetic diet for insects, consideration must be given to the composition of the natural diet, more especially when the insect is a specialized, restricted feeder than when it is omnivorous. Proximate analysis of the natural food often suffices as a basis for formulating an artificial diet. With such information, diets may be prepared with proportions of protein, carbohydrate, lipid and mineral resembling the composition of the natural food. To such ingredients must be added those factors already proved essential for insects generally, such as a sterol and certain vitamins. Initially, such a diet will probably consist of casein; glucose, sucrose or starch; corn, wheat-germ or other oil; a commercially available mineral salt mixture; cholesterol; and a mixture of vitamins. This type of diet was used with insects in the initial investigations of their nutrition (Fraenkel & Blewett, 1942; 1942a; 1943b; 1943c; 1943d; 1946a; 1946b;). Its components permit it to be prepared as a powder; suspended in agar; cooked into a colloidal suspension; or pressed into pellets. Because insects are adaptable, and providing that other factors influencing growth and development have been respected, some growth will probably occur on such a diet. If this growth is suboptimal, development of a satisfactory synthetic diet becomes very much a trial and error procedure, in which further compositional information is applied.

With *Tenebrio molitor*, various protein and carbohydrate sources were tested (Huot & Leclercq, 1958a, 1958b; Leclercq, 1948a, 1948b; Leclercq & Lopez-Francos, 1964a), fats and sterol were eliminated from the diet (Leclercq, 1948c), and trace elements were investigated (Fraenkel, 1958; Leclercq, 1960). At the same time that an adequate diet was being derived, nutritional data on this insect was being gathered. Protein and carbohydrate sources were classified in order of effectiveness, lipid and sterol requirements were defined, and effects of trace elements were determined.

Few insect nutritionists are satisfied to leave the diet at this level of success. Recourse is generally made at this point to published analyses of the natural food or of the components of the artificial diet. Recently (Rock & King, 1967a, 1967b; Davis, 1971), analyses of the pupal stage of the insect under investigation have been made, with the rationale that all necessary metabolic components are present in this closed system, in concentrations necessary for normal development. In animal nutrition, a diet of high protein quality is defined as one supplying essential amino acids in the proportions in which they occur in the protein to be formed (Maynard & Loosli, 1962). Use of analyses of pupal tissue was particularly successful in formulating a diet for *Argyrotaenia velutinana* (Rock & King, 1967a), but not for *Tenebrio molitor* (Davis, 1971). For the latter, an amino acid mixture based on the composition of pupal protein was not much better than one based on an analysis of casein. Both were less effective than inclusion of intact protein in the diet.

Analyses are dependent on the accuracy of the analyst and of the method of analysis. Under optimum conditions, partial or complete destruction of components can take place. Amino acid analyses always leave some doubt concerning true concentrations of cystine, methionine and tryptophan in the protein. By analogy, this situation introduces doubt regarding the completeness of analyses in areas other than amino acids. Such incompleteness can result in the overlooking of important components, especially those occurring in minute quantities, such as essential trace elements and growth factors. On the other hand, with a hypothetically complete analysis of the natural food, no assurance exists that all components may be required for growth or maintenance of the insect. Preparation of the material for analysis may also release components from a bound state in which they are biologically unavailable to the insect when natural products are fed upon. Deterrents and stimulants are known to occur simultaneously in natural foods. Both would be indicated in the analysis, but the deterrents certainly would be unnecessary in the refined diet. Some components would be innocuous and therefore be neither beneficial nor detrimental inclusions of the insect diet and could be omitted without harm.

Such being the case, analyses of natural or of artificial diets cannot be followed dogmatically in refining diets for optimal performance. Although this information can be used as a general guide, it must be supplemented with knowledge of general nutritional requirements of insects. Occasionally, it is best ignored altogether.

Because an insect develops satisfactorily on an artificial diet, that diet is not necessarily capable of inducing optimal development. To determine optimal concentrations of components of synthetic diets, an investigation of the effects of graded doses of a single component must be entered into. Leclercq & Lopez-Francos in early work (1964a), with larvae of *Tenebrio molitor* determined a sensitive growth reaction when the casein level of the diet increased from 0% to 3% (Table 2). Growth was less influenced by concentrations of casein from 3% to 6.5% and was not affected at all as the dietary casein level increased from 6.5% to 20%. For this reason, the 3% protein level has been adopted as the working level for nutrition studies with *Tenebrio molitor.* Recently, work at Gembloux (Davis, 1970) indicated that optimal dietary concentrations of casein were linked to the particular casein being used. Better growth was obtained when the concentration of vitamin-free Difco casein in the diet was 20% than when it was 3%. The reason for this probably lies in the amino acid constitution of the casein compared with that used in the earlier work. Indeed, Leclercq and Lopez-Francos (1964b) had suggested that a 1% or a 2% protein concentration is desirable for larvae of *Tenebrio molitor* when amino acid mixtures are used in place of casein.

Table 2. Average gains in weight of 30 larvae of *Tenebrio molitor* reared on synthetic diets containing 0% to 20% casein (average initial weight = 10.5 mg)[a]

Percentage Casein	Avg wt after 4 wks (mg)	Percentage wt gained
0.0	14.0	33.3
0.5	36.5	247.6
1.0	47.3	350.5
1.5	51.3	388.5
3.0	53.3	407.6
4.5	55.0	423.8
6.5	57.5	447.6
10.0	54.8	421.9
20.0	55.7	430.5

[a]Leclercq, J. & Lopez-Francos, L. 1964a.

Even when other products serve as a protein source in the diet, it is wise to consider the optimal protein level. Work with larvae of *Tenebrio molitor*, in which oilseed meals provided the protein, indicated that 10% was superior to a 3% dietary protein level (Davis & Sosulski, 1972) (Table 3). The necessity of continually assessing the optimal concentrations of dietary components as the sources change therefore cannot be stressed for another. Neither are optimal levels of one protein source the same as those of another protein source for the same insect.

Table 3. Average initial weights and gains in weight of 30 larvae of *Tenebrio molitor,* reared on autoclaved and on unautoclaved oilseed meal diets[a]

Treatment of oilseed meal	Weight		
	Avg initial (mg)	Avg gain (mg)	Percentage gain
3% protein	10.1	26.2	259.4
10% protein	10.2	33.5	328.4

[a]Davis, G. R. F. & Sosulski, F. W. (1972).

Once a synthetic diet of this type has been developed, further modification may be desired. The process is much the same as outlined above. However, reference must be made to chemical analyses of the products included in the synthetic diet. These analyses are the basis for formulating a defined diet, but the information must be accepted or rejected on the basis of past research and experience. To define the protein portion of the diet, the amino acid composition of the component parts must be noted. Components other than the dietary protein source may also contain amino acids. Corn starch has recently been cited (Reiners *et al.*, 1970) as a dietary contributor of some amino acids, which, although present only in minute quantities, were sufficient for maintenance of swine on leucine-free diets (Baker & Allee, 1970). If the synthetic diet is sufficiently chemically pure to be certain that only the protein source contributes amino acids, an analysis of the protein is sufficient for formulating an amino acid mixture to replace the protein.

Immediate success is not always assured by this substitution, because of the many analytical possibilities of misinformation mentioned previously, but usually an acceptable diet does result, which is amenable to further improvement. Such has not been the case with larvae of *Tenebrio molitor* (Leclercq & Lopez-Francos, 1964b, 1966; Davis, 1971). Recently, however (Davis, unpublished), an amino acid mixture has supported as good growth for these larvae as the particular casein used as a control. Comparison of the amino acid diet and the casein diet with the natural control diet indicates, nevertheless, that both synthetic diets are inferior to the natural diet (Table 4).

Table 4. Average gains in weight of 30 larvae of *Tenebrio molitor*, reared on synthetic diets containing vitamin-free casein or an amino acid mixture as the protein source[a]

Protein	Weight		
Source	Avg initial (mg)	Avg gain (mg)	Percentage gain
Vitamin-free Casein	10.0	16.0	160.0
Amino acid Mixture	10.3	20.0	194.2
Natural (control) Diet	10.1	48.0	475.2

[a]Davis, G. R. F. Unpublished data.

In contrast to *T. molitor*, an amino acid mixture based on casein has been satisfactory for rearing larvae of *Oryzaephilus surinamensis* (Davis, 1956). This formulation has served as the basis for further modification of the amino acid content of the diet over a period of more than 15 years. As optimal concentrations of individual amino acids have been determined, the amino acid diet has been changed, so that important differences now exist between that diet and early diets (Table 5). Although casein is an excellent source of protein for these larvae, the amino acid pattern as given by analyses is not, but has served as the point of departure for the development of a satisfactory diet.

Table 5. Major variations of the amino acid pattern (mg/g) of the original and of the present amino acid mixture used in nutrition studies with *Oryzaephilus surinamensis* as compared with that of casein[a]

Amino Acid	Amino Acid Mixture		
	Casein	Early	Present
L-Alanine	51	28	0
L-Cystine	3	3	11
L-Glutamic acid	216	194	160
Glycine	5	18	283
L-Leucine	90	84	43
L-Phenylalanine	46	53	27
L-Proline	65	110	0
L-Serine	71	57	0

[a]Amino acids not included occur in essentially the same amounts in all three products.

[b]After Block, R. J. & Boling, D. (1947).

Too often the failure of a synthetic diet is accepted on the basis that it is synthetic. Statements are made implying that amino acid mixtures are necessarily inferior to natural proteins. In this manner, the importance of other components of the natural diet are discounted. Fraenkel (1958) and Leclercq (1960) improved the nutritional quality of the casein diet for *Tenebrio molitor* through the addition of carnitine. Leclercq (1960) further improved it by adding a minute quantity of $ZnCl_2$ to the vitamin solution. Both compounds can be considered as growth factors, although one is a vitamin and the other an essential trace element. Recently (Davis, 1972), the amino acid diet for *Oryzaephilus surinamensis* has been improved by the addition of a growth factor, which also is effective in this regard for mice and which has been designated as factor G (Table 6). As nutrition science progresses, many more factors may be discovered to dispel the bias against the nutritional efficiency of artificial diets.

Table 6. Effects of adding purified Factor G to diets containing amino acids on survival and on rate of development of *Oryzaephilus surinamensis*[a]

Factor G	Pupation		Emergence	
	Avg surv. (%)	Avg time (hr)	Avg surv. (%)	Age time (hr)
0	64	543	42	639
25	62	507	48	615
50	70	496	58	614

[a]Davis, G. R. F. (1972).

Regardless of the nutritive quality of the diet, it becomes less effective if its physical characteristics do not meet those required by the insect. Glucose could not be used in chemically-defined diets for *Oryzaephilus surinamensis* (Davis, unpublished), because the diet became too hygroscopic under optimal humidity conditions for the insect. Despite recorded occurrences of infestations of this insect in such products as raisins (Back & Cotton, 1940), larvae became caught in the sticky medium which resulted when glucose was included in the artificial diet. Bacteriological dextrin has served as an ideal carbohydrate source, forming a powdery diet acceptable to this insect. Mulkern & Toczek have recently (1970) solved the problem of physical dietary requirements for the grasshopper, *Melanoplus femurrubrum*, by producing pellets of diet on which this insect actively feeds. Personal observations (Davis, unpublished) have likewise indicated that *Melanoplus sanguinipes* will not feed satisfactorily on agar based media, nor on finely powdered diets. In contrast, *Schistocerca gregaria, Locusta migratoria* (Dadd, 1960) and *Melanoplus bivittatus* (Nayar, 1964) have been shown to feed well on powdered diets. Modifications of physical characteristics to improve diets for optimal performance are therefore restricted by the preferences of the insect being studied. Similar modifications have successfully produced diets having appropriate physical characteristics for *Chrysopa* larvae (Hagen & Tassan, 1965) and for the Angoumois grain moth (Chippendale, 1970). Apparently, the success of modifying the physical characteristics of a diet cannot be predetermined and is dependent entirely upon acceptance by the insect.

As with any modification, modification of the dietary composition can create various problems. Modification of dietary components in one area can elicit deficiencies in another, so that the effect of the modification becomes masked. Combined with this aspect, changes in the chemical constitution of a diet may also cause modification of its physical characteristics. The most evident and perhaps the most important problem is the creation of simple toxicity, whereby greater than optimal concentrations of a constituent overload a particular metabolic system. Such overloading can result in increased mortality or in slower rate of development, and becomes most obvious where complete removal of the component results in improvement of survival and rate of development (Table 7).

Table 7. Effect of various dietary concentrations of L-glutamic acid on survival and rate of development of 100 larvae of *Oryzaephilus surinamensis* reared on synthetic diets lacking aspartic acid, guanine, cytosine and putrescine[a]

Glutamic acid in diet (mg/g)	No. pupating	Average time (hr ± S.E.)	No. emerging	Average time (hr ± S.E.)
0.0 (basic)	89	526 ± 9	81 (91%)	646 ± 8
0.1	75	706 ± 27	66 (88%)	806 ± 30
0.2	68	836 ± 19	67 (98%)	960 ± 19
0.5	73	841 ± 21	68 (93%)	954 ± 19
1.0	66	853 ± 17	62 (94%)	951 ± 16
2.0	68	840 ± 16	65 (96%)	952 ± 13
5.0	73	952 ± 31	68 (93%)	1052 ± 32

[a]Davis, G. R. F. (1968).

Closely connected with simple toxicity is the creation of imbalance in the diet as a result of modifying the concentration of a single component. By such action, the effect of a limiting factor may become more pronounced. It may also upset other metabolic systems, interfering with the biosynthesis of products not required in the diet, but essential to optimal performance of the organism. When imbalance is minimal, the organism can develop reasonably well. For this reason, I believe, the efficiency of chemically-defined diets has been discounted both for large animals and for insects. Recent work with *Oryzaephilus surinamensis* (Davis, unpublished) has indicated that once imbalance is eliminated (other factors being equal), performance on a chemically-defined diet can compare favorably with that on a natural diet.

Further problems arise in connection with the concept of optimal diets, mainly because of poor definition and lack of agreement respecting terms used to denote the essentiality of desirability of nutrients. Dietary requirements for growth differ considerably from those for maintenance in other animals. That this situation should hold for insects should also be accepted by the insect nutritionist. Some consideration must be given to restricting the term "essential" to those dietary factors without which growth of immature insects ceases and mortality eventually ensues, either in larval or pre-reproductive adult stages, or in subsequent generations. In instances where the insect is incapable of synthesizing sufficient quantities of a factor so that it must be included in the diet, it should be referred to as: essential for normal growth; essential for pupation; essential for oviposition; or essential for some other physiological function.

In practice, the necessity of dietary factors is seldom cited in relation to function in the insect. This situation arises from ignorance of the interrelationships of nutrition and physiology. The science of insect nutrition has arrived at a stage where much more consideration must be given to the physiological function of required nutrients rather than just listing them.

Once optimal diets have been developed for insects, the rational step becomes a comparative study of quantitative requirements with those of other animals. If practical applications of insect nutrition research are to be forthcoming, these differences and similarities must become known. In this way, insects may be used as test animals to determine general nutritional quality of foods and foodstuffs, or to indicate concentrations of specific dietary factors. They may also be used to determine toxicity of compounds to themselves and to other organisms.

SUMMARY

Refining diets for optimal performance for insects is dependent upon chemical analyses of materials, on a certain empathy with the insect, and on much patience in routine variation of dietary factors. Failure of artificial diets most often occurs because of a lack of an essential factor, rather than of the nature of the diet, itself. Often a fine dietary imbalance is responsible for the apparent failure of synthetic diets. The significance of insect nutrition studies lies in the possibility of applying the knowledge so derived in the fields of general nutrition, general physiology, and toxicology.

Acknowledgements.—This work is contribution No. 481, Canada Agriculture Research Station, Saskatoon, Saskatchewan, Canada.

REFERENCES

BACK, E. A. and COTTON, R. T. (1940). *Stored grain pests.* U.S. Dept. Agr. Farm Bull. 1260.

BAKER, D. H. and ALLEE, G. L. (1970). Effect of dietary carbohydrate on assessment of the leucine need for maintenance of adult swine. *J. Nutr.* **100**:277-280.

BLOCK, R. J. and BOLLING, D. (1947). *The determination of the amino acids.* Burgess Publ. Co., Minneapolis.

CHIPPENDALE, G. M. (1970). Development of artificial diets for rearing the Angoumois grain moth. *J. Econ. Ent.* **63**:844-848.

DADD, R. H. (1960). The nutritional requirements of locusts. I. Development of synthetic diets and lipid requirements. *J. Insect Physiol.* **4**:*319-347.*

DAVIS, G. R. F. (1956). Amino acid requirements *Oryzaephilus surinamensis* (L.) (Coleoptera: Silvanidae) for pupation. *Can. J. Zool.* **34**:82-85.

____________ (1968). Glutamic acid requirements of the saw-toothed grain beetle, **Oryzaephilus surinamensis** (L.) (Coleoptera, Silvanidae). *Comp. Biochem. Physiol.* **24**:395-401.

____________ (1970). Protein nutrition of *Tenebrio molitor* L. XII. Effects of dietary casein concentration and of dietary cellulose on larvae of race F. *Archs. Int. Physiol. Biochem.* **78**:37-41.

____________ (1971). Protein nutrition of *Tenebrio molitor* L. XV. Amino acid mixtures as replacements for protein of the artificial diet. *Archs. Int. Physiol. Biochem.* **79**:11-17.

____________ (1972). A growth factor in Brewer's yeast for the saw-toothed grain beetle, *Oryzaephilus surinamensis* (L.). *Comp. Biochem. Physiol.* **(in press).**

DAVIS, G. R. F. & LECLERCQ, J. (1969). Protein nutrition of *Tenebrio molitor* L. IX. Replacement caseins for the reference diet and a comparison of the nutritional values of various lactalbumins and lactalbumin hydrolysates. *Archs. Int. Physiol. Biochem.* **77**:687-693.

DAVIS, G. R. F. and SOSULSKI, F. W. (1972). Use of larvae of *Tenebrio molitor* L. to determine nutritional value of proteins in six defatted oilseed meals. *Archs. Int. Physiol. Biochem.* **80**: (in press).

FRAENKEL, G. S. (1958). The effect of zino and potassium in the nutrition of *Tenebrio molitor*, with observations on the expression of a carnitine deficiency. *J. Nutr.* **65**:361-395.

FRAENKEL, G. and BLEWETT, M. (1942). Biotin, B_1, riboflavin, nitotinic acid, B_6, and pantothenic acid as growth factors for insects. *Nature, Lond.* **150**:177-178.

_______ & _______ (1943a). Vitamins of the B-group required by insects. *Nature, Lond.* **151**:703-704.

_______ & _______ (1943b). The sterol requirements of several insects. *Biochem. J.* **37**:692-695.

_______ & _______ (1943c). The natural foods and the food requirements of several species of stored products insects. *Trans. R. Ent. Soc. Lond.* **93**:457-490.

_______ & _______ (1943d). The basic food requirements of several insects. *J. Exp. Biol.* **20**:28-34.

FRAENKEL, G. and BLEWETT, M. (1946a). The dietetics of the caterpillars of three *Ephestia* species, *E. Kuhniella, E. elutella,* and *E. cautella*, and a closely related species, *Plodia interpunctella. J. Exp. Biol.* **22**:162-171.

_______ & _______ (1946b). Linoleic acid, vitamin E and other fat-soluble substances in the nutrition of certain insects, *E. kuhniella, E. elutella, E. cautella,* and *Plodia interpunctella. J. Exp. Biol.* **22**:*172-190.*

HAGEN, K. S. and TASSAN, R. L. (1965). A method of providing artificial diets to *Chrysopa* larvae. *J. Econ. Ent.* **58**:999-1000.

HUOT, L. and LECLERCQ, J. (1958a). Nutrition protidique chex *Tenebrio molitor* L. I. Influence de l'etat physiologique des larves sur leur comportement dans des milieux nutritifs carences. *Archs. Int. Physiol. Biochem.* **66**:270-275.

_______ & _______ (1958b). Nutrition protidique chex *Tenebrio molitor* L. II. Notion d'optimum protidique. *Archs. Int. Physiol. Biochem.* **66**:276-281.

LECLERCQ, J. (1948a). Importance de glucides dans la nutrition des larves de *Tenebrio molitor* L. *Archs. Int. Physiol.* **56**:28-34.

_______(1948b). Aspects qualitatifs des besoins en glucides chez *Tenebrio molitor. Archs. Int. Physiol.* **56**:130-133.

_______(1948c). Sur les besoins en sterols des larves de *Tenebrio molitor* L. *Biochem. Biophys. Acta.* **2**:614-617.

_______(1960). Carintine, zinc et potassium dans la nutrition des larves de *Tenebrio molitor* L. *Archs. Int. Physiol. Biochim.* **68**:500-503.

_______(1965). Premiers essais d'utilisation des larves de *Tenebrio molitor* pour comparer la valeur nutritive des proteines. *Annls. Nutr. Aliment.* **19**:47-58.

LECLERCQ, J. and LOPEZ-FRANCOS, L. (1964a). Nutrition protidique chez *Tenebrio molitor* L. V. L'optimum protidique chez les larves de race F. *Archs. Int. Physiol. Biochim.* **72**:95-99.

_______ & _______(1964b). Nutrition protidique chez *Tenebrio molitor* L. VI. Essais de remplacement de la caseine par des melanges artificiels d'acides amines. *Archs. Int. Physiol. Biochim.* **72**:276-296.

_______ & _______(1966). Nutrition protidique chez *Tenebrio molitor* L. VII. Nouveaux essais de remplacement de la caseine par des preparations d'acides amines. *Archs. Int. Physiol. Biochem.* **74**:397-415.

MAYNARD, L. A. and LOOSLI, J. K. (1962). *Animal nutrition,* 5th ed., McGraw-Hill, New York, Toronto, London. p. 102.

MULKERN, G. B. and TOCZEK, D. R. (1970). Bioassays of plant extracts for growth-promoting substances for *Melanoplus femurrubrum* (Orthoptera: Acrididae). *Ann. Ent. Soc. Am.* **63**:272-284.

NAYAR, J. K. (1964). The nutritional requirements of grasshoppers. I. Rearing of the grasshopper, *Melanoplus bivittatus* (Say), on a completely defined synthetic diet and some effects of different concentrations of B-vitamin mixture, linoleic acid and B-carotene. *Can. J. Zool.* **42**:11-22.

REINERS, R. A., MORGAN, R. E. and SHRODER, J. D. (1970). Note on the amino acid composition of the protein in commercial corn starch. *Cereal Chem.* **47**:205-206.

ROCK, G. C. and KING, K. W. (1967a). Quantitative amino acid requirements of the red-banded leaf roller, *Argyrotaenia velutinana* (Lepidoptera: Tortricidae). *J. Insect Physiol.* **13**:175-186.

_______ & _______(1967b). Estimation by carcass analysis of the growth requirements for amino acids in the codling moth, *Carpocapsa pomonella* (Lepidoptera: Oletheutidae). *Ann. Ent. Soc. Am.* **60**:1161-1162.

OPTIMAL PROPORTIONS OF DIETARY AMINO ACIDS

George C. Rock
Department of Entomology
North Carolina State University, Raleigh, 27607

INTRODUCTION

My contribution to this conference will be a discussion of amino acid and protein nutrition in insects in which I emphasize the importance of both qualitative and quantitative amino acid requirements in the development of diets of optimum nutrient value. In particular, studies conducted with the insect *Argyrotaenia velutinana* (Walker), a polyphagous microlepidopteran, will be elaborated since this insect can be reared axenically on chemically defined diets.

QUALITATIVE AMINO ACID REQUIREMENTS:

From a nutritional standpoint the primary value of determining the indispensable amino acids lies in the fact that the nutritive value of proteins depends mainly on the content of the indispensable amino acids, although the pattern of the dispensable amino acids is not altogether without significance. Results of the dietary deletion and radiometric methods to determine the indispensable amino acids for *A. velutinana* and *Heliothis zea* (Boddie), shown in Table 2 are, in general, in agreement with similar studies with other insects and higher animals.

Knowing which amino acids are dietarily indispensable it is then possible to determine the relative effectiveness of nitrogen-containing compounds to promote larval growth and development when added as supplements to the indispensable amino acids. The results of such an experiment are summarized in Table 2. Each dispensable amino acid was added in equal amounts (0.7%) to a diet containing the 10 indispensable amino acids (1.325%); urea, biuret, and diammonium citrate were used in amounts isonitrogenous with the nitrogen supplied by the 8 dispensable amino acids in the control diet. The results demonstrated that the source of nitrogen required for biosynthesis of nitrogen-containing substance is relatively non-specific for various amino acids or diammonium citrate could serve that purpose. Different nitrogeneous supplements, however, are not equally effective, and the relative effectiveness differed with sex. In experiment 1 (Table 2) gluta-

Insect and mite nutrition – North-Holland – Amsterdam, (1972)

mate and serine gave the best larval growth response and proline and cysteine

Table 1. Dietary amino acids requirements for *A. velutinana* and *H. zea* determined by deletion and radiometric techniques (Rock and King, 1968 and Rock and Hodgson 1971).

	A. velutinana		*H. zea*	
Amino acids	^{14}C studies[a]	Deletion studies	^{14}C studies[a]	Deletion studies
Lysine	+	+	+	+
Histidine	+	+	+	+
Arginine	+	+	+	+
Valine	+	+	+	+
Methionine	+	+	+	+
Isoleucine	+	+	+	+
Leucine	+	+	+	+
Phenylanine	+	+	N.D.[c]	+
Threonine	+	+	+	+
Tryptophan	-[b]	+	-[b]	+

[a]Glucose - U -^{14}C
[b]destroyed during acid hydrolysis
[c]N.D. = Not determine because of ^{14}C contaminant eluted along with phenylalanine.

the poorest. Male responses to glutamate, serine and aspartate on the one side, and proline and cysteine on the other, were found to be significantly different. However, females differed in that larval growth on glutamate and serine was significantly faster than that for aspartate, proline and cysteine. Experiment 2 (Table 2) showed that urea and biuret had no effect on larval growth or survival when compared to the unsupplemented diet. However, diammonium citrate significantly increased larval survival and growth rate. Larval growth rates were not dependent upon the total nitrogen content of the diets, as glycine furnished about twice the amount of nitrogen as glutamic acid, but increased growth rates less. The effectiveness of a single nitrogenous compound as a nitrogen source for synthesis of dispensable amino acids depends upon its ability to supply NH_2 groups. The key role of glutamic acid in amino acid metabolism of *A. velutinana* is indicated by the

Table 2. Effect on growth and survival of *A. velutinana* by supplementing the indispensable amino acids with various nitrogenous substances (Rock and King, 1967a).

Substance added[a]	mg N added/ 100g diet	Total larvae	Avg days to pupation		% survival
			Male	Female	
		Experiment 1			
Glutamate	70	56	23.8 [bc]	24.9 [ab]	84
Serine	91	42	23.5 [b]	24.7 [ab]	85
Aspartate	70	47	24.9 [bc]	27.6 [cd]	76
Glycine	133	43	26.1 [cd]	28.8 [cde]	79
Proline	85	45	28.8 [e]	30.7 [de]	80
Cysteine	81	40	28.4 [de]	30.6 [e]	74
Control[c]	95	91	20.8 [a]	23.3 [a]	89
		Experiment 2			
Diammonium citrate	95	63	27.8 [a]	26.6 [a]	89
Biuret	95	67	39.0 [b]	40.3 [b]	30
Urea	95	59	41.8 [b]	44.2 [c]	28
None	0	60	40.1 [b]	42.1 [bc]	36

[a]Substance added to diet containing 1.325% of the 10 indispensable amino acids.
[b]Values not followed by same letter are significantly different ($P < 0.01$).
[c]Basic 18 L-amino acid mixture at 2% dietary level.

fact that it is the dispensable amino acid present in highest concentration in amino acid hydrolysates of the insect carcass (Fig. 2) and helps explain its effectiveness to promote growth. The ineffectiveness of urea and biuret to stimulate larval growth is most likely due to the lack of enzymatic hydrolysis of these substances by the insect. All diets containing the 10 indispensable amino acids plus 1 nitrogenous supplement supported poorer growth rates than the control diet containing 18 amino acids, demonstrating that for optimal growth the number of nutrients required by an organism increases.

An experiment was conducted to determine if *A. velutinana* could develop and reproduce on a diet containing just the 10 indispensable amino acids at an adequate nitrogen level. From growth and survival responses of insects reared on different dietary levels of the 10 indispensable amino acids it was found that a dietary level of 4% was optimal as concentrations above 4% showed no positive effect. Table 3 shows the results of axenic rearing from egg to adult for 3 successive generations on a diet containing the 10 indispensable amino acids as the primary nitrogen source. To check for the

Table 3. Growth for 3 successive generations on a diet containing a mixture of indispensable amino acids (4% dietary level).[a] (Rock and King 1967a).

Generation	Total larvae	Avg days to pupation		% survival
		Male	Female	
1	107	25.1	27.6	86
2	80	28.9	28.6	70
Control[b]	37	27.5	29.8	68
3	64	28.9	30.2	78
Control[b]	24	25.9	27.0	90

[a]Amino acid mixture patterned after that found from carcass analysis.
[b]Larvae from parents reared on semi-synthetic diet.

possibility of an experimental error in preparation of the chemically defined diets, larvae from the stock laboratory strain (control) were reared on portions of the medium used for generations 2 and 3. The prolonged growth of male larvae of the 2nd generation when compared to growth for the control larvae was not significant ($P>0.05$); however, 3rd generation male and female larval growth was significantly slower than growth for the control larvae ($P<0.05$). The prolonged larval growth for generations 2 and 3 when compared with first-generation growth was probably due to a lack of some nutrient reserve present in the first-generation larvae whose parents were reared on a better balanced (semi-synthetic diet containing plant material) diet.

The results of the above experiments show that the minimum qualitative amino acid requirements of *A. velutinana*, perhaps of insects in general, are the indispensable amino acids. Acceleration of larval growth rates and increased survival depends upon the number and amount of additional dispensable amino acids.

QUANTITATIVE AMINO ACID REQUIREMENTS

The importance of amino acid balance in insect nutrition has been well documented. A multitude of studies on amino acid requirements and deficiencies have made it clear that, in order to support optimal growth, a diet must provide each indispensable and dispensable amino acid in the quantity appropriate for the animal. Although amino acid requirements of insects vary with age and other conditions such as the nutritive state of the parent and quantitative relationships between other nutrients, it is possible for a given set of conditions to determine ratios representing the relative requirements of the individual amino acids. The approach I used to determine proper amino acid ratios was analysis of the insect carcass for amino acid composition.

Pattern of dietary amino acids determined by carcass analysis.

A similarity between the amino acid composition of animals and the pattern of amino acid requirements as determined by nutritional studies has been noted in protists (Wu and Hogg, 1952) and in large animals (Williams *et al.* 1954 and King, 1967). Basing the pattern of amino acids in the diet of an animal on that found for its carcass has the main advantage of short-cutting a purely approach to determining quantitative amino acid requirements. Knowing the amino acid composition of the whole egg, larva, pupa and adult should give a reliable estimate of the changes in the amino acid pattern occurring during growth and development. Knowledge of such changes and at which stages in growth they occur is useful in formulating diets containing optimal amino acid balance. The quantitative amino acid requirements of *A. velutinana* were approximated from the information obtained from carcass analysis using an amino acid analyzer. Fig. 1 shows the successive changes in fresh wt. and protein content during development from egg to adult and the results demonstrate that during larval development there was a similar increase in both fresh wt and total protein and that larval growth was exponential. This correlated increase in the mass and protein content of the body with time is characteristic of normal growth. The quantities of 17 amino acids were determined at all stages of development examined, and of those identified all except methionine and cysteine may occur in rather high

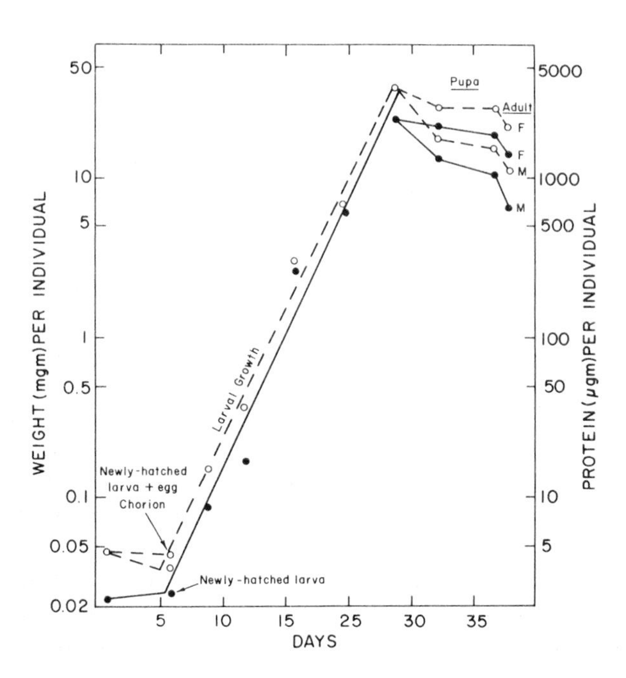

Figure 1. Successive changes in fresh wt (____) and protein (----0) during development of *A. velutinana* from egg to adult at 27°C. Ordinate: right, protein in μg per individual; left, fresh weight in mg per individual. Abscissa: age in days, beginning at egg deposit. Data obtained from amino acid analyzer (Rock and King, 1966).

concentrations (Fig. 2). The amino acid pattern during growth and development was fairly constant, although it is clear that there were variations in percentages of individual amino acids within and between various stages. For example, the quantitative results showed a percentage increase in proline and alanine and a decrease in tyrosine and serine from egg to newly-hatched larvae. Analysis of early (newly hatched) and late (3-day-old) first-instar larvae showed the amino acid pattern to be rather uniform except that lysine, proline and arginine exhibited a slight percentage increase in the older larvae. Assuming that the total amino acid requirements of a rapidly growing animal are determined largely by the proportions of the amino acids contained in the tissues being formed, this change in amino acid pattern within a larval instar would indicate a shifting of amino acid requirements in accordance with tissues being formed. The most pronounced change in the amino acid pattern during larval growth is that the later instars are richer in lysine, arginine, and especially tyrosine and phenylalanine and poorer in glutamate,

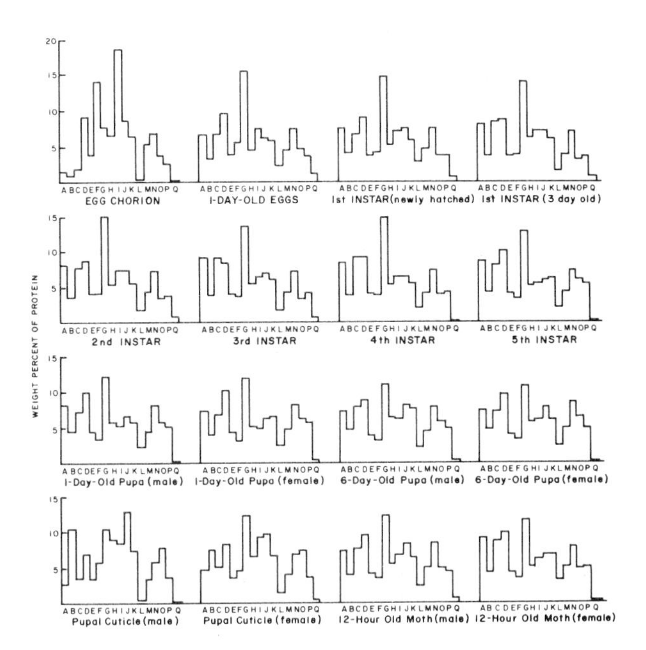

Figure 2. Patterns of amino acids in *A. velutinana* at various stages of development determined by amino acid analyzer (Rock and King, 1966). Ordinate: weight percent of each amino acid. Abscissa: the figures represent the respective amino acids as follows: a, lysine; b, histidine; c, arginine; d, aspartic acid; e, threonine; f, serine; g, glutamic acid; h, proline; i, glycine; j, alanine; k, valine; l, methionine; m, isoleucine; n, leucine; o, tyrosine; p, phenylalanine; q, cysteine.

glycine and alanine. The changes would indicate a shifting in amino acid requirements between different instars. Accordingly, the quantitative amino acid requirements for optimum growth of the later larval stages would be different from those for the earlier stages. The amino acid pattern from the late stage of pupation through adult emergence is fairly constant except for the sharp decrease in tyrosine. However, the increased tyrosine in the cast pupal cuticle accounted for the decrease in tyrosine content from the pupal to adult stage. A comparison of the data concerning the amino acid pattern in male and female pupae showed no large differences with regard to quantitative amino acid requirements. In summary, analysis of the whole larva, pupa, and the adult revealed an amino acid pattern characterized by a high percentage of glutamate, aspartate, and lysine, and a low percentage of serine, histidine and sulfur amino acids. However, entirely different amino acid patterns were revealed when different parts of the insect were analyzed separately, as was the case with hydrolyzed whole egg chorion and cast pupal cuticle. This characteristic amino acid pattern with respect to a particular tissue would be expected to determine the quantitative amino acid requirements of an organism in accordance with the tissues being formed.

Pattern of dietary amino acids determined by growth studies

The primary purpose of this study was to determine if the amino acid composition of whole insects, in this instance, *A. velutinana*, may be used in determining the best amino acid proportionality pattern to support maximum growth and development. For this, four basic amino acid mixtures were used and are listed in Table 4. The amino acid mixture patterned after carcass analysis was that found in 1-day-old pupae of *A. velutinana*. The modified carcass mixture was formulated following experiments to improve the amino acid pattern found by carcass analysis. Sodium bicarbonate was added to the amino acid diets in sufficient amounts to "buffer" the hydrochloride present in lysine-HCL.

Table 5 summarizes the results of growth and survival experiments on two dietary levels of the amino acid mixture patterned after the carcass and on two amino acid mixtures having totally different amino acid patterns. Statistical analysis revealed no significant difference between larval growth rates for male and female on mixtures patterned after carcass analysis and cotton seed protein at the 2 or 3% level ($P>0.05$). However, growth rates on the amino acid mixture resembling casein were significantly slower at the 2 and 3% levels when compared with the carcass and cotton seed mixtures at similar levels ($P<0.01$). Adult survival was lowest on the mixture resembling casein. On all three amino acid mixtures male and female larval growth rates were significantly faster ($P<0.1$) when the dietary amino acid level was increased from suboptimal (2%) to near optimal (3%). These results elucidate the importance of the amino acid composition of diets, both total and relative, on growth and development. Using the amino acid pattern after carcass as a reference standard for the pattern of amino acids utilized most efficiently by *A. velutinana*, it is evident that appreciable deviation in an amino acid pattern is tolerated before deleterious effects on growth occur. For example, although the mixture after cotton seed protein had slight deficiencies of lysine, histidine, aspartic acid, and glutamic acid along with excesses of arginine, phenylalanine and glycine, growth was not significantly slower than that for the carcass pattern. However, the retarded growth rate on a mixture resembling casein, which is very deficient in arginine, histidine, tryptophan, glycine, and cystine, with an excess of isoleucine, leucine, serine and proline, shows this to be a poorly balanced mixture. Growth rates at the 3% dietary level on the unbalanced mixture resembling casein are comparable to that attained at the 2% level on the balanced carcass mixture. This is in line with the observation from large animal nutrition that an organism can tolerate an unbalanced mixture if the amino acid requirements are exceeded.

Table 4. Composition of the basic amino acid mixtures for *A. velutinana*.

Amino acid	Amino acid pattern resembling[a]			
	Cotton seed protein	Carcass analysis	Casein hydrolysate	Modified carcass
Arginine	311	160	94	125
Histidine	75	83	38	100
Isoleucine	113	93	122	92
Leucine	175	156	224	156
Lysine-HCl	113	164	140	190
Methionine	75	60	76	72
Phenylanine	150	107	112	107
Threonine	100	100	76	100
Tryptophan	75	90	38	90
Valine	138	122	132	95
Alanine	150	125	102	125
Aspartic acid	100	184	127	180
Glutamic acid	100	223	218	220
Glycine	150	97	30	97
Proline	50	99	198	99
Serine	50	57	148	57
Cystine	40	35	15	55
Cysteine	10	10	5	10
Tyrosine	50	102	130	102
indispensable	1.325	1.135	1.052	1.127
dispensable	0.700	0.937	0.973	0.945
Total	2.025 g	2.072 g	2.025 g	2.072 g
Mg nitrogen/ 100 g diet	326	298	263	293
Sodium bicarbonate	10	15	15	20

[a]Mg of amino acids per 100 g diet.

To evaluate if indeed the amino acid mixture after carcass analysis was correctly proportioned an experiment was conducted to determine if this mixture could be improved. For this, the level of each indispensable amino acid was either increased or decreased by 20% of the level in the basic mixture. The diets were maintained isonitrogenous by adding to or removing from each diet enough glycine to adjust the nitrogen level provided by the amino acids to 2.51% of total dry wt, approximately 1.7g amino acids/100g diet. This low amino acid nitrogen level was chosen because it is known for insect and large animal nutrition that the deleterous effects of unbalanced nutrients on growth are most pronounced when the nutrient concentration is low. Using the information accumulated from this experiment a new and presumably better balanced basic mixture was formulated, of the composition shown in Table 4 under "modified carcass." To determine if those modified mixture was better balanced than the basic carcass mixture, larval growth and survival were compared on the two mixtures. The results, summarized in Table 6, show that although larval growth on the modified carcass mixture was faster than that attained on the basic carcass mixture the difference was not significant. Survival to the adult stage was comparable for the two diets. These data support the thesis that the amino acid composition of whole insects may be used in achieving an amino acid pattern to support maximum growth. In this study the amino acid pattern found in 1-day-old pupae was the dietary pattern used for larval instars. Feeding the insect on formulations of new dietary amino acid patterns as changes occur during larval growth (see Fig. 2) would presumably produce faster growth rates.

Table 5. Growth on amino acid mixtures differing in total and relative levels of amino acids.

Mixture resembling	Percent amino acids	Total no. larvae	Avg days to pupation		Percent survival[a]
			Male	Female	
Cotton seed	2.0	95	22.4	24.9	87
	3.0	82	19.6	20.9	86
Carcass	2.0	106	22.4	23.8	94
	3.0	80	19.6	20.6	88
Casein	2.0	78	33.3	32.5	55
	3.0	72	22.0	25.2	81

[a]Percent of initial number of larvae to develop to adults.

Table 6. Effect on development of modifications of certain amino acids of the basic carcass amino acid mixture.

Amino acid diet	Total no. insects	Avg time to pupation: Male No.	Male Days	Female No.	Female Days	Percent survival[a]
Basic carcass	103	39	29.6[a]	36	30.0[a]	76
Modified carcass	93	40	28.1[a]	32	28.1[a]	77

[a]Percent of initial number larvae to develop to adults.
[b]Values not followed by same letters are significantly different ($P<0.05$).

NUTRITIVE VALUE OF DIETARY PROTEINS

The ability of dietary proteins to meet the nitrogen requirements of an organism varies, a variation that is an expression of a protein's nutritive value. It is known that the nutritive value of a protein is correlated with the proportions of amino acids in the protein and that this nutritive value may be improved by judicious amino acid supplements. Therefore, a series of experiments were conducted to investigate the nutritive value of proteins for *A. velutinana* based on the amino acid composition of the proteins. For this, the amino acid content of three proteins (casein, soybean and egg albumen) were compared with the amino acid requirements of *A. velutinana* (carcass analysis, Table 4). Estimates of the amino acid composition of the proteins were taken from reference texts. The proteins were supplemented with the predicted most limiting amino acids and evaluated biologically by growth performance. The results of these experiments are summarized in Table 7. The relative order of protein quality (unsupplemented) for larval growth was as follows: egg albumen>soybean protein>casein. All the amino acid supplements, individually or as mixtures, stimulated growth significantly when compared to the unsupplemented proteins, illustrating the practical significance of judicious amino acid supplementation to improve protein quality. The several experiments with casein showed that the individual amino acid supplements may be as effective as mixtures for growth performance. The order of amino acid deficiency from most to least for casein was predicted to be arginine, cystine, and glycine. Experiments 1 and 2 (Table 7) showed that supplemental arginine, cystine, or glycine gave growth responses nearly comparable to that for the three amino acids together. Casein is a rich source of the dispensable amino acids, glutamate and serine; however, if either of these amino acids were added individually when compared to unsupplement casein (table 7, experiment 3). These results indicate that protein quality for *A. velutinana* can be improved by supplementing limiting amino acids and/or increasing total nitrogen intake by supplying excessive dispensable amino acid nitrogen. These results are consistent with chick and human nutritional studies showing that the first limiting nitrogenous factor of a protein can partially or wholly be replaced by dispensable amino acids.

CONSUMPTION AND UTILIZATION OF CHEMICALLY DEFINED DIETS

Waldbauer has pointed out the deficiency of studies concerned with the intake, digestibility and efficiency of conversion of either ingested or digested food into body matter, particularly with diets of known chemical composition. Quantitative work with chemically defined diets usually involves only the measurement of the amount of a particular nutrient required per unit of diet. This defines the relationship between the requirements for particular nutrients, but specifies nothing of food intake (Waldbauer, 1968). An experiment was conducted to obtain information on food consumption and utilization by *A. velutinana* larvae, feeding on a chemically-defined diet. The insects were reared axenically from newly-hatched larvae to the larval-pupal ecdysis on diets similar to the "modified carcass" amino acid diet given in Table 4. Two diets were evaluated, one contained L-methionine (100mg/100g diet) and the second diet contained D-methionine (150mg/100g diet) in lieu of L-methionine. The faecal uric acid method and the gravimetric method (Bhattacharya and Waldbauer, 1969 a,b) were used to determine consumption and utilization of the two diets by the larvae. The approximate digestibility (A.D.), efficiency with which ingested food is converted to body matter (E.C.I.) and the efficiency with which digested food is converted to body matter (E.C.D.) were calculated as described by Bhattacharya and Waldbauer (1969b). The results of this experiment using the uric acid method, summarized in Table 8, show that the digestibility and per cent utilization of the two diets are not significantly different. The gravimetric method gave results nearly identical with those obtained by the uric acid method. During larval development the approximate digestibility (A.D.) of the artificial diet (alfalfa leaf meal) on which *A. velutinana* is routinely reared was shown by the gravimetric method to be 60.4 ± 2.5. These preliminary results indicate that the per cent utilization of chemically defined diets is in close agreement with the per cent utilization of diets containing crude organic matter, at least with regard to approximate digestibility.

Table 7. Effect of Amino Acid Supplementation of Proteins on Development of *A. velutinana*.

Protein (3%)	Supplementation	Total no. larvae	Avg days to pupation: Males	Females	Per cent survival
Egg albumen	None	82	21.4[b]	24.1[a]	81
	0.15 glutamate, 0.05 histidine[a]	70	18.9[b]	21.8[b]	86
Soybean protein	None	69	23.1[a]	24.6[a]	83
	0.10 methionine, 0.4 glycine	69	18.7[b]	20.7[b]	76
Casein					
Expt. 1	None	46	29.2[a]	28.0[a]	65
	0.13 arginine	51	20.9[b]	24.1[b]	87
	0.10 cystine	57	22.3[b]	23.4[b]	89
	0.13 arginine, 0.10 cystine, 0.5 glycine	52	21.2[b]	22.8[b]	96
Expt. 2	None	51	30.4[a]	31.0[a]	68
	0.15 glycine	54	22.8[b]	24.7[b]	83
	0.13 arginine, 0.10 cystine, 0.15 glycine	54	20.0[c]	22.4[c]	89
Expt. 3	None	50	28.5[a]	27.0[a]	78
	0.40 glutamate	49	23.1[b]	23.8[b]	92
	0.40 serine	52	20.0[c]	22.0[b]	83
	0.13 arginine, 0.10 cystine, 0.15 glycine	51	19.3[c]	21.6[b]	89

[a]Number preceding each amino acid supplement is g/100 g diet.
[b]Values.

Table 8. The approximate digestibility (A. D.) efficiency of conversion of ingested food to body substance (E.C.I.) and efficiency with which digested food is converted to body matter (E.C.D.) by *A. velutinana* reared axenically from newly-hatched larvae to larval-pupal ecdysis on amino acid diets containing either L- or D- methionine based on faecal uric acid method.[a]

Methionine isomer	No. larval reared	Survival to larval-pupal ecdysis (%)	No. insects for uric acid method ♂	♀	Dry wt gained (mg)	A.D.	E.C.I.	E.C.D.
L-methionine (100 mg/100g diet)	10	100	6	4	8.9[a]	64.0[a]	13.3[a]	22.7[a]
D-methionine (150 mg/100g diet)	39	79.5	4	6	9.4[a]	61.2[a]	12.0[a]	21.5[a]

[a]Figures not followed by same letter are significantly different at the 5% level of probability.

SUMMARY

The use of amino acid analysis of the whole insect at various developmental stages as a basis for formulating amino acid mixtures is discussed. It is shown that *Argyrotaenia velutinana* larvae grow best on synthetic diet in which the amino acid mixture is modeled on that of pupal carcass analysis rather than on the amino acid pattern of casein or cotton-seed protein. This insect grows well for several successive generations on diet containing only the essential amino acids. Growth is improved by supplements of several individual amino acids or diammonium citrate, and is best with supplement of eight dispensable amino acids.

Acknowledgements.—This investigation was supported in part by research grant A1-98633 from the National Institute of Allergy and Infectious Diseases, National Institutes of Health. This paper is No. 3789 of North Carolina State University Agricultural Experiment journal series.

REFERENCES

BHATTACHARYA, A. K. and WALDBAUER, G. P. 1969a. Quantitative determination of uric acid in insect faeces by lithium carbonate extraction and the enzymatic--spectrophotometric method. *Ann. Ent. Soc. Am.* 62:925-7.

BHATTACHARYA, A. K. and WALDBAUER, G. P. 1969b. Faecal uric acid as an indicator in the determination of food utilization. *J. Insect Physiol.* 15:1129-35.

KING, W. K. 1963. Isotopic evaluation of efficiency of amino acid retention. *Federation Proc.* 22:1115-20.

ROCK, G. C. and HODGSON, E. 1971. Dietary amino acid requirements for *Heliothis zea* determined by dietary deletion and radiometric techniques. *J. Insect Physiol.* 17:1087-97.

ROCK, G. C. and KING, K. W. 1966. Amino acid composition in hydrolysates of the red-banded leaf roller, *Argyrotaenia velutinana* during development. *Ann. Ent. Soc. Am.* 59:273-7.

ROCK, G. C. and KING, K. W. 1967a. Qualitative amino acid requirements of the red-banded leaf roller, *Argyrotaenia velutinana*. J. Insect Physiol. 13:59-68.

ROCK, G. C. and KING, K. W. 1967b. Quantitative amino acid requirements of the red-banded leaf roller, *Argyrotaenia velutinana. J. Insect Physiol.* 13:175-86.

ROCK, G. C. and KING, K. W. 1968. Amino acid synthesis from glucose-U-^{14}C in *Argyrotaenia velutinana* (Lepidoptera:Tortricidae) larvae.*J. Nutr.* 95:369-73.

WALDBAUER, G. P. 1968. The consumption and utilization of food by insects. *Adv. Insect Physiol.* 5:229-88.

WILLIAMS, H. H., CURTIN, L. W., ABRAHAM, J., LOOSLI, J. K. and MAYNARD, L. A. 1954. Estimation of growth requirements for amino acids by assay of the carcass. *J. Biol. Chem.* 208:276-86.

WU, C. and HOGG, J. F. 1952. The amino acid composition and nitrogen metabolism of *Tetrahymena geleu. J. Biol. Chem.* 198:733-56.

AMBIGUITIES IN THE INTERPRETATION OF GROWTH EXPERIMENTS WITH MOSQUITO LARVAE IN SEMI-SYNTHETIC DIETARY MEDIA

R. H. Dadd
Division of Parasitology
University of California, Berkeley, California 94720

In this paper I discuss some problems of interpretation that bear on our general topic of improving synthetic diets. These problems arose while developing a basal dietary medium for the mosquito *Culex pipiens*, and they illustrate how difficult it may be, given a partially satisfactory diet allowing fair growth and development, to distinguish whether suboptimal performances arise from strictly nutritional or physical/mechanical inadequacies of the diet. Such problems might attend the use of any non-homogenous diet if there were uncertainty as to the relative intake of its components or phases. The particular case of non-homogeneity considered here is found in dietary media for filter-feeding animals, media that comprise an aqueous solution of nutrients in which there is a solid component, usually but not necessarily nutritive, and probably needed mechanically for the proper ingestion and passage of material through the gut.

In nature, filter feeders gain most of their nutriment by straining into their mouths solid detritus and small organisms from the water in which they live; the water, though often really a dilute solution of certain nutrients, contributes little to their nourishment, as shown for several phagotrophic crustaceans (Provasoli and D'Agostino, 1969; Provasoli *et al.*, 1971), and in mosquito larvae supposedly is not ingested to any great extent (Clements, 1963). Ideally, a synthetic medium would simulate the phagotrophic situation by providing a complete complement of nutrient chemicals encapsulated into particles of optimal dimensions for rapid ingestion by membranes of some impermeable material that would prevent diffusion of soluble nutrients into the surrounding water. For my first instar larvae, the maximal diameter of particles that can be ingested is about 20μ (Dadd, 1971), and though recent developments in microencapsulation (e.g., of insecticides and viruses [Ignoffo & Batzer, 1971]) in this size range offer future hope, my inquiries into the feasibility of devising an impermeable, microencapsulated, particulate diet have so far drawn a blank. As I had perforce, therefore, to formulate dietary media like those developed for *Aedes aegypti*, I scrutinized the four decades of work on the nutrition of this species to gain insight into those crucial features common to synthetic media that have supported good growth and development.

Although in 1935 Trager reported normal development of *A. aegypti* in sterile media composed of apparently wholly soluble materials (yeast and liver extracts and hydrolysed protein), all subsequent diets that supported good growth included a solid nutrient component: casein, in the later semi-synthetic diets of Trager (1948), and insoluble yeast residue in those of Golberg & DeMeillon (1948a, b). Two highly defined diets with amino acids completely replacing protein allowed slow growth and development of *A. aegypti* to adults. Both contained a nutritionally inert solid phase: in one case, cellulose powder, used as a carrier for lipids (Singh & Brown, 1957); and in the other, agar, holding amino acids and other nutrients in a gel at the bottom of culture tubes with larvae in water above (Lea *et al.*, 1956, 1958). In Akov's hands (1962), the diet of Singh & Brown allowed little growth, and she reverted to a basal diet similar to that of Trager, after finding that insoluble vitamin-free casein gave the best of varied performances when different proteins, sterilized by various procedures (dry baking, autoclaving) were compared; it is to be noted that the same proteins, differently sterilized, gave rise to diets of different physical consistencies and growth supporting potential. Nayar (1966) found casein media a failure with various salt marsh mosquitoes, apparently because his casein developed an unsuitable texture after autoclaving; he then devised a lactalbumen-based diet on which several salt marsh mosquitoes (and also *A. aegypti*, if salts were reduced) developed satisfactorily. This lactalbumen medium contained cellulose powder, without which larvae died in a few days, apparently because of mechanical malfunction of the gut if solids were absent. Besides these studies of *Aedes*, an early note (Lichtenstein, 1948) describes a medium based on hydrolysed casein that supported very slow growth of *Culex molestus* to the adult stage. It contained a salt mixture that was probably incompletely soluble, but otherwise all nutrients were, at face value, in solution. However, it nominally lacked two generally essential nutrients, sterol and choline, probably covert contaminants of the hydrolysed casein, which may therefore have carried other unknowns.

Although studies based on these various diets provided substantial knowledge of the nutrition of *A. aegypti*, there are many inconsistencies between the results obtained by different workers with respect to requirements for sugar, amino acids, vitamins and lipids. Since many of these studies patently encountered difficulties with physical and mechanical properties of the media used, I wondered whether some of the inconsistent, supposedly nutritional findings might more likely reflect unexamined constraints on normal feeding stemming from unsatisfactory physical characteristics of the media. Especially, I speculated that without an appropriate solid phase, adequate ingestion of food might not be possible, for my studies of larval feeding in *C. pipiens* (Dadd, 1968, 1970a, b, 1971) indicated that however much a larva may filter in water devoid of solids, material already ingested within the midgut remains stationary, requiring the provision and ingestion of additional solids to move it rearwards, a process resulting in complete displacement of the original gut contents in about an hour with natural food particles or inert powders of suitable particle size. When the major nutrient in a synthetic medium is an insoluble particulate, as casein is in the diets of Trager and Akov, its ingestion in optimal amounts can readily be understood and the soluble nutrients needed in relatively low concentration (salts and growth factors) may well be carried along with it, absorbed, or as interstitial liquid. However, when essentially all nutrients are soluble, as in amino acid media, if their ingestion were dependent upon being carried in by a nutritionally inert solid, such as cellulose powder or agar gel, intake of nutrients, as distinct from total material, would very likely be grossly suboptimal. This factor possibly accounts for the slow growth observed on such soluble media.

In trying out dietary media for *C. pipiens*, I therefore surveyed a number of insoluble and variously soluble proteins, protein hydrolysates and amino acid mixtures, sometimes augmented with readily ingested, non-nutritive particulates such as kaolin, and all used in an aseptic basal medium otherwise modeled on those of Trager (1948) and Akov (1962). The results of these initial trials may be summarized as follows: (a) Diets with amino acid mixtures, enzymatic casein (or lactalbumen) hydrolysate, or Hammersten casein, all of which appeared to be clear, simple solutions, supported very slow, limited growth; contrary to my expectations, addition of inert particulates generally had no effect or was detrimental. (b) With vitamin-free casein (essentially the same medium that Trager and Akov found adequate for *A. aegypti*), which remained largely an insoluble sediment, growth was slow compared to growth in normal septic culture, and larvae only occasionally completed development to pupae or adults; media with lactalbumen or alpha (soy) protein, both giving insoluble sediments, were poorer than with vitamin-free casein, and none of these sediment-containing media were improved by adding inert particulates. (c) One protein alone of those tested, B.D.H. light white soluble casein (hereafter referred to as LWS casein), allowed larval growth within the normal range with good survival to the adult stage in most experiments. (d) No media were improved by the addition of sugar nor any of the extra vitamins that particular authors have claimed improved the growth of *Aedes*. (e) Very surprisingly, all media were better with cholesterol omitted than with it at any concentration tested in the range of 0.1 to 10mgs per 100ml of medium.

So far as they went, these initial findings showed that although wholly soluble media supported so little growth as to suggest inadequate intake of nutrients, my first hypothesis, that provision of a readily ingestible particulate would alleviate this, was not sustained. But before considering mechanical aspects of media further, I must dispose of a specific nutritional deficiency that was restricting the rate of growth in all media used in the initial trials. As part of the basal medium, I routinely incorporated a small amount (0.1%) of yeast nucleic acid, following studies which show that for *A. aegypti*, as for many Diptera, nucleic acid, though not strictly essential, improves the rate of growth. With *C. pipiens*, however, the requirement for nucleic acid appears more stringent, and for optimal growth is needed at thrice the concentration giving maximal growth acceleration with *A. aegypti* (Dadd *et al.*, in press). With this ribonucleic acid requirement optimally satisfied, growth to the 4th instar in the LWS casein medium was as rapid as in normal culture, but pupation still tended to be delayed, and adult emergence was erratic from experiment to experiment.

Table 1. Basal medium for *Culex pipiens* larvae.

	mg/100 ml		mg/100 ml
Casein, BDH light white soluble	1000.0	K_2HPO_4	60
Ribose nucleic acid, NBC reagent, yeast	300.0	KH_2PO_4	60
Thiamin	0.5	Ca gluconate	5
Riboflavin	0.5	NaCl	2
Nicotinamide	1.0	Fe sequestrene[a]	2
Pyridoxine	0.5	Zn "	2
Ca pantothenate	5.0	Mn "	2
Folic acid	0.1	Cu "	0.5
Biotin	0.1	KOH to pH 7.5	
Choline chloride	10.0		

[a]Geigy products: chelates of the metals with sodium EDTA.

In amplification of some of the aforementioned generalizations, I present graphically in Figure 1 the results of an experiment using synthetic media, all with optimal nucleic acid. The media differ only in the type of casein incorporated, and might be considered essentially the same in a strictly nutritional sense, with one important exception respecting lipid contaminants which I discuss later. The basal composition of these media is given in Table 1, and details of formulation and aseptic methodology are described elsewhere (Dadd *et al.*, in preparation). For comparison with normal development on a natural diet, representative data are included for larvae reared individually on a septic medium of rabbit pellet powder, liver powder and dried yeast in water under identical conditions of culture tube size, liquid volume, temperature, etc.

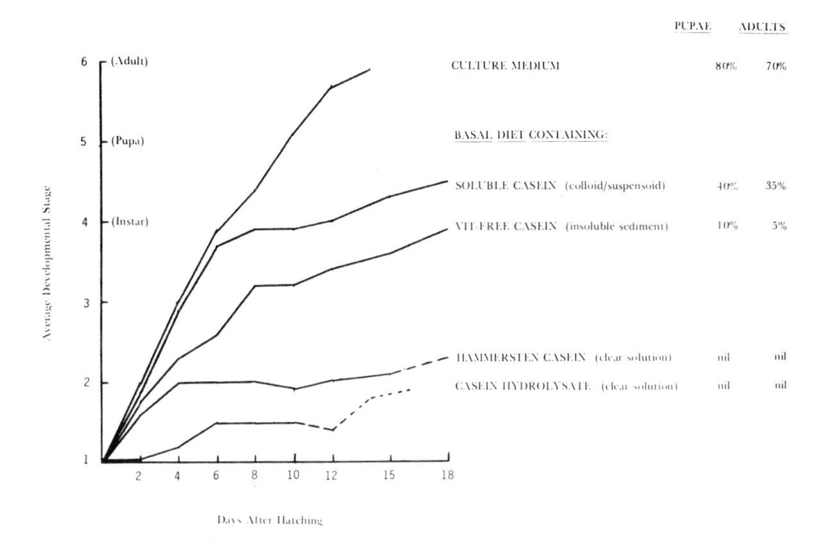

Figure 1. Average developmental stage and final percentage pupation and adult emergence for larvae of *Culex pipiens* reared aseptically and individually in various casein-based media.

The two clear, completely soluble media, those with Hammerstens casein and enzymatic casein hydrolysate respectively, supported slow growth of some larvae to the 2nd instar (most larvae in the case of Hammerstens casein), but little development beyond this stage. The medium containing vitamin-free casein, largely as a sediment, supported rapid growth to the 2nd instar and then moderate growth to the 3rd and early 4th instars, when mortality became so heavy that only two of the original 20 larvae pupated and only one became adult. On the LWS casein medium, growth and development up to the 4th instar were as good as on the crude septic medium, but thereafter development lagged and ultimately only half as many adults emerged.

If we first consider these results in relation to the idea that different developmental rates might reflect variable feeding efficiencies, especially with respect to the hypothesis that intake of nutrients would be inadequate without ingestible nutrient solids, then the medium with LWS casein, which ostensibly lacks a solid phase, presents an apparent anomaly; up to the final instar poor ingestion could not have been a limiting factor in growth, as development was normal. However, LWS casein does not really form a true solution at the pH of these media; when freshly made up, the medium appears opalescant, remains so after autoclaving, and after a day or two deposits a sparse film of fine sediment on the bottom of the culture tubes.

Evidently the casein is present as a colloid, probably of large particle size verging on the suspensoid state.

This prompted me to examine whether simple colloid solutions were well ingested, using my procedures for comparing rates of ingestion of particulate solids (Dadd, 1968, 1970a) in which the relative displacements of solids already in the gut by experimental particulates in water containing the larvae are compared. These experiments, to be described in detail elsewhere, showed that quite low concentrations of colloid materials such as agar, methyl cellulose, soluble starch, etc. displaced charcoal from the gut of 4th instar larvae as well as did kaolin, a rapidly ingested particulate solid. Evidently these colloid solutions could be swept into the gut to act as a ram, more like ingested solid material than like water or watery solutions. Pertinent to this discussion, "solutions" of LWS casein of concentrations from 0.5-2.0% also displaced gut solids very rapidly, and in Table 2 I summarize displacement data obtained for charcoal-glutted 4th instar larvae given access for 30 minutes to the media used in the experiment of Figure 2. It will be seen that the casein hydrolysate and the Hammerstens casein media were as ineffective as water in displacing charcoal from the larval gut, whereas LWS casein medium displaced charcoal as effectively as kaolin.

Table 2. Mean displacement values for groups of charcoal-glutted 4th instar larvae of *Culex pipiens* given access for 30 minutes to various media or water.

Media	Replicate displacement values					Avg
Plain water	0.0	0.0	0.0	0.0	0.0	0.0
Casein hydrolysate (enzymatic)	0.0	0.0	0.0	0.0	0.0	0.0
Hammersten casein	0.0	0.0	0.0	0.0	0.0	0.0
Vitamin-free casein	2.7	2.6	2.6	2.4	2.3	2.5
LWS casein	4.2	4.1	3.8	3.7	3.7	3.9

The results of these displacement studies accounted for the presumptively adequate ingestion of nutrients implicit in the normal development of the first three instars on LWS casein medium, and could well explain the very poor growth in the soluble media. However, two features of the data presented in Figure 1 are not so readily explicable. First, displacement tests showed that vitamin-free casein medium was well ingested, as would be expected in view of its profuse, fine sediment, and yet development, overall, was poor, with nearly all larvae dying before pupation.

In seeking to account for this, I must elaborate on my earlier comment that all media tested have supported better growth when sterol, as a specific component, was omitted. Now, to judge from the fact that with *A. aegypti* sterol omission prevents growth byond the 3rd or early 4th instar (Akov, 1962; Golberg & DeMeillon, 1948a), failure in the 3rd and 4th instars of *C. pipiens* on vitamin-free casein medium without sterol is precisely what would be expected. Why then the nearly normal growth and adult development on LWS casein? Probably because this less refined casein contains sterol: it is stated to contain up to 1.5% lipid contaminants; and whereas a qualitative sterol test on a lipid extract from vitamin-free casein indicated only a doubtful trace of sterol, a similarly made extract from LWS casein was strongly sterol positive (I thank my colleague, E. McClain, for performing these sterol tests). At this time, what remains inexplicable is why the addition of sterols (cholesterol or β-sitosterol) to the media, in amounts that were optimally effective for *A. aegypti*, not only fails to alleviate the growth retardation of *C. pipiens*, but at the higher levels tested further inhibits the already poor growth.

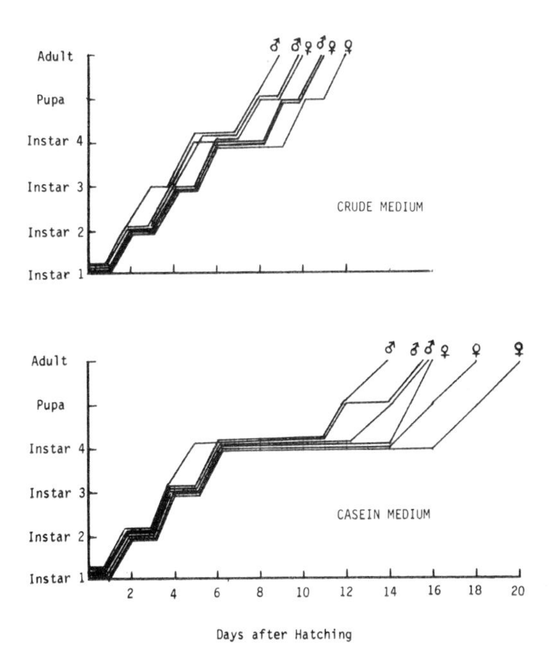

Figure 2. Plots of developmental stage with time for individual larvae of *Culex pipiens* reared in crude medium or LWS casein medium.

The second problematic feature posed by the data of Fig. 1 concerns the retardation on LWS casein medium after development had proceeded normally to the attainment of the 4th instar. That this delay was primarily due to an inordinate extension of the last larval instar is seen most clearly from development curves for individual larvae, examples of which are shown in Figure 2. For LWS casein medium, data for three larvae of each sex that first became perfect adults in the experiment of Figure 1 are used, and are matched with three males and three females from the septic culture group (the fastest developers of each sex were selected, though in fact almost all of both sexes emerged over the span of four days covered by the data given). Though males in septic culture tended to develop to instar 4 in a day less than males in LWS casein medium, the most striking difference between the sets of curves for the two treatments is in the duration of the 4th instar: in septic culture males and females average slightly less and slightly more than three days, respectively, whereas in the synthetic medium, males averaged nearly seven days and females nearly ten.

This stagnation in the last instar might, of course, reflect eventual sterol insufficiency, even though enough sterol was probably supplied by LWS casein for the needs of early larval development. Or it might reflect sub-optimal feeding with increasing size, in spite of my earlier argument for optimal feeding in the early instars. Elaborating this latter possibility, it is conceivable that the "pseudo-solid" properties of colloid solutions, allowing them to act in ingestion and displacement of gut contents as if they were themselves solids, might be a function of size such that they acted like solids less well, the larger the larval dimensions. I tried two approaches to clarify this point. If adequate nutriment in the 4th instar requires a more concentrated ingesta to counteract a relative deterioration in the rate of ingestion of a colloid medium, then perhaps augmentation of the LWS casein medium with some sedimenting vitamin-free casein would provide a concentrated nutritive supplement; but media so augmented proved worse than with LWS casein alone. The other approach hinged on the conjecture that if the high viscosity of colloidal media was crucial in enabling them to be swept into the mouth and to pack back into the midgut, then addition of the various colloids that in low concentration effectively displaced charcoal might increase the viscosity of LWS casein medium so as to significantly increase ingestion rate by large larvae. I therefore compared development on LWS casein diets variously augmented with colloids and obtained the results that are summarized in Table 3. Agar and methyl cellulose, both nutritionally inert, somewhat shortened the delay in the 4th instar, and slightly increased the proportion of larvae that eventually pupated and became adult. Soluble starch, a utilizable nutrient, markedly shortened the 4th instar (though still not to a normal 3-4 days) and also greatly increased the proportions of pupae and adults obtained.

Table 3. Developmental data for groups of twenty 1st instar larvae of *Culex pipiens* reared individually and aseptically on basal LWS casein diet with the colloid supplements indicated.

Supplements	Avg developmental stage[a] at: 4th day	8th day	12th day	Percentage Pupation	Adults	Avg duration of 4th instar (days) ♂♂	♀♀
No additions	3.1	3.9	4.5	64	46	6.0	9.4
plus							
0.02% methyl cellulose	3.1	4.0	4.5	74	61	8.0	9.0
0.02% agar	2.7	3.9	4.7	71	59	4.8	6.7
0.02% soluble starch	3.2	4.1	5.4	85	80	4.8	5.8

[a]Stages 1-4 are larval instars, 5 is the pupa, and 6 the adult. Values given are the averages for all individuals in a treatment alive at the stated times.

Though this latter result could be construed to support my viscosity hypothesis, it still carries interpretational ambiguities. Since starch is a utilizable carbohydrate, if *C. pipiens* had a late-developmental requirement for carbohydrate to metabolize for prepupal fat accumulation, as is known in certain other insects (Beck, 1956; Dadd, 1963), this might account for the improvement. However, we may probably discount this possibility by noting that an equivalent concentration of glucose was no substitute for soluble starch. Among other possibilities, one must still keep in mind the probably marginal sterol in the medium; perhaps starch carried traces of sterol that could improve terminal growth, were sterol demands more imperious in late development. And so long as only natural products, such as starch, are able to improve growth, the possibility remains in the background that the true effective material is some unrecognized contaminating growth factor or trace mineral.

I think it is clear that to further clarify how physical/mechanical features of the medium may be crucially affecting growth it will first be necessary to get all known, purely nutritional requirements under manipulative control. In essence, for *C. pipiens* this means finding some vehicle for incorporating sterol (and perhaps other lipids) in a utilizable form, so that reliance need not be placed on an unrefined protein having a sterol contaminant. For as long as one cannot intentionally add (or withhold), sterol to a sterol-free formulation in amounts demonstrably able to produce and alleviate deficiency, positive results in response to supplements such as soluble starch, added for other conjectural reasons, will carry ambiguous interpretations.

REFERENCES

AKOV, S. (1962). A qualitative and quantitative study of the nutritional requirements of *Aedes aegypti* L. larvae. *J. Insect Physiol.* **8**:319-335.

BECK, S. D. (1956). The European corn borer, *Pyrausta nubilalis* (Hubn.), and its principal host plant. II. The influence of nutritional factors on larval establishment and development on the corn plant. *Ann. Entomol. Soc. Amer.*, **49**:582-588.

CLEMENTS, A. N. (1963). *The Physiology of Mosquitoes.* Pergamon Press, Oxford. 393pp.

DADD, R. H. (1960). The nutritional requirements of locusts - II. Carbohydrate requirements and utilization. *J. Insect Physiol.* **5**:301-316.

DADD, R. H. (1968). A method for comparing feeding rates in mosquito larvae. *Mosquito News*, **28**:226-230.

DADD, R. H. (1970a). Comparison of rates of ingestion of particulate solids by *Culex pipiens* larvae: phagostimulant effect of water-soluble yeast extract. *Ent. Exp. & Appl.*, **13**:407-419.

DADD, R. H. (1970b). Relationship between filtering activity and ingestion of solids by larvae of the mosquito *Culex pipiens*: a method for assessing phagostimulant factors. *J. Med. Entomol.* 7:708-712.

DADD, R. H. (1971). Effects of size and concentration of particles on rates of ingestion of latex particulates by mosquito larvae. *Ann. Entomol. Soc. Amer.*, **64**:687-692.

GOLBERG, L. and DEMEILLON, B. (1948a). The nutrition of the larva of *Aedes aegypti* L. 3. Lipid requirements. *Biochem. J.* **43**:372-379.

GOLBERG, L. and DEMEILLON, B. (1948b). The nutrition of the larva of *Aedes aegypti* L. 4. Protein and amino acid requirements. *Biochem. J.*, **43**:379-387.

IGNOFFO, C. M. and BATZER, O. F. (1971). Microencapsulation and ultraviolet protectants to increase sunlight stability of an insect virus. *J. Econ. Entomol.* **64**:850-853.

LEA, A. O., DIMOND, J. B. and DELONG, D. M. (1956). A chemically defined medium for rearing *Aedes aegypti* larvae. *J. Econ. Entomol.* **49**:313-315.

LEA, A. O. and DELONG, D. M. (1958). Studies on the nutrition of *Aedes aegypti* larvae. *Proc. Tenth Int. Cong. Entomol., Montreal, 1956,* 2:299-302.

LICHTENSTEIN, E. P. (1948). Growth of *Culex molestus* under sterile conditions. *Nature, Lond.*, **162**:999-1000.

NAYAR, J. K. (1966). A method of rearing salt-marsh mosquito larvae in a defined sterile medium. *Ann. Entomol. Soc. Amer.*, **59**:1283-1285.

PROVASOLI, L. and D'AGOSTINO, A. S. (1969). Development of artificial media for *Artemia salina. Biol. Bull. Woods Hole,* **136**:434-453.

PROVASOLI, L., CONKLIN, D. E. and D'AGOSTINO, A. S. (1970). Factors inducing fertility in aseptic Crustacea. *Helgoländer wiss. Meeresunters,* **20**:443-453.

SINGH, K. R. P. and BROWN, A. W. A. (1957). Nutritional requirements of *Aedes aegypti* L. *J. Insect Physiol.*, **1**:199-200.

TRAGER, W. (1935). On the nutritional requirements of mosquito larvae (*Aedes aegypti). Amer. J. Hyg.*, 22:475-493.

TRAGER, W. (1948). Biotin and fat soluble materials with biotin activity in the nutrition of mosquito larvae. *J. Biol. Chem.*, **176**:1211-1223.

INTERACTIONS BETWEEN DIETARY COMPONENTS

T. E. Mittler
Division of Entomology, University of California
Berkeley, 94720

There probably is no one in this audience who has not been plagued with the chore of repeatedly having to make up fresh batches of diet, even when sufficient amounts of previously made-up diet were left to perform a new experiment or to keep a culture going. This stems, of course, from the knowledge or perhaps just the fear that a diet that has been around for some time has deteriorated to a point where the insects we wish to feed it to would perform less well than on a freshly made-up diet. Comparisons of the performances on fresh and old diets are, no doubt, occasionally performed. At other times, one look or sniff at an old diet–even one that has been maintained in the refrigerator or freezer–is enough evidence of its deterioration. Even if no tests are performed, and no color changes, precipitates, turbidity or microbial growths are evident, we mostly play it safe and make up a fresh diet. This may be done from scratch each time, or, more frequently I suspect, stock mixtures–whether of solids or of solutions–are variously combined to give a new diet in a fraction of the time it would take to make it up by measuring out each individual component.

With regard to mixtures of solids, we had the experience some years ago that a dry mix of the 20 amino acids used in our standard aphid diet deteriorated even when maintained in a screw-capped bottle in the refrigerator. While none of the amino acids appeared to be deliquescent, sufficient moisture may have been associated physically or chemically with some of the amino acids to allow decomposition by chemical interaction or by microbes. This was unfortunate as we thought that we had got nearer the day when we could make up an instant diet merely by dissolving some spoonfulls or tablets of the ingredients in a beaker of water. I wonder whether the chemical companies that offered to make up, and perhaps also market, such amino acid mixes, could have overcome this difficulty. We, for our own part, went back to weighing out the amino acids individually each time we made up a diet.

In most cases we do not know what goes on in a diet after it is made up. While microbial contaminations may be the cause of much of the trouble, even sterile diets change. On the physical side, for example, agar based diets lose water; suspensions or emulsions settle out or break; and the most frequent and evident indications of chemical changes occurring in diets is their gradual darkening. This is particularly so when diets containing sugars are autoclaved. The browning of *Drosphila* diets is a well-known example, recently minimized by replacing fructose with sucrose (Cooke & Sang,

1970). However, even without autoclaving, diets containing sugars and amino acids darken—an interaction which occurs in aphid diets and has, no doubt, been observed and recorded on numerous occasions in other studies. Dadd (unpubl. obs.) found that even in the absence of free sugars, recently formulated mosquito diets containing amino acids and nucleic acids sometimes turn brownish when autoclaved.

In the process of providing essential dietary lipids for wax moth larvae, Dadd (1964) previously found that diets containing polyunsaturated acids darken and develop an odor. While either linoleic or linolenic acid were required for normal emergence from the pupa, the insects grew less well on diets containing these fatty acids, probably because of their unpalatable or toxic oxidation products. It was suggested by Gordon (1959) that the rapid deterioration of linolenic acid did not permit it to be used as a replacement for dietary linoleic acid required by *Blattella germanica*. Considerably earlier, Fraenkel & Blewett (1946) had found that vitamin E (i.e., α-tocopherol) and other unrelated anti-oxidants protected unsaturated fatty acids from breaking down. Ascorbic acid and fatty acid monoesters of ascorbic acid are added to food stuffs as anti-oxidants. In their ability to inhibit rancidity, they are synergistic with phospholipids and α-tocopherol. Citric acid too is known to be an effective anti-oxidant synergist (Gortner, 1950).

Precautions that have been taken to reduce oxidation reactions include dissolving the dietary ingredients under a stream of nitrogen (Akey & Beck, 1971), using water from which air has been boiled out to make up the diets, using nitrogen to force diets through filters, and storing made-up diets under nitrogen. While precautions such as these, or the storage of diets in a deep freezer at -20°C., have been shown to delay the decomposition of such labile compounds as ascorbic acid (Dadd *et al.*, 1967), interactions with oxygen or between the dietary constituents themselves are unavoidable when the diets are subsequently used at experimental temperatures. Granted a diet may be frequently renewed, but the nutritional advantages are generally offset by the practical difficulties and harmful effects of too frequently handling the insects.

With regard to storage at sub-zero temperatures, it has been found that the rate of oxidation of ascorbic acid may decrease less than expected or even increase when dilute and simple solutions of the vitamin (particularly at pH>5) are frozen to -17°C (Grant & Alburn, 1965; Thompson & Fennema, 1971). Thus, under certain conditions, freezing of diets or solutions of some of their constituents may not achieve the desired effect of maximally preserving all the dietary nutrients.

To compensate for anticipated losses, it is a general practice to incorporate labile components into diets at concentrations greatly in excess of their required levels. While such high levels of certain trace nutrients may readily be tolerated by some insects, the feeding behavior and growth of others may be adversely affected even by fairly low levels of some trace minerals, nutrients or vitamins.

In the case of ascorbic acid, the 100-200 mg per 100 ml of diet, which we had found optimal for the growth of *Myzus persicae* (Dadd *et al.*, 1967; Mittler *et al.*, 1970), unfortunately makes the diet less acceptable to the insects, at least within the first day of being presented with it (Kunkel & Mittler, unpublished obs.). High levels of riboflavin have been shown to be detrimental to a number of insects, e.g., mosquitoes (Akov, 1962) and silkworm larvae (Horie *et al.*, 1966). While we found that *Myzus persicae* requires dietary riboflavin (Dadd *et al.*, 1967), Ehrhardt (1968a, 1969) was able to maintain *Neomyzus circumflexus* for 5-10 successive generations on diets lacking this vitamin—presumably because this aphid's symbiotes are able to supply the insect with adequate amounts. Ehrhardt moreover demonstrated a detrimental effect of riboflavin at the one level at which he tested it, namely 5 mg per 100 ml. This concentration is ten times greater than the one we used in our vitamin studies on *Myzus persicae* (Dadd *et al.*, 1967) and have routinely been using since 1966. Recent tests showed that the 5 mg per 100 ml of riboflavin is also detrimental to *Myzus persicae* (Mittler & Tsitsipis, unpubl. obs.). However, further experiments with this aphid are needed before its dietary requirement for riboflavin and possibly other vitamins can be considered established, in view of the fact that the 1967 studies employed mineral-deficient and, hence, growth-limiting diets, as pointed out by Ehrhardt (1968a). Nevertheless, these examples show that one can readily overdo things on the basis of playing it safe.

Vitamins are unstable at certain pHs of the medium and in the presence of the ions of certain metals—some of which, however, must be incorporated into diets (although not necessarily in ionizable forms) to meet the mineral requirements of the test animal. Not only are trace amounts of certain metals required, but substantial amounts of other minerals must be available.

Now most diets have their pH adjusted to some value based on preferential acceptability or settling, feeding and growth rates, microbial growth inhibition or the growth of symbiotic microorganisms. However, precipitates tend to form as a result of such pH adjustment, particularly in diets with a pH above neutrality. Should a precipitate result, one must be thankful if it is immediately obvious rather than slow in forming and relatively inconspicuous, for then one is immediately aware of the problem and can make adjustments to avoid getting the precipitate or try to bring it back into solution, if that is possible and deemed necessary. It may be that insoluble metal phosphates or hydroxides, when these are evenly dispersed in, for example, an agar diet ingested by a chewing insect, are as readily available to the insect as, say, the soluble chlorides of the metals. But I do not know of any studies on this. If however, the precipitated minerals are not ingested, or if they are in a chemical form that makes them unavailable for absorption after ingestion, then it becomes a major problem. An aphid can ingest particles small enough to pass up its narrow food canal (Miles *et al.*, 1964), but not particles that, in themselves or due to their aggregation or clumping, are too large to do so; or, even though they may be small enough, do not remain dispersed in the diet but settle out.

Thus, a far greater problem may be created by precipitates that are too fine or sparse to be noticed, particularly if they only form slowly. While not all artificial diets are passed through Millipore or other bacterial filters to sterilize them, when this is done, even inconspicuous precipitates will thereby be removed. It has frequently been our experience that Millipore filters, of the 0.22 micron pore size used to sterilize our aphid diets, get progressively clogged even when apparently clear diets are passed through them in deciliter amounts; but we did not determine whether fine precipitates, bacteria, yeasts, or chemically unchanged solutes at the limit of their solubility were responsible. But it might pay one to examine what it is that one is taking out of a diet in this way.

In our early attempts to rear aphids on artificial diets, we incorporated cholesterol dispersions into the diets. On the whole, these dispersions (made by rapidly stirring acetone solutions of cholesterol into water and then boiling off the acetone) were fairly stable, but did give the diets a milky white look. Occasionally a white precipitate would settle out and the aphids grew poorly on such diets. It took us over six months of experimentation before we realized that it was not the cholesterol that precipitated out, thereby adversely affecting growth, but that it was the dietary magnesium that precipitated as phosphate. That occurred whenever a solution of dietary components including magnesium chloride became strongly alkaline as a result of the addition of the tribasic potassium phosphate used at the time, and was particularly conspicuous if the alkaline mixture was heated to more rapidly dissolve the dietary amino acids and sucrose. It may be emphasized that the precipitate did not go back into solution when neutrality was restored. If, however, the mixture containing the magnesium chloride remained on the acid side due to the amino acids or the use of acid potassium phosphate, and basic phosphate or KOH added to it only in the final stages of bringing the diet to neutrality, no precipitate formed. Dadd (1968) discussed these and other problems connected with inorganic components of aqueous diets at a symposium of the Entomological Society of America on artificial diets at Portland in 1966.

After all the trouble that our attempts to supply cholesterol in the diet had caused us, it was ironical to find that neither cholesterol nor any other sterol or lipid was needed by *Myzus persicae* and by some ten other aphid species in their diet (Dadd & Mittler, 1966; Dadd & Krieger, 1967; Ehrhardt, 1968b; Krieger, 1971; Srivastava & Auclair, 1971). However, the continued use of cholesterol benzoate by Akey & Beck (1971) appears to be justified, for while over 25 generations of pea aphids could be reared on sterol-free diet, aphid weight was increased on diets containing cholesterol (Akey and Beck, personal communication). Although cholesterol benzoate (2.5 mg per 40 gm of diet solutes per 100 ml, as originally used by Auclair & Cartier in 1963 in non-sterile pea aphid diets) may also be performing a bacteriostatic function, the occurrence of similar levels of sterols in young radish plants and in the honeydew excreted by *M. persicae* feeding upon them (Forrest & Knights, 1972) suggests that dietary sterols can be of significance to aphids, even if they can exist without ingesting them, no doubt as a result of sterol synthesis by their symbiotes.

Although the use of cholesterol suspensions literally clouded things for us for some time in the mid-60s, the difficulties we encountered did make us aware of the fact that the sequence in which dietary components are mixed together can be extremely important. While we should have anticipated and avoided the precipitation of magnesium phosphate, not all the interactions between dietary components are as obvious and predictable.

In addition to incorporating and making sure that relatively high levels of P, K and Mg remained in solution in aphid diets, trace amounts of Fe, Mn, Zn and Cu have to be incorporated into the diet, in order to meet the insects' mineral requirements. In fact, it was not until this was accomplished that more than two successive generations of aphids could be reared on artificial diets. The breakthrough in this regard came when Dadd (1967) realized that the addition to diets of lettuce juice, or crude yeast or sperm nucleic acids substantially improved aphid growth by virtue of trace amounts of these metals thereby introduced into the diets as contaminants. Our ability to rear some 20 successive generations of *Myzus persicae* on a completely defined artificial diet (Dadd & Mittler, 1966) resulted from this important discovery.

With the difficulties which we had encountered in maintaining the major mineral constituents in solution, we were most grateful to Dr. H. T. Gordon for suggesting and providing us with the highly soluble and stable sodium EDTA complexes of the trace metals. The concentrations that gave optimal growth were 1.5 mg Fe Na EDTA (229 μg Fe^{+++}), 0.8 mg Mn Na_2 EDTA (113 μg Mn^{++}), 0.4 mg Cu Na_2 EDTA (65 μg Cu^{++}) and 2.4 mg Zn Na_2 EDTA (396 μg Zn^{++}) per 100 ml of diet (Dadd, 1967). Calcium was also incorporated, 5 mg of Ca pantothenate (420 μg Ca^{++}) per 100 ml of diet being provided in the vitamin mix.

It was rather natural to attribute our success in relation to the mineral nutrition of aphids to the use of these metal sequestrenes, particularly in view of the fact that earlier attempts to improve aphid growth by adding the 4 metals to the diet as chlorides did not enhance larval growth (Dadd & Mittler, 1965). Moreover, the suitability of EDTA metal complexes for supplying the trace mineral requirements of 11 other aphid species has been substantiated by a number of recent papers (Cress & Chada, 1971; Dadd & Krieger, 1967; Krieger, 1971. Srivastava & Auclair, 1971). In their paper of last year, Srivastava & Auclair specifically attributed their success in rearing as many as 7 successive generations of the pea aphid to the inclusion of trace amounts of Cu, Mn, Fe and Zn in sequestrene forms, as well as to some other changes in the diet. However, soon after we pointed out the necessity of supplying the trace metals, Ehrhardt (1968a,c) had comparable successes in maintaining one of these 11 aphid species (*Neomyzus circumflexus*) on artificial diets by supplying the trace minerals as chlorides; and recent tests by Akey & Beck (1971) showed that the pea aphid did better on diets to which the metals were added as chlorides rather than as EDTA, DTPA or EDDHA complexes. We too have known for some time that higher than routine levels of the EDTA sequestrenes result in poor settling and larviposition by *Myzus persicae* (Dadd, 1967), and more recently found that even at the routine level these organic complexes make the diets slightly deterrent to the aphids (Mittler & Kleinjan, 1970; Kunkel & Mittler, unpubl. obs.). However, we continued to incorporate the trace metals as EDTA sequestrenes for the same reason, I suspect, that a number of other materials are put in diets - namely, that diets containing them have been found to be satisfactory. Certain formulations tend to be perpetuated over the years, so much so that one becomes almost convinced after a while that certain ingredients are indispensable, while in fact they were put in a diet originally because of some relevant knowledge or sound reasoning rather than to meet a proven requirement. Perhaps the continued use of homoserine in pea aphid diets (e.g., by Akey & Beck, 1971) at the same high concentration (some 20% of all the dietary amino acids) originally used by Auclair & Cartier in 1963, falls into this category; particularly so since Retnakaran & Beck (1968) already in 1968 had shown that homoserine is not required by the pea aphid. Moreover, Srivastava's and Auclair's (1971) recent success in rearing 7 successive generations of the pea aphid was achieved on a diet lacking homoserine. Retnakaran & Beck found that equimolar concentrations of glutamic acid could substitute for the homoserine if the glutamic acid was dissolved together with all the dietary components, but not if it was added only finally to a solution of all the other ingredients. While the authors state that "apparently some type of complexing was involved," no indication was given of the type of interaction envisaged, although solubility and pH considerations come to mind. Nevertheless, this finding is another interesting example of how the sequence of incorporating nutrients into a diet may affect its nutritional value.

Why was it that 8 years ago when we added the trace metals as chlorides we did not get any improvement in aphid growth? Several factors might account for this failure, but one that comes to mind in retrospect was that we may have used a diet lacking ascorbic acid, namely our diet D of

1965 (Dadd & Mittler, 1965) and that the absence of this vitamin might have been rate-limiting for aphid growth. In contrast, in other studies in which chlorides have successfully been used, L-ascorbic acid has been incorporated at the 100 mg per 100 ml level which we originally recommended in 1962 (Mittler & Dadd, 1962). Ehrhardt (1968a) demonstrated a dosage response to L-ascorbic acid with diets containing the trace metals as chlorides, as we had done prior to the adoption of the EDTA complexes but using the USP No 2 salt mixture (Dadd *et al.*, 1967) which, however, only supported suboptimal growth. With both these formulations growth improvements were recorded with L-ascorbic acid concentrations up to 100 mg per 100 ml of diet, and twice that level was found to be optimal in subsequent studies in which the trace metals were incorporated as EDTA complexes (Mittler *et al.*, 1970; Fig. 3). The latter study on *M. persicae* also showed that equivalent amounts of D-araboascorbic acid were as active as L-ascorbic acid. This not only demonstrated that an epimeric specificity is lacking in this aphid—as it is in *Bombyx mori* (Ito & Arai, 1965) but not in *Prodenia litura* (Levinson & Navon, 1969)—but that considerable amounts of ascorbic acid have to be incorporated into aphid diets. The extent to which this is so to meet the metabolic requirements of the insects, to act as an antioxidant or reducing agent, and to compensate for its rapid breakdown in diets at room temperature, has yet to be ascertained. Some insights have, however, been gained from past studies on the effects of freezing and thawing diets (Dadd *et al.*, 1967) and the use of L-dehydroascorbic acid (Mittler *et al.*, 1970).

We were recently surprised to find that optimal growth by some groups of *Myzus persicae* larvae occurred with as little as 10 mg of L-ascorbic acid per 100 ml of a diet in which the trace metals were incorporated in the form of their chlorides, but not when equivalent and routine amounts of these metals were supplied as EDTA complexes. What had made the difference? It was unlikely that it was due to the fact that we had used larvae that were derived from mothers that had up to a day or two previously been feeding on plants and thus had some reserves of ascorbic acid (Dadd *et al.*, 1967; Mittler *et al.*, 1970). Although this might account for a lower dietary requirement for ascorbic acid by these larvae, it should have met the needs of the larvae equally on the two diets.

On examination of our notes regarding the formulation of the diets used in this experiment we found that in making up the ascorbic acid concentration series, appropriate amounts of ascorbic acid were weighed out individually and dissolved, either in aliquots taken from a stock solution of the EDTA trace metals, or in aliquots of a freshly made-up solution of the chlorides. From each of these mixtures an appropriate volume was added to a concentrated mixture of all the other dietary ingredients, and each of the diets individually adjusted to pH 7 with dibasic potassium phosphate. How did this procedure differ from the way we and others had previously made up diets with different levels of ascorbic acid? Sufficient details of the steps taken by others in making up diets are not available to be sure, but we may suppose that in other studies as in our previous work with concentration series of ascorbic acid (or other nutrients) that it was added to a diet after all the other ingredients had been mixed together. In the example before us, however, the ascorbic acid was added in various amounts to each of the trace mineral mixtures before they were added to the diet.

In order to test whether these procedural differences might account for the divergent results, a comparison was made of the growth of aphids on two diets identical in the amounts of nutrients mixed together, but in which one of the diets had the ascorbic acid added to a mixture of metal chlorides before it was added to the diet, whereas the other diet had the ascorbic acid added to it only after the metal chloride mixture had been incorporated in the diet. Again, it was found that, using the former procedure, 10 mg of ascorbic acid per 100 ml of diet are adequate to support optimal growth, whereas more than 100 mg of diet are needed to support comparable growth if the vitamin is added finally. Similarly, 10 mg of D-araboascorbic acid per 100 ml of diet are adequate if added to the mixture of metal chlorides before it is added to the diet. This result too is in sharp contrast with our previous data for this epimer of ascorbic acid.

Two possible roles that L-ascorbic acid or its epimer play in this situation come to mind. The first is that ascorbic acid, because of its reducing properties, effectively reduces the ferric chloride used to the ferrous state, and that it is in this reduced state that the iron might be more readily available to the insect. The complete decoloration of the chloride mix indicates that the iron does in fact get reduced. But this appears not to be the basis for the improvement, for the following reasons: 1) Considerable improvement in growth is obtained when dehydroascorbic acid (which does not have the reducing properties of ascorbic acid) interacts with the trace metal chlorides, although more than 10 mg of dehydroascorbic acid per 100 ml of diet appear to be needed for optimal growth. 2) Citric acid added to the chloride mix prior to adding this to the solution of other dietary ingredients also results in a diet that supports optimal growth, even in the absence of any ascorbic acid. Final dietary concentrations of citric acid of less than 10 mg per 100 ml are adequate. Calcium citrate (10 mg per 100 ml of diet) provides comparable citrate levels in the diet used so successfully by Akey & Beck (1971) for rearing pea aphids, but no data is as yet available as to its effect on aphid growth in the absence of ascorbic acid.

It appears, therefore, that ascorbic acid can play a crucial role in the nutrition of aphids by combining with the trace metals to keep them in a soluble and available form for these suctorial insects to ingest and subsequently to absorb. Citric acid clearly is able to perform the same function. Ascorbic acid, whether in the D, L, or dehydro form, citric acid, and EDTA all are chelating agents that are able to sequester metal cations. As the term sequester implies, these ligands interact with the metal ions to form soluble and stable complexes, thereby preventing the loss of the metals from the system—in this case through precipitation as insoluble phosphates—so that they remain available for subsequent utilization. A stable chelate not only denotes a chemically stable compoud (ascorbic acid in solution is not), but one that has fairly strong binding characteristics for the metals in question. In this regard, EDTA metal complexes have some of the highest stability constants among the chelates. This property, and probably also the relatively large size of the EDTA molecules would make the metals less readily available for intestinal absorption when complexed with EDTA than with ascorbic or citric acid. Since the metals are partitioned between different ligands according to a hierarchy of stability or mass-action equilibrium constants, it is tempting to speculate that the high dietary levels of ascorbic acid are needed in the presence of EDTA to compete for the metals and make them more readily available to the aphids. Comparable concentrations of citric acid do not, however, improve the performance of the insects on diets with EDTA mineral formulations, presumably because of the relatively lower stability constant of the citric acid complexes.

The deleterious effect on the growth of pea aphids of dietary histidine (200 mg per 100 ml of diet; cf. 80 mg per 100 ml used for *M. persicae*) was speculatively linked to the formation by this amino acid of stable chelates with iron and other trace metals, and, by inference, to a mineral deprivation of the aphids (Markkula & Laurema, 1967). The contrary could however be true, since histidine may be a most efficient vehicle for the absorption of metals. Even on normal plant diets, aphids deposit carbonates and phosphates of magnesium and calcium in mid gut cells (Ehrhardt, 1965). While this indicates that aphids have a mechanism for coping with some metals when these are ingested and absorbed in excessive amounts, they may not be able to do so with others. Could it be that the dietary requirement by *M. persicae* for histidine (Dadd & Krieger, 1967) rests, in part at least, on just this property of the amino acid to form these complexes (Provasoli & Pinter, 1960) with minerals—thereby increasing rather than decreasing their availability.

Other dietary amino acids could also play a role in metal transport, as may some of the ketose-amino acids that are formed when sugars interact with amino acids, particularly upon heating. When we dissolve the dietary amino acids and sucrose in hot or even boiling water, do we inadvertently enhance the production of such molecular species? Since fructosyl or glucosyl-amino acids can function as iron transport factors for bacteria (Demain & Hendlin, 1959), similar complexes may be of importance also in making minerals available to the symbiotic microorganisms of aphids, thereby effecting the overall nutrition of the insects.

For interesting discussions of the role of chelators in the mineral nutrition and metabolism of bacteria, algae, and higher animals and plants, see Albert (1965), Hopping & Ruliffson (1966), Kratzer (1965), Pollack *et al.* (1970), Provasoli & Pinter (1960), and Scott (1965).

With regard to the adverse effects of dietary riboflavin reported by Ehrhardt (1968a, 1969) and Markkula & Laurema (1967), it was suggested by these authors that insoluble complexes are formed between this vitamin and several of the trace elements, especially iron, thereby depriving the aphids and their symbiotics of these minerals. The fact that *Neomyzus circumflexus* did not require dietary riboflavin was ascribed by Ehrhardt (1968a) to the aphid's symbiotes, which presumably could synthesize this vitamin (and others) when dietary trace minerals were available, but that this function was impaired as a result of a riboflavin-induced iron deficiency. No direct evidence of the supposed interaction between dietary riboflavin and iron is yet available to support Ehrhardt's contention. There is no doubt, however, that aphids can rely on their symbiotes to synthesize a number of nutrients, e.g., essential amino acids and sterols (Dadd & Mittler, 1966; Dadd & Krieger, 1968; Ehrhardt, 1968a,b).

That the symbiotes of an insect can be sensitive indicators of the mineral nutrition of their host is a phenomenon that was pointed out some time ago by Brooks (1960) in relation to the crucial role of zinc and manganese in the transovarial passage of cockroach symbiotes. Evidently the symbiotes of aphids are affected by the mineral intake of their hosts, since the mycetocytes of *M. persicae* reared for more than a generation on diets deficient in trace metals were found to aggregate into a dense, dark-green cluster in the aphid's abdomen (Dadd, 1968), and Ehrhardt (1966, 1968c) found that the microorganisms themselves degenerate in aphids maintained on a diet lacking these minerals.

The thought of having discovered an important new nutritional role for ascorbic acid made me quite jubilant, until I dug into the vertebrate nutrition literature and found that ascorbic acid has been known for over 30 years to enhance iron absorption in rats and man. That it did so by virtue of its chelating properties, however, has only been elucidated in recent years (Conrad & Schade, 1968). Today, numerous chelating agents are known and in use for enhancing the availability of trace minerals for animals and plants. Let us not neglect their usefulness for insect nutritional purposes.

Lest someone leave this meeting with the impression that ascorbic acid is not needed by aphids, let me say that we have evidence that these insects do have a specific requirement for this vitamin, no doubt for a variety of metabolic purposes. My intention here has been to point up the *important role that ascorbic acid may play in mineral nutrition of insects, and the awareness that one should have or develop for the steps taken in making up diets.* Until we have this awareness, some aspects of diet formulation will remain a mysterious art, only achieved by those who inadvertently or intuitively do things in a certain way—often without realizing the significance of their actions and hence without describing them adequately in the literature.

Acknowledgements—The technical assistance of Miss Joyce E. Kleinjan, the advice and collaboration of Drs. R. H. Dadd, H. T. Gordon and Dr. J. A. Tsitsipis, and the receipt of a Biomedical Sciences Grant (FR 7006) in partial support of these studies, are gratefully acknowledged.

SUMMARY

Some problems encountered in making up and storing artificial diets are discussed. Special consideration is given to non-particulate liquid diets suitable for rearing aphids. A number of organic chelating agents are able to sequester trace minerals and make them available to these insects. L-ascorbic acid can function as a ligand for the trace metals, provided it is allowed to interact with them prior to their addition to the other dietary components. This example stresses the importance of the sequence in which dietary components are combined.

REFERENCES

ALBERT, A. (1965). *Selective Toxicity*. Methuen & Co., London (3d Edn.), 394 pp.

AKEY, D. H. and BECK, S. D. (1971). Continuous rearing of the pea aphid, *Acyrthosiphon pisum*, on a holodic diet. *Ann. Ent. Soc. Amer.* **64**:353-356.

AKOV, S. (1962). A qualitative and quantitative study of the nutritional requirements of *Aedes aegypti* L. larvae. *J. Insect Physiol.* **8**:319-335.

AUCLAIR, J. L. and CARTIER, J. J. (1963). Pea aphid: rearing on a chemically defined diet. *Science* **142**:1068-1069.

BROOKS, M. A. (1960. Some dietary factors that effect ovarial transmissions of symbiotes. *Proc. Helminth. Soc., Wash.* **27**:212-220.

CONRAD, M. E. and SCHADE, S. G. (1968). Ascorbic acid chelates in iron absorption: a role for hydrochloric acid and bile. *Gastroenterology* **55**:35-45.

COOKE, J. and SANG, J. H. (1970). Utilization of sterols by larvae of *Drosophila melanogaster*. *J. Insect Physiol.* **16**:801-812.

CRESS, D. C. and CHADA, H. L. (1971). Development of a synthetic diet for the greenbug, *Schizaphis graminum*. 2. Greenbug development as affected by zinc, iron, manganese, and copper. *Ann. Ent. Soc. Amer.* **64**:1240-1244.

DADD, R. H. (1964). A study of carbohydrate and lipid nutrition in the wax moth,

Galleria mellonella (L.), using partially synthetic diets. *J. Insect Physiol.* 10:161-178.

____________(1967). Improvement of synthetic diet for the aphid *Myzus persicae* using plant juices, nucleic acids, or trace metals. *J. Insect Physiol.* 13:763-778.

____________(1968). Problems connected with inorganic components of aqueous diets. *Bull. Ent. Soc. Amer.,* 14:22-26. Proceedings of Ent. Soc. of Amer. Symposium on Artificial Diets—Current Trends and Challenges, Portland, 1966.

DADD, R. H. and KRIEGER, D. L. (1967). Continuous rearing of the *Aphis fabae* complex on sterile synthetic diet. *J. Econ. Ent.* 60:1512-1514.

DADD, R. H. and KRIEGER, D. L. (1968). Dietary amino acid requirements of the aphid, *Myzus persicae. J. Insect Physiol.* 14:741-764.

DADD, R. H., KRIEGER, D. L. and MITTLER, T. E. (1967). Studies on the artificial feeding of the aphid *Myzus persicae* (Sulzer) -- IV. Requirements for water-soluble vitamins and ascorbic acid. *J. Insect Physiol.* 13:249-272.

DADD, R. H. and MITTLER, T. E. (1965). Studies on the artificial feeding of the aphid *Myzus persicae* (Sulzer) -- III. Some major nutritional requirements. *J. Insect Physiol.* 11:717-743.

DADD, R. H. and MITTLER, T. E. (1966). Permanent culture of an aphid on a totally synthetic diet. *Experientia* 22:832.

DEMAIN, A. L. and HENDLIN, D. (1959). Iron transport compounds as growth stimulators for *Microbacterium* sp. *J. Gen. Microbiol.* 21:72-79.

EHRHARDT, P. (1965). Speicherung anorganischer Substanzen in den Mitteldarmzellen von *Aphis fabae* Scop. und ihre Bedeutung für die Ernährung. *Z. vergl. Physiol.* 50:293-312.

____________(1966). Entwicklung und Symbionten geflügelter und ungeflügelter Virgines von *Aphis fabae* Scop. unter dem Einfluss künstlicher Ernährung. *Z. Morph. Okol. Tiere* 57:295-319.

____________(1968a). Der Vitaminbedarf einer siebröhrensaugenden Aphide, *Neomyzus circumflexus* Buckt. (Homoptera, Insecta). *Z. Vergl. Physiol.* 60:416-426.

____________(1968b). Nachweis einer durch symbiotische Mikroorganismem bewirkten Sterinsynthese in künstlich ernähren Aphiden (homoptera, Rhynchota, Insecta). *Experientia* 24:82.

____________(1968c). Die Wirkung verschiedener Spurenelemente auf Wachstum, Reproduktion und Symbionten von *Neomyzus circumflexus* Buckt. (Aphidae, Homoptera, Insecta) bei künstlicher Ernährung. *Z. vergl. Physiol.* 58:47-75.

____________(1969). Die Rolle von Methionin, Cystein, Cystin und Sulfat bei der künstlichen Ernährung von *Neomyzus (Aulacorthum) circumflexus* Buckt. (Aphidae, Homoptera, Insecta). *Biol. Zbl.* 88:335-348.

FORREST, J. M. S. and KNIGHTS, B. A. (1972). Presence of phytosterols in the food of the aphid, *Myzus persicae. J. Insect Physiol.* 18:723-728.

FRAENKEL, G. and BLEWETT, M. (1946). Linoleic acid, vitamin E and other fat-soluble substances in the nutrition of certain insects, *Ephestia kuehniella, E. elutella, E. cautella* and *Plodia interpunctella* (Lep.). *J. Exp. Biol.* 22:172-190.

GORDON, H. T. (1959). Minimal nutritional requirements of the German roach, *Blattella germanica* L. *Ann. N. Y. Acad. Sci.* 77:290-351.

GORTNER, R. A. (1950). *Outlines of Biochemistry*. III Edn. R. A. Gortner, Jr., Ed. J. Wiley & Sons, N. Y.

GRANT, N. H. and ALBURN, H. E. (1965). Fast reactions of ascorbic acid and hydrogen peroxide in ice, a presumptive early environment. *Science* 150:1589-1590.

HOPPING, J. M. and RULIFFSON, W. D. (1966). Roles of citric and ascorbic acids in enteric iron absorption in rats. *Amer. J. of Physiol.* **210**:1316-1320.

HORIE, Y., WATANABE, K. and ITO, T. (1966). Nutrition of the silkworm *Bombyx mori* -- XIV. Further studies on the requirements for B vitamins. *Bull. Ser. Exp. Sta., Tokyo,* **20**:393-409.

ITO, T. and ARAI, N. (1965). Nutrition of the silkworm, *Bombyx mori* -- IX. Further studies on the nutritive effects of ascorbic acid. *Bull. Seric. Exp. Sta., Tokyo,* **20**:1-19.

KRATZER, F. H. (1965). Chelation–its influence on nutrient utilization. *Feedstuffs* 37:pp 62,64,72.

KRIEGER, D. L. (1971). Rearing several aphid species on synthetic diet. *Ann. Ent. Soc. Amer.* **64**:1176-1177.

LEVINSON, H. Z. and NAVON, A. (1969). Ascorbic acid and unsaturated fatty acids in the nutrition of the Egyptian cotton leafworm, *Prodenia litura. J. Insect Physiol.* **15**:591-595.

MARKKULA, M. and LAUREMA, S. (1967). The effect of amino acids, vitamins, and trace elements on the development of *Acyrthosiphon pisum* Harris (Hom., Aphididae). *Ann. Agr. Fenn.* **6**:77-80.

MILES, P. W., McLEAN, D. L. and KINSEY, M. G. (1964). Evidence that two species of aphid ingest food through an open stylet sheath. *Experientia* **20**:582.

MITTLER, T. E. and DADD, R. H. (1962). Artificial feeding and rearing of the aphid, *Myzus persicae* (Sulzer), on a completely defined synthetic diet. *Nature* **195**:404.

MITTLER, T. E. and KLEINJAN, J. E. (1970). Effect of artificial diet composition on wing-production by the aphid *Myzus persicae. J. Insect Physiol.* **16**:833-850.

MITTLER, T. E., TSITSIPIS, J. A. and KLEINJAN, J. E. (1970). Utilization of dehydroascorbic acid and some related compounds by the aphid *Myzus persicae* feeding on an improved diet. *J. Insect Physiol.* **16**:2315-2326.

POLLACK, J. R., AMES, B. N. and NEILANDS, J. B. (1970). Iron transport in *Salmonella typhimurium*: Mutants blocked in the biosynthesis of enterobactin. *J. Bacter.* **104**:635-639.

PROVASOLI, L. and PINTER, I. J. (1960). *Artificial media for fresh-water algae: problems and suggestions.* Pp 84-96 in *"The ecology of algae" (Pymatuning symposia in ecology,* Spec. Pubn. No. 2) Ed. by C. A. Tyron & R. T. Hartman, Univ. of Pittsburgh Press.

RETNAKARAN, A. and BECK, S. D. (1968). Amino acid requirements and sulfur amino acid metabolism in the pea aphid, *Acyrthosiphon pisum* (Harris). *Comp. Biochem. Physiol.* **24**:611-619.

SCOTT, M. L. (1965). Chelates in animal nutrition. *Proc. 1965 Maryland Nutr. Conf. for Feed Manufacturers* 54-60.

SRIVASTAVA, P. N. and AUCLAIR, J. L. (1971). An improved chemically defined diet for the pea aphid, *Acyrthosiphon pisum. Ann. Ent. Soc. Amer.* **64**:474-478.

THOMPSON, L. U. and FENNEMA, O. (1971). Effect of freezing on oxidation of L-ascorbic acid. *J. Agr. Food Chem.* **19**:121-124.

CHEMICAL FEEDING REQUIREMENTS OF OLIGOPHAGOUS INSECTS

Ting H. Hsiao
Department of Zoology
Utah State University, Logan, Utah 84321

INTRODUCTION

In nutritional studies with insects, the matter of phagostimulation and diet acceptance is not a crucial problem when the tested organisms can be reared on artificial diets of physical and chemical characteristics differing from those of the natural food. Among phytophagous insects, however, especially the monophagous and oligophagous species, feeding requirements are often highly specific and many species cannot be induced to feed readily on artificial diets. For these species, an understanding of these mechanisms of feeding behaviour could enhance successful formulation of artificial diets and facilitate nutritional studies.

Feeding behavior of phytophagous insects has been discussed in several recent reviews (Beck, 1965; Dadd, 1970; Dethier, 1966, 1970; Schoonhoven, 1968). In essence, the process of feeding involves a cyclic sequence of four stereotyped behavioral patterns: host plant orientation and recognition, initiation of biting and feeding, maintenance of feeding, and cessation of feeding, usually followed by dispersal via locomotor activity. Each of these behavioral patterns is manifested only when proper external stimuli (plant characteristics) are combined with internal responses (physiological and behavioral) of the insects. Plant characteristics influence the initiation and completion of these behavioral patterns by providing or not providing the stimuli required for one or more of their components.

Plant chemicals are of major importance in regulating feeding behavior of phytophagous insects. According to their influence on different behavioral components of feeding, chemicals are classified as positive or negative stimuli. Stimuli that elicit positive responses are termed "attractants", "arrestants", "incitants" and "stimulants." Stimuli that evoke negative responses are "repellents" and "deterrents" (Beck, 1965). "Attractants", "arrestants" and "repellents" are olfactory stimuli that affect host plant finding and recognition. "Incitants", "stimulants" and "deterrents" are gustatory stimuli that affect the insects' biting, feeding and maintenance of feeding. The existence of different classes of stimuli illustrates the complicated relationships between plant chemicals and phytophagous insects.

The roles of negative stimuli, repellents and deterrents, in limiting the host ranges of phytophagous insects are well recognized. But the relative importance of different classes of positive stimuli in regulating feeding

Insect and mite nutrition – North-Holland – Amsterdam, (1972)

behavior has been more difficult to assess. Results vary with species and often are influenced by associated environmental conditions. For example, the needs for olfactory stimuli as behavioral releasers diminish when insects are directly placed on their food. Among phytophagous insects, the polyphagous species are less rigid in their requirements for positive stimuli. Many species, when starved, will feed on neutral substrates or diets consisting only of inert nutrients. Some oligophagous insects, in addition to a high degree of sensitivity to negative stimuli, have rigid requirements for several classes of positive stimuli in regulating feeding behavior. These insects provide excellent examples to illustrate the unique chemical feeding requirements of insects.

General aspects of phagostimulation in relation to artificial diets of insects have been discussed by Davis (1968). The scope of this presentation will be restricted to chemical feeding requirements of oligophagous plant feeders. Gustatory stimuli will be emphasized in the discussion, since these stimuli are primarily responsible for regulating insect feeding on artificial diets. The general term "Phagostimulant" will be used in this paper to designate any gustatory stimulus that elicits a feeding response. Some recent findings on the roles of phagostimulants as dietary requirements for the Colorado potato beetle, *Leptinotarsa decemlineata*, are included to illustrate specific points.

THE ROLES OF PLANT CHEMICALS AS PHAGOSTIMULANTS FOR OLIGOPHAGOUS INSECTS

Plant chemicals that serve as phagostimulants for phytophagous insects are conventionally classified into secondary plant substances and primary nutrient chemicals. This division is somewhat arbitrary as pointed out by Herout (1970), but does serve to stress the different roles these chemicals play in regulating feeding behavior of oligophagous insects.

The importance of secondary plant chemicals in host selection by phytophagous insects has been discussed by Fraenkel (1969). The notion that secondary plant substances serve as repellents and deterrents to insects in general is widely accepted. Secondary plant substances also provide important phagostimulants for oligophagous insects. In many instances, these compounds occur sporadically in plants and are essential to elicit proper feeding. The term "token factor or stimulant" has been used to designate this class of phagostimulants, since their presence signals the particular food plants for some monophagous or oligophagous insects. The mustard oil glycosides are classical examples of token stimulants, and have now been shown to serve as phagostimulants for a number of Cruciferae feeders including the diamondback moth, *Plutella maculipennis* (Nayar and

Thorsteinson, 1963); the cabbage butterflies, *Pieris brassicae* and *P. rapae* (David and Gardiner, 1966; Schoonhoven, 1969); the mustard weevil, *Phaedon cochleariae* (Tanton, 1965); the vegetable weevil, *Listroderes obliquus* (Matsumoto, 1970); the flea beetles, *Phyllotreta cruciferae* and *P. striolata* (Feeny *et al.*, 1970) and the aphid, *Brevicoryne brassicae* (Wensler, 1962; Wearing, 1968). Other examples of token stimulants for oligophagous insects include the essential oils (methyl chavicol, carvone etc.) of Umbelliferae for the swallowtail butterfly, *Papilio polyxenes* (Dethier, 1941); cucurbitacins of Cucurbitaceae for the spotted cucumber beetle, *Diabrotica undecimpunctata* (Chambliss and Jones, 1966); catalposides of Bignoniaceae for the catalpa sphinx moth, *Ceratomia catalpae* (Nayar and Fraenkel, 1963a), the cyanogenetic glycosides, phaseolunatin and lotaustrin, of Leguminosae for the mexican bean beetle, *Epilachna varivestris* (Nayar and Fraenkel, 1963b); hypericin of Hypericaceae for *Chrysomela brunsvicensis* (Rees, 1969); and the alkaloid, sparteine, of Leguminosae for an aphid, *Acyrthosiphon spartii* (Smith, 1966). Morin and isoquercitrin of mulberry are biting factors for larvae of the silkworm, *Bombyx mori* (Hamamura, 1970). Several secondary plant substances that are widely distributed in plants have also been shown to serve as phagostimulants for some oligophagous insects, but because of their general distribution, these compounds are not considered as token stimulants. Examples are chlorogenic acid for the Colorado potato beetle, *Leptinotarsa decemlineata* (Hsiao and Fraenkel, 1968a); and shikimic acid and caffeic acid for the spruce budworm, *Choristoneura fumiferana* (Heron, 1965).

Specific chemoreceptors that are sensitive to token stimulants have been demonstrated in a few instances. For example, *Pieris brassicae* has two sensory cells in the maxillae that are sensitive to mustard oil glycosides (Schoonhoven, 1969). One cell in the tarsi of *Chrysomela brunsvicensis* is stimulated by hypericin (Rees, 1969). In *Leptinotarsa decemlineata*, a cell in the maxillae has recently been demonstrated to be sensitive to chlorogenic acid (Schoonhoven, personal communication). These evidences further indicate the importance of secondary plant substances as phagostimulants for oligophagous insects.

Primary nutrient chemicals have now been shown to be important phagostimulants for many oligophagous as well as polyphagous insects. The list of phagostimulative nutrients includes sugars, amino acids, steroids, fatty acids, phospholipids, B-vitamins, ascorbic acid, purines and related substances, and inorganic salts (see reviews of Thorsteinson, 1960; Davis, 1968; Dadd, 1970 for references). Chemoreceptors sensitive to sugars, amino acids, salts and inositol have been recorded in several phytophagous insects (Schoonhoven, 1968, 1969; Ishikawa *et al.*, 1969).

Despite the large number of primary nutrient chemicals present in each plant, most oligophagous insects can respond to only a few compounds.

Phagostimulative nutrients that elicit feeding in one species do not necessarily elicit feeding in other species. For example, amino acids are major phagostimulants for the Colorado potato beetle, but do not affect the alfalfa weevil, *Hypera postica* (Hsiao, 1969a). β-sitosterol is a biting factor for silkworm, but has no such effect on the Colorado potato beetle or the alfalfa weevil (Hsiao, 1969a). Adenine and related substances are specific phagostimulants for the alfalfa weevil (Hsiao, 1969b) and the sweet clover weevil, *Sitona cylindricollis* (Beland, G. L., personal communication), but have no stimulative effects on the Colorado potato beetle (Hsiao, 1969b). Specificity to nutrients can be observed even within a single class of compounds. For example, among 26 sugars and related compounds tested for phagostimulative effects on the Colorado potato beetle, only sucrose elicited feeding (Hsiao and Fraenkel, 1968b). Similarly, only a few aliphatic amino acids were phagostimulative among 27 compounds tested. Chemosensory specificity to nutrient chemicals by oligophagous insects demonstrates the important roles of these compounds in the feeding behavior of phytophagous insects.

Increasing evidence indicates that many phagostimulants are more effective in combination. For example, sugars and amino acids have additive effects on feeding maintenance in the European corn borer, *Ostrinia nubilalis* (Beck and Hanec, 1958) and the grasshoppers, *Melanoplus bivitattus* and *Camnula pellucida* (Thorsteinson, 1960). Davis (1965) found that mixtures of amino acids were more effective than individual amino acids in eliciting feeding of the prairie grain wireworm, *Ctenicera destructor*. Gothilf and Beck (1967) showed that the addition of potassium salts to phospholipids enhanced feeding of the cabbage looper, *Trichoplusia ni*. Hsiao and Fraenkel (1968b) demonstrated additive and synergistic effects between sucrose, amino acids and potassium salts in the feeding of the Colorado potato beetle. Synergistic interactions of phagostimulants at the chemoreceptor level have been reported in the silkworm larvae (Ishikawa *et al.*, 1969). Several studies have shown that a feeding response to secondary plant substances can be facilitated by a nutrient phagostimulant. For instance, the response of diamondback moth larvae to mustard oil glycosides was enhanced by the presence of glucose (Nayar and Thorsteinson, 1963). Silkworm larvae responded to morin, isoquercitrin and β-sitosterol only in the presence of sugars. Most phagostimulants in any green plant are present in relatively low concentrations. Additive and synergistic interactions among such phagostimulants would render a plant more acceptable to insects.

INTERACTIONS OF PHAGOSTIMULANTS IN FEEDING BEHAVIOR OF THE COLORADO POTATO BEETLE

The mechanisms of host selection and feeding behavior of the Colorado potato beetle have been extensively investigated (see review of Hsiao, 1969a). Phagostimulants are essential in regulating the feeding behavior of this species. Among secondary plant chemicals, a phenolic flavonoid is a sign (token) stimulant, while chlorogenic acid is an important phagostimulant (Hsiao and Fraenkel, 1968a). The roles of nutrient chemicals as phagostimulants for this species have been investigated by Hsiao and Fraenkel (1968b). The list of phagostimulative nutrients includes one sugar (sucrose), several amino acids (L-alanine, γ-aminobutyric acid, L-serine, etc), and three phospholipids (lecithin, phosphatidyl L-serine and inositol phosphatide). Inorganic salts (KC1, KH_2PO_4 and NaC1) were inactive alone but acted synergistically with phagostimulants. Further, when the phagostimulants were combined, their additive effects enhanced feeding response of the larvae.

A series of experiments was recently conducted to determine the nature of phagostimulant interactions and their importance in the feeding behavior of this species. For these experiments, a phagostimulant mixture (Table 1) consisting of sucrose, L-alanine, γ-aminobutyric acid, lecithin and KC1 was prepared. The components of this mixture were incorporated into an agar-cellulose medium, and the bioassay was carried out by the method described by Hsiao and Fraenkel (1968b). Fourth instar larvae of 80-100 mg body weight were selected from the culture, starved for 4-6 hours, and then allowed to feed on test media individually for 12 hours. The average number of fecal pellets produced by 20 larvae was the criterion of feeding response.

Table 1. Composition of a phagostimulant mixture for the Colorado potato beetle.

Compound	Molar concentration	% in diet
Sucrose	0.01	0.342
L-alanine	0.005	0.045
γ-aminobutyric acid	0.005	0.053
Lecithin, vegetable	0.25%	0.250
KC1	0.04	0.298
Total		0.988

Since the agar-cellulose medium alone did not elicit feeding, this bioassay was believed to provide an accurate measurement of feeding response of the larvae.

Our first experiment was conducted to determine the relative effects of single vs. mixed phagostimulants on feeding responses of larvae (see Fig. 1, upper graph). Sucrose, alanine and γ-aminobutyric acid, when tested singly, elicited moderate feeding responses. Lecithin and KC1 were non-stimulative at the concentrations tested. However, when two, three, four or five of these phagostimulants were grouped into mixtures, a gradual increase of feeding response was observed, indicating additive and synergistic effects. In comparing the larval responses to different proportions of phagostimulants in mixtures, it was found that additive and synergistic effects were maximized when low concentrations of each phagostimulant were used. When high concentrations of one or more phagostimulants were included in a mixture, the additive and synergistic interactions were reduced or diminished. This result indicates that the concentrations of various phagostimulants in a mixture must be properly balanced to achieve optimal synergisms in feeding activity.

The relative importance of individual phagostimulants in regulating the feeding response of potato beetle larvae was investigated by the deletion method. The removal of any single phagostimulant from the mixture did not reduce the feeding response significantly (Fig. 1, lower graph). Apparently none of the phagostimulants plays a decisive role in regulating feeding, but each compound provides some phagostimulation toward the total feeding response observed.

Feeding responses of the larvae to different concentrations of the phagostimulant mixture (Table 1) were investigated to determine the optimal level of phagostimulation. Various concentrations of an alcohol extract of the potato leaf powder were included in the study for comparison (Fig. 2). A gradual increase of feeding response was evident as increased concentrations of the phagostimulant mixture were added to the tested medium. No drastic decrease of feeding response was observed even when the amount of the mixture was increased to 6%.

The alcohol extract of the potato leaf powder has previously been shown to contain most of the phagostimulants of the potato plant (Hsiao and Fraenkel, 1968a). Larval response to this extract was greater than to the phagostimulant mixture, especially at the lower concentrations tested (Fig. 2). At 0.5%, the alcohol extract elicited more than twice the amount of feeding generated by the phagostimulant mixture. This difference was reduced when higher concentrations of two mixtures were compared. The optimal feeding response was obtained with a 3% level of the alcohol extract. A noticeable reduction in feeding occurred with higher concentrations of the

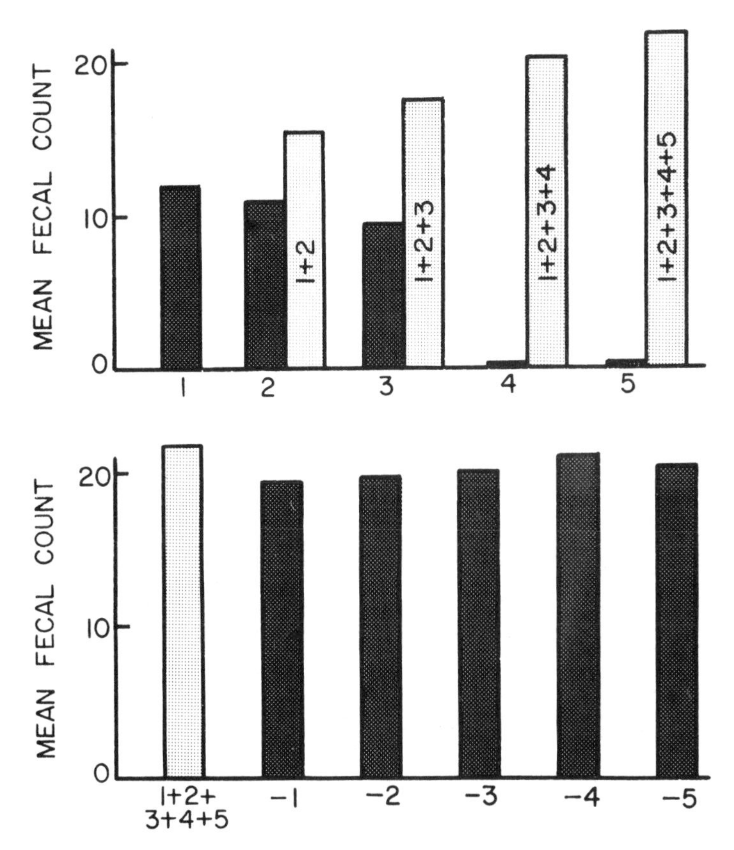

Figure 1. Feeding responses of potato beetle larvae to agar-cellulose media with one or more phagostimulants (upper graph) and different combinations of phagostimulants (lower graph). 1 = 0.01 M sucrose; 2 = 0.005 M L-alanine; 3 = 0.005 M γ-aminobutyric acid; 4 = 0.25% lecithin; 5 = 0.04 M KC1.

extract. The effectiveness of the alcohol extract as compared to the phagostimulant mixture is probably due to more phagostimulants being contained in the extract, for all of the components of the phagostimulant mixture, with the exception of lecithin, were present in it (Hsiao and Fraenkel, 1968a). In addition, the alcohol extract contains phagostimulative secondary plant substances, as mentioned earlier. Since the nutrient phagostimulants, alone, can provide substantial feeding response as compared to the performance of the alcohol extract, no token stimulants are required for feeding and maintenance of feeding behavior patterns in the potato beetle.

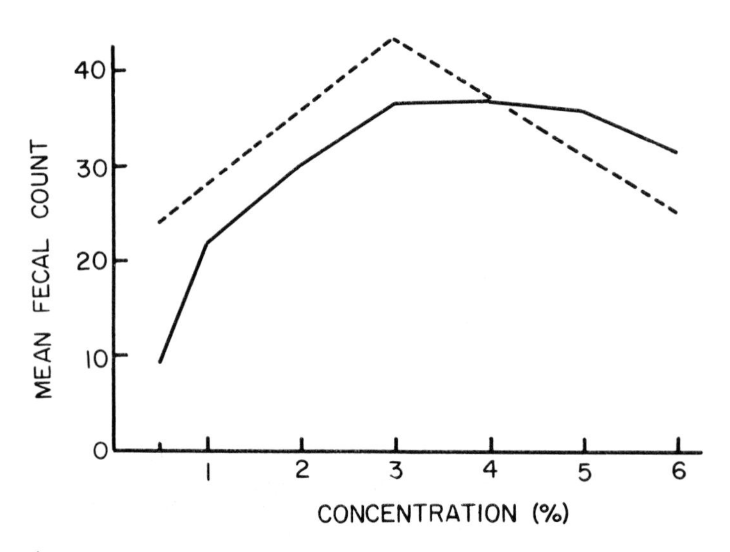

Figure 2. Feeding responses of potato beetle larvae to agar-cellulose media with different concentrations of a phagostimulant mixture (solid line) and the alcohol extract of potato leaf powder (dotted line).

THE ROLE OF PHAGOSTIMULANTS IN ARTIFICIAL DIETS FOR THE COLORADO POTATO BEETLE

The role of phagostimulants is rarely considered in the formulation of artificial diets for phytophagous insects. Consequently, very few studies have been carried out to analyze their significance. The most extensive work has been done with silkworm larvae (Hamamura, 1970; Ishikawa *et al.*, 1969). Several phagostimulants, including secondary and nutrient chemicals, were included in artificial diets fed to this species and were shown to promote feeding and growth. David and Gardiner (1966) reported that the mustard oil glycosides, although not essential for growth of the large white butterfly in artificial diets, would enhance feeding and growth. In his investigation of artificial diets for the Colorado potato beetle, Hsiao (1966) demonstrated the importance of phagostimulants in the diet. For example, replacing sucrose by either glucose or fructose, both non-phagostimulative nutrients drastically reduced larval feeding and weight gain. Hsiao and Fraenkel (1968c) found that potato leaf powder, potato leaf homogenate, or an alcohol extract of the potato leaf powder would improve larval feeding and growth on artificial diets. Wardojo (1969a), while studying artificial diets for the Colorado potato beetle, confirmed the important role of sucrose as a dietary component. His artificial diets for the potato beetle also included several other known phagostimulants to improve feeding and growth of the larvae.

In studying interactions of phagostimulants (described in the previous section), the effects of the test phagostimulant mixture (Table 1) in an artificial diet were investigated. For this investigation a basic diet similar in composition to one used by Hsiao and Fraenkel (1968c) was adopted (Table 2). This diet contains all the essential nutrients for growth, except that the phagostimulative nutrients are replaced or deleted to facilitate evaluation of their roles in feeding and growth. The preparation of the artificial diet and the method of bioassay were the same as described by Hsiao and Fraenkel (1968c). Newly molted 4th instar larvae having body weights of 40-50 mg were used in the experiment. Individual larvae were allowed to feed on the tested diet and weight gains were recorded at intervals of 24 and 48 hours. The average weight gain of 25 larvae was used to evaluate the growth performance of each diet tested.

Table 2. Composition of a basic diet for growth experiments with potato beetle larvae.

Constituents	Grams per 100 ml diet
Casein, vitamin free	4.00
Glucose	4.00
Cellulose	4.00
Agar, bacto	3.00
Corn oil	0.50
Salt mixture[a]	0.50
Ascorbic acid	0.10
Choline chloride	0.05
Cholesterol	0.05
B-vitamin mixture[b]	
Potassium hydroxide	0.15

[a]Beck *et al.* (1968).
[b]Contains 1.0 mg niacin, 1.0 mg calcium pantothenate, 0.24 mg thiamine HCl, 0.50 mg riboflavin, 0.25 mg pyridoxine HCl, 0.25 mg folic acid, 0.02 mg biotin, 20 mg meso-inositol, 0.002 mg vitamin B_{12}.

The upper graph in Fig. 3 shows that the basic diet alone did not elicit feeding, and larvae on that diet actually lost weight during the period of the experiment. When the phagostimulant mixture was added to the basic diet, larval feeding and growth were observed. The amounts of weight gained were proportional to the concentrations of the phagostimulant mixture present in the diet. The greatest larval weight gain was recorded when the phagostimulant mixture constituted 3-4% of the total diet. At higher concentrations, a slight decrease in weight gain was observed. Addition of the alcohol extract of the potato leaf powder generated a greater larval weight gain than had the phagostimulant mixture (Fig. 3, lower graph). With the alcohol extract, the optimal weight gain was reached at 4% concentration. At 5 and 6% extracts, a substantial reduction in weight gain occurred, indicating that the alcohol extract had a deterrent effect on feeding.

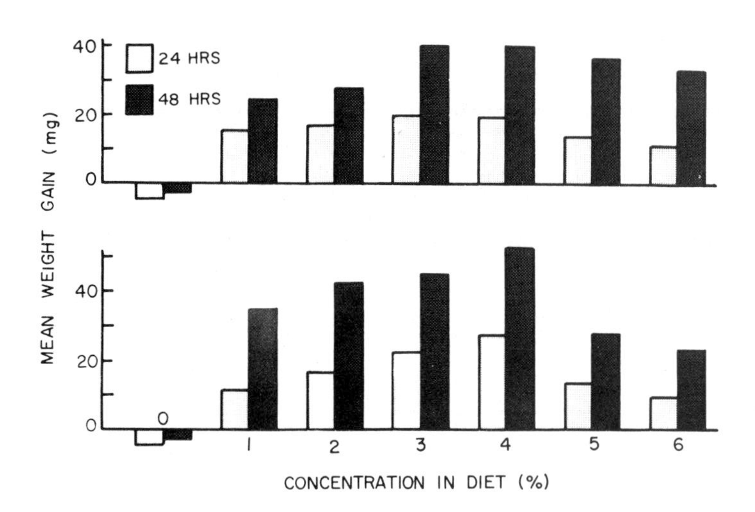

Figure 3. Growth of potato beetle larvae fed on artificial diets with different concentrations of a phagostimulant mixture (upper graph) and the alcohol extract of potato leaf powder (lower graph).

In comparing the results of feeding tests (Fig. 2) and growth experiments (Fig. 3), it is clear that the performance of the basic diet depends upon the relative amounts of phagostimulants that are incorporated. Since most of the phagostimulants used in the artificial diet are also nutrients, it must be assumed that these compounds serve a dual function, although it is necessary to distinguish whether or not they are essential or non-essential nutrients. For the Colorado potato beetle, none of the nutrient phagostimulants has been demonstrated to be essential for larval growth (Hsiao, 1966;

Wardojo, 1969 a & b). Therefore, for this species, their roles as nutrients are likely to be limited to a general improvement of larval feeding and growth. The rather high concentrations of these compounds that are needed to elicit adequate feeding and growth tend to support this view.

The feasibility of improving the ability of the artificial diet to support growth of the potato beetle was investigated by adding plant materials to the basic diet containing 3% of the phagostimulant mixture. Potato leaf powder and an alcohol extract of the leaf powder were incorporated at 1 to 6% concentrations. Larval weight gains on these diets were compared. Diet combinations including either 5% leaf powder or 1% alcohol extract provided the optimal weight gain of the larvae. These findings, along with the data on the weight gain of larvae fed on potato leaves are presented in Fig. 4. Larval feeding and growth were improved by the addition of the alcohol extract. An even greater increase in larval weight gain was obtained with the potato leaf powder. This improvement in growth is probably due to the phagostimulants as well as the nutrients present in the plant materials.

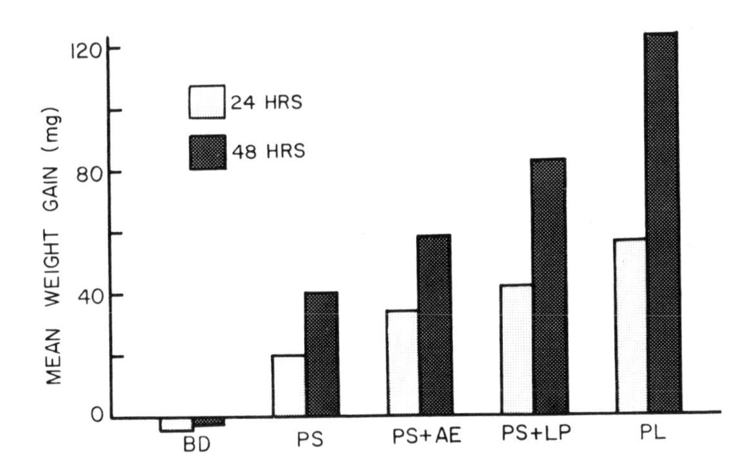

Figure 4. Growth of potato beetle larvae fed on artificial diets with additions of a phagostimulant mixture, the mixture plus alcohol extract, the mixture plus potato leaf powder, or fresh potato leaf. BD = basic diet; PS = phagostimulant mixture (3%); AE = alcohol extract (1%); LP = potato leaf powder (5%); PL = fresh potato leaf.

Larval growth on the best artificial diets, however, was still suboptimal as compared to growth on potato leaves (Fig. 4). This suggests that other feeding requirements have not been met in the artificial diet. Several factors that must be considered are the possible inadequacy of physical feeding requirements provided, the likelihood of insufficient amounts of non-essential nutrients, or any imbalance of nutrients in the diet. Potato beetle larvae, when reared on non-host plants, have required a considerable period of feeding adaptation (Hsiao and Fraenkel, 1968d), a phenomenon also seen in larvae fed on artificial diets. However, a shorter period of feeding adaptation was noted when potato leaf powder or extracts were added to the artificial diets, indicating that phagostimulants from the host plant are important to elicit optimal feeding. The fact that the potato beetle can be reared on artificial diets without plant materials (Wardojo, 1969 a & b) indicates that nutritional requirements of this species do not differ greatly from other phytophagous insects. However, the suboptimal growth of potato beetle larvae fed on artificial diets without plant materials points to the inadequacy of phagostimulation supplied by the diets. Evidence from the present study clearly demonstrates that the level of phagostimulation produced by nutrient chemicals is insufficient to elicit the optimal feeding response of the larvae, for which phagostimulants from the potato plant are vital. The nature of these host specific phagostimulants is currently being investigated.

CONCLUSION

The roles of phagostimulants in regulating feeding behavior vary among phytophagous insects. In polyphagous species, sensory discrimination of phagostimulants has been demonstrated (Haskell and Mordue, 1969). However, phagostimulants are less crucial to these insects and many species can be induced to feed on diets that lack them. Among oligophagous species, phagostimulants are essential in eliciting a feeding response. In many species, the requirements are rigid and highly specific as indicated by the presence of chemosensory receptors that are sensitive to specific phagostimulants. Both secondary plant substances and primary nutrient chemicals serve as phagostimulants for phytophagous insects. Phagostimulants of the nature of secondary plant substances are sporadically distributed in plants and act on only a few insects, suggesting that the insects might have evolved a need for such chemicals to release feeding behavior (Fraenkel, 1959).

Primary nutrient chemicals are widely distributed in plants and serve as phagostimulants for many phytophagous insects. Most oligophagous insects, however, can respond to only a few phagostimulative nutrients, indicating a high degree of chemosensory specificity. The feeding responses of phytophagous insects are often enhanced by additive and synergistic interactions among phatostimulants. Recent investigations with the Colorado potato beetle showed that a balance in the proportion of various phagostimulants is important in order to obtain optimal interactions.

Detailed investigations on the role of phagostimulants in feeding behavior of a few oligophagous insects, i.e. the silkworm, the Colorado potato beetle, the cabbage butterfly, the tobacco hornworm, etc., have made possible the development of artificial diets for these species, which in turn facilitates assessing dietary requirements of phagostimulants. For the Colorado potato beetle, phagostimulants are indispensable components of the artificial diet. A phagostimulant mixture has been shown to elicit moderate feeding response on an agar-cellulose medium and to support growth of larvae on an artificial diet. The addition of potato leaf powder or an alcohol extract of the leaf powder improves feeding and growth, indicating that phagostimulants from the host plant are important to elicit optimal feeding and growth of potato beetle larvae.

Recent success in rearing several oligophagous insects on artificial diets has led to the general opinion that nutritional requirements of phytophagous insects are qualitatively more or less similar, but quantitatively different from species to species (Wardojo, 1969b). The marked differences in the quantitative requirements of phagostimulative nutrients among oligphagous insects illustrate this concept. The fact that oligophagous insects can be reared on artificial diets without plant materials does not necessarily mean that specific phagostimulants from host plants do not play a role in feeding. Results from rearing oligophagous insects on artificial diets invariably showed that the inclusion of plant materials in the diets promoted better growth and survival. Recent findings on the induced feeding preference of lepidopterous larvae (Schoonhoven, 1967; Jermy *et al.*, 1968) further stressed the important role of the host plant in the feeding behavior of phytophagous insects.

SUMMARY

Current knowledge of phagostimulant chemicals in the foodplants of phytophagous insects is reviewed in relation to their importance as components of synthetic diets. Experiments with the oligophagous Colorado potato beetle illustrate that both primary nutrients and secondary plant chemicals acting as sign (token) stimulants are necessary components of synthetic diets to ensure adequate ingestion and hence good growth. It is emphasized that phagostimulant chemicals, both primary nutrients and token stimulants, often act synergistically, so that optimal stimulation of feeding may depend on a proper balance of phagostimulants in the food.

REFERENCES

BECK, S. D. (1965). Resistance of plant to insects. *Ann. Rev. Entomol.* **10**:207-232.

BECK, S. D. and HANEC, W. (1958). Effect of amino acids on feeding behavior of the European corn borer, *Pyrausta nubilalis* (Hubn). *J. Insect Physiol.* **2**:85-96.

BECK, S. D., CHIPPENDALE, G. M. and SWINTON, D. E. (1968). Nutrition of the European corn borer, *Ostrinia nubilalis.* VI. A larval rearing medium without crude plant fractions. *Ann. Ent. Soc. Amer.* **61**:459-62.

CHAMBLISS, O. L. and JONES, C. M. (1966). Cucurbitacins: Specific insect attractants in Cucurbitaceae. *Science* **158**:1392-3.

DADD, R. H. (1970). Arthropod nutrition. In *Chemical Zoology: Arthropoda* Ed. Florkin, M. and Scheer, B. Academic Press, New York, **5A**:35-95.

DAVID, W. A. L. and GARDINER, B. O. C. (1966). Mustard oil glycosides as feeding stimulants of *Pieris brassicae* larvae in a semi-synthetic diet. *Ent. Exp. and Appl.* 9:247-255.

DAVIS, G. R. F. (1965). Phagostimulatory effects of amino acids for larvae of *Ctenicera destructor* (Brown) (Coleoptera, Elateridae). *Arch. Internat. Physiol. Biochim.* 73:610-26.

____________ (1968). Phagostimulation and consideration of its role in artificial diets. *Bull. Ent. Soc. Amer.* **14**:27-30.

DETHIER, V. G. (1941). Chemical factors determining the choice of food plants by *Papilio* larvae. *Amer. Nat.* **75**:61-73.

____________ (1966). Feeding behavior. In *Insect Behaviour.* Ed. P. T. Haskell. *Symposia R. Ent. Soc. Lond.* **3**:46-58.

____________ (1970). Chemical interactions between plants and insects. In *Chemical Ecology* Ed. Sondheimer, E. and Simeone, J. B. Academic Press, New York, pp. 83-102.

FEENY, P., PAAUWE, K. L. and DEMONG, N. J. (1970). Flea beetle and mustard oils: Host plant specificity of *Phyllotreta cruciferae* and *P. striolata* adults (Coleoptera: Chrysomelidae). *Ann. Ent. Soc. Amer.* **63**:832-841.

FRAENKEL, G. (1959). The raison d'etre of secondary plant substances. *Science* 129:1466-70.

____________ (1969). Evaluation of our thoughts on secondary plant substances. *Ent. Exp. and Appl.* **12**:473-486.

GOTHILF, S. and BECK, S. D. (1967). Larval feeding behavior of the cabbage looper, *Trichoplusia ni. J. Insect Physiol.* **13**:1039-54.

HAMAMURA, Y. (1970). The substances that control the feeding behavior and the growth of the silkworm *Bombyx mori* L. In *Control of Insect Behavior by Natural Products.* Ed. Wood, D. L., Silverstein, R. M. and Nakajima, M., Academic Press, New York, pp. 55-80.

HASKELL, P. T. and MORDUE, A. J. (1969). The role of mouthpart receptors in the feeding behavior of *Schistocerca gregaria. Ent. Exp. and Appl.* **12**:591-610.

HERON, R. J. (1965). The role of chemotactic stimuli in the feeding behavior of the spruce budworm larvae on white spruce. *Can. J. Zool.* **43**:247-69.

HEROUT, V. (1970). Some relations between plants, insects and their isoprenoids. In *Progress in Phytochemistry.* Ed. Reinhold, L and Liwschitz, Y., Interscience Publishers, New York, 2:144-202.

HSIAO, T. H. (1966). The host plant specificity of the Colorado potato beetle, *Leptinotarsa decemlineata* (Say). Ph.D. thesis, Univ. of Illinois. Urbana, Illinois. 194 pp.

____________ (1969a). Chemical basis of host selection and plant resistance in oligophagous insects. *Ent. Exp. and Appl.* **12**:777-788.

____________ (1969b). Adenine and related substances as potent feeding stimulants for the alfalfa weevil, *Hypera postica* (Gyllenhal). *J. Insect Physiol.* **16**:1785-1790.

HSIAO, T. H. and FRAENKEL, G. (1968a). Isolation of phagostimulative substances from the host plant of the Colorado potato beetle, *Leptinotarsa decemlineata* (Say). *Ann. Ent. Soc. Amer.* **61**:476-484.

____________ (1968b). The influence of nutrient chemicals on the feeding behavior of the Colorado potato beetle, *Leptinotarsa decemlineata* (Coleoptera: Chrysomelidae). *Ann. Ent. Soc. Amer.* **61**:44-54.

____________ (1968c). The role of secondary plant substances in the food specificity of the Colorado potato beetle, *Leptinotarsa decemlineata* (Say). *Ann. Ent. Soc. Amer.* **61**:485-493.

____________(1968d). Selection and specificity of the Colorado potato beetle for solanaceous and non-solanaceous plants. *Ann. Ent. Soc. Amer.* **61**:493-503.

ISHIKAWA, S., HIRAO, T. and ARAI, N. (1969). Chemosensory basis of hostplant selection in the silkworm. *Ent. Exp. and Appl.* **12**:544-554.

JERMY, T., HANSON, F. E. and DETHIER, V. G. (1968). Induction of specific food perference in lepidopterous larvae. *Ent. Exp. and Appl.* **11**:211-230.

MATSUMOTO, Y. (1970). Volatile organic sulfur compounds as insect attractants with special reference to host selection. In *Control of Insect Behavior by Natural Products.* Ed. Wood, D. L., Silverstein, R. M. and Nakajama, M. Academic Press, New York, pp. 133-160.

NAYAR, J. K. and FRAENKEL, G. (1963a). The chemical basis of host selection in the Catalpa sphinx, *Ceratomia catalpae* (Boisduval) (Lepidoptera: Sphingidae). *Ann. Ent. Soc. Amer.* **56**:119-122.

____________ (1963b). The chemical basis of host selection in the Mexican bean beetle, *Epilachna varivestris. Ann. Ent. Soc. Amer.* **56**:174-178.

NAYAR, J. K. and THORSTEINSON, A. J. (1963). Further investigations into the chemical basis of insect-host relationships in an oligophagous insect, *Plutella maculipennis* (Curtis). *Can. J. Zool.* **41**:923-29.

REES, C. J. C. (1969). Chemoreceptor specificity associated with choice of feeding site by the beetle *Chrysolina brunsvicensis* on its foodplant, *Hypericum hirsutum. Ent. Exp. and Appl.* **12**:565-83.

SCHOONHOVEN, L. M. (1967). Loss of hostplant specificity by *Manduca sexta* after rearing on an artificial diet. *Ent. Exp. and Appl.* **10**:270-272.

____________ (1968). Chemosensory bases of host plant selection. *Ann. Rev. Entomol.* **13**:115-36.

____________ (1969). Gustation and foodplant selection in some lepidopterous larvae. *Ent. Exp. and Appl.* **12**:555-564.

SMITH, B. D. (1966). The role of the plant alkaloid Sparteine in the distribution of *Acyrthosiphon spartii* (Koch) (Homoptera: Aphididae) on broom (*Sarothamnus scoparius* L.). *Nature* **212**:213-214.

TANTON, M. T. (1965). Agar and chemostimulant concentrations and their effect on intake of synthetic food by larvae of the mustard beetle, *Phaedon cochleariae* Fab. *Ent. Exp. and Appl.* **8**:74-82.

THORSTEINSON, A. J. (1960). Host selection in phytophagous insects. *Ann. Rev. Entomol.* **5**:193-218.

WARDOJO, S. (1969a). Some factors relating to the larval growth of the Colorado potato beetle, *Leptinotarsa decemlineata* Say (Coleoptera: Chrysomelidae), on artificial diets. *Meded. Landbouwhogeschool. 69*-**16**:1-71.

____________(1969b). Artificial diet without crude plant material for two oligophagous

leaf feeders. *Ent. Exp. and Appl.* 12:698-702.
WEARING, C. H. (1968). Responses of aphids to pressure applied to liquid diet behind parafilm membrane. Longevity and larviposition of *Myzus persicae* (Sulz.) and *Brevicoryne brassicae* (L.) feeding on sucrose and sinigrin solutions. *New Zealand J. Sci.* 11:105-121.
WENSLER, R. J. D. (1962). Mode of host selection by an aphid. *Nature* 195:830-831.

FEEDING STIMULI AND TECHNIQUES FOR STUDYING THE FEEDING OF HAEMATOPHAGOUS ARTHROPODS UNDER ARTIFICIAL CONDITIONS, WITH SPECIAL REFERENCE TO *RHODNIUS PROLIXUS*

W. G. Friend and J. J. B. Smith
Zoology Department
University of Toronto, Toronto 5, Ontario Canada

INTRODUCTION

In his review of the sensory physiology of blood-sucking arthropods Dethier (1957) pointed out that a considerable number of unrelated species of arthropods had evolved the ability to utilize the blood of vertebrates as food. The process of blood-feeding can be divided into successive steps: (1) location of the host; (2) identification of a site for penetration by the mouthparts; (3) "biting" or penetration of the skin; (4) the production of a haemorrhagic area or the location of a blood vessel; (5) ingestion of the blood meal; (6) cessation of ingestion and withdrawal of the mouthparts. This paper is restricted mainly to a critical appraisal of the research done on the details of step 5, ingestion, and of artificial feeding techniques and diets that were used during this research. Reviews of feeding behaviour giving much broader coverage are available (Dethier, 1957; Galun, 1971; Tarshis, 1956; Downes *et al.*, 1962).

MEMBRANES

Most haematophagous arthropods will not feed from a free surface; consequently, artificial feeding techniques have involved the use of artificial membranes through which feeding takes place. A great many membranes have been used. Tarshis (1956) has extensively reviewed this aspect and concludes . . ." ambient temperature, membrane derivation and thickness, blood temperature and storage and age and degree of starvation of the arthropod were the most critical factors in membrane feeding of arthropods." More recently Galun (1971) has reviewed the subject of membranes from the viewpoint of mass rearing haematophagous arthropods.

She notes that thin membranes may be required for feeding in the case of an arthropod with a short proboscis, such as a louse or a flea, but that tsetse flies require thick membranes. An extensive study of the effects of 4 different membranes on the feeding of 6 species of mosquito has been conducted by Rutledge *et al.* (1964). They found that the synthetic membranes Parafilm M and Saran Wrap were not as effective as the natural membranes chick skin or Baudruche membrane. On the other hand, *Rhodnius prolixus* - and *Triatoma infestans* (Shimamune *et al.*, 1965) - will feed without difficulty through rubber membranes, which have obvious advantages over natural membranes.

ELECTRICAL RECORDING TECHNIQUES

Until quite recently information on the feeding behaviour of blood-feeding arthropods was limited to measurements of nutrient uptake, and to direct observations of the movements of the mouth-parts in certain translucent tissues, such as a mouse's ear, or in artificial feeding chambers. (Friend *et al.*, 1965; Lavoipierre *et al.*, 1959; Dickerson & Laviopierre, 1959; Gordon & Lumsden, 1939). A useful additional method of studying feeding behaviour has recently been made available. Beginning with the work of Kashin and Wakeley (1965), techniques have been developed to measure changes in the electrical resistance of arthropods with sucking mouthparts as they feed. Application of these techniques has provided information on the mosquito *Aedes aegypti* (Kashin 1966), the pea aphid *Acyrthosiphon pisum* (McLean and Kinsey 1964, 1965, 1967, 1968a, b; McLean and Weight 1968), the tsetse fly (Galun 1971) and *R. prolixus* (Smith and Friend, 1970; Friend and Smith, 1971a).

To enhance the usefulness of the electrical recording techniques for differentiating the response of an insect such as *R. prolixus* to various experimental diets we found it necessary to correlate the electrical events with feeding activities such as mouthpart movements, salivation, and ingestion. A method was developed for observing, photographing, and video-taping the mouthparts of *R. prolixus* as its feeding was monitored electrically. By using a split-screen television technique, simultaneous recordings were made of the changes in electrical resistance and the feeding events; the recordings could then be analyzed by slow-speed and stop-frame playback (Smith and Friend, 1971).

The techniques described made it possible to see the action of the mouthparts very clearly and revealed several events not previously described in blood-feeding Hemiptera. These were: the secretion of a 'collar' around the labium; continuous salivation during the initial probing phase; sampling of the diet; regurgitation of diet. These observations are discussed in detail in Friend and Smith (1971a).

The method was particularly useful for the analysis of rapid and transient events. For example, large temporary drops in electrical resistance during the initial feeding phase were linked with a momentary opening of the functional mouth and an intake of diet. Low amplitude but regular variations in resistance were seen to coincide with the thrusting and withdrawal of the maxillae during the initial probing phase; smaller fluctuations in resistance during each cycle corresponded to a characteristic 'flicking' motion of the maxillae. This motion is extremely rapid, taking approximately 0.05 second for each flick.

It is apparent that these techniques could be applied with minor modifications to the study of feeding in any species of insect that can be induced to feed through a membrane on a liquid diet contained in an optically transparent chamber. Furthermore, the method could probably be adapted to the study of insects which do not have piercing mouthparts but still feed on liquids, for example nectar feeders and certain flies.

SAMPLING OF DIET

We have not found reports of sampling of diets prior to feeding in any insects other than aphids (McLean and Kinsey, 1965). Sampling in *R. prolixus* was not reported by Lavoipierre *et al.* (1959); it is, however, such a transient phenomenon that it could easily have been overlooked, particularly under their conditions of observation.

Why blood-feeding insects such as *R. prolixus* should sample the diet prior to gorging is not immediately clear. Sampling may enable the diet to come into contact with certain receptor cells. The repeated sampling seen during the prolonged maxillary probing phase in the absence of a gorging factor is not inconsistent with this suggestion. Alternatively, sampling may represent 'trial activity' of the pharyngeal pump as the animal tests for the availability of fluid. Under normal conditions, *R. prolixus* feeds only from blood vessels and not from a 'pool' of blood resulting from local haemorrhage (Lavoipierre *et al.*, 1959), and consequently there must be some mechanism for signalling that the functional mouth is suitably located.

Whatever the function of sampling, it appears to be an important component of the initial phase of feeding, and is probably associated with the initiation of regular pumping. Feeding in *R. prolixus* involves a sequence of activities, commencing with attraction to a warm surface and culminating in the regular activity of the pharyngeal pump. Each component of this sequence may well be a response to some definite stimulus or pattern of stimuli dependent on previous activity. We have found, using the electrical recording technique, that the initiation of the final phase of feeding, that of prolonged and regular activity of the pharyngeal pump resulting in gorging, is dependent on the presence in the diet of the chemicals previously known to induce gorging. (Smith and Friend, 1970).

CHEMORECEPTION INTERNAL TO THE FOOD CANAL

There has been much speculation concerning the location of the chemoreceptors that respond to chemicals with phagostimulatory activity. These chemoreceptors are probably internal to the food canal since the tarsal and antennal regions do not come in contact with the test diets in most artificial feeding techniques (Friend, 1965; Miles, 1958, Galun, 1967). Hosoi (1954) has shown that mosquitoes will imbibe blood into the stomach after the apical part of the fascicle has been amputated. We have found that *R. prolixus* also will feed after having the tip of the fascicle removed. Suitably located sensilla have been described in a variety of animals. Von Gernet and Buerger (1966) have described sense organs on the cibarial pump of 22 species of mosquitoes and suggest that the epipharyngeal sensilla direct the flow of blood to the stomach. Owen (1963) showed that there were taste receptors which respond to water, sugars, and blood in the cibarium of the mosquito *Culiseta inornata* and he claims that blood drawn into the cibarium stimulates these taste receptors to supply the sensory input for control of sucking. Tawfik (1968) found two types of sensillum in the cibarial regions of the louse *Cimex lectularius*. Wensler and Filshie (1969) have described a sensory structure in the anterior region of the food canal of two species of aphid. In these insects the dorsal wall is innervated by a total of 60 neurons which terminate in groups, at 14 porous papillae on the cuticle, eight of which are in the epipharynx, as in *R. prolixus*. Paired papillae have also been detected in the ventral wall of this region. From their structure, these authors deduce that these sensilla have a chemosensory function. Similar structures are common in Hemiptera (Weber 1930). Kraus (1957) has described a group of innervated structures dorsal and anterior to the pharyngeal pump in *R. prolixus* which he named the epipharyngeal sensilla and has suggested that they are taste receptors. We have studied the epipharyngeal area of *R. prolixus* using both transmission and scanning electron microscopes and our findings support the conclusion that the eight epipharyngeal sensilla are the receptors that respond to chemical feeding factors in the blood and induce feeding.

CHEMICAL FEEDING FACTORS

Hosoi (1959) was the first to identify a specific chemical in blood that would act as a phagostimulant. He showed that adenosine 5' phosphates provide the main stimulus promoting gorging of the mosquito *Culex pipiens* on blood. Since 1959 certain nucleoside phosphates have been shown to act as feeding stimulants for a large diversity of blood feeding arthropods. These feeding stimulants have been tested for effect in two ways: by being presented either in non-nutritive diets such as isotonic NaCl, or as a supplement to blood treated in various ways. Galun (1971) points out that until the phagostimulants of the haematophagous arthropods were recognized it was impossible to evoke optimum uptake of artificial diets. Adenosine triphosphate (ATP) is probably the most important phagostimulant in whole fresh blood. Galun and Rice (1971) present evidence that the blood platelets, which contain more ATP than most tissues, rather than the red blood cells, are the principal source of the phagostimulant. They claim that since the phagostimulatory threshold for ATP is about 10^{-5}M for mosquitoes and tsetse flies, the necessary concentration may be attained only locally at the receptor surface. *R. prolixus* is at least as sensitive to ATP as mosquitoes - 6 x 10^{-6}M will cause 50 per cent gorging (Friend and Smith 1971b) - and so it also could be responding to blood platelets.

CHEMICAL FEEDING FACTORS IN NUTRITIVE DIETS

Rutledge *et al.* (1964) showed that ATP at 5 x 10^{-3}M concentration markedly increased the rather low percentage of mosquitoes feeding on chick erythrocyte extract and on 10 per cent bovine serum, but did not enhance the greater response to fresh blood. They also showed that glucose, lysine and alanine did not stimulate feeding significantly. This work was done with *Aedes aegypti, Aedes togoi, Culex tritaeniorhynchus, Culex pipiens pallens, Armigeres subalbatus* and *Anopheles stephensi.* Galun (1966) found that ATP at 5 x 10^{-3}M concentration would stimulate feeding of the rat flea *Xenopsylla cheopis* on citrated sheep blood plasma. No stimulatory effect was shown by AMP, ADP, the triphosphates of inosine, guanosine or cytidine, or by glutathione or creatine phosphate.

CHEMICAL FEEDING FACTORS IN NON-NUTRITIVE DIETS

The use of phagostimulants such as ATP to enhance the uptake of nutritive diets such as various blood fractions has great practical significance to workers who wish to rear blood-feeding arthropods under artificial conditions. However for the study of specific effects of certain chemicals on phagostimulation there are advantages to having the insects feed on simple non-nutritive diets such as isotonic NaCl. Although these diets do not support growth or development, insects responding to phagostimulants incorporated in these diets demonstrate an apparently normal feeding pattern (Friend and Smith, 1971a).

The mosquitoes *Culex pipiens* (Hosoi 1959), *A. aegypti* (Galun, 1967) and the tsetse fly *G. austeni* (Galun and Margalit, 1970a) will gorge on 0.15M NaCl solutions containing ATP, ADP or AMP. Adenosine tetraphosphate will also stimulate *A. aegypti* (Galun, 1967). Nucleosides and nucleoside phosphates with bases other than adenine had no phagostimulatory effect on these insects.

R. prolixus larvae confined in an artificial feeding apparatus could be induced to gorge on 0.15 M NaCl solutions at pH 7.0 containing 10^{-3}M concentrations of chemicals having phosphate bonds of high energy release. The di- and tri-phosphates of adenosine, guanosine, inosine, cytosine, and uridine all showed high gorging factor activity. Creatine phosphate and tetrasodium pyrophosphate were slightly less active. Riboflavin-5-phosphate, 5'-adenosine monophosphate (AMP) and 3'5'-cyclic adenosine monophosphate (cAMP) also stimulated gorging to a marked degree (Friend, 1965).

The tick *Ornithodoros tholozani* differs from these other animals in that glucose must be included at 1 mg·ml^{-1} concentration in the saline for it to respond to phagostimulants. In the presence of glucose, the following compounds induced feeding: ATP, ITP, GTP, reduced glutathione (GSH), reduced diphosphopyridine nucleotide (DPNH), and the L-amino acids leucine, *allo* leucine, proline, valine, serine, alanine, phenylalanine and glutamine (Galun and Kindler, 1968). These authors suggest that feeding is mediated by at least two kinds of chemoreceptors, one specific for GSH, ATP, and DPNH (based on the observation that glutamate inhibits the response to these three) and the other for amino acids.

These studies in which chemical feeding factors have been added to nutritive and non-nutritive diets reveal a pattern of response. The rat flea *X. cheopis*, the only host-specific insect tested, has a fairly complex response but the only compound with clear phagostimulatory properties was ATP (Galun 1966). Various mosquitoes and the tsetse fly *G. austeni* respond to ATP, ADP and AMP but not to nucleosides or nucleoside phosphates that lack adenine as the nitrogenous base (Hosoi, 1959; Rutledge *et al.*, 1964; Galun, 1967; Galun and Margalit, 1970a). The tick *O. tholozani* responds to a wide variety of molecules, providing glucose is included with the isotonic saline in the diet (Galun and Kindler, 1968) and *R. prolixus* will respond to AMP, cAMP or any of a large number of different molecules that have phosphate bonds of high energy release (Friend, 1965).

Although there is insufficient information at present to erect a hypothesis of broad generality concerning phagostimulation in haematophagous arthropods, it is noteworthy that other non-haematophagous species also respond to nucleosides and nucleoside phosphates. Phosphates of guanosine stimulate the female housefly *Musca domestica* to feed (Robbins *et al.*, 1965). Adenine and adenosine phosphates stimulate feeding in the alfalfa weevil *Hypera postica* (Hsiao, 1969). AMP increases the filtering rate of larvae of *Culex pipiens* (Dadd, 1970) and monophosphates of guanosine and inosine act as flavour enhancers for man (Kuninaka *et al.*, 1964).

DOSE-PERCENT EFFECT STUDIES

Hosoi (1959) and Galun and her associates tested the various adenine-containing nucleotides at concentrations between 10^{-2} and 10^{-6}M (Galun,

1967; Galun and Margalit, 1970a). Galun and Rice (1971) claim that the threshold concentration of ATP which will induce feeding in mosquitoes and the tsetse fly is about 10^{-5} M. Although these authors have all made estimates of the relative potency of chemicals acting as phagostimulants an accurate determination requires adequate dose-percent effect testing and statistical treatment of the resulting data. One good measure of the sensitivity of an insect to stimulating chemicals is the dose that will induce 50 per cent of the insects to respond (median effective dose, ED_{50}). Dose-per cent effect studies are time consuming and use a large number of experimental animals; nevertheless, we feel that they are necessary for further interpretation of the mode of action of nucleoside phosphates as phagostimulants.

We have recently determined the values of ED_{50} for *R. prolixus* of 10 nucleoside phosphates and have attempted to correlate these with the structure of the various molecules - (Friend and Smith, 1971).

Table 1. Comparison of potencies of nucleoside phosphates in inducing the gorging response in third instar larvae of *Rhodnius prolixus.*

Group	Chemical[a]	ED_{50} and confidence limits ($\times 10^{-6}$ M)
I	ATP	6.0 (4.9 - 7.4)
II	ADP	66 (53 - 82)
	CTP	70 (53 - 87)
	GTP	83 (70 - 99)
	CDP	88 (69 - 113)
	ITP	115 (85 - 155)
III	3'5' - cyclic AMP (cAMP)	195 (155 - 245)
IV	IDP	270 (225 - 320)
	GDP	310 (250 - 380)
V	AMP	1000 (780 - 1300)

[a]In 0.15 NaCl, fed through thin rubber membrane.
*P = 0.05

The results were analysed using the rapid graphic method of Litchfield and Wilcoxon (1949). This method includes a test for the significance of any difference between ED_{50} values. Based on the results of this test, the chemicals were divided into five groups (Table 1). All members within any one group differed significantly in potency (P=0.05) from all members of other groups. Members within a group did not differ significantly in potency, except that in Group II ADP and CTP were shown to be significantly more potent than ITP.

This grouping appears to be roughly related to the amount by which the chemicals differ from ATP, the most potent chemical so far tested (See Friend and Smith, 1971). The main exception to the apparent pattern of decreasing potency with greater structural difference from ATP is cAMP. It would be expected that this molecule, which differs structurally from ATP even more than does AMP, would be less potent than AMP.

OTHER DIET COMPONENTS

Although it has been found possible to induce gorging in a variety of blood-feeding arthropods by suitable artificial diets containing stimulating chemicals such as nucleotides and reduced glutathione, in all cases studied it was necessary to include at least one chemical other than water in the diet. Because the nature of the solute appears unimportant in some studies, it has been suggested that an important parameter of the diet is the tonicity or osmolarity. (Hosoi, 1959; Galun and Kindler, 1968; Galun 1966; Galun and Margalit, 1969). For instance *G. austeni* will gorge on 10^{-3}M ATP in 0.15M NaCl, 0.3M sucrose (Galun and Margalit, 1969), or 0.15M KCl (Galun and Margalit, 1970a). The rat flea, *X. cheopsis*, showed significant uptake of artificial diets consisting of isotonic solutions of NaCl, NaBr, NaI, $NaNO_3$, sodium acetate, sucrose and lactose, although no response was as great as that to solutions containing ATP (Galun, 1966). Maximum uptake of a solution of NaCl occurred at a concentration of 0.15M. However, 0.6M lactose produced a greater uptake than 0.3M lactose which is isotonic with 0.15M NaCl.

The mosquito *A. aegypti* appears much more specific in its solute requirements. Galun (1967) obtained greater than 70 per cent response to 0.15M NaCl containing adequate AMP, ADP or ATP. The response was considerably less however to isotonic solutions containing non-electrolytes such as glucose, sucrose or lactose and to isotonic solutions of KCl, $MgCl_2$, $CaCl_2$.

We have found that *R. prolixus*, like *G. austeni* and *X. cheopsis*, is not as specific in its solute requirements as *A. aegypti*. Close to 100 per cent gorging can be obtained on 10^{-3}M ATP in isotonic solutions of NaCl, KCl, LiCl, NaBr, KBr, NaI, KI, sucrose and mannitol. The per cent response to 10^{-3}M ATP declines as a NaCl solution is made more or less concentrated than 0.15M (unpublished data).

It would appear from these studies that at least in *R. prolixus, G. austeni* and *X. cheopsis*, it is the osmotic concentration of the diet rather than the concentration of a specific solute that is important in determining whether or not the insect will gorge. Galun and Kindler (1968) also report that the tick *O. tholozani* is non-specific in its solute requirements although the tonicity of the diet is important. On the other hand, *A. aegypti* seems to require an optimal concentration specifically of sodium chloride. Galun has used this conclusion as supporting evidence for her theory of the mode of action of ATP as a chemical stimulant in mosquitoes (Galun, 1967) and suggests that a different receptive mechanism may be operating in the tsetse fly (Galun and Margalit, 1970b).

It is difficult however to place too much reliance on such conclusions without a more quantitative knowledge of the effect of substituting other solutes for NaCl. It is apparent from our work and the work of others that the gorging response of blood-feeding arthropods is dependent on a variety of factors, some dietary and some non-dietary, such as membrane thickness type and age, degree of starvation, handling, light, humidity and vibration. It is conceivable that several parameters of the diet are monitored prior to commencement of gorging in these arthropods; information concerning these parameters may then interact in complex and variable ways in the decision whether or not to commence gorging. By varying the concentration of the gorging stimulant ATP in diets containing solutes substituting isotonically for NaCl, we have found that even though tests at 10^{-3}M ATP showed no difference between the response to NaCl, KCl, NaBr, and mannitol, there is in fact a significant decrease in sensitivity to ATP as measured by the dose producing a 50 per cent response. For example, Fig. 1 shows the dose-response curves obtained for ATP in 0.15M NaCl and 0.15 M KCl. Preliminary tests have shown that the ED_{50} value for ATP increases from 6 x 10^{-6} in 0.15M NaCl to between 1 x 10^{-4} and 1 x 10^{-3} M in 0.3 M mannitol.

Thus, although with an adequate concentration of a gorging stimulant the tonicity of the diet and not the nature of the principle solute appears important, quantitative testing reveals that the nature of the solute does affect the response and that therefore there may be receptors capable of discriminating between different solutes. We suggest that in the complex stimulus requirements necessary to produce gorging, deficiencies in some parameters can be overcome by suitable stimulation by others, and even by factors other than those of the diet. It might be pointed out that most gorging tests are statistical, and that some insects will in fact gorge completely on a diet of distilled water. An alternative interpretation then to the apparent distinction between the responses of *A. aegypti* and those of *R. prolixus, G. austeni, X. cheopsis* and *O. tholozani* is that the relative importance of different dietary stimuli is not the same in all cases.

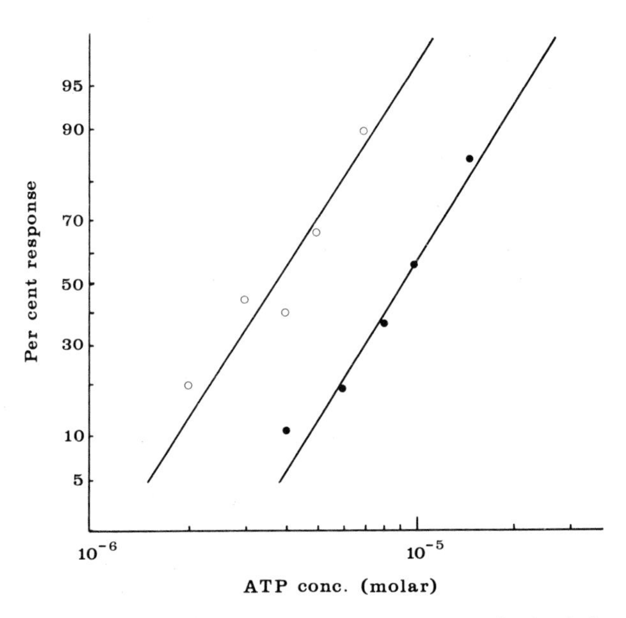

Figure 1. The gorging response of third instar larvae of *Rhodnius prolixus* to ATP fed in solutions of 0.15 M NaCl (open circles) or 0.15 M KCl (solid circles).

MODE OF ACTION OF GORGING STIMULANTS

We have shown previously (Smith and Friend, 1970) that the effect of a gorging stimulant such as ATP on *R. prolixus* is to initiate the regular and prolonged activity of the pharyngeal pump. It seems likely that gorging stimulants are detected by some of the chemoreceptive sensilla in the epipharynx during the sampling activity prior to initiation of ingestion. Some work by ourselves and others has been directed towards determining the mode of action of the gorging stimulants at the receptor sites of the sensilla. Since much is known of the chemistry of ATP, and of its role as a basic metabolic component, it might be hoped that a study of its mode of action as a chemical stimulant will throw light on a fundamental problem of sensory physiology, namely that of describing the way in which a chemical stimulus is transduced by a chemoreceptive cell to give rise to propagated nerve spikes.

The most widely held theory is that a stimulating molecule interacts with a specific molecular site on the surface membrane, thereby inducing some configurational or electronic change to which the cell responds, eventually by production of generator and propagated spike potentials (see, for example, Beidler and Reichardt, 1970).

Our studies have shown that, although *R. prolixus* responds to a wide range of nucleoside phosphates, their potency as gorging stimulants decreases in a manner roughly correlated with the degree of difference between their molecular structure and that of ATP (Friend and Smith, 1971b). Several regions of the molecule appear important. Our interpretation of this finding is that to obtain a sufficient stimulus intensity to elicit gorging enough receptive sites have to be occupied by stimulating molecules enough of the time. As the 'fit' of the stimulating molecule becomes less good, its probability of occupying a receptive site lessens, and thus the smaller is the fraction of time during which that site is occupied. This can, however, be compensated for by increasing the concentration of that molecular species. How much the concentration has to be increased will depend on the details of the interaction between the receptor site and the stimulating molecule; more specific receptors may for instance require a better 'fit' at a larger number of interacting points of the molecule than do less specific receptors. This could account for the apparently lower specificity of response we found for *R. prolixus* as compared to that of *A. aegypti, G. austeni* and *X. cheopsis* as found by Galun and associates.

After the stimulating molecule attaches temporarily to a receptive site, a series of events must occur which culminate in the generation of nerve spikes. In other receptive systems, such as vertebrate Pacinian corpuscles (Gray, 1959), the stimulus is thought to increase the permeability of the membrane locally to ions such as Na^+. The resultant increased influx of cations then partly depolarises the membrane at this region, giving rise to generator currents which in turn initiate spikes in a different region of the cell.

A similar mechanism has been suggested by Galun (1967) for the reception of ATP in mosquitoes. Her hypothesis is that ATP, a fairly strong chelating agent, attaches to a receptor site and then chelates with membrane-bound divalent ions, possibly Zn^{++}; the consequent dislocation of the ion then allows an influx of Na^+. Support for this suggestion comes from observations that the potency of nucleoside phosphates can be correlated with their chelating ability; that prior chelation of ATP with Ni^{++}, Zn^{++} or Co^{++} ions inhibits the gorging response in mosquitoes; that the chelating agent EDTA produces some gorging response, and that substitution of Na^+ as principle cation in the diet by K^+, Ca^{++} or Mg^{++} inhibits gorging.

However, if the gorging response of arthropods other than mosquitoes is considered, some major criticisms of this hypothesis emerge. Galun herself has pointed out that it cannot explain the response of *G. austeni* (Galun and Margalit, 1970b), since prior chelation of ATP with Ca^{++}, Co^{++}, Mn^{++}, Zn^{++}, Ni^{++} or Cu^{++} does not inhibit gorging, and that maximum gorging can be obtained when NaCl in the diet is replaced by either KCl or glucose. We have found that *R. prolixus* is still able to discriminate between diets with and without ATP when NaCl is substituted by a range of other solutes - which indicates that the ATP receptors are working despite their 'abnormal' (in terms of extracellular tissue electrolytes) external environment. In particular, if the receptor cell functions by ionic flux across the receptive membrane, then external KCl would be expected to depolarise the cell and consequently either block reception or mimic a stimulating chemical. And yet neither of these happens. Possibly the cuticle that tightly surround the small exposed dentritic membrane as described in aphids by Wensler and Filshie (1969) acts as an effective insulator preventing the spread of depolarising currents from the specialised receptive regions of the membrane that might occur with external KCl. Whether this is so or not it appears unlikely that the receptive membrane is involved directly in ionic permeability changes. Thus there remains the question: What is the mechanism that links reception of a stimulating molecule at the receptive membrane with the production of the necessary generator currents presumably elsewhere in the cell?

An alternative suggestion for this mechanism has emerged from our most recent work (Smith and Friend, in prep.). The gorging stimulant cAMP has a much greater potency than would be predicted from its molecular structure (Table 1). In other organisms, this compound has recently been shown to act as an internal cell messenger linking reception of a stimulating hormone molecule to events elsewhere in the cell (see for instance Rasmussen, 1971). We tested the effect of adding the drugs theophylline and caffeine to artificial diets presented to *R. prolixus*. These drugs block the action of the phosphodiesterase that breaks down cAMP, and hence enhance the action of the stimulating hormone. When present at a concentration of 10^{-3}M in a diet of 0.15M NaCl and either ATP or ADP these drugs increased the sensitivity of test insects by a factor of from 2-3 as shown by a shift of ED_{50} values. Our hypothesis is then that the stimulating molecule (e.g. ATP) interacts with a receptive site associated with membrane adenyl cyclase, resulting in an increase in internal cAMP synthesis. Diffusion of cAMP to another cell region with a 'normal' extracellular environment then initiates membrane permeability changes with resulting generator currents and spike formation. This hypothesis has the attraction that it could provide a chemoreceptive mechanism with the receptive surface functioning independently of the electrolytes bathing it, a condition that may exist in a wide variety of external chemoreceptors.

CONCLUDING COMMENTS

The use of chemical feeding factors to induce haematophagous arthropods to feed on artificial and non-nutritive diets has provided a technique that can be applied to a broad range of physiological problems. Chemicals that will interfere with the feeding response have been investigated in *A. aegypti* and *R. prolixus* by Salama (1966) who found that both insects would reject diets containing alcohols; Galun *et al* (1969) have demonstrated that methyl phenyldiazenecarboxylate will inhibit the feeding of the tick *O. tholozani*, the tsetse fly *G. austeni*, and the mosquitoes *A. aegypti* and *Culex molestus*. Artificial diets have been used to determine some of the effects of the symbiont *Nocardia rhodnii* on *R. prolixus* (Lake and Friend, 1968) and non-nutritive diets have been used to determine food reserves in *R. prolixus* (Barrett and Friend, 1966) and the relation of abdominal distension and nutrition to molting in this insect (Beckel and Friend, 1964). We have recently developed a technique that allows us to change the diet in less than one second as *R. prolixus* feeds; this technique coupled with the electrical recording system, has revealed that the chemical feeding factor is necessary in a saline diet only at the initiation of feeding and not constantly throughout the feeding period as might be expected. (Smith and Friend, in preparation.)

Research on feeding stimuli and the development of techniques for studying the feeding of haematophagous arthropods are rapidly developing fields and it seems likely that they will provide much information to help both with fundamental problems and with such practical problems as mass rearing for sterile-male control techniques.

SUMMARY

Evidence for the importance of gorging stimulants in the blood meal of haematophagous insects is critically reviewed. All species of ticks and insects studied engorge in response to nucleoside phosphates (nucleotides), in most cases primarily to those containing adenine, with ATP (adenosine triphosphate) usually the most stimulating. Glutathione and certain amino acids also act as phagostimulants for some ticks. Using a split-screen television technique, details of salivatory and gorging behaviour in *Rhodnius prolixus* were correlated with changes in the insect's electrical conductivity as it feeds; complete gorging occurs only after exploratory sips of liquid of appropriate tonicity and containing a suitable phagostimulant arrive in the pharynx, probably because chemoreceptors on the epipharynx are thereby stimulated. Once started, gorging to repletion proceeds in the absence of further phagostimulation. Dose-response studies of percentage engorgement on saline containing various analogues of ATP allowed their efficacies to be ranked in terms of structural divergence from ATP, the most potent phagostimulant. Cyclic AMP ranked anomously with respect to its structure, being more stimulatory than AMP. Since the sensitivity of *Rhodnius* to adenosine phosphates was doubled by caffeine or theophylline, which are phosphodiesterase inhibitors conducive to the maintenance of high levels of cyclic AMP, it is hypothesized that cyclic AMP mediates between external contact of phagostimulant molecules with receptor sites of the sense cell and the eventual generation of nervous impulses.

Acknowledgements.—Our research concerning *Rhodnius prolixus* reported in this paper was supported by the National Research Council and the Defence Research Board of Canada.

REFERENCES

BARRETT, F. M. and FRIEND, W. G. (1966). The use of a non-nutritive diet and uric acid measurements to determine food reserves in *Rhodnius prolixus*. *Can. J. Zool.* **44**:39-46.

BECKEL, W. E. and FRIEND, W. G. (1964). The relation of abdominal distension and nutrition to molting in *Rhodnius prolixus*. *Can. J. Zool.* **42**:71-78.

BIEDLER, L. M. and REICHARDT, W. E. (1970). Sensory transduction. *Neurosciences Res. Prog. Bull.* **8**, No. 5:459-560.

DADD, R. H. (1970). Relationship between filtering activity and ingestion of solids by larvae of the mosquito *Culex pipiens*: A method for assessing phagostimulant factors. *J. Med. Ent.* **7**:708-712.

DETHIER, V. G. (1957). The sensory physiology of blood-sucking arthropods. *Expl. Parisit.* **6**:68-122.

DICKERSON, G. and LAVOIPIERRE, M. M. J. (1959). Studies on the methods of feeding of blood-sucking arthropods II. *Ann. Trop. Med.* **53**:347-357.

DOWNES, J. A., DOWNE, A. E. R., and DAVIES, L. (1962). Some aspects of the behaviour and physiology of biting flies that influence their role as vectors. *Proc. 11th Int. Cong. Entomology* **3**:119-121.

FRIEND, W. G. (1965). The gorging response in *Rhodnius prolixus*. *Can. J. Zool.* **43**:125-132.

FRIEND, W. G., CHOY, C. T. H., and CARTWRIGHT, E. (1965). The effect of nutrient intake on the development and egg production of *Rhodnius prolixus* Stal. *Can. J. Zool.* **43**:891-904.

FRIEND, W. G. and SMITH, J. J. B. (1971a). Feeding in *Rhodnius prolixus*: mouthpart activity and salivation, and their correlation with changes of electrical resistance. *J. Insect Physiol.* **17**:233-243.

FRIEND, W. G. and SMITH, J. J. B. (1971b). Feeding in *Rhodnius prolixus*: potencies of nucleoside phosphates in initiating gorging. *J. Insect Physiol.* **17**:1315-1320.

GALUN, R. (1966). Feeding stimulants of the rat flea *Xenopsylla cheopsis*. *Life Sci.* **5**:1335-1342.

GALUN, R. (1967). Feeding stimuli and artificial feeding. *Bull. Wld. Hlth. Org.* **35**:590-593.

GALUN, R. (1971). Recent developments in the biochemistry and feeding behaviour of haematophagous arthropods as applied to their mass rearing. *Int. Atomic Energy. Agency pub. "Sterility Principle for Insect Control or Eradication." pp. 273-282.*

GALUN, R. and KINDLER, S. H. (1968). Chemical basis of feeding in the tick *Orni-*

thodoros tholozani. J. Insect Physiol. **14**:1409-1421.

GALUN, R., KOSOWER, E. M, and KOSOWER, N. S. (1969). Effect of methyl phenyldiazenecarboxylate (azoester) on the feeding behaviour of blood-sucking invertebrates. *Nature,* Lond. **224**:181-182.

GALUN, R. and MARGALIT, J. (1969). Adenine nucleotides as feeding stimulants of the tsetse fly *Glossina austeni* Newst. *Nature*, Lond. 22:583-584.

GALUN, R. and MARGALIT, J. (1970a). Artificial feeding and feeding stimuli of the tsetse fly *Glossina austeni. Proc. 1st Symp. on tsetse fly breeding under laboratory conditions and its practical application, 22nd and 23 April, 1969, Libson:* 211-220.

GALUN, R. and MARGALIT, J. (1970b). Some properties of the ATP receptors of *Glossina austeni. Trans. Roy. Soc. Trop. Med. Hyg.* **64**:171-174.

GALUN, R. and RICE, M. J. (1971). Role of blood platelets in haematophagy. *Nature New Biology* **233**:110-111.

GERNET, G. VON and BUERGER, G. (1966). Labral and cibarial sense organs of some mosquitoes. *Quaestiones Entomologicae* **2**:259-270.

GORDON, R. M. and LUMSDEN, W. H. R. (1939). A study of the behaviour of the mouth parts of mosquitoes when taking up blood from living tissue; together with some observations on the ingestion of microfilariae. *Ann. Trop. Med.* 33:259-278.

GRAY, J. A. B. (1959). Mechanical into electrical energy in certain mechanoreceptors. *Prog. in Biophys.* 9:285-324.

HOSOI, T. (1954). Mechanism enabling the mosquito to ingest blood into the stomach and sugary foods into the oesophageal diverticula. *Annotnes Zool. Jap.* 27:82-90.

HOSOI, T. (1959). Identification of blood components which induce gorging of the mosquito. *J. Insect Physiol.* 3:191-218.

HSIAO, T. H. (1969). Adenine and related substances as potent feeding stimulants for the alfalfa weevil, *Hypera postica. J. Insect Physiol.* 3:191-218.

KASHIN, P. (1966). Electronic recording of the mosquito bite. *J. Insect Physiol.* 12:281-286.

KASHIN, P. and WAKELEY, H. G. (1965). An insect 'bitometer'. *Nature* Lond. **208**:462-464.

KRAUS, C. (1957). Versuch einer morphologischen und neurophysiolgischen analyse des stechaketes von *Rhodnius prolixus* Stal 1958. *Zool. Anst. Univ. Basel.* 36-84.

KUNINAKA, A., KIBI, M. and SAKAGUCHI, K. (1964). History and development of flavor nucleotides. *Fd. Tech. Champaign,* **18**:287-293.

LAKE, P. and FRIEND, W. G. (1968). The use of artificial diets to determine some of the effects of *Nocardia rhodnii* on the development of *Rhodnius prolixus. J. Insect Physiol.* 14:543-562.

LAVOIPIERRE, M. M. J., DICKERSON, G. and GORDON, R. M. (1959). Studies on the methods of feeding of blood-sucking arthropods-I. The manner in which triatomine bugs obtain their blood meal, as observed in the tissues of the living rodent, with some remarks on the effects of the bite on human volunteers. *Ann. Trop. Med. Parasit.* 53:235-250.

LITCHFIELD, J. T. and WILCOXON, F. (1949). A simplified method of evaluating dose-effect experiments. *J. Pharmac. Exp. Thera.* **96**:99-113.

MCLEAN, D. L. and KINSEY, M. G. (1964). A technique for electronically recording aphid feeding and salivation. *Nature,* Lond. **202**:1358-1359.

MCLEAN, D. L. and KINSEY, M. G. (1965). Identification of electrically recorded curve patterns associated with aphid salivation and ingestion. *Nature,* Lond. **205**:1130-1131.

MCLEAN, D. L. and KINSEY, M. G. (1967). Probing behaviour of the pea aphid

Acyrthosiphon pisum - I. Definite correlation of electronically recorded waveforms with aphid probing activities. *Ann. Ent. Soc. Am.* **60**:400-406.

MCLEAN, D. L. and KINSEY, M. G. (1968a). Probing behaviour of the pea aphid *Acyrthosiphon pisum* - II. Comparison of salivation and ingestion in host and non-host plant leaves. *Ann. Ent. Soc. Am.* **61**:730-739.

MCLEAN, D. L. and KINSEY, M. G. (1968b). Probing behaviour of the pea aphid *Acyrthosiphon pisum* - III. Effect of temperature differences of certain probing activities. *Ann. Ent. Soc. Am.* **61**:927-933.

MCLEAN, D. L. and WEIGT, W. A. (1968). An electronic measuring system to record aphid salivation and ingestion. *Ann. Ent. Soc. Am.* **61**:180-185.

MILES, P. W. (1958). Contact chemoreception in some heteroptera including chemoreception internal to the food canal. *J. Insect Physiol.* **2**:338-347.

OWEN, W. B. (1963). The contact chemoreceptor organs of the mosquito and their function in feeding behaviour. *J. Insect Physiol.* **9**:73-87.

RASMUSSEN, H. (1970). Cell communication, calcium ion and cyclic adenosine monophosphate. *Science* **170**:404-412.

ROBBINS, W. E., THOMPSON, M. J., YAMAMOTO, R. T., and SHORTINO, T. J. (1965). Feeding stimulants for the female house fly *Musca domestica* Linnaeus. *Science,* **147**:628-630.

RUTLEDGE, L. C., WARD, R. A. and GOULD, D. J. (1964). Studies on the feeding response of mosquitoes to nutritive solutions in a new membrane feeder. *Mosquito News* **24**:407-419.

SALAMA, H. S. (1966). Taste sensitivity to some chemicals in *Rhodnius prolixus* Stal and *Aedes aegypti* L. *J. Insect Physiol.* **12**:583-589.

SHIMAMUNE, A., NAQUIRA, C. and MARIN, R. (1965). Artificial feeding of the blood-sucking bug *Triatoma infestans. Boletin Chileno de Parasitologia* **20**:33-42.

SMITH, J. J. B. and FRIEND, W. G. (1970). Feeding in *Rhodnius prolixus*: responses to artificial diets as revealed by changes in electrical resistance. *J. Insect Physiol.* **16**:1709-1720.

SMITH, J. J. B. and FRIEND, W. G. (1971). The application of split-screen television recording and electrical resistance measurements to the study of feeding in a blood-sucking insect (*Rhodnius prolixus). Can. Ent.* **103**:167-172.

TARSHIS, I. B. (1956). Feeding techniques for bloodsucking arthropods. *Proc. 10th Int. Cong. Entomol.* **3**:767-782.

TAWFIK, M. S. (1968). Feeding mechanisms and the forces involved in some blood-sucking insects. *Quaestiones Entomologicae* **4**:92-111.

WEBER, H. (1930). *Biology der Hemipteren.* Berlin: Springer 543 pp.

WENSLER, R. J. and FILSHIE, B. K. (1969). Gustatory sense organs in the food canal of aphids. *J. Morph.* **129**:473-492.

NUTRITIONAL STUDIES: THEIR BEARING ON THE DEVELOPMENT OF PRACTICAL OLIGIDIC DIETS FOR MASS REARING LARVAE OF THE SCREWWORM, *COCHLIOMYIA HOMINIVORAX*

R. E. Gingrich
Entomology Research Division
Agricultural Research Service
United States Department of Agriculture
Kerrville, Texas 78028

INTRODUCTION

The concept of mass producing millions of arthropods daily, unique when it was done for the first time for the sterile-male release program to eradicate first the screwworm, *Cochliomyia hominivorax* (Coquerel), and subsequently other insects, is now envisaged as a vital part of future programs for the control of other insects. However, eradication of localized populations by the release of sterile flies (the screwworm was limited in its distribution in the United States to southern regions) is impractical with more cosmopolitan species or when other confounding circumstances exist. In such cases, the release of sterilized insects must be used in conjunction with other methods of control, some of which also require the mass production of arthropods. For example, when insect parasites, predators, and obligate pathogens such as viruses are to be produced for release or for application on a per-crop-acre basis, large numbers of insects are required as hosts. Also, geneticists interested in releasing strains of insects with domi nant lethal traits are involved in the mass production of arthropods; and insect attractants provided by the insects themselves, or extracts from them, have possible uses in regulatory and control programs against a wide variety of species. Thus, it has been suggested that mass production of arthropods may well become an important industry to support future control programs (Knipling, 1966).

The expansion of mass rearing of insects envisaged will undoubtedly produce a greater need for understanding of the nutritional requirements of the arthropods in question, both in the planning of the initial phases of the work and also when the quality control, which is so vital to continuing success of a program, is to be considered. For example, arthropods, in adapting to artificial rearing, undergo changes in tolerance and/or need for specific ingredients in their diet that can result in an unsuitable final product. However, studies of nutrition have not previously played the role in the rearing of arthropods that one might expect from the apparent relationship between the two subjects. Initially, many insect species were reared in the laboratory on diets containing whole or processed materials similar to their natural food, and, typically, these same ingredients, plus additives found to improve some aspect of development, are the basis for the diets that are used today. Moreover, when such a diet is judged nutritionally adequate to produce arthropods of the desired quality on a small scale, it naturally is the basis for any diet that is to be used for mass rearing. Thus, when the ingredients are easy to obtain and not excessively costly, the role of nutritional studies in the transition from small-scale to large-scale rearing is incidental.

Nevertheless, even when a transition of this type is smooth, complications can arise that will require changes in diet and hence an understanding of the nutritional needs of the arthropod. For example, the cost of labor and materials, one of the most important considerations in maintaining any large-scale rearing program, can and often does become a problem. Although the technical feasibility of developing systems for continuous mass rearing of arthropods has been proved, the commercial feasibility of mass rearing has not, for the most part, and for this, the greatest need has been and will continue to be lower costs of production. In this area, studies of nutrition can play an important role. Also, fluctuations in the quantity and quality of commercially available ingredients may necessitate changes in diet, or it may be advantageous to use diets to intentionally produce biochemical changes in the insects; for instance, the structure of the fat, and thereby possibly the heat tolerance, could be altered by manipulating the fatty acids in the diet (Bracken and Barlow, 1967).

The many aspects of the relationship between nutritional studies, diet, and the development of mass rearing are well-illustrated by the developments that have taken place in the program for rearing larvae of the screwworm. This insect, one of the first to be mass reared, has now been produced in the laboratory in greater numbers than any other insect—about 75 billion screwworms since the beginning of the eradication programs.

In 1936, Bushland (Melvin and Bushland, 1936) first reared larvae of the screwworm, a heretofore obligate parasite of animals, in an artificial medium that contained 3 parts whole milk, 1 part citrated calf blood, 2 parts ground lean beef, and 0.5% formalin (served as a preservative). The puparia produced weighed 40-60 mg and cost about 30¢/1000 to rear. Thus, they weighed more than the puparia of larvae reared in guinea pigs and rabbits (45 mg) but less than pupae reared in calves (75 mg). Later (Melvin and Bushland, 1941), larger insects were produced by omitting the milk and formulating the diet with 2 parts water, 2 parts beef, 1 part blood, and 0.24% formalin. Moreover, this second medium could be used to rear any quantity of larvae in containers of whatever size was appropriate to the amount of medium. For example, a ½-pint wide-mouth mason jar could be used to rear

200 larvae, an aluminum bun pan sufficed to rear 2000 larvae, and a special rearing vat 10 ft long, 4½ ft wide, and 1¼ in. deep has been used to rear 100,000 larvae (A. J. Graham and F. H. Dudley, Entomology Research Division, 1964, unpublished observations). Thus, the tooling-up for mass rearing the screwworm was primarily a matter for the engineers who fabricated the equipment.

In the early days of the screwworm program (the mid-1950's), when the concept of eradication was still somewhat theoretical, consideration of cost was not as important as later. However, once the technique had been tested and proved feasible and interest had increased in attacking the screwworm in wider areas, lower operating costs became an important goal. The first step was automation and modification of procedures and equipment in the rearing plant; still, the ingredients in the rearing medium were a major cost item and now drew attention.

Early in the program, researchers had found that cost could be reduced by using horsemeat instead of beef, and other substitutes for animal and plant protein were also tested (Graham and Dudley, 1959) that were cheaper than horsemeat. However, the larvae produced weighed less, and it was feared that small larvae might produce small flies that would be at a disadvantage in competing with the larger native males. The experimental confirmation of this idea came later (Alley and Hightower, 1966; Gingrich *et al.*, 1971). (To date, horsemeat, whale meat, whole nutria, and pork lungs, which are all cheaper than beef, have been tested in the medium and found to produce flies of an acceptable size.)

To this point in the program, the major dietetic changes, all made to reduce costs, involved substitutions for the meat portion of the medium. However, these substitutions were occasionally influenced to some degree by availability since other industries, for example, dog-food producers and mink raisers, competed for the supply of low-cost meat so that costs went up and supply down. Therefore, a decision was made to try a new type of medium to effect further savings and to enter a less restricted market. Instead of the semisolid, ground meat-blood mixture that had so long been the standby, investigators at the rearing facility at Mission, Texas, then a part of the Agricultural Research Service, tried a fluid rearing medium (A. J. Graham and F. H. Dudley, Entomology Research Division, 1964, unpublished observations). In the first such attempts, dehydrated ingredients such as fish flour, powdered milk, and cottage cheese were combined with dried blood, dissolved or suspended in water, and poured over a layer of cotton. (The cotton fibers provided the support necessary to prevent the larvae from drowning.) This medium had tremendous potential, both because the ingredients were cheaper and because they did not need to be refrigerated; furthermore, the physical properties of the medium enabled most larvae to be reared on the vats. However, again, the small larvae produced were a problem. Nevertheless, the potential savings in costs of this new technique were so great that nutritional studies were initiated to improve the liquid medium so that larger, to 60 mg or more, insects could be reared. Larval weight was the criterion used to judge the adequacy of diets; other criteria such as time required for larvae to complete development were important but secondary to weight; and the fecundity of adults was determined only for diets that met the other criteria for use. Of course, the ultimate evaluation would be in the performance of the flies in the field, but we did not have then and still do not have any convenient way to determine performance except by the ultimate success of field releases as measured by the decrease in incidence of infested animals and the increase in the number of sterile egg masses collected from wounded animals. Also, we have no reliable means of monitoring the populations of adults in the field to determine density.

NUTRITIONAL STUDIES OF LARVAL SCREWWORMS REARED AXENICALLY

As noted, in the early 1960's, after the successful eradication of the screwworm in the southwestern United States and in anticipation of even larger programs in the rest of the U. S. and other countries, a decision was made to determine the individual, nutritional requirements of screwworm larvae in hopes that the information obtained would aid in formulating new diets. First, axenic rearing techniques were used with an oligidic diet containing casein, yeast extract, cholesterol, minerals, agar, and water; the larvae produced averaged 55 mg. Then, the gross ingredients were all replaced in turn with chemically defined ingredients to produce a chemically defined diet except for agar and RNA (Gingrich, 1964). Subsequently, I made further tests with this meridic diet in which I examined the basic qualitative nutritional requirements of the larvae and the effects of any potentially harmful substances that might be present in the foodstuffs that were being considered as prospective ingredients.

The interactions between many ingredients, particularly the amino acids, were judged so complex that a study of the minimum quantitative requirements would not be of value in formulating an oligidic diet. However, I did use deletion to discover that: screwworm larvae required the same well-known 10 essential amino acids required by other animals and also required proline; the vitamins thiamine, riboflavin, niacin, and pantothenate were necessary for growth, and folic acid and biotin were essential for development to adults; pyridoxine was not required but its presence was not harmful; and screwworm larvae required an exogenous source of cholesterol, choline, salts, and RNA. In addition, I found that concentrations of as little as 0.5% of ribose, glucose, and maltose were harmful, but similar concentrations of glycogen and starch were not, though higher concentrations were. Also, histamine, which forms in fish that are improperly preserved after death, was harmful at a concentration of 1% but not at a concentration of 0.02%. Animal and vegetable oils that also occur in many foodstuffs were harmful at 0.5%, the lowest concentration tested in the meridic diet.

The largest larvae produced on the most nutritionally adequate diet tested during this study averaged only about 40 mg (some insects weighed more than 60 mg) and were clearly inferior to the larvae reared on oligidic diets. Still, the information I obtained was valuable later in developing oligidic diets. Also, the results showed that this former obligate parasite, which previously had been reared only in living animals or fresh byproducts of animals had no unique nutritional requirements that precluded rearing on other materials.

NUTRITIONAL STUDIES ON LARVAL SCREWWORMS REARED XENICALLY

The investigations into nutritional requirements were then shifted to larvae reared xenically in oligidic diets; our goal was to improve the nutritional properties of the liquid medium (Gingrich *et al.*, 1971). By adding 0.2% formalin to the media and frequently replacing spent media with fresh, it was possible to prevent excessive microbial effects that would complicate interpretations of the results. (The levels of mercurial and biological antimicrobial agents that were safe for larvae did not retard decomposition of the medium.) In nature, screwworm larvae attack only living flesh and will vacate a dead animal after decomposition advances, so it was unlikely that superior larvae were produced as a result of exposure to microbial byproducts, and, in fact, any important breakdown in control of contamination would probably adversely affect the larvae. However, some imprecision in testing was expected and undoubtedly occurred in calculating the concentrations of nutrients in the formulations because the information was obtained from product labels, correspondence with the manufacturer, and tables in chemical handbooks.

Nevertheless, the studies were productive. In the first place, the xenic rearing caused me to discover that the different developmental stages of screwworm larvae had different nutrient requirements. Also, since I was rearing in open containers, I could alter the composition of the diet to meet the changing requirements of the larvae, something that is very difficult to do when axenic conditions are required. Thus, in subsequent experiments with liquid media, larvae were typically started on small aliquots of cotton-soaked medium and held in a small chamber at 37°C and 100% RH for 48 hr. Then, a sample of larvae was weighed, and the remainder were placed on larger rearing vats in an open room maintained at 35°C. Samples of larvae were weighed again at 72 hr and also when the larvae finished feeding and migrated from the rearing vat in search of a pupation site. (These 3 weighing times roughly corresponded with the 3 larval stages.)

Secondly, areas of study could be determined by comparing the data accumulated from several years of trial and error with gross foodstuff with the data from experiments with the defined diets. For example, dried whole blood and calf suckle were chosen as the basic ingredients because each is inexpensive, dispersible in water, stable without refrigeration, and abundantly available throughout the year, and each had been used without contra-indications in previous media; also, the combination of these 2 ingredients produced a diet that was at least qualitatively adequate for screwworm larvae in all the major classes of nutrients. However, only 6% blood and 4% suckle could be used because of other (limiting) substances in the formulas, so the total concentration of protein was only about 3%, much too low in view of the 11% level in meat diets. Thus, an additional source of protein was necessary. The previous tests with a variety of fresh meats had indicated that the larvae were not especially fastidious in their needs for amino acids, and this adaptability had been confirmed by the results of the study of the defined diet. However, dehydrated fish products proved unsatisfactory, probably because of the loss of essential amino acids that typically occurs during their production (Morrison *et al.*, 1962) since other protein sources added to supplement the fish flour proteins improved the weights of the larvae. (The suitability or lack of suitability of other protein sources, both fresh and processed, was not related to the amino acid content.) The substitute found for fresh meat was whole, dried, chicken egg. When the egg was combined with blood and suckle, larvae were produced that weighed 70-80 mg, which is similar to the weight of larvae from calves.

Third, whole egg, despite its nutritional appeal, was not a desirable major ingredient because of its high cost and instability without refrigeration. However, the other materials tested that were equivalent to egg in amino acid composition were not satisfactory, which suggested that the blood and calf suckle might be qualitatively or more likely quantitatively deficient in essential nutrients other than protein. The possibility therefore existed of a vitamin deficiency since the addition of yeast extract to a protein source such as cottage cheese improved larval weight. Also, the essential ingredient supplied by egg was shown to occur in the yolk fraction. (In the initial tests with blood and suckle, vitamins were not considered because these two products contained all the essential vitamins required by screwworms.) After analyzing the results of these indications and comparing the vitamin composition of meat and the vitamin requirements of screwworms, I concluded that the diets that lacked meat and egg were deficient in choline. Tests with supplemental choline then confirmed the hypothesis. Thus, with diets based on dried whole blood, calf suckle, and some protein source that lacks choline (such as dried cottage cheese, dried milk, or peptone), supplemental choline was necessary though lecithin could also be used to satisfy the requirements. I found, however, that the need for, tolerance to, and interactions of choline in the liquid oligidic diet are complex (Fig. 1). Increasing the amounts of choline chloride in starting diets that lacked egg increased the weights of 48-hr-old larvae; however, the same

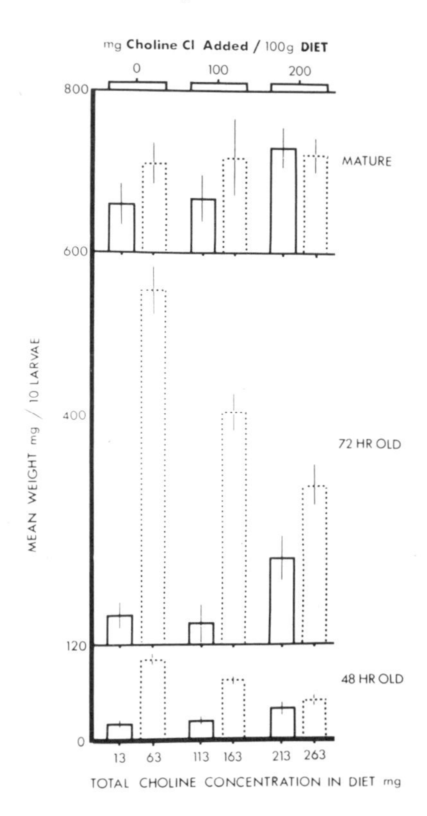

Figure 1. Effects on mean (± SD) weights of larval screwworms of adding amounts of choline chloride to diets. Diets contained 6% whole blood, 4% calf suckle, and 3% cottage cheese; whole chicken egg (2.5%) was added (broken lines) or deleted (solid lines) during the first 48 hr of rearing.

increase in the presence of 2.5% whole egg caused a decrease in the weight of these larvae that persisted in 72-hr-old larvae but not in final-stage larvae. This decrease was not related to the total amount of choline present but to an antagonism between the supplemental choline and substances in the egg. When egg was not included, all stages of larvae responded positively (by higher weights) to increasing levels of choline.

Fourth, the results of the study of defined diets indicated that the screwworm, like all insects, has a requirement for an exogenous source of sterol; and that with these insects the need was specific for cholesterol. Also, in cursory observations, I had noted that larger larvae were produced on an oligidic diet containing pork lung than on diets containing beef or horsemeat. Since the concentration of cholesterol in lung is six times greater than that in muscle, I examined the role of cholesterol in the oligidic diets by increasing the level in the horsemeat-blood mixture 0.1% (supplementation with pure cholesterol) either during the entire rearing (diet 1) or only after the larvae were 48 hr old (diet 2). Then the weights of larvae on these two diets and on the unsupplemented horsemeat-blood mixture (diet 3) were compared with the weights of larvae on pork lung diet (diet 4). The results are shown in Table 1. Cholesterol added to the horsemeat-blood mixture benefited all stages of larvae, but the effect was most striking with 1st-stage larvae: the lower weight gain by these larvae when they were held 48 hr on diet 2 was never recovered even though the concentration of cholesterol was subsequently increased. (After 48 hr, larvae on both diets 1 and 2 had gained about 58% of their total weight, but the larvae that had less gain the first 48 hr were equivalently smaller when mature.) However, no matter when the cholesterol was added, those larvae fed supplemental cholesterol weighed more as mature larvae than when no cholesterol was added. Also, the larvae reared on lung (diet 4) were larger than those reared on diet 3, and this difference too could have resulted because of the greater amount of cholesterol in diet 4 though other nutrients could have been involved. (The smaller weights of the 1st-stage larvae on this diet probably resulted from an excess of fat; we have had rearing problems when lung and other meats were used that have not had the excess fat removed.) I therefore reviewed the concentration of cholesterol in the ingredients used in the liquid oligidic diets containing egg, cottage cheese, calf suckle, and blood. It proved to be 0.12%, a level that was adequate for starting larvae on the meat diets. However, after 48 hr, when egg was no longer included in the diets, the level dropped to 0.06%. However, cholesterol did not increase the weights of these larvae, but since these were not as large as those reared with egg, other deficiencies were probably masking the effects of the cholesterol.

Fifth, carbohydrates have long been recognized as potentially harmful to screwworm larvae. Melvin and Bushland (1941) reported that when 2.6% dextrose or sucrose was added to the meat-blood medium, larvae reached maturity but that the same sugars at 6.7% were lethal. Also, my studies with chemically defined diets showed that the larvae did not require carbohydrates for survival and that the additions of 0.5% ribose, glucose, and maltose were, in fact, harmful. This information was useful in explaining the poor results obtained when fat-free dried milk was used as the protein source

Table 1. Mean weights of larval screwworms reared on media containing pork lung or horsemeat with and without added cholesterol. Test designed to determine the effects of concentration of total cholesterol.

Diet		Supplement of cholesterol	Calculated concentration of cholesterol in diet (mg/g)	Mean wt (mg ± SD) 10 larvae of indicated age	
No.	Basic meat[a]			48 hr old	Mature
1	Horse muscle	1 mg/g (entire rearing)	1.6	348.6±39.5	804.3±30.2
2	Horse muscle	1 mg/g (after 48 hr)	0.6-1.6[b]	303.2±23.8	751.6±19.1
3	Horse muscle	0	.6	318.3± 7.9	690.6±28.0
4	Pork lung	0	3.2	131.9±19.9	708.8±26.1

[a]Horse muscle contained 54.4% lean ground horsemeat, 3.0% dried whole blood, 42.4% water, and 0.2% formalin. Pork lung contained 69.1% lean ground lung meat, 3.8% dried whole blood, 26.9% water, and 0.2% formalin.

[b]Concentration was 0.6 mg/g diet for 48 hr; it increased to 1.6 mg after 48 hr.

instead of cottage cheese: when lactose was added to the cottage cheese diet in an amount equal to its concentration in the milk diet, which is 3X higher than in cottage cheese, the larvae were similarly small. Further testing during which sucrose was added to the liquid diet of blood, calf suckle, and cottage cheese showed that 1st-stage larvae had a low tolerance for carbohydrates; levels of about 2.5% were harmful. However, as the larvae grew older, their tolerance increased, and they grew larger when the level was increased to 3.5%. Thus, sucrose added at the proper time and in the proper amount in these tests improved larval development.

Finally, all the data gained from the many individual experiments were combined in the design of a single test to determine the most expedient method of rearing the heaviest larvae on a liquid oligidic diet. Since screwworm larvae consume only about 1% of their total ration during the first 48 hr of development, we started all larvae on a diet containing whole egg, blood, calf suckle, and cottage cheese (2.5, 6, 4, and 3%, respectively). After 48 hr, egg was omitted, larvae in one group were given 0.5% sucrose for the next 24 hr (diets 4-8, Table 2), and those in the other group received no sugar (diets 1-4, Table 2). Thereafter and until maturity, these "intermediate" larvae were further divided and tested for additional sucrose (0.5 or 1%) and choline (0.2%). The mean weight of 10 larvae after 48 hr on the diets containing egg was 106.6 mg. The mean weight/10 larvae after 24 hr on diet containing 0.5% sucrose (Table 2) was 492.9 mg, significantly higher than the 411.4 mg for larvae on diets without sucrose (t test, $P > 0.01$). Table 2 also shows that the weight advantage of larvae held on the diet with 0.5% sucrose was maintained to maturity and was even increased by additional sucrose. In addition, a slight but nonsignificant benefit was derived from the choline.

DISCUSSION

Examples have been used here from the program for mass rearing screwworms to show how knowledge accumulated from many types of nutritional studies can be applied to solve specific problems as they arise. We obtained useful data from a variety of sources ranging from casual observations during xenic rearing with oligidic diets that contained only a few grossly defined foodstuffs to axenic rearing on diets containing many highly purified chemicals. However, the success of a rearing program also depends on how well data from these various sources can be integrated. Workers in the screwworm eradication program, through their use of such data and because of their continuous efforts to improve the quality and efficiency of their product, reduced the cost of rearing from the initial 30¢/1000 larvae for the first artificial medium to less than 14¢/1000 larvae for the present liquid diet with no apparent adverse effects on the insect. It is anticipated that, with more effort, costs can be even further reduced and insects of better quality produced.

Table 2. Mean weights of mature screwworms when intermediate larvae were fed diets with or without sucrose and then fed finishing diets containing various amounts of choline and sucrose.

Diet no.	% concentration in diet after 72 hr of		Mean weights[a] (mg/10 larvae)
	Choline	Sucrose	
	Intermediate diets contained no sucrose		
1	0	0	465.76[a]
2	.2	0	535.9[b]
3	0	1.0	557.8[b]
4	.2	1.0	501.0[ab]
	Intermediate diets contained 0.5% sucrose		
5	0	.5	622.3[c]
6	.2	.5	636.8[cd]
7	0	1.0	663.58[cd]
8	.2	1.0	676.0[d]

[a]Mean weights of larvae followed by the same letter are not significantly different at the 0.05% level of confidence by Duncan's multiple range test. Values for diet 4 represent average weights of 50 larvae; all other values are average for 100 larvae.

The discovery that different stages of larvae require and tolerate different levels of nutrients was important because it allows much greater latitude in the selection of ingredients. For example, a material that is desirable from the standpoint of cost and abundance but which may be harmful to 1 stage of larva can be used successfully for another stage. Alternatively, a nutritionally desirable ingredient such as egg, which is economically undesirable, can be restricted to use in the diet of 1st-stage larvae, which benefit most from it.

The need for such a broad approach to the solution of dietetic problems stems in part from the crudely defined foodstuffs with which we work. The choice of ingredients for a diet that is to be used in a mass rearing is usually limited to products that were produced for other purposes. Then, as has been the case with the screwworm, these products must be balanced and supplemented with more highly defined ingredients to meet the chemical tolerances and nutritional needs of the specific insect. However, as new programs based on mass rearing procedures are developed and as the press for efficiency intensifies, we can expect that diets will be custom designed and produced in carload quantities for specific needs. Needless to say, extensive studies that will permit simplification of the diet and the proper selection of nutritionally adequate ingredients for each insect will first be necessary.

SUMMARY

The pertinence of nutritional studies using synthetic diets to the modification for economic reasons of mass-rearing procedures is exemplified by the evolution of diets used for the screwworm eradication program. Rising costs of initially satisfactory materials and methods led to a search for cheaper alternatives. Semisolid regimens consisting mainly of meat products requiring refrigeration can be replaced by liquid diets, supported by cotton fibres, based on dried blood, and calf suckle (not requiring refrigeration in storage) and dried whole egg and cottage cheese, variously supplemented with choline, cholesterol and sucrose. Of particular interest is the finding that the nutrient requirements of larvae differ at different stages of development, so that the provision of expensive items such as whole egg may be restricted to only part of the growth period. It is estimated that such liquid diets halve the cost of mass-producing larvae.

REFERENCES

ALLEY, D. A., and HIGHTOWER, B. G. (1966). Mating behavior of the screwworm fly as affected by differences in strain and size. *J. Econ. Entomol.* **59**:1499-1502.

BRACKEN, G. K., and BARLOW, J. S. (1967). Fatty acid composition of *Exeristes comstockii* (Cress) reared on different hosts. *Can. J. Zool.* **45**:57-61.

GINGRICH, R. E. (1964). Nutritional studies on screw-worm larvae with chemically defined media. *Ann. Entomol. Soc. Amer.* **57**:351-60.

GINGRICH, R. E., GRAHAM, A. J., and HIGHTOWER, B. G. (1971). Media containing liquified nutrients for mass-rearing larvae of the screwworm. *J. Econ. Entomol.* **64**:678-83.

GRAHAM, A. J., and DUDLEY, F. H. (1959). Culture methods for mass rearing of screw-worm larvae. *Ibid.* **52**:1006-8.

KNIPLING, E. F. (1966). Introduction, p. 1-12. In Carroll N. Smith (ed.) *Insect Colonization and Mass Production.* Academic Press, New York. 618 p.

MELVIN, R., and BUSHLAND, R. C. (1936). A method of rearing *Cochliomyia americana* C. & P. on artificial media. *USDA Bur. Entomol. Pl. Quar. ET-88.* 2 p.

MELVIN, R., and BUSHLAND, R. C. (1941). The nutritional requirements of screw-worm larvae. *J. Econ. Entomol.* **33**:850-2.

MORRISON, A. B., SABRY, Z. I., and MIDDLETON, E. J. (1962). Factors influencing the nutritional value of fish flour. I. Effects of extraction with chloroform or ethylene dichloride. *J. Nutr.* **77**:97-104.

CELLULAR INTERACTIONS

INTRODUCTION
by

Marion A. Brooks, Section Editor
Department of Entomology, Fisheries, and Wildlife
University of Minnesota
St. Paul, Minnesota 55101

After considerable discussion of the variable needs for specific nutrients, the importance of phagostimulants in feeding behavior, and the optimal formulation of diets, we arrived at the consensus that there must be a wide array of factors as yet unidentified which affect the feeding and growth of insects and mites. These factors may be molecules known to biochemistry but not isolated and manipulated in nutritional work. Alternatively, there is the possibility that hitherto unidentified components might be present, unrecognized, and remaining to be identified chemically. More likely, in laboratory rearing, it is the absence of these components, or an imbalance of components, which is troublesome.

As an example of the role of balance in the control of growth and differentiation, we have before us a description of the presumed significance of changing proportions of water and other nutrients in caste determination of honeybees. The next paper emphasizes the control that the honeybee has over its own body protein components, by regulatory mechanisms completely unknown. The ability of *Tribolium* to metabolize carbohydrates by unconventional pathways is another example of apparent uniqueness in insect biochemistry.

Earlier reports in this conference argued the relative significance of studying the gnotobiotic versus the "natural" animal. In the second group of papers in this section, we attempt to assess the exact contributions made not by chance, ubiquitous contaminants, but by constantly associated mutualists or even by microorganisms so intimately a part of the insect that they are regarded as akin to cell organelles. Growth rate, viability, metamorphosis and reproduction are all affected by these interactions. Very real possibilities of applying insect nutrition strategies in integrated pest control programs will be discussed. These strategies entail the use of food sprays where both honeydew feeding chrysopids and tephritids occur together in the ecosystem.

Finally, with increasing development of insect cell culture, we find ourselves asking whether the requirements at the cellular level, as determined *in vitro*, have any bearing on requirements of the same animal *in toto*. Two papers reveal the similarity of many of the requirements, but emphasize the inadequacy of our present means of culture for precise assays because we lack completely defined media. Even factors which are relatively controllable, such as pH and osmotic pressure, have not been defined for insect cells, and it is surprising to find how much a population of single cells can regulate their environment. Finally, a method for simplifying analysis of culture media is offered as an aid in determining the utilization of nutrients by cultured cells.

Thus the collection of papers in this section bring together a group of fringe areas in nutrition which are basic to clarifying many of the problems of insects and their relationship to the environment.

THE NUTRITIONAL BASIS OF CASTE DETERMINATION IN HONEY BEES

Alfred Dietz
Department of Entomology
University of Georgia
Athens, Georgia 30601

INTRODUCTION

Dimorphism in the female honey bee is a function of nutrition. On the basis of chemical composition and quantity, the larval foods of honey bees have been termed, respectively, royal jelly, worker jelly and modified worker jelly (Haydak, 1959; Townsend and Shuel, 1962). Worker larvae are provisioned with the grayish-white worker jelly from the time of hatching until they are 2½ to 3 days old. The glandular secretions given to worker larvae over 2½ to 3 days old is modified by the addition of pollen and honey, resulting in the yellowish modified worker jelly (Haydak, 1970). Queen larvae are supplied throughout their larval life with an abundance of a white, jelly-like secretion known as royal jelly.

COMPOSITION OF THE LARVAL FOOD OF HONEY BEES

The first chemical analysis of the brood food of the queen, worker and drone was done by von Planta (1888-1889). Subsequent analyses of brood food were reported by Elser (1929), Haydak (1943) and others. Haydak (1957) found that carbohydrates present in the food of 1 to 2 day old drone larvae was 7.53% (fresh matter basis) and it increased to 24.94% for larvae 3 to 5 days old. von Rhein (1951) was able to rear normal drones on worker food, thus indicating that both foods are physiologically equivalent.

Jung-Hoffman (1966) has recently shown that young worker larvae and queen larvae are provided with two types of food components, a white and a clear secretion. She also observed that nurse bees averaging 12 ± 2 days of age secrete the white food component and those of 17 ± 2 days of age produce the clear food component. The ratio of these two components is approximately 1:1. In general, queen larvae are provided with more of the white component and young worker larvae more of the clear component. The clear component is a mixture of honey and hypopharyngeal gland secretion (Haydak, 1970). The white component, however, is a mixture of the secretion of the hypopharyngeal and mandibular gland (Jung-Hoffmann, 1966).

Rembold (1965) made a very extensive and the most recent analysis of

Insect and mite nutrition – North-Holland – Amsterdam, (1972)

royal jelly collected from 3 day old queen larvae, and found it contained 60% water and 40% dry matter (Table 1).

The vitamin composition of royal jelly and the larval food of workers show considerable differences, the most striking being 10-fold more pantothenic acid in royal jelly (Rembold, 1965). Pantothenic acid, along with folic acid, which is approximately 2-fold concentrated in royal jelly, are considered by some investigators to be responsible for the tremendous metabolic changes in honey bee larvae. This assumption could not be substantiated by Rembold (1965). Biopterin, isolated by Butenandt and Rembold (1958), averages 25 μg per gram of royal jelly and 2 μg per gram of worker jelly. The effect of biopterin on caste determination is still unknown.

Table 1. Composition of royal jelly (Rembold, 1965).

Component	Percentage		
A. Water:			60%
B. Dry matter:			40%
1. Lipids:		10%	
a. Strongly acid fraction:	90%		
b. Weakly acid fraction:	2%		
c. Neutral fraction:	8%		
2. Dialyzable material (sugars, amino acids, vitamins, etc.):		52%	
3. Nondialyzable material:		38%	
a. Water-soluble	55%		
b. Water-insoluble	45%		

The strongly acidic lipid fraction of royal jelly consists of a number of free fatty acids, of which 10-hydroxy-2-decenoic acid is the major constituent. Not very much is known at present about the physiological significance of this hydroxydecenoic acid, except that it is important in the preservation of royal jelly. Since it is present in equal concentration in the larval food of workers, it cannot play a role in caste differentiation.

Pain *et al.* (1962) identified adipic, pimelic and suberic acid, and 24-methylene cholesterol in royal jelly, and worker jelly as well. What effect these lipids have in the differentiation of queens is also unknown.

The presence of a remarkably high content of acetylcholine in both foods (1.7 mg/gm in royal jelly and 1.1 mg/gm in worker jelly) (Henschler and von Rhein, 1960), is of interest since acetylcholine is responsible for transmission of nerve impulses.

The protein constituents of royal jelly and worker jelly were investigated by Habowsky and Shuel (1959) on a preparative basis, and on an analytical scale by Patel *et al.* (1960). Qualitative differences were not evident, even though small quantitative differences could not be excluded. Both royal jelly and worker jelly proteins split into 5 electrophoretic bands; while in the modified worker jelly, 2 of these bands are absent. The larval food proteins are also present in the hypopharyngeal gland of nurse bees. Takahashi *et al.* (1963) described biological activity of 2 glycoprotein fractions, one of which showed a slight gonadotrophic effect in mice. A protein fraction with insulin-like activity in royal jelly and worker jelly has been isolated by Dixit and Patel (1964). A highly labile, trophogenic queen determining substance was initially isolated by Rembold and Hanser (1964) from the low molecular, water-soluble components of royal jelly, but so far this has not been characterized (Rembold, 1969).

Analyses of the amino acids have been presented by a number of researchers (see Rembold, 1965). Although there is no considerable quantitative difference between royal and worker jelly, there is a possibility that these small differences are real and not based on biological variation.

In summary, there are no known differences in the major components of royal jelly and worker jelly which can be held responsible for caste determination. However, there are large quantitative differences in pantothenic acid, neopterin and biopterin; and to what extent these 3 substances, or others which are still unknown, bring about the change of a worker larva to a queen larva can only be determined by biological tests.

THEORIES OF EFFECT OF NUTRITION ON CASTE DETERMINATION

Since the classical experiment of von Rhein (1933), who was the first one to rear successfully bee larvae outside the colony on royal jelly, but who failed to produce a queen, several workers such as Weaver (1955-1956), Michael and Abramovitz (1955), Hoffman (1956), Smith (1959), Jay (1959), and Dietz (1966) have been able to produce queens and workers in the laboratory.

Hormonal deficiency during sensitive period

A report that the labile queen determining substance is also present in extracts of male and female silkworms (Rembold, 1969) is of interest. The amount of determining substance present in 1000 silkworms is equivalent to the amount present in 1 kg of royal jelly. Based on this finding Rembold (1969) proposed the following working hypothesis: Caste determination is a consequence of hormonal deficiency during a sensitive developmental period. The purpose of the determining substance is to correct this deficiency, so that queens will develop; workers, on the other hand, are the result of hormonal deficiency.

Increasing moisture content of royal jelly

Haydak (1943) pointed out that larvae which eat more food should get more of the "essential nutrients" and develop into queens, while larvae which consume less should turn out to be workers or intermediates. The effect of larval food moisture on queen differentiation has been proposed by Dietz and Haydak (1967, 1971).

In a normal colony of bees, queens and workers are reared from bipotent female larvae. During larval growth, the food supplied to queen larvae does not undergo any striking changes, as previously pointed out, while the worker larvae receive a considerably modified food in the latter period of development. The sequential feeding of worker larvae is thought to play a role in caste differentiation of honey bees. The actual differentiation mechanism, however, has remained obscure.

In all of our previous rearing experiments, in which laboratory reared larvae were fed royal jelly of unchanged consistency throughout the growth period of the larvae, we obtained a roughly 1:1:2 ratio of queens, intermediates and workers.

However, analysis of larval foods (Elser, 1929; Gontarski, 1949; Smith, 1959; Dietz, 1966) showed that royal jelly supplied to queen larvae (1 to 3 days old) consistently had a low moisture content (50% to 65%) which increased (62% to 69%) in the food of older larvae. The opposite trend was demonstrated in the case of worker larvae, where a rather high moisture content (73% to 75%) was evident for the food of young larvae (1 to 3 days old) but a considerably lower one (64% to 62%) for that of older larvae. If the combined moisture contents of royal jelly in subsequent days of larval development, measured by some of these investigators, are plotted against time of development, then a linear relationship is evident indicating a gradual increase in the moisture content of the food during the growth of queen larvae.

Studies by Stabe (1930), Wang (1965), and Dietz (1966) on the growth of queen and worker larvae in a normal colony showed that during the early period of life (1 to 3 days) the queen larvae are generally restricted in their growth, while the worker larvae grow at a faster rate; in the later period of their development the growth of worker larvae is restricted or reduced, while the queen larvae now grow at a greatly increased pace. Under normal colony conditions, the growth curves for the queen and the worker larvae and the corresponding increase or decrease respectively in the moisture content of the larval foods consistently show the same correlation. It was reported (Weaver, 1955; Smith, 1959; Petit, 1963) that the addition of water, lowering the total solids of royal jelly, appeared to improve larval growth.

The relationship of moisture content to weight and survival of transition larvae (prepupae) was illustrated by Dietz and Haydak (1971). Based on the number of queens, intermediates and workers obtained, there was no evidence that food with a constant high or low moisture content has any influence on caste determination.

On the other hand, a high percentage of queens was produced by gradually increasing the moisture content of royal jelly, either fresh or stored in a refrigerator for a considerable length of time. No labile substance seemed to be involved in queen differentiation (Dietz and Haydak, 1970). It seems possible that nurse bees may initiate the mechanism of queen differentiation by regulating the moisture content of the food and thus the food intake of the growing queen larvae.

Food consumption at specific periods during larval development

It is impossible to study honey bee larval food consumption by gravimetric methods, particularly because of extensive evaporation of the exposed food and because the larvae do not defecate until their feeding period is almost completed. A method of measuring the intake of ^{32}P-labeled royal jelly in live larvae suspended within the liquid scintillator in a sealed inner chamber was developed (Dietz and Lambremont (1970a).

With this technique, it was possible to obtain data to support another explanation, advanced by Dietz and Lambremont (1970b), dealing with the amount of royal jelly consumed by individual larvae at specific periods of development. The difference in growth and weight gain of queen and worker larvae seems to indicate that food consumption of individual larvae, at specific periods of development, plays an important role in caste differentiation. Haydak (1943), Jay (1964) and Dietz (1966) found that some queen larvae removed from the food before completion of their feeding period could develop into queens, although some individuals became intermediates and workers.

In general, the female larvae developing into queens had their largest food intake on the 6th day of the feeding period. However, larvae destined to become queens ate 13% more food than workers during the first 3 days of larval life. This difference increased to approximately 40% after 6 days of larval life. The mean rate of ingestion was 8% less in intermediates as compared to queen larvae during the first 3 days after hatching, and 16% less after 6 days of larval life. Queen larvae consumed an average of 5% more royal jelly than intermediates, and 19% more than workers. In the case of intermediates and workers, maximum food intake was observed at the 168-hr time period. The mean prepupal weights, however, were rather similar between queens and intermediates; while the prepupal weights of workers were approximately 25% lower than those of queens. There is very little difference in the total food intake and prepupal weight between queens and intermediates. The difference between workers and queens, however, is statistically significant.

EFFORTS TO USE ARTIFICIAL DIETS

A major critical problem in laboratory rearing experiments is the inability to produce consistently a large number of adults. In very few instances is it possible to rear all test larvae to the adult stage. As a matter of fact, a consistent success rate of 50% would be a considerable improvement. Shuel and Dixon (1968) recently showed that the addition of sugar to worker jelly to make the total sugar concentration equal to that of royal jelly permits the development of normal honey bee pupae.

However, in addition to an improvement in the rearing procedure, a partial or complete synthetic diet could be a great asset not only in elucidating the nutritional requirements of honey bee larvae, but also in shedding some light on other factors possibly important in queen differentiation.

The nutritional effects of feeding various artificial diets were studied by Bertholf (1927), von Rhein (1933), Michael and Abramovitz (1955), Weaver (1956), Gontarski (1958), Patel and Gochnauer (1959) and Hoffmann (1960). The results of these studies indicated that rearing of 2 to 3 day old honey bee larvae on such diets is only partially successful. Our own results to this date have also shown only limited success in rearing larvae on partial and complete synthetic ciets (Dietz, 1972), but efforts are continuing along these lines. In general, it can be said that it is at present very difficult to rear 1 to 3 day old larvae, and impossible to rear newly hatched larvae, to the adult stage on chemically defined diets.

The findings of Luscher (1969) that soldier development in the termite *Kalotermes flavicollis* can be induced by injecting or feeding synthetic juvenile hormone to larvae and pseudergates is of considerable interest. Chai and Shuel (1970) reported that injections of 2.2 or 4.4 μg of farnesol or farnesyl methyl ether into 4 day old worker larvae did not affect the number of ovarioles but resulted in the development of oocytes. Larger doses, however, killed the larvae within a few days. Bowers *et al.* (1965) demonstrated that amounts in excess of 5000 μg of farnesol and farnesyl methyl ether are required to induce gonadotropic effects in *Periplaneta.*

The suggestion advanced by Haydak (1943) that the anatomical and physiological differences between the queen and the worker honey bee are due to hormones, and a modification of this concept by Shuel and Dixon (1960), plus the above mentioned findings, warrants an investigation of the interrelationship of the morphogenetic hormones during caste determination. In this connection, the report by Roller and Dahm (1968) that juvenile hormone must act prior to the action of edcysone in order to exert its morphogenetic activity in *Galleria mellonella* is of special interest to us. In view of these results and various other findings, we recently have initiated feeding experiments with various combinations of morphogenetic hormones. At present, however, our data are insufficient to warrant specific conclusions.

Acknowledgement.–I am most grateful to Dr. Karl Weiss, Bavarian Honey Bee Research Institute, Erlangen, West Germany, for our stimulating discussions on caste determination during his stay in Athens, Georgia.

REFERENCES

BERTHOLF, L. M. (1927). Utilization of carbohydrates as food by honey bee larvae. *J. Agric. Res.* 35:429-452.

BOWERS, W. S., THOMPSON, M. J. and UBEL, E. C. (1965). Juvenile and gonadotropic hormone activity of 10,11-epoxyfarnesenic acid methyl ester. *Life Sci.* 4:2323-2331.

BUTENANDT, A. and REMBOLD, H. (1958). Uber den Weiselfuttersaft der Honigbiene. II. Isolierung von 2-amino-4-hydroxy-6 (dihydroxypropyl) pteridin. *Hoppe-Seyler's Z. Physiol. Chem.* 311:78-83.

CHAI, B. I. and SHUEL, R. W. (1970). Effects of supernumerary corpora allata and farnesol compounds on ovary development in the worker honeybee. *J. Apic. Res.* 9:19-27.

DIETZ, A. (1966). The influence of environmental and nutritional factors on growth and caste determination of female honey bees. Ph.D. Thesis, Univ. of Minnesota, 134 pp.

DIETZ, A. (1972). Longevity and survival of honey bee larvae on artificial diets. *J. Georgia Entomol. Soc.* **(In Press).**

DIETZ, A. and HAYDAK, M. H. (1967). Caste determination in honey bees. The significance of moisture in larval food. *Proc. Int. Apicult. Congr.*, College Park, Md. 21:470 (Abst.).

DIETZ, A. and HAYDAK, M. H. (1970). The effect of refrigerated royal jelly stored for several years on larval growth and development of female honey bees. *J. Georgia Entomol. Soc.* 5:205-206.

DIETZ, A. and HAYDAK, M. H. (1971). Caste determination in honey bees. I. The significance of moisture in larval food. *J. Exp. Zool.* 177:353-358.

DIETZ, A. and LAMBREMONT, E. N. (1970a). A method for studying food consumption of live honey bee larvae by liquid scintillation counting. *Ann. Entomol. Soc. Amer.* 63:1340-1342.

DIETZ, A. and LAMBREMONT, E. N. (1970b). Caste determination in honey bees. II. Food consumption of individual honey bee larvae determined with ^{32}P-labeled royal jelly. *Ann. Entomol. Soc. Amer.* 63:1342-1345.

DIXIT, P. and PATEL, N. G. (1964). Insulin-like activity in larval foods of the honey bees. *Nature* 202:189-190.

ELSER, E. (1929). Die chemische Zusammensetzung der Nahrungsstoffe der Biene. *Märkische Bienenzeitung* 19:211-215; 232-235; 248-252.

GONTARSKI, H. (1949). Mikrochemische Futtersaftuntersuchungen und die Frage der Königinentstehung. *Die Hessische Biene Nr.* 9:89-92.

GONTARSKI, H. (1958). Die Larvennahrung und die Koniginenentstehung bei Honigbienen. *Proc. Int. Beekeep. Congr.* 17:49 (Abst.).

HABOWSKY, J. E. J. and SHUEL, R. A. (1959). Separation of the protein constituents of the larval diets of the honeybee by continuous flow electrophoresis. *Can. J. Zool.* **37**:957-964.

HAYDAK, M. H. (1943). Larval foods and development of castes. *J. Econ. Entomol.* **36**:778-792.

HAYDAK, M. H. (1957). The food of the drone larvae. *Ann. Entomol. Soc. Amer.* **50**:73-75.

HAYDAK, M. H. (1959). Amer. Comm. Bee Research Assoc., 2nd Meeting, Detroit.

HAYDAK, M. H. (1970). Honey bee nutrition. *Annu. Rev. Entomol.* **15**:143-156.

HENSCHLER, D. and RHEIN, W. VON. (1960). Änderungen des Acetylcholingehaltes von Bienenfuttersäften in der Madenentwicklung. *Naturwissenschaften* **47**:326-327.

HOFFMAN, I. (1956). Die Aufzucht weiblicher Bienenlarven (*Apis mellifica* L.) ausserhalf des Volkes. *Z. Bienenforsch.* **3**:134-138.

HOFFMAN, I. (1960). Rearing worker honeybee larvae in an incubator. *Bee World* **43**:119-122.

JAY, S. C. (1959). Factors affecting the laboratory rearing of honeybee larvae (*Apis mellifera* L.). M.S. Thesis, Univ. of Toronto, 61 pp.

JAY, S. C. (1964). Rearing honeybee brood outside the hive. *J. Apic. Res.* **3**:51-60.

JUNG-HOFFMANN, I. (1966). Die Determination von Königin und Arbeiterin der Honigbiene. *Z. Bienenforsch.* **8**:296-322.

LÜSCHER, M. (1969). Die Bedeutung des Juvenilhormons fur die Differenzierung der Soldaten bei der Termite *Kalotermes flavicollis. Proc. Congr. IUSSI, Bern* **6**:165-170.

MICHAEL, A. S. and ABRAMOVITZ, M. (1955). A method of rearing honeybee larvae *in vitro. J. Econ. Entomol.* **48**:43-44.

PAIN, J., BARBIER, M., BOGDANOVSKY, D. and LEDERER, E. (1962). Chemistry and biological activity of the secretions of queen and worker honeybees *(Apis mellifica* L.). *Comp. Biochem. Physiol.* **6**:233-241.

PATEL, N. G. and GOCHNAUER, T. A. (1959). The toxicity of extracts from foul-brood scale residues for honey bee larvae maintained *in vitro. J. Insect Pathol.* **1**:190-192.

PATEL, N. G., HAYDAK, M. H. and GOCHNAUER, T. A. (1960). Electrophoretic components of the proteins in the honeybee larval food. *Nature* **186**:633-634.

PETIT, I. (1963). Étude in vitro de la croissance des larves d'abeilles *(Apis mellifera* L.). *Ann. Abeille.* **6**:35-52.

PLANTA, A. VON. (1888-1889). Über den Futtersaft der Bienen. *Hoppe-Seyler's Z. Physiol. Chem.* **12**:327-354; 552-561.

REMBOLD, H. (1965). Biologically active substances in royal jelly. In *Vitamins and Hormones* 23:359-382. (Transl. from German by A. DIETZ), Academic Press, Inc., N. Y.).

REMBOLD, H. (1969). Biochemie der Kastenentstehung bei der Honigbiene. *Proc. Congr. IUSSI, Bern* **6**:239-246.

REMBOLD, H. and HANSER, G. (1964). Uber den Weiselzellenfuttersaft der Honigbiene, VIII. Nachweis des determinierenden Prinzips im Futtersaft der Königinenlarven. *Hoppe-Seyler's Z. Physiol. Chem.* **339**:251-254.

RHEIN, W. VON. (1933). Über die Entstehung des weiblichen Dimorphismus in Bienenstaate. *W. Roux Arch. Entwicklungsmech. Organ.* **129**:601-655.

RHEIN, W. VON. (1951). Über die Ernährung bei Dronenmaden. *Z. Bienenforsch.* **1**:63-66.

RÖLLER, H. and DAHM, K. H. (1968). The chemistry and biology of juvenile hormone. *Recent Prog. in Hormone Res.* 24:651-680.

SHUEL, R. W. and DIXON, S. E. (1960). The early establishment of dimorphism in the female honeybee, *Apis mellifera* L. Insectes sociaux 7:265-282.

SHUEL, R. W. and DIXON, E. (1968). The importance of sugar for the pupation of the worker honey bee. *J. Apic. Res.* 7:109-112.

SMITH, M. V. (1959). Queen differentiation and the biological testing of royal jelly. Mem. Cornell Agric. Exp. Sta. No. 356.

STABE, H. A. (1930). The rate of growth of worker, drone and queen larvae of the honey bee *Apis mellifera* Linn. *J. Econ. Entomol.* 23:447-453.

TAKAHASHI, K., KIUCHI, K., ENDO, M., FURNO, Y., SHINOZUKA, T. and HASE-GANA, S. (1963). Active fractions of royal jelly. Chem. Abstr. 59:13246.

TOWNSEND, F. G. and SHUEL, R. W. (1962). Some recent advances in apicultural research. *Annu. Rev. Entomol.* 7:481-500.

WANG, D. I. (1965). Growth rates of young queen and worker honeybee larvae. *J. Apic. Res.* 4:3-5.

WEAVER, N. (1955). Rearing of honey bee larvae in the laboratory. *Science* 121:509-510.

WEAVER, N. (1956). Rearing honeybee larvae in the laboratory. *X. Int. Congr. Entomol.* 4:1031-1036.

LACK OF EFFECT OF DEFICIENT DIETS ON HEMOLYMPH PROTEINS OF ADULT WORKER HONEYBEES

Martha Gilliam
Entomology Research Division, Agricultural Research Service, U.S.D.A., Tucson, Arizona 85719, U.S.A.

INTRODUCTION

The present investigation was therefore made to examine more closely the effect of diet on hemolymph proteins in adult worker bees and to determine whether diet affects the electrophoretic patterns of hemolymph proteins. Since Herbert *et al.* (1970) had found that dandelion pollen was deficient in L-arginine, an amino acid essential for bees (Lue and Dixon, 1967), and that cages of confined newly emerged bees were unable to sustain brood rearing when fed a diet of pure dandelion pollen (unless the diet was fortified with L-arginine), I decided to feed this deficient diet to one colony of bees and to feed only 50% sucrose solution to another colony. A deficiency in hemolymph protein might then be expected with either diet.

Recently, when the proteins of all stages of adult worker honeybees, *Apis mellifera* L., were separated by polyacrylamide gel disc electrophoresis (Gilliam and Jackson, 1972a), the protein patterns of developing worker bees (eggs, larvae, and pupae) were found to change during development and aging. However, hemolymph proteins remained unchanged in the mature adult bee, and no electrophoretic differences were detected. Subsequently, the electrophoretic patterns of enzymes (lactate dehydrogenase, α-glycerophosphate dehydrogenase, esterase, and malate dehydrogenase) in the hemolymph of adult worker honeybees were also found to remain constant throughout the life of the adult (Gilliam and Jackson, 1972b). It therefore appeared that hemolymph proteins remained unchanged in the adult worker bee regardless of age or diet though adult worker bees change from nursing to foraging as they grow older. In general, protein concentration in insect hemolymph can be greatly influenced by nutrition, and hemolymph protein serves as an important reserve (Chen, 1966). Then since changes in electrophoretic patterns of adult worker bees of different ages receiving different diets or working different plants were never observed, the possibility existed that bees differed in this respect from other insects.

MATERIALS AND METHODS

Approximately 4000 worker bees were placed in each of two small hive nuclei indoors (as soon as they had emerged and before feeding) and provided with water and food. One colony was fed 7.5% protein dandelion pollen diet as candy, water, and 50% (w/w) sugar solution. The other colony was given 50% sugar solution and water only. Both colonies were provided with a caged queen.

Bees were collected daily at 2:00 p.m. from each of the hives. The hemolymph was collected from the thorax after decapitation of the bees as previously described (Gilliam and Jackson, 1972a). Then this freshly collected hemolymph from each bee was immediately subjected to separation by a Canalco®* Model 12 polyacrylamide gel disc electrophoresis apparatus. The procedures were essentially those of Davis (1964) and were described previously (Gilliam and Jackson, 1972a). The hemolymph from 12 bees from each colony was analyzed before feeding and on each of the following 26 days and again after 31, 32, 89, 90, and 91 days. All electrophoresis gels were analyzed with a Photovolt® densitometer and automatic integrator.

RESULTS

Fig. 1 shows the electrophoretic pattern of hemolymph proteins obtained for all bees which were analyzed. These banding patterns were identical to those obtained from normal adult worker bees (Gilliam and Jackson, 1972a). Thus, no electrophoretic differences were detected in any bees, whether they received a complete diet, a deficient protein + carbohydrate diet, or only a carbohydrate diet. In addition, the protein patterns were identical in adult bees of all ages. Densitometry confirmed visual results. This experiment was terminated after 91 days since no changes in hemolymph protein patterns were evident.

Electrophoretic patterns of hemolymph proteins obtained from the 12 bees from each colony before feeding were the same as those found in newly emerging worker honeybees (Gilliam and Jackson, 1972a).

---

*Mention of a proprietary product does not constitute an endorsement of this product by the USDA.

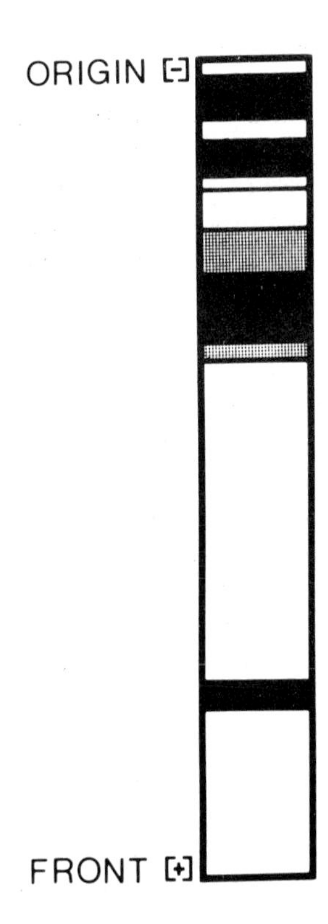

Figure 1. Electropherogram of the hemolymph of adult worker honeybees.

DISCUSSION

Since bees from both test groups had electrophoretic patterns identical to those of bees receiving complete diets, the test diets apparently had no effect on hemolymph proteins. This result is in contrast to the general condition that protein concentration in insect hemolymph is greatly influenced by nutrition (Chen, 1966). This discrepancy might be explained by the fact that hemolymph proteins which change during aging in other female adult insects are involved in egg production as in *Schistocerca* (Hill, 1962), *Hyalophora* (Telfer and Williams, 1953; Telfer, 1954), *Calliphora* (Strangways-Dixon, 1962), and *Phormia* (Orr, 1964a,b). Furthermore, Van der Geest (1968) found no differences in the hemolymph protein concentration of female larvae of *Pieris brassicae* L. reared on three different diets.

Also, Maurizio (1965) observed that the sugar spectrum of the hemolymph of adult worker bees was not influenced by protein food. Halberstadt (1966) found that if worker bees were reared with sucrose but without protein so the hypopharyngeal glands were undeveloped, the electropherogram of the hypopharyngeal gland secretion appeared lighter, but the arrangement of the bands and the quantitative ratio of the proteins were not influenced.

Therefore, the electrophoretic patterns of hemolymph proteins of adult worker bees apparently remain unchanged by nutritional differences and deficiencies. In addition, aging has no effect on electrophoretic patterns. (Perhaps I should note that the bees used in these experiments were not dying at an abnormally high rate. In fact, Haydak (1937a) maintained bees on a pure carbohydrate diet for 189 days).

Bees fed only carbohydrate can probably convert sugar to non-essential amino acids and amino acids to protein. However, the reason or the mechanism for this maintenance of protein stability in the hemolymph is not known. It is possible that proteins produced elsewhere in the body of the bee can be transported to the hemolymph when needed. Chippendale and Kilby (1970) suggested that the midgut of *Pieris brassicae* L. might synthesize proteins which would be released into the haemolymph protein pool. Proteins from other places in the body of the bee may serve as reserves instead of those in the hemolymph. Feir and Krzywda (1969) suggested that the milkweed bug, *Oncopeltus fasciatus* (Dallas) used its tissue mass to maintain its haemolymph protein level. Also, Haydak (1935, 1937a,b) found that the dry weight and nitrogen content of the bodies of bees maintained on carbohydrate diets diminished. The concentration of hemolymph protein in adult worker bees has been reported to vary with diet (Poltev and Karbaskova, 1969) and age (Foti *et al.*, 1969). Perhaps, electrophoretic techniques are not sensitive enough to detect small differences.

SUMMARY

Hemolymph proteins of adult worker honeybees, *Apis mellifera* L., from two colonies were separated daily for 26 days and again at days 31, 32, 89, 90, and 91. One colony of bees was given water, 7.5% protein dandelion pollen diet (deficient in L-arginine), and 50% (w/w) sucrose solution. The other colony received only water and 50% sucrose solution. All bees examined after feeding contained identical electrophoretic patterns of hemolymph proteins regardless of age or diet. These banding patterns were identical to those obtained from normal adult worker bees receiving complete diets (Gilliam and Jackson, 1972a).

Acknowledgements—I thank Mr. Keith Doull, who was on leave from the Waite Agricultural Research Institute of the University of Adelaide in South Australia, for assistance with the bees in the earlier part of this work. Mrs. Karen Jackson provided excellent technical assistance. The investigation was conducted in cooperation with the Arizona Agricultural Experiment Station and was supported in part by Cooperative Agreement Grant No. 12-14-100-9062 (33) from the Entomology Research Division of the USDA.

REFERENCES

CHEN, P. S. (1966). Amino acid and protein metabolism in insect development. *Adv. Insect Physiol.* 3:53-132.

CHIPPENDALE, G. M. and KILBY, B. A. (1970). Protein biosynthesis in larvae of the large white butterfly, *Pieris brassicae. Comp. Biochem. Physiol.* 34:241-243.

DAVIS, B. J. (1964). Disc electrophoresis. II. Method and application to human serum proteins. *Ann. N. Y. Acad. Sci.* 121:404-427.

FEIR, D. and KRZYWDA, L. (1969). Concentration of insect hemolymph proteins under various experimental conditions. *Comp. Biochem. Physiol.* 31:197-204.

FOTI, N., POPA, L. and CRISAN, I. (1969). Variabilité du composant protéique de l'hémolymphe chex les abeilles *(Apis mellifica* L.) par rapport á leur age, á la saison et á leurs activités. *Bull. Apic. Doc. Sci. Tech. Inf.* 12:121-129.

GILLIAM, M. and JACKSON, K. K. (1972a). Proteins of developing worker honey bees, *Apis mellifera. Ann. Entomol. Soc. Amer.* 65:516-517.

GILLIAM, M. and JACKSON, K. K. (1972b). Enzymes in honey bee (*Apis mellifera* L.) hemolymph. *Comp. Biochem. Physiol.* **In press.**

HALBERSTADT, K. (1966). Über die Proteine der Hypopharynxdrúse der Bienenarbeiterin. I. Electorphoretischer Vergleich von Sommer-, Winter- und gekäfigten Bienen. *Annls Abeille* 9:153-163.

HAYDAK, M. H. (1935). Brood rearing by honeybees confined to a pure carbohydrate diet. *J. Econ. Entomol.* 28:657-660.

HAYDAK, M. H. (1937a). Changes in weight and nitrogen content of adult worker bees on a protein-free diet. *J. Agric. Res.* 54:791-796.

HAYDAK, M. H. (1937b). The influence of a pure carbohydrate diet on newly emerged honeybees. *Ann. Entomol. Soc. Amer.* 30:258-262.

HERBERT, E. W., BICKLEY, W. E. and SHIMANUKI, H. (1970). The brood-rearing capability of caged bees fed dandelion and mixed pollen diets. *J. Econ. Entomol.* 63(1):215-218.

HILL, L. (1962). Neurosecretory control of haemolymph protein concentration during ovarian development in the desert locust. *J. Insect Physiol.* 8:609-619.

LUE, P. F. and DIXON, S. E. (1967). Studies on the mode of action of royal jelly in honeybee development. VIII. The utilization of sugar uniformly labelled with ^{14}C and aspartic-1-^{14}C acid. *Can. J. Zool.* 45:595-599.

MAURIZIO, A. (1965). Untersuchungen über das Zuckerbild der Hämolymphe der Honigbiene (*Apis mellifera* L.). I. Das Zuckerbild des Blutes erwachsener Bienen. *J. Insect Physiol.* 11:745-763.

ORR, C. W. M. (1964a). The influence of nutritional and hormonal factors on egg development in the blowfly *Phormia regina* (Meig.). *J. Insect Physiol.* 10:53-64.

ORR, C. W. M. (1964b). The influence of nutritional and hormonal factors on the chemistry of the fat body, blood and ovaries of the blowfly, *Phormia regina* Meig. *J. Insect Physiol.* 10:103-119.

POLTEV, V. I. and KARBASKOVA, V. B. (1969). [Experimental protein insufficiency in bees (*Apis mellifera* L.) and the influence of certain protein foods on its prevention.] *Vses. Akad. Sel'skokhoz. Nauk. V. I. Lenina, Dokl.* **4**:34-36.

STRANGWAYS-DIXON, J. (1962). The relationship between nutrition, hormones and reproduction in the blowfly *Calliphora erythrocephala* (Meig.). III. The corpus allatum in relation to nutrition, the ovaries, innervation and the corpus cardiacum. *J. Exp. Biol.* **39**:293-306.

TELFER, W. H. (1954). Immunological studies of insect metamorphosis. II. The role of a sex-limited blood protein in egg formation by the *Cecropia* silkworm. *J. Gen. Physiol.* **37**:539-558.

TELFER, W. H. and WILLIAMS, C. M. (1953). Immunological studies of insect metamorphosis. I. Qualitative and quantitative description of blood antigens of the *Cecropia* silkworm. *J. Gen. Physiol.* **36**:389-413.

VAN DER GEEST, L. P. S. (1968). Effect of diets on the haemolymph proteins of larvae of *Pieris brassicae*. *J. Insect Physiol.* **14**:537-542.

THE CARBOHYDRATE NUTRITION OF *TRIBOLIUM*

Shalom W. Applebaum
Faculty of Agriculture, The Hebrew University
Rehovot, Israel

The rust-red flour beetle, *Tribolium castaneum* Herbst., exhibits a relatively wide host distribution: Other than its common infestation of wheat flours and products, it can successfully develop in various legume meal products and in whole groundnuts, for it is able to adequately utilize lipid and protein in lieu of carbohydrate as a source of energy. In previous work, using defined diets, we had demonstrated that *Tribolium* develops well on as little as 5% and up to 60% protein, and equally well on high glucose, maltose, dextrin or starch diets. It seemed appropriate at the time to conclude that *Tribolium* is capable of utilizing excess protein for energy production in all cases, and on the other hand is not particularly exacting as to the molecular size of available carbohydrate, being metabolically equipped for degrading oligo- and polysaccharides to glucose residues. I intend in this report to present data which necessitate the qualification of these metabolic characteristics of *Tribolium*. These reservations arose in the process of an attempt to reassess the developmental response of *Tribolium* to different dietary hexoses and pentoses on the basis of data available at the time (Pant & Dang, 1965).

Preliminary experiments indicated that galactose, mannose and xylose, individually substituting for 10% of the rice starch in standard experimental diets, had no apparent effect on larval development of *Tribolium*. The standard experimental diet contained 80% carbohydrate (rice starch), 5% dried yeast, 7% gluten, 6% cellulose, 1% cholesterol and 1% salts.

Further experiments with galactose showed that higher concentrations markedly depressed development, and that this adverse effect was significantly more acute when glucose was the supplementary carbohydrate, than in the presence of rice starch (Fig. 1).

The toxicity of galactose to higher animals has been attributed to biochemical lesions (e.g. Tygstrup & Keiding, 1969; Mays *et al.*, 1970) and to hyperosmolar dehydration (Malone *et al.*, 1971). Galactose is detrimental to adult *Anthonomus grandis* and it has been shown that this hexose is predominantly reduced in this insect to the sugar alcohol galactitol, which is a dead-end metabolite for this organism (Nettles, Jr. & Burks, 1971).

There was no obvious reason why development on galactose/starch diets should be better than on galactose/glucose diets. The combined action of amylases and glucosidases should result in substantial hydrolysis of starch

Insect and mite nutrition – North-Holland – Amsterdam, (1972)

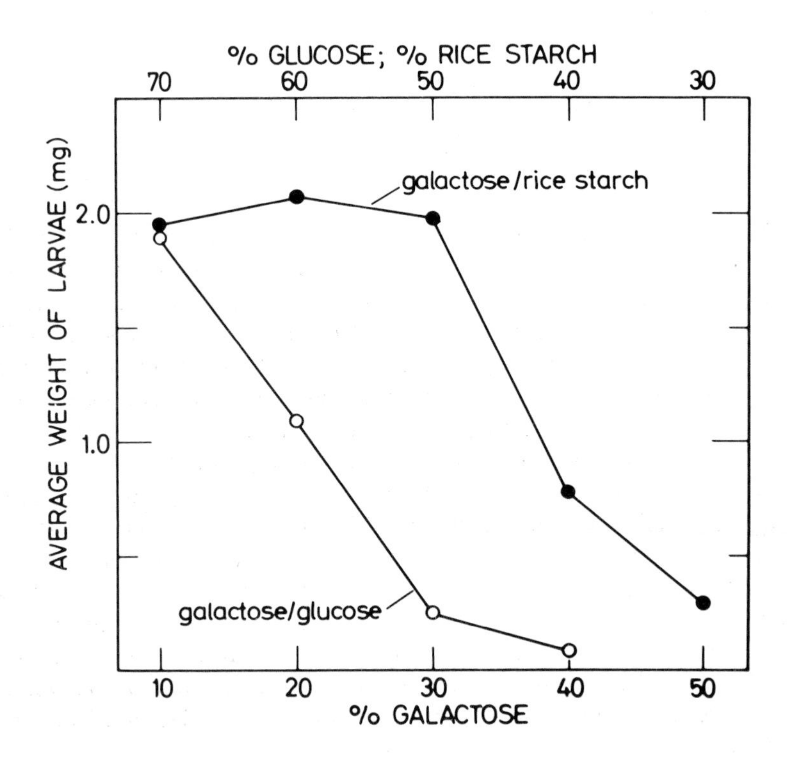

Figure 1. The effect of galactose/rice starch and galactose/glucose combinations on larval development of *Tribolium castaneum*. Total carbohydrate was 80% of the diet. Average larval weights are from 5 replicates of 10 larvae each, reared at 32°C and weighed 12 days after hatching.

to glucose, and this partial release of glucose would not be expected to exceed the energetic value of a similar diet in which glucose is *a priori* the major source of energy. Several further observations are pertinent to this problem:

(1) Purified α-amylase inhibitors, isolated from wheat (Shainkin & Birk, 1965), inhibit *Tribolium* larval midgut amylase *in vitro* at very low concentrations—less than 1 μg per reaction mixture effectively inhibits 50% of the amylolytic activity. Yet relatively high levels (0.2% and 0.4%) of these inhibitors in diets based on 80% rice starch had no apparent effect on larval development. We know that α-amylase inhibitor activity is retained after prolonged digestion *in vitro* by intestinal proteases of *Tribolium*. This seems to suggest that starch can be digested by *Tribolium* in some manner unrelated to conventional amylolysis.

(2) Glycogen, which is inferior to starch as a substrate *in vitro* for *Tribolium* larval amylase (but is the preferred substrate for glycogen phosphorylase), sustains excellent growth of *Tribolium*. The optimal growth obtained contrasts with what we formerly reported (Applebaum & Konijn, 1965) and was obtained with a preparation of glycogen free of trace nucleotide contamination which is commonly present in commercial preparations.

(3) Larval development of *Tribolium* on a 30% galactose diet supplemented with α-amylase-limit-dextrin of potato starch (Applebaum, 1966) is depressed to the same extent as on a similar galactose/glucose diet. This limit dextrin contains much organic phosphorus bound to glucose residues at the 6-carbon (Posternak, 1951), a composition amenable to α-amylase but not to phosphorylase.

These observations are all compatible with phosphorylytic degradation of dietary polysaccharide, whereby hexose phosphate esters are produced and are degraded in a manner and at a stage which is not inhibited by free galactose or galactose metabolites. Galactose enters into the general metabolism of carbohydrates before the main junction of hexose metabolism. It has been suggested that galactose 1-phosphate may exert its effect by inhibiting phosphoglucomutase, which converts glucose 1-phosphate to glucose 6-phosphate, leading to glycolytic degradation, or by inhibiting glucose 6-phosphate dehydrogenase (Lerman, 1959) leading to the pentose phosphate cycle. Galactose, or galactose-1-phosphate, could also conceivably interfere in the initial phosphyorylation of glucose. It has been also suggested (Chefurka, 1965) that insects may be capable of metabolizing glucose via the glucuronic acid pathway, and hence by decarboxylation to the pentose phosphate pathway (for review see Burns & Conney, 1966). In this scheme glucose is phosphorylated to gluceose-6-phosphate, converted by phosphoglucomutase to glucose-1-phosphate and hence to UDP-glucose. Galactose or galactose-1-phosphate could inhibit the first two stages of this pathway with glucose as a source of energy. If however starch were degraded by phosphorylysis to glucose-1-phosphate residues, this would evade inhibition by galactose and could explain the difference in developmental response of galactose/glucose and galactose/starch diets. Under normal conditions the glucuronic acid pathway appears to have little quantitative importance in carbohydrate metabolism. However, inhibition of the two other pathways might well shift metabolism in this direction. We are investigating these possibilities.

Tribolium cannot efficiently utilize high levels of dietary protein in the presence of 10% galactose and supplementary glucose (Fig. 2). The depression of growth at high protein concentrations did not differ significantly from development on high cellulose diets (Fig. 3). This suggests that protein is not available for energy production in the presence of a low level of galactose, and is in contrast to the efficient utilization in a similar situation but with supplementary rice starch (Fig. 2).

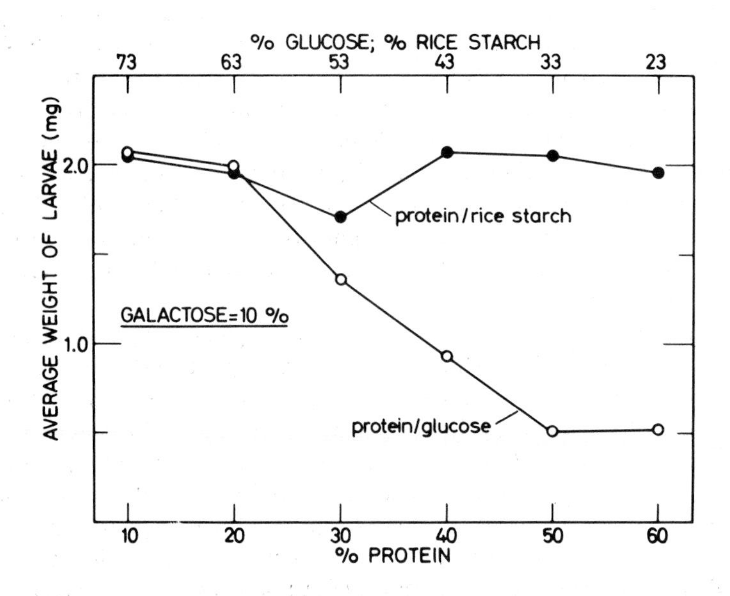

Figure 2. The effect of protein/rice starch and protein/glucose combinations, in the presence of 10% galactose, on larval development of *Tribolium castaneum*. The protein used was egg albumen (coagulated), previously shown to be optimal for development.

If dietary protein is to serve efficiently as substrate for production of energy in lieu of carbohydrate, it must first be hydrolysed by appropriate proteolytic enzymes and then penetrate the intestinal mucosa. Galactose does not interfere with proteolysis either by directly inhibiting the enzymes involved or by indirectly depressing enzyme synthesis. This was shown firstly by assaying the effect of galactose (0.1% and 1%) on *in vitro* proteolysis where it had no inhibitory effect. In a second experiment *Tribolium* larvae were adapted for 2 days on 50% casein diets, with or without 10% galactose and with either glucose or rice starch as supplementary carbohydrate. The diets contained in addition 5% dried yeast, 1% cholesterol and 1% salts. Relative larval proteolytic activity *in vitro* was then measured by the casein digestion method (Kunitz, 1947). The relative activity was: galactose/glucose>galactose/starch>glucose>starch. Proteolytic activity on the galactose/glucose diet was twice that obtained on the starch diet. This activity is not in correlation to growth on the various diets.

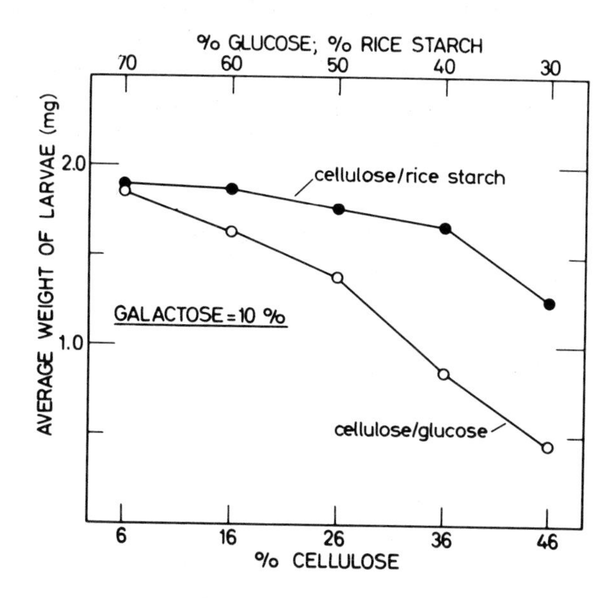

Figure 3. The effect of replacing dietary carbohydrate with cellulose in the presence of 10% galactose, on larval development of *Tribolium castaneum*.

We intend to investigate the possibility of active transport of amino acids in the larval midgut of *Tribolium*. A mechanism of this sort, requiring energy expenditure, is amenable to inhibition by galactose in the presence of glucose, but not in the presence of starch, assuming that starch is degraded directly to glucose-phosphate residues as detailed above. In chicks, amino acid excretion in cloacal contents is increased during galactose feeding (Wells & Segal, 1969).

Mannose was found to be an adequate source of dietary carbohydrate for *Tribolium*. Growth of larvae was optimal in the range of 10 - 60% mannose (supplemented to 80% with rice starch) and slightly suboptimal at 80%. Mannose/glucose diets were too hydroscopic to assay reliably. In the intestine of higher animals, where active transport of carbohydrates against a concentration gradient is the rule, luminal mannose does not effectively penetrate the intestinal mucosa. However, mannose does possess hexokinase specificity and is an effective source of energy to the extent that it can enter those cells via their basal membrane. The energetic availability of dietary mannose presumably stems from the passive transport of luminal carbohydrates in insects (Treherne, 1958 a,b; Shyamala & Bhat, 1965; Gelperin, 1966) and from its subsequent phosphorylation, a process which would assure a concentration gradient in the direction of uptake.

The developmental response of *Tribolium* to dietary xylose is similar to that obtained with galactose. The relative toxicity of xylose to *Tribolium* is dependent on the supplementary carbohydrate, and is more pronounced in the presence of glucose than in the presence of starch (Fig. 4). Here again the possibility is that xylose, which possesses hexokinase specificity, is competing with glucose.

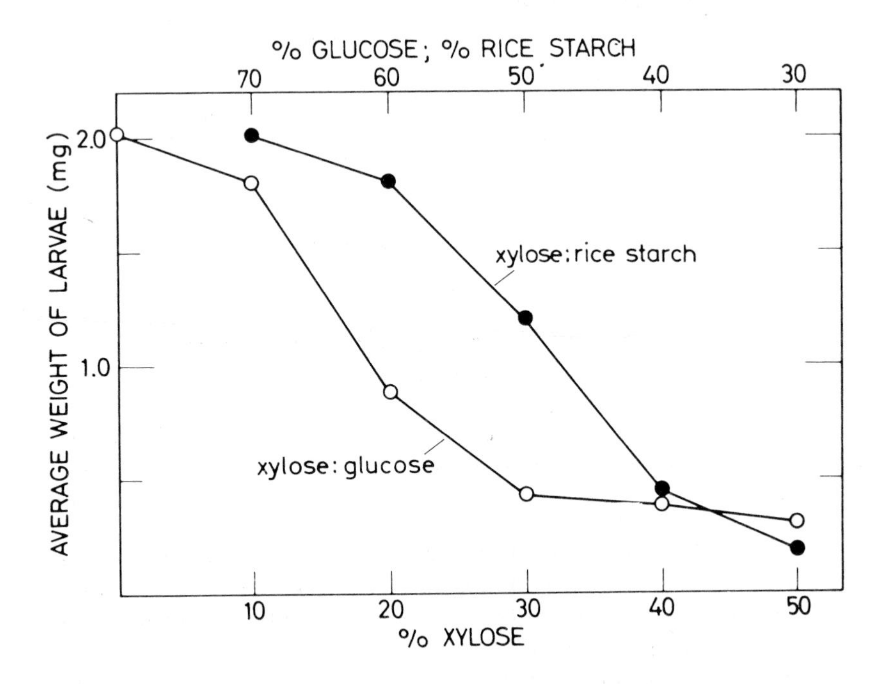

Figure 4. The effect of xylose/rice starch and xylose/glucose combinations on larval development of *Tribolium castaneum*.

A similar situation is encountered with dietary arabinose. The toxicity of this pentose is more pronounced in the presence of glucose than when rice starch supplements the diet, with maltose and dextrin exhibiting an intermediate position in this respect (Table 1).

Both xylose and arabinose have been shown to severely depress growth and derange the normal levels of plasma constituents in chicks (Wagh & Waibel, 1966; 1967).

Substitution of cellulose for either glucose or starch, in the presence of 10% xylose, resulted in a similar decreased larval growth rate (Fig. 5) which in the case of the cellulose/starch diet was equivalent to that obtained in the presence of 10% galactose (Fig. 3).

Table 1. The effect of increasing levels of arabinose in the presence of various carbohydrates on the larval development of *Tribolium castaneum*.

Arabinose (%)	Supplementary carbohydrate (%)	Average weight of larvae (mg)[a]			
		+Glucose	+Maltose	+Dextrin	+Rice starch
4	76	2.06	2.17	2.17	2.12
8	72	1.15	2.15	1.97	2.14
12	68	0.45	1.23	1.26	1.90
16	64	-	0.54	0.27	1.70
20	60	-	0.30	0.18	1.06
30	50	-	0.11	-	0.26

[a]After 12 days at 32°C.

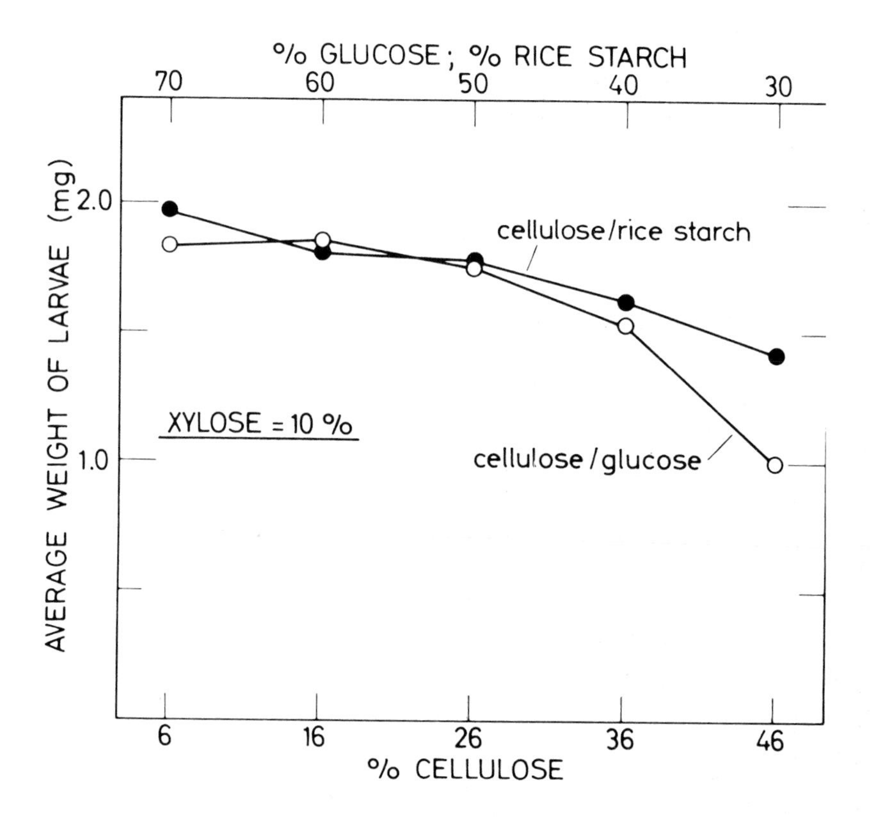

Figure 5. The effect of replacing dietary carbohydrate with cellulose, in the presence of 10% xylose, on larval development of *Tribolium castaneum.*

Tribolium cannot efficiently utilize high levels of dietary protein in the presence of 10% xylose (Fig. 6) or 12% arabinose (Table 2).

Whereas this effect is evident only in the presence of glucose on 10% galactose diets (Fig. 2), xylose and arabinose exert this negative effect on starch-containing diets as well. This difference may well be fundamental. In contrast to galactose, xylose and arabinose or metabolites presumably inhibit at some stage in the pentose phosphate pathway. This pathway has been demonstrated in some insects (see review by Chefurka, 1965). At the pentose level of metabolism, no benefit accrues to the organism from direct production of glucose-1-phosphate by phosphyorylysis of starch.

Xylose may enter the pentose phosphate pathway by first being reduced to xylitol and then dehydrogenated to xylulose, as demonstrated in rat lens extract (van Heyningen, 1959). This initial dehydrogenation is presumably not efficient in *Tribolium*, since xylitol is a much better dietary component than is xylose: On a diet containing 30% xylitol and 80% glucose the average larval weight (after 12 days at 32°C) was 1.60 mg, as compared to 0.43 mg on 30% xylose (Fig. 4). In contrast, arabitol (as 8% in a glucose-supplemented diet) was, if anything, a bit worse than arabinose.

Table 2. The effect of protein/rice starch combinations, in the presence of 12% arabinose, on larval development of *Tribolium castaneum.*

Protein (%)	Rice Starch (%)	Avg wt of larvae (mg)
10	65	1.88
20	55	1.98
30	45	1.62
40	35	1.13
50	25	0.60
60	15	0.43

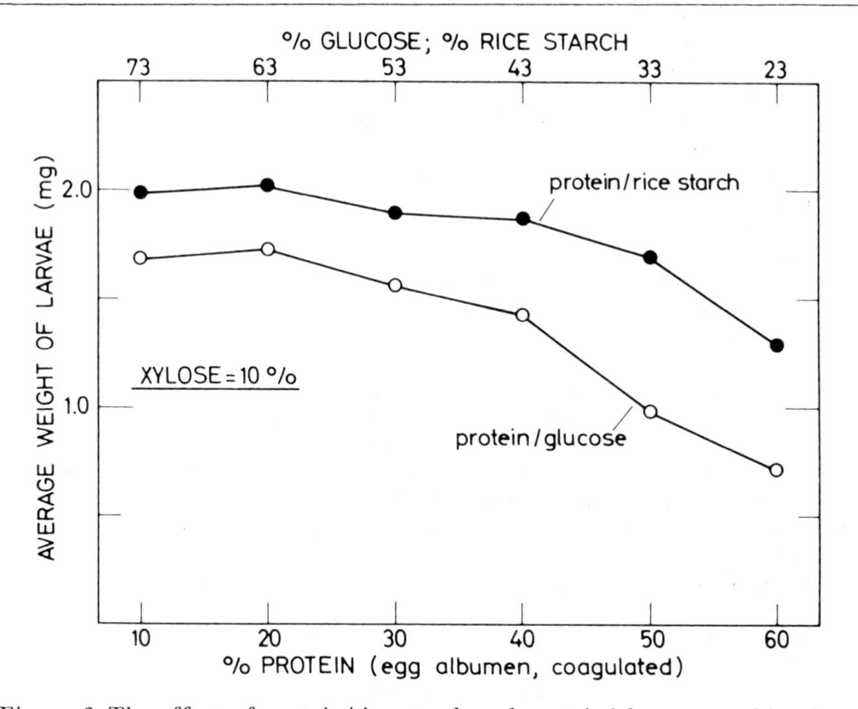

Figure 6. The effect of protein/rice starch and protein/glucose combinations, in the presence of 10% xylose, on larval development of *Tribolium castaneum.*

In conclusion, it is suggested that the ability of *Tribolium* to metabolically cope with various hexoses and pentoses, and to efficiently utilize protein for energy production, is dependent on the potential of this insect to degrade starch in a manner other than by conventional amylolysis, and by its ability to shunt carbohydrate metabolism to alternate pathways in the event of inhibition of glycolysis.

Acknowledgement—I wish to acknowledge the skillful technical assistance of Mrs. N. Levin in the growth experiments.

REFERENCES

APPLEBAUM, S. W. (1966). The digestion of potato starch by larvae of the flour beetle, *Tribolium castaneum*. *J. Nutrition* **90**:235-239.

APPLEBAUM, S. W. and KONIJN, A. M. (1965). The utilization of starch by larvae of the flour beetle, *Tribolium castaneum*. *J. Nutrition* **85**:275-282.

BURNS, J. J. and CONNEY, A. H. (1966). "Metabolism of glucuronic acid and its lactone." In: Glucuronic Acid (G. J. Dutton, ed.) Academic Press, pp. 365-384.

CHEFURKA, W. (1965). "Intermediary metabolism of carbohydrates in insects." In: Physiology of Insecta (M. Rockstein, ed.) Vol. II, Academic Press, pp. 581-667.

GELPERIN, A. (1966). Control of crop emptying in the blowfly. *J. Insect Physiol.* **12**:331-345.

VAN HEYNINGEN, R. (1959). Metabolism of xylose by the lens. 2. Rat lens *in vivo* and *in vitro*. *Biochem. J.* **73**:197-207.

KUNITZ, M. (1947). Crystalline soybean trypsin inhibitor - II. General properties. *J. Gen. Physiol.* **30**:291-310.

LERMAN, S. (1959). Enzymatic factors in experimental galactose cataract. *Science* **130**:1473-1474.

MALONE, J. I., WELLS, H. J. and SEGAL, S. (1971). Galactose toxicity in the chick: Hyperosmolality. *Science* **174**:952-955.

MAYS, J. S., MILLER, L. R. and MYERS, F. K. (1970). The relationship of galactose-1-phosphate accumulation and uridyl transferase activity to the differential galactose toxicity in male and female chicks. *Biochem. Biophys. Res. Comm.* **39**:661-665.

NETTLES, JR. W. C. and BURKS, M. L. (1971). Absorption and metabolism of galactose and galactitol in *Anthonomus grandis*. *J. Insect. Physiol.* **17**:1615-1623.

PANT, N. C. and DANG, K. (1965). Effect of different carbohydrates on the growth and development of *Tribolium castaneum* Herbst. *Indian J. Entomol.* **27**:432-441.

POSTERNAK, T. (1951). On the phosphorus of potato starch. *J. Biol. Chem.* **188**:317-325.

SHAINKIN, R. and BIRK, Y. (1965). α-amylase inhibitors from wheat. *Israel J. Chem.* **3**:96 p.

SHYAMALA, M. B. and BHAT, J. V. (1965). Intestinal transport of glucose in the silkworm, *Bombyx mori* L. *Indian J. Biochem.* **2**:101-104.

TREHERNE, J. E. (1958a). The absorption of glucose from the alimentary canal of the locust, *Schistocerca gregaria* (Forsk.). *J. Exp. Biol.* **35**:297-306.

TREHERNE, J. E. (1958b). The absorption and metabolism of some sugars in the locust, *Schistocerca gregaria* (Forsk.). *J. Exp. Biol.* **35**:611-625.

TYGSTRUP, N. and KEIDING, S. (1969). Lethal effect of feeding rats on galactose-ethanol. *Nature (Lond.)* 222:181.

WAGH, P. V. and WAIBEL, P. E. (1966). Metabolizability and nutritional implications of L-arabinose and D-xylose for chicks. *J. Nutrition* 90:207-211.

WAGH, P. V. and WAIBEL, P. E. (1967). Metabolism of L-arabinose and D-xylose by chicks. *J. Nutrition* 92:491-496.

WELLS, H. J. and SEGAL, S. (1969). Galactose toxicity in the chick: Tissue accumulation of galactose and galactitol. *FEBS Letters* 5:121-123.

DEPENDENCE OF FERTILITY AND PROGENY DEVELOPMENT OF *XYLEBORUS FERRUGINEUS* UPON CHEMICALS FROM ITS SYMBIOTES

Dale M. Norris
Department of Entomology
University of Wisconsin
Madison, Wisconsin 53706

INTRODUCTION

Ambrosia beetles (Coleoptera: Scolytidae) live in obligatory symbiosis with mutualistic fungi and bacteria. The beetles derive nutrients and other chemicals from their symbiotes, and these microbes, at least, derive nutrients from, and are disseminated by, the beetles. These types of symbioses were recognized by Hartig (1884). Possible nutritional interdependencies among ambrosia beetles and their microbial symbiotes have been discussed for many years (Francke-Grosmann, 1963, 1965 and 1967; Baker, 1963; Batra, 1963 and 1966; Norris, 1965; and Graham, 1967); however, experimental demonstration of specific nutritional interdependencies among species in such symbiotic complexes has only occurred recently (e.g., Norris and Baker, 1967; Norris *et al.*, 1969; Abrahamson and Norris, 1970; Chu *et al.*, 1970; Kok *et al.*, 1970).

This paper is a combined elaboration on previously published or "in press" findings, and presentation of new data especially on the mutualistic contributions of symbiotic fungi and bacteria to the beetle, *Xyleborus ferrugineus*.

LABORATORY MAINTENANCE OF THE SYMBIOTIC COMPLEX

The progress which our research group has made on interpreting this symbiosis in terms of chemical interactions among component species has been possible because a series of diets for continuous laboratory cultivation of the complex or component species were developed. From our early studies, Saunders and Knoke (1967) published data on the successful continuous maintenance of the insect with its symbiotic complex on several synthetic diets. This was the first published continuous culture of a wood-inhabiting insect on such diets. Our laboratory cultures of these types of complexes are now routinely sustained on the diet given in Table 1.

Table 1. Composition of the diet for mass rearing of symbiotic *Xyleborus ferrugineus* in the laboratory.

Constituent	Quantity
Sucrose[a]	15.0 g
Yeast extract (Difco, Bacto)	10.0 "
Casein[b]	10.0 "
Starch[a], soluble	10.0 "
Wheat germ[b]	5.0 "
Salts, Wesson's[b]	1.3 "
Agar (Difco, Bacto)	40.0 + 40.0[h] g
Sawdust (elm)	150.0[f] "
Corn oil[c]	5.0 ml
Water, distilled	1,000.0 "
Additional ingredient[i]	
Hydrochloric acid (conc)[d]	2.0 ml
Sorbic acid[e]	2.0 g
Streptomycin[b]	0.7 "

[a]Mallinckrodt Chemical Works.
[b]Nutritional Biochemical Co.
[c]Magnus, Mabee & Reynard, Inc., New York.
[d]Baker Chemical, Allied Chemical Co.
[e]Eastman Organic Chemicals.
[f]Amount varies slightly with moisture content of sawdust.
[g]Weight in grams.
[h]Additional agar added to mixture after initial autoclaving to maintain cohesiveness of diet.
[i]Any one of the three below.

Life history of the symbiotic insect in laboratory culture

The distribution of individuals of a progeny population of symbiotic *X. ferrugineus* among the developmental stages of larva, pupa and adult; and the sexes over time (3 to 9 weeks), when reared as indicated on the diet in Table 1, is shown in Table 2. Both unfertilized (which yield haploid males) and fertilized (which yield diploid females) eggs (see Figure 1) have been laid and have resulted in progeny adults within 3 weeks. Egg laying has largely ceased by the 7th week; at this time most of the progeny are adults. Under the given experimental conditions, about two-thirds as many sampled progeny were observed at 3 weeks as were seen in the 9th week. It is because of these kinds of findings on progeny production and development that our standard experimental period on a batch of diet is limited to 6 weeks. Beyond this period, the diet will usually shrink away from the walls of the container so that progeny perish in the cracks, and the diet is then otherwise not readily acceptable by adult females for tunnelling and nutrition.

Table 2. The distribution of individuals of a progeny population of symbiotic *X. ferrugineus* among the developmental stages of larva, pupa and adult; and the sexes, over time.

Time of harvest (wk)	No. of living progeny[a]					
	Adult		Pupa		Larva	Total
	Male	Female	Male	Female		
3rd	11	112	15	220	653	1,011
4th	0	96	18	222	910	1,246
5th	7	496	12	300	509	1,324
6th	33	900	0	55	21	1,009
7th	42	1,649	0	1	0	1,692
8th	23	1,294	0	3	1	1,321
9th	27	1,556	0	6	5	1,594

[a]Reared in three jars; each with 30 mother females and 5 father males per jar on the diet of Table 1 in 32-oz glass breeding jars incubated at 28°C, 70% R.H. in darkness.

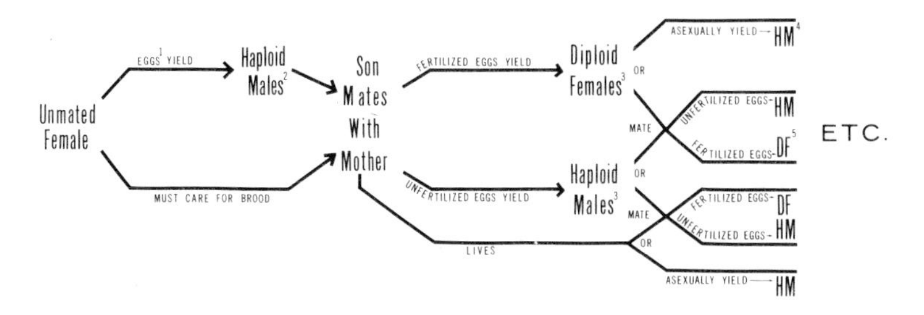

Figure 1. Flow diagram depicting the arrhenotokous reproduction, inbreeding, multiple brood fecundity, and the long life of female *X. ferrugineus*. (1) Unfertilized eggs; (2) males are much smaller than females and normally die in the brood chamber where they formed; (3) mating normally is sister x brother or son x mother; (4) HM means haploid male; (5) DF means diploid female.

Freeing individual insects of ectosymbiotic fungi

Xyleborus insects can be conveniently freed of ectosymbiotic fungi by either surface sterilization of eggs or pupae. The basic technique for eggs is given in Table 3. In this method, the concentration of mercuric chloride is sometimes increased to 0.13%.

Pupae are similarly sterilized by submerging them in 0.13% mercuric chloride for 4 minutes, and then rinsing in 3 changes of sterile water (each rinse is for 1 minute). The pupae are then placed on the surface of 2% agar-water medium in petri dishes. The transforming adults from these ectosymbiotic fungus-free pupae are subsequently transferred aseptically to the chosen diet. These insects still have bacterial symbiotes (Norris and Baker, 1967).

Table 3. A technique for surface sterilization of eggs of *X. ferrugineus* without destroying their viability.

Technique step	Time period
Wash in sterile, distilled water	1 min
Wash in sterile, distilled water	1 "
Submerge in 0.1% $HgCl_2$	4 "
Rinse in sterile, distilled water	1 "
Rinse in sterile, distilled water	1 "
Wash in 70% ethanol	30 sec
Rinse in sterile, distilled water	1 min
Transfer aseptically to the surface of nutrient-agar medium, and observe for microbial growth	2-3 days

Diets for continuous rearing of ectosymbiotic fungi-free insects

A meridic diet for continuous rearing of the ectosymbiotic fungi-free insects is given in Table 4. This meridic diet was refined to a holidic state by replacing the vitamin-free casein in the former diet with a regime of amino acids. A mixture (Table 5) of the 10 commonly essential amino acids supported continued reproduction and progeny development of the ectosymbiotic fungi-free insect. Minimal amino acid requirements of this insect have not been established.

Table 4. Composition of a meridic diet adequate for continuous rearing of the ectosymbiotic fungus-free insect

Component	Quantity
Sucrose	10.0 g
Casein, vitamin-free[a]	10.0 "
Salts, Wesson's[a]	2.5 "
Vitamin solution[b]	20.0 ml
Ergosterol[a] (g)	0.4 g
Linolenic acid, methyl ester[c]	50.0 mg
Ethanol, 95%	2.5 ml
Cellulose, pure powdered[a]	15.0 g
Cellulose, fibrous[e]	4.0 "
Sorbic acid	0.4 "
Agar, Bacto	10.0 "
Streptomycin, 20% wettable	0.35 g
Water, deionized	230.0 ml

[a]Nutritional Biochemical Co.

[b]A solution containing the following gram quantities of each vitamin per liter: choline chloride, 10.3; inositol, 4.6; calcium pantothenate, 0.85; niacin, 0.33; p-aminobenzoic acid, 0.155; riboflavin, 0.048; pyridoxine HCL, 0.037; thiamine HCL, 0.028; folic acid, 0.011; biotin, 0.008 and cobalamine B_{12}, 0.002.

[c]Hormel Institute.

[d]Cellucotton®, Kimberly-Clark Corp.

DEPENDENCE OF THE INSECT UPON ITS ECTOSYMBIOTIC FUNGI FOR AN ADEQUATE STEROL SOURCE

Regardless of whether the initial stage of *X. ferrugineus* on test diets is virgin females from ectosymbiotic fungi-freed pupae or is surface-sterilized eggs, the insects have pupated in the second progeny brood or generation (Figure 2) only on synthetic diets containing fungus mycelium or a Δ^7 sterol (e.g., ergosterol or 7-dehydrocholesterol). The only other studied insect that has not used cholesterol as a sole sterol source is *Drosophila pachea* (Heed and Kirscher, 1965).

When virgin ectosymbiotic fungi-free females were placed on the diet shown in Table 4, but with 0.4 g of cholesterol or lanosterol substituted for 0.4 g of ergosterol, first-brood progeny pupated, but not second-brood

Table 5. The regime of the 10 commonly essential amino acids used in the holidic diet for ectosymbiotic fungi-free insects.

Amino acid[a]	Quantity[b] (mg)
L-Arginine	1,111
L-Histidine	610
L-Isoleucine	1,320
L-Leucine	2,653
L-Lysine	1,776
L-Methionine	533
L-Phenylalanine	1,560
L-Threonine	899
L-Tryptophan	446
L-Valine	1,563
Total	12,471

[a]Obtained from Nutritional Biochemical Co.

[b]Based on our analysis of the amino acids in the casein used with meridic diets for *X. ferrugineus* (Norris *et al.*, 1969); and present in the mutualistic fungus, *Fusarium solani.*

progeny from the same females. Lumisterol as the sterol source in the same situation supported (allowed) first-brood progeny to develop into the larval stage, but all died before 6 weeks. Vitamin D_2 or D_3 at 0.4 g in place of ergosterol prevented progeny production (no eggs). It is interesting, and probably significant, that the tested chemicals which were detrimental to even first-brood progeny, lumisterol and vitamin D_2 and D_3, differ from all other tested molecules at the highly important C_{10} position.

Passage of pupation factor(s) from mother beetle to progeny

In a sterol-depletion study starting with ectosymbiotic fungi-freed, virgin *X. ferrugineus* female adults, evidence was obtained that such a mother beetle can pass, apparently transovarily, to her asexual progeny sterol or sterol-dependent metabolites necessary for pupation. In these investigations, 80 females initially were placed individually in test tubes containing the diet in Table 4, except the sterol source was omitted. Of the 80, 77 became established (tunnelled) in the diet (Figure 3). After 6 weeks, the progeny per female was determined and the numbers of original females indicated in Figure 3 were transferred to a new batch of the same diet with the indicated sterol condition. There thus were 33 original females now on a

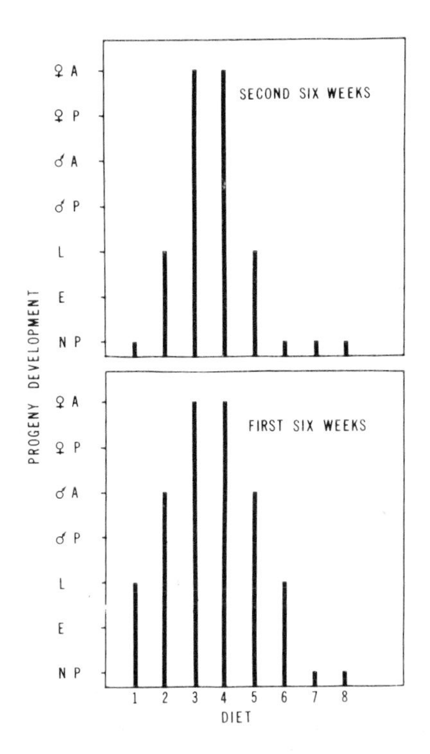

Figure 2. Development stages of progeny of fungus-free *X. ferrugineus* observed in each diet treatment during two consecutive 6-week periods (original females were transferred to a fresh batch of their respective diet for the second 6 weeks). NP, no progeny; E, eggs; L, larvae; P, pupae; and A, adults. 1, no-sterol control; 2, cholesterol; 3, ergosterol; 4, 7-dehydrocholesterol; 5, lanosterol; 6, lumisterol; 7, vitamin D_2; 8, vitamin D_3.

cholesterol sterol source; 14, on no sterol; and 19, on ergosterol. At the end of the second 6 weeks, the progeny per female during this second period was determined, and the indicated number of original females placed on yet a new (third) batch of the diet for 6 weeks with the indicated sterol condition (Figure 3).

Progeny at the end of the first 6 weeks on the no-sterol diet had not pupated; normally pupae are present after 3 weeks on the diet containing a proper sterol. However, when original females were than transferred to a fresh batch of diet with cholesterol or ergosterol, progeny produced during this second 6 weeks included pupae and adults. Because our previous research has established that cholesterol as the sole sterol will not allow pupation, the original mother female must have provided the necessary sterol

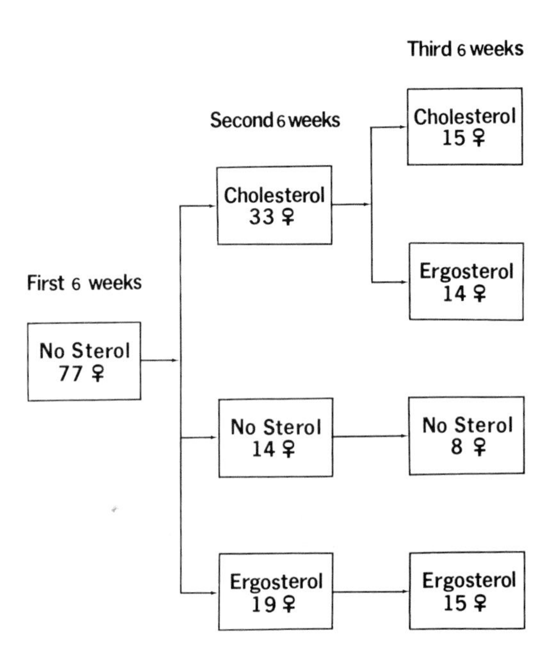

Figure 3. Sequences of 6-week no-sterol or sterol exposures of the indicated numbers of ectosymbiotic fungi-free female *X. ferrugineus* adults.

or sterol-dependent metabolite(s) required for pupation of progeny. This mother-derived factor(s) only functioned in progeny pupation when a sterol source which would support egg and larval development (i.e., cholesterol), but not pupation, was included in the second batch of diet (Figure 3). Progeny produced by original females during the third 6 weeks and on cholesterol-containing diet (Figure 3) did not pupate; this indicates that the pupational factor(s) provided to progeny by the mother female during the second 6 weeks had been depleted or reduced to an inadequate level by the third 6 weeks.

TRANSOVARIALLY-TRANSMITTED BACTERIAL SYMBIOTE OF *X. FERRUGINEUS*

Baker and Norris (1968) reported that the microbial symbiotic complex associated with *X. ferrugineus* included bacteria. In recent studies, histological sections of male and female adults, pupae and all stages of larvae have revealed the presence of gram-positive cocci, usually in pairs, in the gut lumen of the insects. Serial sections of both virgin and mated females showed that these cocci migrate (penetrate) into the developing oocytes in the ovarioles and become intimately associated with each oocyte nucleus. Smearing and gram staining of the internal contents of surface-sterilized eggs established the presence of these cocci in the laid eggs. Serial sections of first-instar larvae, collected upon hatching, from surface-sterilized eggs revealed the cocci apparently confined initially to the gut lumen.

These symbiotes were isolated and pure cultured on Nutrient-Agar medium (Difco) from the guts of adults which transformed under aseptic conditions from surface-sterilized pupae. Following the methods given by Baird-Packer in Gibbs and Skinner (1966), the coccus was placed in the genus *Staphylococcus.*

Dependency of oocyte maturation in X. ferrugineus upon the bacterial symbiote

When ectosymbiotic fungi-free, virgin female *X. ferrugineus* adults were reared for 6 weeks on the sodium benzoate- or sorbic acid-supplemented diet given in Table 6 with one of the following additions of antibiotic per batch of diet: (1) 2417 units/g of Streptomycin sulfate: (2) 3031 μg/g of Chlorotetracycline-HC1; (3) a combination of 1208 units/g of Streptomycin sulfate and 2121 μg/g of Chlorotetracycline-HC1; or (4) 5178 units/g of "G" Potassium Penicillin, the number of ovipositing females was significantly (0.0005 level) reduced as compared to those on antibiotic-free diet. With the antibiotic-free control diet, 58.3% of the females oviposited during the 6 weeks; whereas, with the various antibiotic treatments, the mean oviposition ranged from 17.9 to 25.7%.

The mean number of progeny/female in the above diets with antibiotics also was significantly reduced. In diets containing Streptomycin sulfate or Chlorotetracycline-HC1 (or both), the mean number of progeny/female was 3.06 to 4.17 times smaller than in the antibiotic-free control diet. Penicillin only reduced the mean progeny/female to a level 2.56 times smaller than in the positive control. Larvae only completed development, and yielded male pupae and adults during the 6 weeks in the antibiotic-free control diet.

Microscopic study of serial histological section of females, reared for 6 weeks on antibiotic-containing diet, that did not lay eggs, revealed some initiation of primary oocytes; however, none matured. In these oocytes yolk deposition was noticed, but the nucleus of such oocytes remained static; not showing divisional figures. Oocyte maturation had progressed beyond these points in females that had not yet laid eggs in the no-antibiotic positive control.

In all antibiotic-treated females that did not lay eggs, the symbiotic cocci were not found in their oocytes. This does not necessarily mean that these microorganisms were completely eliminated from these insects. However, as indicated previously, all maturing oocytes and laid eggs contain very

Table 6. The basal meridic diet for rearing ectosymbiotic fungi-free *Xyleborus ferrugineus* female adults into which one of several antibiotic treatments was added.

Ingredient	Amount
Water, deionized and distilled	230.0 ml
Vitamin solution[a]	20.0 "
Ethanol, 95%	2.5 "
Sucrose	10.0 g
Casein, vitamin-free[b]	10.0 "
Inorganic salts, Wesson's[b]	2.5 "
Ergosterol[b]	0.4 "
Sorbic acid[c]	0.4 "
or	
Sodium benzoate[d]	1.5 "
Methyl linolenate[e]	0.1 "
Cellulose, powdered[b]	15.0 "
Cellulose, fibrous[f]	4.0 "
Agar, Bacto[g]	10.0 "

[a] See Table 4.
[b] Nutritional Biochemical Co.
[c] Eastman Kodak.
[d] Allied Chemical Co.
[e] Hormel Institute.
[f] Cellucotton®, Kimberly-Clark Corp.
[g] Difco Laboratories.

obvious population levels of this coccus. When antibiotic-treated non-laying females were subsequently transferred aseptically to a new batch of the diet with antibiotic treatment plus an aliquot of an autoclaved pure culture of Staphylococcus they unfailingly began to lay viable eggs within 4 to 5 days. This surely indicates that some chemical entity or entities produced by the *Staphylococcus* allow(s) oocyte maturation.

Bacterial symbiote supercedes sperm in activation of embryonic development and genome decoding in X. ferrugineus

As previously indicated oocytes do not develop beyond the primary oocyte stage in the absence of the bacterial symbiote. Thus, sexual as well as asexual reproduction is primarily dependent upon the *Staphylococcus.* Beyond this, in the presence of an adequate population of the bacterium, embryonic development is initiated while eggs are still enclosed in the follicular walls of the ovarioles. Therefore, the usual sperm role of mature-oocyte activation in sexual reproduction is indeed assumed by the bacterial symbiote. The remaining major role of the sperm thus seems to be the creation of the diploid progeny state which yields females.

DISCUSSION AND SUMMARY

Reported progress in interpreting certain chemical interdependencies among the species components of the involved symbiotic complex required continuous culturing of the entire complex, and its individual components, on synthetic diets including holidic ones. The series of reported investigations demonstrate that the natural pupation of the insect is dependent upon a sterol source, ergosterol, provided by the three mutualistic ecto-symbiotic fungi (*Fusarium solani, Cephalosporium* sp. and *Graphium* sp.) associated with *X. ferrugineus.* Aspects of the sterol studies also showed that the virgin mother beetle can pass, apparently transovarily, to its progeny the sterol or sterol-dependent metabolites required for progeny pupation.

A bacterial symbiote in the complex was isolated, pure cultured and classified in the genus *Staphylococcus.* It is associated saprophytically with the fungi and beetle, but is especially important to the beetle as a facultative endosymbiote. In this latter mutualistic role, it controls the fertility of the female insect and has superceded the sperm in the primary activation of mature oocytes. The chemical entity or entities involved in this bacterial control of maturation of oocytes and primary activation of embryonic development in virgin female beetles were stable to extensive autoclaving.

Acknowledgements.–This research was supported by the College of Agricultural and Life Sciences, University of Wisconsin, Madison; in part by research grant No. AI-06195-06 from The Institute of Allergy and Infectious Diseases, N.I.H.; and in part by funds from the Wisconsin Department of Natural Resources, The Schoenleber Foundation, and the Cancer Institute, N.I.H. The author is indebted to the following colleagues who contributed to the research findings discussed: Drs. J. L. Saunders, J. K. Knoke, G. Moya, L. P. Abrahamson, L. T. Kok and B. Peleg; Ms. J. M. Baker; and Mr. W. O. Bishop and Mr. H. M. Chu.

REFERENCES

ABRAHAMSON, L. P., and NORRIS, D. M. (1970). Symbiontic interrelationships between microbes and ambrosia beetles. V. Amino acids as a source of nitrogen to the fungi in the beetle. *Ann. Entomol. Soc. Amer.* **63**:177.

BAKER, J. M. (1963). Ambrosia beetles and their fungi with special reference to *Platypus cylindrus* Fab. *Sympos. Soc. Gen. Microbiol.* **13**:232.

BATRA, L. R. (1963). Ecology of ambrosia fungi and their dissemination by beetles. *Trans. Kansas Acad. Sci.* **66**:213.

BATRA, L. R. (1966). Ambrosia fungi: extent of specificity to ambrosia beetles. *Science* **153**:193.

CHU, H. M., NORRIS, D. M., and KOK, L. T. (1970). Pupation requirement of the beetle, *Xyleborus ferrugineus*: sterols other than cholesterol. *J. Insect Physiol.* **16**:1379.

FRANCKE-GROSMANN, H. (1963). Some new aspects in forest entomology. *Ann. Rev. Entomol.* **8**:415.

FRANCKE-GROSMANN, H. (1965). Uber Symbiosen von Xylo-Mycetophagen und Phloeophagen, Scolytoidea mit Holzbewhnenden Pilzen. *Holz und Organisman* **1**:503.

FRANCKE-GROSMANN, H. (1967). Ectosymbiosis in Wood-Inhabiting Insects. In *Symbiosis* (S. M. Henry, ed) Vol. 2, pp. 141-205 (Academic Press, New York).

GIBBS, B. M., and SKINNER, F. A. (1966). *Identification Methods for Microbiologists, Part A* (Academic Press, New York).

GRAHAM, K. (1967). Fungal-insect mutualism in trees and timber. *Ann. Rev. Entomol.* **12**:105.

HARTIG, T. (1844). Ambrosia des *Bostrychus dispar. Allgem. Forst-u. Jagdztg.* **13**:73.

HEED, W. B., and KIRSCHER, H. W. (1965). Unique sterol in the ecology and nutrition of *Drosophila pachea. Science* **149**:758.

KOK, L. T., NORRIS, D. M., and CHU, H. M. (1970). Sterol metabolism as a basis for a mutualistic symbiosis. *Nature* **225**:661.

NORRIS, D. M. (1965). The complex of fungi essential to *Xyleborus sharpi. Holz und Organismen* **1**:523.

NORRIS, C. M., and BAKER, J. M. (1967). Symbiosis: Effects of a mutualistic fungus upon the growth and reproduction of *Xyleborus ferrugineus. Science* **156**:1120.

NORRIS, D. M., BAKER, J. M., and CHU, H. M. (1969). Symbiontic interrelationships between microbes and ambrosia beetles. III. Ergosterol as the source of sterol to the insect. *Ann. Entomol. Soc. Amer.* **62**:413.

PELEG, B., and NORRIS, D. M. (1972a). Bacterial symbiote activation of insect parthenogenetic reproduction. *Nature New Biology* **236**:111.

PELEG, B., and NORRIS, D. M. (1972b). Symbiotic interrelationships between microbes and ambrosia beetles. VII. Bacterial symbionts associated with *Xyleborus ferrugineus. J. Invertebrate Pathol.*, in press.

SAUNDERS, J. L., and KNOKE, J. K. (1967). Diets for rearing the ambrosia beetle, *Xyleborus ferrugineus* (Fabr.). *Science* **157**:460.

PHYSIOLOGY AND ELIMINATION OF INTRACELLULAR SYMBIOTES IN SOME STORED PRODUCT BEETLES

N. C. Pant
Himachal Pradesh University
Simla, India
and
K. Dang
Indian Agricultural Research Institute
New Delhi-12, India

INTRODUCTION

Various forms of endosymbiosis are known in insects which harbour a variety of microorganisms in special organs called mycetomes, located in different organs of insects. The information on the location, identity, mode of transmission and function of symbiotes has been reviewed by several authors (Koch, 1956, 1960, 1967; Buchner, 1965; Richards and Brooks, 1958; Steinhaus, 1946; Toth, 1952; Wigglesworth, 1952; Francke-Grosmann 1967). Although in the majority of cases, function of symbiotes are either imperfectly known or not known at all, the relationship is logically considered as symbiotic because of the very intimate relationship. Special transmission mechanisms evolved by the host ensure that each and every progeny is infected with the parental symbiotes. The transmission mechanism is simple in *Coptosoma* (Schneider, 1940) where symbiote-filled capsules are laid among aposymbiotic egg masses by the mother. Hatching nymphs receive the symbiotes by sucking bacteria-containing fluid of the capsule. A more complicated arrangement exists in *Lasioderma, Stegobium* and some other anobiid beetles where special transmission organs smear the symbiotes on the external surface of the egg at the time of oviposition. The larvae upon hatching acquire the infection where they eat the shell as a matter of their normal behaviour (Breitsprecher, 1928). The mechanisms become highly complicated in insects like *Oryzaephilus, Sitophilus* and in those forms where transmission is through ovarial infection. The symbiotes in several cases enter the egg during early oogenesis in the ovary. In such cases the symbiotic cycle is difficult to break even through the use of antibiotics, heat, deficient diets, ultra violet rays or centrifugal force. The literature contains conflicting reports on the efficacy of these methods for making insects aposymbiotic for conducting physiological investigations. Where the transmission cycle can be interrupted as in *Lasioderma* and others, it has been possible to obtain aposymbiotic individuals. It is in these cases that we have a much better understanding of the physiology of symbiosis (Blewett and Fraenkel, 1944; Fraenkel and Blewett, 1943; Pant and Fraenkel 1950, 1954a, 1954b; Kolya and Pant, 1962; Pant *et al.*, 1959).

Insect and mite nutrition – North-Holland – Amsterdam, (1972)

In this communication an attempt is made to review the earlier work on our understanding of the physiology of symbiotes in *Lasioderma, Stegobium, Oryzaephilus* and *Sitophilus oryzae* and also the problems involved in making insects aposymbiotic.

SYMBIOSIS IN SOME STORED PRODUCT BEETLES

Lasioderma serricorne and Stegobium paniceum

The mycetome which harbours yeast-like symbiotes in *Stegobium* was first described by Karawaiew (1899). Buchner (1930) provisionally identified the microorganism as *Saccharomyces anobii.* The mycetome of *Stegobium* and *Lasioderma symbiotes* and their mode of transmission from one generation to another have been investigated by several workers (Karawaiew, 1899; Buchner, 1912; Heitz, 1927; Breitsprecher, 1928, Pant and Fraenkel, 1954a; Grabner, 1954; Foeckler, 1961).

The first success in the cultivation of symbiotes of *Lasioderma* was reported by Pant and Fraenkel (1954a). Later Grabner (1954) confirmed and extended the investigations to offer an explanation of the manner in which the symbiotes could be a source of supply of vitamins of B complex to the host. The precise taxonomic identity of the symbiotes in the two species of insects has never been settled. Contrary to Buchner (1930), Grabner (1954) named the symbiotes of Stegobium as *Torulopsis buchnerii* but Kuhlwein and Jurzitza (1961) suggested them to be *Symbiotaphrina buchnerii.* The symbiotes of *Lasioderma* closely resemble *Taphrina* (Milne, 1963; Foeckler, 1961).

The symbiotes, although yeast-like, are morphologically and functionally different in the two insects. They differ markedly in size and shape. In *Lasioderma* the cells are broadly oval, with only one bud attached to a cell except in pupae where two buds to a cell are common. In *Stegobium* the yeasts are elongate, pear-shaped and pointed at one end to which a single bud may be attached. The point of attachment is very narrow. These insects can be made aposymbiotic by surface sterilization of the eggs with a suitable germicidal fluid. The resulting larvae which are devoid of symbiotes behave differently in their nutritional requirements. Symbiotes of *Lasioderma* are superior to those of *Stegobium* which are unable to supply enough biotin and folic acid to the host. This superiority is further seen by the growth of aposymbiotic *Stegobium* larvae infected artificially with *Lasioderma* yeast (Pant and Fraenkel, 1954a). Exchange of symbiotes has lately been accomplished also in these two insect species by Jurzitza (1964).

The symbiotes of *Stegobium* have further been shown to supply several essential and at least one nonessential amino acid, i.e. glycine, in the absence of which aposymbiotic larvae completely failed to develop. Table 1 shows

Table 1. Growth of Stegobium larvae (with and without symbiotes) on diets lacking in one of the amino acids (Pant *et al.*, 1959).

Diet lacking in	With symbiotes		Without symbiotes	
	No. of adults out of 30 larvae	Developmental period (days)	No. of adults out of 30 larvae	Developmental period (days)
'Nonessential'				
D1-Alanine	29	31-36	10	72-118
D1-Aspartic acid	24	35-57	18	43-118
D1-Cystine	26	32-57	17	52-188
L-Glutamic acid	24	33-57	16	52-118
Glycine	14	64-94	0	-
L-Hydroxyproline	27	33-57	17	35- 72
L-Proline	28	31-47	23	40-118
DL-Serine	28	31-50	18	43-108
L-Tyrosine	29	33-50	18	43-108
'Essential'				
L-Arginine	2	51-78	0	-
L-Histidine	13	51-93	7	43-108
DL-Isoleucine	11	54-84	0	-
L-Leucine	9	54-88	0	-
L-Lysine	10	42-84	0	-
DL-Methionine	14	35-52	2	72-103
DL-Phenylalanine	18	35-86	0	-
DL-Threonine	7	57-78	0	-
L-Tryptophane	24	46-73	0	-
DL-Valine	13	42-75	0	-
No amino acids	26	38-57	-	-
Casein diet	27	27-37	29	28- 40

that in the absence of nonessential amino acids growth was very much prolonged in insects without their symbiotes. This probably was the first instance where intracellular symbiotes in an insect have been shown to help in the nitrogen metabolism of the host.

The symbiotes of *Lasioderma* and *Stegobium* also act as sources of sterols. The aposymbiotic individuals failed to become adults in a diet lacking a sterol (Blewett and Fraenkel, 1944; Fraenkel and Blewett, 1943; Pant and Fraenkel, 1954a, b).

Oryzaephilus surinamensis

In the fat body of this insect are present two pairs of mycetomes which contain bacterium-like microorganisms which are transmitted transovarially (Pierantoni, 1929, 1930; Koch, 1931). Due to the nature of the association it has always proved difficult to obtain aposymbiotic colonies of this insect. Koch (1936) reported that a thermal treatment of *Oryzaephilus* at 36°C renders the host aposymbiotic, but later only partial (Pant and Fraenkel, 1954a) or no elimination (Kolya and Pant, 1962) was obtained. The trophic role assigned to the symbiotes by Pant and Fraenkel (1954b) could not be supported by Kolya and Pant (1962).

The efficacy of upper limits of temperature tolerance as a means of eliminating microorganisms in the insect tissue is probably not a very reliable method in view of the conflicting reports about its success. Further, the 'thermal resistant' residue of the symbiotes are liable to repopulate the partially or completely empty mycetomes, as has been reported by Huger (1956) in *Rhyzopertha dominica.* The microorganisms are highly pleomorphic and under unfavourable conditions of high temperature, they may react differently to histological strains used for their detection. It is important then, to confirm the aposymbiotic condition by the microbiological methods for cultivating the symbiotes in artificial media. Unfortunately this technique was not employed in these investigations.

Recently these improved techniques have been used by Dang-Gabrani (1970) and Dang-Gabrani and Pant (unpublished data) which establishes the inefficacy of thermal treatment for the elimination of intracellular microorganisms in *Sitophilus oryzae.*

Sitophilus oryzae and Sitophilus granaria

Among Curculionidae, special attention has been paid to the microbial symbiosis in *Sitophilus granaria.* During embryonic and postembryonic stages of *S. granaria* and *S. oryzae*, the mycetomes are represented by the accessory cell mass which has no connection with the gut (Mansour, 1927, 1930). The infection of a new generation takes place transovarially. A very comprehensive account of the embryological development of mycetomes and the fate of symbiotes during metamorphosis has been given by several authors (Murray and Tiegs, 1935; Tiegs and Murray, 1938; Schneider, 1954, 1956). In a series of interesting papers Musgrave and his co-workers gave the results of their investigations on the ultrastructure of symbiotes, their development, location and susceptibility to high temperatures and chemicals like methyl bromide (Musgrave *et al.*, 1963); Musgrave *et al.*, 1962; Musgrave and Miller, 1951; Musgrave *et al.*, 1965).

EFFECT OF SUPRA OPTIMAL TEMPERATURES AND CHEMOTHERAPY ON SYMBIOTES

Thermal treatment

Due to the transovarial transmission of symbiotes from one generation to the other, it is difficult to make the insect aposymbiotic. Several methods reported to be successful in closely related insects were applied to *S. oryzae* with a view to making it aposymbiotic but without success. The insects were subjected to high temperatures, methyl bromide, and oral administration of antibiotics with food.

Even after 4 generations the mycetome contained symbiotes. These findings do not support the observations made by earlier workers (Koch, 1936; Schneider, 1956; Musgrave and Miller, 1956) that by holding *Sitophilus* to 32°C or more, the symbiotes are eliminated. Failure of thermal treatment has, however, been reported in *O. surinamensis* (Pant and Fraenkel, 1954a; Kolya and Pant, 1962) *Blattella germanica* (Brooks and Richards, 1955; Selmair, 1962) and *Sitophilus* (Musgrave *et al.*, 1963).

The thermal sensitivity of symbiotes is not very well understood and awaits further clarification. Exposure to 33°C. proved fatal to *S. oryzae.* However, the reported tolerance to high temperature in adult *Sitophilus* reared at 33°C aborted their symbiotes. In almost all these cases absence of symbiotes was decided only histologically and not microbiologically. Before any method for eliminating the symbiote is considered effective it would be necessary to establish the following:

(a) that the mycetomes of a successfully treated individual must give a negative reaction when inoculated into a culture medium suitable for normal growth of the symbiotes,

(b) that the symbiotes have not changed towards the usual reaction to the stains and thus escape detection when examined in a histological preparation,

(c) that the treated population when returned to normal conditions will keep on showing an aposymbiotic condition. In none of the reports so far available in the literature on the successful utilization of thermal or chemotherapy, have these tests been performed.

Chemotherapy

Musgrave, Monro and Upitis (1965) reported that by subjecting *S. granaria* to sub-lethal dosages of methyl bromide, symbiotes gradually disintegrate leading to partially or totally aposymbiotic conditions. *S. oryzae*, exposed to methyl bromide 3-4 times in each of 4 generations, failed to show any effect on its microflora. The survivors of each generation were allowed to oviposit to obtain the succeeding generation for further treatment.

Use of antibiotics for eliminating symbiotic microorganisms at first appears to have a bright prospect because the symbiotes belong to that class of microbes which are susceptible to antibiotics of one or the other types. The effect of antibiotics in several cases failed to produce aposymbiotic insects (Glaser 1946; Brooks and Richards, 1955). Schneider (1956), on the contrary, claimed destruction of mycetomal microorganisms of *Sitophilus* when terramycin or Aureomycin were orally administered. In our experiments administration of antibiotics was orally administered to *S. oryzae* with the artificial pellets made of wheat flour to which were added Aureomycin, Hostacyclin, penicillin, polymixin-B, sulphadiazine or sulphanalimide at concentrations ranging from 75 to 500mg/16 gm of artificial pellets. Except for Hostacylin insects died in 3rd of 4th generations when the amount of antibiotics was more than 125 mg/16 gm of food. In none of the cases were symbiotes adversely affected.

When the antibiotics failed to eliminate the intracytoplasmic symbiotes *in vivo*, these were tested against the cultivated symbiotes *in vitro*. Cultures were made on nutrient broth or agar of the organisms isolated from the mycetome, ovaries and eggs. Table 2 gives the area of inhibition zone developed by different antibiotics.

Table 2. Effect of different antibiotics on symbiotes of *Sitophilus oryzae* cultured from different sources (Dang, 1971).

	Inhibition zone in cms		
Antibiotic	Mycetome	Ovary	Egg
Aureomycin	2.2 x 2.5	1.7 x 2.2	3.1 x 3.3
Penicillin (Crystapan)	2.7 x 3.5	1.7 x 1.5	1.5 x 1.9
Polymixin	0.8 x 0.8	1.5 x 1.6	1.3 x 1.3
Sulfadiazine	No inhibition	1.5 x 1.5	1.5 x 1.5
Sulfanalimide		No inhibition	
Hostacyclin	2.0 x 2.2	2.1 x 2.4	3.6 x 3.5

In vitro all antibiotics except sulfanalimide were able to inhibit only the symbiotes cultured from ovaries and eggs, but not those from mycetomes, indicating differential responses in normal and infective forms. The symbiotes cultured from the egg were more susceptible than those from the mycetomes and ovaries to Aureomycin and Hostacyclin. On the other hand, penicillin affected the mycetomal microorganisms more, as compared to those derived from eggs or ovaries. Polymixin produced much less zone of inhibition in organisms isolated from mycetome.

The results suggest a possibility of the existence of some unknown mechanism in insects which protects the symbiotes from the adverse effect of antibiotics. The symbiotes appear to be resistant to these chemicals, but the host is not. The larvae feeding on pellets with antibiotics produced adults with soft pale brown cuticle, except in the case of Hostacyclin and polymixin. An important question arises that while most of the bacteria associated with higher animals are susceptible to antibiotics, how is it that the symbiote of *S. oryzae*, which has been identified as a bacterium very closely resembling *Bacillus circulans* (Dang, 1971; Dang-Gabrani, 1970) resists the action of potent chemotherapeutics? Similar results have also been obtained in *Cletus signatus* (Coreidae, Heteroptera) (Singh 1971) in which repeated injections of several antibiotics failed to eliminate the symbiotes present in mycetome, ovar, malipighian tubes or even in haemolymph. Malke (1964) argued that the lysozyme in the cockroach could kill symbiotic and pathogenic bacteria alike but symbiotes remain well protected in the mycetomes. Another explanation can be suggested on the basis of the work done on the mode of action of antibiotics on microbial infection in human beings. It has been demonstrated that bacteria in multiplying phase are more susceptible than those in non-multiplying phase (Willett, 1964). Probably a similar situation exists in *S. oyrzae*; the symbiotes are generally not in a multiplying phase intracellularly, but when cultured *in vitro* they are in active phase and have no barriers in the form of cell membrane or other mechanism present in the host body. The tests with *S. oryzae* further show that in spite of a prolonged chemotherapy for 4 generations, the symbiotes remain unaffected, suggesting the presence of a far more complicated mechanism, probably involving some enzyme system in the host or in the symbiotes which neutralizes the effect of antibiotics. Certain penicillin-resistant strains of pathogens are known to produce penicillinase (Moyed, 1964). However, no clear evidence can be given from our experiments or from the available information in the literature for the nature of the mechanism which imparts resistance to symbiotes *in vivo*.

FUNCTION OF SYMBIOTES

Vitamin production

All attempts to render *S. oryzae* aposymbiotic having failed, the trophic function was investigated indirectly. *S. oryzae* were grown in artificial pellets consisting of casein, corn starch, McCollum's salt, cholesterol and vitamins of B complex. Unlike stored grain insects without microbial symbiosis, this insect can develop in pellets lacking in nicotinic acid, riboflavin, pantothenic acid, pyridoxin, etc. These vitamins are very vital to all insects and must be available. Only exceptions are *Lasioderma, Stegobium* and others which have symbiotes supplying one or more of these substances to the hosts (Pant and Fraenkel, 1954a). By this analogy, we must presume that the bacterium-like symbiotes of *S. oryzae* may be compensating for the dietary deficiency of vitamins for the host.

To supplement this analogy, synthesis of nicotinic acid, pantothenic acid and riboflavin by cultured symbiotes was demonstrated in vitro microbiologically by using *Lactobacillus arabinosus* and *L. casei* as test organisms. It was found that the symbiotes are capable of synthesizing 0.39 microgram riboflavin, 1.70 microgram of nicotinic acid and 0.14 microgram of pantothenic acid in 10 ml of the medium containing 1.5 ml. of mycetomic suspension. No tests could be performed with respect to other vitamins because of the non-availability of test organisms; but one can presume that the symbiotes may synthesize other vitamins also. This assumption gathers further support from the fact that a few larvae became adults even in no-vitamin diet.

Fixation of atmospheric nitrogen

In *S. oryzae*, as in almost all insect species, mycetomes wherever present are richly supplied by tracheae. This suggests aerobic nature of symbiotes. Toth (1946, 1953) demonstrated that symbiotes in aphids are capable of fixing atmospheric nitrogen. Smith (1948), however, failed to support nitrogen fixation in aphids. Owing to the successful isolation, identification and cultivation of symbiotes in artificial medium, tests were carried out to see if symbiotes of *S. oryzae* can fix nitrogen from air. Isolates from mycetome, ovary and egg could grow on nitrogen free media and fix up to 0.28 mg/gm of sucrose in 14 days. This amount is much smaller than the 2-10 mg fixed by *Azotobacter* and 5.6-7.87 mg by *C. signatus* (Narula 1969, Singh, 1971). Symbiotes have often been supposed to be playing an active role in the nitrogen metabolism of the host (Sulc, 1910, Buchner, 1912). These ideas were based mainly on anatomical evidence such as the presence of a profuse tracheal network. Toth (1951-53) demonstrated that cultured microorganisms of *Carpocoris* and *Pyrrhocoris* fix nitrogen in culture.

The significance and the extent of the role the symbiotes play in the nitrogen economy of the host can only be fully appreciated if aposymbiotic individuals can be obtained to compare their behaviour with that of normal insects. This naturally draws our attention again to the need for successful elimination of symbiotes without which several aspects of endosymbiosis will continue to lack positive experimental support. The failure of antibiotics in eliminating intracellular symbiotes is a problem which requires our immediate attention.

CONCLUSION

For a proper understanding of the physiology of symbiosis in insects, a thorough understanding of the symbiotes *in vivo* and *in vitro* is essential. Enough information is available in the literature on the location, development and transmission of symbiotes. These are based mainly on the histological evidence which itself suffers from certain shortcomings. In most instances, common stains like haematoxylin or Giemsa have been used. These, not being specific to bacteria, stain several other inclusions in the tissue rendering the observations less accurate. A method needs to be evolved by which symbiotes and their various pleomorphic forms could be differentially stained within the insect tissue.

The discussion will not be complete without referring to *in vitro* culture of symbiotes. Brooks (1963) has raised strong, yet valid doubts that "It appears unlikely that intracytoplasmic microorganisms can be cultured on simple nutrient broths but instead will require complex media like those used for tissue culture . . ." She has pointed out that Koch's postulates must be fulfilled which of course can be attempted if the host can be made aposymbiotic.

A successful reaction to simple medium like nutrient broth by the symbiotes of *S. oryzae* or *C. signatus* (Singh, 1971) is bound to raise doubts if we presume that the symbiotes because of their long associations with the hosts have become obligatory partners. But the results reported in this communication and also in literature indicate that the symbiotes in well-studied insects have retained their power of independent existence and have remained autotrophic for several nutrients - a property so common with all the microorganisms. The success of culturing the symbiotes of *Sitophilus* (Dang, 1971; Dang-Gabrani and Pant, unpublished data) and those of *Lasioderma* and *Stegobium* (Pant and Fraenkel, 1954a,b) cannot be doubted as these have been achieved in several replicated tests following the usual microbiological procedures. Further, the nutritional experiments in all these three cases show that the symbiotes are able to synthesize nutrients in the host as well as they do *in vitro*. The usual biochemical tests carried out with the cultivated symbiotes of *S. oryzae* show their close resemblance to *B.*

circulans. Serological tests have further confirmed that the cultured symbiotes are similar to those found in insect tissues. It will not be correct to suppose that symbiotes undergo functional and anatomical degeneration, a characteristic of parasites. It is this virtue of symbiotes which forms the foundation of insect-microbe relationships and in at least many instances influences the nutrition of the host.

REFERENCES

BLEWETT, N. and FRAENKEL, G. (1944). Intracellular symbiosis and vitamin requirements of two insects, *Lasioderma serricorne* and *Sitodrepa panicea*. *Proc. Rov. Soc.* B. **132**:212-221.

BREITSPRECHER, E. (1928). Betrage zur Kenntnis der Anobiidensymbiose. *Z. Morphol. Okol. Tiere* **11**:

BROOKS, M. A. (1963). Symbiosis and aposymbiosis in arthropods. *Symp. Soc. Gen. Microbiol.* **13**:200-231.

BROOKS, M. A. and RICHARDS, A. G. (1955). Intracellular symbiosis in cockroaches. I. Production of aposymbiotic cockroaches. *Biol. Bull.* **109**:22-39.

BUCHNER, P. (1912). Studien and intrazellularen Symbioten, 1: Die Symbionten der Hemipteren. *Arch. Protistenk.* **26**:1-116.

________(1930). "Tier und Pflanze in Symbiose." *Verlag Gebruder Born-Traeger, Berlin.* pp. 900.

________(1965). "Endosymbiosis of Animals with Plant Microorganisms" (English edition). Wiley (Interscience) New York, pp. 901.

DANG, K. (1971). Studies on the intracellular symbiotes in *Sitophilus oryzae* Linn. Doctoral thesis, Bihar University, Muzaffarpur, India. pp. 87.

DANG-GABRANI, K. (1970). On the function of intracellular symbiotes of *Sitophilus oryzae*. *Experientia*, **27**:107.

FOECKLER, F. (1961). Reinfektions versuche steriler larven von *Stegobium* paniceum L. mit Fremdhefen und die Beziehungen zwischen der Entwicklungsdauer ler Larven und dem Vitamingehalt des Futters und der Hefen. *Z. Morphol. Okol Tiere.* **50**:119-162.

FRANCKE-GROSMANN, H. (1967). Ectosymbiosis in Wood - Inhabiting Insects, pp. 141-205. In "Symbiosis" vol. 2 ed. by S. Mark Henry, pp. 443. Academic Press, New York and London.

FRAENKEL, G. and BLEWETT, M. (1943). Intracellular symbionts of insects as sources of vitamins. *Nature* **152**:506-507.

GLASER, R. W. (1946). Vergleichende morphologische und physiologische Studien and osis. *J. Parasitol.* **32**:483-489.

GRABNER, K. E. (1954). Vergleichende morphologische und physiologische Studien and Anobiiden-und Çerambyciden-symbionten. *Z. Morphol. Okol Tiere.* **41**:471-528.

HEITZ, E. (1927). Über intrazellulare Symbiose bei holzfressenden Kaferlarven. *Z. Morphol. Okol Tiere.* **7**:279-305.

HUGER, A. (1956). Experimentelle Untersuchungen uber die künstliche Symbionten-elimination bei Vorratsschadlingen: *Rhizopertha dominica* F. (Bostrychidae) und *Oryzaephilus surinamensis* L. (Cucujidae). *Z. Morphol. Okol. Tiere.* **44**:626-701.

JURZITZA, G. (1964). Studien an der Symbiose der Anobiiden. II. Physiologische Studien am Symbioten von *Lasioderma serricorne* F. Arch. *Mikrobiol.* **49**:331-340.

KARAWAIEW, W. (1899). Anatomie und Metamorphosis des Darmkanals der Larve von *Anobium paniceum*. *Biol. Zentr.* 19:122-130, 161-171, 196-220.

KOCH, A. (1931). Uber die symbiose von *Oryzaephilus surinamensis* L. *Z. Morphol. Okol. Tiere.* 23:389-424.

_________(1936). Symbiosestudien, II: Experimentelle Untersuchungen an *Oryzaephilus surinamensis* L. *Z. Morphol. Okol. Tiere.* 32:137-180.

___________(1956). The experimental elimination of symbionts and its consequences. *Expl. Parasit.* 5:481-518.

_________(1960). Intracellular symbiosis in insects. *Ann. Rev. Microbiol.* 14:121-140.

_________(1967). Insects and their endosymbionts, pp. 1-106. In "Symbiosis" ed. by S. Mark Henry, pp. 443. Academic Press, New York and London.

KOLYA, A. K. and PANT, N. C. (1962). Effect of ultraviolet irradiation and centrifugation on the mycetomes and symbiotes of *Oryzaephilus surinamensis*. *Indian J. Ent.* 24:191-198.

KUHLWEIN, H. and JURZITZA, G. (1961). Studien an der Symbiose der Anobiiden I. Mitt.: Die Kulter der Symbionten von *Sitodrepa panicea* L. *Arch. Mikrobiol.* 40:247-260.

MALKE, H. (1964). Production of aposymbiotic cockroaches by means of lysozyme. *Nature* 204:1223-1224.

MANSOUR, K. (1927). The development of the larval and adult midgut of *Calandra oryzae* (L). The rice weevil. *Quart. J. Microscop. Sci.* 71:313-352.

MANSOUR, K. (1930). Preliminary studies on the bacterial cell-mass (accessory cell-mass) of *Calandra oryzae* (L.): The rice weevil. *Quart. J. Microscop. Sci.* 73: (New Series) 421-436.

MILNE, D. L. (1963). A study of the nutrition of the cigarette beetle, *Lasioderma serricorne* F. (Coleoptera:Anobiidae) and a suggested new method for its control. *J. Entomol. Soc. S. Africa* 26:43-63.

MOYED, H. S. (1964). Biochemical mechamisms of drug resistance. *Ann. Rev. Microbiol.* 18:347-366.

MURRAY, F. V. and TIEGS, O. W. (1935). The metamorphosis of *Calandra oryzae*. *Quart. J. Microscop. Sci.* (N.S.)77:405-495.

MUSGRAVE, A. J., ASHTON, G. C. and HOMAN, R. (1963). Quantitative and qualitative effects of temperature and type of grain on population of *Sitophilus* (Coleoptera: Curculionidae) and on their mycetomal microorganisms. *Can. J. Zool.* 41:1245-1261.

MUSGRAVE, A. J., GRINYER, T. and HOMAN, R. (1962). Some aspects of the fine structure of the mycetomes and mycetomal microorganisms in *Sitophilus* (Coleoptera:Curculionidae). *Can. J. Microbiol.* 8:747-751.

MUSGRAVE, A. J. and MILLER, J. J. (1951). A note on some preliminary observations on the effect of the antibiotic Terramycin on insect symbiotic microorganisms. *Can. Entomologist* 83:343-345.

__________&_________(1956). Some microorganisms associated with the weevils *Sitophilus granarius* (L.) and *Sitophilus oryza* (L.) (Coleoptera). II. Population differences of mycetomal microorganisms in different strains of *S. granarius*. *Can. Entomologist* 88:97-100.

MUSGRAVE, A. J., MONRO, H. A. U. and UPITIS, E. (1965). Apparent elimination of symbiotes in successive generations of *Sitophilus* (Coleoptera) fumigated with methyl bromide. *J. Invert. Pathol.* 7:506-511.

NARULA, M. (1969). Studies on intracytoplasmic microorganisms in *Cletus signatus* Walker (Coreidae:Heteroptera). Doctoral thesis, Indian Agricultural Research Institute, New Delhi, pp. 80.

PANT, N. C. and FRAENKEL, G. (1950). The function of the symbiotic yeasts of two insect species, *Lasioderma serricorne* F. and *Stegobium (Sitodrepa) paniceum* L. *Science* 112:498-500.

________ & ________(1954a). Studies on the symbiotic yeasts of two insect species, *Lasioderma serricorne* F. and *Stegobium paniceum* L. *Biol. Bull.* **107**:420-432.

________ & ________(1954b). On the function of the intracellular symbionts of *Oryzaephilus surinamensis* L. (Cucujidae:Coleoptera). *J. Zool. Soc. India* **6**:173-177.

PANT, N. C., GUPTA, P. and NAYAR, J. K. (1959). Physiology of intracellular symbionts of *Stegobium paniceum* L. with special reference to amino acid requirements of the host. *Experientia* **16**:311-312.

PIERANTONI, U. (1929). L'organo simbiotico di *Silvanus surinamensis* (L.). *Rc. R. Accad. Lincei* **(6)9**:451-455.

________(1930). Origine e sviluppo degli organi simbiotici d' *Oryzaephilus (Silvanus) surinamensis. Att. R. Accad. Sci. fis. mat. Napoli* (2)**18**:1-16.

RICHARDS, A. G. and BROOKS, M. A. (1958). Internal symbiosis in insects. *Ann. Rev. Entomol.* **3**:37-56.

SCHNEIDER, H. (1940). Beitrage zur Kenntnis der symbiontischen Einrichtungen der Heteropteren, *Z. Morphol. Okol. Tiere* **36**:595-644.

________(1954). Künstlich symbiontenfrei gemachte Kornkäfer (*Calandra granaria* L.) *Naturwissenschaften.* **41**:147-148.

________(1956). Morphologische und experimentelle Untersuchungen über die Endosymbiose der Korn-und Reiskäfer (*Calandra granaria* L. und *Calandra oryzae* L.). *Z. Morphol. Okol Tiere.* **44**:555-625.

SELMAIR, E. (1962). Beitrage zur Wirkung wachstumsfordernder Stoffe auf die Entwicklung der Blattiden. *Z. Parasitenk.* **21**:321-362.

SINGH, G. (1971). Studies on the endosymbiotic microorganisms in *Cletus signatus* Walker (Coreidae:Heteroptera). Doctoral thesis, Indian Agricultural Research Institute, New Delhi. pp. 143.

SMITH, J. D. (1948). Symbiotic microorganisms of aphids and fixation of atmospheric nitrogen. *Nature, Lond.* **162**:930-931.

STEINHAUS, E. A. (1946). 'Insect Microbiology', pp. 763. Comstock Publishing Co., Ithaca, N.Y.

SULC, K. (1910). "Pseudovitellus" und ähnliche Gewebe der Homopteren sind Wohnstätten symbiontischer Saccharomyceten. *Sitzber. Bohm. Ges. Wiss.*

TIEGS, O. W. and MURRAY, F. V. (1938). The embryonic development of *Calandra oryzae. Quart. J. Microscop. Sic.* **(N.S.) 80**:159-284.

TOTH, L. (1946). The biological fixation of atmospheric nitrogen. *Monographs Hung. Mus. Natl. Sci.* **5**.

________(1950). Die Rolle der Mikroorganismen in den Stickstoff - Stoffwechsel der Insekten. *Zool. Anz.* **146**:8191-196.

________(1952). The role of nitrogen - active microorganisms in the nitrogen metabolism of insects. *Tijdschr. Entomol.* **95**:43-59.

________(1953). Nitrogen active microorganisms living in symbiosis with animals and their role in the nitrogen metabolism of the host animal. *Arch. Mikrobiol.* **18**:242-244.

WIGGLESWORTH, V. B. (1952). Symbiosis in blood-sucking insects. *Tijdschr Ent.* **95**:63-68.

WILLETT, H. P. (1964). The action of chemical agents and chemotherapeutic drugs on bacteria, pp. 139-171. In "Zinsser Microbiology" ed. by Smith, Conant, Willett, pp. 1281. Appleton Century - Crofts, Division of Meredith Corporation (ACC), N.Y.

EXPLORING NUTRITIONAL ROLES OF EXTRACELLULAR SYMBIOTES ON THE REPRODUCTION OF HONEYDEW FEEDING ADULT CHRYSOPIDS AND TEPHRITIDS

K. S. Hagen and R. L. Tassan
Department of Entomological Sciences
Division of Biological Control
University of California, Berkeley 94706

Homopterous honeydews attract and are eaten by adults of certain Chrysopidae and Tephritidae. Adult green lacewings, whose larvae are predaceous, and adult fruit flies, whose larvae are phytophagous, depend mainly on honeydews for oogenesis. Although honeydews vary greatly in quality, rarely are they nutritionally complete enough to satisfy chrysopid and tephritid requirements for egg production. Yet these insects respond with high fecundities by feeding upon certain honeydews. It seems it is the extracellular yeast symbiotes in *Chrysopa* and extracellular bacterial symbiotes in tephritids that often provide the missing nutrients in honeydews essential for high fecundity and fertility. Furthermore, bacterial symbiotes are required by at least some fruit fly larvae for plant substrate utilization. We were unable, as would be expected, to find symbiotes in the predaceous *Chrysopa* larvae.

We will show the similarity in the adult nutritional requirements between the green lacewing, *Chrysopa carnea* Stephens and the tephritids, the oriental fruitfly, *Cacus dorsalis* Hendel, the melon fly, *D. cucurbitae* Coquillett, and the Mediterranean fruit fly, *Ceratitis capitata* (Wiedemann). Additionally, both *C. carnea* and *D. dorsalis* produce many fertile eggs in the laboratory when fed honeydew or an artificial diet simulating honeydew.

Artificial honeydews attract both *Chrysopa* and tephritids, and have been used to increase *Chrysopa* oviposition in the field. Baits made by mixing a poison with an artificial honeydew (protein hydrolysates) have been used in different parts of the world to control fruit fly adults. In ecosystems where honeydew-feeding chrysopids and tephritids occur together it is conceivable that a single food spray (artificial honeydew) could be used to attract beneficial green lacewings and increase their populations, at the same time attracting fruit flies and causing decrease of their populations. Incorporating selective antibiotics in artificial honeydews could achieve this.

It is generally accepted that it is mostly the phytophagous sap sucking insects and blood sucking insects that depend upon symbiotes to provide required missing factors in their substrates for growth and reproduction, while parasitoids and predators usually do not need internal symbiotes. This presents another approach that should be more thoroughly explored in integrated control strategies. Disrupting a specific symbiotic association provides

Insect and mite nutrition – North-Holland – Amsterdam, (1972)

a selective way of attacking only target pests without disrupting natural enemies or other components of the ecosystem. Furthermore, since these food sprays attract the potential predators and pests, the diet could be applied to restricted areas.

When biological control workers import or export predators, they must keep in mind the rare cases wherein the predator relies on mutualistic micro-organisms for growth or reproduction. If the predator is imported in a stage without its symbiote and is cultured on artificial diets, it may grow and reproduce normally, but, upon release, the progeny may not be able to utilize food and subsequently will not become established in the new environment.

MATERIALS AND METHODS

Chrysopidae culture

Adults were from F_3 to F_{30} collected originally from a Kerman, California alfalfa field. Larvae were fed eggs and coddled larvae of *Phthorimaea operculella* (Zeller) (Finney, 1948). A male and female were kept in an ovipositional unit constructed from a pint cardboard carton (Hagen, 1950), or in a clear polystyrene plastic container (90 diam x 35 mm) with a 37 mm hole cut in its removable lid. Every 2 days a duplicate set of ovipositional units were prepared. Fresh liquid diet was brushed on the wax paper liner and a water soaked wad of cotton wool was placed on the bottom of the cage; the top was covered with cloth. In the plastic cages, diet was brushed on the wall, and toilet tissue paper was used as the ovipositional site on the top. Usually 10 units were used per test. Where 30 females were used there were 10 males and 10 females per unit. Laboratory tests were conducted at 27 ± 1°C, 65 ± 2% R.H. and under continuous fluorescent light.

Diet- The two standard or control diets used to obtain eggs were (1) 4g enzymatic protein hydrolysate of brewers' yeast (Type PH yeast autolysate from Yeast Products Inc., Patterson, N. J.) + 7g fructose + 10 ml of distilled water (Hagen and Tassan, 1966), and (2) 4.8g Wheast® (Knudsen Creamery Co., Los Angeles, Calif.) + 5.8g sucrose + 10 ml water (Hagen and Tassan, 1970).

Some chemically defined diets used here were similar to those described by Hagen and Tassan (1966b). The composition of the diets provided to *C. carnea* adults to determine the influence of omission of major nutritional groups upon fecundity, as well as the basal diets are shown in Table 1. The detailed B vitamin and amino mixtures used are shown in Tables 2 and 3 respectively. In order to maintain the same total N value in all the diets involving amino acid deletion studies arginine was varied (Table 4) to compensate for the amount of N omitted when a particular amino acid was

Table 1. Composition of chemically defined diets provided to *Chrysopa carnea* adults (lettered diets) and for 3 tephritid spp. (numbered diets) to determine influence of major nutritional group deficiencies on fecundity and fertility.

Ingredients[a] Diets:	A	B	C	D	E	F-U	1	2	3	4	5	6	7	8	9
Fructose (g)	3.5	3.5	3.5	3.5	3.5	3.5	4	4	4	4	5	4	4	4	4
B Vit. + water (Table 2) (ml)	–	5	5	–	5	5	–	–	–	5	–	5	5	–	–
Salt mixture (g)	–	0.16	–	0.16	0.16	0.16	–	–	0.15	–	0.15	–	0.15	–	–
19 Amino acids (Table 3) (g)	–	–	1.5	1.5	1.5	–	–	1.5	–	–	1.5	1.5	1.5	–	–
10 Amino acids (Table 3) (g)	–	–	–	–	–	0.926	–	–	–	–	–	–	–	–	–
Cholesterol (mg)	–	3	3	3	3	3	–	–	–	–	–	–	–	–	–
Water (ml)	5	–	–	–	–	–	5	5	5	–	5	–	–	5	5
Yeast hydrolysate[b] (g)	–	–	–	–	–	–	–	–	–	–	–	–	–	2	–
Soy hydrolysate, trypsin (g)	–	–	–	–	–	–	–	–	–	–	–	–	–	–	2

[a] Vitamins, amino acids and salt mixtures (W for *Chrysopa* and No. 2 (U.S.P XIII) for tephritids) from Nutritional Biochemical Corp. were used. Cholesterol, from Eastman Kodak Co.

[b] Enzymatic protein hydrolysate of brewers yeast.

deleted. Vitamin omission diets are shown in Table 5.

The stock diets were refrigerated. No axenic technique was practiced other than using microorganism inhibitors in certain diets. High sugar concentration in the diets greatly precluded contamination by bacterial activity.

The various honeydews tested were collected on wax paper, and a honeydew surplus was provided every 2 days in cages accompanied by a water-soaked cotton wool.

Table 2. Composition of vitamin mixture used in diets listed in Tables 1 and 5.

Vitamin[a]	mg per ml water	
	Chrysopa	Tephritids
Nicotinic acid amide	0.50	0.10
Calcium pantothenate	0.25	0.05
Thiamine (HC1)	0.25	0.05
Riboflavin	0.25	0.05
Pyridoxine (HC1)	0.25	0.05
Folic acid	0.25	0.05
Biotin	0.01	0.01
Inositol	1.00	-
Choline chloride	5.00	10.0
Para-aminobenzoic acid	-	0.05
B_{12}	-	-

[a] 1 mg B_{12} in 5 liters distilled water; the above vitamins were added to 100 ml of this solution.

Tephritidae culture

In tests where 15 flies of each sex were used, 5 adults of each sex were confined in a 125 cm^3 cage. In later tests only 5 pairs of flies were used; each pair was held in the 125 cm^3 cage. The 3 species of fruit flies used were progeny from an original 100 flies collected in 1949 in Honolulu, Hawaii and reared several generations under laboratory conditions. Flies in each test emerged on the same day and came from the same batch of medium. In the majority of tests, *D. dorsalis* and *C. capitata* adults were from larvae cultured in an artificial diet (Maeda *et al.* 1953): In 100 ml water add, in g, 1.3 agar, 0.12 n-butyl p-hydroxybenzoate, 4.9 dextrose, 0.35 salt mixture (U.S.P. XIII), 0.175 wheat germ oil, 0.175 cholesterol, 3.0 brewers' yeast and 0.07 choline chloride. The pH was adjusted to 4.5; once set, it was blended. *D. cucurbitae* larvae were reared on slices of yellow squash. It appeared later than this melon fly could be cultured on the above artificial diet if methyl

Table 3. Composition of amino acid mixtures used in diets listed in Tables 1, 4 and 5.

Amino acid, L form	Proportions in grams in basal diet mixtures *Chrysopa* C-E (19 AA)	*Chrysopa* F-ZZ (10 AA)	Tephritids 1-7 (19 AA)
alanine	.075	-	0.120[b]
arginine (HC1)	.060	.374[a]	0.045
aspartic	.090	-	0.142[b]
cystine	.021	-	0.017
glutamic	.329	-	0.225
glycine (ethyl ester)	.045	-	0.033
histidine (HC1)	.030	.030	0.021
hydroxyproline	.030	0	0.021
isoleucine	.081	.081	0.120[b]
leucine	.099	.099	0.070
lysine (HC1)	.099	.099	0.070
methionine	.036	.036	0.054[b]
phenylalanine	.060	.060	0.045
proline	.126	-	0.093
serine	.111	-	0.165[b]
threonine	.060	.060	0.090[b]
tryptophane	.021	.021	0.017
tyrosine	.066	-	0.052
valine	.066	.066	0.099[b]

[a]Arginine was varied in the amino acid deletion studies (Table 4).
[b]DL form doubled to provide ample L form.

p-hydroxybenzoate was used instead of the butyl form.

Diets were exposed to adult flies as droplets on wax paper and were changed daily. Axenic practices were not followed, but bacterial contamination of adult food was essentially precluded by high sugar concentrations employed. Composition of the diets are shown in Table 1. The B vitamin and amino acid mixtures used are shown in Tables 2 and 3 respectively. Racemic mixtures of DL amino acids were used at 2x to provide ample L form.

Ovipositional substrate for *C. capitata* and *D. dorsalis* has been used by USDA workers. The technique utilizes one tangential section of orange rind, pulp removed, sealed to a glass slide with paraffin. For *D. cucurbitae* a similar section of melon rind was used. These substrates were changed daily; eggs were counted and incubated on damp filter paper to determine fertility. The flies were held at room temperature ($26.7 \pm 2.7°C$) and relative humidity of $55\% \pm 10$ and normal diurnal light (Honolulu).

Table 4. Amino acid mixtures and AA deletions (-) incorporated in chemically defined diets exposed to *Chrysopa carnea* adults to determine amino acid requirements.

Ingredients[a]		Quantity (g)																
	Diets:	E	F	G	H	I	J	K	L	M	N	O	P	Q	R	S	T	U
19AA (Table 3)		1.5	-	-	-	-	-	-	-	-	-	-	-	-	-	-	-	-
10AA (Table 3)		-	.926	-	-	-	-	-	-	-	-	-	-	-	-	-	-	-
Arginine[b]		.20	.374	.399	.383	.433	.374	.402	.405	.383	.374	.390	.396	.399	.374	-	.399	.433
(-) Histidine		-	-	.921	-	-	-	-	-	-	-	-	-	-	-	-	-	-
(-) Tryptophane		-	-	-	.914	-	-	-	-	-	-	-	-	-	-	-	-	-
(-) Lysine		-	-	-	-	.886	-	-	-	-	-	-	-	-	-	-	-	-
10AA (Table 3)		-	-	-	-	-	.926	-	-	-	-	-	-	-	-	-	-	-
(-) Isoleucine		-	-	-	-	-	-	.874	-	-	-	-	-	-	-	-	-	-
(-) Leucine		-	-	-	-	-	-	-	.858	-	-	-	-	-	-	-	-	-
(-) Methionine		-	-	-	-	-	-	-	-	.889	-	-	-	-	-	-	-	-
10AA (Table 3)		-	-	-	-	-	-	-	-	-	.926	-	-	-	-	-	-	-
(-) Phenylalanine		-	-	-	-	-	-	-	-	-	-	.885	-	-	-	-	-	-
(-) Threonine		-	-	-	-	-	-	-	-	-	-	-	.888	-	-	-	-	-
(-) Tryptophane		-	-	-	-	-	-	-	-	-	-	-	-	.914	-	-	-	-
(-) Valine + symbiotes		-	-	-	-	-	-	-	-	-	-	-	-	-	.885	-	-	-
(-) Arginine[c]		-	-	-	-	-	-	-	-	-	-	-	-	-	-	1.177	-	-
(-) Valine + sorbic acid[d]		-	-	-	-	-	-	-	-	-	-	-	-	-	-	-	.926	-
(-) Lysine + sorbic acid[d]		-	-	-	-	-	-	-	-	-	-	-	-	-	-	-	-	.886

[a] See Table 1 for total composition of diets.
[b] Arginine was varied in each diet where an amino acid was deleted in order to maintain equal amount of total N in each diet.
[c] 0.625 g of L lysine (HCL) was added to equalize total N when arginine was omitted.
[d] 50 mg of sorbic acid added to diets.

Table 5. Composition of chemically defined diets provided to *Chrysopa carnea* adults to determine the influence of certain B vitamin deletions (-) on fecundity and fertility.

Components	Diets: V	W	X	Y	Z	ZZ
Fructose (g)	3.5	3.5	3.5	3.5	3.5	3.5
10 Amino acids (Table 3) (g)	0.926	0.926	0.926	0.926	0.926	0.926
Salt mixture (W) (g)	0.16	0.16	0.16	0.16	0.16	0.16
Cholesterol (mg)	3	3	3	3	3	3
Ribose nucleic acid (mg)	50	50	50	50	50	50
Sorbic acid (mg)	50	50	50	50	50	50
Vitamin mixture (ml)						
Table 2	5	–	–	–	–	–
(-) Nicotinic acid	–	5	–	–	–	–
(-) Thiamine	–	–	5	–	–	–
(-) Folic acid	–	–	–	5	–	–
(-) Riboflavin	–	–	–	–	5	–
(-) Choline	–	–	–	–	–	5

RESULTS

Major nutritional group requirements

C. carnea larvae transfer sufficient metabolites from their larvae to the adults to support production and deposition of some eggs. On a carbohydrate alone, a female may deposit a total of ca 30 eggs, which may hatch if the female was mated. On a sugar diet, ca 80% of the eggs are deposited in the first 15 days after adult emergence. Exposed to a "complete" adult chemically defined diet, *C. carnea* lays ca 10 eggs/day (Fig. 1, Table 6). By omitting 1 major nutritional group at a time from the complete adult diet, the relative quality and quantity of metabolites possibly transferred to adults could be determined. Scanning data shown in Fig. 1 and Table 6 indicates that fewer eggs were deposited by females when the 19 amino acid mixture or salt mixture was omitted compared to the egg count obtained from providing only fructose. Thus, it appears that only small quantities of amino acids (free and protein-bound) and minerals were incorporated in the adult following larval feeding. Minerals must also be present in the diet in order for amino acids to be utilized; omission of the B vitamin mixture reduces fecundity compared to the egg count obtained from the complete diet. But omission of the B vitamin mixture permitted a higher egg deposition compared to that obtained from a carbohydrate-only diet. This further verifys the minimal transfer of metabolites from the larvae. Therefore a rather complex diet must be available from extrinsic sources for high fecundity.

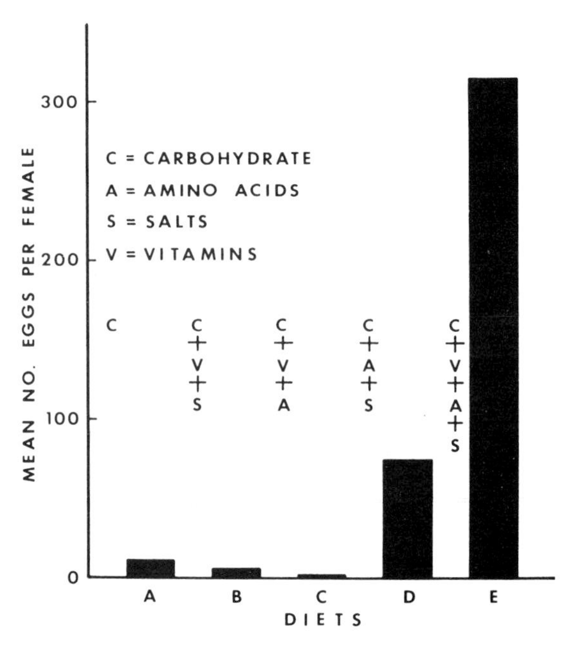

Figure 1. Mean total number of eggs deposited by *Chrysopa carnea* in 28 days exposed to various adult chemically defined diets from which major nutritional groups were omitted.

Table 6. Fecundity of *Chrysopa carnea* resulting from diets deficient of major nutritional groups over 28 days plus a 5-day preoviposition period.

Diet (Table 1)	$\bar{x}$ no. eggs[a] per ♀/day	% Mortality ♀	% Mortality ♂
A. Fructose	0.9 ± 1.2	26.6	26.6
B. Complete - 19 Amino acids	0.2 ± 0.2	10	30
C. Complete - Salt mixture	0.06 ± 0.09	80	60
D. Complete - Vitamins	3.8 ± 3.2	70	10
E. Complete (19AA)	11.5 ± 3.8	20	20
F. Complete (10AA)	9.0 ± 4.4	50	10

[a]Diet A: 15♀♀, 15♂♂; diets B-F: 10♀♀, 10♂♂.

Fruit fly data on fecundity and fertility obtained from diets with major nutritional group omissions are shown in Table 7 for *D. dorsalis* (Fig. 2), for *D. cucurbitae* (Fig. 3), and for *C. capitata* (Fig. 4). Adults of these 3

Table 7. Influence of major nutritional group deficiencies on the fecundity and fertility on *Dacus dorsalis, D. cucurbitae* and *Ceratitis capitata* over the first 30 days.[a]

Carbohydrate Diet,[b] plus	*Dacus dorsalis*			*Dacus cucurbitae*			*Ceratitis capitata*		
	$\bar{x}$ eggs/♀ per day	% Hatch	$\bar{x}$ no. days ♀alive	$\bar{x}$ eggs/♀ per day[c]	% Hatch	$\bar{x}$ no. days ♀ alive	$\bar{x}$ eggs/♀ per day[c]	% Hatch	$\bar{x}$ no. days ♀alive
Carbohydrate	0	0	26.4	0	0	26.4	0.05	0	27.0
Amino acids	.05	0	17.4	0.5	0	25.6	0.97	4.7	21.2
Salt mixture	0	0	28.6	0.2		29.0	0.17	20.0	30.0
B vitamin	0	0	21.4	0	0	25.8	0.13	40.0	26.0
Amino acids Salt mixture	24.2[d]	0	28.4	6.2	0	26.4	13.68	19.7	30.0
Amino acids B vitamin	0.16	0	27.2	0.2	0	22.0	0.17	0	24.6
Amino acids Salt mixture B vitamin	34.2	63	30.0	20.0	70.8	30.0	14.96	65.7	26.2
Yeast hydrolysate	56.4	68	29.0	16.0	72.6	30.0	26.12	78.0	30.0
L.S.D. .05	11.3	n.s.	5.7	7.0	n.s.	2.7	4.12	22.15	4.5
L.S.D. .01	13.7		7.6	9.3		3.6	5.42	29.15	5.9

[a] Based on 5 males and 5 females, 1 pair per cell; *D. dorsalis* males were fed separately on trypsin soy hydrolysate for 10 days then introduced to females.
[b] Composition of diets shown in Table 1.
[c] Based on total egg production divided by number of egg laying days.
[d] Majority of eggs visibly malformed, possibly lacking chorion.

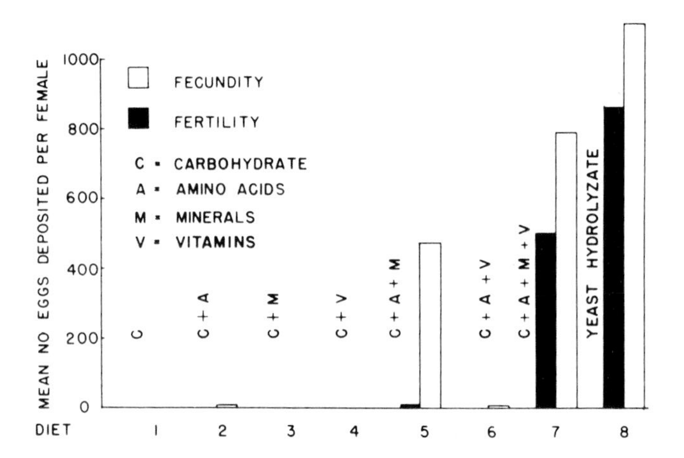

Figure 2. Mean total number of eggs and their fertility deposited in 30 days by the oriental fruit fly, *Dacus dorsalis*, exposed to various chemically defined diets from which major nutritional groups were omitted (from Hagen, 1958).

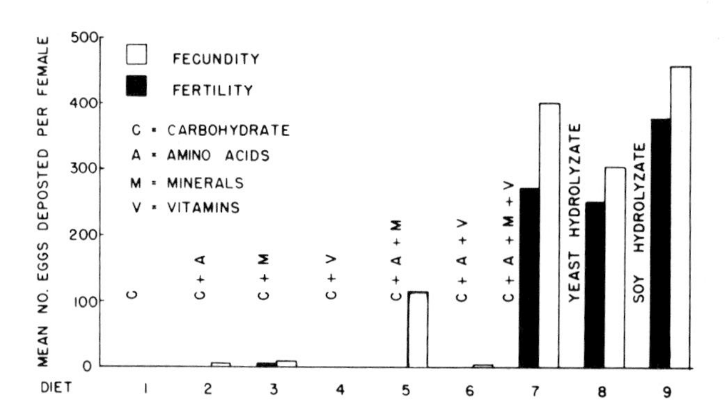

Figure 3. Mean total number of eggs and their fertility deposited in 30 days by the melon fly, *Dacus cucurbitae*, exposed to various chemically defined diets from which major nutritional groups were omitted.

tephritids have gross nutritional requirements for egg production comparable to those of *C. carnea* adults. Neither *Dacus* species oviposits when fed a diet of only sugar, but *Ceratitis* like *C. carnea* is able to produce and deposit a few eggs when fed fructose alone. Hanna (1947) obtained a few eggs when *C. capitata* was fed only a sucrose solution. Metabolite reserves are evidently

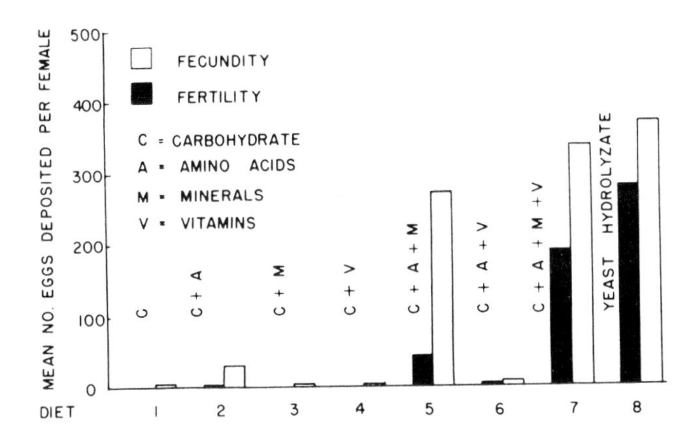

Figure 4. Mean total number of eggs and their fertility deposited in 30 days by the Mediterranean fruit fly, *Ceratitis capitata*, exposed to various chemically defined diets from which major nutritional groups were omitted.

transferred from larval feeding through the pupae to adults via the fat (Langley *et al.*, 1972). *D. dorsalis* fed a carbohydrate diet alone or plus any of the following nutritional groups: amino acids, a salt mixture or B vitamins were not able to produce eggs, but a diet of sugar, amino acids and a salt mixture did permit oogenesis, but with poor fertility (Hagen, 1958). *D. cucurbitae* responded to gross nutritional group omissions like *D. dorsalis* in egg production, and unlike *C. capitata* did not produce eggs on a fructose diet alone. Matsumoto and Nishida (1962) used a sucrose diet and obtained no eggs. *C. capitata* was quite similar in its fecundity responses, and as in *Chrysopa*, reveal some metabolite transfer from immature stages.

For high fecundity and fertility the tephritid adults, like the *C. carnea* adults, require at least a carbohydrate, amino acids, a salt mixture and a B vitamin mixture. These required nutrients must come from an extrinsic origin. If any of the nutrients are lacking in the ingested diet symbiotes would have to be present to provide the missing nutritional elements.

Chemically defined adult diets were not as effective as the enzymatic protein hydrolysates which they were formulated to emulate. There are many explanations for the lower responses, but one likely reason was the absence of fat soluble factors in the chemically defined diets. However, the males of *D. dorsalis* which were fed a trypsin soy hydrolysate did obtain some fat soluble factors including a sterol and vitamin E. Males fed only the chemically defined diet mated but the resulting fertility was poor indeed.

Nitrogen levels in adult diets

C. carnea exposed as adults to an amino N level of 2.3% in an adult chemically defined diet deposited ca 300 eggs. Yet when fed a diet of enzymatic protein hydrolysate of brewers yeast + sugar, with a calculated amino N level of 2.5%, the green lacewing species each deposited ca 400 eggs over a 21 day period. A 3.8% amino N level (=total organic N) in the defined diet permitted ca 400 eggs/female to be laid, but there was a great variability between individuals. Higher egg production from the yeast hydrolysate, which had a lower amino N level, could be accounted for by the other sources of N that it contained for the total organic N was 4%. In the undiluted yeast hydrolysate containing 11.2% total organic N, 6.7% is amino N, 1.5% peptone, 0.3% proteose N, and 2.6% polypetides, purines, pyrimidines and amide N by difference.

A diet of enzymatically hydrolyzed protein of brewers yeast + sugar permits *C. carnea* to oviposit 2x the number of eggs than obtained from a diet of unhydrolyzed brewers yeast + sugar (Hagen and Tassan, 1966b). Up to 3000 eggs can be deposited by a female *C. carnea* in her lifetime (Hagen, 1950). Although pure amino acids are readily utilized for egg production perhaps certain polypetides found in the enzymatically hydrolyzed protein are more easily metabolized and therefore are desirable in the diet.

D. dorsalis increased its egg deposition as a 19 amino acid mixture was increased to 3.8% amino N in chemically defined diets, and the peroviposition period was 9 or 10 days when fed 2.8 and 3.8% amino N levels as compared to 15 to 19 days on a diet at 1.2% amino N level. Thus, at the 3.8% amino N concentration it appears that the oriental fruit fly may be reaching its upper tolerable level of amino N intake. As in *C. carnea*, egg production was greater on an enzymatically hydrolyzed brewers yeast with a calculated 2.5% amino N than the fecundity obtained on a higher amino N in the chemically defined diets. However, here again total protein was greater in the yeast product.

Amino acid requirements

C. carnea is able to deposit a total of ca 30 eggs in its lifetime if fed only a sugar solution. The adults fed a "complete" chemically defined diet containing 19 amino acids (Diets F, J, N, R in Table 8) deposited ca 10 eggs/female/day over a 28 day period. Therefore, any daily egg deposition of about 1 egg/day or less was considered to reflect a deficiency in the diet. In Table 8, the data indicate that when any of 8 essential amino acids were omitted from the adult defined diet, oviposition dropped below a mean of 1 egg/day. Diets O, P, and Q in Table 8 which did not contain phenylalanine, threonine or valine respectively, but permitted deposition of about 5 eggs/day. After the 28 days tests, the adults were dissected; it was found that 3, 6 and 7 individuals of the adults fed diets O, P, and Q respectively contained a *Torulopsis* yeast in their crops. By adding sorbic acid to suppress

Table 8. Fecundity and fertility of *Chrysopa carnea* adults exposed to chemically defined diets with single amino acid (AA) deletions (-) over initial 28 ovipositional days (n=10♀♀, 10♀♀).

	Diet, 10AA Table 3[a]	$\bar{x}$ eggs/♀/day	$\bar{x}$ % fertile egg	Females		Males	
				% mortality	$\bar{x}$ no. days live	% mortality	$\bar{x}$ no. days live
F.	10 AA (Table 3)	9.0±4.4	55.6±11.2	50	23.4±6.2	10	27.4±1.8
G.	(-) Histidine	0.9±0.7	54.8±10.4	60	16.2±10.5	20	26.4±3.7
H.	(-) Tryptophane	0.6±0.6	62.2±11.7	70	16.8±9.5	30	26.2±4.1
I.	(-) Lysine	0.02	50.0	90	12.2±7.0	20	27.1±2.5
J.	10 AA (Table 3)	11.0±2.3	79.3±9.9	40	25.8±3.0	0	28±0.0
K.	(-) Isoleucine	0.4±0.1	17.9±19.1	80	19.0±6.6	60	23±5.4
L.	(-) Leucine	0.2±0.2	89.8±11.5	10	27.6±1.2	20	24±7.9
M.	(-) Methionine	0.6±0.4	76.8±16.1	90	20.4±4.7	30	26.8±2.0
N.	10 AA (Table 3)	9.5±1.4	65.6±26.7	60	24.6±3.8	30	26.8±3.8
O.	(-) Phenylalanine[b]	2.3±2.2	45.3±37.6	60	21.6±7.1	50	23.0±6.9
P.	(-) Threonine[c]	5.1±3	71.4±30.2	20	26.4±3.5	30	25.6±4.2
Q.	(-) Valine[d]	5.3±3.2	59.9±19.6	0	28.0±0.0	20	25.6±6.3
R.	10 AA (Table 3) + Sorbic acid	8.4±3.2	71	40	23.8±6.9	0	28.0±0.0
S.	(-) Arginine	0.7±0.5	17.2±26.2	40	22.0±9.4	20	26.4±4.4
T.	(-) Valine + Sorbic acid	0.1±0.16	0	40	24.6±5.7	30	25.4±4.2
U.	(-) Lysine + Sorbic acid	0.2±0.2	0	90	16.4±5.7	40	24.4±6.4

[a]Composition of diets shown in Table 1.
[b,c,d]3, 6, and 7 females with yeast symbiotes respectively.

Torulopsis to another valine deficient diet (Diet T, Table 8) it became clear that valine is indeed required as very few eggs were produced. Thus the *Torulopsis* yeast made up the missing valine and probably also phenylalanine and threonine. However, we did not test the omission of these latter 2 amino acids in the presence of sorbic acid. It appears that the 10 amino acids usually considered essential are also required by *C. carnea* for high fecundity. The accumulated egg deposition for *C. carnea* fed 10 amino acids for 28 days was $<$ 300 eggs; with sorbic acid added, oviposition decreased to 200; the 10 amino acids with valine deleted caused an oviposition decrease to 150 eggs. Adding sorbic acid to a valine-deficient diet (10 amino acids) resulted in no oviposition.

Tephritid amino acid requirements apparently have not been specifically determined. Some research done on *D. dorsalis* using a 13 amino acid mixture in a "complete" chemically defined diet revealed that the omitted amino acids - alanine, aspartic acid, glycine, hydroxyproline, proline and serine - were not essential for egg production.

Adding tryptophane, arginine, histidine and cystine to an acid hydrolysate of casein produced better results with *Rhagoletis pomonella* adults than with casein as the sole protein source (Neilson and McAllan, 1965).

A chemically defined diet, containing a 19 amino acid mixture, fed to the apple maggot, *R. pomonella*, stimulated egg production. However, when tryptophane was omitted, no larvae or puparia were produced. (Boush *et al.*, 1969).

B vitamin requirements

C. carnea fecundity and fertility were not greatly influenced by omission of several B vitamins from a chemically defined diet. Sorbic acid was used to suppress microorganisms in these chemically defined diets and in the test adults sorbic acid was used. Data in Table 9 indicate that omitting either nicotine acid, riboflavin or choline chloride had some influence in reducing egg production, but because of variability in fecundity there were no significant differences between diets. Absence of nicotinic acid strikingly reduced fertility and longevity. Apparently there is some B vitamin transfer from immatures as indicated in Fig. 1. Had the experiment been extended, perhaps the deficiencies would have produced a greater impact.

D. dorsalis adults were much more sensitive to B vitamin omissions than *C. carnea* (Table 10). Omitting choline chloride had the most striking effect in egg production and egg hatch. A choline deficiency is manifested after 15 days by production of many malformed eggs. Thus this lipogenic factor may in small degree be coming from the metabolites accruing to the adult from larval feeding. There appears to be no chorion formation on the malformed eggs. Omission of nicotinic acid also influences fecundity and fertility signifi-

Table 9. Influence of single B vitamin deletions (-) in a chemically defined diet on the fecundity and fertility of *Chrysopa carnea* over a 28 day period, (n = 10♀♀, 60♂♂).

Complete Diet, Table 5	$\bar{x}$ eggs/♀ per day[a]	% Hatch	% Mortality ♀	% Mortality ♂
V. Complete	9.4±4.4	53.8	70	10
W. (-) Nicotinic acid	4.8±2.7	21.9	90	60
X. (-) Thiamine	7.9±5.8	50.8	40	25
Y. (-) Folic acid	10.9±2.9	63.5	40	25
Z. (-) Riboflavin	5.7±2.9	42.2	30	15
ZZ. (-) Choline cl.	4.1±2.4	50.8	50	25

[a]No significant difference at 5% between complete diet and any vitamin omissions, but there was significant difference between Diet Y and Diets W and ZZ.

Table 10. Influence of single B vitamin omissions on the fecundity, fertility and longevity of *Dacus dorsalis* over entire life of females (n = 5♀♀, 5♂♂).

Diet[a]	$\bar{x}$ eggs/♀ per day	$\bar{x}$ fertility[b]	$\bar{x}$ life ♀ days
Vitamin omitted			
Thiamine	34.3	32.6*	43.6
Riboflavin	39.2	65.2	53.6
Nicotinic acid	22.9*	25.0*	56.5
Pyridoxin	34.5	70.0	41.6
Ca Pantothenate	37.9	75.4	47.2
Choline Cl	21.4*	5.0**	49.8
Folic acid	36.2	28.2**	48.6
p-Amino benzoic acid	41.2	61.4	49.4
Complete diet	36.7	63.7	49.2
Complete diet + biotin	32.7	60.2	55.2
L.S.D. .05	9.6	23.7	n.s.
L.S.D. .01	12.6	31.2	

[a]Composition of diets in Table 2.
[b]Males were fed trypsin soy hydrolysate with sucrose for 10 days before introducing to females.

cantly. Fertility is a more sensitive indicator of the influence of nutritional factors than just the no. of eggs deposited. Fertility responses showed that a diet deficient in thiamine or folic acid significantly reduced the no. of eggs hatching. Males of *D. dorsalis* were fed a trypsin hydrolysate of soy + sugar for 10 days and then were introduced to the females. Therefore even though there were no fat soluble factors in the female diet, males had received the required factor to insure effective spermatozoan insemination from the soy hydrolysate. Vitamin E was implicated as the required male factor (Hagen, 1952).

Apparently the B vitamin requirements in the diet vary between insects. If no or only a few B vitamins are required in the adult diet the adults probably depend upon metabolite transfer from their immatures or upon the ability of associated microorganisms in the substrate or in the host to provide the missing required B vitamins for oogenesis. In these honeydew feeding groups *Chrysopa* and *Dacus*, choline chloride and nicotinic acid stand out as being required. *D. dorsalis* is similar in its egg producing requirements to the ichneumonid *Exeristes comstockii* (Cresson), a pollen feeder and host blood feeder, as Braken (1966) found that this ichneumonid needed thiamine, folic acid and pantothenate. Pantothenic acid, however, apparently was not required in the diet of *D. dorsalis*.

Honeydews and artificial honeydews

C. carnea adults respond with high fecundity when fed only citrus mealybug honeydew, or the same honeydew with honey provided separately (Table 11). When other natural honeydews from the walnut aphid, the pear psylla and the albizzia psyllid were provided to *C. carnea* adults, the resulting fecundity was at best only slightly greater than when honey alone was used as the adult food. "Maximum" egg productivity occurred when artificial honeydews were used as adult food.

An artificial honeydew made up of an enzymatic protein hydrolysate of brewers' yeast (*Saccharomyces cerevisiae*), which is water soluble, with honey provided separately gives a slightly higher fecundity than mealybug honeydew, Table 11 (Hagen, 1950). Mixing yeast hydrolysate with sugar achieved an even higher egg production (Hagen and Tassan, 1966b). Similarly high egg production occurred when an inexpensive commercial dairy byproduct Wheast was mixed with sugar (Hagen and Tassan, 1970). Wheast is composed of a yeast (*Saccharomyces fragilis*) plus its whey substrate. The Wheast product is not completely water soluble, and the particulate nature of the mixture is readily eaten by many *Chrysopa* species. Wheast may thus not only simulate honeydews but pollens as well.

Thus, using fecundity as a criterion, artificial honeydew made from either protein hydrolysate of brewers' yeast or Wheast + sugar most closely

Table 11. Fecundity of *Chrysopa* and *Dacus dorsalis* exposed to various honeydews, artificial honeydews and chemically defined diets.

Species and diet (water also provided separately)	x̄ no. eggs/day/♀ 28 days	(n ♀)
Chrysopa carnea		
Honeydew (mealybug)[a]	18.7± 4.9	30
Honeydew, honey[a]	19.7± 1.9	30
Honeydew (aphid)[b]	2.3± 1.5	10
Honeydew (psyllid)[c]	1.1± 0.7	3
Honey	1.1± 0.1	30
Defined diet E (Table 1)	11.5± 3.8	10
Yeast hydrolysate, honey[a]	20.1± 4.1	30
Yeast hydrolysate ± sugar[e]	26.3± 5.6	10
Wheast ± sugar[f]	27.7± 5.5	10
Chrysopa sp. (from Australia)		
Honeydew (psyllid)[d]	2.1± n.a.	3
Wheast ± sugar[f] (21 days)	10.4± n.a.	3
Honey (21 days)	0.0	3
Dacus dorsalis		
Honeydew (mealybug[1])[g]	21.1± 5.6	15
Defined diet	25.6± 4.2	5
Yeast hydrolysate ± sugar[g]	36.9±11.8	15

[a] Data from Hagen (1950); *Planococcus citri*, cultured on potato tuber sprouts.
[b] *Chromaphis juglandicola* cultured on English walnut trees.
[c] *Psylla pyricola* cultured on winter nellis pear trees.
[d] *Psylla uncatoides* (21 days) cultured on Acacia.
[e] 4 g yeast hydrolysate Type pH (Yeast Products, Inc., Patterson, N.J.), + 7g fructose ±10 ml water.
[f] 4.8 Wheast®(Knudsen Creamery Co., Los Angeles, Calif.) + 5.8 fructose + 10 ml water.
[g] Hagen (1958).

simulates natural mealybug honeydew and far surpasses any of the other natural honeydews fed to *C. carnea.*

Fecundity obtained from a "complete" chemically defined diet was about 1/2 of that obtained from mealybug honeydew or from yeast products; as complex as the defined diet is, there may be additional factors required. Addition of nucleic acids to the defined diet has improved the fecundity obtained from *C. carnea.* Thus an effective honeydew as a diet for *Chrysopa* probably contains nucleic acids.

A *Chrysopa* sp. (*C. dispar* group) from Australia found to be associated with the albizzia psyllid (*Psylla uncatoides* Tuthill), deposited 1/2 as many eggs as *C. carnea* when fed the Wheast-sugar diet (Table 11). Oviposition was relatively poor when it was fed albizzia psyllid honeydew and it did not produce eggs on honey alone. It must be stressed here that this species was introduced into California without its symbiotic microorganisms. With the presence of its symbiote the albizzia honeydew probably would induce high egg production. Adults of this species did not feed on psyllid eggs, nymphs or adults, but its larvae are voracious predators of all stages of albizzia psyllid.

Dacus dorsalis fecundity was quite similar to that of *Chrysopa carnea* when fed similar diets (Table 11). Oviposition was ca 20 eggs/day when fed mealybug honeydew, ca 35 eggs/day on the enzymatic protein hydrolysate of yeast and an intermediate no. of eggs on chemically defined diet.

The data clearly show (Table 11) that all honeydews are not of equal nutritional quality, and that rather complex diets must be provided to both *C. carnea*, the Australian *Chrysopa* sp., and *D. dorsalis* to obtain high fecundity. Diets made of yeast products with a carbohydrate added simulate the citrus mealybug honeydew in fecundity and fertility levels obtained from both *C. carnea* and fruit flies. In accumulated mean eggs/female, yeast hydrolysate plus sucrose and Wheast plus sucrose gave higher oviposition, $>$ 800, than mealybug honeydew or yeast hydrolysate and honey, $>$ 550.

DISCUSSION

Food habits of the adult green lacewings and tephritid fruit flies in the field are little known. The few field observations and dissections of *Chrysopa* adults reveal, depending upon the species, that honeydews, pollens, nectars and insects are used as food. Adult tephritids have been observed to feed on honeydews, fruit exudations, bird droppings and nectars.

Using chemically defined adult diets in the laboratory has shown that the green lacewing, *C. carnea*, and 3 fruit fly species, *D. dorsalis, D. cucurbitae* and *C. capitata*, require complex diets in the adult stage for high fecundity and fertility, as there is little if any transfer to the adults of metabolites incorporated during larval feeding. For the adult it has thus been established that most of the nutrients must come from extrinsic sources, and that honeydew, even though often nutritionally incomplete, is one of these sources. Nutrients missing in honeydew, however, can apparently be provided by extracellular microorganisms in both *C. carnea* and tephritids.

Knowledge derived from these and other nutritional studies has led to field manipulation of both certain green lacewing species and tephritid fruit fly species. An inexpensive complete artificial honeydew has been used in the field to attract *C. carnea* and induce oviposition. An incomplete artificial honeydew for various tephritid fruit flies has been used to attract and kill them.

Chrysopidae

Recent reviews of the feeding habits of adult *Chrysopa* indicate that about 1/2 of the 33 species studied are predaceous in both adult and larval stages. The remaining 15 or so species are predaceous as larvae but not as adults (Hagen, Tassan and Sawall, 1970; Ickert, 1968; Sheldon and MacLeod, 1971). The foods of the non-predaceous *Chrysopa* adults are mainly honeydews and pollens or mixtures of both.

C. carnea adults fed the honeydew of the citrus mealybug, *Planococcus citri* (Risso), were not only sustained for weeks but also responded with high fecundity (Finney, 1948; Hagen, 1950). Honeydew from aphids in Israel was found to be an effective diet for *C. carnea* by Neumark (1952). Moreover, honeydew from the soft brown scale, *Coccus hesperidum* L., was an effective adult diet for *C. carnea* (Elbadry and Fleschner, 1965). By dissecting *C. carnea* Sheldon and MacLeod (1971) found that leaf scrappings in the gut contained pollens and spores but these items were in the honeydew and were ingested along with honeydew. But as we have shown in Table 11 not all honeydews produce similar fecundity levels in *Chrysopa*.

All honeydews are comprised of mostly sugars and this alone usually makes them important sustaining foods for many insects. In addition most honeydews have some amino N present. Apparently only 2 honeydews analyzed, both from aphids, contain all 10 of the essential amino acids (Auclair, 1963). Recent analyses of various honeydews confirm most earlier reports that 1 or more essential amino acids are absent (Adomako 1972; Boush *et al.*, 1969; Salama and Rizk, 1969; Saleh and Salama 1971; Sidhu and Patton, 1970). Histidine, lysine, phenylalanine and tryptophane are consistently lacking in most honeydews; however, the absence of tryptophane may be a result of the techniques used.

Pollens found in the gut of *C. carnea* Stephen (=*C. vulgaris* Schneider, *C. plorabunda* Fitch, *C. californica* Coquillett) have come from diverse plants e.g., monocots, confiers, and composits (Grinfeld, 1959; Ickert, 1968; Sheldon and MacLeod, 1971; Sundby, 1966). According to Ickert (1968), Ponisch (1964 thesis) could not induce *C. carnea* to eat hazelnut, birch or poplar pollen. Apparently, however, pollens eaten by *C. carnea* are utilized for egg production if the pollens are mixed with sugars (Elbadry and Fleschner, 1965).

Pollens vary considerably interspecifically in amino acid content (Haydak, 1970; Herbert *et al.*, 1970) and vary in the ratio of carbohydrate to protein (Todd and Bretherick, 1942; Standifer *et al.*, 1970). Since a sugar, nectar of honeydew seems to be required along with the pollen, and since the carbohydrate in many pollens is mainly starch, it may be that *Chrysopa* adults cannot utilize complex carbohydrates but require simple sugars, as suggested by Shelden and MacLeod (1971). Gilbert (1972) suggested that the free amino acids in pollen are leached out by mixing the pollen in nectar or sugar and thus are made available for egg production. The possibility exists, however, that the carbohydrate must be in the form of simple sugars for the yeast symbiotes in *C. carnea*, and it is the yeast symbiotes that make up the possible nutrient deficiencies in the pollens that *C. carnea* may ingest. Yeasts require fewer amino acids for growth than insects and have the ability to synthesize most essential amino acids (Harris 1958).

C. carnea is the first species among the Neuroptera found to contain mutualistic microorganisms (Hagen and Tassan, 1966; Hagen, Tassan and Sawall, 1970). The extra-cellular symbiotes are yeasts of the genus *Torulopsis.* At first we thought we had found an exception to Buchner's 1930 postulate (Buchner, 1965; Brooks, 1963; and Koch, 1967) that predaceous insects should not require internal mutualistic symbiotes. Although *C. carnea* is a member of an Order which is predominantly predaceous, *C. carnea* and some other related species are not predaceous in the adult stage, therefore, do not violate Buchner's rule.

Torulopsis yeasts are the symbiotes and they reside in the oesophageal diverticulum (crop) of adult *C. carnea.* The yeasts can be seen budding in the crop and apparently some honeydews provide an excellent substrate for the osmophilic yeasts. *Torulopsis* sp. found in *C. carnea* is easily cultured on Difco's yeast morphology agar. This medium contains only 4 amino acids (histidine, methionine, tryptophane and asparagine). We suspect the important role of *Torulopsis* is provisioning *C. carnea* with the essential amino acids that may be missing in the honeydews and pollens they eat.

The morphology of *C. carnea* does not differ greatly from *Chrysopa* that are predaceous in the adult stage. Mandibles of the predaceous species are slightly larger than in the non-predaceous species (Ickert, 1968), but there is a significant difference in the size of the tracheal trunks that serve the crop. Non-predaceous *Chrysopa* that contain symbiotes have much larger tracheal trunks than the predaceous species, and crop tracheal trunks in females of the non-predaceous species are also of much greater diameter than those of their males (Hagen, Tassan and Sawall, 1970).

The symbiotic relationship of *Torulopsis* and *C. carnea* seems to be vital, but the association appears rather loose since we believe yeasts occur only in the adult stage of *C. carnea.* Therefore, it is necessary for each generation of *C. carnea* to obtain *Torulopsis* in its food or by trophallaxis. Before mating, adult *Chrysopa* often exchange a droplet of regurgitated liquid by mouth and this could spread the yeast symbiotes in the *Chrysopa* population. The ubiquitous *Torulopsis* may exist in an emphemeral environment of honeydew or on flowers during the growing season, but may have to survive the winter in the crops of overwintering adults of non-predaceous *Chrysopa,* as *Torulopsis* do not form spores. Hence the adults of the non-predaceous *Chrysopa* may provide the winter reservoir of the yeasts. Predaceous *Chrysopa* species which do not have symbiotes mainly over-winter as prepupae in cocoons.

Attempts were made in recent years to import *C. carnea* from California, into New Zealand and several shipments of eggs were sent. In New Zealand, *C. carnea* larvae were fed aphids and the adults were fed the Wheast-sugar diet before release.[1] No establishment occurred. Since no *Chrysopa* species exist in New Zealand, establishment would have been easily detected. The failure was probably the result of *C. carnea* being shipped without its yeast symbiotes as we sent only eggs, and in the absence of any honeydew- or pollen-feeding *Chrysopa* in New Zealand there may not have been any type of suitable symbiotic yeast available to the *Chrysopa* in the field.

Recently we imported eggs of a *Chrysopa* sp. (*dispar* group) into California from Australia. We have successfully cultured several generations in the laboratory by feeding the adults the Wheast-sugar diet. This species is obviously a honeydew feeder in the adult stage; the crop trachae are similar in size to *C. carnea*, the females trachae being 4 times greater than those found in predaceous *Chrysopa* spp. We realize that this species too may require a specific yeast symbiote and that our releases may fail. We are now also trying to introduce the *Torulopsis* from *C. carnea* of central California into the adults of the Australian species to determine whether or not the new association will permit the alien *Chrysopa* to survive on its own in California. By applying a complete adult diet in the field, we can maintain the Australian species, but we do not know yet if it will encounter a suitable yeast symbiote in the wild.

Some biological control attempts in the past may have failed as the result of importing an insect that at some stage depended upon a symbiote, and inadvertantly excluding the symbiote by importing the wrong stage. It is also possible some factor in the new environment inhibited the symbiote. Thus in the future when entomologists import honeydew-feeding *Chrysopa*, the symbiote should be included either *in vivo* or *in vitro*.

After obtaining high fecundity from *C. carnea* in the laboratory in response to various yeast hydrolysates + sugar, (Hagen, Sawall and Tassan, 1970; Hagen and Tassan, 1966a) we recognized the possibility of attracting naturally occurring *C. carnea* adults by spraying an artificial honeydew in the field to simulate the presence of high homopterous population densities. However, the expense and phytotoxicity of the yeast hydrolysate to cotton prompted further research in the laboratory leading to the discovery of Wheast, an excellent candidate for field testing (Hagen and Tassan, 1970). Field applications of this inexpensive whey yeast product + sugar revealed no phytotoxicity to alfalfa, cotton, or green peppers and *Chrysopa carnea* adults were not only attracted the Wheast-sugar diet but were induced to feed and lay significantly more eggs in the food-sprayed field plots (Fig. 5). Furthermore, other natural enemies of aphids and lepidopterous eggs and larvae were also beneficially influenced (Hagen, Tassan and Sawall, 1970). We have found there is little value in applying artificial predator diets in crops to control certain pests if the natural enemies to be increased are not present in the area.

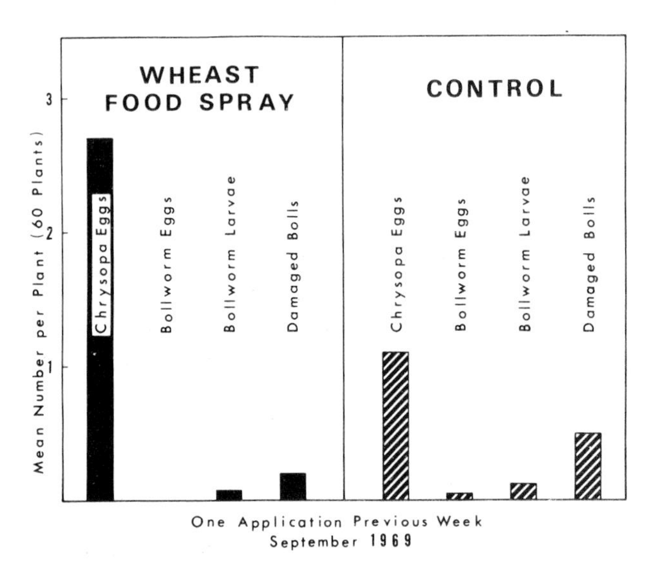

Figure 5. Mean number of *Chrysopa carnea* eggs, bollworm eggs and larvae and damaged bolls sampled from cotton plants in a plot sprayed with Wheast® + sucrose + water (4.8:5.8:20) compared to the number of insects and damaged bolls sampled in a control plot in the same field. (From Hagen, Sawall and Tassan, 1970.)

A relatively heavy deposit of the food spray is necessary on a crop if one hopes to do more than just to attract *C. carnea.* To obtain *Chrysopa* egg deposition a mixture of at least 20 pounds Wheast + 20 pounds sucrose + 20 gallons of water should be applied per acre but actually only an area of 1/2 acre receives this rate of application. Since the artificial honeydew attracts *C. carnea* from a considerable distance, only the alternate row of a crop or aircraft spray swath need to be covered by the food spray. One application costs ca $5/acre.

Tephritidae

Tephritid fruit flies are phytophagous as larvae and are among the most important fruit pests. As adults many species are attracted to honeydews and may depend mainly upon this type of diet for sustenance and egg production (Bateman, 1972; Boush *et al.*, 1969; Christensen and Foote, 1960; Hagen, 1953, 1958; Matsumoto and Nishida, 1962; Middlekauff, 1941; Neilson and Wood, 1966). Honeydews, as already discussed, are usually deficient in certain essential amino acids and probably often lack other nutritional elements known to be required by tephritid adults for high fecundity and fertility. It is suspected that missing required nutrient factors in honeydew are provided by bacterial symbiotes. Petri (1910) discovered the unique morphological adaptations in both adult and larvae of the olive fly, *Dacus oleae*, that harbored bacteria, and he believed the bacterium *Pseudomonas savastanoi* played a mutualistic role with the olive fly. Most tephritid species studied revealed presence of bacteria believed to have a mutualistic association in at least some of these species (Bateman, 1972; Baerwald and Boush, 1968; Boush *et al.*, 1969; Boush *et al.*, 1972; Buchner, 1965; Christensen and Foote, 1960; Fytizas, 1970; Fytizas and Mazomenos, 1971; Fytizas and Tzanakakis, 1966; Hagen, 1966; Hagen *et al.*, 1963; Mryazaki *et al.* 1968; Yamvrias *et al.* 1970).

Certainly the bacterial symbiotes in the larvae of the monophagous olive fly *D. oleae*, play an obligatory role, for without their symbiotic bacteria the larvae are unable to utilize the olive. With the addition of Streptomycin to a complete adult diet (enzymatic protein hydrolysate of yeast + sugar and water), of *D. oleae*, the adults deposited symbiote-free eggs. When such eggs were deposited into green olives, the larvae hatched but all died in the 1st instar (Fytizas, 1970; Fytizas and Tzanakakis, 1966; Hagen *et al.*, 1963; Hagen, 1966). One role the mutualistic bacteria may play is digesting protein; perhaps certain amino acids deficient or low in olives are also being synthesized by the symbiotic bacteria (Hagen, 1966). Hurpin (1966) reviewed the digestive function of various insect symbiotes.

Bacterial symbiotes in the Tephritidae also play a role in the adult fly, an area only recently explored. Boush *et al.*, (1969) using a chemically defined diet, demonstrated that *R. pomonella* adults required the usual 10 essential amino acids for egg production. An analysis of a honeydew normally eaten by the apple maggot adults indicated an absence of tryptophane, one of the essential amino acids. When fed the honeydew, eggs were deposited; thus Boush *et al.* (1969) speculated that either microbial growth on the honeydew or the bacterial symbiote *Pseudomonas* melophthora synthesized it, and such synthesis was shown to be possible in an earlier work by Mryazaki *et al.* (1968).

Enzymatic protein hydrolysates of yeast + sugar and water, which were used so successfully in stimulating high egg production in *C. carnea*, were also found to be an excellent diet for several species of tephritidae (Hagen and Finney, 1950; Hagen, 1953). In addition, a trypsin hydrolysate of soy with sugar was found to be an effective adult diet producing high fecundity and fertility (Hagen, 1953, 1958). These enzymatic protein hydrolysates with sugar are used as laboratory adult diets for many fruit fly species in such genera as *Anastrepha, Ceratitis, Dacus* and *Rhagoletis* (Christensen and Foote, 1968).

Steiner (1952) developed an effective bait-spray formula containing yeast hydrolysate to control tephritids in the field, and the use of contact poisonous protein hydrolysate-bait sprays is currently being used for control purposes in most parts of the world (Bateman, 1972).

Christensen and Foote (1960) visualized the potential of controlling fruit flies by killing their symbiotes. Laboratory research has demonstrated the possibility of either preventing tephritid larvae from developing or precluding oogenesis in the adults with antibiotics. Development of selective antibiotics to be included in protein hydrolysate-bait sprays which would disrupt the bacterial symbiotes of tephritids without disrupting symbiotes of *Chrysopa* would be a further development in the strategies of integrated pest control.

Some applied insect nutrition strategies in integrated pest control programs

In agroecosystems where both honeydew feeding green lacewings are co-existing with honeydew feeding tephritid fruit flies, an interesting problem is posed if either of these insect populations is to be manipulated by using protein hydrolysates with sugar. If only the green lacewing is considered, and a food spray is applied to attract and induce oviposition of *Chrysopa* adults so that hatching predaceous larvae could potentially suppress an aphid, mite or bollworm population, a population explosion of fruit flies may simultaneously be created. This may occur because the adult fruit flies in the area would also be attracted to the honeydew simulating food spray and adults not only would be sustained but egg production would be increased. Conversely, should the decision be to apply a poison bait spray with the attractant being a protein hydrolysate, fruit flies would be attracted and killed but so would the green lacewing adults because they are also attracted to the protein hydrolysates. This could result in an increase in aphids or bollworms in surrounding areas.

There are several other strategies that possibly could be employed if sufficient nutritional knowledge is available. The food sprays in Table 12 are considered attractive and possess the phagostimulants necessary to elicit good feeding response by the insects involved. Protein hydrolysates + a sugar usually provide both these qualities in honeydew feeding insects. The following reactions from *C. carnea* and several tephritid fruit fly species should occur if the adults encounter any of the following diets shown in Table 12.

Other insects occurring in the agroecosystem and feeding on the above diets would not be killed or disrupted unless they relied on symbiotic bacteria or yeasts. Since most predators and parasitoids do not rely on symbiotes, the effect of using antibiotics in food sprays would be mainly a reduction in numbers of phytophagous insects. Insects feeding on incomplete diets would be sustained as the carbohydrate alone would provide some energy. As more knowledge of insect nutrition becomes available, it will be possible to aim the impact of a nutritional spray at the specific target predator we wish to increase or at the specific target pests we wish to decrease.

Table 12. The development expected from several honeydew-feeding insects exposed to diets containing attractants and phagostimulents applied in the field. The resulting larvae from *Chrysopa* are predaceous, while those from the fruit flies are pests of fruits. (+ = yes, - = no).

Species and stage	Complete[a]	Complete + bacterial antibiotic	Incomplete[b]	Incomplete + bacterial antibiotic	Incomplete + yeast antibiotic
Chrysopa carnea					
egg	+	+	+	+	-
larval	+	+	+	+	-
Dacus oleae					
egg	+	+	+	-	+
larval	+	-	+	-	+
Dacus dorsalis & D. cucurbitae					
egg	+	+	+	-	+
larval	+	?	+	-	+
CeritITIS capitata					
egg	+	+	+	+[c]	+
larval	+	?	+	?	+
Rhagoletis pomonella					
egg	+	+	+	-	+
larval	+	?	+	-	+

[a] Enzymatic protein hydrolysate of yeast + sugar or Wheast® + sugar.
[b] Any protein hydrolized diet including sugar, but missing certain essential nutrients.
[c] A few eggs are produced on only a carbohydrate diet; some progeny possibly could survive.

SUMMARY

The neuropteron *Chrysopa carnea* is a honeydew feeder in its adult stage and predaceous in its larval stage. The dipteron family Tephritidae comprises many species that are honeydew feeders in adult stage and phytophagous in larval stage. Both certain *Chrysopa* and certain tephritids rely upon honeydew as food for egg production.

It has been shown that in *C. carnea* an extracellular yeast symbiote, residing in the adult crop, is able to provide essential nutrients that may be lacking in honeydews. A similar role is extrapolated for extracellular bacterial symbiotes residing in special gut crypts in the Tephritidae. In both *C. carnea* and in the tephritid species very little metabolite transfer from larvae to adults occurs. Adults must obtain most nutrients required for high fecundity from extrinsic sources and symbiotes.

Artificial honeydew, to attract and induce *C. carnea* to oviposit, can be applied in field situations where pest densities are low. The resulting increase in predaceous *Chrysopa* larvae can prevent a pest from reaching damaging levels. In ecosystems where tephritid pests are present, they too will be attracted and feed upon artificial honeydew. It may be possible to suppress the tephritids by incorporating selective antibiotics in the attractant food so that *Chrysopa* populations can be increased and tephritids decreased.

Acknowledgements.—We thank Dr. Minora Tamashiro of the University of Hawaii for the statistical analyses of the tephritid data, Dr. Michael Tansey of the University of Indiana for the identification of the *Torulopsis* yeast, and Mr. Irwin Sawall, Jr. who assisted in developing the predator food sprays in the field. We also wish to thank those California growers who provided field plots for our food spray experiments. This research was supported in part by grants: USDA Cooperative Regional Project W84; U.C. Berkeley Hatch Act Project 1748; California Division of Highways Project RTA 13945–A 13308 UCB.

REFERENCES

ADOMAKO, D. (1972). Studies on mealybug *Planococcoides njalensis* (Laing) nutrition: a comparative analysis for the free carbohydrate and nitrogenous compounds in cocoa bark and mealybug honeydew. *Bull. Ent. Res.* **61**:523-531.

AUCLAIR, J. L. (1963). Aphid feeding and nutrition. *Ann. Rev. Ent.* **8**:439-490.

BAERWALD, R. J. and BOUSH, C. M. (1968). Demonstration of the bacterial symbiote *Pseudomonas melophthora* in the apple maggot, *Rhagoletis pomonella*, by fluorescent antibody technique. *Invert. Path.* **11**:251-259.

BATEMAN, M. A. (1972). The ecology of fruit flies. *Ann. Rev. Ent.* **17**:493-518.

BOUSH, G. M., BAERWALD, R. J. and MIYAZAKI, S. (1969). Development of a chemically defined diet for adults of the apple maggot based on amino acid analysis of honeydew. *Ann. Ent. Soc. Amer.* **62**:19-21.

BOUSH, G. M., SALEH, S. M. and BARANOSKI, R. M. (1972). Bacteria associated with the Caribbean fruit fly. *Environ. Ent.* **1**:30-33.

BRACKEN, G. K. (1966). Role of ten dietary vitamins on fecundity of the parasitoid *Exeristes comstockii* (Cresson). *Can. Entomol.* **98**:918-922.

BROOKS, M. A. (1963). Symbiosis and aposymbiosis in arthropods. In: R. Dubos and A. Kessler editors. Symbiotic Associations 30th Symp. Soc. Gen. Microbiol. London, pp. 200-231. University Press, Cambridge.

BUCHNER, P. (1965). Endosymbiosis of animals with plant microorganisms. English translated edition. Interscience Publ. pp. 909. N. Y., London, Sydney.

CHRISTENSON, L. D. and FOOTE, R. H. (1960). Biology of fruit flies. *Ann. Rev. Ent.* **5**:171-192.

ELBADRY, E. A. and FLESCHNER, C. A. (1965). The feeding habits of adults of *Chrysopa californica* Coquillett. *Bull. Soc. Ent. Egypte.* **49**:359-366.

FINNEY, G. L. (1948). Culturing *Chrysopa californica* and obtaining eggs for field distribution. *J. Econ. Ent.* **41**:719-721.

FYTIZAS, E. (1970). Action de quelques antibiotique sur les adultes de *Dacus oleae* et leur descendance. *Zeit. Angew. Ent.* **65**:453-458.

FYTIZAS, E. and MAZOMENOS, B. (1971). Development dans les olives des larves de *Dacus oleae* issues de parents eleves au stade larvaire sur an substrat nutritif artificiel. *Ann. Zool. Ecol. Anim.* 3:217-223.

FYTIZAS, E. and TZANAKAKIS, M. E. (1966). Some effects of streptomycin, when added to the adult food, on the adults of *Dacus oleae* and their progeny. *Ann. Ent. Soc. Amer.* **59**:269-273.

GILBERT, L. E. (1972). Pollen feeding and reproduction and reproductive biology of *Heliconias* butterflies. *Proc. Nat. Acad. Sci. USA* **69**:1403-1407.

GRINFELD, E. K. (1959). Die Fütterung von Neuroptera mit pollen und ihre offenbare Miturkung bei der Entwicklung entomophiler Pflanzen. Vestn. Leningrad Univ. 9. *Ser. Biol.* 2:48-55. (Russian); from Ickert (1968).

HAGEN, K. S. (1950). Fecundity of *Chrysopa californica* as affected by synthetic foods. *J. Econ. Ent.* **43**:101-104.

HAGEN, K. S. (1952). Influence of adult nutrition upon fecundity, fertility and longevity of three tephritid species. Ph.D. thesis, Univ. of Calif., Berkeley.

HAGEN, K. S. (1953). Influence of adult nutrition upon the reproduction of three fruit fly species. pp. 72-76. Special Rep. on Control of Oriental fruit fly (*Dacus dorsalis*) in Hawaiian Islands, 3rd Senate of the State of California.

HAGEN, K. S. (1958). Honeydew as an adult fruitfly diet affecting reproduction. *Proc. Int. Congr. Ent.* **10**:25-30.

HAGEN, K. S. (1966). Dependence of the olive fly, *Dacus oleae,* larvae on symbiosis with *Pseudomonas sarastanoi* for the utilization of olive. *Nature* (London) **209**:423-424.

HAGEN, K. S. and FINNEY, G. L. (1950). A food supplement for effectively increasing the fecundity of certain tephritid species. *J. Econ. Ent.* **43**:735.

HAGEN, K. S., SANTAS, L. and TSECOURAS, T. (1963). A technique of culturing the olive fly, *Dacus oleae*, on synthetic media under xenic conditions. In radiation and radioisotopes applied to insects of agricultural importance. Inter. At. Energy Agency, Vienna. pp. 333-356.

HAGEN, K. S., SAWALL, E. F., JR. and TASSAN, R. L. (1970). The use of food sprays to increase effectiveness of entomophagous insects. Proc. Tall Timb. conf. ecol. animal control by Habitat Management, No. 2:59-81. Tallahassee, Fla.

HAGEN, K. S. and TASSAN, R. L. (1966a). Artificial diet for *Chrysopa carnea* Stephens. In: Hodek, I., Ecology of aphidophagous insects. pp. 83-87. Academia, Prague.

HAGEN, K. S. and TASSAN, R. L. (1966b). The influence of protein hydrolysates of yeasts and chemically defined diets upon the fecundity of *Chrysopa carnea*

Stephens. *Vest. Cs. Spol. Zool.* **30**:219-227.

HAGEN, K. S. and TASSAN, R. L. (1970). The influence of food Wheast and related *Saccaromyces fragilis* yeast products on the fecundity of *Chrysopa carnea. Can. Entomol.* **102**:806-811.

HAGEN, K. S., TASSAN, R. L. and SAWALL, E. F. JR. (1970). Some ecophysiological relationships between certain *Chrysopa,* honeydews and yeasts. *Boll. Lab. Ent. Agr., Portici.* **28**:113-134.

HAYDAK, M. H. (1970). Honey bee nutrition. *Ann. Rev. Ent.* **15**:143-156.

HARRIS, G. (1958). Nitrogen metabolism. Ch. 9, pp. 437-533. In: Cook, A. H. ed. The Chemistry and Biology of Yeasts. Academic Press, Inc., N. Y. 763 pp.

HANNA, A. D. (1947). Studies on the Mediterranean fruit fly, *Ceratitis capitata* Wied. II. *Bull. Soc. Soc. Fouad. ter Ent.* **31**:251-285.

HERBERT, E. W., BICKLEY, W. E. and SHIMANUKI, H. (1970). The brood-rearing capability of caged honey bees fed dandelion and mixed pollen diets. *J. Econ. Ent.* **63**:215-218.

HURPIN, B. (1966). Role des microorganismes dans les processus digestifs chez les Insectes. *Rev. Zool. Agr. et Appliq. No.* **4-6**:63-77.

ICKERT, G. (1968). Beiträge zur Biologie einhemischer Chrysopiden. *Entomol. Abhandl.* (Dresden) **36**:123-192.

KOCH, A. (1967). Insects and their endosymbionts. Ch. 1, pp. 1-106. In: Henry, S. M. ed., Symbiosis Vol. 2, 443 pp. Academic Press, N. Y. and London.

LANGLEY, P. A., MALY, H. and RUHM, F. (1972). Application of the sterility principle for the control of the Mediterranean fruit fly (*Certatitis capitata*): pupal metabolism in relation to mass-rearing techniques. *Ent. Exp. & Appl.* **15**:23-34.

MAEDA, S., HAGEN, K. S. and FINNEY, G. L. (1953). Artificial media and the control of microorganisms in the culture of tephritid larvae. *Proc. Hawaii. Entomol. Soc.* **15**:177-185.

MATSUMOTO, B. and NISHIDA, T. (1962). Food preference and ovarian development of the melon fly, *Dacus cucurbitae* Coquillett, as influenced by diet. *Proc. Hawaii. Entomol. Soc.* **28**:137-144.

MIDDLEKAUFF, W. W. (1941). Some biological observations of the adults of the apple maggot and the cherry fruit flies. *J. Econ. Ent.* **34**:621-624.

MRYAZAKI, S., BOUSH, G. M. and BAERWALD, R. J. (1968). Amino acid synthesis by *Pseudomonas melopthora* bacterial symbiote of *Rhagoletis pomonella. J. Insect. Physiol.* **14**:513-518.

NEILSON, W. T. A. and MCALLAN, J. W. (1965). Artificial diets for the apple maggot III. Improved defined diets. *J. Econ. Ent.* **58**:542-543.

NEILSON, W. T. A. and WOOD, F. A. (1966). Natural source of food of the apple maggot **59**:997-998.

NEUMARK, S. (1952). *Chrysopa carnea* St. and its enemies in Israel. *Forest Res. Sta. Ilanoth* **1**:125 pp.

PETRI, I. (1910). Untersuchungen uber die Darmbakterien der olivenfliege. *Zentralbl. Bakteriol. II* **26**:357-367.

SALEH, M. and SALAMA, H. S. (1971). Main chemical components of the honeydew excreted by the vine mealy bug *Planococcus vitis. J. Insect. Physiol.* **17**:1661-1663.

SALAMA, H. S. and RIZK, A. M. (1969). Composition of the honeydew in the mealy bug, *Saccharicoccus sacchari. J. Insect Physiol.* **15**:1873-1875.

SHELDON, J. K. and MACLEOD, E. G. (1971). Studies on the biology of the Chrysopidae II. The feeding behavior of the adult of *Chrysopa carnea. Psyche* **78**:107-121.

SIDHU, H. S. and PATTON, R. L. (1970). Carbohydrates and nitrogenous compounds in the honeydew of the mustard aphid, *Lipaphis erysimc. J. Insect. Physiol.* **16**:1339-1348.

STANDIFER, L. H., MACDONALD, R. H. and LEVIN, M. D. (1970). Influence of the quality of protein in pollens and of a pollen substitute on the development of the hypopharyngeal glands on honey bees. *Ann. Ent. Soc. Amer.* **63**:909-910.

STEINER, L. F. (1952). Fruit fly control in Hawaii with poison-bait sprays containing protein hydrolysates. *J. Econ. Ent.* **45**:838-843.

SUNDBY, R. A. (1966). A comparative study of the efficiency of three predatory insects *Coccinella septempunctata* L. *Chrysopa carnea* St. and *Syrphus ribesii* at two different temperatures. *Entomophaga* **22**:395-404.

TODD, F. E. and BRETHERICK, O. (1942). The composition of pollens. *J. Econ. Ent.* **35**:312-317.

YAMVRIAS, C., PANAGOPOULOS, C. G. and PSALLIDAS, P. G. (1970). Preliminary study of the internal bacterial flora of the olive fruit fly (*Dacus oleae* Gmelin). *Ann. Inst. Phytopath. Benaki, Greece. n.s.* **9**:201-206.

POLYPEPTIDES AND PROTEINS AS GROWTH FACTORS FOR *APOSYMBIOTIC BLATTELLA GERMANICA* (L.)

Marion A. Brooks
Department of Entomology, Fisheries, and Wildlife
University of Minnesota
St. Paul, Minnesota 55101
and
Wendell B. Kringen
Department of Biology
Union College
Barbourville, Kentucky 40906

INTRODUCTION

The intracytoplasmic bacterial symbiotes of *Blattella germanica* will not be transmitted transovarially if the mother insect is fed 0.1% or 0.2% Aureomycin(R) (chlortetracycline HC1) in her diet throughout life, starting on the day she hatches (Brooks, 1970; Brooks and Richards, 1955). An insect treated in this way regularly produces offspring lacking symbiotes for all practical purposes. The "empty" mycetocytes can be observed, in histological sections, extremely small and atrophied, although occasional bacteria can be detected by diligent search of these abnormal mycetocytes. The aposymbiotic numphs are very weak and frail. The treated mother does not lose her symbiotes; the bacteria may show some degenerative changes and some decrease in numbers, but they are not eliminated. If the antibiotic treatment is stopped, the insect again transmits symbiotes to her eggs. For these reasons we do not use the parental generation, i.e., the first treated insects, of *B. germanica* for nutritional studies, but use their offspring. While the aposymbiotic colony can be maintained by feeding them food containing yeast, Aureomycin® is also employed to suppress the residual symbiotes. The most consistent results are obtained by using filial generations of insects through about the 6th. After that, the symbiotes which are refractory to antibiotic treatment gradually increase and eventually the symbiotic hosts dominate the population, since symbiotic cockroaches are much more vigorous than aposymbiotic ones. Before this stage is reached, a new colony is started by feeding Aureomycin® to a fresh supply of normal insects.

Surprisingly, a number of quite ordinary dietary components also inhibit symbiote transmission. A deficiency of manganese or zinc, an excess of calcium (which antagonizes manganese and zinc), a high level of urea, or fatty acids which are rancid, all are deleterious to bacterial-transmission through the oocytes (Brooks, 1960, 1962).

MATERIALS AND METHODS

Nutrition of aposymbiotic B. germanica on dog biscuit diet

The newly-hatched aposymbiotic nymphs of *B. germanica*, which are weak and light-colored, have difficulty in standing on their feet and in feeding. If they are offered a standard food, such as dog biscuit which has been finely ground, they live for many months without growing. Normal control insects fed the dog biscuit grow much faster, maturing in 60 days at 28°C.

Groups of aposymbiotic nymphs were fed finely ground dog biscuit plus supplements and, as shown in Table 1, only yeast permitted growth.

Table 1. Growth of aposymbiotic *B. germanica* on supplements added to dog biscuit diet.

	Supplement	Growth
a)	5.0% cysteine	no growth - - - death
b)	5.0% ascorbic acid	no growth - - - death
c)	2.5% L-methionine	no growth - - - death
d)	2.5% L-tyrosine	no growth - - - death
e)	2.5% cholesterol	no growth - - - death
f)	5.0% ascorbic acid + 20.0% brewers' yeast	slow growth, maturing in 180 days, poor reproduction
g)	25.0% brewers' yeast	slow growth, maturing in 180 days, poor reproduction

Since yeast is vital to the life of aposymbiotic cockroaches, we undertook a determination of which growth factor(s) in yeast substitute for the bacteria. In addition, other sources of growth factors were sought, and attempts made to characterize the growth factors as far as possible.

Nutrition of aposymbiotic B. germanica on casein diets

Females carrying ripe egg capsules were collected from stock colonies of either dog biscuit or dog biscuit plus Aureomycin® chlortetracycline HCl (donated by American Cyanamid Co.). Each female was confined in a clean glass fruit-canning jar until her offspring hatched. The nymphs were maintained on distilled water and sucrose for up to 3 days. When an adequate number of numphs had hatched, 4 or 5 from each parent were distributed to each test diet. Pooled in this way, the batches of 16 to 32 nymphs were comparable with respect to parentage for every set of nutritional variables. The nymphs were reared on the experimental diets in the glass jars with filter paper discs on the bottom, and with silk or nylon mesh covers.

The casein diet used (Gordon's diet) has the following composition [its formula is given in more detail and expressed in molar concentrations elsewhere (Brooks, 1960)]:

casein (vitamin-free)	300 g
dextrose	360 g
salt mixture	28 g
vitamin mixture	7 g
cholesterol	10 g
corn oil	30 g
non-nutritive celluflour	265 g
	1000 g

Substitutions and deletions are possible if the proper compensation in the amount of non-nutritive celluflour is made, so that all other nutritive factors are held in constant proportion. The finished product, a dry, slightly oily powder, was stored in a deep freezer to deter oxidation of the corn oil.

An amount of food just sufficient to feed the cockroaches for one week was placed in a chemically-clean glass vial. The food was replenished weekly with all residue discarded. Distilled water was placed in chemically-clean glass vials stoppered with absorbent cotton, and replaced with fresh vials weekly. The insects thus had no access to crude organic substances such as wood shavings, nor did they have contact with metallic substances. Although aseptic technique was not employed, mold did not grow on the food or water containers. The normal gut flora was probably not disturbed, but this question was not examined. The temperature was maintained at 25 + 1°C.

RESULTS

On the above regime, fed Gordon's Casein Diet, our normal stock *B. germanica* matured in 43 to 59 days, with a median of 48 days, the actual times varying with different lots of insects. Survival was not perfect, with about 19/20, or 95%, reaching maturity. Each mature female produced an egg capsule in 10 days, with a total of 3 or 4 egg capsules, averaging 28-30 nymphs in each, during her lifetime. There was an incidence of about 15% of spontaneous abortion of egg capsules on this diet. Life span of the adults was 6 to 42 weeks, the males being the first to die. The offspring on the same diet survived as well as the parents had done, and grew approximately as well.

On the other hand, the survival rate of the aposymbiotic cockroaches on the casein diet was only about 33%. Presumably the ones most deficient in symbiotes were the first to die, but there was no simple way to verify this since dead insects are not satisfactory for histological preparations. The surviving insects reached maturity over a long period extending anywhere from 48 to 122 days. They usually died without producing any viable egg capsules; some females produced abnormal-appearing egg capsules which soon shriveled and dropped off.

The growth rates and the differences in performance between normal and aposymbiotic insects was checked repeatedly and verified to hold true over minor variations in the mineral composition, vitamin composition, choline level, and source of fatty acids in the diet.

In contrast to casein, amino acids were very poorly utilized by the aposymbiotic nymphs. A diet was made in which the casein was replaced by its constituent amounts of essential amino acids, and nymphs fed this died in a few weeks. The amino acid diet was then supplemented as shown in Table 2 with casein, or the additional amino acids which are deficient in casein or most likely involved in cuticle sclerotization.

Table 2. Growth of aposymbiotic *B. germanica* on supplements added to amino acid diet.

	Supplement	Result
a)	10.0% casein	very poor growth
b)	2.0% L-methionine + 0.5% L-cysteine	no growth - - - death
c)	2.0% L-tyrosine + 2.0% DL-phenylalanine	no growth - - - death
d)	1.0% L-methionine + 0.25% L-cysteine + 1.0% L-tyrosine + 1.0% DL-phenylalanine	no growth - - - death
e)	same as d) + 2.0% L-asparagine	no growth - - - death
f)	same as e) + 20% brewers' yeast	poor growth

On a diet made with acid hydrolyzed casein, the nymphs died within a few weeks, but on diet made with enzymatically hydrolyzed casein a slightly better performance was observed (i.e., they survived longer and a few matured but then died).

In all of the above diets, only when yeast was present did any significant growth occur. Yeast was either added to an already complete diet, or in some cases, it was substituted for the vitamin mixture. When yeast replaced the vitamin mixture, it also replaced part of the celluflour and constituted 25% of the total diet. Regardless of whether yeast was a supplement or a substitute, it did not alter the growth rate of normal cockroaches very much compared to their growth on a dog biscuit diet or a casein diet without yeast. The median time for maturity was 44 days instead of 48 on stock diet; survival rate was 95%.

The aposymbiotic cockroaches, however, were much more viable on yeast-supplemented or yeast-substituted casein diets, their time for maturity ranging from 48 to 92 days, with a median of 60 days, and their survival was improved to about 84%.

Calculations based on the vitamin components of yeast show that the growth factor is not any of the known B-vitamins. When yeast is substituted for the vitamin mixture at the level of 25% of the diet, the resulting B-vitamin content is less than what it was in the standard casein diet, and yet growth of the insects was dramatically improved by the feeding of yeast.

In addition to yeast, a few other products were tried for sources of growth factor. Liver was superior to yeast. White mice were sacrificed, bled, and their livers were removed and lyophilized. This liver powder, when added to a casein diet in proportions of 1:3, caused the aposymbiotic nymphs to grow as quickly as normal ones. Since liver is a rich source of vitamin B-12, this vitamin by itself was tested in casein diet, but it was not a growth factor under these conditions. Growth factor was not found in alcohol-insoluble liver extract, dried egg yolk, fresh apples, fresh wheat germ, or a dehydrated leaf factor (Cerophyll). A petroleum ether extract of the feces of normal, symbiotic *B. germanica* was prepared, dried, and added to dog biscuit diet, but this was ineffective also. It was fed at a level of extract from 10 g of feces added to 40 g of dog biscuit, and at 1/5 this amount, but in both cases a few insects survived for 6 months without growing.

The B-vitamins are only a small fraction of the contents of the intact yeast cell (White, 1954). Various species of yeasts consist of 45-52% protein, thus ranking between high quality animal protein and the best plant proteins in biological value. Yeasts also contain up to 2.6% fats, including sterols; a high purine content; and various minerals, especially calcium, iron, and phosphorus.

Fractionation of yeast was carried out to try to determine which component carries the growth factor. As a preliminary to this, we undertook a comparison of several species and strains of yeasts to ascertain which one, if any, is the most effective. Samples of several different types of Saccharomyces were donated by Standard Brands, Inc., and Anheuser-Busch, Inc. These were compared with commercially available brewers' yeast, *Torula* yeast, yeast hydrolysate, yeast extract, and yeast nucleic acid. The growth of the aposymbiotic insects was rated by a scale of 0, 1, 2, 3, with 0 indicating indefinite survival with little growth; while 3 indicated maximum growth with maturity in about 8 weeks (which compares favorably with 6-7 weeks for normal control insects). As shown in Table 3 all of the insects on whole yeast supplements grew well. Those on yeast-extract supplement (water soluble) grew poorly; while those on yeast-nucleic acid supplement failed to grow at all. Yeast nucleic acid was therefore eliminated from further consideration. Since brewers' yeast is the most conveniently and economically obtainable, we restricted our fractionation efforts to that one yeast.

Table 3. Growth of aposymbiotic *B. germanica* on various yeast supplements added to casein diet

Yeast supplement	Proportion in diet	Growth
Casein[a] diet (control	---	0
Yeast nucleic acid[b]	1:50	0
Yeast nucleic acid[b]	1:3	0
Yeast extract (water soluble)[a]	1:18	1
Yeast hydrolysate[a]	1:3	3
Brewers' yeast[a]	1:3	3
Torula yeast[a]	1:3	3
Saccharomyces yeast[c] (2 strains)	1:3	3
Saccharomyces yeast[d] (4 strains)	1:3	3

[a]Nutritional Biochemicals Corporation
[b]California Corporation for Biochemical Research
[c]Anheuser-Busch, Inc.
[d]Standard Brands, Inc.

Lots of brewers' yeast were extracted with the solvents listed in Table 4. The extracts from 25 to 50 g of yeast were evaporated to dryness and each added to 10 g of the casein diet. As indicated in the table the only fraction which showed any growth effect was that extracted with 70% ethanol. Presumably this extract contained some proteins and a little ergosterol. Ergosterol was probably also present in the extracts made with ethyl ether and chloroform, which showed no growth effect. A water extract was not made because we did not have facilities for evaporating large volumes quickly enough to prevent growth of microbial contaminants. However, yeast water-extract was obtained commercially and this was not a good growth factor, as shown in Table 3.

Table 4. Growth of aposymbiotic *B. germanica* on various extracts of brewers' yeast added to casein diet.

Solvent	Amount of yeast extracted and added to 10 g diet	Growth
95% ethanol	50 g	0
70% ethanol	25 g	2
100% isopropanol	50 g	0
60% isopropanol	25 g	0
acetone	50 g	0
benzene	50 g	0
chloroform	50 g	0
dioxane	50 g	0
ethyl acetate	40 g	0
ethyl ether	50 g	0
toluene	50 g	0
trichlorotrifluoroethane	50 g	0

With this indication that the growth factor is contained in a protein or polypeptide fraction, a number of peptides were obtained and added individually to the casein diet, at 10%. This made the protein equal to about 37% in all of these diets and in the diet which contained 20% yeast. Growth performance of the aposymbiotic insects is shown in Table 5. The addition of yeast or hydrolyzed protein improved growth over that which occurred on the casein control diet. All the peptides were not equally efficacious; both enzymatic lactalbumin hydrolysate (from milk) and proteose-peptone (from meat infusion) were as good as yeast for sources of the growth factor.

When milk is acidified to pH 4.6, the first fraction which precipitates is casein (Jenness and Patton, 1959). When the filtrate is 0.5 saturated by $(NH_4)_2SO_4$, a lactoglobulin fraction is precipitated. If the filtrate (lactalbumin) is exhaustively dialyzed, β-lactoglobulin crystallizes out. Thus, milk yields the following fractions:

1. casein		(fractionation not considered here)
2. lactoglobulin	a. b.	(two fractions not considered here)
3. lactalbumin	a.	β-lactoglobulin
	b.	blood albumin
	c.	α-lactalbumin

Table 5. Growth of aposymbiotic *B. germanica* on various proteins or polypeptides added to casein diet.

Supplement	Proportion in diet	Growth
Casein[a] diet (control)	--	0
Brewers' yeast[a] (control)	1:4	3
Casein hydrolysate, enzymatic[a]	1:9	2
Lactalbumin hydrolysate, enzymatic[a]	1:9	3
Proteose peptone, enzymatic[b]	1:9	3
Protone, enzymatic[b]	1:9	2
Tryptose, enzymatic[b]	1:9	1
Tryptone, enzymatic[b]	1:9	1
Peptone, enzymatic[b]	1:9	1
Neopeptone, enzymatic[b]	1:9	1
Lactalbumin hydrolysate, enzymatic[a]	1:9	
+ yeast extract[a]	3%	3
Carnitine, DL[c]	10 mg/kg	0

[a]Nutritional Biochemicals Corporation
[b]Difco Laboratories
[c]Courtesy Biochemistry Dept., U. of Minn.

β-lactoglobulin and blood albumin are available commercially. Blood albumin accounts for 50-60% of the lactalbumin, and is considered to be identical to bovine blood serum, which has been characterized. β-lactoglobulin has also been characterized. It consists of two very similar proteins called β-lactoglobulin A and β-lactoglobulin B, and its amino acid content is known. α-lactalbumin is not readily obtainable.

Blood albumin, β-lactoglobulin and lactalbumin hydrolysate were compared with yeast in the diets as shown in Table 6. In these diets, the casein was reduced to 25%; the proteins were substituted for celluflour in the diet at 10% or yeast was substituted at 20%, so that the total protein in all the diets was approximately 35%.

The results of this experiment showed first, that either lactalbumin hydrolysate or one of its fractions, β-lactoglobulin, supported growth of the aposymbiotic cockroaches as well as does yeast; second, that another fraction, that is, blood albumin (from blood) does not contain a growth factor. Blood albumin may contain a toxic factor, or it may simply be so distasteful that the insects did not eat properly. The normal control insects grew on the blood albumin diet, but mortality was high, with only 3/20 surviving a month past maturity.

Although the emphasis in the above experiments was on growth performance, it became apparent that growth and reproduction were not dependent on the same factor. All of the aposymbiotic insects which grew at a rate of 3 and matured, also laid egg capsules; but all the eggs aborted, except those which received a small amount of yeast water-extract. About 25% of the eggs from this group were viable, which is comparable to the results when the diet includes whole yeast. Thus, apparently something in the aqueous yeast extract is necessary for egg viability.

Table 6. Growth of aposymbiotic *B. germanica* on various protein supplements added to diet.

Amount of casein in diet	Supplement[a]	Growth
35% casein (control)	------------------------	0
25% casein	20% brewers' yeast	3
25% casein	10% lactalbumin hydrolysate (enz.)	3
25% casein	10% lactalbumin hydrolysate (enz.) + 3% yeast extract (water soluble)	3
25% casein	10% β-lactoglobulin	3
25% casein	10% blood albumin (from bovine blood serum)	0
30% casein	20% brewers' yeast	3
0% casein	50% brewers' yeast	2
0% casein	35% lactalbumin hydrolysate (enz.)	2

[a]Nutritional Biochemicals Corporation for all products.

DISCUSSION

The essential results of this work were that the addition of a mixture of certain peptides, or particular proteins, to a diet made with a presumably complete protein (casein), furnished a growth factor for aposymbiotic cockroaches which was lacking in the casein diet alone. Brewers' yeast, lactalbumin hydrolysate, β-lactoglobulin, and proteose peptone Difco were equally effective. One of these, lactalbumin hydrolysate, was fed as the sole source of nitrogen, in which case it was not so good as when it was fed in addition to casein. Since β-lactoglobulin is one of the more refined fractions of milk, it was considered in more detail. It is a complete protein and its amino acid composition is known (Jenness and Patton, 1959). On the other hand, proteose peptone Difco, which equaled β-lactoglobulin, presumably is a mixture of polymers of amino acids of lower molecular weight than proteins; but its source and analysis are unavailable.

If one compares β-lactoglobulin with whole casein, the only particularly arresting difference is that β-lactoglobulin contains twice as much sulfur as casein contains and 7 times as much cystine plus some additional cysteine. We checked the possible need for additional cystine as a growth factor by adding 0.2% to the casein diet, but this did not support growth of the aposymbiotic nymphs.

Thus while the identity of the growth factor common to all of the effective substances is unknown, the most likely candidate seems to be a combination of partially hydrolyzed protein fragments or peptides. This is not surprising since it is accepted that (mammalian) cells take up macromolecules; and there are suggestions that some of these exogenous proteins may function in the control of growth and differentiation (Ryser, 1968). Also, strepogenin, which is a variable assemblage of peptides, is a known growth factor for *Lactobacillus casei* (Woolley and Merrifield, 1963). Strepogenin is effective in minute quantities, like those required of vitamins, when added to a purified medium in which the nitrogen source is made up only of amino acids.

Woolley and his colleagues have formulated a general picture of the biological activity of peptides. Strepogenin activity resides in several dissimilar peptides, and individual amino acids can be replaced by others of similar structure without loss of activity. Thus no one amino acid is indispensable, and even the sequence of amino acids in the peptide is not always essential. Single, pure substances with biological activity can be obtained from natural products, for instance, trypsinized casein or trypsinized insulin. If minor substitutions are made, such as replacing serine with threonine (replacing one H with $-CH_3$), the resulting peptide is inactive. Similarly, the introduction of unnatural amino acid residues, such as cysteic acid for cysteine residues, renders the peptide inactive. Two positive statements can be made, however, regarding strepogenin activity. First, all active peptides contain either serine or cysteine; and second, the size of the peptide is important—at least 5 amino acids are required but large molecules such as intact proteins are inactive.

In the microbiological assays used for studying the peptides, the amounts required are too small to account for all of the amino acid requirement of the stimulated cells. This fact seems to refute the argument suggesting that peptides are merely efficient means of getting the required amino acids into the bacterial cells. Another possibility is that some molecules are inactive because they are degraded or excreted too easily before they reach the site of action, particularly in a whole animal. A compound such as a peptide would then give protection to the essential residue incorporated in it.

Woolley and Merrifield (1963) have shown that in addition to strepogenin, the activities of several peptide hormones and peptide vitamins also are independent of a single unique structure.

Peptide growth factors, hormones and vitamins are commonly assayed by their effects on growth of a bacterium, darkening of a toad skin, or contraction of a rat uterus. If we extrapolate to cover the observations reported here on growth of a cockroach, we have to consider the interaction of a second system, *viz.*, the bacterial symbiotes. Since on a minimal diet, only those cockroaches with symbiotes can grow, we must grant the premise that the symbiotes produce a growth factor, either through synthesizing activities or metabolic by-products, which is released from the mycetocytes and reaches the target tissues. The growth factor for aposymbiotic cockroaches provided by the peptide or protein supplements must then either penetrate the gut to reach the target tissues, or alter the permeability of the gut so that the essential requirements can be absorbed, or bind some trace impurities unrecognized as yet.

SUMMARY

The German cockroach, *B. germanica*, normally contains numerous symbiotic bacteria in specialized cells of the fat body, called mycetocytes. The bacteria are transmitted to the offspring internally in the egg cytoplasm. Constant feeding of Aureomycin® prevents egg infection, resulting in a generation of symbiote-deficient offspring in which empty mycetocytes can be found. The aposymbiotic nymphs die if fed a stock diet such as dog biscuit, or they survive for long periods without growing if they are maintained on a casein stock diet. However, the addition of one part of brewers' yeast to 3 parts of the casein diet, or 1 part of lyophilized mouse liver to 9 parts of casein diet, enabled the aposymbiotic nymphs to grow. Their growth on the liver supplement was equal to that of normal symbiotic nymphs.

Growth factor is not in any of the following: B-vitamins, fresh apples, dehydrated egg yolk, wheat germ, leaf factor (Cerophyll), alcohol-insoluble liver extract, yeast sterols, yeast nucleic acid, choline, carnitine, and the amino acids tryptophan, phenylalanine, tyrosine, methionine, and cysteine or cystine. Growth factor is present in a milk fraction, hydrolyzed lactalbumin, and in β-lactoglobulin, a constituent which can be crystallized out of lactalbumin. Growth factor is also present in a meat infusion preparation, proteose peptone Difco.

The action of these protein fragments or polypeptides seems comparable to that of strepogenin, which is a peptide growth factor for certain bacteria. It is not known whether the size of the molecule or its constituent amino acids are the controlling factors. Furthermore it is not known how the symbiotic bacteria supply the growth factor normally.

Acknowledgement.—This is Paper No. 8019, Scientific Journal Series, Minnesota Agricultural Experiment Station, St. Paul. The work was supported also by U.S. Public Health Service Research Grant No. AI-09914 from the National Institute of Allergy and Infectious Diseases.

REFERENCES

BROOKS, M. A. (1960). Some dietary factors that affect ovarial transmission of symbiotes. *Proc. Helm. Soc. Wash.* 27:212-220.

BROOKS, M. A. (1962). The relationship between intracellular symbiotes and host metabolism. *Symposia genetica et biologica Italica* 9:456-463.

BROOKS, M. A. (1970). Comments on the classification of intracellular symbiotes of cockroaches and a description of the species. *J. Invert. Pathol.* **16**:249-258.

BROOKS, M. A. and RICHARDS, A. G. (1955). Intracellular symbiosis in cockroaches. I. Production of aposymbiotic cockroaches. *Biol. Bull.* **109**:22-39.

JENNESS, R. and PATTON, S. (1959). *Principles of Dairy Chemistry.* John Wiley & Sons, Inc., New York, 446 pp.

RYSER, H. J. P. (1968). Uptake of protein by mammalian cells: An underdeveloped area. *Science* **159**:390-396.

WHITE, J. (1954). *Yeast Technology.* Chapman & Hall, London, 432 pp.

WOOLLEY, D. W. and MERRIFIELD, R. B. (1963). Anomalies of the structural specificity of peptides. *Ann. N. Y. Acad. Sci.* **104**:161-171.

NUTRITION OF *IN VITRO* CULTURED INSECT CELLS

W. Fred Hink
Department of Entomology
The Ohio State University
Columbus, Ohio 43210

INTRODUCTION

Nutrition at the cellular level will be investigated by discussing growth requirements and metabolic utilization of nutrients *in vitro*. Most insect cell culture media consist of chemically defined minimal media supplemented with insect hemolymph, vertebrate sera, vertebrate extracts, protein fractions, hydrolysates, yeast extracts, or tissue culture media designed for vertebrates. The chemically defined compounds usually present in media are amino acids, sugars, vitamins, salts, and organic acids. These substances are often used at concentrations approximating those in the hemolymph of the insect from which the cells were obtained. An example of a medium with this general composition is given in Table 1.

My discussion of nutritional requirements includes both defined and undefined compounds. Some media constituents are not necessarily of nutritional value to cultured insect cells but are nevertheless beneficial in terms of cell proliferation and survival.

REQUIREMENTS OF CULTURED INSECT CELLS FOR COMPLEX MEDIA CONSTITUENTS

Insect Hemolymph

The media used for early experiments in insect tissue culture consisted of hemolymph, saline solutions supplemented with hemolymph, or complex mixtures of chemicals supplemented with insect tissue extracts or hemolymph. Hemolymph from one insect species may be beneficial to cells from other species (Sen Gupta, 1964). Intestinal tissues of *Galleria mellonella*, *Antheraea pernyi* and *Melolontha melolontha* were cultured in media supplemented with homologous hemolymph and heterologous hemolymphs from the other two species. Growth in these cultures was not significantly different, which indicates a lack of specificity in hemolymph requirement. Established insect cell lines, *Aedes aegypti*, *Bombyx mori*, and *Spodoptera frugiperda*, also grown in the presence of heterologous hemolymph (Grace, 1966, 1967; Vaughn, personal communication). However, in some systems heterologous hemolymph may be toxic or inadequate. The *Antheraea eucalypti* cell line was usually cultured in a medium that contained 3% *A. pernyi* hemolymph and Rahman *et al.* (1966) assayed hemolymph from six species of Lepidoptera as possible substitutes. All hemolymphs, except that from *Antheraea polyphemus*, were toxic. The generalization derived must be that heterologous hemolymph will support cell growth, but each cell culture must be evaluated individually.

Table 1. Insect tissue culture medium (TNM-FH) used to culture cells from *Trichoplusia ni*, *Carpocapsa pomonella*, and *Heliothis zea*[a]

Salts	mg/100 ml		mg/100 ml
$NaH_2PO_4 \cdot 2H_2O$	114	L-tryptophan	10
$NaHCO_3$	35	L-glutamine	60
KCl	224	L-threonine	17.5
$CaCl_2$	100	L-valine	10
$MgCl_2 \cdot 6H_2O$	228	Sugars	
$MgSO_4 \cdot 7H_2O$	278	Sucrose (in g.)	2.668
Amino acids		Fructose	40
L-arginine HCl	70	Glucose	70
L-aspartic acid	35	Organic acids	
L-asparagine	35	Malic	67
L-alanine	22.5	Alpha-ketoglutaric	37
B-alanine	20	Succinic	6
L-cystine HCl	2.5	Fumaric	5.5
L-glutamic acid	60	Vitamins	
L-glycine	65	Thiamine HCl	0.002
L-histidine	250	Riboflavin	0.002
L-isoleucine	5	Ca Pantothenate	0.002
L-leucine	7.5	Pyridoxine HCl	0.002
L-lysine HCl	62.5	p-aminobenzoic acid	0.002
L-methionine	5	Folic acid	0.002
L-proline	35	Niacin	0.002
L-phenylalanine	15	Isoinositol	0.002
DL-serine	110	Biotin	0.001
L-tyrosine	5	Choline chloride	0.02

To 90.0 ml of above medium add: 8.0 ml fetal bovine serum, 8.0 ml chicken egg ultrafiltrate, 0.3 gm yeastolate, 0.3 gm lactalbumin hydrolysate, and 0.5 gm crystallized bovine plasma albumin.

[a]This is Grace's (1962) insect tissue culture medium modified from Wyatt (1956) with added supplements of albumin, yeastolate, lactalbumin hydrolysate, serum, and egg ultrafiltrate.

Sen Gupta (1961) was among the first to suggest that hemolymph may not be required for growing insect cells *in vitro*. Various tissues of the *G. mellonella*, grew in the absence of hemolymph, and variation in the hemolymph supplement between 5 and 20% did not affect growth significantly. In cultures of silkworm ovarian tissue, the growth promoting hemolymph factor or factors reside in the hemolymph dialysate (Aizawa and Sato, 1963).

Vertebrate Sera

Currently, insect hemolymph is seldom used in insect cell culture media. Sera from vertebrate animals, usually bovine, is often employed in place of hemolymph. Sen Gupta (1961) and Vago and Chastang (1962a) demonstrated that insect cells will grow in media supplemented with 15-20% calf serum. In minimal medium without calf serum, ovaries of *Culex pipiens* survive for more than three months without cell division (Kitamura, 1965). Addition of calf serum to the medium stimulated cell migration and maintenance of healthy cell sheets. Calf serum was more beneficial than ovine serum or chicken plasma.

Fetal bovine serùm (FBS) is used at levels of 4-20% in media for culturing established cell lines. Those cell lines that were originally established in media that contained insect hemolymph may be adapted to grow in hemolymph-free media that is supplemented with FBS (Nagle *et al.*, 1967; Yunker *et al.*, 1967; Mitsuhashi and Grace, 1969; Sohi, 1969). The *A. eucalypti, A. aegypti,* and *B. mori* lines established by Grace (1962, 1966, 1967) and subsequently adapted to hemolymph-free media require FBS for sustained growth (Sohi and Smith, 1970). Maximum growth of *A. aegypti* cells occurred in medium with 10% FBS while the *A. eucalypti* and *B. mori* cells multiplied most rapidly in the presence of 5% FBS. Kuno *et al.* (1971) also reported that *A. aegypti* cells require FBS. Dialysed FBS retained about 50% of its growth-promoting activity, thus suggesting that 1) the growth-promoting molecules were either of large molecular weight, 2) they are found to large molecules, or 3) several growth-stimulating substances of different molecular weights are present in FBS.

Extracts and Hydrolysates

Sera along with extracts, filtrates, and hydrolysates stimulate cell proliferation and multiplication. Medium containing chick embryo extract and no hemolymph supported mitosis in cultured gonadal tissues from three species of Lepidoptera (Vago and Chastang, 1962b). However, the cultures in extract supplemented medium did not live as long as those in hemolymph supplemented medium. Chick embryo extract was thought to be the growth stimulating constituent in medium for culturing tissues of the blow fly, *Calliphora erythrocephala*, and *Drosophila melanogaster* (Demal and Leloup,

1963). Whole chicken egg ultrafiltrate is used as a component of several media that support growth of cell lines (Yunker *et al.*, 1967; Hink, 1970).

A medium containing only lactalbumin hydrolysate, yeast extract, sugars, salts, and amino acids supported the migration of cells from cultured tissues of *B. mori* (Aizawa and Sato, 1963). Kuno (1970) also observed the importance of these two compounds by demonstrating that *A. aegypti* cells, routinely grown in medium that contains lactalbumin hydrolysate and yeast extract, will not multiply in their absence.

REQUIREMENTS AND UTILIZATION OF CHEMICALLY DEFINED COMPOUNDS

Amino Acids

Amino acid utilization has been studied in insect cell lines and tick primary cultures. The following amino acids decreased in the medium that supports growth of cells from the tick, *Rhipicephalus sanguineus:* aspartic acid, glutamic acid, leucine, methionine, phenylalanine, proline, and threonine (Rehacek and Brzostowski, 1969). Cystine, leucine, and isoleucine remained fairly constant while the concentrations of seven other amino acids increased as the cultures grew. As the *A. eucalypti* cell line multiplied, 14 of the 21 amino acids in the medium decreased (Grace and Brzostowski, 1966). L-alanine was the only amino acid that increased in the medium, and the remaining six stayed at their original levels. A cell line (EPa) from the cockroach, *Periplaneta americana* requires 16 amino acids for continued cell division (Landureau and Jolles, 1969). All but one of the amino acids in the medium decreased as the cultures increased in age. After a seven day growth period, the α-alanine value had risen to about five times its original concentration. There was transamination between aspartic acid, glutamic acid, and α-alanine.

Two cell lines (CP-1268 and CP-169) from the codling moth, *Carpocapsa pomonella*, may require fewer amino acids than the previously mentioned cultured cells (Hink *et al.*, 1972). Only five amino acids (aspartic acid, glutamic acid, 1/2 cystine, methionine, and tyrosine) disappear from the medium of both lines during cell growth (Figures 1 and 2). Proline metabolism of the two lines must be different because in CP-169 culture media the proline level remains constant and in CP-1268 media it drops to about 50% of its original value after a ten day culture period (Figure 3). Most amino acids remain relatively constant throughout the ten day period of growth. Like other cultured insect cells, alanine accumulated in the media of both lines (Figures 1 and 2). It could not be proven that the amino acids that decreased as cultures multiplied were growth limiting. These results suggest that there are specific differences and similarities in the amino acid metabolism of cultured arthropod cells.

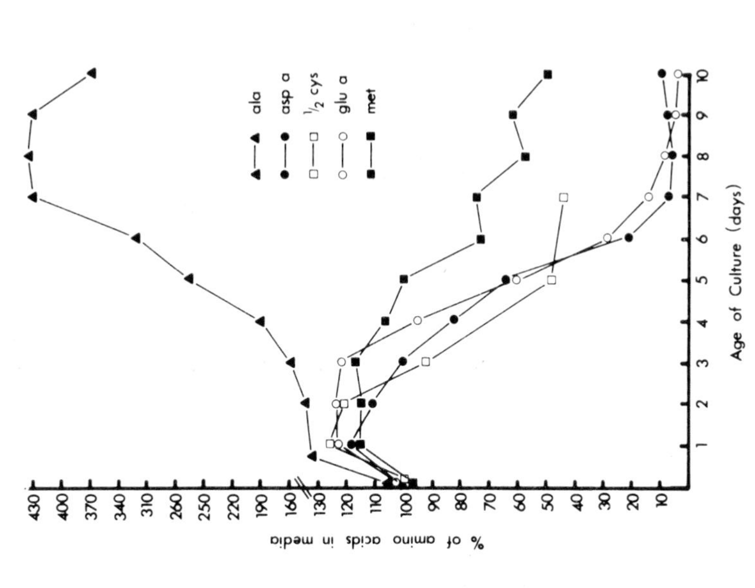

Figure 1. The amino acid values in media from CP-1268 cells cultured for 10 days *in vitro*.

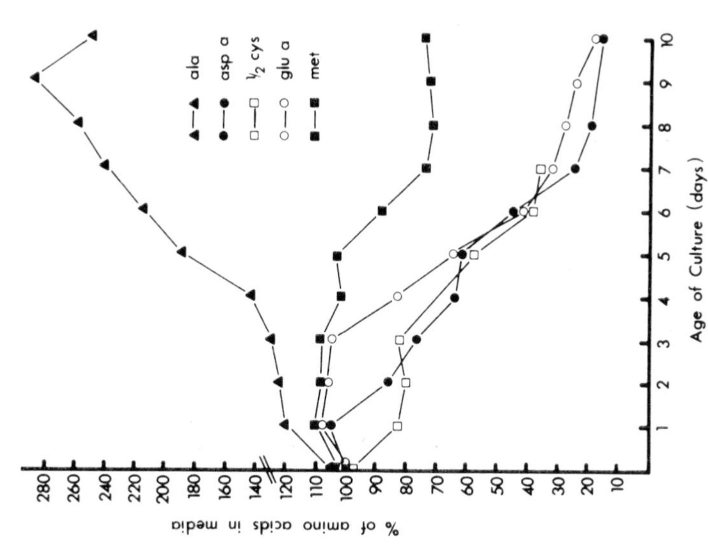

Figure 2. The amino acid values in media from CP-169 cells cultured for 10 days *in vitro*.

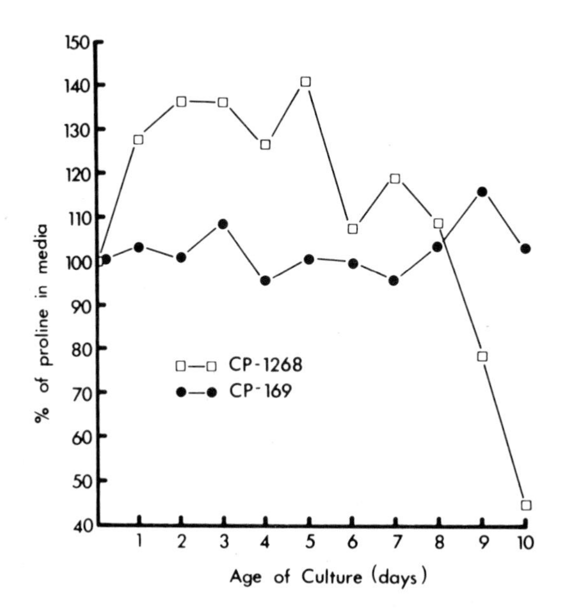

Figure 3. The proline values in media from CP-1268 and CP-169 cells.

Sugars

The *A. eucalypti* cell line metabolizes trehalose at approximately the same rate as glucose (Clements and Grace, 1967). Glucose disappears from the media by the fifth day of culture and is utilized more rapidly than fructose. After the glucose level has fallen to approximately one half of its original value, sucrose utilization begins.

In contrast, the two *C. pomonella* lines, grown in media with similar amounts of sugar, do not use sucrose (Hink *et al.*, 1972). Fructose and glucose are depleted from the media after eight days of cell growth. The single omission of either sugar reduces the rate of cell division and produces cytopathic effects on both cell lines. In sugar deficient media the cells become vacuolated, rounded up, clumped, and then lysed. A two-fold increase in the amounts of fructose and glucose results in higher cell numbers and an increased period of active multiplication.

Cultured tick cells use glucose and inositol (Rehacek and Brzostowski, 1969). During the first five days of culture, these two sugars are utilized at similar rates. After this period, glucose is used more extensively.

Vitamins

Grace (1958) reported that incorporation of B vitamins improved cell morphology and stimulated cellular migration from cultured *B. mori* tissues. The rate of mitosis was not increased. The vitamins used were: thiamine, riboflavin, pyridoxine, niacin, pantothenic acid, biotin, folic acid, p-aminobenzoic acid, choline, and mesoinositol. The *P. americana* (EPa) cell line requires Ca pathothenate, cyanocobalamine, riboflavin, inositol, thiamine, choline, pyridoxine, folic acid, nicotinic acid, and biotin for optimum cell growth. However, the line requires only the first five vitamins for cell survival (Landureau, 1969). Cyanocobalamine is a very important vitamin, and EPa cells growing in serum-free medium are stimulated to more active growth when this compound is added (Landureau and Steinbuch, 1969). Folate is also beneficial in medium for primary cultures of wax moth and silk worm ovaries (Zielinska and Saska, 1972). Increased amounts of choline produced higher populations of *A. aegypti* cells (Nagle, 1969). The cell numbers were increased about 57% by increasing the choline level from 1 mg/liter to 50 mg/liter.

Proteins

Fetal bovine serum (FBS) often supplies the protein in insect tissue culture media. About 40-50% of the FBS proteins are represented by fetuin, an alpha-glycoglobulin (Spiro, 1960). Starch gel electrophoresis revealed as many as eight protein fractions in FBS (DiDomizio and Minoccheri, 1964). FBS was fractionated by gel filtration and the fractions with high levels of protein were growth promoting for cultured *A. aegypti* cells (Kuno, 1970). When treated with 6M urea or pronase, these fractions lost their stimulatory activity. The serum proteins were isolated by starch block electrophoresis, chemical fractionation methods, or obtained from commercial sources, and only albumin was growth promoting. Disc electrophoresis of fresh tissue culture medium and medium from seven-day-old cultures revealed that 13% of protein at the albumin region and 67% of protein at the fetuin region had disappeared (Figure 4). This study suggests that specific proteins or substances bound to them are required and metabolized by *A. aegypti* cells *in vitro*.

The cockroach cell line (EPa) also exhibits a requirement for protein (Landureau and Steinbuch, 1969). Human serum fraction V had a growth promoting effect. The activity of this protein was attributed to substances, such as vitamins, that are bound to this macromolecule. The serum proteins may also be beneficial because they function as antiproteases and thus protect the cultured cells from proteolytic enzymes (Landureau and Steinbuch, 1970).

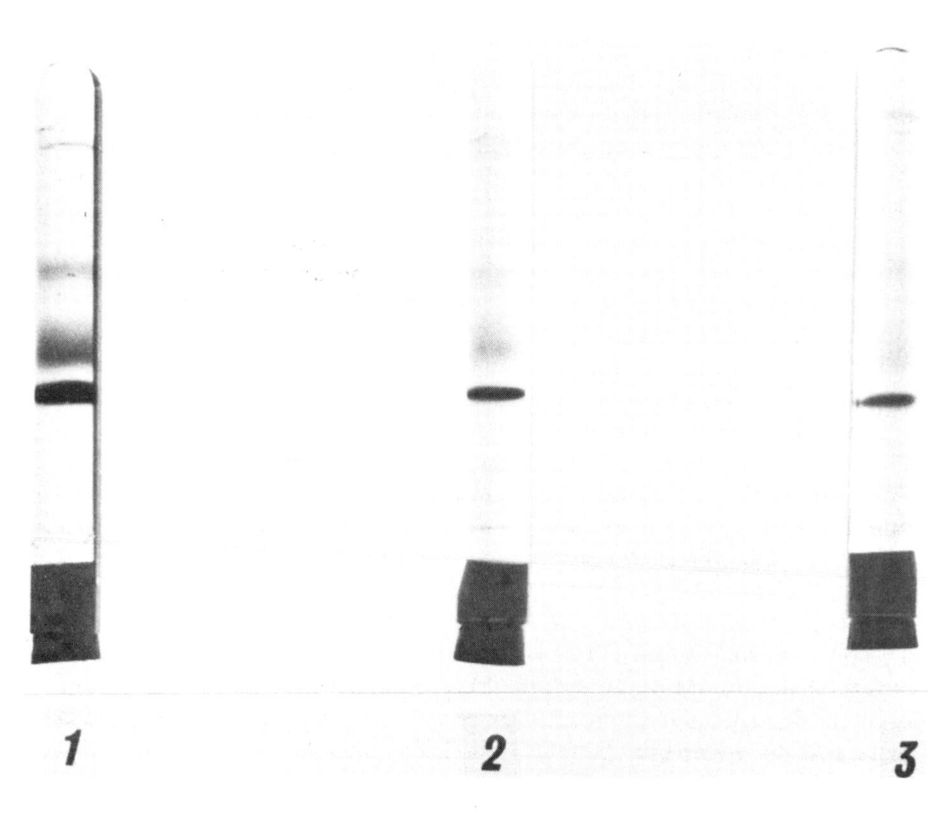

Figure 4. Disc electrophoresis of fresh and 7-day-old culture media from growing *A. aegypti* cells. The bottom band is albumin and the broader band immediately above albumin is the fetuin region. A-FBS; B-fresh medium; C-medium from a 7-day-old culture. From Kuno, 1970.

REFERENCES

AIZAWA, K. and SATO, F. 1963. Culture de tissus de ver a soie, *Bombyx mori* L. dans un milieu sans hemolymphe. *Ann. Epiphyties.* 14:125:126.

CLEMENTS, A. N. and GRACE, T. D. C. 1967. The utilization of sugars by insect cells in culture. *J. Insect Physiol.* 13:1327-1332.

DEMAL, J. and LELOUP, A. M. 1963. Essai de culture *in vitro* d'organes d'insectes. *Ann. Epiphyt.* 14:91-93.

DIDOMIZIO, G. and MINOCCHERI, F. 1964. Sieroproteine fetali ed evoluzione post-natale del ferogramma sierico bovino. *Arch. Vet. Ital.* 15:9-20.

GRACE, T. D. C. 1958. Effects of various substances on growth of silkworm tissues *in vitro*. *Aust. J. Biol. Sci.* 11:407-417.

GRACE, T. D. C. 1962. Establishment of four strains of cells from insect tissues grown *in vitro*. *Nature* **195**:788-789.
GRACE, T. D. C. 1966. Establishment of a line of mosquito (*Aedes aegypti* L.) cells grown *in vitro*. *Nature* **211**:366-367.
GRACE, T. D. C. 1967. Establishment of a line of cells from the silkworm *Bombyx mori*. *Nature* **216**:613.
GRACE, T. D. C. and BRZOSTOWSKI, H. W. 1966. Analysis of the amino acids and sugars in an insect cell culture medium during cell growth. *J. Insect Physiol.* **12**:625-633.
HINK, W. F. 1970. Established insect cell line from the cabbage looper, *Trichoplusia ni*. *Nature* **226**:466-467.
HINK, W. F., RICHARDSON, B. L., SCHENK, D. K., and ELLIS, B. J. 1972. Utilization of amino acids and sugars by two codling moth, *Carpocapsa pomonella*, cell lines (CP-1268 and CP-169) cultured *in vitro*. Proc. Third Intern. Coll. Invertebr. Tissue Culture, Smolenice.
KITAMURA, S. 1965. The *in vitro* cultivation of tissues from the mosquito, *Culex pipiens* var *molstus* II. An improved culture medium useful for ovarian tissue culture. *Kobe J. Med. Sci.* **11**:23-30.
KUNO, G. 1970. Studies of growth-promoting proteins in fetal bovine serum using *Aedes aegypti* cells cultured *in vitro*. Ph.D. Dissertation, The Ohio State University.
KUNO, G., HINK, W. F., and BRIGGS, J. D. 1971. Growth-promoting serum proteins for *Aedes aegypti* cells cultured *in vitro*. *J. Insect Physiol.* **17**:1865-1879.
LANDUREAU, J. C. 1969. Etude des exigences d'une lignee de cellules d'insectes (Souche EPa) II. Vitamines hydrosolubles. *Exptl. Cell Res.* **54**:399-402.
LANDUREAU, J. C. and JOLLES, P. 1969. Etude des exigences d'une lignee de cellules d'insectes (Souche EPa) I. Acides amines. *Exptl. Cell Res.* **54**:391-398.
LANDUREAU, J. C. and STEINBUCH, M. 1969. Cyanocobalamine as a support of the *in vitro* cell growth-promoting activity of serum proteins. *Experientia* **25**:1078-1079.
LANDUREAU, J. C. and STEINBUCH, M. 1970. *In vitro*, cell protective effects by certain antiproteases of human serum. *Z. Naturforsch.* **25**:231-232.
MITSUHASHI, J. and GRACE, T. D. C. 1969. Adaptation of established insect cell lines to a different culture medium. *Appl. Ent. Zool.* **4**:121-125.
NAGLE, S. C. 1969. Improved growth of mammalian and insect cells in media containing increased levels of choline. *Appl. Microbiol.* **17**:318-319.
NAGLE, S. C., CROTHERS, W. C., and HALL, N. L. 1967. Growth of moth cells in suspension in hemolymph-free medium. *App. Microbiol.* **15**:1497-1498.
RAHMAN, S. B., PERLMAN, D., and RISTICH, S. S. 1966. Growth of insect cells in tissue culture. *Proc. Soc. Exptl. Biol. Med.* **123**:711.
REHACEK, J. and BRZOSTOWSKI, H. W. 1969. The utilization of amino acids and sugars by tick cells cultivated *in vitro*. *J. Insect Physiol.* **15**:1683-1686.
SEN GUPTA, K. 1961. Studies on insect tissue culture I. Culture of tissues from the wax moth, *Galleria mellonella*, L. *in vitro*. *Folia Biol.* **7**:400-408.
SEN GUPTA, K. 1964. Studies on the specificity of haemolymph from different insects for the culture of tissues from other insects. *Proc. Indian Acad. Sci.* **59**:103-109.
SOHI, S. S. 1969. Adaptation of an *Aedes aegypti* cell line to hemolymph-free medium. *Can. J. Microbiol.* **15**:1197-1200.
SOHI, S. S. and SMITH, C. 1970. Effect of fetal bovine serum on the growth and survival of insect cell cultures. *Can. J. Zool.* **48**:427-432.

SPIRO, R. C. 1960. Studies on fetuin, a glycoprotein of fetal serum. I. Isolation, chemical composition, and physicochemical properties. *J. Biol. Chem.* 235:2860-2869.

VAGO, C. and CHASTANG, S. 1962a. Cultures de tissus d'insectes a l'aide de serum de mammiferes. *Entomophage* 7:175-179.

VAGO, C. and CHASTANG, S. 1962b. Culture de tissus d'insectes a l'aide d'extrait d'embryon de poulet en l'absence d'hemolymphe. *C.R.H. Acad. Sci., Paris.* 255:3226-3228.

WYATT, S. S. 1956. Culture *in vitro* of tissue from the silkworm, *Bombyx mori* L. *J. Gen. Physiol.* 39:841-852.

YUNKER, C. E., VAUGHN, J. L., and CORY, J. 1967. Adaptation of an insect cell line (Grace's Antheraea cells) to medium free of insect hemolymph. *Science* 155:1565-1566.

ZIELINSKA, Z. M. and SASKA, J. 1972. Effect of folate and some of its analogues on insect ovaries *in vitro*. *Proc. Third Intern. Coll. Invertebr. Tissue Culture, Smolenice.*

THE ROLE OF STEROLS AND FATTY ACIDS IN THE NUTRITION OF INSECT CELLS IN CULTURE

J. L. Vaughn and S. J. Louloudes
Insect Pathology Laboratory
Agricultural Research Service
United States Department of Agriculture
Beltsville, Maryland 20705

INTRODUCTION

The amount of discussion at this conference concerning the role of lipids in insect nutrition adequately demonstrates their importance for insect growth and development. The importance of sterols for the synthesis of structural cell components and as precursors for various compounds which act as metabolic regulators recently has been discussed in many original papers and reviews. The fact that all insects require an exogenous source of sterols is now generally accepted (Robbins *et al.*, 1971). In addition, certain fatty acids are needed for the normal growth and maturation of insects and some of these too must be obtained from exogenous sources. In Lepidoptera for example, a dietary source of either linoleic or linolenic acid is required for normal larval growth and for the development and emergence of the adults (Gilbert, 1967). Also, the importance of fatty acids, triglycerides and other neutral lipids as energy sources and in the storage and transportation of energy in insects is likewise well known (Fast, 1964).

Lipids, therefore, are almost certainly of importance in the nutrition of insect cells in culture and it is of practical as well as academic interest to know whether or not the cell lines derived from Lepidoptera have requirements for exogenous sources of sterols and fatty acids similar to those of the insects from which they were obtained.

Several different cell lines have been developed from insects, but few studies have been made to determine the requirements of these cell lines for sterols and fatty acids. Of those compounds related to lipid synthesis, only choline, a precursor for phospholipid synthesis, has been routinely added to insect cell culture media. In these media the level of choline was between 0.02 mg/100 of medium (Marks and Reinecke, 1965) and 25 mg/100 ml of medium (Pudney and Varma, 1971). Also a few media have contained added cholesterol (Larson, 1967; Horikawa and Kuroda, 1959; and Schneider, 1969), and Hirumi and Maramorosch (1964) added alphatocopherol-phosphate (Na_2) to their medium for the culture of leafhopper cells. None of these investigators attempted to evaluate the effect of added lipids on cell growth; however, a few such studies have been reported. In 1968 Grace used

Insect and mite nutrition – North-Holland – Amsterdam, (1972)

cholesterol in a medium for primary cultures of cells of the silkworm, *Bombyx mori* (L.), but found that a concentration of 3 mg/100 ml of medium had no effect on the growth of these cells. Nagle (1969) increased the growth of a mosquito cell line by increasing the choline in his medium to a concentration of 5 mg/100 ml of medium. At a concentration of 20 mg/100 ml of medium the choline had a toxic effect. Choline at a concentration of 0.04 mg/100 ml of medium has been reported as necessary for the optimum growth of cockroach cells in culture, but not essential for cell survival (Landureau, 1969).

In studies of the metabolism of cholesterol by insect cell cultures, Vaughn *et al.* (1971) found that cholesterol added to the culture medium bound to serum proteins was taken up by insect cells in culture and that added ^{14}C labeled cholesterol was incorporated into the cells in preference to the cholesterol that occurred naturally in the various serums used to supplement the minimal medium. However, further examination of the cell sterols showed that only 6.5% of the label in the cell was still in the cholesterol. The remainder was in more polar compounds, indicating that the cells were capable of altering the cholesterol molecule.

Studies of the role of fatty acids in the nutrition of insect cells in culture have been even more limited in number than those on sterols. Jenkin *et al.* (1971) reported that in one insect cell line the total lipids contributed 44% of the dry weight of the cells and that of the total lipids, 49.3% were neutral lipids and the remainder were phospholipids. No glycolipids were found in the cell. They also found that of the classes of neutral lipids present in the cell, nearly half (49.5%) were free fatty acids compared with only 8.6% free fatty acids in the serum supplement. The cells also contained higher levels of monoglycerides and triglycerides than did the serum. However, the percentages of neutral lipids present in the cells as steroids and sterol esters were less than the percentages of these compounds in the serum. Their analysis of the constituent fatty acids of the neutral lipids revealed higher percentages of the 16:1, 18:0 and 18:1 fatty acids in the cells than in the serum supplement. The cells also contained 10% C_{22} fatty acids although none were detected in the serum. In addition, the phospholipids of the insect cells contained high levels of phosphatidylcholine (42.0%) and phosphatidylethanolamine (31.1%). The major constituent fatty acids of the phospholipids were 16:1 (21.0%), 18:0 *20.3%) and 18:1 (33.1%); and C_{22} fatty acids again accounted for about 9% of the total fatty acids in the phospholipids.

Thus, several classes of lipids are clearly important for the growth of insect cells *in vitro*; and the data of Jenkin *et al.* (1971) also provides evidence that insect cells in culture have some ability to produce fatty acids by synthesis, chain lengthening or interconversion of fatty acids in the medium. Previously the only source of these compounds in insect cell culture media has been the serum supplements added to basal medium, so the wide differences in the percentages of fatty acids in the cells and in the supplements may indicate shortages of some critical fatty acids. We are interested in the role of the various lipids for several reasons: One, the shortage of one or more lipids may be limiting either the rate of growth or the maximum growth of insect cell cultures. Two, we hope to develop a chemically defined medium and presumably such a medium will require some essential lipids or suitable precursors. Three, the shortage of some lipids, though not limiting to the growth of healthy cells, may be essential for the replication of certain insect pathogenic viruses in the established cell lines.

We therefore attempted to identify the critical lipids by supplementing normal medium with additional sources of lipid and also by depleting the culture medium of essential fatty acids.

MATERIALS AND METHODS

Supplementation of insect cell culture medium with additional lipids.

Cholesterol and Cholesterol Esters. There are no reports available on the effect of additional sterols on the growth of insect cell lines. Although Grace found that added sterols had no effect on primary cultures, it seemed possible that such cultures, particularly those from ovarian tissue, may carry sufficient pools of sterol to permit growth of the cells in culture for short periods. However, with established lines the cells should have exhausted such reserves and therefore should have a greater requirement for exogenous sterols.

The sterols and sterol esters to be tested were added to the culture medium by two procedures. In the early studies, the stock solution of each sterol was prepared in Tween 80 and then added to the complete medium at the desired test concentration. In later studies the sterols or sterol esters were combined with part of the normal fetal calf serum supplement, and the test medium was then supplemented with the sterol-enriched fetal calf serum to give the desired concentration of sterol. Also sufficient additional non-enriched fetal calf serum was added to each test medium to achieve the normal level of serum supplementation.

The sterols were tested for their effect on the growth of RML-10 cells, a strain of the Grace *Antheraea* cell line adapted to hemolymph-free medium (Yunker *et al.*, 1967) as follows: Three 3-ml cultures were prepared in 30 ml Falcon disposable tissue culture flasks and the cultures were inoculated with 150,000 cells/ml. The flasks were incubated at 26°C for seven days and then the concentration of cells in each flask was determined by counting on an electronic counter. From this the mean of the three cultures was obtained and the effect of the test sterol was expressed as the ratio of the mean cell count in the test medium to the mean cell count in the normal, non-enriched control medium. The results are shown in Table 1. The first part of the table contains the data from those tests in which the sterol was prepared as a stock solution in Tween 80. In each test the concentration of Tween 80 incorporated with the sterol is also shown. An apparent stimulation of cell growth

Table 1. Effect of sterols and sterol esters on the growth of insect cells.

Additive	Concentration (μg/ml)	Ratio[a]
	Added in Tween 80	
Cholesterol	1.5	1.42
Tween 80	25.0	
Cholesterol oleate	1.2	1.94
Tween 80	25.0	
Methyl oleate	5.2	1.39
Tween 80	25.0	
Cholesterol	1.5	
Methyl oleate	5.2	1.64
Tween 80	25.0	
Tween 80	25.0	1.22
	Added in serum	
NH_4 cholesterol, SO_4	1.9	1.01
K cholesterol, SO_4	1.9	1.09
Glycine cholesterol SO_4	2.1	1.05
Cholesterol	25.0	1.01
Cholesterol oleate	2.6	1.03

[a]Ratio = mean cell concentration in test medium/mean cell concentration in non-enriched medium.

occurred as a result of the addition of either cholesterol at a concentration of 1.5 μg/ml or with cholesterol oleate at a concentration of 1.25 μg/ml. However, as shown in line 5 of the table, similar stimulation of growth was obtained with Tween 80 alone. When the test sterols were added as part of an enriched serum, none of the sterols tested produced an increase in cell growth (the latter part of Table 1). These results are in agreement with our earlier findings (Vaughn *et al.*, 1971) that the sterols present in the medium from the normal serum supplements were not completely depleted by the growing cells.

Later we attempted to improve the growth of a second established cell line, IPRL-21, from the fall armyworm, *Spodoptera frugiperda* (J. E. Smith), by adding cholesterol to the medium. However, concentrations as high as 4.2 μg cholesterol/ml of medium also gave negative results.

Table 2. The fatty acid composition of naturally occurring oils tested for their effect on insect cell growth.[a]

	Percent fatty acids in			
Fatty acid	Linseed oil	Safflower oil	Coconut oil	Menhaden oil
Lauric	-	-	51	-
Myristic	-	-	16	12
Palmitic	6	8	9	22
Palmitoleic	-	-	-	14
Stearic	4	2	4	3
Oleic	20	13	5	17
Linoleic	18	76	2	1
Linolenic	52	1	-	1
Misc.	-	-	13	30

[a]From Zielinski, *et al.* (1961).

Fatty Acids. The observation that Tween 80 stimulated the growth of the insect cell cultures led us to study the effect of fatty acids on cell growth. Further investigation of the effect of Tween 80 and other related compounds indicated that only Tween 80 at concentrations between 25 and 50 μg/ml of medium gave some stimulation of cell growth. However, the increase was small and the effect could not be demonstrated consistently, therefore, other sources of fatty acids were tested for their effect.

Stock of the fatty acids used in the tests were prepared by dissolving weighed amounts of the lipid materials in benzene and sterilizing them by filtration through a Gelman membrane filter (pore size 0.2μ). Test solutions were prepared by coating sterile glass beads with the desired amount of stock solution and evaporating the benzene with sterile nitrogen. The lipid materials were removed from the glass surface by immersing them in a measured amount of fetal calf serum. The fatty acid-enriched serum was then used in place of non-enriched serum to supplement the culture medium.

The composition of the 4 naturally occurring oils, tested for their effect on cell growth, is shown in Table 2. Those oils that had as the major constituent fatty acid either linoleic acid (safflower seed oil) or linolenic acid (linseed oil) had the most stimulating effect on cell growth (Table 3).

Table 3. The effect of various fatty acid sources on the growth of insect cells.

Fatty acid source	Concentration (μg/ml)	Ratio[a]
	100.0	1.10
	50.0	1.30
Linseed Oil	25.0	1.39
	12.5	1.39
	6.2	1.40
	100.0	1.38
	50.0	1.14
Safflower Oil	25.0	1.38
	12.5	1.12
	6.2	1.04
	100.0	0.81
	50.0	0.99
Coconut Oil	25.0	0.94
	12.5	1.23
	6.2	1.06
	100.0	0.51
	50.0	1.03
Menhaden Oil	25.0	1.02
	12.5	0.97
Trilinolenin	9.5	1.39
Trilinolenin	0.9	1.27
Trilenolein	9.5	1.27
Trilenolein	0.9	1.47

[a]Ratio = mean cell concentration in test medium/mean cell concentration in non-enriched medium.

Coconut oil, whose major fatty acid constituent is a 12 carbon, saturated fatty acid, had some stimulatory effect. Menhaden oil, which contains several fatty acids of 16 carbons or less, had no stimulatory effect. It was observed that all of the oils had an inhibitory effect at high concentrations.

The results of tests with two triglycerides, trilenolein and trilenolenin, containing 3 linolenic acid residues and 3 linolenic acid residues, respectively, are also shown in Table 3. Both compounds stimulated cell growth considerably, but a concentration of trilenolein ten times that of trilinolenin was required to produce an equivalent effect on cell growth.

Depletion of fatty acid from the medium during cell growth

As noted, Jenkin *et al.* (1971) reported a high lipid content for insect cells in culture (44% of the dry weight) and particularly a high level of free fatty acids (49.5%) of the neutral lipids) in comparison with that found in the fetal calf serum supplement (4.5% and 8.6% respectively). These findings plus our own, to the effect that high levels of linolenic or linoleic acid increased the growth of our cell line, seemed to indicate that some fatty acid or acids might be growth limiting factors. Thus, we investigated the fatty acid content of our medium before and after cell growth. To accentuate any reduction in the fatty acid levels, the medium was reused for three successive cultures. Cells from the previous culture was used as inoculum for each new culture to avoid the introduction of additional essential fatty acids via the intracellular pool. The details of this study are reported in a paper by Louloudes, Vaughn, and Dougherty which is in preparation so only a synopsis of those findings as they relate to the present subject will be reported here.

The cultures used in this study were grown in glass flasks with 50 ml of medium per flask. The medium was supplemented with 5.6% fetal bovine serum, 8.1% whole egg ultrafiltrate, 2.4% of a 35% aqueous solution of bovine serum albumin and 2.4% *Bombyx mori* hemolymph. Prior to extraction, the cells and medium from each culture were separated, and the cells washed twice with Locke's saline solution. Saponified and unsaponified lipids were extracted separately, and the saponified fractions were treated with diazomethane in ether to prepare the fatty acid methyl esters. The methyl esters of fatty acids were determined qualitatively and quantitatively on a Barber Coleman Model 10 gas chromatograph.

Table 4 shows the level of the principal fatty acids in the control medium and in the medium after each of the 3 successive cultures. Even after the third passage none of these fatty acids was completely depleted from the medium. Of the several fatty acids present in the control medium in trace amounts only one, 16:2, was completely removed from the medium after the third culture. Since this fatty acid was never found in the cells, it presumably was metabolized and not incorporated intact into some cell component. Thus, the absence of an essential fatty acid did not appear to be a factor in preventing cell growth in the third culture.

Further evidence supporting our observation is given in Table 5 which shows the amount of the individual fatty acids present in the cells at the end of each passage. Cells examined after the third incubation contained as much or more of each fatty acid, except 20:3 and 20:4, as did cells examined after the first incubation. In addition, the total amount of fatty acids per million cells was higher in the cells harvested after the third incubation than in cells harvested after the first incubation.

Table 4. Levels of the principal fatty acids in culture medium before and after cell growth.

		Concentration (μg/ml)			
Fatty acid	Carbon number	control	First incubation	Second incubation	Third incubation
Myristic	14:0	0.333	0.454	0.462	0.508
Palmitic	16:0	6.914	7.692	4.926	3.973
Palmitoleic	16:1	1.583	2.660	2.053	1.903
Stearic	18:0	4.058	6.108	3.221	2.426
Oleic	18:1	7.463	10.555	6.229	5.040
Linoleic	18:2	4.484	2.891	2.067	1.685
Linolenic	18:3	9.245	3.670	3.317	2.004
Eicosatrienoic	20:3	0.600	0.218	0.086	0.102
Arachidonic	20:4	2.030	0.597	0.633	0.148
TOTAL		36.710	34.845	22.994	17.789

DISCUSSION

From the data gathered in our studies and those of Jenkin *et al.* (1971), two general conclusions can be drawn: First, the insect cell culture media as currently supplemented with various sera contain more than adequate amounts of the sterols and fatty acids essential for the growth of the insect cell lines studied. The depletion studies showed that neither cholesterol nor any of the fatty acids in the media were completely utilized during growth of the cultures. However, most cell cultures have a requirement for "non-essential" nutrients which are not required for the survival of a culture but do improve some aspect of cell growth. For example, the rate of growth of cells in culture can sometimes be increased by increasing the concentration of exogenous nutrients. However, the failure of additional sterols and sterol esters in our culture medium to produce improved growth, indicates that the existing levels of these materials were in excess of the cell needs.

On the other hand, the growth of the cells was consistently improved by the addition of higher levels of exogenous linoleic and linolenic acids. The medium as normally supplemented for these cells contained 4.48 μg of linoleic and 9.24 μg of linolenic acid/ml of complete medium. Since linolenic acid was the most abundant fatty acid in our normal medium, it would appear that the cells have a substantial requirement for these fatty acids. This result, of course, agrees with the general requirement for either or both linoleic and linolenic acid by the Lepidoptera.

Table 5. Levels of the principal cellular fatty acid after three incubations in the same medium.

		Concentration (μg/10^6 cells)		
Fatty acid	Carbon number	First incubation	Second incubation	Third incubation
Myristic	14:0	0.074	0.053	0.125
Palmitic	16:0	0.789	0.950	1.279
Palmitoleic	16:1	0.964	1.948	3.107
Stearic	18:0	1.383	1.702	2.091
Oleic	18:1	3.990	7.133	10.290
Linoleic	18:2	0.536	0.822	1.256
Linolenic	18:3	1.051	1.695	2.458
Eicosatrienoic	20:3	0.095	0.041	0.071
Arachidonic	20:4	0.208	0.107	0.112
TOTALS		0.090	14.451	20.789

The second conclusion is that the cells can modify the sterols and fatty acids in the medium to fulfil their requirements for other related compounds. For example, the small percentage of radioactive carbon recovered as cholesterol from cells grown in medium containing ^{14}C labeled cholesterol established that the cells could metabolize that sterol, but we did not determine what changes were made in the molecule. Also, similar conversions occurred with the fatty acids in the medium. Table 5 reveals a continual shift in the levels and relative proportions of the individual fatty acids in the cells from one culture to the next.

The one aspect of the lipid metabolism by insect cells in culture for which there is no information is the ability to synthesize the required lipids from simple precursors. Presumably the cells are capable of making some lipid compounds and thus the determination of the truly essential lipids must be made after this information is available.

SUMMARY

The addition of sterols and sterol esters to the culture medium for two insect cell lines, RML-10 (a strain of the Grace *Antheraea* line) and IPRL-21 (from *Spodoptera frugiperda*), did not improve the growth of either of these cell lines. Therefore, the basal medium (supplemented with fetal bovine serum and insect hemolymph) probably contains more than an adequate level of the sterols needed for the growth of these two cell lines.

Studies on the utilization of fatty acids by the cells of the IPRL-21 line revealed that even when the medium was used repeatedly for cultures, no fatty acids were completely depleted. However, when natural oils or triglycerides high in linoleic and linolenic acids were added to the culture medium growth was improved. Thus, the cells of this line require an exogenous source of these fatty acids like that of many insects in the order Lepidoptera.

REFERENCES

FAST, P. G. (1964). *Insect Lipids: A Review.* Memoirs Entomol. Soc. Can. No. 37, 50 pp.

GILBERT, L. I. (1967). Lipid metabolism and function in insects. *Adv. Insect Physiol.* **4**:69-211.

GRACE, T. D. C. (1968). Effects of various substances on growth of silkworm tissues *in vitro. Aust. J. Biol. Sci.* **11**:407-417.

HIRUMI, H., and MARAMOROSCH, K. (1964). Insect tissue culture: Use of blastokinetic stage of leafhopper embryo. *Science* **144**:1465-1467.

HORIKAWA, M., and KURODA, Y. (1959). *In vitro* cultivation of blood cells of *D. melanogaster* in a synthetic medium. *Nature* (London) **184**:2017-2018.

JENKIN, H., TOWNSEND, D. and MAKINO, S. (1971). Comparative lipid analysis of *Aedes aegypti* and monkey kidney cell (MK-2) cultivated *in vitro. Curr. Topics Microbiol. Immunol.* **55**:97-102.

LANDUREAU, J. C. (1969). Étude des exigences d'une lignie de cellules d'insectes (Sauche EPa) II. Vitamines hydrosolubles. *Exp. Cell Res.* **54**:399-402.

LARSON, W. P. (1967). Growth in an insect organ culture. *J. Insect Physiol.* **13**:613-619.

MARKS, E. P. and REINECKE, J. P. (1965). Regenerating tissue from the cockroach leg: Nutrient media for maintenance *in vitro. J. Kans. Entomol. Soc.* **38**:179-182.

NAGLE, S. C., JR. (1969). Improved growth of mammalian and insect cells in media containing increased levels of choline. *Appl. Microbiol.* **17**:318-319.

PUDNEY, M. and VARMA, M. G. R. (1971). *Anopheles stephensi* var. *mysorensis*: Establishment of a larval cell line (Mos. 43). *Exp. Parasitol.* **29**:7-12.

ROBBINS, W. E., KAPLANIS, J. N. SVOBODA, J. A. and THOMPSON, M. J. (1971). Steroid metabolism in insects. *Ann. Rev. Entomol.* **16**:53-72.

SCHNEIDER, I. (1969). Establishment of three diploid cell lines of *Anopheles stephensi* (Diptera: Culicidae). *J. Cell Biol.* **42**:603-606.

VAUGHN, J. L., LOULOUDES, S. J. and DOUGHERTY, K. (1971). The uptake of free and serum-bound sterols by insect cells *in vitro*. *Curr. Topics Microbiol. Immunol.* 55:92-97.

YUNKER, C. E., VAUGHN, J. L. and CORY, J. (1967). Adaptation of an insect cell line (Grace's *Antheraea* cells) to medium free of insect hemolymph. *Science* 155:1565-1566.

ZIELINSKI, W. L., JR., MOSELEY, W. V., JR., and BRICKER, R. C. (1961). A critical examination of the use of gas chromatography for the qualitative determination of oil content in organic coatings. *Official Dig. Fed. Soc. Paint Technol.* 33:622-634.

THE EFFECT OF OSMOTIC PRESSURE AND pH ON THE GROWTH OF *HELIOTHIS ZEA* CELLS *IN VITRO*

T. J. Kurtti and Marion A. Brooks
Department of Entomology, Fisheries, and Wildlife
University of Minnesota
St. Paul, Minnesota 55101

INTRODUCTION

Animal cells growing *in vitro* are sensitive to certain changes in the environmental (physical) characteristics of the medium, such as osmotic pressure and pH. These cannot be ignored in determinations of the nutritional requirements of cultured cells as they may have dramatic effects on overall cellular activity.

The Wyatt-Grace medium (Wyatt, 1956; Grace, 1962) patterned after the hemolymph of the silkworm, *Bombyx mori*, is used to culture cells of various lepidopterous insects. The organic constituents of this medium (amino acids, sugars and organic acids) are important effectors of the physical attributes, particularly osmotic pressure. Nutritional studies requiring alterations in the levels of organic molecules will thus cause changes in the physical characteristics.

In this study the osmotic pressure and pH of the Wyatt-Grace medium were assessed for their influence on growth of *Heliothis zea* (Boddie) cells. These factors were chosen to establish a general framework of the basic environmental requirements of these cells.

MATERIALS AND METHODS

Cell line

The cell line (IMC-HZ-1), derived from adult *Heliothis zea* ovaries by Hink and Ignoffo (1970), was obtained from Dr. C. M. Ignoffo. These cells grow in suspension and at 26°C have a generation time of approximately 28 hours.

Media

Basic medium: The unsupplemented medium was made as outlined by Grace (1962) (final volume 100 ml). The medium used to maintain the cell line consisted of 90 ml Wyatt-Grace medium (WGM), 10 ml heat inactivated fetal bovine serum (FBS) (Grand Island Biological Co.), 10 ml whole egg ultrafiltrate (WEUF) (Grand Island Biological Co.) and 1 g bovine serum albumin (BSA) A grade (Pentex) (Yunker, Vaughn and Cory, 1967; Hink and Ignoffo, 1970). A medium consisting of 90 ml WGM and 20 ml FBS will also support the growth of this cell line without a reduction in the growth rate.

Variations in osmotic pressure: The substitutions made in the medium to achieve the various osmotic pressures are shown in Table 1 along with the control media. The medium was made hypotonic by reducing the salts and sugars (#1, 2, 3). Additional glucose was added to the medium when sucrose and fructose were omitted. Appropriate controls (#8, 9, 10, 11) were made to establish that a reduction of growth was caused by hypotonicity and not the reduction of salts or sugars. Cellobiose was added to increase the osmotic pressure (#4 through #11).

All the media used in this section were supplemented with FBS, WEUF, and BSA.

Variations in pH: This *H. zea* cell line can grow in a medium (#12) with reduced calcium and magnesium (i.e. reduced to 0.9 mM calcium and 2.25 mM magnesium) (unpublished); these levels were used in the pH experiments to prevent the precipitation which occurs at higher pH values. Osmotic deficits were corrected by cellobiose (56 mM). The medium was adjusted with 2N KOH and 1N HCl to pH values of 6.0, 6.5, 7.0, 7.5 or 8.0. These media were supplemented with FBS only.

Saline solution: A saline solution patterned after the basic medium was used to wash and dilute the cells. In grams per liter the contents were: $NaH_2PO_4 \cdot H_2O$, 1.010; $NaHCO_3$, 0.350; KCl, 6.606; NaCl, 1.310; sucrose, 26.68; glucose, 0.700; $MgCl_2 \cdot 6H_2O$, 0.228; $MgSO_4 \cdot 7H_2O$, 0.278; and $CaCl_2$, 0.100. The pH was adjusted to 6.5 with 1N KOH. The osmotic pressure of this solution was 315 milliosmols per kilogram.

Culture methods

Falcon Flasks (30 ml) (Falcon Plastics, Div. of Bioquest) were used in all experiments. Each flask was inoculated with 5 ml of cell suspension. Five replicate cultures of each experimental medium were set up.

To set up experimental cultures, a suitable number of cells were taken from stock cultures in the logarithmic phase of growth (population density of approximately 6 to 7 x 10^5 cells/ml). The use of cells growing in the logarithmic phase as inocula helped to minimize the lag phase. The cells were washed, diluted and centrifuged in saline solution. The desired number of cells (2.5 x 10^5 to 5 x 10^5) for each Falcon Flask were pipetted into a 15 ml centrifuge tube and centrifuged at 275 r.c.f. for 5 minutes. The pellet of cells was then resuspended in 5 ml of the culture medium and inoculated into the culture flask.

All solutions were sterilized by membrane filtration. All experiments were run at 26°C.

Table 1. Salt and sugar mixtures in millimoles used to achieve the desired physical characteristics of the culture media.

Medium Code	WGM[a]	1	2	3	4	5	6	7	8	9	10	11	12
$NaH_2PO_4 \cdot H_2O$	7.32	1.41	1.41	1.41	7.32	7.32	7.32	7.32	1.41	7.32	1.41	7.32	7.32
KH_2PO_4	-----	5.90	5.90	5.90	----	-----	-----	-----	5.90	-----	5.90	-----	-----
$NaHCO_3$	4.17	4.17	4.17	4.17	4.17	4.17	4.17	4.17	4.17	4.17	4.17	4.17	4.17
KCl	30.05	9.12	9.12	9.12	30.05	30.05	30.05	30.05	9.12	30.05	9.12	30.05	30.05
$CaCl_2$	9.01	4.51	9.01	9.01	9.01	9.01	9.01	9.01	9.01	9.01	4.51	4.51	0.9
$MgCl_2 \cdot 6H_2O$	11.21	5.61	11.21	11.21	11.21	11.21	11.21	11.21	11.21	11.21	5.61	5.61	1.12
$MgSO_4 \cdot 7H_2O$	11.28	5.64	11.28	11.28	11.28	11.28	11.28	11.28	11.28	11.28	5.64	5.34	1.13
Sucrose	77.94	-----	77.94	77.94	77.94	77.94	77.94	77.94	-----	-----	-----	-----	77.94
Fructose	2.22	-----	2.22	2.22	2.22	2.22	2.22	2.22	-----	-----	-----	-----	2.22
Glucose	3.89	50	3.89	3.89	3.89	3.89	3.89	3.89	50	50	50	50	3.89
Cellobiose	-----	-----	-----	-----	50	100	150	200	69	34	117	84	56
Osmotic pressure[b]	316	230	251	286	360	411	467	515	314	314	335	335	316

[a]Organic acids, amino acids and vitamins as in original WGM medium.

[b]Milliosmols/kg of culture medium after addition of FBS, WEUF and BSA.

Quantitative methods

Cell counts: Cells were counted daily, directly in the Falcon Flasks using an inverted phase microscope fitted with an ocular grid. Ten fields were counted in each flask. Since the volume of medium introduced into each flask was known, the value of cells per unit area was converted to cells per ml. Cells not appearing to be viable under phase microscopy were not scored. In some experiments nonviable cells were determined by the use of trypan blue (0.4%) (Grand Island Biological Co.).

Osmotic pressure determinations: Osmotic pressures were determined by the use of an Osmette Precision Osmometer (Precision Systems). Determinations were expressed in milliosmols per kilogram.

Parameters of growth

Two parameters of growth were measured in these studies, the growth rate and overall cell increase. Generally, logarithmic growth was maintained for 4 or 5 days. A best fit line was drawn through these points (fits of 0.98 to 0.99 were obtained). The growth rate was determined from the slope of this line $R = \frac{\log_{10} Y_2/Y_1,}{t_2 - t_1}$ where Y = log cells/ml; t = time in days. The overall cell increase was calculated using the formula:

$$\text{Cell Increase} = \frac{\text{Final number of cells/ml in stationary phase.}}{\text{Initial number of cells/ml}}$$

This calculation was advantageous in cases where a linear phase of growth was not obtained or where the attainment of a maximum cell population was delayed.

RESULTS AND DISCUSSION

Influence of osmotic pressure on growth

The *H. zea* cell line shows an ability to grow in a wide range of osmotic pressures (Tables 2 and 3) (Fig. 1). The maximum growth rate was obtained at 320 milliosmols per kg. If the osmotic pressure was increased or decreased from this, the growth rate decreased. However, in the 275 to 380 milliosmols/kg range the cells were still able to grow at 90% or better of the maximum growth rate. The overall cell increase was also affected, maximum increase being obtained at 286 and 316 milliosmols/kg. Between 260 and 375 milliosmols/kg, 90% or better of the maximum cell yield was obtained. The control media for the hypotonic series supported the normal growth of these cells in reduced sugars and salts, provided the osmotic deficit was corrected by cellobiose.

Animal cells in general are able to grow in a wide range of osmotic pressures (Wilmer, 1927; Paul, 1959; Pirt and Thackeray, 1964a, 1964b).

Table 2. The influence of osmotic pressure on growth rate and cell yield compared to original medium.

Medium Code	±	WGM	1	2	3	4	5	6	7	8	9	10	11
Growth	a	0.22		0.15						0.20	0.19		
Rates[a]	b	0.27	0.11	0.16							0.26		
	c	0.26	0.13									0.24	0.24
	d	0.23			0.22	0.28	0.19	0.11	ind.[b]				
	e	0.23			0.22	0.21	0.16	ind.	ind.				
Cell	a	13.6		10.4						12.1	13.5		
	b	11.3	5.7	9.8							12.0		
Increase[a]	c	11.6	7.5									11.7	10.5
	d	15.6			15.6	15.6	8.3	3.1	1.4				
	e	15.5			16.2	14.6	7.9	2.9	1.3				
Osmotic pressure		316	230	251	286	360	411	467	515	314	314	335	335

[a] See text for definitions.

[b] Ind. = growth rate indeterminable; no linear growth phase.

± The value for WGM on line a serves as a control for the other cultures on that line, done at the same time. Likewise, cultures on line b, line c, etc., were done simultaneously.

Table 3. The influence of osmotic pressure on the per cent maximum growth compared to original medium.

Medium Code	WGM	1	2	3	4	5	6	7	8	9	10	11
Per cent Maximum	100	43	68	95	98	81	49	ind.	93	89	92	92
Growth Rate	100	48	60	99	92	71	ind.	ind.	--	98	--	--
Per cent Maximum	100	50	87	100	100	53	20	9	89	106	100	90
Cell Increase	100	64	77	104	94	51	19	9	--	99	--	--
Osmotic Pressure	316	230	251	286	360	411	467	515	314	314	335	335

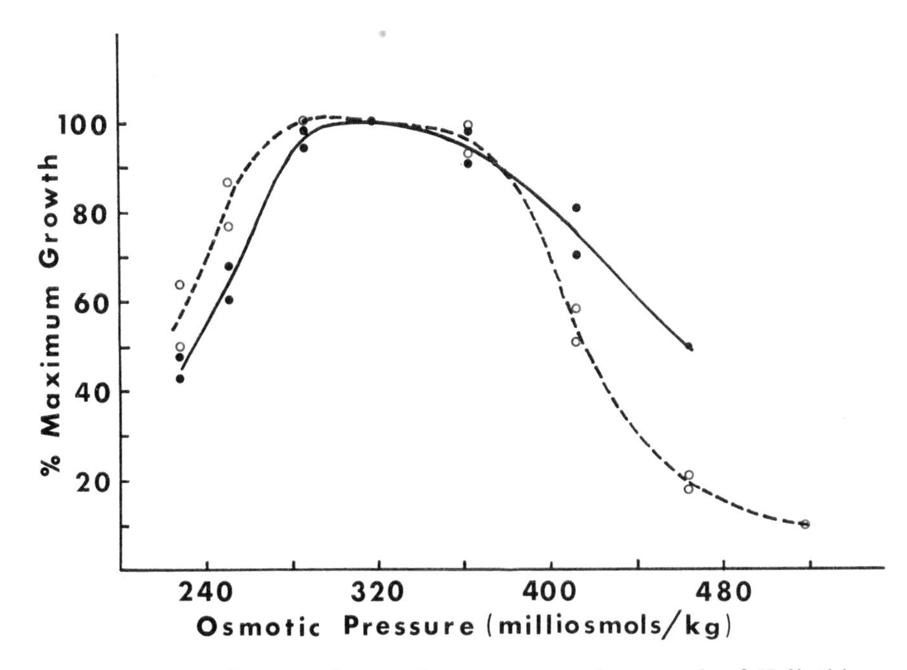

Figure 1. The influence of osmotic pressure on the growth of *Heliothis zea* cells *in vitro*. Percent maximum cell increase (broken lines). Per cent maximum growth rate (solid lines).

Pirt and Thackeray (1964a) found that ERK cells (human HeLa type) growing as a monolayer show 90% maximum growth in the 235 to 340 milliosmols (ideal)/1 range, with a midpoint of 288. The same cells growing in suspension were more sensitive to tonicity but had the same midpoint (Pirt and Thackeray, 1964b). Thus the range of osmotic tolerances between the ERK and the *H. zea* cell lines is similar, with the midpoint (318 to 328 milliosmols/kg) for the *H. zea* cells being higher than that for the ERK cells. However, a basic difference should be noted between these studies and ours, as we used cellobiose to effect increases in osmotic pressure; whereas additional salts were used by Pirt and Thackeray to increase the osmotic pressure.

Influence of pH on growth

The optimum pH for growing the *H. zea* cells was found to be between 6.5 and 7.0 (Table 4) (Fig. 2). Reductions in the growth rate occurred at pH values of 6.0, 7.5 and 8.0. The cell yield was not affected by the pH values of 6.0 and 7.5 to the same extent as was the growth rate. A long lag phase (3 days) occurred with an initial pH of 8.0, after which the cells started to grow

Table 4. Influence of pH on growth in medium #12.

Initial pH	Growth rate	% Max. growth rate	Cell increase	% Max. cell increase
6.0	0.20	80	12.3	97
6.5	0.25	100	12.4	98
7.0	0.24	97	12.7	100
7.5	0.20	82	12.3	97
8.0	0.17	71	9.5	75

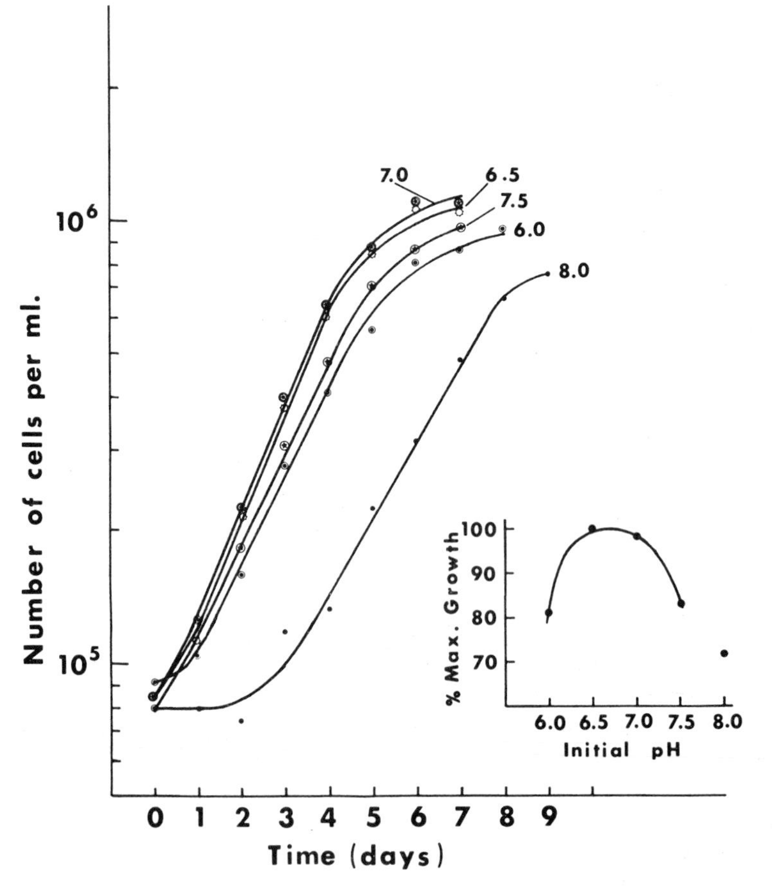

Figure 2. The influence of pH on the growth of *Heliothis zea* cells *in vitro*. Inset graph shows the effect of initial pH on the per cent maximum growth rate.

at a reduced rate and yielded fewer cells upon entering the stationary phase. The pH determinations of used culture media were made on the 7th day. A slight pH shift was found to have occurred if the initial pH had been 6.5 (to 6.6). When the initial pH had been 6.0, the pH on the 7th day was 6.3. A decrease in the pH was found when the initial pH values had been 7.0 (to 6.7), 7.5 (to 6.9), and 8.0 (to 7.2). The drop in pH from the initial pH of 8.0 probably explains the ability of the cells to grow in an apparently abnormally high pH. Paul (1959) [using L cells (mouse subcutaneous, fibroblast), HeLa (human), lymphoma (mouse)] found that for many cell types the growth rate was affected by pH, whereas the maximum population was not. Rubin (1971) and Ceccarini and Eagle (1971), however, found that the maximum population in certain animal cell cultures was affected by pH.

Acknowledgements.—This is Paper No. 8013, Scientific Journal Series, Minnesota Agricultural Experiment Station, St. Paul. The work was supported also by U. S. Public Health Service Research Grant No. AI 09914 from the National Institute of Allergy and Infectious Diseases. It is a pleasure to acknowledge the technical assistance of Miss Fay Freeman and Mr. Surendra Chaudhary.

REFERENCES

CECCARINI, C. and EAGLE, H. (1971). pH has a determinant of cellular growth and contact inhibition. *Proc. Natl. Acad. Sci.* **68**:229-233.

GRACE, T. D. C. (1962). Establishment of four strains of cells from insect tissues grown *in vitro*. *Nature* **195**:788-789.

HINK, W. F. and IGNOFFO, C. M. (1970). Establishment of a new cell line (IMC-HZ-1) from ovaries of cotton bollworm moths, *Heliothis zea* (Boddie). *Exptl. Cell Res.* **60**:307-309.

PAUL, J. (1959). Environmental influences on the metabolism and composition of cultured cells. *J. Exptl. Zool.* **142**:475-505.

PIRT, S. J. and THACKERAY, E. J. (1964a). Environmental influences on the growth of ERK mammalian cells in monolayer culture. *Exptl. Cell Res.* **33**:396-405.

PIRT, S. J. and THACKERAY, E. J. (1964b). Environmental influences on growth of L and ERK mammalian cells in shake-flask cultures. *Exptl. Cell Res.* **33**:406-412.

RUBIN, H. (1971). pH and population density in the regulation of animal cell multiplication. *J. Cell Biol.* **51**:686-702.

WILLMER, E. N. (1927). Studies on the influence of the surrounding medium on the activity of cells in tissue culture. *Brit. J. Exptl. Biol.* **4**:280-291.

WYATT, S. S. (1956). Culture *in vitro* of tissue from the silkworm *Bombyx mori* L. *J. Gen. Physiol.* **39**:841-852.

YUNKER, C. E., VAUGHN, J. L. and CORY, J. (1967). Adaptation of an insect cell line (Grace's Antheraea cells) to medium free of insect hemolymph. *Science* **155**:1565-1566.

APPLICATION OF CENTRIFUGAL CHROMATOGRAPHY TO THE SEPARATION OF SUGARS IN ARTHROPOD HEMOLYMPH AND TISSUE CULTURE MEDIUM

Reno Parker, Edgar Ribi and Conrad E. Yunker
USHEW - PHS - NIH
National Institute of Allergy and Infectious Diseases
Rocky Mountain Laboratory
Hamilton, Montana 59840

INTRODUCTION

Recently, specific classes of compounds have been successfully separated on beds of microparticulate silica by the use of centrifugal chromatography (Ribi *et al.*, 1970a,b). Anacker *et al.* (1971) employed this technique to separate mixtures of mono-, di- and tri-saccharides. It was concluded that sugar molecules under the influence of centrifugal force migrate through the silica gel and appear in the gel as narrow bands according to their molecular structure and size. These chromatographic separations could be altered by the application of different solvent systems.

The present investigation entailed the application of centrifugal chromatography to the separation of naturally occurring sugars in the hemolymph of specifically selected arthropods including ticks, *Dermacentor andersoni* Stiles, beetles, *Leptinotarsa juncta* (Germer) and grasshoppers, *Melanoplus bivittatus* (Say). The resolution quality of centrifugal chromatography was compared to that of conventional thin-layer chromatography with these naturally occurring sugars. Also, sugars were analyzed in tissue culture medium used in the *in vitro* maintenance of cells of *Aedes albopictus* (Skuse) and *Culex quinquefasciatus* (Say).

METHODS AND MATERIALS

Hemolymph was obtained from field collected *M. bivittatus* females by excising a hind leg, from laboratory reared *L. juncta* females by puncturing the wing vein at the base of the elytra, and from laboratory reared *D. andersoni* females by puncturing the upper body wall. All samples were collected in 50 μl pipettes and pooled in 200 μl amounts. Total sugar content of the hemolymph was monitored according to the method of Dische (1962).

Insect tissue culture medium (HLH, Grand Island Biological Co., Grand Island, N. Y.), which contained 1 μg/μl of glucose, received like amounts of

Insect and mite nutrition – North-Holland – Amsterdam, (1972)

trehalose or fructose. These sugars were assayed in freshly prepared medium and then in medium that had been used in cell cultures for 7 and 14 days. Total sugar content was monitored by the Anthrone method of Dische (1962). Cells used were the lines established by Singh (1967) of *Aedes albopictus* and by Hsu (1971) of *Culex quinquefasciatus*. Cells were grown at 27°C in 30 ml plastic tissue culture flasks (BioQuest, Oxnard, Calif.) according to the methods of Yunker and Cory (1969) (but only 10% fetal calf serum was used in the medium). After monolayers of *A. albopictus* or near-monolayers of *C. quinquefasciatus* were produced (approximately 3 days), growth medium was replaced by medium to be tested.

Protein material was precipitated from the hemolymph and tissue culture medium with equal volumes of 10% trichloroacetic acid in 15 ml centrifuge tubes. The mixture was centrifuged at 2500 X g. Supernatants were collected and washed with a chloroform-methanol (2:1) mixture to rid each sample of lipid material (Folch *et al.*, 1957). The aqueous fractions, containing mostly amino acids and sugars, were applied to a mixed ion exchange column of Dowex 50W X 12 (H+) and Dowex 2W X 12 (C1-) resins (Wyatt and Kalf, 1957). The sugar fractions were then eluted with water, collected and dried under nitrogen. Each fraction was dissolved in 100 μl of dimethylsulfoxide (DMSO) to which was added an equal volume of chloroform. Standard solutions of trehalose, glucose, fructose and rhamnose were prepared and analyzed in parallel.

Microparticulate silica (Sorvall 48312 with 2% moisture) was obtained from Ivan Sorvall, Inc., Hamilton, Montana. The slurry consisted of 3 g of 2% gel mixed with 80 ml of solvent. The following solvent system was employed: chloroform:methanol:water (65:24:4). All gels were packed at 1500 X g for 10 min and developed with the same solvent at 2500 X g for approximately 13 min. The gels were extruded, dried for 10 min and divided into two groups. One group was sprayed with a solution of sulfuric acid saturated with sodium dicromate and charred for detection of the sugar compounds as suggested by Anacker *et al.* (1971). The other group was sprayed with an anisaldehyde chromogenic reagent which produces sugar specific colors (Stahl, 1969).

Thin-layer chromatography plates were prepared by applying a suspension of 30 g Kieselguhr G (E. Merck, Darmstadt, Germany) per 80 ml of H_2O in a 25 μ thick layer to glass plates. The developing solvent system was identical to that used for centrifugal chromatography.

RESULTS

Various concentrations of total sugar were obtained from the hemolymph of the three arthropods. *Melanoplus bivittatus* hemolymph contained the highest total concentration and *D. andersoni* hemolymph contained the lowest total concentration of sugar material (Table 1).

Table 1. Comparison of total sugars in the hemolymph of three species of arthropods. Sugar concentrations were based on the anthrone reaction utilizing glucose for standard concentrations.

Arthropod	Sugar in mg/100 ml of hemolymph
Melanoplus bivittatus	415 ± 10
Leptinotarsa juncta	240 ± 12
Dermacentor andersoni	190 ± 5

Centrifugal chromatography adequately separated natural sugars and their respective standards (Fig. 1). *Melanoplus bivittatus* hemolymph displayed a predominance of trehalose with smaller concentrations of glucose and fructose. *Leptinotarsa juncta* hemolymph contained a predominance of glucose with smaller concentrations of fructose and rhamnose. Glucose was the only detectable sugar in *D. andersoni* hemolymph.

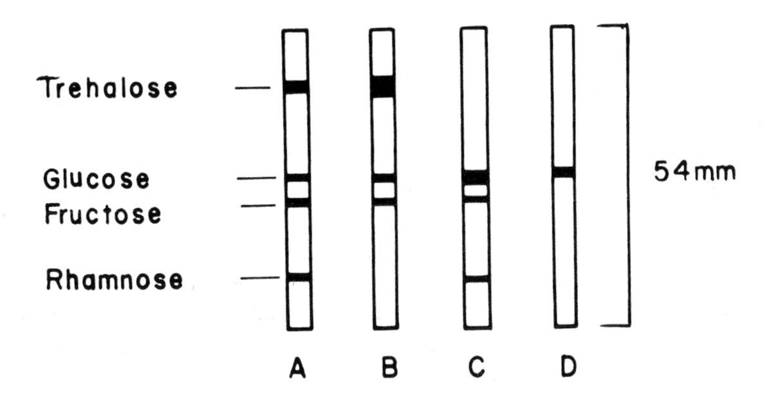

Figure 1. Centrifugal chromatography chromatograms of standard and hemolymph sugars. A.–Standard sugar mixture (5 μg each); B.–*Melanoplus bivittatus* hemolymph; C.–*Leptinotarsa juncta* hemolymph; D.–*Dermacentor andersoni* hemolymph.

When decreasing concentrations of sugar, from 10 μg to 1 μg, were separated by centrifugal chromatography it was found that 2-5 μg were ample concentration for reproducible results and most sugars could be detected at concentrations of approximately 1 μg (Fig. 2). Two to 5 μg concentrations are comparable to those used in thin-layer chromatography separations.

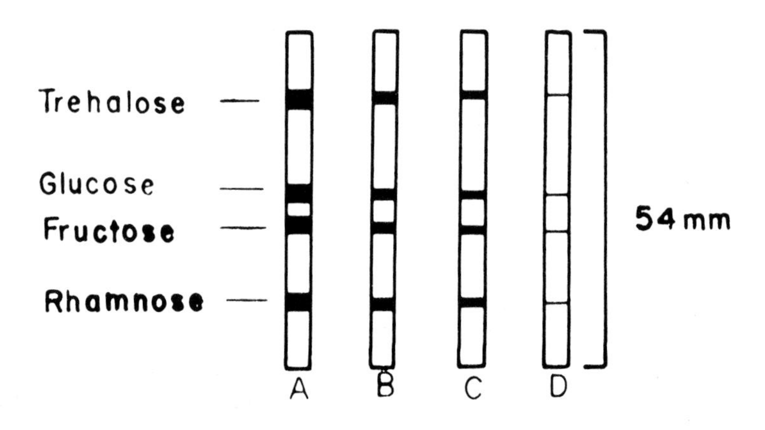

Figure 2. Separation of sugars at different concentrations. Quantities in μg per sugar are: A.—10 μg; B.—5 μg; C.—2.5 μg; D.—1.25 μg.

Use of thin-layer chromatography to separate hemolymph sugars was comparable to use of centrifugal chromatography except that glucose and fructose could not be spacially resolved (Fig. 3). In such instances detection of individual sugars usually depends on characteristic colors produced by spraying with stains such as anisaldehyde (Stahl, 1969). However, when this strain was applied to microparticulate gels we found that most sugars stained only greenish-purple and rhamnose stained a greenish-yellow.

Sugar concentrations in tissue culture medium showed substantial decreases over the 2-week period. Trehalose and fructose added to the medium of *A. albopictus* cells were not as rapidly utilized as was glucose (Figs. 4 & 5). Here, the concentration of trehalose at day 7 was approximately the same as that found in unused medium whereas glucose substantially disappeared. By day 14 only a trace of glucose remained and, in addition, trehalose began to disappear. Fructose appears to play a similar role to trehalose in the tissue culture medium (Figs. 5 & 8). By microscopic observation, cells grown for 14 days in the presence of fructose or trehalose appeared no different from control cells grown in medium lacking these sugars.

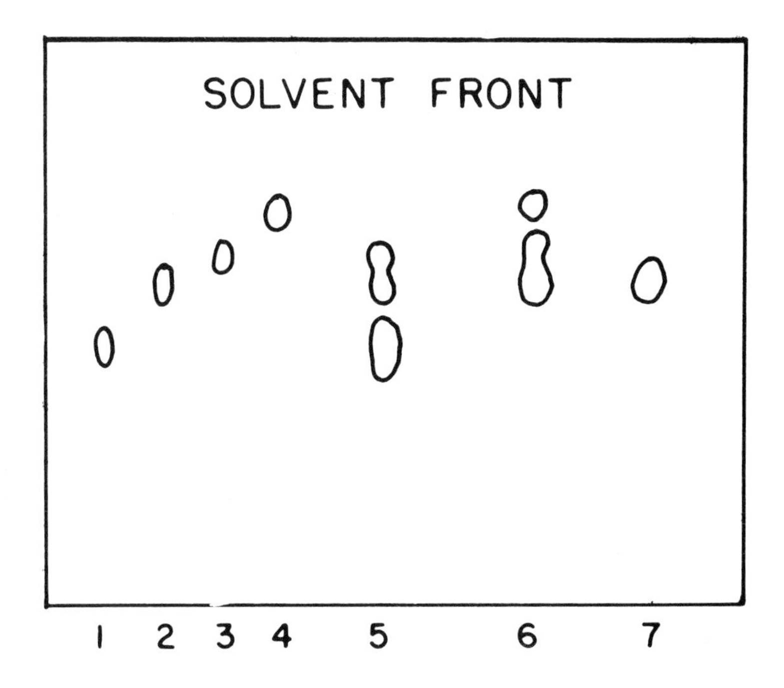

Figure 3. Thin-layer chromatogram separation of standards and hemolymph sugars on Kieselguhr G with the solvent system chloroform:methanol:water (65:25:4); developed for approximately 90 min. 1.–Trehalose; 2.–Glucose; 3.–Fructose; 4.–Rhamnose; 5.–*Melanoplus bivittatus* hemolymph; 6.–*Leptinotarsa juncta* hemolymph; 7.–*Dermacentor andersoni* hemolymph.

A somewhat different utilization of sugars was apparent in *C. quinquefasciatus* cells. Here, constituent sugars diminished faster than in cultures of *A. albopictus* (Figs. 6 & 7). However, in contrast to the latter cultures, trehalose and fructose were apparently being utilized to a greater extent by *C. quinquefasciatus* cells inasmuch as total sugar concentrations were virtually depleted from this system by day 14 (Fig. 9). Also, *C. quinquefasciatus* cells grown in the presence of glucose as the only sugar were in extremely poor condition from day 7 to day 14. In contrast, the cell cultures containing the added trehalose or fructose appeared to be in a more viable condition.

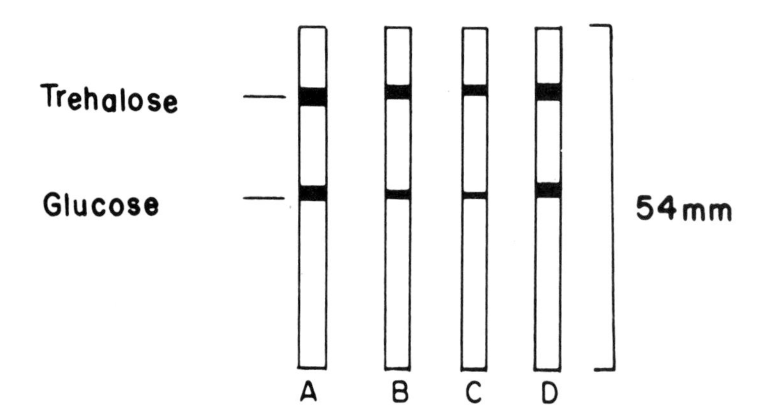

Figure 4. Centrifugal chromatography chromatograms of sugar separated from *Aedes albopictus* tissue culture medium. A.—Sugars from unused medium; B.—Sugars from medium used in cells for 7 days; C.—Sugars from medium used in cells for 14 days; D.—Standard sugars at 5 μg/sugar.

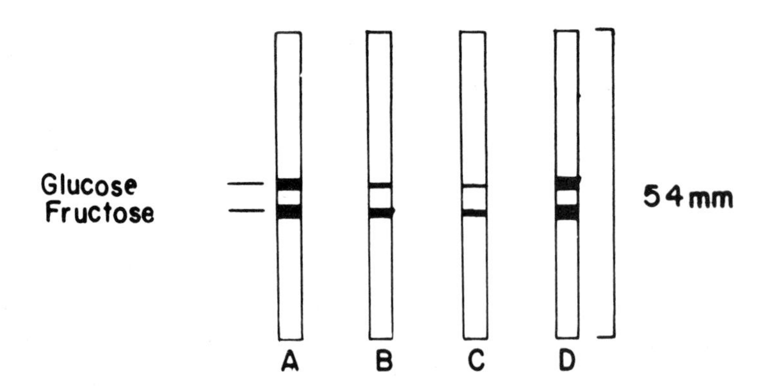

Figure 5. Centrifugal chromatography chromatograms of sugars separated from *Aedes albopictus* tissue culture medium. A.—Sugars from unused medium; B.—Sugars from medium used in cells for 7 days; C.—Sugars from medium used in cells for 14 days; C.—Standard sugars at 5 μg/sugar.

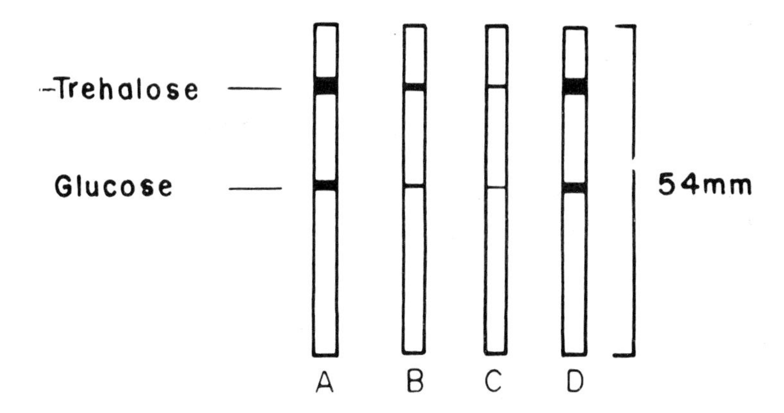

Figure 6. Centrifugal chromatography chromatograms of sugars separated from *Culex quinquefasciatus* tissue culture medium. A.—Sugars from unused medium; B.—Sugars from medium used in cells for 7 days; C.—Sugars from medium used in cells for 14 days; D.—Standard sugars at 5 μg/sugar.

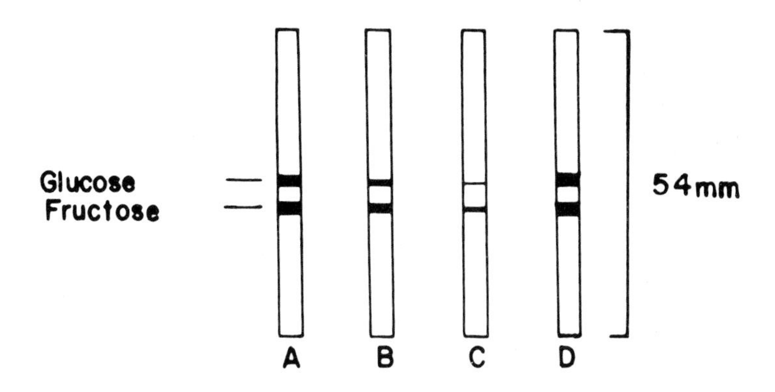

Figure 7. Centrifugal chromatography chromatogram of sugars separated from *Culex quinquefasciatus* tissue culture medium. A.—Sugars from unused medium; B.—Sugars from medium used in cells for 7 days; C.—Sugars from medium used in cells for 14 days; D.—Standard sugars at 5 μg/sugar.

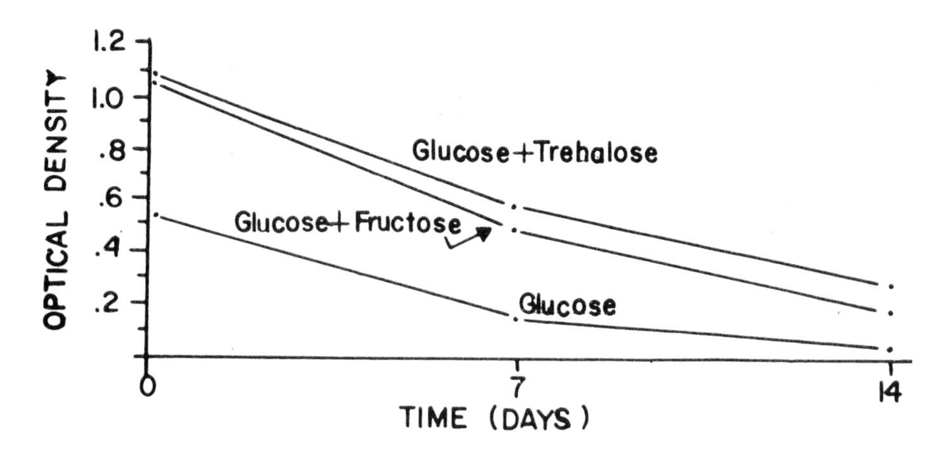

Figure 8. Changes of sugar concentration in *Aedes albopictus* tissue culture medium. Sugar concentrations analyzed using the anthrone reaction; color density read at 620 mμ.

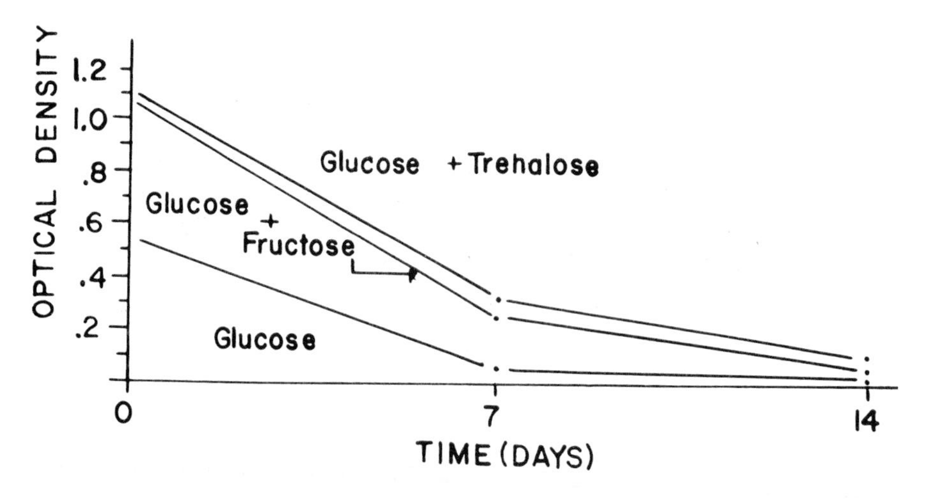

Figure 9. Changes of sugar concentration in *Culex quinquefasciatus* tissue culture medium. Sugar concentrations analyzed using the anthrone reaction; color density read at 620 mμ.

DISCUSSION

The particular methods used in the separation of compounds from biological fluids are of primary interest to the insect physiologist. The use of thin-layer chromatography or paper chromatography often entails chromatographing closely moving compounds in a second direction to obtain accurate resolution. However, this technique produces a tendency for the separated concentrations to become diluted and difficult to detect. Although still in a primary developmental stage, centrifugal chromatography appears to have substantial advantages over the previous methods. Excellent results may be produced in one step with no need for a two dimensional separation. Anacker *et al.* (1971) also reported centrifugal chromatography to be less time consuming than thin-layer chromatography and also relatively inexpensive when compared to gas chromatography in the separation of sugars. Superior resolutions with the use of centrifugal chromatography were noted by Ribi *et al.* (1970a) in the separation of steroid isomers and other lipid related compounds.

We conclude that as with all types of analytical methods, more exact quantitation is needed to verify amounts of separated compounds. Further analysis is presently being carried out through a process of pressure elution from the gel columns. The eluted sugars can then be quantified in a similar manner as suggested by Mopper and Degans (1972). This elution process will make it feasible to determine the exact amount of each sugar utilized in tissue culture medium or other biological fluids.

SUMMARY

Centrifugal chromatography was applied to the separation of naturally occurring sugars in the hemolymphs of ticks, *Dermacentor andersoni* Stiles; beetles, *Leptinotarsa juncta* (Germer); and grasshoppers, *Melanoplus bivittatus* (Say). In addition, sugars were analyzed in tissue culture medium used in the maintenance of Singh's *Aedes albopictus* (Skuse) and Hsu's *Culex quinquefasciatus* (Say) cell-lines. The resolution quality of centrifugal chromatography, when compared to conventional thin-layer methods, was found to be less time consuming and gave superior separation of the sugars.

Acknowledgment–The authors wish to thank Jack Cory for his valuable technical assistance in the maintenance of the insect cells. This investigation was supported in part by Public Health Service Fellowship No. 1-FO2-A1-49521-01.

REFERENCES

ANACKER, R. L., SIMMONS, J. H. and RIBI, E. (1971). Separation of sugars by centrifugal microparticulate bed chromatography. *J. Chromatography* **62**:93-97.

DISCHE, Z. (1962). Section V - Color reactions of carbohydrates. In *Methods in Carbohydrate Chemistry* (WHISTLER and WOLFROM Eds.), pp. 477-514, Academic Press, New York.

FOLCH, J., LEES, M. and SLOAN-STANLEY, G. H. (1957). A simple method for the isolation and purification of total lipids from animal tissue. *J. Biol. Chem.* **226**:497-509.

HSU, S. H. (1971). Growth of arboviruses in arthropod cell cultures: Comparative studies. I. Preliminary observations on growth of arboviruses in a newly established line of mosquito cell (*Culex quinquefasciatus* Say). In *Current Topics in Microbiology and Immunology* (EMILIO WEISS Ed.), pp. 140-148, Springer-Verlag, New York/Heidelberg/Berlin.

MOPPER, K. and DEGENS, E. T. (1972). A new chromatographic sugar autoanalyzer with a sensitivity of 10^{-10} moles. *Analytical Biochemistry* **45**:147-153.

RIBI, E., FILZ, C. J. GOODE, G., STRAIN, M. S., YAMAMOTO, K., HARRIS, S. C. and SIMMONS, J. H. (1970a). Chromatographic separation of steroid hormones by centrifugation through columns of microparticulate silica. *J. Chromatographic Sci.* **8**:577-580.

RIBI, E., FILZ, C. J., RIBI, K., GOODE, G., BROWN, W., NIWA, M. and SMITH, R. (1970b). Chromatography of microbial lipids by centrifugation through microparticulate gel. *J. Bacteriol.* **102**:250-260.

SINGH, K. R. P. (1967). Cell cultures derived from larvae of *Aedes albopictus* (Skuse) and *Aedes aegypti* (L.). *Curr. Sci.* **36**:506-508.

STAHL, E. (Ed.) (1969). *Thin-Layer Chromatography*. pp. 811-831. Springer-Verlag, New York.

WYATT, G. R. and KALF, G. F. (1957). The chemistry of insect haemolymph. II. Trehalose and other carbohydrates. *J. Gen. Physiol.* **40**:833-847.

YUNKER, C. E. and CORY, J. (1969). Colorado tick fever virus: growth in a mosquito cell line. *J. Virol.* **3**:631-632.

METABOLIC FATE OF NUTRIENT COMPOUNDS

INTRODUCTION
by

Ernest Hodgson, Section Editor
Department of Entomology
North Carolina State University
Raleigh, North Carolina 27607

The consideration of metabolic aspects of nutrition differed in some respects from the remainder of the conference, simultaneously showing enough unity of goal and interaction with other sections, that the entire conference emerged as a single worthwhile enterprise.

The call to relevance was frequent, however, it was frequently too narrow, a call to devote the attention of wide fields of scientific endeavor exclusively to the immediate problems of pest control. The metabolic section served this need in certain obvious respects, its relation to mass rearing for example. More important, by its ties to comparative biochemistry, it tended to expand the criteria by which relevance should be measured. The concept of the unity of biochemistry, broadly true and essential for a time in the development of biochemistry itself, has tended to inhibit the development of comparative biochemistry, the study of rich diversity apparent in the chemistry of living creatures. Proper understanding of the biochemistry of target organisms, such as insects, and its contrast to non-target organisms, such as vertebrates, is an aspect of comparative biochemistry relevant to insect control and is essential to the fundamental understanding of comparative toxicology necessary for the rational development of selective biocides. No less important is the business of comparative biochemistry itself—a modular basis for the theory of evolution.

Lambremont's opening presentation on radioisotope technique again illustrated that science is often no better than the tools it uses by bringing into focus the pitfalls that await the unwary—an important point in an era which takes radioisotopes for granted. The following papers by Chippendale and Barlow dealt with an area of current interest and related it to parasitism, again relevant to pest management. Presentations covering choline and choline analogs, transmethylation and sterols demonstrated the significance of insect nutrition to other fields, such as genetics and endocrinology, moreover, demonstrating significant biochemical differences between insects and vertebrates. Ito's paper on silkworm nutrition differed from the others by focusing on the improvement of man's lot by encouraging rather than discouraging insects.

In summary, the section concerning metabolism of nutrient compounds added much to our appreciation of the differences between insects and other animals and clarified the necessary relationships between insect and mite nutrition, the fundamental areas of genetics, endocrinology, comparative biochemistry and toxicology, and additionally to important applied fields such as pest management.

RADIOISOTOPES IN INSECT NUTRITION AND METABOLISM: THE DESIGN AND INTERPRETATION OF RADIOTRACER EXPERIMENTS

Edward N. Lambremont
Nuclear Science Center
Louisiana State University
Baton Rouge, Louisiana 70803

INTRODUCTION

The discovery and production of radioisotopes, particularly those of hydrogen, carbon, phosphorus and sulfur, have contributed one of the most significant advances in research technology of the past 30 years. Physiology and biochemistry have undergone an especially enormous expansion during this period, a veritable explosion of information, resulting largely from a coupling of isotope techniques with numerous other advances in laboratory practice. Volumes have been written describing in great detail the advantages of using radioactive tracers to explore rapidly and with great sensitivity problems that often cannot be investigated in any other way. When properly employed, the radiotracer method gives the researcher an indespensible tool for following step by step some of the most intricate biochemical processes that take place in living systems.

In setting out to prepare a paper for this conference, I felt that it would be redundant to add yet another paper on uses of isotopes in nutritional research. Numerous fine textbooks, reviews, comprehensive journal articles and training programs are already available for those who wish to learn radiotracer technique and apply it to their work. All too often, the advantages of isotopic tracers are stressed while their limitations are either overlooked or not given proper emphasis. My objective will be to highlight some of the problems inherent in the design and execution of tracer experiments so that data derived from them will be interpreted properly. It is my firm belief that once one understands the pitfalls that exist in this method (or any other), and knows how to avoid making mistakes, he is then free to take full advantage of all the opportunities the technique has to offer.

In the space and time available it will not be possible to present a comprehensive coverage of all aspects of design and interpretation of tracer experiments. Therefore, I shall emphasize two areas namely, (1.) general assumptions and limitations associated with radiotracer experiments, and in keeping with the theme of the conference, (2.) special considerations in designing isotope experiments to study insect nutrition and metabolism. It will be necessary to omit some other basic features of experimental design such as the nature of the problem, limits of precision and accuracy, sampling techniques, and statistical treatment of radioassay data. All of these points are amply covered in one or more of the works listed in the selected bibliography at the end of this paper.

GENERAL ASSUMPTIONS AND LIMITATIONS

There are several basic assumptions underlying the validity of radiotracer experiments. With some exceptions these can generally be dismissed in most work dealing with insect nutrition and metabolism. However an awareness of possible problems arising from these basic assumptions is essential during the planning of experimental work.

Tracer levels and biological compartments. First, it must be determined that the administered compound can be obtained or synthesized with sufficiently high specific radioactivity so that one is truly dealing with "tracer" levels. If the labeled compound is to be diluted with unlabeled compound (carrier) before it is administered to the organism, care must be taken to avoid reaching concentrations above the normal physiological range. Most ^{14}C- and ^{3}H-labeled materials can be obtained commercially at high specific radioactivities so that the endogenous pool of a metabolite in question is not altered during the experiment.

In *in vivo* experiments, further complications may arise because of biological compartmentalization. Side reactions and cyclization can also be expected to occur. Although we assume that an injected or fed labeled compound is in equilibrium with the total pool of that metabolite in the organism, often it probably is not. The influence of such factors on the outcome of an experiment must be determined or recognized on an individual basis, and the data must be carefully interpreted. This is especially true with kinetic data. *In vitro* experiments with whole tissue preparations, such as midgut or fat body, could also be influenced by these drawbacks. Generally, kinetic data about a specific biochemical process are difficult to assess *in vivo*, and often are investigated with cell-free preparations. When derived *in vitro*, kinetic data are usually more reliable, barring possible isotope effects (to be considered below). Such cell-free systems, however, suffer the limitation of being very artificial, and one cannot always equate the findings with what takes place in the intact organism. By contrast, measurements of a physiological nature in the living organism, such as uptake, transport and distribution of nutrients, are well-suited to isotopic labeling, because the multicompartment processes will affect the tagged compound like the natural compound in most instances.

Radiobiological effects. A second factor one must consider in obtaining meaningful results with radioactive tracers is that there is no radiobiological damage to the organism. Compounds labeled with the soft beta-emitting isotopes commonly used in biological work will not produce such a problem.

The ease of detection of trace amounts of ^{3}H, ^{14}C and ^{35}S by liquid scintillation assay minimizes the possibility of the radiation affecting the organism abnormally. Nevertheless, one should use the minimum amount of labeled compound necessary to achieve the desired results. One should further consider the possibility of some types of labeled compound being concentrated in certain tissues or subcellular structures to the degree that local radiation damage might result. Excessive use of labeled-thymidine might produce an intense local beta-irradiation in the nuclei of cells undergoing mitosis. This could produce undesirable effects, especially in many adult insects, in which only certain tissues undergo cell division (midgut and testis, for example). The malphigian tubules also could concentrate certain labeled metabolites and be selectively irradiated to unphysiological levels.

Whole-body radiation damage becomes more probably with the more energetic beta emitters such as ^{32}P (1.71 MeV E max). It also may occur with many of the metallic nuclides such as ^{22}Na, ^{45}Ca, ^{59}Fe, ^{60}Co, ^{64}Cu, ^{65}Zn and others which are mixed beta-gamma emitters. Damaging whole-body radiation can occur because of the high interaction of their energetic beta particles, and from a small fraction of the gamma photon interactions.

Two precautions can be taken in experimental design to minimize the possibility of radiobiological damage. Often one can choose a different isotope of the same element with lower disintegration energy or shorter physical half-life. For example, the relatively soft beta emitting ^{33}P (0.248 MeV E max) could be used in place of ^{32}P, and ^{55}Fe, which emits soft x-rays and no beta particles or gamma photons, instead of ^{59}Fe. If a local source of neutrons is available to the investigator, he might prefer to produce small quantities of a short-lived nuclide instead of using the commercially available one with a longer physical half life. Nuclear reactors, cyclotrons, Cockroft-Walton accelerators and other neutron sources are located at many universities, hospitals and research centers for the production of research quantities of many such nuclides having properties that minimize the chance of radiobiological damage. Although the shorter-lived nuclides generally decay with greater disintegration energies, the total-body dose is delivered over a shorter time span. Such isotopes could be given to the insect during periods when radiation damage is least likely to occur, particularly when mitotic rate is at a minimum.

A second adjustment in design to avoid radiation damage would be to use short-duration experiments. If the results can be achieved quickly, there is no need to lengthen the experimental period to a point where the accumulated radiation effects might begin to distort the results.

Physical and chemical state of labeled precursor. The third fundamental assumption in employing tracers is that the physical and chemical state of the labeled compound is exactly the same as the unlabeled one in the organism. Except in certain circumstances, differences in the physical state are unimportant in the present context. Certain cations, particularly in extremely dilute solution, behave as colloids and not as ions. Loss of dosing solutions of very high specific radioactivity by sorption or colloidal accumulation in certain tissues could lead to a physiological state other from that intended by the investigator.

The chemical state, however, is a matter of the utmost importance. First, it is the investigator's responsibility to determine that the administered compound is truly what he intended to use in the experiment. Thus if a sulfate is required, he should be sure he did not administer a sulfite. In biochemical applications, the correct biologically-active isomer should be used. Any label contained in the inactive isomer of the compound before use would seriously affect the interpretation of all resulting data.

The single most serious problem involved with chemical state is the question of radiochemical purity. The quality of a labeled compound can never be assumed, and the investigator must be sure that the compound is free of spurious labeled contaminants that might seriously affect the interpretation of his data. Often the compound supplied by the manufacturer will contain a trace of the very metabolite being sought in the experiment. For example, in our studies of the ω-oxidation of 1-^{14}C-hexadecane to palmitic acid in the house fly and boll weevil (Joiner and Lambremont, 1969) a preliminary purity check of the hexadecane revealed a trace of labeled palmitic acid. It was necessary to remove the labeled palmitic acid by preparative thin-layer chromatography before proceeding with the experiment. In other studies of the interconversion of fatty acids and fatty alcohols (Lambremont, in press), it was found that 1-^{14}C-hexadecanol and octadecanol contained tracers of labeled hexadecanoic and octadecanoic acid. Similarly-labeled fatty acids contained ^{14}C-labeled fatty alcohols as impurities. These precursors had to be carefully purified, and rechecked at frequent intervals to avoid misleading results. Once this was done, it was possible to draw the proper conclusions.

Always assume the labeled compound is impure, and is becoming more so as each day passes. The very fact that it does contain radioactive atoms means that it will have a shelf life far shorter than its unlabeled counterpart. Radioactive impurities arise from two principal sources. First, by radiolysis (the compound is irradiating itself), highly reactive ions and free radicals are produced that can lead to self-decomposition, even in materials stored in the frozen state. This is especially significant with the soft-beta emitting nuclides in which virtually all of the disintegration energy is absorbed in the sample itself. Compounds of high specific radioactivity, which have a relatively large unstable atom ratio with very little actual material, are particularly susceptible to radiation decomposition. Secondly, with all particle-emitting or electron capture radioisotopes a different chemical element forms at the instant of disintegration. For example, when ^{14}C emits its beta particle it becomes stable ^{14}N; ^{3}H becomes stable ^{3}He, ^{32}P becomes stable ^{32}S, and ^{35}S becomes stable ^{35}Cl. A compound purchased with ^{14}C in a specific location in the molecule is constantly changing to the N-analog. These latter products are often highly unstable and react quickly with the parent compound to produce further chemical degradation. Uniformely-labeled ^{14}C compounds pose a special problem, because they may break apart at almost any point in the carbon chain to produce a wide variety of labeled fragments. With UL ring compounds, labeled open-ring products will accumulate in the sample.

In addition to checking the radiochemical purity of the materials you purchase, and repurifying them yourself to your own specifications, there are further precautions to reduce excessive decomposition. Keep all labeled materials dispersed in solution or spread over a glass or filter paper surface to minimize self-irradiation. Further chemical breakdown can be avoided by storing compounds at low temperature (-20°C or lower), in the dark, O_2 free, and under sterile conditions, and in organic solvent, preferably benzene. Contaminating microorganisms could alter the compound before you use it in the experiment, giving misleading results about the metabolic capability of the insect.

Since accumulation of most of the radiolytic and other degradative products is time-dependent, it is highly recommended that supplies of labeled materials be expended quickly or repurified frequently. Suggestions specified by the manufacturer for storing the compound should be carefully followed. Several excellent publications are available (see Bibliography) describing the preparation and purification of radiolabeled compounds, and these should be consulted as necessary.

Although most investigators scrupulously adhere to these precautions with their labeled compounds before using them, many completely overlook the fact that the changes are also taking place in their biological samples. Avoid keeping labeled biological materials or metabolic fractions for long periods before analysis of the final results. Chromatographic, electrophoretic and other analytical procedures should be done in a reasonably short time after samples are obtained to avoid the formation of undesirable labeled products. If samples are stored too long, various labeled impurities might arise and be attributed to biological processes when actually they did not originate in the living organism at all.

Metabolism and randomization of label. A fourth basic assumption is that only the labeled atom is being traced. Radioactivity associated with biological samples after introduction of the labeled compound can be attributed only to the labeled atom. The original compound may not be intact, and a whole series of labeled metabolites usually will be formed during the course of the experiment. This feature of the dynamic aspects of metabolism is usually clearly understood by present-day investigators. A less obvious feature, however, is the randomization of labeled atoms that takes place during metabolism. Because of biochemical cycles and interconnected pathways of intermediary metabolism, the labeled atom can be rearranged to a new location in the same compound. A particularly good example of this occurs with labeled glucose when it enters the hexosemonophosphate pathway. For each six molecules of glucose entering the cycle, five are generated anew, and one net glucose molecule is lost as 6 moles of CO_2. All the CO_2 carbon atoms come from the 1-position of glucose. Of the five newly generated glucoses, two will be formed from the original 2-3-2-4-5-6 carbon atoms, two from carbons 2-3-3-4-5-6, and one from triose condensation will have a pattern 6-5-4-4-5-6. Suppose, for example, one fed or injected 2-^{14}C-glucose into an insect. After a single passage thru the cycle four of the five glucoses would have their label in the 1 position. If 6-^{14}C-glucose were used, one of the newly formed glucoses would be 1-labeled. Thus, randomization in the pentose cycle favors a moving of the label toward the 1-position and away from the 6-position (Lambremont and Bennett, 1966; Katz and Wood, 1960). Similar randomization can be expected from other metabolic cycles.

Exchange of label between the tagged incoming compound and other unlabeled molecules in the organism is known to take place. This is an especially serious problem with tritium-labeled compounds, particularly those that are labeled by exposing them to ^{3}H gas (random or general labeling). Many compounds labeled by this technique (Wilzbach exchange method) can just as easily loose their ^{3}H to surrounding molecules by the same exchange reactions. When attached to inactive carbon atoms, ^{3}H labels are relatively permanent. However, on active molecular sites, such as -O^{3}H or -COO^{3}H, -N^{3}H$_2$, etc., the label could rapidly appear in the general metabolic pool of the insect.

Isotope effects. A final basic assumption in utilizing isotopes is that the chemical behavior of the unstable atom will be exactly the same as that of the stable one it is representing. For most of the nuclides above mass 40, the isotopic mass difference contributes an error of negligible significance. With the lighter elements, the difference in bond strength (and hence chemical behavior) of the stable and radioactive atoms becomes progressively greater. This phenomenon, known as the *isotope effect*, results from the fact that the stability of a chemical bond formed between two atoms is directly proportional to the masses of the participating atoms. Thus, in an experiment with a ^{14}C-labeled compound, bonds of ^{12}C-^{14}C are being used experimentally to represent the normal ^{12}C-^{12}C. In uniformly labeled compounds, the masses of the participating atoms are ^{14}C-^{14}C. A ^{14}C-^{14}C bond will be ruptured slower than a ^{14}C-^{12}C which in turn is broken slower than the normal ^{12}C-^{12}C. With other elements as well, the isotope of a greater mass reacts at a slower rate. The most pronounced isotope effects, therefore, occur with the isotopes of hydrogen: stable ^{2}H and radioactive ^{3}H have two and three times, respectively, the masses of ^{1}H that they represent in the experiment. Deuterium and tritium, therefore, cannot be expected to have the same chemical behavior as ^{1}H.

The most serious drawback of isotope effect occurs in designing kinetic experiments. Depending upon the molecular location of the ^{2}H or ^{3}H, cleavage of such bonds may proceed as much as 10 to 20% slower than the unlabeled compound. In extreme situations, tritium isotope effects retarded some reaction rates by as much as 1/2 (Weinberger and Porter, 1953) and carbon isotope effects producing up to 20% reduction in reaction rate have been reported (Rabinowitz *et al.*, 1960).

Two general classes of isotope effect are recognized: intermolecular and intramolecular. In the intermolecular isotope effect, one is actually dealing with two different populations of a compound, i.e., a ^{12}C population and another population of molecules that contain ^{14}C. Unless one is using carrier-free compounds, or the reaction under study is carried to completion, the results would be biased toward early cleavage of the ^{12}C and an enrichment of ^{14}C in the unreacted material. Specific radioactivities calculated for the labeled product of an incomplete reaction would be low. Conversely, calculations based on the unreacted precursor would give relatively higher specific radioactivities.

The intramolecular isotope effect occurs in symmetrical molecules in which only one of the bonds or planes of symmetry contains the label of different mass. The isotope effect renders the compound asymmetrical. Even if the reaction is carried to completion, the isotopic label will tend to concentrate in one of the metabolic fractions, thereby distorting the results. The intramolecular effect could be partially compensated by labeling equally both ends or bonds in the symmetrical molecule so that differences in activity from one end to the other were not introduced. In so doing, however, the intermolecular isotope effect would be created.

Several alternatives are open to the investigator in dealing with isotope effects. When possible, compounds should be selected that are labeled with an element of heavier mass than carbon. If one cannot avoid using ^{3}H or ^{14}C for some reason, care should be taken to see that the label is at some distance from the active part of the molecule. In no instance should it be a part of the chemical bond under study unless the experimental intent is to make use of the isotope effect to establish the point of reaction by specific site labeling.

ISOTOPES IN EXPERIMENTAL INSECT NUTRITION AND METABOLISM

The two most significant techniques developed during the past twenty years in utilizing radioisotopes in studies of insect nutrition and metabolism are: (1.) the labeled-pool technique of Winteringham (1956) and (2.) the broad-spectrum indirect determination of nutrient requirements first described by Black *et al.* (1952) and later adapted to insect nutrition by many workers (see Kasting and McGinnis, 1966).

The labeled-pool technique. The metabolic pool, a circulating quantity of chemical substances in partial or total equilibrium with those derived either from the diet or tissues, is a fundamental biochemical concept. The labeled-pool technique involves labeling various groups of biochemically related metabolites *in vivo* for later analysis of the many metabolic fractions formed in the course of intermediary metabolism. By feeding or injecting an insect with a suitably labeled precursor, the whole range of compounds in its metabolic pathway become labeled. These may be separated and determined qualitatively and quantitatively by a combination of techniques. The labeled-pool technique and numerous variations of it have been used successfully to explore many aspects of insect nutrition, metabolism and the toxic action of insecticides.

The investigator is seeking answers to two fundamental questions when he uses the labeled-pool technique: (1.) what kinds of molecules are utilized in the structural elements of cells, and (2.) what mechanisms are required to mobilize stored energy reserves and incoming nutrients for synthesis reactions. Several difficulties or limitations are encountered in interpretating data from this technique. A prime requirement is that the various labeled intermediates must be isolated in a pure state so that their specific radioactivity may be determined accurately. Failure to do this would lead to serious misinterpretations. It is especially critical to avoid the error of partial separations of compounds that are closely related or that behave similarly during the various purification procedures. Traces of labeled contaminants with high specific radioactivity mixed with a non-labeled metabolite under study contribute such errors. The accepted criterion of purity in the tracer method is constancy of specific labeled atom content under all conditions of chemical purification. Although this criterion is satisfactory in most instances, Kamen (1957) has shown specific instances in which labeled metabolic products still contained labeled impurities after exhaustive purification.

Secondly, it is generally assumed that the administered labeled compound is in equilibrium with the same unlabeled endogenous material already present in the organism. As biochemical knowledge advances, it is clear that this assumption in many instances simply is not true. Although all metabolic pools are in a dynamic state, their rates of flux are not always equal, and many compounds exist in a biological compartment that may turnover faster or slower than the same compound in another compartment. The various compounds of any given metabolic pool participate at different and characteristic rates. Nitrogenous compounds, for example, may participate in the hypothetical nitrogen pool differently, and may vary between proteins and amino acids. Often only a small fraction of the pool of any given compound is actively involved in the dynamic state.

Determining nutritional requirements with labeled compounds. The indirect radioisotopic method of determining insect nutritional requirements is a specialized adaptation of the labeled-pool technique. When used in conjunction with enzymatic methods, nutrient balance experiments, and nutritional deletion technique, the radioisotopic method offers a procedure for rapid confirmation of nutritional requirements. The procedure usually is carried out as follows: A radioactive substrate (usually ^{14}C-labeled) is given the insect in its diet, or by injection or cuticular absorption, and a period of metabolism is allowed to pass. The insects are sacrificed or selected tissues are obtained, the metabolites of interest are isolated and purified, and their specific radioactivity is determined. Nutritionally non-essential components contain radioactivity, whereas those that do *not* become labeled through the metabolic processes in the insect are considered to be nutritionally essential. Presence of the radioactivity in any given compound after metabolism is generally easier to assess and interpret provided it was isolated pure; absence of radioactivity, however, may be more difficult to interpret.

Although the indirect radioisotopic method has been utilized numerous times to determine the nutritional essentiality of amino acids, fatty acids, and sterols in the insect diet, it is not without its limitations. Once these are clearly understood, and every precaution taken as necessary, the method can be applied successfully.

Since the radioisotopic determination of nutritional compounds is an adaptation of the labeled-pool technique, the same requirements exist for isolating all the labeled and unlabeled metabolites in a high state of purity. There are several well-documented examples in the literature of insect nutrition in which investigators reported the biosynthesis of compounds (such as sterols and polyunsaturated fatty acids) usually recognized as essential in other higher animals. More careful investigation often revealed that the supposedly synthesized compounds contained labeled impurities that were isolated with the compound under study (Gilbert, 1967). For example, it was recently shown (Lambremont, 1971) that radioactivity is often associated with the 18-carbon polyunsaturated fatty acids in insects injected with or fed a ^{14}C-labeled precursor. Close examination revealed that the radioactivity was due to labeled components that co-chromatographed with the polyunsaturated acids. Identification and careful removal of the labeled impurities revealed that the insect under study did not incorporate ^{14}C labeled acetate into the polyunsaturated fatty acids, and the acids were more than likely required by the insect in its diet.

Equilibration of the labeled precursor with a significant portion of the active metabolic pool also is an important consideration. The choice of which type of precursor to use may make a significant difference, because in some insects it is possible that certain substrates may not be available or metabolized. For this reason, formate or acetate or an obvious acetate precursor such as glucose, sucrose, pyruvate, glutamate, glycine and other similar compounds are often selected. Kasting and McGinnis (1966) have wisely recommended that a mixture of several of these compounds (generally labeled) be used. While this would make it difficult to decide the relative contribution of any one precursor, or the specific labeling positions of the products, this is an excellent point of design for the preliminary experiments to determine the probable overall capacity for biosynthesis of a particular class of compounds. Some investigators use several different precursors in individual experiments to assess biosynthetic capacity, and this procedure is highly recommended as a more advanced alternative. The choice of label position in the precursor molecule must also be carefully selected to make sure that the label is not lost in metabolism, which would render the final desired metabolic product non-radioactive.

Another important factor to consider is the method of administering the candidate precursor. Injection is the most rapid technique, and allows one to give accurately known doses if bleeding is controlled. Most commonly a single pulse or injection of the compound under study is given during what is hoped will be an active period of metabolism. If the investigator chooses the wrong period he could conclude erroneously that the substrate was not being metabolized. More continuous perfusion of the labeled material through serial injections might be preferred, particularly if several different developmental stages were chosen.

Injection is not, however, the normal route of entry of nutrients into animals. Feeding the labeled compound overcomes this objection, but is limited in that it is often difficult to know precisely what portion of the label in the diet was eaten, and what portion was finally absorbed and assimilated. Contamination of the diet by bacteria, molds or yeasts could seriously affect the outcome of the experiment due to their metabolic action on the substrate. Sterile rearing conditions are mandatory to avoid this source of error, but these are sometimes difficult to achieve. Stronger and more meaningful evidence for the essentiality of a given metabolite would be obtained if a variety of substrates, administration techniques and metabolic periods still failed to produce incorporation of isotope into the compound in question.

The physical and chemical state of the compound before it is administered also must be examined and considered. Many water-imiscible substances when given without pretreatment could agglomerate to form a relatively inactive mass of small surface area. Most labeled materials should be dispersed in the diet as uniformly as possible. When injected, such materials should be pretreated either chemically or physically to bring them into the molecular range so that they can enter cellular compartments, or be available for attachment to enzymes. Many lipoidal substrates can be dispersed into fine stable suspensions by chemical emulsifiers or by sonication, or both. Some lipids are more easily metabolized when they are given as a salt or complexed with a protein such as albumin. Without these precautions, the metabolic availability of many compounds is seriously curtailed.

After the period or periods of metabolism have passed, one should consider the method by which the labeled sample will be recovered. The usual procedure in insect nutrition is to homogenize and extract the whole insect or groups of insects of a like developmental stage. While this is usually acceptable in most preliminary work, it destroys any chance to determine the specific tissue location of the newly synthesized material. A further limitation is the possible distortion of final specific radioactivity in its relation to that of the *de novo*-synthesized compound. If the endogenous pool is low, the net final specific activity will appear to be high, suggesting that the compound is a major metabolite. The reverse situation, dilution of the *de novo* compound by a large endogenous pool would give the reverse image. One should not draw conclusions from a single parameter of isotope incorporation without examining all the factors that could influence the final interpretation of results. These factors include, among others, total isotope incorporation, size of the endogenous pool, turnover, alternate routes of metabolism, and competing metabolic pathways. It is often wise to assess the overall metabolism of the substrate in the early stages of the experiment by radioassaying one or more obvious metabolic end-products such as tritiated water or $^{14}CO_2$. Such information can give insight into the amount of original substrate still available for metabolism during later experimental periods.

In this discussion I have attempted to highlight some of the general and special limitations inherent in using radioisotopes to study insect nutrition and metabolism. In addition to good radioisotope technique, sound, basic biological knowledge about the organism is one of the best guarantees that large mistakes will not go undetected. A vigorous pursuit of knowledge employing tracer technique will add much new information. I advocate, therefore, not a dismissal of radioisotope methodology, but its continued and proper application, so that each investigator may take full advantage of all the advantages the technique has to offer.

REFERENCES

BLACK, A. L., KLEIBER, M. and SMITH, A. H. (1952). Carbonate and fatty acids as precursors of amino acids in casein. *J. Biol. Chem.* **197**:365-370.

GILBERT, L. I. (1967). Lipid metabolism and function in insects. *Adv. Insect Physiol.* **4**:69-211.

JOINER, R. L. and LAMBREMONT, E. N. (1969). Hydrocarbon metabolism in insects: Oxidation of hexadecane-1-^{14}C in the boll weevil and the house fly. *Ann. Ent. Soc. Amer.* 62:891-894.

KAMEN, M. D. (1957). *Isotopic Tracers in Biology* Academic Press, New York, N. Y. 474 pp.

KASTING, R. and McGINNIS, A. J. (1966). Radioisotopes and the determination of nutrient requirements. *Ann. N.Y. Acad. Sci.* **139**:98-107.

KATZ, J. and WOOD, H. G. (1960). The use of glucose-^{14}C for the evaluation of the pathways of glucose metabolism. *J. Biol. Chem.* 235:2165-2177.

LAMBREMONT, E. N. (1971). Synthesis and metabolism of long chain fatty acids during late developmental stages of *Heliothis zea* (Lepidoptera: Noctuidae). *Insect Biochem.* 1:14-18.

LAMBREMONT, E. N. (1972). Lipid metabolism of insects: Interconversion of fatty acids and fatty alcohols. *Insect Biochem.* (In press).

LAMBREMONT, E. N. and BENNETT, A. F. (1966). Lipid biosynthesis in the boll weevil: Formation of the acetate precursor for lipid synthesis from glucose and related carbohydrates. *Can. J. Biochem.* **44**:1597-1606.

RABINOWITZ, J. L., CHASE, G. D., STRAUSS, H. D. and KLINE, G. G. (1960). Studies on isotope effects with carbonic anhydrase using ^{14}C sodium bicarbonate. *Atompraxis* **6**:432-435.

WEINBERGER, D. and PORTER, J. W. (1953). Incorporation of tritium oxide into growing *Chlorella pyrenoidasa* cells. *Science* **117**:636-638.

WINTERINGHAM, F. P. W. (1956). Labelled metabolic pools for studying quantitatively the biochemistry of toxic action. *Int. J. Appl. Radiat. Isotopes* **1**:57-65.

SELECTED BIBLIOGRAPHY

ARONOFF, S. 1956. *Techniques of Radiobiochemistry.* The Iowa State College Press, Ames. 228 pp.

BINGGELI, M. H. *Radioisotopes and Ionizing Radiations in Entomology.* Bibliographical Series No. 9 (1950-1960, 414 pp); No. 15 (1961-1963, 564 pp); No. 24 (1964-1965), 454 pp); No. 36 (1966-1967, 805 pp). International Atomic Energy Agency, Vienna (Austria).

BRANSOME, E. D. (Editor) 1970. *The Current Status of Liquid Scintillation Counting.* Grune and Stratton, New York, 394 pp.

CHASE, G. D. and RABINOWITZ, J. L. 1967. *Principles of Radioisotope Methodology.* 3rd Ed. Burgess Publ. Co. Minneapolis, Minn. 633 pp.

International Atomic Energy Agency, Vienna (Austria) and Food and Agriculture Organization of the United Nations, Rome (Italy). 1965. *Proceedings of FAO/IAEA Training Course on Use of Radioisotopes in Entomology,* Gainesville, Florida. 4 Oct. - 26 Nov., 1965. Gainesville, University of Florida, 171 pp.

International Atomic Energy Agency, Vienna (Austria) 1966. *Laboratory Training Manual on the Use of Isotopes and Radiation in Entomology.* Tech. Reports Series No. 61, IAEA, Vienna, 144 pp.

KAMEN, M. D. 1957. *Isotopic Tracers in Biology.* Academic Press, New York, 474 pp.
O'BRIEN, R. D. and WOLFE, L. S. 1964. *Radiation, Radioactivity and Insects.* Academic Press, New York, 211 pp.
ROTHCHILD, S. (Editor). *Advances in Tracer Methodology.* Vol. 1 (1963, 332 pp); Vol. 2 (1965, 319 pp); Vol. 3 (1966, 333 pp); Vol. 4 (1968, 293 pp). Plenum Press, New York.
Use of Radioisotopes in Animal Biology and the Medical Sciences. 1962. Proceedings of a Conference held in Mexico City, 21 November - 1 December, 1961. Vol. 1 (565 pp), Vol. 2 (328 pp). Academic Press, New York. Published for the IAEA.
WANG, C. H. and WILLIS, D. L. 1965. *Radiotracer Methodology in Biological Science.* Prentice-Hall, Englewood Cliffs, New Jersey. 382 pp.
WILSON, B. J. (Editor). 1966. *The Radiochemical Manual.* The Radiochemical Centre, Amersham. 327 pp.

INSECT METABOLISM OF DIETARY STEROLS AND ESSENTIAL FATTY ACIDS

G. M. Chippendale
Department of Entomology, University of Missouri
Columbia, Missouri 65201

INTRODUCTION

Research on insect lipid requirements gained prominence with the discovery that cholesterol and linoleic acid were essential nutrients for two species (Hobson, 1935; Fraenkel and Blewett, 1946). Subsequent nutritional research has shown that sterols are indispensable nutrients for all insects while several species also require dietary polyunsaturated fatty acids for normal development. Studies examining sterol requirements have primarily focused on the amounts needed for optimal growth and development, the nutritional effect of cholesterol analogues and the role of symbiotic microorganisms in sterol biosynthesis (Clayton, 1964). Several species of Lepidoptera and Orthoptera have now been shown to also require polyunsaturated fatty acids for optimum larval growth and normal wing development (Dadd, 1963). Results from many studies have shown that the fatty acid requirements of lepidopterous larvae generally fall into one of three categories: (1) those which require linoleate or linolenate for normal wing development, (2) those which have a specific linolenate requirement, or (3) those which do not require exogenous fatty acids (Chippendale and Reddy, 1972a).

While this nutritional research has been valuable in revealing the stringent lipid requirements of insects, it has only been able to provide limited information about essential lipid metabolism. However other investigators studying insect biochemistry have been able to provide some insights into sterol and fatty acid metabolism. For example, much has recently been learned about sterol interconversions, fatty acid biosynthesis, and lipid transport in the insect body (Gilbert, 1967). In fact the recent discovery of the lipid nature of the juvenile hormones and ecdysones has given new impetus to the study of lipid metabolism which may soon bring a clearer understanding of lipid biosynthesis, transport, storage, and function (Dahm and Roller, 1970; Robbins *et al.*, 1971). Investigators are also starting to combine nutritional and metabolic studies and are thereby beginning to establish relationships between essential dietary lipids and those synthesized *in vivo* (Beenakkers and Scheres, 1971).

The present article examines the metabolism of dietary lipids in insects. Much of the discussion is illustrated with data obtained from the southwestern corn borer, *Diatraea grandiosella*, and the Angoumois grain moth, *Sitotroga cerealella*. Lipid requirements and the metabolism of these two plant- and grain-feeding Lepidoptera are currently under investigation in my laboratory. Nutritional results have shown that while dietary sterols are essential for both species, plant sterols promote the larval growth rate of *Diatraea*. An examination of the fatty acid requirements of both species revealed that dietary linoleate or linolenate was required for normal wing formation of *Diatraea*, but not for *Sitotroga* (Chippendale and Reddy, 1972a; Chippendale, 1971a). Data on the lipid metabolism in their mudguts, hemolymph, and fat body have also been obtained during development and following experimental treatments (Chippendale, 1971b, 1971c; 1973). Although attention will focus on the metabolism of essential lipids, that of non-essential and synthesized lipids will also be considered. The fate of all three lipid categories is interrelated, and they share some common metabolic processes. The effect of lipids in initiating and maintaining feeding as well as their digestion, absorption, storage, transport, and ultimate function are briefly examined.

Effect of lipids on feeding behavior

The influence of dietary lipids on insect feeding behavior has not been sufficiently investigated. Yet ideally nutritional and behavioral studies should complement each other since the detrimental effect of a test lipid could be due to a negative influence on feeding behavior rather than an inability to digest or absorb the lipid or some other endogenous metabolic effect. Without adequate behavioral data it is therefore possible to incorrectly classify growth responses in nutritional experiments.

The few behavioral experiments to date have shown that free and esterified sterols, fatty acids, glycerides, and phospholipids directly influence insect feeding behavior. For example, β-sitosterol and stigmasterol served as feeding stimulants for the silkworm, *Bombyx mori*, (Ito *et al.*, 1964), phospholipids increased the feeding activity of two species of grasshoppers (Thorsteinson and Nayar, 1963), and free and esterified fatty acids served as arrestants for the confused flour beetle, *Tribolium confusum* (Starratt and Loschiavo, 1971).

Recent experiments using *Diatraea* larvae have shown that while the C_{29} plant sterols, β-sitosterol and stigmasterol, were feeding stimulants, cholesterol, 7-dehydrocholesterol, ergosterol and β-sitosterol acetate had neutral effects and cholesterol esters served as feeding deterrents (Chippendale and Reddy, 1972b). Further results showed that those diets containing the stimulatory plant sterols also yielded the highest growth rate

(Chippendale and Reddy, 1972a). Behavioral data obtained for *Sitotroga* have shown that the fatty acids, oleic and linoleic, and triglycerides, triolein and trilinolein, served as arrestants. Sterols, on the other hand, had an entirely neutral effect on feeding behavior (Chippendale and Mann, 1972). These studies reveal that both essential and non-essential dietary lipids play a central role in insect-host orientation and in feeding maintenance. The effect of lipids on feeding behavior should therefore not be overlooked in nutritional and metabolic studies.

Lipid digestion and absorption

Most studies to date have been limited to demonstrating the existence of lipid hydrolytic enzymes without investigating the dynamics of digestion and absorption. The hydrolysis of ingested lipids and the mechanisms by which they are absorbed have only been systematically examined in a few species. These studies have focused on the role of esterification in sterol absorption and the hydrolysis of triglycerides (Clayton, 1964; Treherne, 1962; 1967). Specific data about the digestion and absorption of other lipid classes, including phospholipids, have yet to be obtained (Gilbert, 1967). Recent analyses of midgut lipids of *Diatraea*, along with related nutritional results, have provided new data on the role of sterols and sterol esters in the absorptive processes. The digestion and absorption of free and esterified fatty acids were also investigated (Chippendale, 1971b). Lepidopterous larvae are particularly suitable for these studies because their midgut makes up the major portion of the alimentary canal and serves as the main site for digestive enzyme secretion and nutrient absorption.

The effect of sterol esterases on sterol digestion and absorption has been examined in several species. For example, an examination of cholesterol absorption in the German cockroach, *Blattella germanica*, revealed that free cholesterol was readily absorbed and that esterification occurred in the ventricular cells following absorption (Robbins *et al.*, 1961). A related study on the cockroach, *Eurycotis floridana*, (Clayton *et al.*, 1964) showed that while the lumen esterification of sterols was not essential it probably facilitated sterol absorption at low dietary concentrations. Cholesterol absorption was found to occur in the crop and gastric caeca. Studies with *Diatraea* further support the conclusion that free sterols are readily absorbed. Sterols were the major lipid class in the midguts of larvae fed cholesterol as their sole dietary lipid (Chippendale, 1971b). *Sitotroga* also appeared to readily absorb free sterols (Chippendale, 1971a). Lumen esterification of sterols does not therefore appear to be a prerequisite for their absorption.

Since triglycerides probably form the major portion of consumed lipids, their digestion and absorption have received special attention (Eisner, 1955; Treherne, 1958). Essential linoleic and linolenic acids are usually ingested in the form of triglycerides. A single ingested triglyceride molecule may contain both essential and non-essential fatty acids. The effect of lumen hydrolysis on the rate of triglyceride absorption and the sites of absorption have been examined. Results have shown that the midgut and caecal epithelium of the cockroach, *Periplaneta americana*, secreted a lipase which hydrolyzed triglycerides. Since some of this lipase was also transferred to the lumen of the foregut, hydrolysis occurred in both gut sections. However, the lipase only partially hydrolyzed the ingested triglycerides. The accumulation of fatty acids within the partially-hydrolyzed triglyceride aggregates appeared to prevent complete hydrolysis. Although intact triglycerides were absorbed they were taken up at a much lower rate than the partially hydrolyzed di-mono-glyceride and fatty acid mixtures. The polar fatty acids in the mixture apparently facilitated transport across the ventricular cell wall. The caeca and anterior portion of the midgut were the main sites of absorption of these partially-hydrolyzed glycerides.

Analytical and nutritional data obtained for *Diatraea* also suggest that dietary glycerides do not require extensive hydrolysis for their absorption (Chippendale, 1971b). The neutral lipids of the midguts of actively-feeding fifth stage larvae contained about 48% di- and tri-glycerides, while fatty acids could barely be detected. The virtual absence of fatty acids in the ventricular cells therefore suggests that glycerides are not extensively hydrolyzed in the lumen then reesterified following absorption. Nutritional results also showed that *Diatraea* developed normally on artificial diets containing cholesterol and methyl linoleate as the only lipid components (Chippendale and Reddy, 1972a). Midgut cells are therefore able to absorb fatty acids or their methyl esters in the absence of dietary glycerides. In this case, absorbed linoleic acid is probably rapidly incorporated into glycerides and sterol esters in the midgut.

Lipid metabolism in and release from ventricular cells

Only limited information is available about lipid metabolism in ventricular cells or in what form lipids are released into the hemolymph (Gilbert, 1967). After their absorption lipids may be transported directly to the hemolymph, held in storage for later release, or serve as structural components of the ventricular cells. Efforts to discover their precise fate are complicated by several factors. For example, a lipid class may be absorbed in a different form from the one in which it is released into the hemolymph. Furthermore, since ventricular cells have been found to synthesize many lipid classes from simple precursors (Bade, 1964), their total lipid content in feeding insects is made up of both *in vivo* synthesized and absorbed lipids which may be either structural or transient components.

Analyses of the lipid composition of larval midgut cells and hemolymph have provided information about ventricular lipid metabolism in *Diatraea*

(Chippendale, 1971b,c). Free sterols were found to form a major lipid class in all larval midguts examined (15-25% neutral lipids) and one of their functions may be to serve as an integral structural component of the cell. However, since *Diatraea*, like other insects, cannot synthesize the sterol nucleus, it is totally dependent upon the efficient transfer of absorbed sterols to sites of extraintestinal utilization. Results suggested that free sterols themselves were released into the hemolymph, where they form a major lipid class. Sterol esters, on the other hand, formed only a minor component of the hemolymph. The midgut titer of the sterol esters fluctuated sharply while the sterol titer remained fairly constant. The highest sterol ester concentration was found in midguts of actively-feeding fifth stage larvae (29% of neutral lipids) and the lowest was in midguts of larvae fed oil-free diets (<1% neutral lipids). Sterol esters, therefore, are probably not structural components of the midgut cells. Since sterol esters do not appear to be the preferred molecular form for sterol absorption and form only a minor component of the hemolymph, they are probably not involved in sterol transport. The titer of sterol esters in the midgut was found to be directly proportional to larval feeding activity. It may therefore be concluded that sterol esters serve mainly as storage molecules in midgut cells during periods of excessive nutrient intake. Dutky *et al.* (1963) suggested that sterol esters also served as storage molecules for embryogenesis in the egg of the housefly, *Musca domestica*. They found that most of the sterol ester reserves were converted to free sterols during embryonic development.

Glyceride and fatty acid metabolism in midgut cells of *Diatraea* larvae was also examined during the active-feeding period of the fifth instar (Chippendale, 1971b). Results showed that fatty acids were only a minor component (<1% neutral lipids), while di- and tri-glycerides were present in higher concentrations and comprised 6% and 42%, respectively, of the midgut neutral lipids. No monoglycerides were detected. Lipid analyses of larval hemolymph have shown that most of the fatty acids occurred as 1,2 diglycerides and only small amounts of free fatty acids and triglycerides could be detected (Chippendale, 1971c). These data suggest that no significant amounts of free fatty acids are released from the midgut cells into the hemolymph. It seems more likely that midgut di- and tri-glycerides are involved in this phase of fatty acid transport. Both of these lipid classes were found in the midgut cells and diglycerides were always present in lower titers than triglycerides. Midgut diglycerides may therefore represent partially hydrolyzed triglycerides which are selectively transferred from the midgut cells to form the major glyceride class in the hemolymph.

Lipid metabolism in and release from fat body cells

A principal function of the insect fat body is to synthesize polar and neutral lipids and to store both synthesized and absorbed lipid molecules

(Kilby, 1965). The fat body of mature holometabolous larvae, for example, is packed with lipids, protein, and glycogen which form the reserves for adult differentiation, survival, and reproduction (Butterworth *et al.*, 1965). Although little is known about the mechanism of lipid uptake by fat body cells, several investigators have examined their lipid biosynthetic capabilities and specific lipid composition. Research has focused on neutral rather than phospholipid metabolism. It has been clearly shown that fat body homogenates, in the presence of appropriate co-factors, rapidly incorporate acetate into fatty acids and glycerides (Zebe and McShan, 1959; Tietz, 1961). Recent results have demonstrated that the rate of fatty acid synthesis in *B. mori* larvae is controlled by the dietary intake of fatty acids and carbohydrates (Horie and Nakasone, 1971).

Several studies have shown that triglycerides form the major fat body lipid class. They are likely to contain essential, non-essential, and *in vivo* synthesized fatty acids as molecular constituents. Investigators have shown that triglycerides form 95% of the total fat body lipids in diapause pupae of the silkmoth, *Hyalophora cecropia*, and 90% of the neutral lipids in the fat body of *P. americana* (Chino and Gilbert, 1965; Cook and Eddington, 1967). Quantitative analyses of fat body lipids of *Diatraea* and *Sitotroga* agreed with these findings and also provided information on minor neutral lipid components (Chippendale 1971c; 1973). Lipids comprised about 75% of the fat body dry weight of mature larvae of both species. Thin layer chromatographic analyses showed that these fat bodies contained the following neutral lipid classes (%): in *Diatraea*, triglycerides 90, sterols 5, 1,2-diglycerides 4, sterol esters 1; in *Sitotroga*, triglycerides 85, sterols 8, sterol esters 3, 1,2-diglycerides and hydrocarbons 2. These data show that while triglycerides formed the major lipid class no free fatty acids and monoglycerides were detected. Furthermore, sterols were only a minor components and were primarly present in the free form. Sterol esters therefore do not appear to serve as sterol storage molecules in the fat body. Since triglycerides are the principal reserve, they or their hydrolyzed derivatives most likely represent the molecular form in which lipids are released from the fat body.

The method by which neutral lipids are released from fat body cells into the hemolymph has recently been investigated. Most success has been achieved by examining the dynamics of the process in *in vitro* fat body - hemolymph incubation systems. Tietz (1967) and Beenakkers and Gilbert (1968) worked with the locust, *Locusta migratoria*, and *H. cecropia* and concluded that lipids are released from the fat body mainly as diglycerides. On the other hand Wlodawer *et al.* (1966) found that lipids were released from the larval fat body of the wax moth, *Galleria mellonella*, primarily as free fatty acids which were then esterified and exchanged with fatty acids of hemolymph glycerides. Cook and Eddington (1967) proposed that both triglycerides and fatty acids were released from the fat body of *P. americana*, suggesting that this species has a fairly nonspecific lipid release mechanism. More recently Chang and Friedman (1971) conducted a developmental study of lipid release from the fat body of the tobacco hornworm, *Manduca sexta*. They found that while little lipid release occurred from larval fat body, the pupal or pharate adult fat body selectively released free fatty acids into an *in vitro* incubation medium. Clearly, the dynamics of lipid release from the fat body need to be studied in many more species.

While lipid analyses on *Diatraea* tissues were not specifically aimed at this problem, they do provide some additional clues about lipid release from the fat body. A low titer of 1,2-diglycerides was found in the lipids of larval and pupal fat bodies. These diglycerides may represent partially hydrolyzed triglycerides *en route* to the hemolymph to form the major glyceride class. Since no free fatty acids were detected in the fat body, they are probably not involved in lipid release mechanisms. Triglycerides may also be released from the fat body of *Diatraea*, but they were only present in small amounts in the hemolymph (Chippendale, 1971c).

Lipid composition of the hemolymph

Lipid analyses of the hemolymph are especially valuable for providing information on lipid metabolism since most absorbed and synthesized lipids must be transported in the hemolymph from the alimentary canal or fat body to sites of utilization. Recent analyses of hemolymph lipids have begun to reveal in what molecular form lipids are transported. Although metamorphic changes in hemolymph lipid concentration have been detected, the lipid titer in the hemolymph generally remains low. For example, the hemolymph of mature larvae of *Diatraea* contained only 10 mg lipid while that of *Sitotroga* contained 17 mg lipid per ml. (Chippendale, 1971c; 1973).

The few analyses to date suggest that glycerides and sterols form the principal lipid components of the hemolymph. Essential fatty acids are therefore present in the hemolymph predominantly in the esterified form. Wlodawer *et al.* (1966) worked with the larval hemolymph of *G. mellonella* and found phospholipids as well as seven neutral lipid classes. However glycerides formed the major component and comprised 50% of the total lipids. Chino and Gilbert (1965) found that di- and triglycerides made up 84% of the hemolymph lipids of diapause pupae of *H. cecropia*. In related studies sterols have been shown to comprise about 8% of the hemolymph lipids of the house cricket, *Acheta domesticus*, and queens of the termite, *Macrotermes natalensis* (Wang and Patton, 1969; Cmelik, 1969).

Studies of the hemolymph lipids of *Diatraea* and *Sitotroga* indicate that variations in the relative titers of hemolymph lipid classes exist even in identical developmental stages of related species. While the hemolymph neutral lipids of mature larvae of *Diatraea* contained triglycerides (7%)

diglycerides (41%) and sterols (33%), those of *Sitotroga* contained triglycerides (38%), diglycerides (15%), and sterols (11%) (Chippendale, 1971c; 1973). These results suggest that the molecular forms in which lipids are released into and transported within the hemolymph vary among species.

The mode of lipid transport in insect hemolymph has received some attention. While hemolymph lipids may occur free, conjugated to protein, or in both states depending on the developmental stage, lipoproteins are probably a principal agent of lipid transport. Several lipoproteins have been isolated and purified from the pupal hemolymph of the silkmoths *Philosamia cynthia* and *H.scecropia* (Chino *et al.*, 1968; Thomas and Gilbert, 1968). Analyses of the lipid portions of the molecules showed that glycerides, phospholipids, sterols, and carotenoids were conjugated to the carrier proteins. It is likely that these lipoproteins, either as intact molecules or only their lipid portion, are transferred from the hemolymph to sites of utilization. Major sites of utilization include muscles and developing oocytes. Evidence exists that oocytes of *H. cecropia* take up hemolymph lipoproteins by pinocytosis (Hausman *et al.*, 1971).

The mode of transport of ecdysones in insect hemolymph from their site of synthesis, believed to be the ring and prothoracic glands, to target cells has also been investigated. Since no evidence so far has implicated carrier proteins in this process, ecdysones may be transported in the free or esterified form or as glycosides (Chino *et al.*, 1970; Willig *et al.*, 1971).

Function of essential lipids

More information has been obtained about the function of dietary sterols than is yet available about essential fatty acids. Currently available data show that sterols function in two capacities (1) as structural components of insect membranes, especially the myelin sheath of nerves and (2) as precursors of the molting hormones, ecdysones (Clayton, 1964). Sterols apparently are integral components of insect membranes even though they are only minor components compared with the associated phospholipids, proteins, and carbohydrates (Singer and Nicolson, 1972). The membrane location of the majority of tissue cholesterol has been inferred from its association with particular cellular fractions following ultracentrifugation and its low turnover rate (Gilbert, 1967). For example, Lasser and Clayton (1966) studied the intracellular distribution of sterols in several tissues of *E. floridana*. From analyses of the sterol contents of nuclear, mitochondrial, and microsomal fractions they suggested that sterols were components of common repeating structural units of subcellular membranes. The precise role of sterols in membrane organization and regulation of permeability, however, has yet to be determined.

Recent investigations have focused on the *in vivo* conversion of ingested sterols. For example, several plant-feeding insects have been found to transform absorbed β-sitosterol and stigmasterol into cholesterol (Ritter and Wientjens, 1967; Thompson *et al.*, 1972). It is now clear that cholesterol is converted to humorally-active molecules within the insect body. Several investigations have shown that insects convert labelled cholesterol to labelled ecdysones. Larvae of the blowfly, *Calliphora erythrocephala,* for instance, converted injected ^{14}C-cholesterol into ^{14}C-ecdysone and 20-hydroxy-ecdysone. Further study revealed that isolated brain-ring gland complexes converted cholesterol into products which co-chromatographed with these two ecdysones. The ring gland is presumably the principal site of this transformation in fly larvae (Willig *et al.*, 1971).

The function of essential polyunsaturated fatty acids in insects is not yet entirely clear. Recent findings suggest that insects which require linoleate or linolenate are unable even to synthesize suboptimal levels of these chemicals (Lambremont, 1971). Since these indispensable lipids are required in relatively small amounts they do not form a metabolic reserve for normal energy demands. Insects which require polyunsaturated fatty acids can efficiently synthesize others which serve as energy reserves for development, diapause, flight, and reproduction. Essential polyunsaturated fatty acids may have a structural function or serve as precursors for humoral agents which have yet to be discovered. They may then be incorporated into more complex lipid molecules such as structural phospholipids. Much recent biochemical research has focused on the phospholipid and protein content of membranes. Evidence now exists that globular proteins associated with a phospholipid bilayer provide membranes with their typical permeability properties (Singer and Nicolson, 1972). Furthermore, it has been suggested that membrane permeability may in part be regulated by the degree of unsaturation of the fatty acid components of the phospholipids (Wilson and Fox, 1971). Essential linoleic and linolenic acids may therefore be preferentially incorporated into membrane phospholipids of insects.

To date the most thorough study of the metabolism of essential fatty acids has been conducted on the cabbage looper, *Trichoplusia ni*. Nutritional results showed that while either linoleate or linolenate promoted larval growth, linolenate was essential for normal wing development (Chippendale *et al.*, 1964). Later experiments showed that the larvae could not synthesize linoleate or linolenate from ^{14}C-acetate. Radioactive acetate, however, was rapidly incorporated into other tissue fatty acids, and palmitic and oleic acids were the major components (Nelson and Sukkestad, 1968). Subsequently Grau and Terriere (1971) examined the fatty acid composition of the major lipid classes during growth and metamorphosis of *T. ni.* They found that linoleate and linolenate acids were preferentially incorporated into phospholipids. Their results showed that these polyunsaturated fatty acids made up 64% of the fatty acids of phospholipids and only 22% of those of triglycerides. A structural role of linoleate and linolenate in membrane phospholipids were postulated. Since dietary deficiencies of essential fatty acids usually result in deformed wings, linoleic and linolenic acids may be essential components of phospholipids in membranes of choracic hypodermal cells of pharate adults.

CONCLUSION

Insect metabolism of dietary lipids was examined. Special attention was placed on these processes in plant- and grain-feeding Lepidoptera since nutritional and analytical data obtained for the southwestern corn borer, *Diatraea grandiosella*, and the Angoumois grain moth, *Sitotroga cerealella*, were used to illustrate much of the discussion. The fate of dietary sterols and essential linoleate and linolenate was considered in relation to that of other non-essential and *in vivo* synthesized lipids. The current status of our knowledge of the effect of dietary lipids on feeding behavior and mechanisms of lipid digestion and absorption by ventricular cells was briefly reviewed. The storage of absorbed lipids in ventricular and fat body cells, their transport in the hemolymph, and ultimate function were also examined.

SUMMARY

The present review shows that investigators are beginning to look beyond the more descriptive aspects of insect lipid requirements and are examining the role of essential lipids in life processes. Although only a start has been made in understanding the metabolic fate of essential lipids, recent results indicate that the use of modern biochemical techniques will soon uncover many new facets of their metabolism. Fruitful areas for future research include lipid absorption and metabolism within ventricular cells, the mechanism of transport to sites of storage and utilization, the site of synthesis of ecdysones from cholesterol, the relationship between essential and *in vivo* synthesized fatty acids, and the incorporation of essential fatty acids into phospholipids. Once this information becomes available for many insect species, unifying principles of insect lipid metabolism may emerge.

Acknowledgements.—This work is a contribution from the Missouri Agricultural Experiment Station, as journal series No. 6372.

REFERENCES

BADE, M. L. (1964). Biosynthesis of fatty acids in the roach *Eurycotis floridana*. *J. Insect Physiol.* **10**:333-341.

BEENAKKERS, A. M. T., and GILBERT, L. I. (1968). The fatty acid composition of fat body and haemolymph lipids in *Hyalophora cecropia* and its relation to lipid release. *J. Insect Physiol.* **14**:481-494.

BEENAKKERS, A. M. T., and SCHERES, J. M. J. C. (1971). Dietary lipids and lipid composition of the faty body of *Locusta migratoria. Insect Biochem.* **1**:125-129

BUTTERWORTH, F. M., BODENSTEIN, D., and KING, R. D. (1965). Adipose tissue of *Drosophila melanogaster.* I. An experimental study of larval fat body. *J. Exp. Zool.* **158**:141-154.

CHANG, F., and FRIEDMAN, S. (1971). A developmental analysis of the uptake and release of lipids by the fat-body of the tobacco hornworm, *Manduca sexta. Insect Biochem.* **1**:63-80.

CHINO, H., and GILBERT, L. I. (1965). Lipid release and transport in insects. *Biochem. Biophys. Acta* **98**:94-110.

CHINO, H., GILBERT, L. I., SIDDALL, J. B., and HAFFERL, W. (1970). Studies on ecdysone transport in insect haemolymph. *J. Insect Physiol.* **16**:2033-2040.

CHINO, H., MURAKAMI, S., and HARASHIMA, K. (1968). Diglyceride-carrying lipo-proteins in insect hemolymph, isolation, purification and properties. *Biochem. Biophys. Acta* **176**:1-26.

CHIPPENDALE, G. M. (1971a). Lipid requirements of the Angoumois grain moth, *Sitotroga cerealella. J. Insect Physiol.* **17**:2169-2177.

CHIPPENDALE, G. M. (1971b). Metamorphic changes in midgut lipids of the southwestern corn borer, *Diatraea grandiosella. Insect Biochem.* **1**:283-292.

CHIPPENDALE, G. M. (1971c). Fat body and haemolymph lipids of the southwestern corn borer, *Diatraea grandiosella*, during metamorphosis. *Insect Biochem.* **1**:39-46.

CHIPPENDALE, G. M. (1973). Metabolic reserves of larvae and pupae of the Angoumois grain moth, *Sitotroga cerealella,* manuscript in preparation.

CHIPPENDALE, G. M. BECK, S. D., and STRONG, F. M. (1964). Methyl linolenate as an essential nutrient for the cabbage looper, *Trichoplusia ni* (Hubner). *Nature* **204**:710-711.

CHIPPENDALE, G. M., and MANN, R. A. (1972). Feeding behaviour of Angoumois grain moth larvae. *J. Insect Physiol.* **18**:87-94.

CHIPPENDALE, G. M., and REDDY, G. P. V. (1972a). Polyunsaturated fatty acid and sterol requirements of the southwestern corn borer, *Diatraea grandiosella. J. Insect Physiol.* **18**:305-316.

CHIPPENDALE, G. M., and REDDY, G. P. V. (1972b). Dietary sterols: role in larval feeding behaviour of the southwestern corn borer, *Diatraea grandiosella. Experientia* **28**:485.

CLAYTON, R. B. (1964). The utilization of sterols by insects. *J. Lipid Res.* **5**:3-19.

CLAYTON, R. B., HINKLE, P. C., SMITH, D. A., and EDWARDS, A. M. (1964). The intestinal absorption of cholesterol, its esters and some related sterols and analogues in the roach, *Eurycotis floridana. Comp. Biochem. Physiol.* **11**:333-350.

CMELIK, S. H. W. (1969). Composition of the neutral lipids from termite queens. *J. Insect Physiol.* **15**:1481-1487.

COOK, B. J., and EDDINGTON, L. C. (1967). The release of triglycerides and free fatty acids from the fat body of the cockroach, *Periplaneta americana. J. Insect Physiol.* **13**:1361-1372.

DADD, R. H. (1963). Feeding behaviour and nutrition in grasshoppers and locusts. *Adv. Insect Physiol.* **1**:47-109.

DAHM, K. H., and RÖLLER, H. (1970). The juvenile hormone of the giant silkmoth, *Hyalophora gloveri* (Strecker). *Life Sciences* **9**:1397-1400.

DUTKY, R. C., ROBBINS, W. E., KAPLANIS, J. N., and SHORTINO, T. J. (1963). The sterol esters of housefly eggs. *Comp. Biochem. Physiol.* **9**:251-255.

EISNER, T. (1955). The digestion and absorption of fats in the foregut of the cockroach,

Periplaneta americana (L.). *J. Exp. Zool.* 130:159-176.
FRAENKEL, G., and BLEWETT, M. (1946). Linoleic acid, vitamin E and other fat soluble substances in the nutrition of certain insects (*Ephestia kuehniella, E. elutella, E. cautella* and *Plodia interpunctella (Lep.)*). *J. Exp. Biol.* 22:172-190.
GILBERT, L. I. (1967). Lipid metabolism and function in insects. *Adv. Insect Physiol.* 4:69-211.
GRAU, P. A., and TERRIERE, L. C. (1971). Fatty acid profile of the cabbage looper, *Trichoplusia ni*, and the effect of diet and rearing conditions. *J. Insect Physiol.* 17:1637-1649.
HAUSMAN, S. J., ANDERSON, L. M., and TELFER, W. H. (1971). The dependence of cecropia yolk formation *in vitro* on specific blood proteins. *J. Cell Biol.* 48:303-313.
HOBSON, R. P. (1935). On a fat-soluble growth factor required by blow-fly larvae. II. Identity of the growth factor with cholesterol. *Biochem. J.* 29:2023-2026.
HORIE, Y., and NAKASONE, S. (1971). Effects of the levels of fatty acids and carbohydrates in the diet on the biosynthesis of fatty acids in larvae of the silkworm, *Bombyx mori. J. Insect Physiol.* 17:1441-1450.
ITO, T., KAWASHIMA, K., NAKAHARA, M., NAKANISHI, K., and TERAHARA, A. (1964). Effects of sterols on feeding and nutrition of the silkworm, *Bombyx mori* L. *J. Insect Physiol.* 10:225-238.
KILBY, B. A. (1965). Intermediary metabolism and the insect fat body. *Biochem. Soc. Symp.* No. 25:39-48.
LAMBREMONT, E. N. (1971). Synthesis and metabolism of long-chain fatty acids during late developmental stages of *Heliothis zea* (Lepidoptera: Noctuidae). *Insect Biochem.* 1:14-18.
LASSER, N. L., and CLAYTON, R. B. (1966). The intracellular distribution of sterols in *Eurycotis floridana* and its possible relation to subcellular membrane structures. *J. Lipid Res.* 7:413-426.
NELSON, D. R. and SUKKESTAD, D. R. (1968). Fatty acid composition of the diet and larvae and biosynthesis of fatty acids from ^{14}C-acetate in the cabbage looper, *Trichoplusia ni. J. Insect Physiol.* 14:293-300.
RITTER, F. J., and WIENTJENS, W. H. J. M. (1967). Sterol metabolism in insects. *TNO-Nieuws* (Delft, Netherlands) 22:381-392.
ROBBINS, W. E., KAPLANIS, J. N., MONROE, R. E., and TABOR, L. A. (1961). The utilization of dietary cholesterol by German cockroaches. *Ann. Ent. Soc. Amer.* 54:165-168.
ROBBINS, W. E., KAPLANIS, J. N., SVOBODA, J. A., and THOMPSON, M. J. (1971). Steroid metabolism in insects. *Ann. Rev. Ent.* 16:53-72.
SINGER, S. J., and NICOLSON, G. L. (1972). The fluid mosaic model of the structure of cell membranes. *Science* 175:720-731.
STARRATT, A. N. and LOSCHIAVO, S. R. (1971). Aggregation of the confused flour beetle, *Tribolium confusum*, elicited by mycelial constituents of *Nigrospora sphaerica. J. Insect Physiol.* 17:407-414.
THORSTEINSON, A. J. and NAYAR, J. K. (1963). Plant phosphlipids as feeding stimulants for grasshoppers. *Can. J. Zool.* 41:931-935.
TIETZ, A. (1961). Fat synthesis in cell-free preparations of the locust fat body. *J. Lipid Res.* 2:182-187.
TIETZ, A. (1967). Fat transport in the locust: the role of diglycerides. *European J. Biochem.* 2:236-242.
THOMAS, K. K., and GILBERT, L. I. (1968). Isolation and characterization of the

hemolymph lipoproteins of the American silkmoth, *Hyalophora cecropia*. *Arch. Biochem. Biophys.* 127:512-521.

THOMPSON, M. J., SVOBODA, J. A. KAPLANIS, J. N., and ROBBINS, W. E. (1972). Metabolic pathways of steroids in insects. *Proc. Royal Soc. London B.* **180**:203-221.

TREHERNE, J. E. (1958). The digestion and absorption of tripalmitin in the cockroach, *Periplaneta americana* L. *J. Exp. Biol.* 35:862-870.

TREHERNE, J. E. (1962). The physiology of absorption from the alimentary canal in insects. In *Viewpoints in Biology* (ed. Carthy, J. D. and Duddington, C. L.) 1:201-241. Butterworths, London.

TREHERNE, J. E. (1967). Gut absorption. *Ann. Rev. Ent.* 12:43-58.

WANG, C. M., and PATTON, R. L. (1969). Lipids in the haemolymph of the cricket, *Acheta domesticus*. *J. Insect Physiol.* 15:851-860.

WILLIG, A., REES, H. H., and GOODWIN, T. W. (1971). Biosynthesis of insect moulting hormones in isolated ring glands and whole larvae of *Calliphora*. *J. Insect Physiol.* 17:2317-2326.

WILSON, G., and FOX, C. F. (1971). Biogenesis of microbial transport system: evidence for coupled incorporation of newly synthesized lipids and proteins into membranes. *J. Mol. Biol.* 55:49-60.

WLODAWER, P., LAGWINSKA, E., and BARANSKA, J. (1966). Esterification of fatty acids in the wax moth haemolymph and its possible role in lipid transport. *J. Insect Physiol.* 12:547-560.

ZEBE, E. C., and McSHAN, W. H. (1959). Incorporation of ^{14}C acetate into long chain fatty acids by the fat body of *Prodenia eridania* (Lep.). *Biochem. Biophys. Acta* 31:513-518.

SOME HOST-PARASITE RELATIONSHIPS IN FATTY ACID METABOLISM

J. S. Barlow
Pestology Centre, Department of Biological Sciences
Simon Fraser University
Burnaby 2, B. C.

The recent accumulation of abundant evidence of the danger inherent in distributing large amounts of toxic materials for the control of insect pests is prompting a great resurgence of interest in biological control or at least in integrated biological and chemical control procedures. A great deal of knowledge of the nutritional and biochemical characteristics of both host and parasite and of the interaction between them is needed to assure maximum success of such efforts. Indeed, in reviewing the present state of knowledge concerning the physiology and biochemistry of endoparasitic hymenoptera, Fisher (1971) has pointed out, "Proposals to use parasitic insects more widely in the biological control of pests will have to take account of the full physiology of the host-parasitoid relationship, the cellular defence reactions of insects towards new or alien parasitoids as well as the physiological ecology of both species and the integration of life cycles."

This paper deals chiefly with the relationship between two hymenopterous parasites, *Exeristes comstockii* (Cresson) and *Itoplectus conquisitor* (Say) and several host insects with respect to their fatty acid nutrition and metabolism.

The study of this relationship began in an attempt to determine some factor that restricted *E. comstockii* to lepidopterous hosts. This insect, found as an external parasitoid of the pine shoot moth, *Ryaconia bouliana* (Schiff), was reared in the laboratory on the wax moth, *Galleria mellonella* (L.). By transferring newly hatched larvae from *G. mellonella*, some growth on a hymonopteron, the larch sawfly *Neodiprion sertifer* (Geoffroy) was obtained. The parasite grew rapidly on *N. sertifer* during the first instar but growth finally stopped about halfway through the second. In a search for a possible reason for the arrest of growth, the fatty acids of the parasite larvae and the hosts upon which they had been reared were examined. Figure 1, which is also figure 1 of the ensuing paper by Bracken and Barlow (1967), shows the fatty acid distribution in these insects.

The first major peak in the chromatogram of the fatty acids in the wax moth is palmitic acid. This is followed by a very small peak for palmitoleic acid; stearic acid is present in a very small amount which is hardly detectable on this chromatogram and the following major peak is that of oleic acid. The last two peaks shown are linoleic and linolenic acids respectively. The chromatogram of the fatty acids in the sawfly shows a small quantity of

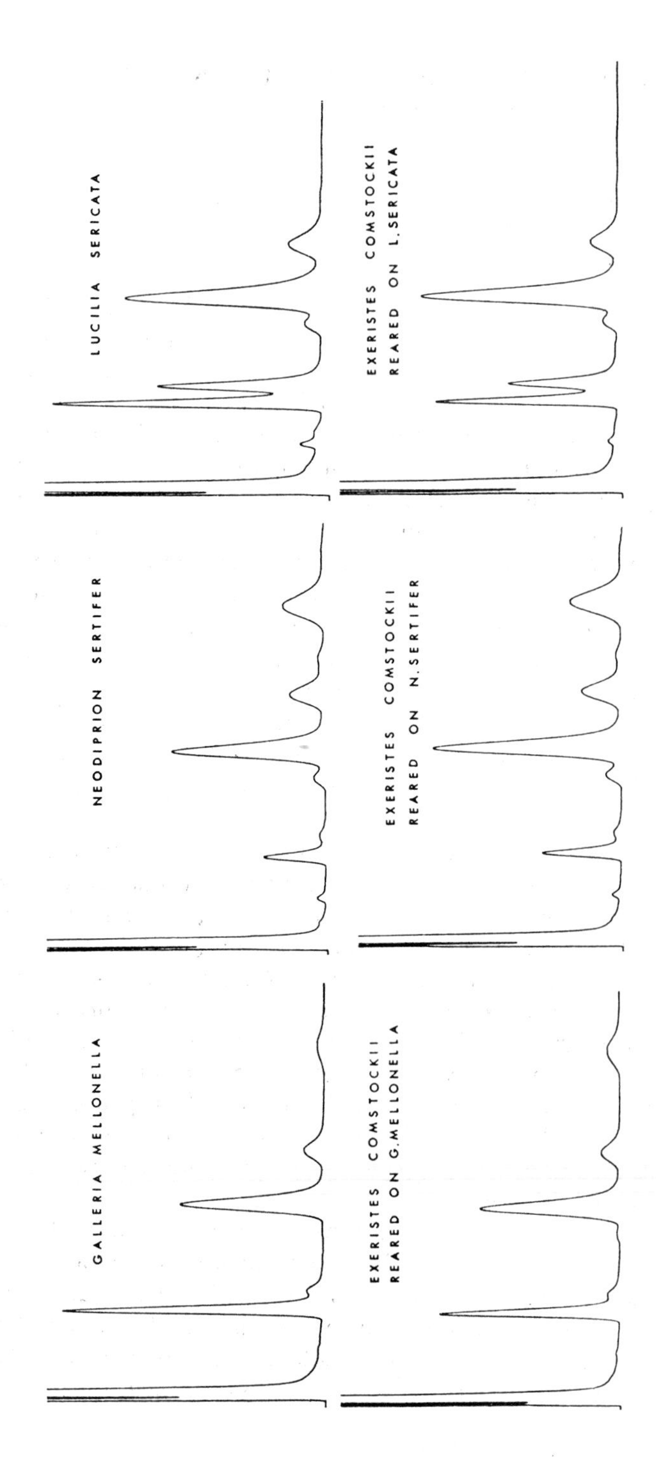

Figure 1. Gas chromatographs of the fatty acids in *Exercistes comstockii* larvae compared with those in three host species.

fourteen carbon fatty acids prior to the palmitic acid peak which in this case is much lower than in the wax moth. Small amounts of palmitoleic and stearic acid and a major peak produced by oleic acid are again evident in this chromatogram and the obvious subsequent peaks are linoleic and linolenic. It is quite easy to see that the fatty acids of the parasite, shown below its hosts, are considerably influenced by the fatty acids in the host and indeed the chromatograms suggest that the parasite mimics the fatty acid composition of the host. For a time it was thought that this phenomenon might explain the failure of the parasite to complete its growth on the hymenopterous host and a search was made for a host whose fatty acid pattern departed further from that of *G. mellonella* than did the fatty acid pattern of *N. sertifer*. Such was the blow fly, *Lucilia sericata* (Meigen). The chromatogram (Fig. 1) of its fatty acids shows some fourteen carbon fatty acids and an amount of palmitic acid similar to that found in the wax moth. However, palmitoleic acid is present in much higher amounts than in either of the previous insects. Stearic, oleic and linoleic acids are evident as they were in the previous two chromatograms but there is no linoleic acid because these insects were reared axenically on chemically defined diets lacking this acid and the insects do not synthesize polyunsaturates.

The first attempts to raise the parasite on this insect were not successful but fortunately Bucher and Williams (1967) had discovered that the parasites were susceptible to bacterial infections in the host. Therefore, parasite larvae were transferred to *L. sericata* reared axenically on a chemically defined diet. The *L. sericata* larvae deteriorated quickly and had to be changed several times, but the parasites grew rapidly and a few pupated and formed adults. The successful growth of the parasite on the dipterous larvae discouraged the belief that fatty acids were involved in the host specificity of the parasite, but fat analysis showed that again the fatty acids in the parasite were strongly influenced by those of the host. Moreover, the fatty acid pattern in the parasite larvae was largely unchanged during pupation and retained the characteristically high palmitoleic acid content in the newly emerged adult even though these had not fed for some time.

Work with three species of dipterous insects *Pseudosarcophaga affinis, Musca domestica* and *Lucilia sericata* reared axenically on chemically defined diets had shown that the lipids of these insects retained their characteristic fatty acid pattern over a wide variety of dietary fat including diets which contained no fatty acids and diets which contained oleic acid only. Barlow, 1965; 1966a; 1966b). Similarly, Table 1 (below) shows the result of a recent experiment in which we have added large quantities of palmitoleic acid to wax moth diets. Although an increase in palmitoleic acid in the larvae is quite noticeable, it is by no means as great as that observed in the parasite in Fig. 1. Similarly adding a large quantity of stearic acid to the wax moth diet produced little or no change in this acid in wax moth larvae as shown by

Table 2. Apparently the fatty acids of the parasite are much more susceptible to changes in the diet than are fatty acids in the dipterons or in the wax moth.

Table 1. The fatty acid composition of wax moth larvae reared on palmitoleic acid supplemented diet.

Fatty acid	Per Cent Composition					
	Control diet			Palmitoleate supplemented diet		
	Diet	Larvae	Frass	Diet	Larvae	Frass
C14:0	0.2	0.5	0.7	0.3	Tr	0.2
C16:0	15.0	36.1	15.7	13.4	36.4	14.3
C16:1	Tr	2.7	2.4	23.6	6.4	21.4
C18:0	1.9	3.5	1.6	2.2	2.1	1.0
C18:1	37.7	47.9	51.1	29.1	45.8	41.7
C18:2	44.0	9.3	28.4	30.6	9.4	21.4
C18:3	1.3	-	-	0.8	-	-
C20:1	Tr	-	-	Tr	-	-

Table 2. The fatty acid composition of wax moth larvae reared on stearic acid supplemented diet.

Fatty acid	Per Cent Composition					
	Control diet			Stearate supplemented diet		
	Diet	Larvae	Frass	Diet	Larvae	Frass
C14:0	0.2	0.5	0.7	0.3	Tr	0.7
C16:0	15.0	36.1	15.7	12.7	34.4	15.3
C16:1	Tr	2.7	2.4	Tr	3.2	2.3
C18:0	1.9	3.5	1.6	34.1	2.2	28.8
C18:1	37.7	47.9	51.1	25.6	50.9	39.6
C18:2	44.0	9.3	28.4	26.5	9.4	13.4
C18:3	1.3	-	-	0.7	-	-
C20:1	Tr	-	-	Tr	-	-

At this stage, two main problems were posed: what is the generality of the tendency to reproduce in such detail the fatty acids pattern of the host? and what is the mechanism by which the similarity in fat composition between the parasite and its host is achieved?

To effectively answer the first question, one should feed each parasite on at least two hosts with distinctive fatty acid patterns. Fortunately, Bracken and Harris (1969) had shown that several Lepidoptera including the corn borer, *Ostrinia nubilalis* (Hubner), contained large amounts of palmitoleic acid in their fats and there was an available supply of the ichneumonid parasite *Itoplectis conquisitor* (Say) which would attack both the *O. nubilalis* and *G. mellonella.* Analysis of the fats of *I. conquisitor* on these two hosts showed that this parasite too adopted a fatty acid composition similar to that of its host. The chromatograms are shown in Fig. 2 (Thompson and Barlow, 1970). A more quantitative view of the similarities and differences between the parasite and its hosts can be obtained from Table 3. *I. conquisitor* reared on *O. nubilalis* differs very little from its host but the difference between parasite and host is somewhat greater when *I. conquisitor* is raised on *G. mellonella.*

Table 3. The per cent composition of the individual fatty acids of *I. conquisitor* reared on different hosts and that of corresponding hosts.

	Per cent composition of individual fatty acids			
Fatty acid[a]	*G. mellonella*	*I. conquisitor* reared on *G. mellonella*	*O. nubilalis*	*I. conquisitor* reared on *O. nubilalis*
C12's	0.2	0.1	0	0
C14:0	trace	trace	0.6	0.6
C16:0	29.7	38.4	22.5	20.6
C16:1	2.5	1.4	43.6	42.0
C18:0	1.2	1.2	1.0	1.0
C18:1	51.5	44.2	22.6	25.6
C18:2	13.5	10.9	9.6	10.1
C20:1	1.3	3.7	0	0

[a]The first number represents the number of carbon atoms, the second, the number of double bonds.

Among other differences, the parasite has some 9% more palmitic acid and 7% less oleic acid than its host. The phospholipids were separated from the neutral lipids of *I. conquisitor* and analysis gave the chromatograms in Fig. 3 showing a considerable change in phospholipid fatty acids. Table 4 shows the detail of the analyses. Regardless of the host, the phospholipids

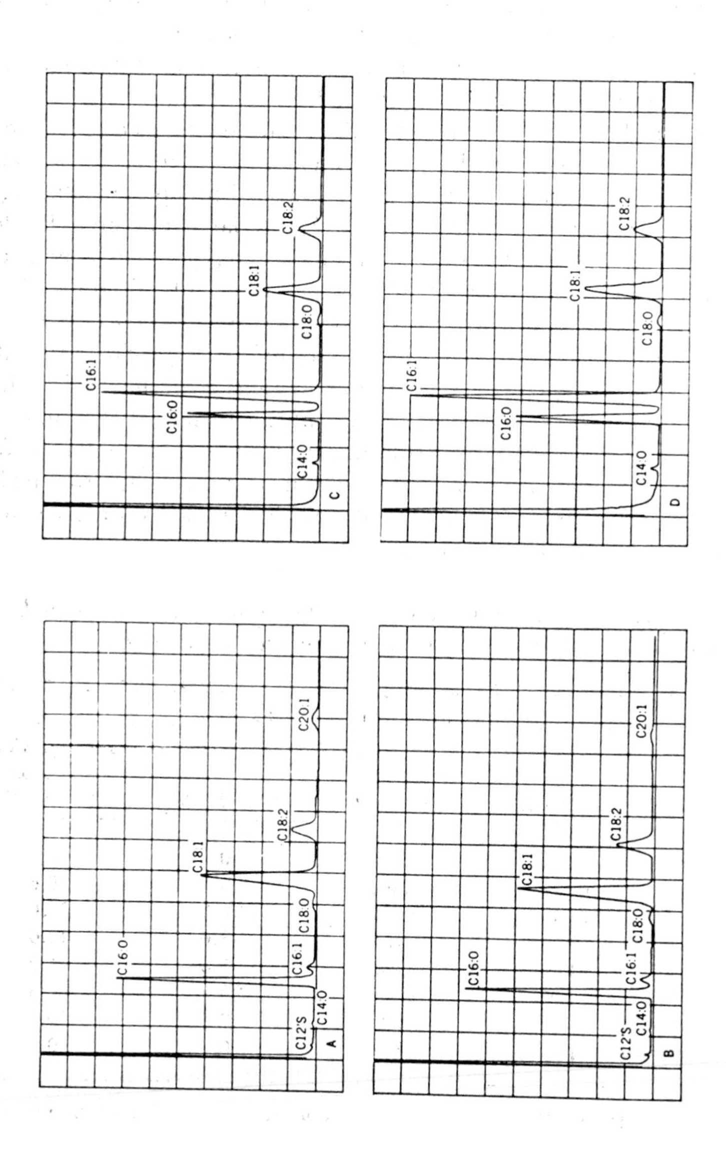

Figure 2. The fatty acid patterns of *I. conquisitor* and its corresponding hosts. A. *G. mellonella*, B. *I conquisitor* reared on *G. mellonella*, C. *O. nubilalis*, D. *I. conquisitor* reared on *O. nubilalis*.

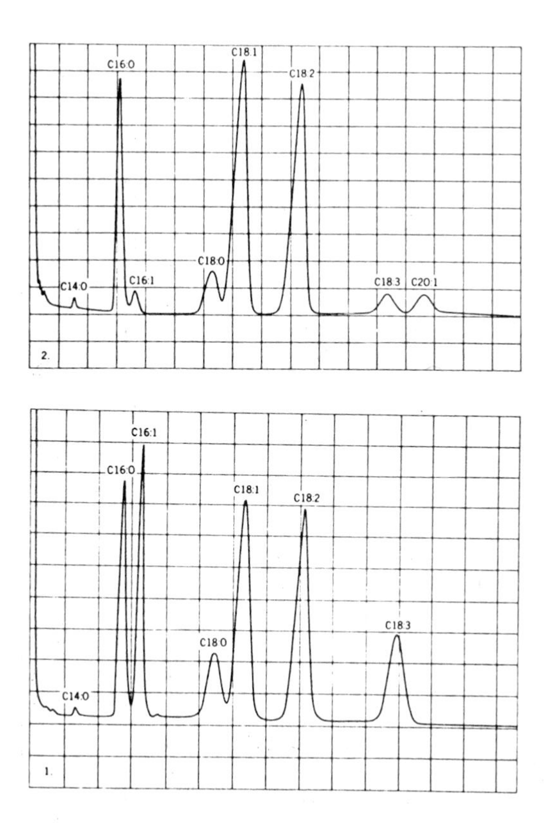

Figure 3. The phospholipid fatty acids of *Itoplectus conquisitor* reared on *Galleria mellonella*, above, and *Ostrinia nubilalis*, below.

contain less palmitic acid and more stearic and polyunsaturated acids than do the neutral fats.

Moreover, the changes in the phospholipids though considerable are not as great as the changes in the neutral fats. For example, palmitoleic acid in the neutral fats of the parasite reared on corn borer is 35% compared with 2% when the parasite is reared on the wax moth, while in the content of this fatty acid in the phospholipids is only 15% and 2% respectively. Some ratios between classes of fatty acids for the phospholipids are shown in Table 5. It is evident that such ratios as saturated to poly-unsaturated and mono-unsaturated to poly-unsaturated are similar regardless of the host. The parasite apparently maintains certain characteristics in its phospholipids. This accounts to some extent for the fact that the patterns of fatty acid composition of the parasite do not completely reflect those of the host.

Table 4. The fatty acid composition of the phospholipids and neutral lipids of *I. conquisitor* reared on two hosts.

	Percent composition							
Source	C14:0	C16:0	C16:1	C18:0	C18:1	C18:2	C18:3	C20:1
I. conquisitor reared on *G. mellonella*								
Phospholipids	Tr	14.1	1.6	6.5	36.1	34.9	3.3	3.4
Neutral lipids	Tr	31.0	1.7	1.5	43.4	9.9	Tr	12.4
I. conquisitor reared on *O. nubilalis*								
Phospholipids	Tr	13.6	15.4	8.2	24.8	25.6	12.2	---
Neutral lipids	Tr	25.9	3.5	0.9	26.9	6.5	4.8	---

Table 5. The ratio of fatty acid classes of the phospholipids of *I. conquisitor* reared on two hosts.

	Ratios			
Phospholipid Source	Saturates / Unsaturates	Saturates / Monounsaturates	Saturates / Polyunsaturates	Monounsaturates / Polyunsaturates
I conquisitor reared on *G. mellonella*	0.26	0.50	0.54	1.08
I. conquisitor reared on *O. nubilalis*	0.28	0.54	0.58	1.06

Other host parasite systems have not been available but heuristic, though much less definitive information is contained in Fig. 4. Several host-parasite couples, including parasites of the families Aphidiidae, Brachonidae, Ichneumonidae, Pteromalidae, and Eulophidae were analysed.

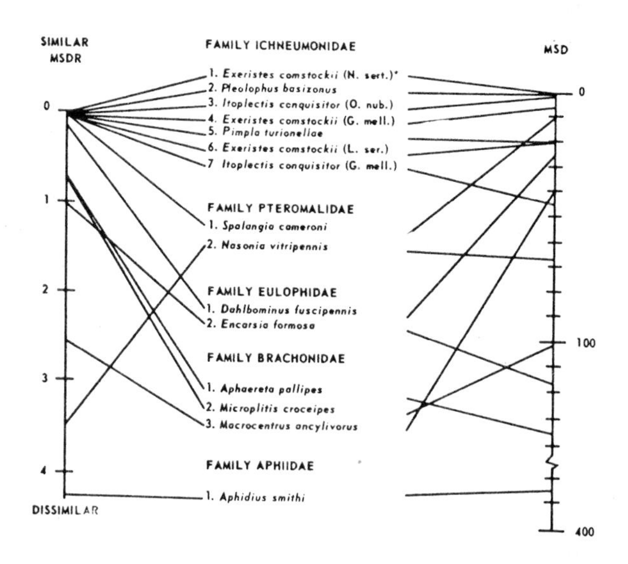

Figure 4. The mean square distance (MSD) and the mean square distance of the ratios (MSDR) between the fatty acid composition of various parasites and their hosts.

The prominence of ichneumonids in the figure is caused by the fact that the five examples previously discussed are included, in these cases the host is indicated in parenthesis. The additional ichneumonids are *Pleolophus basizonus* parasitizing *Neoiprion sertifer* and *Pimpla turionellae* parasitizing *G. mellonella*. The pteromalids, *Spalangia cameroni* and *Nasonia vitripennis* were parasitizing *Musca domestica*. The eulophids *Dahlbominus fuscipennis* and *Encarsia formosa* were parasitizing *Neodiprion lecontei* and *Trioleurodes vaporariorum* respectively. Among the brachonids, *Aphaereta pallipes* was parasitizing *Pseudosarcophaga affinis; Microplitis croceipes, Heliothis zea;* and *Macrocentrus ancylivorus, Gnorimoshema operculella.* The lone aphid, *Aphidius smithi* was parasitizing *Acyrthosiphon pisum.*

If the fatty acid pattern of the parasite does not resemble that of its host, the phenomenon certainly does not exist. If on the other hand, the pattern of the parasite does resemble that of its host, the phenomenon may exist but is not proven.

Boyce (1969) has suggested a measure of diversity which he calls the mean square distance (MSD). This, as the name implies, is the mean of the differences between the parasite and its host for each fatty acid squared. It tends to be least for the ichneumonids including *E. comstockii* and *I. conquisitor* (MSD = 1 to 45) and greatest for *Aphidius smithi* (MSD = 385). This measure is influenced as much by a small proportional difference between major components, for example the 1.2 fold difference between 25 and 30%, as by a proportionally large difference between lesser components, for example the 2 fold difference between 5 and 10%. Therefore, we have calculated a second measure of similarity which in keeping with Boyce's nomenclature may be called the mean square distance of the ratios (MSDR). This is the mean of the squares of the differences from unity of the ratios of the individual fatty acids (highest percentage / over lowest percentage).

Again the ichneumonids exhibit the lowest MSDR's (0 - 0.06) and *A. smithi* the greatest (4.29). It is evident that a close relationship between parasite and host fatty acid pattern is quite common particularly but not exclusively among the ichneumonids. On the other hand, many parasites certainly do not have fatty acid patterns at all resembling the patterns of their hosts.

Turning now to a study of the mechanisms, the larvae of *E. comstockii* were transferred from *G. mellonella* to *L. sericata* and *vice versa*, and their fatty acid compositions were analysed 12, 36 and 72 hours after the transfer. The change observed in the proportion of palmitoleic acid in the fat of the parasite is illustrated in Fig. 5 taken from Barlow and Bracken (1971). Its

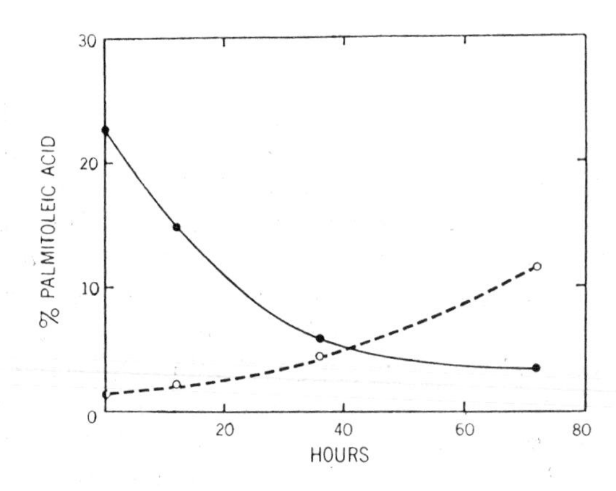

Figure 5. Changes in the concentration of palmitoleic acid (C16:1) when *Exeristes comstockii* larvae are transferred from *Lucilia sericata* to *Galleria mellonella* (o) and vice versa (o).

concentration falls rapidly when the parasite is transferred from a host with a high concentration. It is reduced by 50% in about 20 hours. Probably deposition is reduced immediately and decomposition continues until equilibrium at the lower level is reached. On the other hand, when the parasite is transferred from low to high concentration of palmitoleic acid the build up of this fatty acid in its fats takes place at a rate considerably slower than the reduction observed in the former case. It did not reach 50% of its probably final concentration until 60 hours had elapsed.

It can be hypothesized that the mimicry is achieved by the parasite not synthesizing fatty acids but incorporating into its body fats the fatty acids in its diet. A preliminary experiment in which ^{14}C-sodium acetate was injected into the parasite dispelled this notion. The radio-activity was incorporated into the fatty acids of the parasite. In a more rigorous experiment (Barlow and Bracken 1971), parasite larvae were injected with radio-active acetate (2 uCi per 50 mg of larvae) and after 24 hours the fats were extracted and analysed. The results in Table 6 show the relatively high concentration of palmitoleic acid in *E. comstockii* reared on *L. sericata.* The relatively large standard deviations of the specific activity data reflect the difficulty of injecting precise amounts of tracer into small insect larvae. However, several differences statistically significant at the 5% level are apparent. In *E. comstockii* regardless of the host on which it is reared, the specific activity of stearic acid is high and the specific activity of oleic acid is low compared with the specific activities of 16 carbon fatty acids. The differences among the latter are not statistically significant. There are also differences between *I. conquisitor* and *E. comstockii* reared on *G. mellonella* aside from the fact that both duplicate the fatty acid pattern of their hosts; these will be recalled below. Looking at *E. comstockii* only, it is apparent that regardless of the tenfold difference in the amount of palmitoleic acid, the specific activity is nearly the same. In experiments of this type, the specific activity at a given time for any chemical entity, must be related to the amount of the entity which has been synthesized and into which the radio-active label has been incorporated, less the amount of the entity which has been degraded, divided by the total amount of that same entity in the animal. If this ratio is the same in two instances in which the total quantities differ by a factor of ten, it is apparent that the numerator of the ratio, the amount of synthesis less the amount of degradation must also differ by a factor of ten. The conclusion is that in *E. comstockii* reared on *L. sericata*, the high concentration of palmitoleic acid is not the result of direct deposition of this fat from the diet into the fat depot of the parasite but is maintained by active metabolic processes.

Since it is obvious that the first hypothesis is not tenable and that the parasite does indeed synthesize fatty acids, a second hypothesis was tested; this was that the parasite synthesized fatty acids by the same mechanism

Table 6. Percentage of various fatty acids in the fat of parasites reared on different hosts and the specific activity of the fatty acids 24 h after the injection of Na-1-^{14}C-acetate into the parasite.

Parasite and host	No. samples	(%/specific activity) of fatty acid* (±S.D.) C16:0	C16:1	C18:0	C18:1
I. conquisitor / *G. mellonella*	4	32+2 / 41±25	1.0±0.3 / 85±17	1.0±0.7 / 186±75	50±2 / 8±3
E. comstockii / *G. mellonella*	4	29+2 / 58±18	2.0±0.4 / 50±12	2.0±0.3 / 318±62	53±1 / 7±2
E. comstockii / *L. sericata*	3	19±2 / 111±37	21±1 / 44±13	2.0±0.5 / 260±57	42±5 / 22±7

*Symbols for fatty acids give number of carbon atoms: number of double bonds.
Significantly different (P 0.05) from similar result in column.
Significantly different (P 0.05) from 53±1 only in column.
Significantly different (P 0.05) from specific activities in row.

Table 7. Incorporation of ^{14}C-1-acetate into fatty acids of *G. mellonella, L. sericata* and *E. comstockii* reared on *G. mellonella,* after 24 hours exposure.

Hosts (2) and parasite on host	Specific activity of fatty acid fractions (dpm/g)						
	C14:0*	C16:0	C16:1	C18:0	C18:1	C18:2	C20:1
G. mellonella	160 ± 15	8.9 ± 1.4	5.5 ± 0.5	140 ± 12	1.7 ± 0.1	0	3.94 ± 0.89
L. sericata	19.5 ± 1.8	10.1 ± 1.2	5.36 ± 0.15	11 ± 0.55	2.71 ± 0.1	0	0
E. comstockii reared on *G. mellonella*	795 ± 22	106 ± 4	77.3 ± 6.6	487 ± 45	11.2 ± 0.3	0	0

*The first number represents the carbon chain length, the second, the number of double bonds.

used by the host. To test this, the rates of incorporation of radio-active carbon from acetate into the fatty acids of *G. mellonella, L. sericata* and *E. comstockii* reared on *G. mellonella* was measured. The results in Table 7 show that there is little relationship if any among them.

This together with the differences between *E. comstockii* and *I. conquisitor* both reared on *G. mellonella* alluded to above (Table 6) indicates that the parasites have their own specific metabolic processes and only the level at which the processes operate is dictated by the concentration of the end product in the host.

It is interesting to speculate that such host control of parasite metabolism might be a factor in some examples of obligate parasitism.

At this point *in vivo* experiments have been suspended and examination of the synthetic mechanisms of the larvae begun. Crude preparations of fatty acid synthetase from *I. conquisitor, G. mellonella* and *L. sericata* incorporated the carbon of ^{14}C-Acetyl-CoA into palmitic and stearic acids. Subsequent purification of the synthetase from *L. sericata* developed the techniques which will be used in a comparative study of the other species.

The synthetase behaves somewhat differently in the fractionation procedures than does synthetase from mammalian systems. A somewhat higher level of ammonium sulphate is required for its precipitation and the complex is completely excluded by Sephadex G 200 but not by Sepharose 4 B. Attempts to elute the synthetase complex from DEAE cellulose were unsuccessful. However, fractionation was obtained on ECTEOLA cellulose. In the crude preparation about 40% of the incorporated activity is in palmitic acid, 60% in stearic. After 40 fold purification, the ratio of palmitate to stearate produced remains the same. This matter remains to be resolved. Further purification of the synthetase system from *L. sericata* and the comparative study of this system among the species of host insects is being carried out at present.

REFERENCES

BARLOW, J. S. (1965). Effects of diet on the composition of body fat in *Agria affinia* (Fallen). *Can. J. Zool.* 43:337-340.

_____________ (1966a). Effects of diet on the composition of body fat in *Musca domestica* L. *Can. J. Zool.* 44:775-779.

_____________ (1966b). Effects of diet on the composition of body fat in *Lucilia sericata* (Meigen). *Nature, Lond.* 212:1478-1479.

BARLOW, J. S. and BRACKEN, G. K. (1971). Incorporation of Na-1-^{14}C-acetate into the fatty acids of two insect parasites (Hymenoptera) reared on different hosts. *Can. J. Zool.* 49:1297-1300.

BRACKEN, G. K. and BARLOW, J. S. (1967). Fatty acid composition of *Exeristes comstockii* (Cress) reared on different hosts. *Can. J. Zool.* 45:57-61.

BRACKEN, G. K. and HARRIS, P. (1969). High palmitoleic acid in Lepidoptera. *Nature, Lond.* 224:84-85.

BOYCE, A. J. (1969). *Mapping diversity: a comparative study of some numerical methods. Proc. Colloguium in Numerical Taxonomy,* Academic Press, London.

BUCHER, G. E. and WILLIAMS, R. (1967). The microbial flora of laboratory cultures of the greater wax moth and its effect on rearing parasites. *J. Invert. Path.* 9:367-473.

FISHER, R. C. (1971). Aspects of the hysiology of endoparasitic Hymenoptera. *Biol. Rev.* 46:243-278.

THOMPSON, S. N. and BARLOW, J. S. (1970). The changes in fatty acid pattern of *Itoplectus conquisitor* (Say) reared on different hosts. *J. Parasit.* 56:845-846.

NUTRITION AND METABOLISM OF CERTAIN METHYL-CONTAINING COMPOUNDS IN INSECTS

Ernest Hodgson, Ewald Smith and Karl D. Snyder
Department of Entomology
North Carolina State University
Raleigh, North Carolina 27607

INTRODUCTION

Compounds containing methyl groups are important in the nutrition and metabolism of animals. Two of these, methionine and choline, are important in protein and phospholipid metabolism respectively, while, at the same time, being the primary nutritional source, in many phyla, of labile methyl groups.

Both the nutrition and metabolism of these compounds have been of considerable interest to us for a number of years, as well as to a number of other laboratories and it has become apparent that there are several distinct differences between insects and vertebrates in these areas.

Limitations of space prohibit the writing of an exhaustive review, the following emphasizes the investigations carried out in our own laboratories, citing the literature necessary to place this work in proper perspective. Where possible reviews are cited as a source of references to earlier work.

METHIONINE REQUIREMENTS

An examination of previous findings reveals that methionine is required for growth and development of many insect species (Table 1), from several different orders. Although it is apparent that insects exhibit a general requirement for this particular nutrient one insect, *Phormia regina,* appeared to have an alternative requirement for either methionine or cystine (Hodgson *et al.*, 1956). At the time this work was carried out adequate growth of *Phormia regina* on chemically defined diets was difficult to obtain. In the subsequent search for a suitable diet we found (Hodgson *et al.*, 1960) that the addition of a small amount of purified casein (=16% of the total nitrogen) provided a much more satisfactory growth rate. This enabled a great deal of work to be carried out on choline and its analogs (see below) but unfortunately complicated investigations on methionine. More recently, by careful attention to a number of details, particularly the pH of the medium, and the availability of all L amino acid mixtures, we have been able to dispense with added casein. The results of recent investigations are as follows: 1. A requirement for methionine can be demonstrated (Table 2); 2. Neither cystine nor cysteine can substitute for methionine but both can

Table 1. Insect in which dietary methionine is known to be required for normal growth and development

Order	Species	References[a]
Coleoptera	*Anthonomus grandis*	Altman & Dittmer, 1968
	Attagenus sp.	Altman & Dittmer, 1968; House, 1965
	Oryzaephilus surinamensis	Altman & Dittmer, 1968
	Tribolium confusum	Altman & Dittmer, 1968; House, 1965
	Trogoderma granarium	Altman & Dittmer, 1968; House, 1965
Diptera	*Aedes aegypti*	Altman & Dittmer, 1968; House, 1965
	Calliphora vicina	Altman & Dittmer, 1968; House, 1965
	Ctenicera destructor	Kasting & McGinnis, 1966
	Drosophila melanogaster	Altman & Dittmer, 1968; House, 1965
	Hylemya antiqua	Altman & Dittmer, 1968; House, 1965
	Hypoderma bovis	Alikhan, 1963; Kasting & McGinnis, 1966
	Phormia regina	Hodgson *et al.*, 1956; House, 1965; this paper
	Agria (=Pseudosarcophage) affinis	Altman & Dittmer, 1968; House, 1965
Hemiptera	*Acyrthosiphon pisum*	Retnakaran & Beck, 1968
	Aphis gossypii	Turner, 1971
	Dysdercus fasciatus	Alikhan, 1968
	Myzus persicae	Dadd & Krieger, 1968; Kasting & McGinnis, 1966
Hymenoptera	*Apis mellifera*	House, 1965
Lepidoptera	*Agrotis orthogonia*	Altman & Dittmer, 1968; House, 1965; Kasting & McGinnis, 1966
	Argyrotaenia velutinana	Altman & Dittmer, 1968; Rock & King, 1967; Sharma *et al.*, 1972
	Bombyx mori	Altman & Dittmer, 1968; Ito & Arai, 1967
	Chilo suppressalis	Altman & Dittmer, 1968; House, 1965
	Heliothis zea	Rock & Hodgson, 1971
	Pectinophora gossypiella	Altman & Dittmer, 1968; House, 1965
Orthoptera	*Blattella germanica*	Altman & Dittmer, 1968

[a]Altman & Dittmer (1968), House (1965) and Kasting & McGinnis (1966) are reviews which may be consulted for the original references.

spare the methionine requirement (Tables 3 and 4). Sparing of methionine by cysti(e)ne has also been shown in *Tribolium confusum* (Taylor & Medici, 1966) *Bombyx mori* (Ito & Arai, 1967) and *Argyrotaenia velutinana* (Sharma *et al.*, 1972).

Table 2. Effect of methionine on the growth of *Phormia regina* larvae.

Methionine added (mg.)	Avg. wt./larva (mg.)
0	0
0.5	0
1.0	21.9
1.5	45.7
2.0	49.0
2.5	39.2
3.0	33.6

Table 3. Sparing action of cysteine on the methionine requirements of *Phormia regina* larvae.

Methionine (mg.)	Cysteine (mg.)	Avg. wt./larva (mg.)
0.1	4.0	1.3
0.2	4.0	7.8
0.3	4.0	22.7
0.4	4.0	33.9
0.5	4.0	33.5
0.6	4.0	32.7
0.7	4.0	33.1
0.8	4.0	31.6

It is of interest that homocysteine, the demethylated analog of methionine, will not substitute for methionine in the diet of several insects (Table 5) (Sharma *et al.*, 1972; Geer & Vovis, 1965) but in some cases will spare part of the methionine requirement (Sharma *et al.*, 1972).

Table 4. Sparing action of cystine and cysteine on the methionine requirement of *Phormia regina* larvae.

Complete diet (18 amino acids) 16 Amino acid diet plus Methionine (mg.)	Cystine (mg.) or	Cysteine (mg.)	Avg. wt./larva (mg.) Expt. I 47.2	Expt. II 46.6
0	0	0	0	0
0.5	0	0	1.0	0
0.5	4.0	0	23.0	37.0
0.5	10.0	0	25.0	45.5
0.5	20.0	0	40.9	43.7
0	0	4.0	0	0
0.5	0	4.0	35.6	47.0
0.5	0	10.0	41.3	38.2

Table 5. Substitution of dietary methionine by homocysteine in *Phormia regina.*

16 Amino acid diet plus methionine (mg.) or	homocysteine (mg.)	Avg. wt./larvae (mg.)
0	0	0
5.0	0	26.5
0	1.0	0
0	5.0	0
0	10.0	0
0	20.0	0

Determination of amino acid requirements by the use of ^{14}C-labeled precursors (Kasting & McGinnis, 1966) have in general confirmed the essentiality of methionine but in some cases have demonstrated incorporation of ^{14}C into methionine (Kasting & McGinnis, 1962; Kasting *et al.*, 1962) including some insects in which dietary deletion techniques have demonstrated a methionine requirement (Sharma, *et al.*, 1972; Rock & Hodgson, 1971). Several possible reasons have been advanced, including synthesis at a level not adequate to supply normal needs and the possibility of labeled contaminants in the protein hydrolysate not resolved from methionine.

METHIONINE AS A SOURCE OF METHYL GROUPS

In vertebrates methionine is converted to a high energy intermediate, S-adenosylmethionine, which is important in methyl transfer (Greenberg, 1963), particularly in the stepwise methylation of phosphatidylethanolamine to phosphatidylcholine. Recent studies by Willis and Hodgson (1970) using several insect species, namely *Phormia regina, Heliothis zea* and *Hylobius pales*, have shown that methionine is not a source of labile methyl groups, at least insofar as the formation of choline is concerned.

In these studies methyl-^{14}C labeled methionine was fed for extended periods and the insects were examined for possible transmethylation products. No labeled choline was found in the watersoluble fraction or in the lipid hydrolysate. The principal products were methionine sulfoxide or methionine sulfone although some metabolism of the methyl group was apparent from the presence of unknown labeled products in both the lipid and protein hydrolysates.

This is in agreement with the work of Moulton *et al.* (1970) who showed that in *Musca domestica* the methyl group of methionine was incorporated into serine, ethanolamine, fatty acids, unknown neutral lipids and RNA but not choline. Thus the methyl group appears to be metabolized, but not generally transferred intact. These studies confirmed earlier indirect studies which suggested that transmethylation might be impaired in *Phormia regina* (House, 1965; Hodgson & Dauterman, 1964; Hodgson *et al.*, 1964), *Drosophila melanogaster* (Geer *et al.*, 1968), *Musca domestica* (Bridges & Rocketts, 1965; 1967) and *Bombyx mori* (Shyamala & Bhat, 1959).

Following administration of ^{14}C-formate (Zielinska & Dominas, 1967) to *Acantholyda nemoralis* ^{14}C appeared in phosphatidylethanolamine, phosphatidylmonomethylethanolamine and phosphatidyldimethylethanolamine but not phosphatidylcholine. Although the authors suggest methylation from methionine the experiments are not definitive. The dried fruit moth, *Vitula edmondsae serratilineela,* did, however, appear to methylate phosphatidylethanolamine to phosphatidylcholine (Blankenship & Miller, 1971).

CYSTATHIONINE PATHWAY FOR METHIONINE METABOLISM

A well known pathway for methionine metabolism involves its conversion to cysteine (Fig. 1). The intermediate, cystathionine, has been isolated from larvae and pupae of *Bombyx mori* and pupae of *Antheraea pernyi*

(Kondo, 1959; Rao *et al.*, 1967). The conversion of methionine to cysteine via the cystathionine pathway has been shown in *Musca domestica* (Cotty *et al.*, 1958), *Blattella germanica* (Block & Henry, 1961; Henry & Block, 1961), *Phormia regina* (Henry & Block, 1962) and *Antheraea pernyi* (Kondo, 1962), although evidence for the complete pathway was lacking in *Bombyx mori* (Kondo, 1962), *Acyrothsiphon pisum* (Retnakaran & Beck, 1968) or *Aphis gossypii* (Turner, 1971). This pathway has been found to be irreversible in all insects studied (Hilchey *et al.*, 1957; Cotty *et al.*, 1958; Block & Henry, 1961; Henry & Block, 1961; 1962).

Nutritional investigations on *Argyrotaenia velutinana* have shown that methionine is essential (Rock & King, 1967) but that approximately 75% of the methionine requirement could be spared by cysteine or by other members of the cystathionine pathway (cystathionine and homocysteine) (Sharma *et al.*, 1972). As you may recall, methionine was also spared by cysteine and cystine in the blowfly, *Phormia regina*.

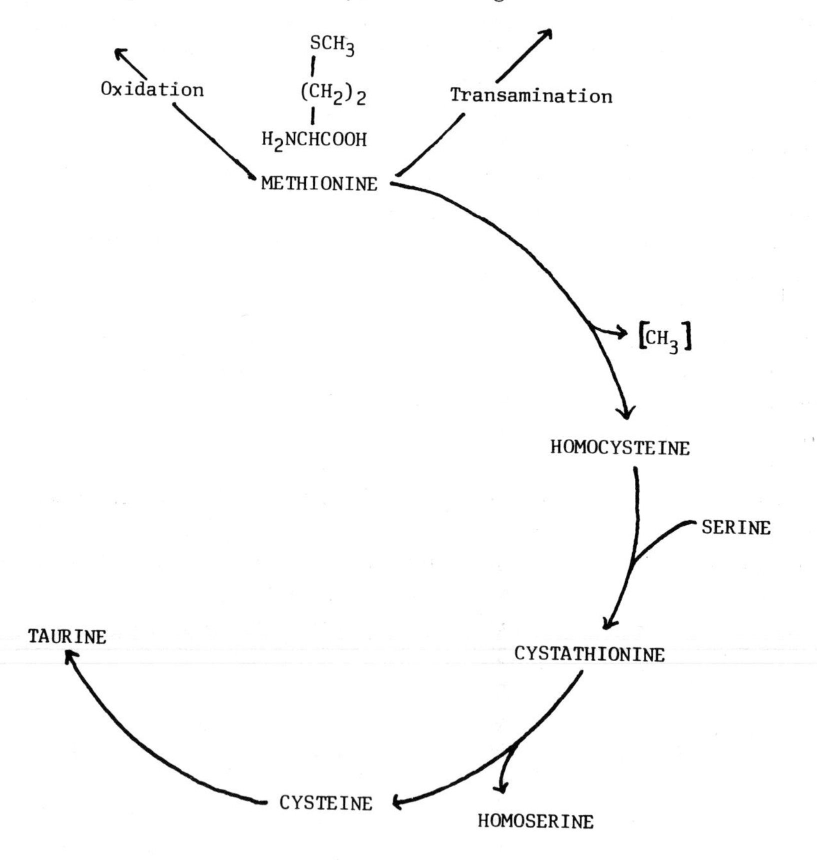

Figure 1. Cystathionine pathway.

Recently Sharma, Rock and Hodgson (in press) using ^{35}S-methionine and ^{35}S-cysteine demonstrated the cystathionine pathway in *Argyrotaenia velutinana.* The pathway was irreversible, no methionine being synthesized from cysteine. This is of importance in two regards: 1. Since the pathway is irreversible it explains why neither cystine nor cysteine can substitute for methionine, and 2. It explores how the spring action of homocysteine is consistent with the lack of transmethylation since it may function not as a methyl acceptor, but in the same way as cystine or cystathionine to spare that portion of the dietary methionine which would normally be metabolized via the cystathionine pathway. Presumably some type of feedback control reduces the activity of the pathway in the presence of these intermediates leaving the reduced amount of methionine available for indispensible methionine needs.

OTHER PATHWAYS FOR METHIONINE METABOLISM AND THEIR RELATION TO METHIONINE NUTRITION

Methionine is known to be involved in transamination reactions in insects and to be oxidized by their amino acid oxidases (Chefurka, 1965), both of these reactions resulting in the formation of a keto acid. Since keto acids lack the asymmetry involved in the L- and D-configuration, the deamination of D-methionine either oxidatively or by transamination and the subsequent reamination of the keto acid formed could explain the ability of D-methionine to substitute, in part, for L-methionine requirement (Sharma *et al.*, 1972; Rock, 1971) in the red banded leaf roller. Other examples of substitution of L-methionine by D-methionine have been reported (Dadd, 1970; Retnakaran & Beck, 1968).

A number of other analogs have been tested for their ability to replace methionine in the diet of insects. We have recently (Rock *et al.*, in preparation) tested a number of methionine analogs in the diet of the red banded leaf roller. L-methionine ethyl ester was as effective as L-methionine, presumably due to hydrolysis to methionine although it is interesting to note that this compound is not utilized in the rate (Miller & Samuel, 1968). The effective utilization of methionine sulfoxide is similar to that reported for the rat and the german cockroach (Gordon, 1959) although this analog was not utilized by another insect, *Ctenicera destructor* but had antimetabolite activity (Davis, 1965). Methionine sulfoxide is one of the principal metabolites of methionine in insects (Willis & Hodgson, 1970) and has been shown to be a normal constituent of *Phormia regina* (Levenbook & Dinamarca, 1966). The methionine-D-hydroxy analog was approximately 60% as effective as methionine, presumably via amination to methionine. Miller and Samuel (1968) showed this analog to be as effective as methionine for the rat. L-ethionine, L-methionine-DL-sulfoximine, L-methionine sulfone and seleno-DL-methionine were not utilized by the red banded leaf roller. When fed with methionine, ethionine had no effect, methionine sulfone was slightly inhibitory while L-methionine-DL-sulfoximine and seleno-DL-methionine were toxic. These results are similar to those reported for the roach and many vertebrates (Gordon, 1959; Miller & Samuel, 1968; Meister, 1965).

CHOLINE REQUIREMENTS

Choline requirements have been demonstrated in a larger number of insects (Table 6). In those few species which do not appear to require choline the investigations were carried out either on poorly defined diets which may have contained choline or a suitable substitute, or in the presence of micro-organisms. It is probably an accurate generalization to say that choline is a requirement for the normal growth and development of insects. Choline may also be required for egg development in adult females (Geer, 1968; Geer *et al.*, 1970; Hagen, 1956; House, 1965; Vanderzant & Richardson, 1964) and sperm motility in adult males (Geer, 1967).

SUBSTITUTION OF CHOLINE BY RELATED COMPOUNDS

Investigations on the substitution of dietary choline by related compounds started with the observation that *Drosophila melanogaster* (Fraenkel *et al.*, 1955) and *Phormia regina* (Hodgson *et al.*, 1956) could utilize carnitine *in lieu* of choline in the larval diet. Subsequent studies (Hodgson *et al.*, 1960) showed that γ-butyrobetaine and 2,2-dimethylaminoethanol would also substitute for choline in the diet of *Phormia regina* larvae and that they would reverse the action of the choline inhibitor 2-amino-2-methyl propanol, as would choline and carnitine. Carnitine is itself a specific requirement for a number of tenebrionid beetles (Fraenkel & Friedman, 1957) and has functions in animals unrelated to its ability to substitute for choline in certain insects. These will not be discussed further. Hodgson and Dauterman (1964) investigated the relationship between the chemical structure of a large series of choline analogs and the ability to replace choline in the larval diet of *Phormia regina*. We found that at least two methyl groups must be present on the nitrogen atom and that a hydroxyl group must be present on the second carbon atom from the nitrogen. The third N-alkyl can be missing or it can vary from methyl to n-butyl.

Geer and his co-workers (Geer & Vovis, 1965; Geer & Ricker, 1965) investigated the effects of dietary substitutes for choline on growth and development of *Drosophila melanogaster* and found that a hydroxyl group both terminal and on the second carbon was necessary for maximum growth

Table 6. Insects in which dietary choline is known to affect growth and development.

Order	Species	References[a]
Coleoptera	*Anthonomus grandis*	Altman & Dittmer, 1968; Hodgson, 1960
	Attagenus sp.	Altman & Dittmer, 1968; House, 1965
	Carpophilus hemipterus	Hodgson, 1960
	Dermestes maculatus	Altman & Dittmer, 1968; Hodgson, 1960
	Dermestes vulpinus	Hodgson, 1960
	Lasioderma serricorne	Altman & Dittmer, 1968; Hodgson, 1960
	Palorus ratzeburgi	Hodgson, 1960
	Ptinus tectus	Altman & Dittmer, 1968; Hodgson, 1960
	Stegobium panaceum	Altman & Dittmer, 1968; Hodgson, 1960
	Tenebrio molitor	Altman & Dittmer, 1968; Hodgson, 1960; House, 1965
	Tribolium castaneum	Applebaum & Lubin, 1967
	Tribolium confusum	Altman & Dittmer, 1968
	Trogoderma granarium	Altman & Dittmer, 1968
Diptera	*Aedes aegypti*	Altman & Dittmer, 1968; House, 1965
	Agria affinis	Altman & Dittmer, 1968; House, 1965
	Calliphora vicina	Altman & Dittmer, 1968; Hodgson, 1960; House, 1965
	Drosophila ambigua	Royes & Robertson, 1964
	Drosophila funebris	Royes & Robertson, 1964
	Drosophila immigrans	Royes & Robertson, 1964
	Drosophila melanogaster	Altman & Dittmer, 1968; House, 1965
	Drosophila obscura	Royes & Robertson, 1964
	Drosophila simulans	Erk & Sang, 1966
	Drosophila subobscura	Royes & Robertson, 1964
	Hylemya antiqua	Altman & Dittmer, 1968; House, 1965
	Musca domestica	Altman & Dittmer, 1968; House, 1965
	Musca domestica vicina	Altman & Dittmer, 1968; House, 1965
	Phaenicia (Lucilia) sericata	Altman & Dittmer, 1968; House, 1965
	Phormia regina	Altman & Dittmer, 1968; House, 1965
Hemiptera	*Acyrthosiphon pisum*	Auclair, 1965
	Myzus persicae	Altman & Dittmer, 1968; Dadd & Krieger, 1967
Lepidoptera	*Argyrotaenia velutinana*	Rock, 1969
	Bombyx mori	Altman & Dittmer, 1968; Horie & Ito, 1965
	Ephestia elutella	Hodgson, 1960
	Ephestia kuhniella	House, 1965
	Pectinophora gossypiella	Hodgson, 1960
	Pyrausta nubilalis	Hodgson, 1960
	Tineola bisselliella	House, 1965
Orthoptera	*Acheta domestica*	Altman & Dittmer, 1968; House, 1965
	Blattella germanica	Altman & Dittmer, 1968; House, 1965
	Schistocerca gregaria	Altman & Dittmer, 1968; House, 1965

[a] Altman & Dittmer (1968), Hodgson (1960) and House (1965) are reviews which may be consulted for the original references.

effect. The nitrogen atom and the three N-methyl groups also appeared to be important. Carnitine, deoxycarnitine and β-methylcholine appeared to be exceptions to the requirement that the hydroxyl group be terminal.

Both ethanolamine and choline can be replaced, in part, in the phospholipids of vertebrates by compounds related, and the incorporation appears to take place via the cytidine diphosphate derivatives (see Hodgson *et al.*, 1969, for references).

Although phospholipids containing ethanolamine predominate in *Phormia regina* (Hodgson *et al.*, 1960; Kasting *et al.*, 1962), those containing choline form a significant part. When choline is replaced in the larval diet by carnitine or γ-butyrobetaine, the phospholipids containing choline are almost completely replaced by phospholipids containing β-methylcholine (Bieber *et al.*, 1963). Phosphatidyl carnitine is probably an intermediate in the metabolism of carnitine to phosphatidyl β-methylcholine (Mehendale *et al.*, 1966). Bieber and Newburgh (1963) have shown that the dimethyl analog of β-methyl choline, 2-dimethylaminoisopropanol and 2-dimethylaminoethanol are incorporated into the lipids of larval *Phormia regina* in place of choline.

Although choline is not an absolute requirement for larval growth in *Musca domestica*, in its absence carnitine, γ-butyrobetaine and β-methylcholine are incorporated into the larval lipids as β-methylcholine (Bridges *et al.*, 1965). Dimethyl and monomethylaminoethanol (Bridges & Ricketts, 1965), as well as a number of other amino alcohols and their alkyl derivatives (Bridges & Ricketts, 1967), are also incorporated into the phospholipids of *Musca domestica*. However, unlike *Phormia regina*, they may replace ethanolamine as well as choline. Diacylglyceryl-2-(aminoethyl)-phosphonate is synthesized in *Musca domestica* when 2-aminoethylphosphonate is included in the diet (Bridges & Ricketts, 1966). The above investigations on *Phormia regina* were concerned with larval growth, incorporation into the phospholipids being considered for only five compounds. Subsequent studies considered a wide range of compounds, their effects on post-larval development and their incorporation into the larval phospholipids (Hodgson *et al.*, 1969).

In these studies, twenty-six compounds were tested for their ability to support growth and development when fed, *in lieu* of choline, to *Phormia regina*. Sixteen supported larval growth, and previous findings on structure-function relationships in choline substitutes were confirmed (Gordon, 1959).

Only ten of these compounds permitted appreciable formation of puparia and only four of these permitted emergence of normal adults. The four compounds were dimethylethyl choline, dimethylisopropyl choline, α- and β-methylcholine. Thus the structural requirements for post-larval development are much more rigorous than those for larval growth. No compound was as effective as choline.

All of the compounds which support larval growth are incorporated into phospholipids analogous to phosphatidylcholine, indicating that the cytidine pathway for phospholipid synthesis is relatively non-specific.

i-Propylaminoethanol is a growth inhibitor when fed to *Phormia regina* larvae.

We then compared the metabolism of dimethylethylcholine and dimethylisopropylcholine, two compounds which permit growth and development of *Phormia regina* to the adult stage when fed *in lieu* of choline with the metabolism of dimethyl-n-propylcholine and dimethyl-n-butycholine, two compounds which permit only larval growth under the same conditions.

All four compounds are incorporated into phospholipids, analogous to phosphatidyl choline, without further modification.

The predominant water-soluble metabolite, in each case, is the phosphate monoester. Although CDPcholine can be detected when choline is fed, the CDP derivatives of the above choline analogs cannot.

The effectiveness of choline analogs as choline substitutes has been compared by measuring their ability to compete with choline as a precursor of phospholipids. Those analogs which permit superior growth and development when fed *in lieu* of choline are the most effective competitors for phospholipid synthesis.

The synthesis of phospholipids from unnatural bases by the CDP pathway has also been demonstrated in adult houseflies (Bridges & Holden, 1969). Recently we have concluded that insects other than the higher diptera may have much more restrictive requirements for choline substitutes. In *Argyrotaenia velutinana* (Table 7) only dimethylethylcholine permitted the emergence of adults while other substitutes permitted little or no growth. Similar findings have been made for *Heliothis virescens. Aedes aegypti* (Akov, 1962) cannot utilize carnitine or γ-butyrobetaine *in lieu* of choline, although 2,2-dimethylaminoethanol was partially effective. Similarly, *Anthonomus grandis* cannot use carnitine, γ-butyrobetaine or ethanolamine *in lieu* of choline.

CHOLINE AS A SOURCE OF LABILE METHYL GROUPS

Biochemical investigations on mammals (Greenburg, 1963) have shown that choline is oxidized to betaine and that betaine can donate a methyl group to homocysteine to form methionine. As discussed above methionine can then be converted to a high energy compound, S-adenosyl methionine, itself a methyl donor. Choline metabolism in vertebrates is shown in Fig. 2. Early fragmentary nutritional studies raised the possibility that transmethylation was impaired in insects (see compilation in House, 1965). Experiments in which enzyme preparations from a number of invertebrates were screened

Table 7. Substitution of choline by related compounds in the diet of *Argyrotaenia velutinana.*

	Larvae (total no.)	Stage at death					Survivor to	
		L1	L2	L3	L4	L5	pupa	adult
-	84	82	2	0	0	0	0	0
Choline	80	4	2	4	2	0	68(28)[a]	68
Dimethylethyl choline	78	31	0	11	4	0	32(43)	14
β-methyl choline	77	12	28	27	8	0	2(62)	0
Dimethyl isopropyl choline	68	28	15	20	4	0	1(54)	0
Choline + isopropyl ethanolamine	79	77	2	0	0	0	0	0

[a]Average no. days to pupation.

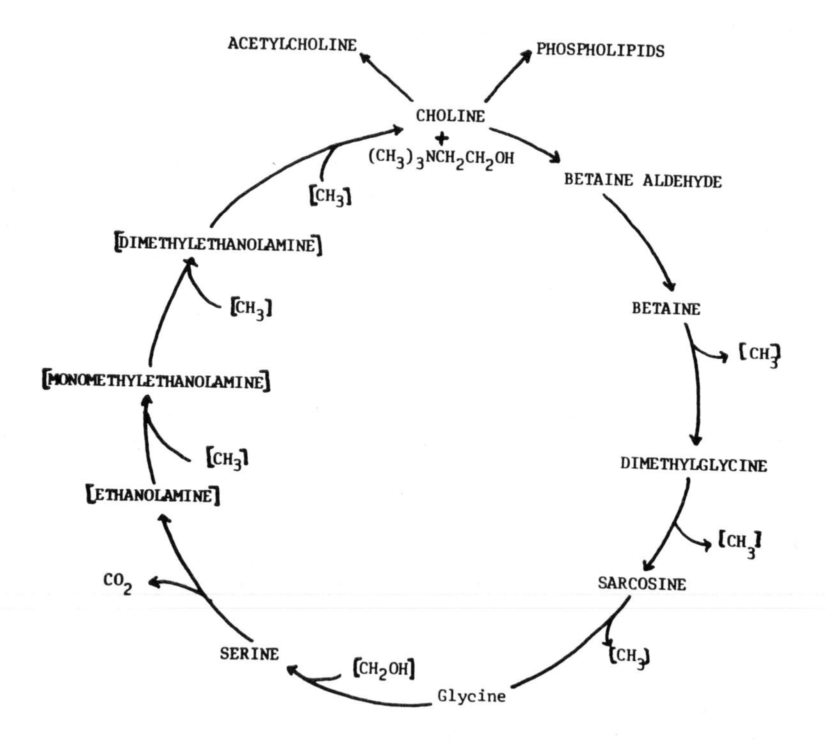

Figure 2. Generalized pathway for choline metabolism in animals.

for betaine-homocysteine transmethylase included a limited number of preparations from *Formica, Apis* and *Drosophila* (Ericson, 1960). All of these preparations were inactive. In none of the choline substitution experiments on *Phormia regina* (Hodgson & Dauterman, 1964; Hodgson *et al.*, 1960; 1969; Mehendale, 1970), *Drosophila melanogaster* (Geer *et al.*, 1968), or *Musca domestica* (Bridges & Ricketts, 1967; Bridges *et al.*, 1965) was there any evidence of transmethylation. In order to obtain direct evidence on this point we (Willis & Hodgson, 1970) fed methyl-^{14}C-labeled choline and betaine to *Phormia regina, Heliothis zea* and *Hylobius pales* larvae for extended periods and then examined the insects for possible transmethylation products. In all cases, when choline was fed, only intermediates of the cytidine pathway for phospholipid synthesis were found. In the case of betaine there was no metabolism, in all three insects the betaine added to the diet could be quantitatively recovered, either from the uneaten diet, the excreta, or the insects. This was true even though the insects were fed on labeled betaine for periods up to seven days.

The above investigations provide strong evidence that transmethylation from choline is impaired in insects and that betaine is without a role in insect metabolism. This leaves unexplained the early observation (Noland & Baumann, 1949) that betaine could substitute for choline in the diet of *Blattella germanica*. However, the equimolar equivalence observed would probably require the presence of a unique pathway in this organism.

OTHER METABOLIC PATHWAYS FOR CHOLINE METABOLISM

The formation of acetylcholine and its role in the nervous system are well known and will not be discussed further.

The principal pathway for choline metabolism is its incorporation into phospholipids, primarily phosphatidylcholine. The cytidine pathway for phosphatidylcholine synthesis first described by Kennedy and Weiss (1956) is well known in mammals and enough fragmentary evidence has been available to indicate its presence in insects, although it was not well known in the Insecta (for references see introduction to Shelley & Hodgson, 1970).

Shelley and Hodgson (1970) demonstrated the entire pathway in the larval fat body of the blowfly, *Phormia regina*. Biosynthesis of phosphatidylcholine from choline was demonstrated in both intact fat body and cell free preparations. The intermediates phosphocholine and cytidine diphosphate choline were identified and phosphorylcholine synthesis from them confirmed. Choline kinase and phosphorylcholine cytidyltransferase were located in the soluble part of the cell while phosphorylcholine glyceride transferase is located in the particulate fractions. More recent studies have studied the characteristics of partially purified choline kinase from *Peri-*

planeta americana (Kumar & Hodgson, 1970) and *Phormia regina* (Shelley & Hodgson, 1971a) and the role of choline analogs as substrates and competitive inhibitors of choline kinase (Shelley & Hodgson, 1971b). There appear to be separate choline and ethanolamine kinases, and compounds which can substitute for choline in the diet are readily phosphorylated and act as competitive inhibitors of choline phosphorylation.

If we consider the overall pathway for choline metabolism we can see that the stepwise methylation of ethanolamine to choline is probably missing insects (above). From the work of Taylor and Hodgson (1965) we know that at least in *Phormia regina* ethanolamine can be formed from glycine via serine and formate. There appears to be no metabolism of betaine in several insects. Betaine aldehyde, however, can substitute for choline in the larval diet of *Drosophila* (Geer & Vovis, 1965) and *Phormia regina* (this paper). In the latter case it is reduced to choline and appears in the phospholipids as such. It might be noted that we have never been able to demonstrate choline oxidase activity in *Phormia regina* either *in vivo* or *in vitro* in spite of many attempts by several different methods. It appears that choline oxidase is also missing in this insect.

CONCLUSIONS

A consideration of the evidence presented above suggests that the following may be true of insects in general:

1. They require a dietary source of both choline and methionine
2. The cystathionine pathway is operative in insects
3. Transmethylation reactions are impaired, and several reactions involving methionine and choline as a source of labile methyl groups may be missing completely.
4. Other restrictions of choline metabolism may also occur such as a lack of choline oxidase.

Since there are only an extremely small number of reports which conflict with these generalities, and in no case is the conflicting investigation extensive, it is possible that these apparent conflicts will be resolved. Other points of interest in the area of methionine and choline metabolism, such as the unequalled ability of some insects to synthesize 'unnatural' phospholipids may be characteristic, not of insects in general, but of the higher diptera.

REFERENCES

AKOV, S. (1962). Antimetabolites in the nutrition of *Aedes aegypti* L. larvae. The substitution of choline by related substances and the effect of choline inhibitors. *J. Insect Physiol.* **8**:337-348.

ALIKHAN, M. A. (1968). Amino acid requirements of the cotton stainer, *Dysdercus*

fasciatus (Pyrrhocoridae, Hemiptera). *Entomologia Exp. Appl.* 11:43-50.
ALTMAN, P. L. and DITTMER, D. S. (1968). *Metabolism.* Committee on Biological Handbooks. Fed. Am. Soc. Exp. Biol. Washington, D.C.
APPLEBAUM, S. W. and LUBIN, Y. (1967). The comparative effects of vitamin deficiency on development and on adult fecundity of *Tribolium castaneum. Entomologia Exp. Appl.* 10:23-30.
AUCLAIR, J. L. (1965). Feeding and nutrition of the pea aphid, *Acyrthosiphon pisum* (Homoptera:Aphidae), on chemically defined diets of various pH and nutrient levels. *Ann. Ent. Soc. Am.* 58:855-875.
BIEBER, L. L., CHELDELIN, V. H. and NEWBURGH, R. W. (1963). Studies on a β-methylcholine-containing phospholipid derived from carnitine. *J. Biol. Chem.* 238:1262-1265.
BIEBER, L. L., HODGSON, E., CHELDELIN, V. H., BROOKES, V. J. and NEWBURGH, R. W. (1961). Phospholipid patterns in the blowfly, *Phormia regina* (Meigen). *J. Biol. Chem.* 236:2590-2595.
BIEBER, L. L. and NEWBURGH, R. W. (1963). The incorporation of dimethylaminoethanol and dimethylaminoisopropyl alcohol into *Phormia regina* phospholipids. *J. Lipid Res.* 4:397-401.
BLANKENSHIP, J. W. and MILLER, G. J. (1971). *In vivo* synthesis of phosphatidyl choline and ethanolamine in larvae of the dried-fruit moth. *J. Insect Physiol.* 17:2061-2068.
BLOCK, R. J. and HENRY, S. M. (1961). Metabolism of the sulfur amino acids and of sulfate in *Blattella germanica. Nature,* Lond. 191:392-393.
BRIDGES, R. G. and HOLDEN, J. S. (1969). The incorporation of amino-alcohols into the phospholipids of the adult housefly, *Musca Domestica. J. Insect Physiol.* 15:779-788.
BRIDGES, R. G. and RICKETTS, J. (1965). Incorporation of N-methylaminoethanol and N-diemthylaminoethanol into the phospholipids of the housefly, *Musca domestica. Biochem. J.,* 95:41p.
_____ & _____ (1966). Formation of a phosphonolipid by larvae of the housefly, *Musca domestica. Nature,* Lond. 211:199-200.
_____ & _____ (1967). The incorporation, *in vivo*, of aminoalcohols into the phospholipids of the larva of the housefly, *Musca domestica. J. Insect Physiol.* 13:835-850.
BRIDGES, R. G., RICKETTS, J. and COX, J. T. (1965). The replacement of lipid-bound choline by other bases in the phospholipids of the housefly, *Musca domestica. J. Insect Physiol.* 11:225-236.
CHEFURKE, W. (1965). Intermediary metabolism of nitrogenous and lipid compounds in insects. In *The Physiology of Insecta, Vol. 2.,* Academic Press, N. Y., pp. 669-768.
COTTY, V. F., HENRY, S. M. and HILCHEY, J. D. (1958). The sulfur metabolism of insects. III. The metabolism of cystine, methionine, taurine, and sulfate by the housefly, *Musca domestica* L. *Contr. Boyce Thompson Inst. Pl. Res.* 19:379-392.
DADD, R. H. (1970). Arthropod nutrition. In *Chemical Zoology, Vol. 5.* Academic Press, N. Y. pp. 35-38.
DADD, R. H. and KRIEGER, D. L. (1968). Dietary amino acid requirements of the aphid, *Myzus persicae. J. Insect Physiol.* 14:741-764.
DADD, R. H., KRIEGER, D. L. and MITTLER, T. E. (1967). Studies on the artificial feeding of the aphid *Myzus persicae* (Sulzer)-IV. Requirements for water-soluble vitamins and ascorbic acid. *J. Insect Physiol.* 13:249-272.
DAVIS, G. R. F. (1965). The effect of structural analogs of methionine and glutamic acid on larvae of the prarie grain wireworm, *Ctenicera destructor* (Brown). *Archs. Int.*

Physiol. Biochim. 73:177-188.

ERICSON, L. E. (1960). Betaine-homocysteine-methyl-transferases I. Distribution in nature. *Acta. Chem. Scand.* 14:2102-2112.

ERK, F. C. and SANG, J. H. (1966). The comparative nutritional requirements of two sibling species *Drosophila simulans* and *D. melanogaster*. *J. Insect Physiol.* 12:43-51.

FRAENKEL, G. S. and FRIEDMAN, S. (1957). Carnitine. *Vitamins and Hormones.* 15:73-118.

FRAENKEL, G. S., FRIEDMAN, S., HINTON, T., LASZLO, S. and NOLAND, J. L. (1955). The effect of substituting carnitine for choline in the nutrition of several organisms. *Archs. Biochem. Biophys.* 54:432-439.

GEER, B. W. (1967). Dietary choline requirements for sperm motility and normal mating activity in *Drosophila melanogaster*. *Biol. Bull. Mar. Biol. Lab.*, Woods Hole. 133:548-566.

_________(1968). Modification of the larval nutritional requirements of *Drosophila melanogaster* by maternally inherited choline. *Archs. Int. Physiol. Biochim.* 76:797-805.

GEER, B. W., OLANDER, R. M., and SHARP, P. L. (1970). Quantification of dietary choline utilization in adult *Drosophila melanogaster* by radioisotope methods. *J. Insect Physiol.* 16:33-43.

GEER, B. W. and RICKER, J. G. (1965). The growth effects of carnitine, deoxycarnitine, and sulfocholine for *Drosophila, Neurospora,* and *Saccharomyces*. *Growth.* 29:405-413.

GEER, B. W. and VOVIS, G. F. (1965). The effects of choline and related compounds on the growth and development of *Drosophila melanogaster*. *J. Exp. Zool.* 158:233-236.

GEER, B. W., VOVIS, G. F. and YUND, M. A. (1968). Choline activity during the development of *Drosophila melanogaster*. *Physiol. Zool.* 41:280-292.

GORDON, H. T. (1959). Minimal nutritional requirements of the German roach, *Blattella germanica* L. *Ann. N.Y. Acad. Sci.* 77:290-352.

GREENBERG, D. M. (1963). Biological methylation. *Adv. Enzymol.* 25:395-431.

HAGEN, K. S. (1956). Honeydew as an adult fruit fly diet affecting reproduction. *10th Int. Congr. Ent. Montreal.* 3:25-30.

HENRY, S. M. and BLOCK, R. J. (1961). The sulfur metabolism of insects. VI. Metabolism of the sulfur amino acids and related compounds in the German cockroach, *Blattella germanica* (L.) *Contr. Boyce Thompson Inst. Pl. Res.* 21:129-144.

HENRY, S. M. and BLOCK, R. J. (1962). The sulfur metabolism of insects. VII. The metabolism of the sulfur amino acids and of sulfate in the blowfly, *Phormia regina* (Meig). *Contr. Boyce Thompson Inst. Pl. Res.* 21:447-452.

HILCHEY, J. D., COTTY, V. F. and HENRY, S. M. (1957). The sulfur metabolism of insects. II. The metabolism of cystine-S^{35} by the house fly, *Musca domestica* L. *Contr. Boyce Thompson Inst. Pl. Res.* 19:189-200.

HODGSON, E. (1960). Ph.D. Thesis, Oregon State University.

HODGSON, E., CHELDELIN, V. H. and NEWBURGH, R. W. (1956). Substitution of choline by related compounds and further studies on amino acid requirements in nutrition of *Phormia regina* (Meig). *Can. J. Zool.* 34:527-532.

_____, ______ & ______(1960). Nutrition and metabolism of methyl donors and related compounds in the blowfly, *Phormia regina* (Meigen). *Archs. Biochem. Biophys.* 87:48-54.

HODGSON, E. & Dauterman, W. C. (1964). The nutrition of choline, carnitine, and related compounds in the blowfly, *Phormia regina* Meigen. *J. Insect. Physiol.* 10:1005-1008.

HODGSON, E., DAUTERMAN, W. C., MEHENDALE, H. M., SMITH, E. and KHAN, M. A. Q. (1969). Dietary choline requirements, phospholipids and developments in *Phormia regina. Comp. Biochem. Physiol.* 29:343-359.

HORIE, Y. and ITO, T. (1965). Nutrition of the silkworm, *Bombyx mori-x.* Vitamin B requirements and the effects of several analogues. *J. Insect. Physiol.* 11:1585-1593.

HOUSE, H. L. (1965). Insect nutrition. In *The Physiology of Insecta, Vol. 2*, Academic Press, N. Y., pp. 769-813.

ITO, T. and ARAI, N. (1967). Nutritive effects of alanine, cystine, glycine, serine, and tyrosine on the silkworm, *Bombyx mori. J. Insect Physiol.* 13:1813-1824.

KASTING, R., DAVIS, G. R. F. and MCGINNIS, A. J. (1962). Nutritionally essential and non-essential amino acids for the prairie grain wireworm, *Ctenicera destructor* Brown, determined with glucose-U-C^{14}. *J. Insect Physiol.* 8:589-596.

KASTING, R. and MCGINNIS, A. J. (1962). Nutrition of the pale Western cutworm, *Agrotis orthogonia Morr.* (Lepidoptera, Noctuidae). IV. Amino acid requirements determined with glucose-U-C^{14}. *J. Insect Physiol.* 8:97-103.

_____&_____(1966). Radioisotopes and the determination of nutrient requirements. *Ann. N. Y. Acad. Sci.* 139:98-107.

KENNEDY, E. P. and WEISS, S. B. (1956). The function of cytidine coenzymes in the biosynthesis of phospholipides. *J. Biol. Chem.* 222:193-213.

KONDO, Y. (1959). The isolation of L-cystathionine from the body fluids of silkworm larvae, *Bombyx mori. Nippon Sanshigaku Zasshi.* 28:1-9.

_____(1962). The cystathionine pathway in the silkworm larva, *Bombyx mori. J. Biochem., Tokyo.* 51:188-192.

KUMAR, S. S. and HODGSON, E. (1970). Partial purification and properties of choline Kinase from the cockroach, *Periplaneta americana. Comp. Biochem. Physiol.* 33:73-84.

LEVENBOOK, L. and DINAMARCA, M. L. (1966). Free amino acids and related compounds during metamorphosis of the blowfly, *Phormia regina. J. Insect Physiol.* 12:1343-1362.

MEHENDALE, H. M., DAUTERMAN, W. C. and HODGSON, E. (1966). Phosphatidyl carnitine: a possible intermediate in the biosynthesis of phosphatidyl β-methylcholine in *Phormia regina* (Meigen). *Nature,* Lond. 211:759-761.

_____,_____&_____(1970). The incorporation of choline analogues into the phospholipids of *Phormia regina. Int. J. Biochem.* 1:429-437.

MEISTER, A. (1965). *Biochemistry of Amino Acids, Vol. 1,* Academic Press, N. Y., 592 p.

MILLER, D. S. and SAMUEL, P. Methionine sparing compounds. *Proc. Nutr. Soc.* 27:21A.

MOULTON, B., ROTTMAN, F., KUMAR, S. S. and BIEBER, L. L. (1970). Utilization of methionine, choline, and β-methylcholine by *Musca domestica* larvae. *Biochim. Biophys. Acta.* 210:182-185.

NOLAND, J. L. and BAUMANN, C. A. (1949). Requirement of the German cockroach for choline and related compounds. *Proc. Soc. Exp. Biol. Med.* 70:198-201.

RAO, D. R., ENNOR, A. H. and THORPE, B. (1967). The isolation and identification of L-lanthionine and L-cystathionine from insect haemolymph. *Biochemistry.* 6:1208-1216.

RETNAKARAN, A. and BECK, S. D. (1968). Amino acid requirements and sulfur amino acid metabolism in the pea aphid *Acyrthosiphon pisum* (Harris). *Comp. Biochem. Physiol.* 24:611-619.

ROCK, G. C. (1969). Sterol and water-soluble vitamin requirements for larvae of the red-banded leaf roller, *Argyrotaenia velutinana. Ann. Ent. Soc. Am.* **62**:611-613.

_____(1971). Utilization of D-isomers of the dietary indispensable amino acids by *Argyrotaenia velutinana* larvae. *J. Insect Physiol.* **17**:2157-2168.

ROCK, G. C. and HODGSON, E. (1971). Dietary amino requirements for *Heliothis zea* determined by dietary deletion and radiometric techniques. *J. Insect Physiol.* **17**:1087-1097.

ROCK, G. C. and KING, K. W. (1967). Quantitative amino acid requirements of the red-banded leaf roller, *Argyrotaenia velutinana* (Lepidoptera:Tortricidae). *J. Insect Physiol.* **13**:175-182.

ROCK, G. C., LIGON, B. G. and HODGSON, E. (in preparation).

ROYES, W. V. and ROBERTSON, F. W. (1964). The nutritional requirements and growth relations of different species of *Drosophila. J. Exp. Zool.* **156**:105-115.

SHARMA, G. K., HODGSON, E. and ROCK, G. C. (1972). Nutrition and metabolism of sulfur amino acids in *Argyrotaenia velutinana* larvae. *J. Insect Physiol.* **18**:9-18.

SHARMA, G. K., ROCK, G. C. and HODGSON, E. (in press). *J. Insect Physiol.*

SHELLEY, R. M. and HODGSON, E. (1970). Biosynthesis of phosphatidylcholine in the fat body of *Phormia regina* larvae. *J. Insect Physiol.* **16**:131-139.

_____&_____(1971a). Choline kinase from the fat body of *Phormia regina* larvae. *J. Insect Physiol.* **17**:545-558.

_____&_____(1971b). Substrate specificity and inhibition of choline and ethanolamine kinases from the fat body of *Phormia regina* larvae. *Insect Biochem.* **1**:149-156.

SHYAMALA, M. D. and BHAT, J. V. (1959). Antimetabolites in the nutrition of the silkworm, *Bombyx mori* L.: Part II-Ethionine as antagonist to methionine. *J. Scient. Ind. Res.* **18C**:242-245.

TAYLOR, J. F. and HODGSON, E. (1965). The origin of phospholipid ethanolamine in the blowfly, *Phormia regina* (Meig). *J. Insect Physiol.* **11**:281-285.

TAYLOR, M. W. and MEDICI, J. C. (1966). Amino acid requirements of grain beetles. *J. Nutr.* **88**:176-180.

TURNER, R. B. (1971). Dietary amino acid requirements of the cotton aphid, *Aphis gossypii*:the sulphur-containing amino-acids. *J. Insect Physiol.* **17**:2451-2456.

VANDERZANT, E. S. (1963). Nutrition of the boll weevil larva. *J. Econ. Ent.* **56**:357-362.

VANDERZANT, E. S. and RICHARDSON, C. D. (1964). Nutrition of the adult boll weevil: lipid requirements. *J. Insect Physiol.* **10**:267-272.

WILLIS, N. P. and HODGSON, E. (1970). Absence of transmethylation reactions involving choline, betaine, and methionine in the insecta *Int. J. Biochem.* **1**:659-662.

ZIELINSKA, Z. M. and DOMINAS, H. (1967). The origin of phospholipid ethanolamine and choline in a sawfly, *Acantholyda nemoralis. J. Insect Physiol.* **13**:1769-1779.

THE FUNCTIONS OF CHOLINE AND CARNITINE DURING SPERMATOGENESIS IN *DROSOPHILA*

B. W. Geer
Department of Biology, Knox College
Galesburg, Illinois 61401

INTRODUCTION

The quaternary amines, choline and carnitine, are important constituents of the male reproductive tract. The primary lipids of spermatozoa are choline-containing phospholipids, the principal phospholipid of invertebrate spermatozoa being lecithin (Mohri, 1958; Hartree and Mann, 1959; Barnes and Dawson, 1966) with choline plasmalogen and phosphatidylcholine both being prominent in vertebrate spermatozoa (Lovern *et al.*, 1957; Hartree and Mann, 1959; Gray, 1960; Scott *et al.*, 1963; Hartree, 1964; Bratanov *et al.*, 1965; Minassian and Terner, 1965). Though mammalian spermatozoa preferentially derive energy via carbohydrate degradation, under aerobic carbohydrate-free conditions mammalian spermatozoa metabolize either phosphatidylcholine or choline plasmalogen (Lardy and Phillips, 1941a, 1941b; Carlson and Wadstrom, 1958; Hartree and Mann, 1959, 1961; Hartree, 1964). Phospholipid may also serve as the source of energy for mammalian spermatozoa during the maturation period in the epididymis (Scott *et al.*, 1967). In contrast to the capacity of mammalian spermatozoa to acquire energy by either phospholipid or carbohydrate degradation, marine invertebrate spermatozoa, which are shed into an aquatic environment without the benefit of an accessory fluid, acquire energy by oxidation of intracellular phospholipid (Rothschild and Cleland, 1952; Mohri, 1957, 1964; Gonse, 1962).

Tissues of the mammalian male reproductive system contain high levels of carnitine (Marquis and Fritz, 1965; McCaman *et al.*, 1966); and carnitine acetyltransferase, the enzyme that metabolizes carnitine, is very active in late stage rat spermatocytes, spermatids, and mature spermatozoa (Go *et al.*, 1971; Vernon *et al.*, 1971). Although carnitine metabolism appears to be important to mammalian spermatozoan development, its exact function remains undetermined.

Drosophila related genera are relatively unique in that they can synthesize little, if any, choline (Bieber *et al.*, 1963; Bridges *et al.*, 1965; Geer *et al.*, 1968) and are dependent on dietary choline for growth and adult maintenance. Thus, *Drosophila* is conducive to studies of quaternary amine metabolism because the choline contents of *Drosophila* tissues can be altered by dietary manipulation and because lecithin-type phospholipid synthesis in most *Drosophila* tissues permits the substitution of structurally-related

Insect and mite nutrition – North-Holland – Amsterdam, (1972)

compounds for choline (Geer and Vovis, 1965). Developmental processes may, consequently, be studied in animals under essentially choline-free conditions or at different choline concentrations. Choline-specific processes may be identified and metabolic relationships between different quaternary amines established. This article reviews the nutritional and biochemical investigations of choline- and carnitine-dependent processes during *Drosophila* spermatogenesis.

DISCUSSION

Drosophila nutritional requirements

In general, the nutritional requirements of *Drosophila* for growth are like those of other higher organisms. The ten essential amino acids for vertebrate growth are needed (Schultz and Rudkin, 1949; Geer, 1966a) plus an adequate source of nonessential nitrogen such as glutamic acid (Geer, 1966b). Seven B vitamins - folic acid, nicotinic acid, riboflavin, thiamine, pyridoxine, pantothenic acid, and biotin - plus choline and cholesterol are essential for development (Hinton *et al.*, 1965; Sang, 1956). For optimal larval growth, adenylic acid and either cytidylic or uridylic acid (Sang, 1957; Geer, 1963) are needed along with a carbohydrate source. Although contaminating inorganic ions make dietary requirements for inorganic substances difficult to assess, magnesium, manganese, calcium, sodium, iron and phosphorus are known requirements for larval growth (Sang, 1956). The absence of a dietary fatty acid requirement is notable, *Drosophila melanogaster* larvae being able to synthesize both unsaturated and saturated fatty acids (Keith, 1967; Goldin and Keith, 1968).

Several studies in this and other laboratories have dealt with the dietary choline requirement of *Drosophila* (Hinton *et al.*, 1951; Sang, 1956; Geer and Vovis, 1965; Geer *et al.*, 1968; Geer, 1968; Geer and Dolph, 1970; Geer *et al.*, 1970). The status of the *Drosophila* choline requirement is:

1. *Drosophila* has little, if any, capacity to synthesize choline.
2. Choline is required for phospholipid synthesis. It is incorporated into membrane phosphatidylcholine and lysophosphatidyl choline.
3. The choline requirement for growth during the third larval instar is critical, the demand for phospholipid being greatest during this stage.
4. No structurally-related compound can completely replace choline during larval growth. Except for carnitine, choline-related compounds that can substitute for choline are incorporated into phospholipid without modification.
5. Choline is required for *Drosophila* adults for both spermatogenesis and egg production.

The choline requirement for male reproduction

The first indication of the dependence of *Drosophila* male reproduction upon dietary quaternary amines came with the observation that adult *D. melanogaster* males are sterile when raised on a diet supplemented with DL-carnitine in place of choline (Geer, 1967), a condition that is correctable by feeding choline. Onset of fertility, however, is dependent upon the dietary concentration of choline, the length of the feeding period, and the time that elapses before the fertility test. For example, a diet containing 5.7 x 10^{-4} M choline is sufficient to induce fertility in carnitine-raised adult males if fed for 5 days and if a total of 7 days elapses before the fertility test. A lower concentration of choline lengthens the period of choline feeding and total time required before males become fertile. Feeding of choline at any time during the larval period results in fertile adult males, but no choline-related compounds are capable of inducing fertility in carnitine-raised males.

Electron microscope observations of spermiogenesis in carnitine-raised *D. melanogaster* males have shown that the sterility is due to aberrant development of the sperm mitochondrion (Geer and Newburgh, 1970). The most striking abnormality is the presence of an atypical number of mitochondrial elements within spermatids, although nebenkern formation and elongation of the mitochondrion during spermiogenesis are also aberrant (Figs. 1 and 2). Formation of the axial filament complex, on the other hand, is normal.

Utilization of carnitine for phospholipid synthesis

When *Drosophila* utilize carnitine as a choline substitute, carnitine is decarboxylated to β-methylcholine and the β-methylcholine is subsequently incorporated into phospholipid (Geer *et al.*, 1971). The incorporation of carnitine derivatives into *D. melanogaster* phospholipids is dependent upon a choline-deficient physiological state, with the affinity of the *in vivo* biosynthetic mechanism being over 100 times greater for choline than for DL-carnitine. However, when starved for choline, approximately 85 per cent of the dietary DL-carnitine is metabolized to β-methylcholine (Geer and Newburgh, 1970).

(-)-Carnitine is much more effective than (+)-carnitine as a choline substitute for *D. melanogaster*. Larvae fed (-)-carnitine synthesize 2.4 times as much phosphatidyl-β-methylcholine as larvae fed (+)-carnitine though both are deficient in lecithin-type phospholipid compared to their choline-fed counterparts. (-)-Carnitine-fed larvae have only 62 per cent and (+)-carnitine-fed larvae 25 per cent as much lecithin-type phospholipid as choline-fed larvae. Furthermore, when fed (+)-carnitine or (-)-carnitine, larvae grow at rates which correspond to the capacities of the carnitine isomers to sponsor the biosynthesis of lecithin-type phospholipid (Table 1). These observations

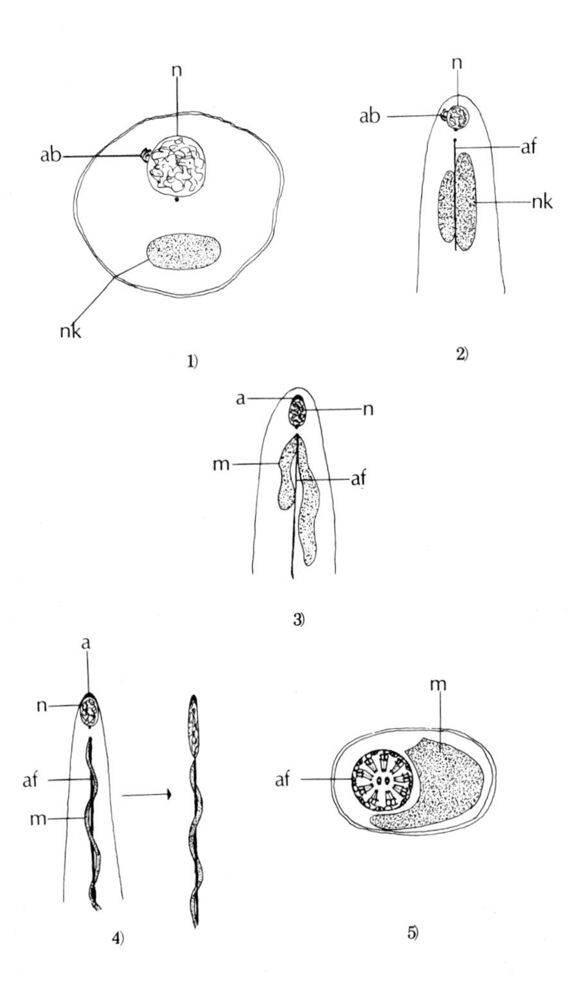

Figure 1. A diagram of spermiogenesis in fertile choline-raised *Drosophila melanogaster*. (1) Early spermatid with acroblast and nebenkern formation complete; (2) initial elongation, formation of the axial filament complex has begun and the nebenkern has divided; (3) elongation of the mitochondrion with degeneration of the other nebenkern derivative, elongation of the nucleus; (4) formation of the mature sperm, elongation completed, the hypothetical spiraling of the mitochondrion about the axial filament complex is shown; (5) a cross-section of a normal mature sperm tail showing the ultrastructure. During spermiogenesis the nebenkern divides giving rise to two mitochondrial elements (nebenkern derivatives). In *D. melanogaster* one mitochondrial element differentiates into the mature sperm mitochondrion; the other mitochondrial element degenerates. Explanations: a, acrosome; ab, acroblast; ax, axial filament complex; m. mitochondrion; n, nucleus; nk, nebenkern.

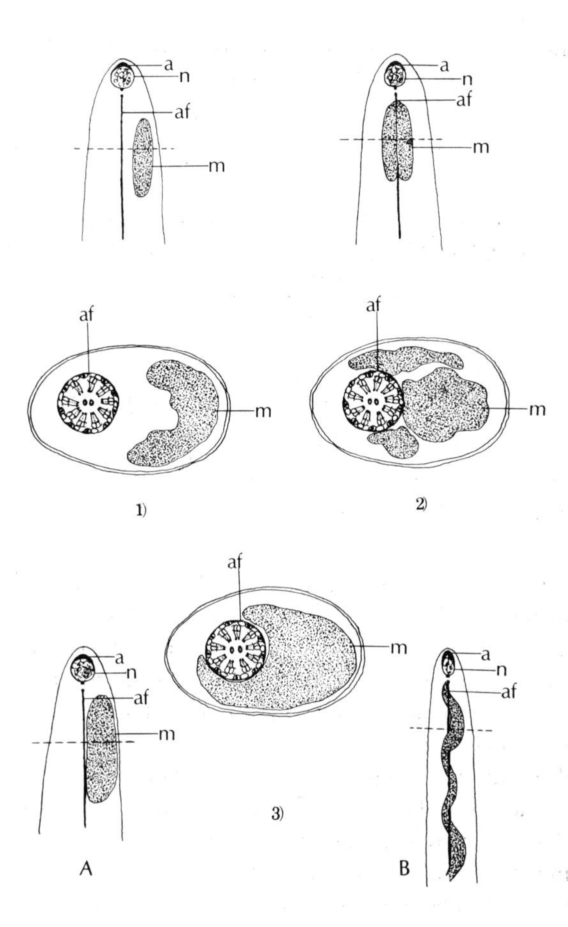

Figure 2. Diagrams showing the developmental abnormalities that occur during spermiogenesis in *Drosophila melanogaster* males raised on a synthetic diet in which DL-carnitine has been substituted for choline. (1) In many developing sperm the mitochondrion is not properly associated with the axial filament complex; (2) in other spermatids there is an atypical number of mitochondrial elements; (3) many spermatids exhibit a greatly enlarged mitochondrion in cross-section. Either (A) the entire mitochondrion is enlarged, or (B) the mitochondrion has improperly elongated so that it is enlarged only in certain regions. The explanations are as given for Fig. 1. From the observations of Geer and Newburgh (1970).

Table 1. Growth and phospholipid content of *Drosophila melanogaster* larvae fed a synthetic diet containing choline or carnitine isomers[a]

			Phospholipid	
				Phosphatidyl-
	Larvae	Growth	Lecithin-type	ethanolamine
Supplement	to enclose	period	% in choline	% in choline
5.7×10^{-4} M	%	days	fed larvae	fed larvae
Choline	82.8	7.5	100	100
DL-carnitine	71.9	8.3	55	80
(-)-carnitine	78.4	8.2	62	87
(+)-carnitine	2.9	12.6	25	88

[a]From Geer *et al.* (1971).

contrast with those in *Phormia regina* (Hodgson *et al.*, 1969) and *Musca domestica* (Bieber *et al.*, 1963) where (-) and (+)-carnitine were equally adept in replacing dietary choline. Observation of label turnover from *D. melanogaster* adults raised on labelled choline or carnitine-containing diets indicates that tissue-bound phosphatidyl-β-methylcholine is less stable than tissue-bound phosphatidylcholine.

These observations suggest that the sterility of carnitine-raised *D. melanogaster* males may be due to 1) inadequate amounts of lecithin-type phospholipid for sperm development, or 2) the inadequacy of phosphatidyl-β-methylcholine to replace phosphatidylcholine in specific developmental events during spermatogenesis, or 3) a deficiency of intact carnitine during choline starvation because of the decarboxylation of β-methylcholine to meet the quaternary amine demands for phospholipid synthesis. Current experimental results do not allow us to choose between these possibilities. If the sterility is due to a deficiency of lecithin-type phospholipid, it is not correctable by feeding high concentrations of DL-carnitine to larvae or adults.

The effect of carnitine substitution on locomotor activity

The capacity of adult *Drosophila* males to mate is dependent on diet-influenced locomotor activity. *D. melanogaster* adults raised on diet containing 5.7×10^{-4} M DL-carnitine, instead of choline, court and mate less readily than adults raised on choline (Geer, 1967) because the locomotor activity of adults raised at this concentration of dietary carnitine is much less than that of choline-raised adults (Geer *et al.*, 1971). Increasing the level of DL-carnitine above 5.7×10^{-4} M in the larval diet results in increased adult activity

(Fig. 3) with the greatest adult activity being observed at 45.6 x 10^{-4} M DL-carnitine, 8 times the amount giving maximum larval growth. Adults raised on a diet with this level of DL-carnitine are as active as their choline-raised counterparts. The greatest adult activity is observed in *D. melanogaster* raised on a diet supplemented with both choline and DL-carnitine, 5.7 x 10^{-4} M choline to 1.4 x 10^{-4} M DL-carnitine being optimal. Administration of choline to adult *D. melanogaster* adults who are inactive for dietary

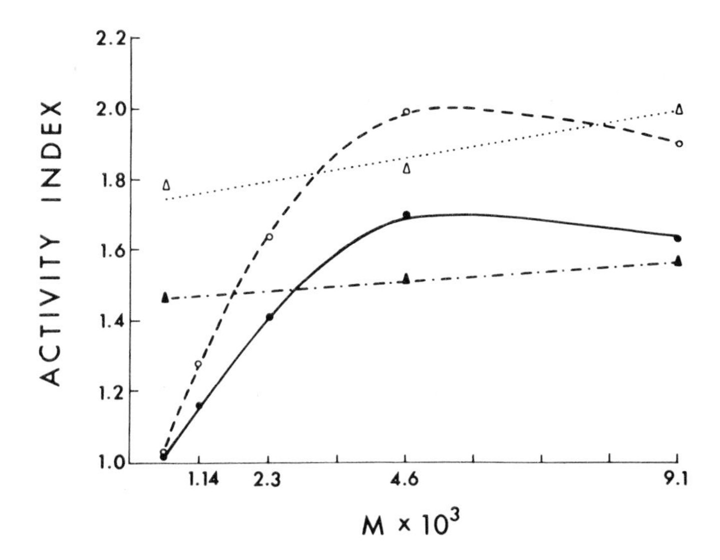

Figure 3. The activity indices of *D. melanogaster* males (o--o) and females (o--o) raised on diets supplemented with concentrations of DL-carnitine and males (······) and females (------) raised on diets supplemented with different concentrations of choline. The activity index for a group of animals may range from 1, indicating no measurable activity, to 4. The figure is from Geer *et al.* (1971).

reasons increases their locomotor activity but never to the activity level of adults raised on a choline-supplemented diet. All compounds metabolized to β-methylcholine by *D. melanogaster* - carnitine, acetyl-β-methylcholine, deoxy-carnitine, and 0-acetyl-DL-carnitine - result in low level adult activities when substituted for dietary choline at concentrations equivalent to the choline level sufficient to support maximum larval growth, 5.7 x 10^{-4} M.

The correlation between adult locomotor activity and dietary carnitine concentration indicates that physiological processes other than spermatogenesis are affected by the tissue levels of choline and/or carnitine. Any of the three hypotheses presented in the previous section may be advanced to explain the low levels of locomotor activity when choline is absent from the diet and dietary carnitine is present in low concentrations. In addition, it is conceivable that alterations of dietary quaternary amine concentration alter the levels of transmitter substances in the nervous system. Connolly, Tunnicliff, and Rick (1971) recently observed that adult locomotor activity can be altered by diet-influenced changes in brain quaternary amine concentrations.

Carnitine acetyltransferase in sperm and the male reproductive tract

Several investigations indicate an important role for carnitine in male reproduction. The function of carnitine is believed to be linked to the action of the enzyme, carnitine acetyltransferase. Carnitine acetyltransferase (EC 2.3.1.7) catalyzes the reaction

$$(CH_3)\overset{+}{N}CH_2\underset{\underset{\underset{O}{\|}}{O\text{-}CCH_3}}{CH}CH_2COOH + CoA\text{-}SH \rightleftharpoons CoA\text{-}S\text{-}CCH_3 + (CH_3)\overset{+}{N}CH_2\underset{OH}{CH}CH_2COOH$$

and is hypothesized to function in the transfer of acetyl groups generated by fatty acid or pyruvate degradation to sites containing Krebs cycle enzymes within the mitrochondrion (Fritz, 1967). As a consequence, translocation of acetyl groups via carnitine acetyltransferase could be important to the maintenance of basal metabolism. Carnitine also stimulates long-chain fatty acid oxidation by facilitating the transport of acyl groups into the mitochondrion as acylcarnitine derivatives.

Carnitine acetyltransferase is very active in the male reproductive tract. Marquis and Fritz (1965) traced carnitine quantity and carnitine acetyltransferase activity during rat development and found carnitine and carnitine acetyltransferase levels in the testis to increase strikingly during the period of sperm maturation in the rat. Carnitine acetyltransferase activity is linked to the formation of functional sperm in *Drosophila* (Geer and Newburgh, 1970). In *D. melanogaster* males raised on a choline-containing diet, labeled carnitine accumulated in the testis to a level constituting more than 25 per cent of the total body quantity. In contrast, the amount of labeled carnitine in the testes of sterile carnitine-raised males was less than one-third of their fertile counterparts.

D. melanogaster males that are sterile for genetic or nutritional reasons exhibit low levels of carnitine and carnitine acetyltransferase as compared to their fertile counterparts (Table 2). The quantity of carnitine that accumulates in the testes of sterile carnitine-raised males is less than one-third and the enzyme level about 55 per cent that of choline-raised males. However, when sterile carnitine-raised males are fed choline the quantity of testicular carnitine and level of carnitine acetyltransferase increases (Fig. 4) at the time

Table 2. Carnitine acetyltransferase activity in testes of sterile and fertile adult *Drosophila melanogaster* males[a]

Larval diet	Sex chromosome composition	Fertility	Enzyme activity moles/mg/protein/hr
Synthetic + choline	XY	Fertile	3.55 + 0.44
Synthetic + carnitine	XY	Sterile	1.54 + 0.26
Cornmeal	XY	Fertile	3.54 + 0.48
Cornmeal	XO	Sterile	1.76 + 0.39

[a]From Geer and Newburgh (1970).

that the males become fertile. The low activity of carnitine acetyltransferase in the testes of XO males as compared to XY males indicates that the Y-chromosome is needed for the developmental events leading to high enzyme activity.

There is evidence from our studies with *Drosophila hydei* that carnitine acetyltransferase activity is critical to spermiogenesis as well as to mature spermatozoan metabolism. *D. hydei* is a slow maturing *Drosophila* species, males being unable to successfully inseminate females until 8-10 days after eclosion. During the period of sexual maturation testis size and gametic content change markedly (Geer *et al.*, 1972). Testis length increases steadily from 0.81 cm at eclosion to 2.45 cm at 8 days of age, the time that motile sperm first appear in the testis. At eclosion the gametic content in terms of volume is approximately 37 per cent spermatogonia and spermatocytes, and 63 per cent early spermatids. At sexual maturation spermatogonia and spermatocytes constitute about 11 per cent and spermatids and mature sperm 89 per cent of the total testis gametic content. Carnitine acetyltransferase activity peaks in *D. hydei* testes at an age before mature functional sperm are formed (Table 3), whereas glycolytic and Krebs cycle enzymes are most active in sexually mature testes.

Developing gametes are arranged in the mature *D. hydei* testis so that gametes of different stages of spermatogenesis may be isolated by dissection of the testis into different parts. Gametes of the middle region containing 100 per cent mid and late stage spermatids have the highest carnitine acetyltransferase activity; the posterior region possessing mature sperm have a lower, but still high, activity; and the enzyme is least active in the anterior region which contains 70 per cent spermatogonia and spermatocytes, and 35 per cent early spermatids (Table 4).

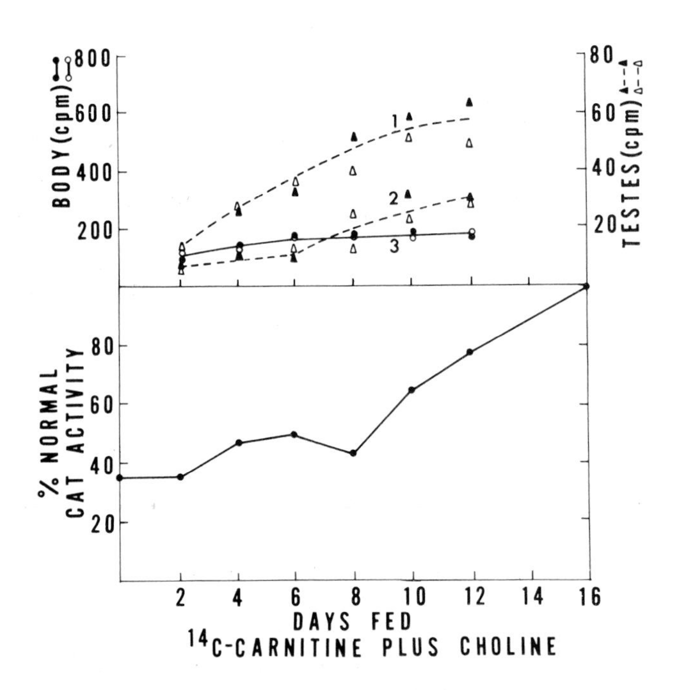

Figure 4. The upper half of the figure shows the incorporation of label from dietary DL-carnitine-methyl-^{14}C (solid circles and triangles) and DL-carnitine-carboxyl-^{14}C (open circles and triangles) into the testes and whole body of adult *D. melanogaster* males. Curve 1 represents testes of males grown previously on a choline diet; Curve 2 represents testes of males grown previously on a carnitine diet; Curve 3 represents the bodies of males grown previously on choline or carnitine. The horizontal axis indicates the number of days that males were fed a diet containing 0.05 Ci of ^{14}C-carnitine per ml with sufficient unlabeled DL-carnitine to raise the concentration of 5.7×10^{-4} M. In addition, the diet contained 5.7×10^{-4} M unlabeled choline.

The lower half of the figure indicates the relative testicular carnitine acetyltransferase (CAT) activity of carnitine-raised males fed a diet of the same composition as indicated above for different time periods. The activity of the enzyme in the testes of choline-raised males was used as the normal level. The figure is modified from Geer and Newburgh (1970).

Table 3. Enzyme activities at different ages in the testis of male *Drosophila hydei*[a]

Enzyme	Age			
	1 Day	4 Days	6 Days	10 Days
Hexokinase[b]	21.7 ± 1.5[c]	23.6 ± 3.6	37.7 ± 9.1	63.6 ± 9.0
3-Phosphoglycerate kinase	52.6 ± 8.3	46.9 ± 7.6	46.6 ± 8.4	86.0 ± 12.1
Malate dehydrogenase	485.0 ± 56.2	99.8 ± 7.8	212.9 ± 13.2	599.2 ± 26.5
Malic enzyme	220.7 ± 18.0	235.9 ± 59.0	228.8 ± 34.9	219.2 ± 39.0
Carnitine acetyltransferase	45.6 ± 4.7	100.3 ± 7.7	80.0 ± 6.4	80.5 ± 6.6

[a]From Geer *et al.* (1972).
[b]Glucose was employed as the substrate.
[c]Nanomoles of cofactor oxidized or reduced/mg of homogenate protein/minute of incubation at 30°C. Carnitine acetyltransferase activity is given in nanomoles of acetylcarnitine formed/mg of homogenate protein/min of incubation at 37°C. The mean ± the standard deviation is given.

In general, glycolytic and Krebs cycle enzymes are most active in the region containing mature sperm, the posterior region, and least active in the anterior region where gametes of the early stages of spermatogenesis are concentrated (Table 4). *Drosophila* spermatozoa apparently possess a much higher glycolytic-Krebs cycle capacity than early stage gametes. This is confirmed by observations of enzyme activities in the testes of *D. hydei* males at different adult ages (Table 3). Enzymes associated with metabolic functions other than carbohydrate degradation, however, exhibit different testis developmental patterns. Malic enzyme, often associated with NADPH production for lipid synthesis (Wise and Ball, 1954; Olson, 1966; Horie, 1968), is very active in the testis at all maturation stages, but is most prominent in testis regions possessing early stage gametes. β-Hydroxyacyl dehydrogenase, a fatty acid oxidation enzyme, declines in activity during *D. hydei* maturation and is of little consequence in mature sperm metabolism. The aminotransferases are very active in the testis at all maturation stages, but L-aspartate aminotransferase is less active in elongating spermatids than at earlier and later stages of spermatogenesis, whereas L-alanine aminotransferase is like the glycolytic and Krebs cycle enzymes in that it is most active in mature sperm.

Table 4. Enzyme activities in *Drosophila hydei* testis regions[a]

Enzyme	Posterior	Middle	Anterior
Hexokinase	151.7 ± 12.4[b]	38.2 ± 2.5	10.4 ± 2.4
Citrate synthase	203.9 ± 40.8	150.0 ± 26.3	77.3 ± 23.6
Malate dehydrogenase	494.1 ± 54.6	430.2 ± 63.4	553.2 ± 48.6
Malic enzyme	64.6 ± 13.4	130.1 ± 19.0	136.5 ± 12.7
Carnitine acetyltransferase	112.7 ± 7.8	141.9 ± 7.1	94.7 ± 9.0

[a]From Geer *et al.* (1972).
[b]Enzyme activities and substrate as given for Table 3. Mean activity ± standard deviation.

Because of these observations we have proposed (Geer *et al.*, 1972) that early in spermatogenesis gametes depend on glycolysis, the Krebs cycle, and, to a lesser extent, the pentose shunt cycle, for energy. We postulated that the capacity of sperm to degrade carbohydrate and amino acids via glycolysis and the Krebs cycle increases 2-4 fold in *Drosophila* spermatogenesis, and we speculated that acetylcarnitine and L-alanine are temporary glycolytic end-products in mature sperm. Both L-alanine and acetylcarnitine, once formed, can be metabolized by conversion to acetyl CoA and subsequent degradation by the Krebs cycle reactions (Fig. 5).

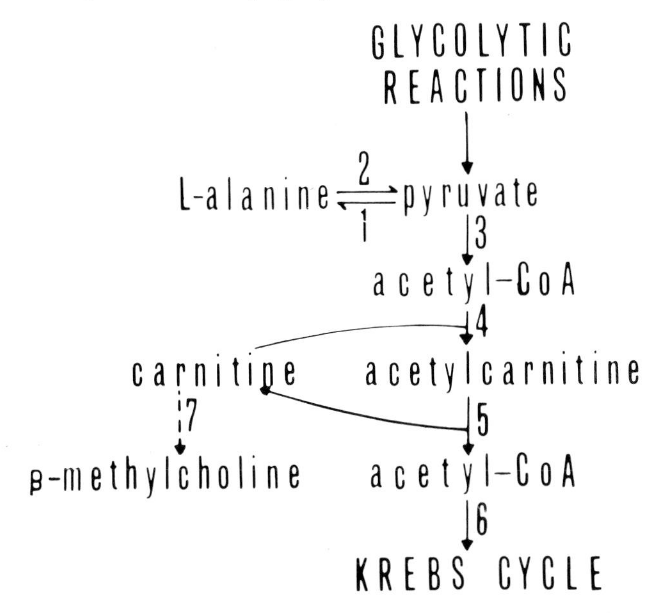

Figure 5. The possible relationships of L-alanine and acetylcarnitine to glycolysis and the Krebs cycle in *Drosophila* spermatozoan metabolism. L-alanine (reaction 1) and acetylcarnitine (reactions 3 and 4) may serve as temporary endproducts of glycolysis. However, L-alanine (reactions 2, 3, 4, 5 and 6) and acetylcarnitine (reactions 5 and 6) can be metabolized after formation by the Krebs cycle reactions. Reactions 4 and 5 may be important in the translocation of "acetyl" groups within the mitochondrion with acetylcarnitine serving as the transport molecule. Under choline-starvation conditions carnitine is metabolized to β-methylcholine (reaction 7). This may result in a carnitine deficiency that limits carbohydrate metabolism.

Carnitine stimulation of substrate degradation

Recently, we tested the metabolic model for *Drosophila* spermatozoan energy-yielding metabolism by examining the influence of exogenous DL-carnitine on ^{14}C-labeled substrate degradation. Homogenates of testes and thoraces from sexually immature one-day old and ten-day old *D. hydei* males were assayed (Table 5). The most striking characteristic of mature *D. hydei* testis metabolism is its high pyruvate degradative activity. The specific activity for [1-^{14}C] pyruvate decarboxylation in the mature testis and twice as great as the mature thoracic tissue specific activity. The rate of [1-^{14}C]

Table 5. Decarboxylation of labeled substrates by *Drosophila hydei* testis and thoracic tissue.

Substrate	Thoracic tissue		Testis	
	No supplement	2.4 M DL-carnitine	No supplement	2.4 M DL-carnitine
	Immature			
[1-^{14}C] Pyruvate	191[a]	315	167	170
[2-^{14}C] Pyruvate	52	56	54	31
[1-^{14}C] L-alanine	111	163	342	382
[U-^{14}C] L-alanine	70	91	64	153
[1-^{14}C] Acetate	43	55	43	64
[1-^{14}C] Palmitate	1	1	2	2
	Mature			
[1-^{14}C] Pyruvate	1030	1263	2332	2707
[2-^{14}C] Pyruvate	122	104	63	92
[1-^{14}C] L-alanine	241	332	516	607
[U-^{14}C] L-alanine	152	228	199	251
[1-^{14}C] Acetate	209	231	66	74
[1-^{14}C] Palmitate	4	4	3	1

[a]Measured radiometrically by capturing $^{14}CO_2$ emitted from the reaction mixture with Hyamine and determining the radioactivity by liquid scintillation counting methods. The reaction mixture contained 100 moles KH_2PO_4-Na_2HPO_4 buffer, pH 7.4; 4.0 moles ADP; 1.2 moles DL-carnitine (in supplemented reactions), 66 moles NaCl; 2.5 moles KCl; 0.6 mole Mg 504; 2.15 moles ^{14}C-substrate (2.5 million dpm per assay), and 1-1.5 mg of homogenate protein in a total reaction volume of 0.5 ml. The specific activities were calculated as picomoles of substrate decarboxylated per minute of incubation at 30°C per mg of homogenate protein using the known specific activities of the substrates.

L-alanine decarboxylation in the mature testis was about twice as rapid for that in mature thoracic tissue, but [2-^{14}C] acetate was metabolized three times more rapidly in the mature thorax as in the mature testis. Butyrate and palmitate were degraded very poorly, if at all, by *Drosophila* tissue. Exogenous DL-carnitine stimulated pyruvate, L-alanine, and acetate degradation 20 - 30 per cent in both *D. hydei* testis and thoracic tissue.

That mature testes degrade [1-^{14}C] pyruvate much more readily than mature thoracic tissue, but not [2-^{14}C] pyruvate indicates that the degradation process in the testis is not directly dependent upon Krebs cycle activity.

This is consistent with our model for sperm metabolism which contends that pyruvate may be converted to the temporary endproducts, L-alanine and acetylcarnitine. Thus, the enzymes catalyzing the reactions yielding these compounds, L-alanine aminotransferase and carnitine acetyltransferase, may control the rate of *Drosophila* spermatozoan glycolysis. Although L-alanine is readily decarboxylated in the mature testis, earlier observations (Geer and Downing, 1972) indicated that L-alanine formation from pyruvate occurs more rapidly in the mature testis than the conversion of L-alanine to pyruvate. Furthermore, the low level of acetate degradation in the mature testis, as compared to the mature thorax, evidences a moderate level of testis Krebs cycle activity, since acetate presumably is degraded via the Krebs cycle. In conclusion, *Drosophila* late stage developing gametes and spermatozoa appear to depend upon carnitine acetyltransferase and L-alanine aminotransferase activities, as well as Krebs cycle activity, for maximum carbohydrate degradation.

Protein and lipid synthesis during spermatogenesis

The formation and accumulation of late stage gametes in the testis during the maturation of *D. hydei* is accompanied by more than 4-fold increases in both lipid and protein synthesis (Geer and Downing, 1972). The rates of protein and lipid synthesis in the mature testis are high compared to the rates of synthesis in other tissues. Insect flight muscle is known to exhibit a high level of protein synthesis (Wyatt, 1968), and the rate of protein in the *D. hydei* mature testis is as high as that in mature thoracic tissue. Also, the rate of lipid synthesis from acetate in the *D. hydei* mature testis is comparable to that found in the desert locust fat body (Walker and Bailey, 1970), which presumably possesses a well developed lipid synthesis system. In contrast, lipid synthesis in *D. hydei* thoracic tissue occurs at a very low rate (Geer and Downing, 1972). Undoubtedly, this is related to the observation that glycogen, and not lipid as in the desert locust, is the primary energy source for *Drosophila* flight (Wigglesworth, 1949). In any event, the capacity to synthesize lipid varies markedly between tissues at different developmental stages in *D. hydei*.

Exogenous DL-carnitine has no detectable effect on *D. hydei* testis lipid and protein synthesis. Perhaps there is sufficient carnitine present in the testis to meet the requirements for these two processes or, more likely, compounds such as pyruvate or citrate are the vehicles of transport of substrate materials to protein and lipid synthesis sites rather than carnitine derivatives.

Relation of nutritional and genetic factors that affect spermatogenesis

The research reported in this study was designed to examine the developmental metabolic pattern of spermatogenesis and to gain some insight about factors controlling spermatogenesis. Among other things, these studies suggest that the demands for phospholipid synthesis in developing sperm are more precise than in other *Drosophila* tissues. In contrast to the development of other *Drosophila* tissues, no quaternary amine can replace choline in sperm development, making spermatogenesis susceptible to dietary deficiencies. In addition, the quaternary amine demands for phospholipid synthesis are related to the requirement for carnitine, which is involved in the regulation of testis metabolism. This appears to be a facet of a general sensitivity of *Drosophila* spermatogenesis to genetic and environmental factors. Meyer (1969) has shown that several chemicals influence testis development and the number of motile sperm formed when injected into males. Furthermore, high temperature is known to be capable of sterilizing *Drosophila* males.

Genetically, it is known that the Y-chromosomes of *D. melanogaster* and *D. hydei* (Brosseau, 1960; Meyer and Hess, 1968) possess genes that function in spermiogenesis, 5 genes located on the Y-chromosome of *D. hydei* and 7 genes on the Y-chromosome of *D. melanogaster*. Mutant genes that block spermiogenesis have been induced by mutagens at many different loci on the X-, second, and third chromosomes of *D. melanogaster* (Edmundson, 1952; Meyer, 1969; Romrell *et al.*, 1972a, b; Bowman, unpublished; Geer, unpublished). Furthermore, the segregation-distorter gene of the second chromosome and modifier genes of the X and second chromosomes influence the dysfunction of developing sperm bearing a non-distorter allele (Zimmering *et al.*, 1970). Many other mutant genes exerting phenotypic effects seemingly unrelated to male sterility result in male sterility in *D. melanogaster*. Although mutation studies have not advanced sufficiently to accurately determine the number of genes whose primary action is in *D. melanogaster* spermatogenesis, estimates (Bowman and Geer, unpublished) indicate the number of gene loci to be greater than 100 and perhaps as high as 300. Thus, many genes must function, ostensibly in a highly regulated pattern, to maintain functional spermatogenesis.

The time of gene activity that supports spermatogenesis is subject to debate. Lindsley and Grell (1969) have shown that sperm lacking any of the chromosomes of *D. melanogaster* may develop into functional sperm provided that genes of the chromosomes are represented in the genome of the male. Two alternate models have been presented to explain these observations: 1. Genes that support spermiogenesis are active in long-lived RNA synthesis with the long-lived RNA functioning during sperm maturation after meiosis (Beermann, 1965, 1967; Hess, 1966, 1967; Hess and Meyer, 1968). 2. The spermatids are interconnected by cytoplasmic bridges after meiosis and thus represent a 64-nucleus, 64-ploid heterokaryon, which forms 64 spermatozoa as a unit under the control of the combined genotype of the 64 nuclei (Lindsley and Grell, 1969). The models differ basically in that genes supporting spermatogenesis would be active either only before meiosis as indicated in the first model, or both before and after meiosis according to the latter model. At present, the absence of detectable RNA synthesis in the spermatid of *Drosophila* (Olivieri and Olivieri, 1965),) would seem to necessitate the inclusion of long-lived RNA into a workable model.

Spermatogenesis in *Drosophila*, and in other animals, is very susceptible to genetic and environmental influences. Evolutionists have long known that hybrid male offspring derived from the mating of members of different species are sterile much more often than their sisters. This may reflect, as Meyer (1969) has suggested, the phylogenetically primitive regulation mechanisms that still control the differentiation of the flagellate-like gametes of higher organisms. If this is true, in light of the number of genes that must act and number of gene products and substrate materials that must be coordinated for functional spermatogenesis, the susceptibility of spermatogenesis to environmental and genetic variations is not surprising. On the other hand, why have the controlling mechanisms escaped the forces of natural selection?

Perhaps, the disadvantages of a relatively unstable sperm-forming mechanism are outweighed by advantages. Failure of hybrid offspring to reproduce prevents the interbreeding and thus genetic intermixing of different species. Similarly, the sensitivity of spermatogenesis to nutritional, temperature, and other environmental factors limits the ecological niche of the species. In brief, the susceptibility of spermatogenesis to genetic and environmental disruption may represent an important species barrier. The disadvantages of a reduced reproduction capacity due to environmental and genetic conditions may be counterbalanced by the restriction of the species to optimal environmental conditions and preservation of a favorable gene pool for growth and reproduction.

CONCLUSION

Experimentation in this laboratory has shown that mitochondrion development during spermiogenesis in *Drosophila* is aberrant when carnitine is substituted for choline in the diet, probably due to the specific need for phosphatidylcholine. Phosphatidyl-β-methylcholine, a derivative of carnitine, is an inadequate replacement. (-)Carnitine is important to developing and mature spermatozoa because it stimulates carbohydrate degradation. Acetylcarnitine apparently serves as a temporary endproduct of glycolysis, facilitating pyruvate degradation by effecting the removal of acetyl-CoA and thus preventing inhibition of the pyruvate dehydrogenase complex.

Choline and carnitine are related metabolically in *Drosophila* in that under choline-deprivation conditions carnitine is decarboxylated to β-methylcholine to replace choline. As a result, carnitine may be present in suboptimal concentrations when *Drosophila* is starved for choline, even though sufficient carnitine is synthesized by *Drosophila* for metabolic purposes when dietary choline is present in adequate quantities. Choline-starvation conditions may thus disrupt spermatogenesis in *Drosophila* by blocking phosphatidylcholine synthesis, needed for spermatozoan mitochondrion development, and limiting the supply of carnitine required for the energy-yielding metabolism of developing sperm.

Studies of nutritional factors that influence spermatogenesis complement genetic studies. Manipulation of environmental and genetic factors has good potential for future use in the control of animal reproduction. Eradication of the screw-worm from southeastern United States is an example where this research has already been put to use. Needless to say, control of reproduction by genetic disruption is not possible, nor desirable, in all animals; consequently, knowledge of metabolic control mechanisms governing spermatogenesis that are subject to external manipulation is of great interest. A nutritional method of reproduction control would have the advantage over other methods because it would be easier to administer to the animal.

Acknowledgements.—I am indebted to the many students at Knox College who have actively participated in the research reported here and without whose assistance this work would have been impossible. I would like to thank Miss Deborah Neil for making the drawings in Figures 1 and 2. I would also like to acknowledge the financial support of National Science Foundation Grant GB-13393.

REFERENCES

BARNES, H. and DAWSON, R. M. C. (1966). Lipids of *Balanus balanus* spermatozoa. *J. Mar. Biol. Ass. U.K.* 46:263-265.

BEERMANN, W. (1965). Operative Gliederung der Chromosomen (Von W. Beermann). *Naturwissenschaften.* 52:365-375.

________(1967). Gene action at the level of the chromosome pp. 179-201. in *Heritage from Mendel* (R. A. Brink, Ed.) University of Wisconsin Press, Madison, Wisconsin.

BIEBER, L. L., CHELDELIN, V. H. and NEWBURGH, R. W. (1963). Studies on a β-methyl choline-containing phospholipid derived from carnitine. *J. Biol. Chem.* 238:1262-1265.

BIEBER, L. L. and MONROE, R. E. (1969). The relation of carnitine to the formation of phosphatidyl-β-methylcholine by *Tenebrio molitor* L. larvae. *Lipids.* 4:293-298.

BRIDGES, R. G., RICKETTS, J. and COX, J. T. (1965). The replacement of lipid-bound choline by other bases in the phospholipids of the housefly, *Musca domestica. J. Insect Physiol.* 11:225-236.

BRATONOV, K., KIKOV, V. and ANGELOVA, Z. (1965). Fluorescent-cytochemical investigations into the lipids of the spermatozoa of certain farm animals. *Vet. Med. Nauki.,* 2:731-737.

BROUSSEAU, G. E., JR. (1960). Genetic analysis of the male fertility factors in the Y. chromosome of *Drosophila melanogaster. Genetics.* **45**:257-274.

CARLSON, L. A. and WADSTROM, L. B. (1958). Determination of unesterified fatty acid in plasma. *Scand. J. Clin. Lab. Invest.* **10**:407-414.

CONNOLLY, K., TUNNICLIFF, G. and RICK, J. T. (1971). The effects of -Hydroxybutyric acid on spontaneous locomotor activity and dopamine level in a selected strain of *Drosophila melanogaster. Comp. Biochem. Physiol.* **40B**:321-326.

EDMUNDSON, M. (1952). New mutants. Drosophila Information Service, **26**:61.

FRITZ, I. B. (1967). An hypothesis concerning the role of carnitine in the control of interrelations between fatty acid and carbohydrate metabolism. *Perspect. Biol. Med.* **10**:643-677.

GEER, B. W. (1963). Ribonucleic acid-protein relation in *Drosophila* nutrition. *J. Exp. Zool.* **154**:353-364.

_________(1966a). Utilization of D-amino acids for growth by *Drosophila melanogaster. J. Nutr.* **90**:31-39.

GEER, B. W. (1966b). Comparison of some amino acid mixtures and proteins for the diet of *Drosophila melanogaster. Trans. Ill. St. Acad. Sci.* **59**:3-10.

_________(1967). Dietary choline requirements for sperm motility and normal mating activity in *Drosophila melanogaster. Biol. Bull. Mar. Biol. Lab.,* **Woods Hole.** 133:548-566.

_________(1968). Modification of the larval nutritional requirements of *Drosophila melanogaster* by maternally inherited choline. *Archs Int. Physiol. Biochem.* **76**:797-805.

GEER, B. W. and DOLPH, W. W. (1970). A dietary choline requirement for egg production in *Drosophila melanogaster. J. Reprod. Fert.* **21**:9-15.

GEER, B. W., DOLPH, W. W., MAGUIRE, J. A. and DATES, R. J. (1971). The metabolism of dietary carnitine in *Drosophila melanogaster. J. Exp. Zool.* **176**:445-460.

GEER, B. W. and DOWNING, B. C. (1972). Changes in lipid and protein synthesis during spermatozoan development and thoracic tissue maturation in *Drosophila hydei.* Wilhelm Roux' Archiv., **170**:83-89.

GEER, B. W., MARTENSEN, D. V., DOWNING, B. C. and MUZYKA, G. S. (1972). Metabolic changes during spermatogenesis and thoracic tissue maturation in *Drosophila hydei. Dev. Biol.,* **28**:390-406.

GEER, B. W., OLANDER, R. M. and SHARP, P. L. (1970). Quantification of dietary choline utilization in adult *Drosophila melanogaster* by radioisotope methods. *J. Insect Physiol.* **16**:33-43.

GEER, B. W. and VOVIS, G. F. (1965). The effects of choline and related compounds on the growth and development of *Drosophila melanogaster. J. Exp. Zool.* **158**:223-236.

GEER, B. W., VOVIS, G. F. and YUND, M. A. (1968). Choline activity during the development of *Drosophila melanogaster. Physiol. Zool.* **41**:280-292.

GO, V. L. W., VERNON, R. G., and FRITZ, I. B. (1971). Studies on spermatogenesis in rats: III. Effects of hormonal treatment on differentiation kenetics of the spermatogenic cycle in regressed hypophysectomized rats. *Can. J. Biochem.* **49**:768-775.

GOLDIN, H. H. and KEITH, A. D. (1968). Fatty acid biosynthesis by isolated mitochondria from *Drosophila melanogaster. J. Insect Physiol.* **14**:887-900.

GONSE, P. H. (1962). Respiration and oxidative phosyhorylation in relation to sperm motility. pp. 99-132. in *Spermatozoan Motility* (D. W. Bishop, Ed.). American Association for the Advancement of Science, Washington, D. C.

GRAY, G. M. (1960). The presence of lecithin in whole ram semen. *Biochem. J.* **74**:1P-2P.

HARTREE, E. F. (1964). Metabolism of plasmalogens. in *Metabolism and Physiological Significance of Lipids.* (R. M. C. Dawson and D. N. Rhodes, Eds.). Wiley, New York. pp. 207-218.

HARTREE, E. F. and MANN, T. (1959). Plasmalogen in ram semen, and its role in sperm metabolism. *Biochem. J.* 71:423-434.

_____&_____(1961). Phospholipids in ram semen: metabolism of plasmologen and fatty acids. *Biochem. J.* **80**:464-476.

HESS, O. (1966). Structural modifications of the Y-chromosome in *Drosophila hydei* and their relation to gene activity. in *Chromosomes Today* (C. D. Darlington and K. R. Lewis, Eds.). Vol. 1. Oliver and Boyd, Edinburgh.

_________(1967). Complementation of genetic activity in translocated fragments of the Y-chromosome in *Drosophila hydei. Genetics.* **56**:283-295.

HESS, O. and MEYER, G. F. (1968). Genetic activities of the Y-chromosome in *Drosophila* during spermatogenesis. *Adv. Genet.* **14**:171-223.

HINTON, T., NOYES, D. T. and ELLIS, JR. (1951). Amino acids and growth factors in a chemically defined medium for *Drosophila. Physiol. Zool.* **24**:335-353.

HODGSON, E., DAUTERMAN, W. C., MEHENDALE, H. M., SMITH, E. and KHAN, M. A. Q. (1969). Dietary choline requirements, phospholipids and development in *Phormia regina. Comp. Biochem. Physiol.* **29**:343-359.

HORIE, Y. (1968). The oxidation of NADPH by the soluble fraction of the fat body of the silkworm, *Bombyx mori* L. *J. Insect Physiol.* **14**:417-424.

KEITH, A. D. (1967). Fatty-acid metabolism in *Drosophila melanogaster.* Interaction between dietary fatty acids and *de novo* synthesis. *Comp. Biochem. Physiol.* **21**:587-600.

LARDY, H. A. and PHILLIPS, P. H. (1941a). The interrelation of oxidative and glycolytic processes as sources of energy for bull spermatozoa. *Am. J. Physiol.* **133**:602-609.

_____&_____(1941b). Phospholipids as a source of energy for motility of bull spermatozoa. *Am. J. Physiol.* **134**:542-548.

LINDSLEY, D. L. and GRELL, E. H. (1969). Spermiogenesis without chromosomes in *Drosophila melanogaster. Genetics.* **61** (supplement), 69-78.

LOVERN, J. A., OLLEY, J., HARTREE, E. F. and MANN, T. (1957). The lipids of ram spermatozoa. *Biochem. J.* **67**:630-643.

MCCAMAN, R. E., MCCAMAN, M. W. and STAFFORD, M. L. (1966). Carnitine acetyltransferase in nervous tissue. *J. Biol. Chem.* **241**:930-934.

MANN, T. (1964). *The Biochemistry of Semen and of the Male Reproductive Tract.* Wiley, New York, 1964. pp. 433.

MARQUIS, N. R. and FRITZ, I. B. (1965). Effects of testosterone on the distribution of carnitine, acetylcarnitine, and carnitine acetyltransferase in tissues of the reproductive system of the male rat. *J. Biol. Chem.* **240**:2197-2200.

MEYER, H. U. (1969). New mutants. Drosophila Information Service, 33:97.

MEYER, G. F. (1968). Spermiogenese in normalen und Y-defizienten Mannchen von *Drosophila melanogaster* und *D. hydei. Z. Zellforsch. Mikrosk.* **84**:141-175.

_________(1969). Experimental studies on spermiogenesis in *Drosophila. Genetics.* **61** (supplement): 79-92.

MINASSIAN, E. S. and TERNER, C. (1966). Biosynthesis of lipids by human and fish spermatozoa. *Am. J. Physiol.* **210**:615-618.

MOHRI, H. (1957). Endogenous substrates of respiration in sea-urchin spermatozoa. *J. Fac. Sci. Tokyo Univ.* (section 4). **8**:51-63.

_________(1964). Phospholipid utilization in sea-urchin spermatozoa. *Publ. Staz. Zool. Napoli.* **34**:53-58.

OLIVIERI, G. and OLIVIERI, A. (1965). Autoradiographic study of nucleic acid synthesis during spermatogenesis in *Drosophila melanogaster*. *Mutat. Res.* 2:366-380.
OLSON, J. A. (1966). Lipid metabolism. *A. Rev. Biochem.* 35:559-598.
ROMRELL, L. J., STANLEY, H. P. and BOWMAN, J. T. (1972a). Genetic control of spermiogenesis in *Drosophila melanogaster*: an autosomal mutant (ms(2)3R) demonstrating failure of meiotic cytokinesis. *J. Ultrastruct. Res.* 38:563-577.
________(1972b). Genetic control of spermiogenesis in *Drosophila melanogaster*: an autosomal mutant (ms(2)10R) demonstrating disruption of the axonemal complex. J. Ultrastruct. Res. 38:578-590.
ROTHSCHILD, LORD and CLELAND, K. W. (1952). The physiology of sea-urchin spermatozoa, the nature and location of the endogenous substrate. *J. Exp. Biol.* 29:66-71.
SANG, J. H. (1956). The quantitative nutritional requirements of *Drosophila melanogaster*. *J. Exp. Biol.* 33:45-72.
________(1957). Utilization of dietary purines and pyrimidines by *Drosophila melanogaster*. *Proc. R. Soc. Edinb.* (section B). 64:339-359.
SCHULTZ, J. and RUDKIN, G. T. (1949). Nutritional requirements and the chemical genetics of *Drosophila melanogaster*. *Proc. 8th Int. Congr. Genetics.* 657-658.
SCOTT, T. W., DAWSON, R. M. C. and ROWLANDS, I. W. (1963). Phospholipid interrelations in rat epididymal tissue and spermatozoa. *Biochem. J.* 87:507-512.
SCOTT, T. W., VOGLMAYR, J. K. and SETCHELL, B. P. (1967). Lipid composition and metabolism in testicular and ejaculated ram spermatozoa. *Biochem. J.* 102:456-461.
VERNON, R. G., GO, V. L. W. and FRITZ, I. B. (1971). Evidence that carnitine acetyltransferase is a marker enzyme for the investigation of germ cell differentiation. *Can. J. Biochem.* 49:761-767.
WALKER, P. R. and BAILEY, E. (1970). Metabolism of glucose, trehalose, citrate, acetate, and palmitate by the male desert locust during adult development. *J. Insect Physiol.* 16:499-509.
WIGGLESWORTH, V. B. (1949). The utilization of reserve substances in *Drosophila* during flight. *J. Exp. Biol.* 26:150-163.
WISE, E. M., JR. and BALL, E. G. (1964). Malic enzyme and lipogenesis. *Proc. Natn. Acad. Sci. U.S.A.* 52:1255-1263.
WYATT, G. R. (1968). Biochemistry of insect metamorphosis. In *Metamorphosis* (W. Etkin and L. I. Gilbert, Eds.). Appleton-Century-Crofts, New York. pp. 143-184.
ZIMMERING, S., SANDLER, L. and NICOLETTI, B. (1970). Mechanisms of meiotic drive. *A. Rev. Genetics.* 4:409-436.

UTILISATION OF LIPIDS BY *DROSOPHILA MELANOGASTER*

J. H. Sang
Sussex University, Brighton, England

I want to consider two of the nutrients which insects (and possibly all arthropods) cannot synthesise for themselves, although many other higher organisms can: namely, sterols and choline. Of course, both must be provided in axenic diets, and it is amusing to note that Guyenot (1913) provided only lecithin in the first germ-free *Drosophila* food. Presumably, as now, his lecithin was contaminated with a utilisable sterol, and his experiment was saved from failure by that piece of good luck. Contamination is still a problem for anyone working with lipids, but it is now usually a curse, not a blessing. Progress has its disadvantages.

In his "Molecular Biology of the Gene" (1971), Watson reckons that auxotrophy in higher organisms is secondary; i.e., in the present example, that ancestral arthropod forms had the ability to synthesise choline and sterols which they lost when they became adapted to lipid-rich diets. There would then be no selective advantage in preserving essentially useless genes, and some gain in losing them, or in modifying them for other tasks. The alternative is that insects were auxotrophs from the beginning (which is my guess). But, either way, we are now confronted with handling organisms which are highly adapted to their natural diets. Two things follow: (a) we might expect a diversity in lipid utilisation among insects such that one lipid may be best for the growth of one species whereas it is a poor nutrient for another species, and (b) insofar as particular lipids may have special, and essential, metabolic roles, different species will have varied ways of converting their common dietary lipids to these ends. The conveniences of supply, if they are to be exploited, may demand an originality in metabolism. The classic example in *Drosophila* is *D. pachea*, which cannot survive on cholesterol alone (Goodnight and Kirchner, 1971).

As usual, there are two sides to the coin of adaptation; some gain and some loss. Consequently, we might expect to find differences in the utilisation of the wide range of available sterols by different species, as has been found (Clayton, 1964), but perhaps also differences in response to mixtures, which seems to have been largely ignored, although it probably represents the natural dietary situation. Certainly there is evidence that a non-utilisation sterol can fulfill a "bulk" or macro-role, if supplemented with a small amount (micro-role) of a utilisable sterol (Lasser, Clayton and Edwards, 1963). This evident dual function of sterols has variously been related to structural and metabolic functions (Clayton, 1964), but we know so little about the selective utilization of sterols that it seems premature to do more than define what we are doing, and to leave interpretation to more

Insect and mite nutrition – North-Holland – Amsterdam, (1972)

detailed studies.

As in the preceding paper (Sang, 1972) in this Symposium, I shall refer to results obtained with the melanotic tumor strain of *Drosophila melanogaster*, using tumor frequency as an indicator of some (unknown) metabolic function of sterols which is sensitive to the treatments described. This sensitivity appeared when the strain was being studied from a genetic viewpoint (Sang and Burnet, 1967), and it is important to remember that other treatments have similar phenotypic consequences, as we have seen. However, what was important in this instance was that tumor penetrance increased as the amount of dietary sterol (cholesterol routinely) was reduced. The converse did not apply: massive provision of cholesterol did not "cure" the genetically tumorous condition. However, a change to another sterol (ergosterol) virtually did. It was therefore clear that the kind (as well as the amount) of dietary sterol was important. This was the starting point for the work.

METHODS

Technical details will be found in Cooke and Sang (1970 and 1972). No lecithin was used in the cultures, since all samples tested were found to be contaminated with sterols. The choline requirement was provided as choline chloride. Perhaps it is worth noting that differences between experiments were due to two causes; growth varied a) with the casein source used, and b) with the lighting regimen in the 25°C culture room. This last cause of variation has only recently been discovered, but it is perhaps not unexpected in view of Winfree's (1972) work. Controls within experiments were always run, and infected cultures discarded.

RESULTS

Cooke and Sang (1970) have already surveyed the sterols used by *Drosophila melanogaster*, and their findings will not be repeated here. They also examined, by pair comparisons, the significance of modifications of the cholesterol molecule insofar as they affected larval growth rate. Table 1 also gives similar data for tumor penetrance, from which it will be evident that while there is a rough correlation between growth and tumor incidence, this is not always so. In short, particular modifications of the molecule may result in one effect on growth and another on tumor frequency. It follows that tumors are not induced only by slow growth but depend also on the character of the dietary sterol: tumor incidence can be used as an index of sterol activity.

Unfortunately the available samples of sterols did not permit a full exploration of all possible combinations of sterol provision, so attention was directed to cholesterol (as standard), and to the three sterols which differed from it by single structural changes; dihydrocholesterol (which slowed

Table 1. Comparison of the effects of structural changes of the cholesterol molecule on development rate and on tumor penetrance.

Structural change	Comparison	Development rate	Tumor percentage
C5 double bond saturated	Dihydrocholesterol–Cholesterol	+	++
	Lathosterol–7-Dehydrocholesterol	+	+
C7 double bond added	7-dehydrocholesterol–Cholesterol	0	----
	Lathosterol–Dihydrocholesterol	0	----
C22 double bond added	Stigmasterol–β-sitosterol	0	+
C24 double bond added	Desmosterol–Cholesterol	++	----
C24 ethyl added	β-sitosterol–Cholesterol	--	----
C24 methylene added	Ostreasterol–Cholesterol	+	----

Data summarised from Cooke and Sang (1972). + indicates a significant slowing of development or an increase of tumor penetrance, -, an improvement of development or tumors, and 0 shows no significant effect. All values are for the difference of the second from the first of the pair.

growth and increased tumor penetrance), 7-dehydrocholesterol, and β-sitosterol (both lower tumors, but only the latter improves growth). Desmosterol is examined separately. The major characteristic differences were therefore covered, although incompletely.

Since there was no guarantee that uptake of each sterol is equal, dose responses were run in order to find if they behaved similarly, and to find the sensitive levels for subsequent work (Fig. 1). The responses to three of the sterols differ only in level, and it is interesting that a small amount of β-sitosterol gives as good growth as a large amount of cholesterol, or of dihydrocholesterol. Such a result could be due to differences in assimilation, and we were interested to note that larvae reared on the highest dihydrocholesterol diet gave adults which contained crystal masses which were soluble in lipid solvents. What is taken in is not necessarily what is used. However, this was not the case at lower amounts, and it seems reasonable to assume that the different levels of the three curves represent the functional competence of the three molecules. In each case, tumor frequency was proportional to development rate, confirming the previous finding that slowed development resulted in increased tumors. 7-dehydrocholesterol is quite different; growth is similar to that with cholesterol only when more than 0.008 per cent is provided. Less than this gives very slow growth and

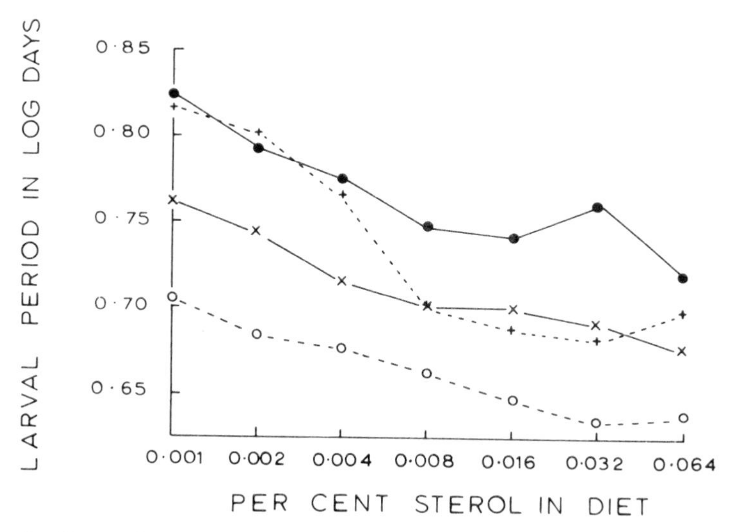

Figure 1. Rate of growth in relation to sterol supply. Dihydrocholesterol = o, β-sitosterol = 0, Cholesterol = X and 7-dehydrocholesterol = +. Growth rate is measured in terms of log·larval period for reasons already given (Sang, 1956). The responses are from separate experiments.

very high tumors. A similar result is found with ergosterol (Cooke and Sang, 1972) indicating that this peculiarity is a consequence of their common 7. Both β-sitosterol and 7-dehydrocholesterol are metabolised in part to cholesterol (Cooke and Sang, 1970), but their activities are almost certainly not a consequence of this. Finally, it is clear that feeding levels around 0.008 per cent sterol are likely to be responsive to treatments both for growth and for tumors.

Mixtures of sterols were next tested for their effects on development rate and on tumor penetrance (Table 2a). Except for the nil addition controls (0.004 per cent), the total sterol was provided at 0.008 per cent and was composed of one (sole sterol) or an equal mixture of two sterols. Development was significantly superior to that with the sole sterol only when the mixture contained β-sitosterol. In fact, all the additions were more or less equivalent when half the provision was β-sitosterol. So β-sitosterol has a significant and important role in improving growth which is not matched by the other sterols. The situation is not the same with respect to tumor penetrance (Table 2b), although there is a high correlation between developmental rate and tumor percentage (i.e., the faster development the fewer the tumors, including the nil addition classes). As the correlation implies, the β-sitosterol combinations have the lowest tumor percentages, but

Table 2. The effects of feeding mixtures of sterols on development rate and on tumors.

(2a)	Larval development time (log days)			
Addition	Cholesterol	Dihydrocholesterol	7-dehydrocholesterol	β-sitosterol
Nil	0.776	0.792	0.807	0.747
Cholesterol	0.786	0.783	0.760	0.729
Dihydrocholesterol	0.783	0.787	0.763	0.743
7-dehydrocholesterol	0.760	0.763	0.776	0.711
β-sitosterol	0.729*	0.743*	0.711*	0.727

(2b)	Tumor frequency (percent)			
Addition	Cholesterol	Dihydrocholesterol	7-dehydrocholesterol	β-sitosterol
Nil	72.5	97.6	94.4	26.8
Cholesterol	55.2	82.8	61.3*	32.6*
Dihydrocholesterol	82.8*	92.4	65.2*	33.8*
7-dehydrocholesterol	61.3	65.2*	82.1	13.1
β-sitosterol	32.6*	33.8*	13.1*	12.9

Each sterol was fed at the sensitive level of 0.004%, and the additions were also of this proportion. Comparisons (within columns) are with the double quantities (underlined) since this is the relevant total sterol provision (0.008%).

the 7-dehydrocholesterol/β-sitosterol mix is as good as β-sitosterol alone. Except when mixed with cholesterol, dihydrocholesterol has a similar action to cholesterol: both lower tumors when fed along with 7-dehydrocholesterol, but have the converse effect if fed with β-sitosterol. Together, they are not distinguishable from the average of their two effects when fed as sole sterol. It follows that the tumor system is a more sensitive indicator of sterol function than measurement of development rate, and that not only is βsitosterol functionally different from the other sterols but so also is 7-dehydrocholesterol, whereas cholesterol and dihydrocholesterol are not certainly distinguishable from one another by these criteria.

Since there is good evidence that 7-dehydrocholesterol and β-sitosterol are converted to cholesterol, interpretation of the data must be subject to this reservation. Dihydrocholesterol is not subject to this change, and it is interesting that larvae reared on dihydrocholesterol develop to small adults with defective ovaries. Such flies lay no viable eggs (and preliminary experiments suggest that feeding them with cholesterol as adults makes no difference to this). Kirchner (personal communication) and his students have found the same result, which suggests that larvae reared on a dihydrocholesterol diet survive only because some other sterol has been passed to them from their yeast-fed mothers, *via* the egg. We have shown that the quantity of sterol in eggs is very small (Sang and King, 1961), about 35 μg/m dry weight, but presumably this is sufficient for the necessary synthesis of such important products as ecdysone, and possibly other unknown sterols. Clearly this area of metabolism merits further study, since it implies an additional complication of the simple assumption that what is fed is what is used.

Since the evidence is against any conversion of dehydrocholesterol, it seemed worth testing the effects of adding small amounts of the other sterols to a sensitive level of dietary dehydrocholesterol (Table 3). In this case the addition was kept as low as seemed practicable, being a fifth of the total. As expected, such an addition of dihydrocholesterol or of cholesterol had no

Table 3. Sterol interactions with dihydrocholesterol

		Development time	Tumors
Dihydrocholesterol 0.008%		0.721 log days	82.3
"	+ dihydrocholesterol 0.002%	0.729 " "	90.2
"	+ cholesterol 0.002%	0.691 " "	75.2
"	+ 7-dehydrocholesterol 0.002%	0.684* " "	57.6^{xx}
"	+ β-sitosterol	0.711 " "	51.2^{xx}

Data from Cooke and Sang (1972)

effect on growth or on tumor penetrance. 7-dehydrocholesterol improved growth and lowered tumors, whereas β-sitosterol only lowered tumors. This confirms the previous findings, since it implies that neither 7-dehydrocholesterol or β-sitosterol function only because they can be converted to cholesterol. Further, since the amount of these sterols un-converted must be less than that fed, it follows that their effects on development and/or tumor penetrance must be a consequence of some role they play not in macro-sterol usage, but in micro-usage. It is tempting to follow Thompson *et al.*, (1972) and to suggest that the growth stimulating function of 7-dehydrocholesterol derives from its easy conversion to ecdysone, but the fact that dihydrocholesterol fed larvae can pupate warns us against such a simple assumption.

Cooke and Sang (1972) have produced even more convincing evidence that the kind of sterol involved in the micro-function of sterols is of great importance for growth and for tumor penetrance. If desmosterol is fed as sole sterol, larval growth is slow but tumors are infrequent, as might be expected from this relative of β-sitosterol (Table 4). At a level of cholesterol which gives the same growth rate, tumors are common. The kind of sterol

Table 4. The effects of a small addition of cholesterol to desmosterol

	Development time	Tumors
1. Desmosterol 0.06%	0.921 log days	22.0 per cent
2. Cholesterol 0.002%	0.911 " "	95.0 " "
3. Desmosterol 0.06% + cholesterol 0.002%	0.859 " "	41.5 " "
4. Control - cholesterol 0.03%	0.847 " "	61.0 " "

Data from Cooke and Sang, (1972).
Development times of the pairs 1 and 2, and 3 and 4 are not significantly different, but one pair is significantly different from the other. All tumor differences are significant.

fed can therefore influence growth and tumor penetrance differentially. However, if a mixture of desmosterol and cholesterol is fed - the cholesterol being only 1/31st of the total - growth is dramatically improved, and is as good as if the total diet was made up of cholesterol. That is, if the sterol microfunction is fulfilled (with cholesterol) the provision of a bulk sterol (desmosterol) permits more or less optimal growth. Tumor frequency with the mixture is more or less intermediate, indicating that the small amount of cholesterol also plays a significant part in determining tumors, possibly out of proportion to its amount.

It will also be obvious from these data that the hypothesis that in insects β-sitosterol is converted to cholesterol *via* desmosterol is not supported by these results. However, we should be ignoring the complication introduced by this demonstration of the significance of the micro-role of sterols for growth, if we tried to deduce more than this. It is sufficient to conclude at this stage that more work is needed using sterol mixtures.

It would be useful if we could move on to a discussion of macro- and micro-roles of sterols, but there are so many possibilities that we have first to find some way into this area. What follows was originally conceived as a contribution to this search for an opening, on the assumption that manipulation of the other lipids, the phospholipids, might be instructive. In other words, it was tempting to exploit the work of Geer *et al.* (1968). Table 5 shows that lowering dietary choline slows development more or less consistently, except if β-sitosterol is provided when the effect is less obvious. There are no significant changes in tumor penetrance, again with the exception of the β-sitosterol set. In this instance, tumor penetrance is notably raised. Altering the choline provision has an effect on the utilisation of β-sitosterol in the tumor system. As is known (Geer and Vovis, 1965), carnitine can be substituted for choline, but again to the detriment of development rate. The effect is more or less consistent and equivalent to the low choline diet. Once more only the β-sitosterol set shows a significant tumor increase, but in this case only the male flies are affected. When carnitine is supplemented with choline, development is normal for all four sterols, but β-sitosterol fed males are again more tumorous than females. There is a sex difference which, apparently, depends on carnitine-replacing choline (as it can do according to Bridges and Ricketts, 1970) in some function.

The functions of choline in tumorigenesis, more particularly with respect to the action of a tumor suppressing gene, is another complex story (Sparrow and Sang, unpublished). The point here is that we can conclude that three of the sterols are not linked with phospholipid metabolism in their action, although manipulating phospholipids affects development, as expected. On the other hand, β-sitosterol metabolism in males is dependent on the kind of phospholipids formed in major amount. This rules out the acetyl-choline function of choline, but unfortunately leaves us with the remaining complexities of choline metabolism.

CONCLUSIONS

In nature, *Drosophila melanogaster* feeds on the yeasts and bacteria of rotting fruits. Its sterol diet is mixed, and mostly phytoserols. Not surprisingly, it grows best when these are provided as sole sterol. It has the ability to convert at least some of them (β-sitosterol) to cholesterol, and it

Table 5. The effects of reducing choline and of replacing it by carnitine.

Sterol provided at 0.06 per cent	Choline 8 x 10^{-4}M		Carnitine 8 x 10^{-4}M			Choline 1 x 10^{-4}M + Carnitine 8 x 10^{-4}M		
	d.t.	tu	d.t.	tu	d.t.	tu	d.t.	tu
Cholesterol	0.813	56.0	0.886*	64.1	0.864*	53.8	0.784	64.6
Dihydrocholesterol	0.846	84.8	0.915*	92.1	0.893*	73.1	0.854	79.2
7-dehydrocholesterol	0.790	8.6	0.868*	17.7	0.852*	20.4	0.761	15.4
β-sitosterol	0.779	9.9	0.857*	46.8*/17.1	0.814	50.7*	0.737	69.8*/9.8

Development time (d.t.) is in log. days and tumors (tu) in percentages, as before. Comparisons are with the normal choline diet (8 x 10^{-4}M) and within the sterol row. Where two tumor percentages are shown in the β-sitosterol row, the first is for male and the second for female flies: in the other cases there is no significant sex difference.

must do this for some purpose. However, larvae will grow on dihydrocholesterol, but only because a sufficiency of some other sterol(s) is passed through the egg. Perhaps this small source is cholesterol, but it might equally well be 7-dehydrocholesterol, or some other sterol which can be used as an ecdysone precursor. However, it some phytosterol is provided, the bulk of the dietary sterol may be any one of the three major sterols considered above (and possibly others) without serious detriment to development (or survival). It is therefore suggested that future studies of sterol usage in insects should note that the minor supply of a particular sterol can alter the apparent function of another (bulk) sterol. Similarly, the provision of a small, but necessary, amount of choline allows other aminoalcohols to give normal growth. Mixtures seem important in more than one area of metabolism.

Use of the melanotic tumor mutant has been useful in confirming conclusions reached from growth measurements alone. It has further shown that particular sterols have roles in metabolism which could not easily be made obvious from growth data. In particular, it has shown that β-sitosterol functions in the context of the phospholipid supply, particularly in males. Substitution of choline by carnitine inhibits this function with respect to tumors, but not for growth. The further study of the relationships of sterols to other lipids should certainly be rewarding.

Acknowledgements—Support of this work by the Science Research Council is gratefully acknowledged, as is the competent technical assistance of Mrs. June Atherton.

REFERENCES

BRIDGES, R. G. and RICKETTS, J. (1970. The incorporation of analogues of choline into the phospholipids of the housefly, *Musca domesticae*. *J. Insect Physiol.* 13:835-850.

CLAYTON, R. B. (1964). The utilisation of sterols by insects. *J. Lipid Res.* 5:3-19.

COOKE, J. and SANG, J. G. (1970). Utilisation of sterols by larvae of *Drosophila melanogaster*. *J. Insect Physiol.* 16:801-812.

_______ & _______(1972). Physiological genetics of melanotic tumors in *Drosophila melanogaster*. The influence of dietary sterols on tumor penetrance in the *tu bw* strain. *Genet. Res.* (in press).

GEER, B. W. and VOVIS, G. F. (1965). The effects of choline and related compounds on the growth and development of *Drosophila melanogaster*. *J. Exp. Zool.* 158:223-236.

GEER, B. W., VOVIS, G. F. and YUND, M. A. (1968). Choline activity during the

development of *Drosophila melanogaster*. *Physiol. Zool.* **41**:280-292.
GOODNIGHT, K. C. and KIRCHNER, H. W. (1971). Metabolism of lathosterol by *Drosophila Pachea*. *Lipids*. **6**:166-169.
GUYENOT, E. (1913). Etudes biologiques sur une mouche, *Drosophila amelophila* Lou. I. Possibilite de la vie aseptique pour 1 ' individu et la ligne. *C.r., Seanc. Soc. Biol.* 74:97.
LASSER, N. L., CLAYTON, R. B. and EDWARDES, A. M. (1963). The dynamic state of sterols in the cockroach. *Fedn. Proc. Fedn. Am. Socs. Exp. Biol.* 22:590.
SANG, J. H. and BURNET, B. (1967). Physiological genetics of melanotic tumors in *Drosophila melanogaster*. IV. Gene-environment interactions of *tu bw* with different third chromosome backgrounds. *Genetics*. **56**:743-754.
SANG, J. H. and KING, R. C. (1961). Nutritional requirements of axenically cultured *Drosophila melanogaster* adults. *J. Exp. Biol.* **38**:793-809.
THOMPSON, M. J., SVOBODA, J. A., KAPLANIS, J. N. and ROBBINS, W. E. (1972). Metabolic pathways of steroids in insects. *Proc. R. Soc. Lond. B.* **180**:203-221.
WATSON, J. D. (1970). *Molecular Biology of the Gene* 2nd Ed. W. A. Benjamin, Inc., New York.
WINFREE, A. (1972). Acute temperature sensitivity of the circadian rhythm in Drosophila. *J. Insect Physiol.* **18**:181-185.

PHYTOSTEROL UTILIZATION AND METABOLISM IN INSECTS: RECENT STUDIES WITH *TRIBOLIUM CONFUSUM*

J. A. Svoboda, W. E. Robbins, C. F. Cohen, and T. J. Shortino
Insect Physiology Laboratory, Entomology Research Division,
Agricultural Research Service
U. S. Department of Agriculture
Beltsville, Maryland 20705

INTRODUCTION

In their classic work, Clark and Bloch (1959) were the first to conclusively demonstrate the dealkylation of a plant sterol in an insect. They showed that ^{14}C-ergosterol was converted to ^{14}C-22-dehydrocholesterol by the German cockroach *Blattella germanica* (L.) through the dealkylation of the 24-methyl group and reduction of the $\Delta 7$-bond. The first report of the dealkylation and conversion of the phytosterol to cholesterol was made three years later by Robbins *et al.* (1962). In this research, the German cockroach was fed ^{3}H-β-sitosterol, and ^{3}H-cholesterol was subsequently found to be the major metabolite, thus showing that this insect can also remove a 24-ethyl group from the sterol side chain. Since then, the biochemical mechanism for conversion of β-sitosterol to cholesterol has been shown to be present in a number of species of insects (Schaefer *et al.*, 1965; Ikekawa, *et al.*, 1966; Earle *et al.*, 1967; Svoboda *et al.*, 1967; Ritter & Wientjens, 1967); however, not all insects are able to dealkylate phytosterols. Kaplanis *et al.* (1963) have conclusively demonstrated that both the larva and adult of the house fly *Musca domestica* (L.) are unable to convert ^{3}H-β-sitosterol to cholesterol and that this insect relies entirely on a dietary source of "essential" cholesterol (Kaplanis *et al.*, 1963; 1965).

We have been interested not only in the utilization and the conversion of phytosterols such as β-sitosterol to cholesterol but also in the intermediate steps involved in these processes in insects. To this end, an in-depth study was initiated to elucidate the pathway(s) of plant sterol metabolism in the tobacco hornworm, *Manduca sexta* (L.) and parallel studies have been carried out with several additional species of insects that convert phytosterols to cholesterol, including representatives of each of the three types of metamorphosis (Robbins *et al.*, 1971).

Our research with the tobacco hornworm has shown this insect to be extremely versatile in its ability to metabolize C_{28} and C_{29} phytosterols to cholesterol (Svoboda & Robbins, 1968). The hornworm converts each of the sterols shown in Fig. 1 to cholesterol. These conversions include the removal of a 24-R-ethyl group from β-sitosterol or stigmasterol, a 24-R-methyl from campesterol, a 24-S-methyl from brassicasterol or 22,23-dihydrobrassicasterol, a 24-methylene from 24-methylenecholesterol and a 24-ethylidene from fucosterol. In addition, the hornworm is able to reduce the Δ^{22}-bond of stigmasterol and in this respect differs from the German cockroach (Clark & Bloch, 1959).

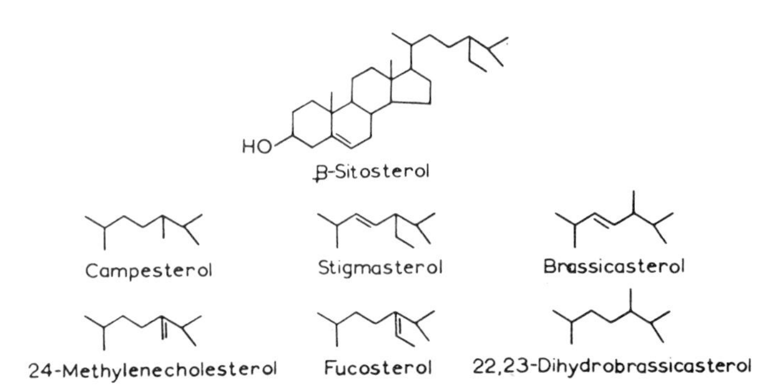

Figure 1. C_{28} and C_{29} Phytosterols.

The first known intermediate in the conversion of phytosterols to cholesterol in insects was discovered when ^{3}H-desmosterol (24-dehydrocholesterol) was isolated and positively identified from the minor sterols present in hornworms fed ^{3}H-β-sitosterol (Svoboda *et al.*, 1967). Extensive studies demonstrated that desmosterol was indeed an intermediate in the conversion of β-sitosterol to cholesterol (Fig. 2) in this insect and these results have been verified by the subsequent finding that desmosterol is an intermediate in the conversion of β-sitosterol to cholesterol in a number of insects (Svoboda & Robbins, 1971).

Certain hypocholesterolemic agents have been shown to exert their effect by inhibiting the Δ^{24}-sterol reductase in vertebrates and thus blocking the conversion of desmosterol to cholesterol (Avigan *et al.*, 1960; Thompson *et al.*, 1963). Since desmosterol is an intermediate in the conversion of β-sitosterol to cholesterol in the tobacco hornworm, two of these inhibitors, 22,25-diazacholesterol and triparanol (Fig. 3), were tested in the hornworm (Svoboda & Robbins, 1967). Both of these compounds, when fed in combination with β-sitosterol, inhibited Δ^{24}-sterol reductase of this insect and caused an accumulation of desmosterol and unmetabolized β-sitosterol at the expense of cholesterol formation. In addition, larval development was disrupted by these inhibitors, particularly by the diazasterol. These results demonstrated for the first time that sterol utilization in an insect can be blocked through the inhibition of specific enzymes involved in sterol metabolism and that this block may be accompanied by the disruption of normal development and metamorphosis.

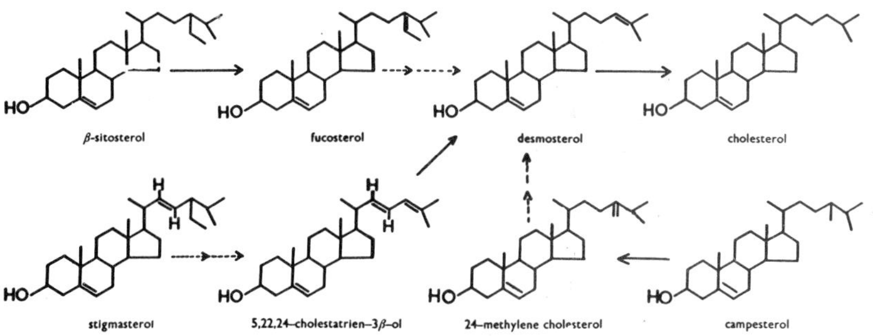

Figure 2. Summary of phytosterol metabolism in the tobacco hornworm.

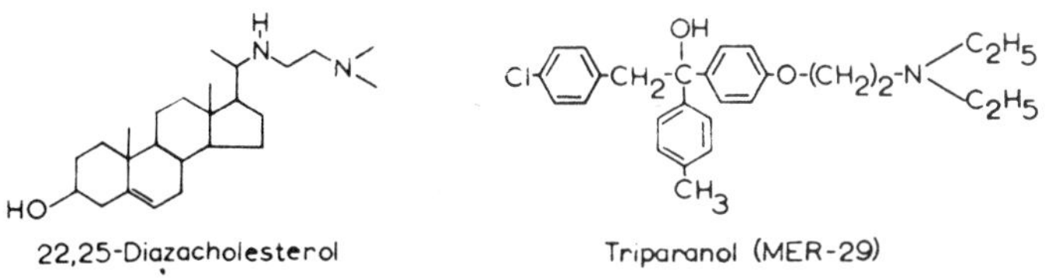

Figure 3. Inhibitors of $\Delta 24$-sterol reductase.

These inhibitors have also proven valuable for studying the utilization and metabolism of plant sterols in insects. The 20, 25-diazacholesterol was used to demonstrate that desmosterol is a common intermediate in the conversion of each of the C_{28} and C_{29} plant sterols in Fig. 1 to cholesterol in the hornworm (Svoboda & Robbins, 1968). The usefulness of the azasterols in metabolic studies again became apparent when a new sterol, 22-trans-5,-22,24-cholestatrien-3β-ol (Fig. 2), was isolated and identified as an intermediate in the conversion of stigmasterol to cholesterol in several insects (Svoboda *et al.*, 1969; Hutchins *et al.*, 1970). This sterol, first found to accumulate in tobacco hornworms when a diazasterol was fed in combination with stigmasterol, was later shown, however, to be a normal intermediate that precedes desmosterol in the conversion of stigmasterol to cholesterol.

Another step in the biochemical conversion of β-sitosterol to cholesterol was elucidated when ^{3}H-fucosterol was isolated and identified as a metabolite of ^{3}H-β-sitosterol in the hornworm (Fig. 2) (Svoboda *et al.*, 1971). An analogous pathway was found to exist in the case of the conversion of ^{3}H-campesterol to ^{3}H-cholesterol in the hornworm. We recently reported on the isolation and positive identification of ^{3}H-24-methylenecholesterol as a constant metabolite of campesterol and presented evidence that it is an intermediate in the dealkylation and conversion of campesterol to cholesterol in this insect (Fig. 2) (Svoboda *et al.*, 1972).

Generally, similar pathways for plant sterol metabolism have been found for several omnivorous and phytophagous insects (Robbins *et al.*, 1970); however, distinct specific differences do occur in the sterol utilization and metabolism of certain insects. For example, in the confused flour beetle, *Tribolium confusum* Jaquelin du Val, one-half or less of the sterol present in the tissues is cholesterol and the remainder is 7-dehydrocholesterol (Beck & Kapadia, 1957). Since there have been no detailed biochemical studies on sterol metabolism in this insect, it was deemed of interest to determine what effect, if any, the presence of the high titer of 7-dehydrocholesterol in this insect would have on phytosterol utilization and metabolism. This paper provides information on the pathways of plant sterol metabolism in *Tribolium* and describes the isolation and identification of two new sterol metabolites from this insect.

EXPERIMENTAL PROCEDURES

Newly hatched confused flour beetle larvae, from a laboratory colony, were reared to mature larvae or prepupae at 30°C and 55% RH on a semidefined diet at the rate of 40 larvae per gram of diet. The insects were weighed and held frozen until they were homogenized and the lipids were extracted. The semidefined diet was composed of glucose, casein, and Wesson salt mixture (48:50:2), supplemented with a B-vitamin mixture (Monroe, 1960), D,L-carnitine (2.5 mg/100 g) and zinc chloride (11 mg/100 g). The casein was made sterol deficient by three extractions with chloroform-methanol. The dietary sterols and azasterol inhibitors were coated on the diet at concentrations of 0.1% and 0.05% (dry wt.), respectively. The metabolism of cholesterol, ^{14}C-desmosterol, ^{3}H-β-sitosterol, ^{3}H-campesterol, ^{3}H-7-dehydrocholesterol acetate, and ^{3}H-stigmasterol was examined when each of these sterols was utilized as the sole dietary sterol and, with the exception of ^{3}H-7-dehydrocholesterol acetate, with each of these labeled sterols fed in combination with an azasterol. ^{14}C-desmosterol, ^{3}H-β-sitosterol and ^{3}H-campesterol were tested in combination with 25-azacholesteryl methyl ether (Aza-1) and the ^{3}H-stigmasterol with 22,25-diazacholesterol (Aza-2). All radiolabeled sterols were purified and analyzed for radiochemical purity by gas liquid chromatography (GLC) and thin layer chromatography (TLC) prior to use. The labeled sterols added to the diets had specific activities of 800 to 1,000 dpm/pg.

The insect sterols were isolated and purified as previously described (Svoboda *et al.*, 1967). Since *Tribolium* normally has a high percentage of

7-dehydrocholesterol and since sterols with a 5,7-diene system are more stable as the acetates, the sterols were acetylated immediately after isolation. In order to preserve the labile sterols isolated from these insects, the processes of extraction, purification, and analysis were all carried out, as far as possible, under conditions of subdued light. As a further precaution, sterol samples were stored under nitrogen in benzene-methanol solution in a freezer compartment when not being processed. The sterol acetates were chromatographed on columns of 20% $AgNO_3$-impregnated silicic acid (Unisil)[1] as previously described (Svoboda *et al.*, 1972). Column fractions were monitored by GLC and TLC on $AgNO_3$-impregnated Silica Gel H. Quantitation of insect sterols and their acetate derivatives were carried out by GLC analysis. Desmosterol acetate and 7-dehydrocholesterol acetate are not completely separated on either of the GLC systems employed (Table 1),

Table 1. GLC analyses of sterols isolated from the confused flour beetle and their acetates.[a]

Compound	RRT[b]			
	0.75% SE-30		1.0% OV-17[c]	
	Free sterol	Sterol acetate	Free sterol	Sterol acetate
Desmosterol	1.95	2.79	3.18	4.40
7-Dehydrocholesterol	1.98	2.82	3.18	4.23
5,22,24-Cholestatrien-3β-ol[d]	2.15	3.00	3.88	5.37
5,7,24-Cholestatrien-3β-ol	2.15	3.00	3.86	5.37
5,7,22,24-Cholestatetraen-3β-ol	2.28	3.21	4.62	6.24

[a]The sterols isolated from *Tribolium* were analyzed by comparison of their RRTs with those of authentic standards with the exception of 5,7,24-cholestatrien-3β-ol, for which no authentic standard was available.

[b]Retention time relative to cholestane.

[c]SE-30 column conditions as previously described (6). OV-17 column 6 ft X 4 mm ID, on 100-140 mesh Gas Chrom P, 30 Psi, 230°C, Cholestane time 4.0 min.

[d]5,22,24-Cholestatrien-3β-ol and 5,7,24-cholestatrien-3β-ol are separable as their acetates on $AgNO_3$-impregnated TLC chromatoplates.

---

[1]Mention of a company name or a proprietary product does not imply endorsement by the U. S. Department of Agriculture.

but these acetates are readily separable by both column and thin layer argentation chromatography, thus facilitating comparisons of relative distribution of radioactivity in the *Tribolium* sterols. Unknown compounds were identified by GLC and TLC analysis, NMR spectroscopy, gas chromatography-mass spectrometry (GC-MS), and UV spectroscopy. Distribution of radioactivity was determined by radioassay of fractions from column chromatography, TLC, and GLC effluent on a Packard Tri-Carb Scintillation Spectrometer.

RESULTS AND DISCUSSION

In previous studies, both omnivorous and phytophagous insects that dealkylate phytosterols were found to accumulate cholesterol as the major tissue sterol when fed either β-sitosterol or a number of other plant sterols (Robbins *et al.*, 1962; Schaefer *et al.*, 1965; Ikekawa *et al.*, 1966; Earle *et al.*, 1967; Svoboda *et al.*, 1967; Ritter & Wientjens, 1967; Svoboda & Robbins, 1968). In the tobacco hornworm, for example, when β-sitosterol is the sole added dietary sterol, cholesterol comprises about 85% of the neutral sterols isolated from prepupae (Svoboda & Robbins, 1968) and is also the major tissue sterol of hornworms when either campesterol, stigmasterol, fucosterol, or 24-methylenecholesterol is the dietary sterol (Svoboda & Robbins, 1968). In the present study, the confused flour beetle was found to efficiently dealkylate and utilize ^{3}H-β-sitosterol, ^{3}H-campesterol, and ^{3}H-stigmasterol. This insect grew as well on diets containing either of these phytosterols as on a diet containing cholesterol. However, although over 40% of the total sterols was found to be ^{3}H-cholesterol in the insects fed each of these ^{3}H-phytosterols, in contrast to the hornworm, ^{3}H-7-dehydrocholesterol was present in greater quantity than ^{3}H-cholesterol in the sterols from mature *Tribolium* larvae. In the test insects, the cholesterol to 7-dehydrocholesterol ratios, respectively, were as follows: follows: for ^{3}H-β-sitosterol (42:50), ^{3}H-campesterol (41:48), and ^{3}H-stigmasterol (41:53). Furthermore, desmosterol metabolism in *Tribolium* paralleled that found in a number of other insects (Svoboda & Robbins, 1968) in that ^{14}C-desmosterol was largely converted into the major tissue sterol(s): ^{14}C-cholesterol and ^{14}C-7-dehydrocholesterol (47:53). There exists an equilibrium between cholesterol and 7-dehydrocholesterol in this insect as indicated by experiments with partially grown larvae (avg. wt. 0.61 mg) fed for a week on a diet containing ^{3}H-7-dehydrocholesterol acetate. Approximately 19% of the radioactive sterols isolated from these larvae was ^{3}H-cholesterol, indicating that 7-dehydrocholesterol is readily converted to cholesterol in *Tribolium* larvae even when the insects already possess a considerable pool of sterol. In addition, control insects fed cholesterol were found to contain approximately equal quantities of cholesterol and 7-dehydrocholesterol, further substantiating the equilibrium between these sterols in *Tribolium* tissues.

Growth and development were inhibited in each instance when *Tribolium* larvae were fed diets containing an azasteroid in combination with any of the labeled sterols. However, in *Tribolium* a characteristic effect in inhibition of the larval to pupal molt, whereas in the hornworm, the primary effect is on the 4th instar larva causing formation of a high percentage of precocious "4th instar prepupae" (Robbins *et al.*, 1971). In addition, as might be expected, GLC analysis of the sterols from these inhibited insects revealed that the azasterol also disrupted neutral sterol metabolism. A primary effect was a substantial decrease in the amount of cholesterol present in the inhibited larvae. The relative percentage of cholesterol in the sterols from these insects was: ^{3}H-β-sitosterol + Aza-1 (5%), ^{3}H-campesterol + Aza-1 (5%), ^{14}C-desmosterol + Aza-1 (5%), and ^{3}H-stigmasterol + Aza-2 (17%). The higher level of cholesterol present in the ^{3}H-stigmasterol + Aza-2 test insects is probably due to the fact that the diazasterol (Aza-2) is in general a much less potent inhibitor of insect sterol metabolism than is Aza-1 (Svoboda & Robbins, 1971; Robbins *et al.*, 1971). The 7-dehydrocholesterol-desmosterol GLC peak (Table 1) comprised the following percentages of the total sterols: ^{14}C-desmosterol + Aza-1 (50%), ^{3}H-β-sitosterol + Aza-1 (32%), ^{3}H-campesterol + Aza-1 (9%), and ^{3}H-stigmasterol + Aza-2 (23%). Each of these peaks was shown, by argentation chromatography of the sterol acetates, to contain 7-dehydrocholesterol and a greater than normal quantity of desmosterol. In every case, the 7-dehydrocholesterol content was considerably less than that found in normal larvae. The decrease in cholesterol production plus the accumulation of desmosterol indicate that the azasteroids effectively inhibit the Δ^{24}-sterol reductase enzyme system in *Tribolium* just as they do in other insects (Svoboda & Robbins, 1971).

The disruption of sterol metabolism was further evidenced by the presence of an unknown component in the sterols from the inhibited larvae observed as a major GLC peak in both systems. This peak was found to contain a single ^{3}H-compound in all the samples except those from the ^{3}H-stigmasterol + Aza-2, which were found to be mixtures. The major unknown sterol from the inhibited larvae comprised about 41% of the total sterols from the ^{14}C-desmosterol + Aza-1 test, 37% from the ^{3}H-β-sitosterol + Aza-1 test, and 11% from the ^{3}H-campesterol + Aza-1 test. This compound was quite easily separated from the other sterols after acetylation and argentation chromatography by column and thin layer chromatography. The polarity of this unknown, as evidenced by column chromatography, its Rf value (TLC), and its RRT (GLC) (Table 1) indicated it to be a triene. A UV spectrum of this material, taken in methanol, was typical for a sterol with a $\Delta^{5,7}$-diene system (shoulder at 264 nm, peaks at 272, 282, and 294 nm). Its mass spectrum showed a M^+ at m/e 382 and additional prominent peaks at m/e 364, 349, 271, and 253, indicating loss of H_2O, $H_2O + CH_3$, C_8H_{15}

(side chain), and C_8H_{15} + H_2O, respectively. The M^+ peak at m/e 382 indicated that the compound had two less hydrogens than 7-dehydrocholesterol or an additional ring system. However, the side chain fragment of C_8H_{15} places a double bond in the side chain, and the compound's NMR spectrum which showed a far downfield shift for the C-26 and C-27 methyl signals as a doublet at 1.65 ppm readily positioned the double bond at C-24 (Hutchins *et al.*, 1970). The C-18, C-19, and C-21 methyl resonances appeared in similar regions as those of 7-dehydrocholesterol. These data provide conclusive proof that this sterol from *Tribolium* is 5,7,24-cholestatrien-3β-ol. This new insect sterol would fit into the scheme of sterol metabolism in *Tribolium* as shown in Figs. 4 and 5. That this sterol also occurs as a normal metabolite in *Tribolium* was verified by the presence of the triene in very small quantities (1 - 2%) in the sterols from non-inhibited larvae fed diets containing ^{3}H-β-sitosterol, ^{3}H-campesterol, or ^{3}H-stigmasterol.

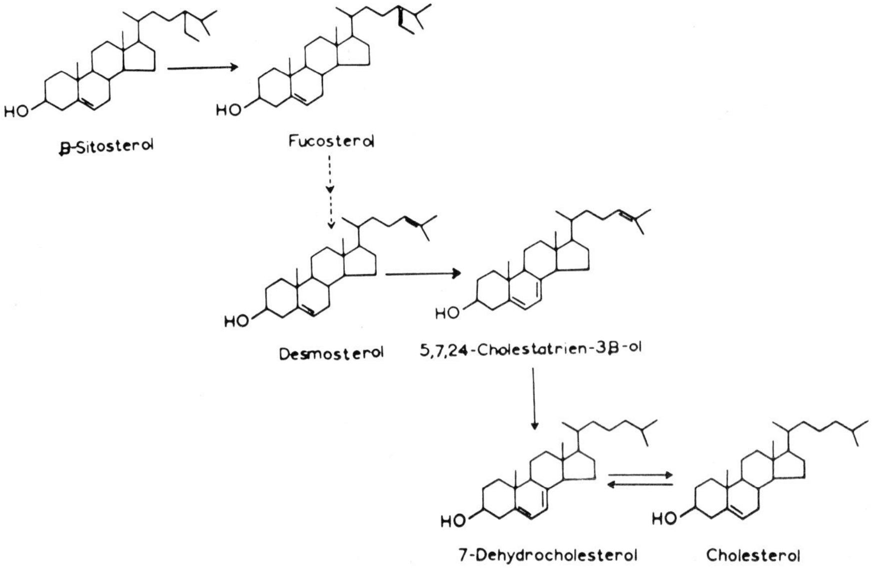

β-Sitosterol
Fucosterol
Desmosterol
5,7,24-Cholestatrien-3β-ol
7-Dehydrocholesterol
Cholesterol

Figure 4. Conversion of β-sitosterol to cholesterol in *Tribolium*.

Examination of sterol acetates from the sterols from larvae fed ^{3}H-stigmasterol revealed the presence of ^{3}H-5,22,24-cholestatrien-3β-ol acetate in fractions from insects that had been reared both with and without diazasterol in the diet. This sterol was first isolated from the hornworm (Svoboda *et al.*, 1969), and its presence in *Tribolium* indicates a further

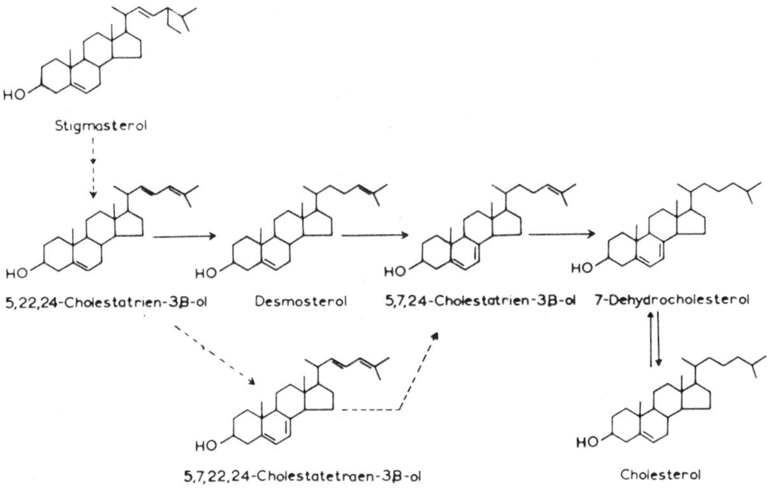

Figure 5. Conversion of stigmasterol to cholesterol in *Tribolium.*

similarity in the metabolic pathways of the hornworm and *Tribolium* (Figs. 2 and 5). However, the presence of an azasteroid in the diet does not bring about the accumulation of this metabolite in *Tribolium* to the same extent that it does in the hornworm (Svoboda *et al.*, 1969).

A major fraction of the more polar sterol acetates that exceeded 15% of the total sterols from inhibited larvae fed ^{3}H-stigmasterol + Aza-2 was identified as 5,7,22,24-cholestatetraen-3β-ol acetate. It was more polar than 5,7,24-cholestatrien-3β-ol acetate by $AgNO_3$-TLC analyses, was ^{3}H-labeled, and had RRTs both as the free sterol and its acetate identical to those of the authentic sterol and its acetate. The UV spectrum of the sample from the insect containing this fraction has absorption bands typical of both a Δ5,7-diene and a Δ22,24-diene (232, 239 nm and shoulder at 248 nm) system of sterols (Svoboda *et al.*, 1969). The GC-MS of the acetate derivatives of authentic 22-trans-5,7,22,24-cholestatetraen-3β-ol and unknown from *Tribolium* showed identical fragmentation patterns. Their mass spectra exhibited M^+ peaks at m/e 422, with additional prominent peaks at m/e 362, 347, 313, 280, and 253, indicating loss of CH_3COOH, $CH_3COOH + CH_3$, C_8H_{13} (side chain), $CH_3COOH + C_6H_9$, and $CH_3COOH + C_8H_{13}$, respectively. The base peaks at m/e 109 (side chain fragment) were also observed for other Δ22,24-sterols (Hutchins *et al.*, 1970) and further support this side chain conjugated diene system. This tetraene accumulates when the 24-reductase system is blocked, but is present only in trace amounts in normal larvae. As indicated in Fig. 5, it is not known whether it is in a normal pathway, or whether it may only be produced when the insect is unable to reduce the Δ^{24}-bond of the 5,7,24-cholestatrien-3β-ol. The accumulation of the latter sterol may activate a feedback mechanism that triggers the introduction of the Δ^{7}-bond into the 5,22,24-cholestatrien-3β-ol, even though it is not a normal substrate.

The fact that feeding desmosterol with Aza-1 results in a large accumulation of 5,7,24-cholestatrien-3β-ol and a greatly reduced production of 7-dehydrocholesterol and cholesterol lends support for the metabolic sequence for these sterols as shown in Fig. 5. No Δ^{22}-sterols are produced from stigmasterol except for 5,22,24-cholestatrien-3β-ol, indicating that the Δ^{22}-bond is reduced before the Δ^{24}-bond, as was found for the tobacco hornworm (Svoboda *et al.*, 1969).

In comparing the metabolic pathways of sterols for the hornworm (Fig. 2) with those outlined for *Tribolium* (Figs. 4 and 5), it is obvious that the major difference in these pathways is the involvement of an intermediate containing a Δ^{7}-bond in the routes from the phytosterols to cholesterol in the confused flour beetle. A similar pathway may exist in the hornworm and other insects that contain only very small amounts of 7-dehydrocholesterol (Robbins *et al.*, 1971), but as a minor route serving to provide only that 7-dehydrocholesterol needed for specific purposes such as a precursor for the steroid molting hormones or ecdysones (Robbins *et al.*, 1971).

The equilibrium between cholesterol and 7-dehydrocholesterol and the high percentage of 7-dehydrocholesterol that occur in the tissues of *Tribolium* is not currently understood. 7-Dehydrocholesterol has been shown to be the major sterol of *Drosophila pachea* Patterson and Wheeler, but cholesterol was not found in this insect, and a dietary source of a sterol with a Δ^{7}-bond is essential to this insect (Goodnight & Kircher, 1971). Other insects have been shown to introduce a Δ^{7}-bond into the steroid nucleus to form 7-dehydrocholesterol (Robbins *et al.*, 1971), but the 7-dehydrocholesterol usually constitutes but a small percentage of the total insect sterols, except in certain special examples such as the prothoracic glands of the American cockroach (Robbins *et al.*, 1971) or housefly eggs (Robbins, 1963).

The present study adds interesting comparative information to our growing body of knowledge of phytosterol metabolism in insects particularly on the intermediary metabolism of plant sterols in the confused flour beetle. Further studies on the new metabolites or intermediates identified from *Tribolium* could aid us in determining whether or not similar or identical pathways are present in those phytophagous insects that contain very small quantities of 7-dehydrocholesterol as well as perhaps eventually providing us with information on the biological and biochemical significance of these pathways.

SUMMARY

Three radiolabeled plant sterols (β-sitosterol, campesterol, stigmasterol) were dealkylated and converted to cholesterol by *Tribolium confusum* and desmosterol was found to be a common intermediate in the conversion of each of these sterols to be cholesterol. As was previously found with the tobacco hornworm, fucosterol is an intermediate in the conversion of β-sitosterol to cholesterol and 5,22,24-cholestatrien-3β-ol is an intermediate in the conversion of stigmasterol to cholesterol by *Tribolium*. A new intermediate, 5,7,24-cholestatrien-3β-ol, was isolated from *Tribolium* and identified as being involved in the conversion of each of the three phytosterols to cholesterol and a second steroid, 5,7,22,24-cholestatetraen-3β-ol, was identified as a metabolite of stigmasterol. The pathway(s) of phytosterol metabolism in *Tribolium* and the tobacco hornworm are compared.

REFERENCES

AVIGAN, J., STEINBERG, D., THOMPSON, M. J., and MOSETTIG, E. (1960). Mechanism of action of MER-29, an inhibitor of cholesterol biosynthesis. *Biochem. Biophys. Res.Commun.* 2:63-65.

BECK, S. D. and KAPADIA, G. G. (1957). Insect nutrition and metabolism of sterols. *Science,* N. Y. **126**:258-259.

CLARK, A. J. and BLOCH, K. (1959). Function of sterols in *Dermestes vulpinus. J. Biol. Chem.* **234**:2583-2588.

EARLE, N. W., LAMBREMONT, E. N., BURKS, M. L., SLATTEN, B. H. and BENNETT, A. F. (1967). Conversion of β-sitosterol to cholesterol in the boll weevil and the inhibition of larval development by two aza-sterols. *J. Econ. Ent.* **60**:291-293.

GOODNIGHT, K. C. and KIRCHER, H. W. (1971). Metabolism of lathosterol by *Drosophila pachea. Lipids.* **6**:166-169.

HUTCHINS, R. F. N., THOMPSON, M. J. and SVOBODA, J. A. (1970). The synthesis and the mass and nuclear magnetic resonance spectra of side chain isomers of cholesta-5, 22-dien-3-β-ol and cholesta-5, 22, 24-trien-3β-ol. *Steroids.* **15**:113-130.

IKEKAWA, N., SUZUKI, M., KOBAYASHI, M. and TSUDA, K. (1966). Sterols of *Bombyx mori.* IV. Sterol conversion in the silkworm. *Chem. Pharm. Bull,* Tokyo. **14**:834-836.

KAPLANIS, J. N., MONROE, R. E., ROBBINS, W. E. and LOULOUDES, S. J. (1963). The fate of dietary H^3-β-sitosterol in the adult house fly. *Ann. Entomol. Soc. Amer.* **56**:198-201.

KAPLANIS, J. N., ROBBINS, W. E., MONROE, R. E., SHORTINO, T. J. and THOMPSON, M. J. (1965). The utilization and fate of β-sitosterol in the larva of the housefly, *Musca domestica* L. *J. Insect Physiol.* **11**:251-258.

MONROE, R. E. (1960). Effect of dietary cholesterol on house fly reproduction. *Ann. Entomol. Soc. Amer.* **53**:821-824. The B-vitamin mixture was used at one-half the concentration used for the adult house fly diet.

RITTER, F. J. and WIENTJENS, W. H. J. M. (1967). Sterol metabolism of insects. *T.N.O.-Nieuws* **22**:381-392.

ROBBINS, W. E. (1963). *Radiation and Radioisotopes Applied to Insects of Agricultural Importance.* International Atomic Energy Commission, Athens.

ROBBINS, W. E., DUTKY, R. C., MONROE, R. E. and KAPLANIS, J. N. (1962). The metabolism of H^3-β-sitosterol by the German cockroach. *Ann. Entomol. Soc. Amer.* 55:102-104.

ROBBINS, W. E., KAPLANIS, J. N., SVOBODA, J. A. and THOMPSON, M. J. (1971). Steroid metabolism in insects. *Annu. Rev. Ent.* 16:53-68.

ROBBINS, W. E., KAPLANIS, J. N., THOMPSON, M. J. and SVOBODA, J. A. Chemistry and biological activity of the ecdysones and ecdysone analogs. *1971 Symposium on Chemistry and Activity of Insect Hormones.* 2nd International Congress of Pesticide Chemistry, Tel Aviv, Israel. (In press)

SCHAEFER, C. H., KAPLANIS, J. N. and ROBBINS, W. E. (1965). The relationship of the sterols of the Virginia pine sawfly, *Neoprion pratti* Dyar, to those of two host plants, *Pinus virginiana* Mill and *Pinus rigida* Mill. *J. Insect Physiol.* 11:1013-1021.

SVOBODA, J. A., HUTCHINS, R. F. N., THOMPSON, M. J. and ROBBINS, W. E. (1969). 22-Trans-cholesta-5, 22, 24-trien -3β-ol: An intermediate in the conversion of stigmasterol to cholesterol in the tobacco hornworm, *Manduca sexta* (Johannson) *Steroids.* 14:469-476.

SVOBODA, J. A. and ROBBINS, E. W. (1967). Conversion of beta sitosterol blocked in an insect by hypocholesterolemic agents. *Science*, N. Y. 156:1637-1638.

SVOBODA, J. A. and ROBBINS, W. E. (1968). Desmosterol as a common intermediate in the conversion of a number of C_{28} and C_{29} plant sterols to cholesterol by the tobacco hornworm. *Experientia.* 24:1131-1132.

_____&_____(1971). The inhibitive effects of aza-sterols on sterol metabolism and growth and development in insects with special reference to the tobacco hornworm. *Lipids.* 6:113-119.

SVOBODA, J. A., THOMPSON, M. J. and ROBBINS, W. E. (1967). Desmosterol, an intermediate in dealkylation of β-sitosterol in the tobacco hornworm (*Manduca sexta*). *Life Sci.* 6:395-404.

_____,_____&_____(1971). 24-Methylenecholesterol. Isolation and identification as an intermediate in the conversion of campesterol to cholesterol in the tobacco hornworm. *Nature New Biology.* 230:57-58.

_____,_____&_____(1972). Identification of fucosterol as a metabolite and probable intermediate in conversion of β-sitosterol to cholesterol in the tobacco hornworm. *Lipids.* 7:156-158.

THOMPSON, M. J., DUPONT, J. and ROBBINS, W. E. (1963). The sterols of liver and carcass of 20, 25-diaza-cholesterol-fed rats. *Steroids.* 2:99-104.

NUTRITION AND METABOLISM OF AMINO ACIDS IN THE SILKWORM, *BOMBYX MORI*

Toshio Ito and Tamio Inokuchi
Sericultural Experiment Station, Suginami-ku
Tokyo, Japan

AMINO ACID REQUIREMENTS IN GENERAL

In the field of sericultural science the nutrition of the silkworm, *Bombyx mori*, has been of primary importance because cocoon production is influenced by the nutritive value of foodstuffs, i.e. the quality of mulberry leaves. However, quantitative analyses of amino acids of mulberry leaves, larval hemolymph and tissues, fibroin, and sericin as well as investigations on amino acid metabolism have revealed a limited aspect of amino acid nutrition in this insect. Development of artificial diet for this insect began some years ago, and since then much information has been accumulated on the nutritional requirements of the silkworm, including those for amino acids (Ito, 1967, 1970).

It was at first reported that the silkworm requires proline in addition to the same ten amino acids known to be essential for other species of insects; that is, arginine, histidine, isoleucine, leucine, lysine, methionine, phenylalanine, threonine, tryptophan, and valine (Arai and Ito, 1964; Ito and Arai, 1965). Subsequent experiments showed, however, that proline is semi-essential since it was revealed that the larva can grow and develop, though slowly, on a proline-deficient diet as a result of improvement in the basal composition of an amino-acid diet (Arai and Ito, 1967; Inokuchi *et al.*, 1967). The omission of any one of the ten essential amino acids did not allow growth at all, even on the improved diet.

A single omission of aspartate or glutamate from the complete diet had a slight effect, whereas a double deletion strongly suppressed growth. Their addition to a diet containing the ten essential amino acids plus proline resulted in nearly normal growth (Ito and Arai, 1965), and it was finally shown that the silkworm requires either aspartate or glutamate (Ito and Arai, 1966). It is well known that they play a central role in amino acid metabolism, including transamination (Koide *et al.*, 1955; Fukuda, 1957; also see a review by Gilmour, 1961) and respiration (Ito *et al.*, 1958) in the silkworm.

The deletion of either alanine, glycine, or serine does not bring about any adverse effect, and they are thus classed as nonessential amino acids for the silkworm. However, the addition of each of them to a diet containing the ten essentials plus proline evidently results in some improvement in larval growth. This is also true for the diet containing added aspartic and glutamic acid (Ito and Arai, 1966). Although those amino acids are readily synthesized by the silkworm, it seems to be necessary to add them to maintain the highest efficiency of the amino-acid diet.

Neither cystine nor tyrosine are considered to be essential for the silkworm, except under limited dietary conditions (Ito and Arai, 1967). The nutritional effect of these amino acids will be discussed below, in relation to the metabolism of methionine and phenylalanine.

REQUIREMENT FOR PROLINE AND 'ORNITHINE CYCLE'

The minimum optimum level of proline required for the normal growth and development is 4 mg per g of the dry diet, and a somewhat detrimental effect is observed at a level of 12 mg (Arai and Ito, 1967).

The addition of proline brought about different effects according to the composition of the diet used. Almost no improvement was obtained when the diet contained only the ten essential amino acids. The efficiency of the diet containing ten essentials plus both aspartate and glutamate was largely improved by their inclusion, but such an improvement was still limited to a certain level. For further improvement it was necessary to add five non-essential amino acids in addition to proline (Arai and Ito, 1967) as shown in Fig. 1. Subsequently, the requirement for proline was shown to be spared partly by glutamic acid, ornithine, or arginine, but not by aspartic acid (Inokuchi, 1969b). The use of labeled compounds revealed the metabolic pathway involved (Inokuchi, 1969b; Inokuchi *et al.*, 1969). By injecting several radioisotopes, it was finally demonstrated that the silkworm is able to convert citrulline to arginine, arginine to proline, and ornithine to proline (Table 1). The formation of urea was confirmed simultaneously with the ornithine formation.

It is interesting that the rate of conversion of these compounds - the members of 'ornithine cycle' - is dependent on varying dietary conditions. As shown in Table 1, a higher radioactivity in larval arginine (1.44 times control) is found in larvae fed on the arginine-deficient diet after injection of labeled citrulline-ureido. In the diet lacking proline, a higher rate of incorporation of ^{14}C from both labeled arginine and ornithine into proline is obtained (3.29 and 1.71 times, respectively). In the silkworm proline is apparently synthesized by means of the ornithine cycle, but the amount biosynthesized seems to fulfill only a part of the quantity required for normal growth and metabolism, and the rest must be supplied from dietary sources. The acceleration of proline formation in the absence of dietary proline is one of the examples showing the control of metabolic rates by nutrition.

The incorporation of ^{14}C into citrulline after the injection of radioactive ornithine is only observed at a low level, if at all (Inokuchi, 1969b),

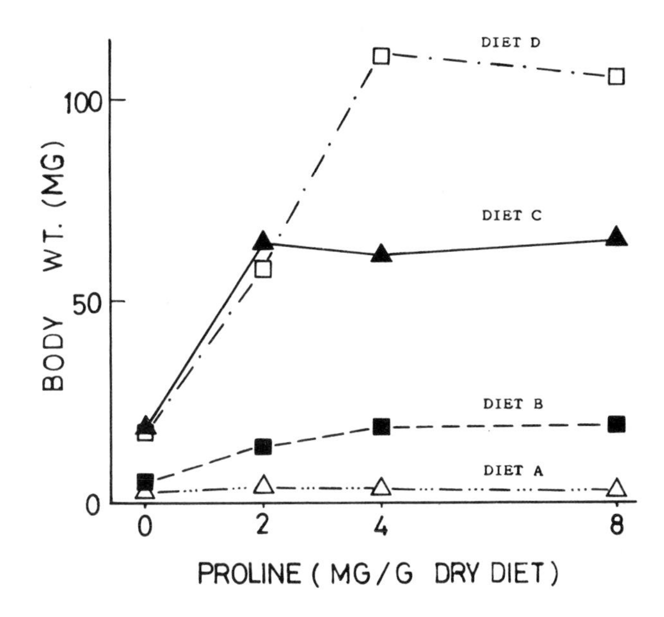

Figure 1. Effect of graded doses of proline on growth of the silkworm, when the diet contains varying amino acid mixture. Mean weight after 15-day rearing is shown.

Diet A: ten usual essential amino acids alone.

Diet B: (A) plus five non-essential amino acids (alanine, cystine, glycerine, serine, and tyrosine.

Diet C: (A) plus aspartic and glutamic acids.

Diet D: (C) plus five non-essential amino acids. (Modified from the data of Arai and Ito, 1967).

suggesting the occurrence of an incomplete ornithine cycle in the silkworm. This may be related to the fact that urea is not the main end product of nitrogen metabolism in the silkworm, although the formation of small amounts of urea was reported for this insect by several authors previously (see the reviews by Gilmour, 1961, and by Chefurka, 1965).

When the larva are feeding on the diet lacking proline, they become reddish in color (Arai and Ito, 1967). The red coloration is not retained, however, when the proline-deficient diet is supplemented with a sufficient amount of arginine or ornithine (Inokuchi, 1969b).

Arginine-HCl must be added at a level of at least 8 mg per g of the dry diet (Arai and Ito, 1967), but the requirement is partly spared by citrulline (Inokuchi, 1969b). As far as the diet contained a certain, but insufficient amount of arginine, the sparing effect of citrulline was very remarkable. In the absence of dietary arginine, a slight larval growth was obtained by an addition of excess citrulline. Thus, arginine requirements seem to depend partly on whether or not dietary citrulline is supplied from the diet. It is unlikely that citrulline is readily synthesized by the silkworm.

Table 1. Effect of dietary condition on incorporation of ^{14}C into amino acids of larval protein of the silkworm.

Isotope injected	Amino acid	Radioactivity (cpm/mg protein) (Complete diet)	(Arginine-deficient diet)[a]
^{14}C-L-citrulline-ureido	Proline	28 ± 3.9	31 ± 4.4
	Arginine	1539 ± 27.8	2221 ± 33.3
		(Complete diet)	(Proline-deficient diet)[b]
^{14}C(U)-L-arginine	Proline	362 ± 13.5	1192 ± 24.4
	Arginine	6195 ± 55.1	4739 ± 48.6
		(Complete diet)	(Proline-deficient diet)[b]
^{14}C-5-DL-ornithine	Proline	2591 ± 22.8	4441 ± 28.2
	Arginine	50 ± 3.2	70 ± 3.7

[a,b]Larvae were kept on the deficient diet for 24 hours, (a) or for about 3 days (b), then analyzed.

From Inokuchi (1969) and Inokuchi, Horie, and Ito (1969).

REQUIREMENT FOR SULFUR AMINO ACIDS AND THEIR METABOLISM

The minimum optimum level of methionine required is 4 mg per g of the dry diet of the silkworm (Arai and Ito, 1967). An addition of cystine to the methionine-deficient diet did not bring about any growth, but cystine showed a conspicuous growth-promoting effect in a diet containing insufficient methionine, (2 mg per g of the dry diet), and cystine was nutritionally almost equivalent to methionine. The single omission of cystine from the complete diet scarcely affected larval growth, but later experiments showed that cystine still possessed a slight nutritive effect even in the presence of sufficient methionine (Ito and Arai, 1967). The nutritional study is strongly suggestive that the silkworm is capable of methionine-cystine conversion, and the requirement for cystine is concluded to be met usually by the presence of sufficient methionine, though the supplementation of cystine seems to be still necessary. Both amino acids are important precursors in further metabolic conversions, as will be shown below.

A large amount of free cystathionine is present in the larval and pupal hemolymph of the silkworm (Kondo, 1959; 1960) and it was reported that neither cystine nor any other sulfur amino acids were recovered from labeled methionine, although cystathionine was recovered from it (Kondo, 1962). Thus, the composition of the free amino acids of the hemolymph was re-investigated recently with larvae injected with labeled methionine, and the results obtained were mostly in accord with the previous reports. However, when the tissue protein of larvae was analyzed, ^{35}S of the labeled methionine was found to be incorporated actively into cystine (Inokuchi, 1969a).

The rate of conversion from methionine to cystine was influenced by dietary conditions; that is, a higher recovery of the free cystathionine of the hemolymph and a much higher recovery of cystine of tissue protein from the labeled methionine were obtained on the cystine-deficient diet than on the control diet (Inokuchi, 1969a).

There occurs also a large amount of free L-lanthionine in the hemolymph of the silkworm, but ^{35}S from methionine is incorporated more rapidly into cystathionine than into lanthionine (Rao *et al.*, 1967). According to these authors, the injection of ^{35}S-L-cystine is followed by a rapid incorporation of isotopes into these two compounds, and the presence of lanthionine is associated with a complete absence or, at least, a barely detectable trace of methionine and cystine in *Bombyx* pupae. A recent study by Inokuchi (1972b), however, revealed that the levels of cystathionine, lanthionine, methionine, and cystine in a free state in the hemolymph are variable according to the dietary levels of methionine and cystine, and appreciable amounts of both methionine and cystine were present with an accumulation of large amounts of both cystathionine and lanthionine (Table 2). Thus, the concentration of free lanthionine was dependent only on the level of dietary cystine, not that of dietary methionine, and it is considered that the dietary cystine is readily converted to lanthionine by the silkworm. In the absence of dietary methionine the accumulation of free cystathionine in the hemolymph was small, and seemed to be independent on the dietary cystine. On the contrary, it was dependent on the level of dietary cystine, as far as the diet contained methionine. Thus, it is concluded that free cystathionine accumulates in the hemolymph in the presence of both methionine and cystine in the diet. The physiological role of the accumulation of these sulfur amino acids in the hemolymph is not well understood.

The occurrence of lanthionine has already been reported for the egg shell chorion on the silkworm (Inoue and Kawaguchi, 1943). Recently, it was found that the concentrations of both cystathionine and lanthionine of the hemolymph of the silkworm vary conspicuously during development, and differ significantly between male and female, being higher in the latter. Free lanthionine is considered to be related to egg formation. There is also a significant difference in the level of free lanthionine of the hemolymph between Japanese and Chinese races, being higher in the former group

(Inokuchi, 1972a; Ito and Inokuchi, in preparation).

Table 2. Effect of dietary methionine and cystine on the concentration of free sulfur amino acids of hemolymph of the silkworm.[a]

Methionine (mg/g dry diet)	Cystine (mg/g dry diet)	Sex	Mean body wt of larvae analyzed	Amino acids (μg/ml)			
				Cystathionine	Lanthionine	Methionine	Cystine
0	0	Male	1.06	trace	34.7	trace	trace
		Female	1.18	37.3	27.2	16.6	4.3
	1	Male	1.14	50.0	353.0	17.9	trace
		Female	1.23	53.3	278.0	13.4	7.2
	2	Male	1.19	56.7	446.7	11.2	10.8
		Female	1.24	56.7	449.8	17.9	10.8
4	0	Male	1.77	30.0	6.2	64.9	32.4
		Female	1.78	27.0	trace	105.4	32.4
	1	Male	1.70	93.3	131.2	73.9	68.5
		Female	1.84	160.0	218.6	94.0	28.8
	2	Male	1.74	206.7	346.7	64.9	54.1
		Female	1.81	336.7	462.3	91.8	50.5

[a]Newly ecdysed fifth-instar larvae were kept on the experimental diet for 48 hours, then analyzed. The rate of larval growth differed in the presence or absence of methionine.
From Inokuchi (1972b).

REQUIREMENT FOR PHENYLALANINE AND TYROSINE AND THEIR METABOLISM

The minimum optimum amount of phenylalanine required by the silkworm is 8 mg per g of the dry diet (Arai and Ito, 1967). Tyrosine cannot replace phenylalanine, in a basal diet containing little or no phenylalanine. However, tyrosine improves growth and development markedly in combination with insufficient amounts of phenylalanine, and nutritionally is almost equivalent to phenylalanine in such sparing action (Ito and Arai, 1967), similar to the case of methionine-sparing by cystine. The omission of tyrosine from the complete diet results in a slight drop in larval growth, but the addition of some tyrosine is still necessary for full growth even in the presence of sufficient phenylalanine. The nutritional study is in accord with the previous reports on the conversion from phenylalanine to tyrosine (Bricteux-Gregoire *et al.*, 1956; Fukuda, 1956), and it will be concluded that tyrosine requirement is usually met by the supply of sufficient phenylalanine.

With the addition of phenylpyruvate to the phenylalanine-deficient diet normal growth of the silkworm resumed to some degree (Horie *et al.*, 1970). This compound must be transaminated to phenylalanine by the silkworm. This direct precursor might occur very slightly or not at all in the larva of the silkworm.

REQUIREMENT FOR TRYPTOPHAN AND ITS METABOLISM

The minimum optimum amount of tryptophan required by the silkworm is 2 mg per g of the dry diet (Arai and Ito, 1967). With the addition of excess tryptophan to the diet the larva becomes strongly reddish in color, perhaps from the deposition of pigments in the epidermis (Arai and Ito, 1967). The conversion of tryptophan to pigments via kynurenine and 3-hydroxykynurenine has been established in the silkworm (Kikkawa, 1941), but tryptophan cannot be a precursor of niacin in this insect (Ito, 1952; Kikkawa and Kuwana, 1952). Requirement for niacin as a vitamin by the silkworm is well demonstrated (Horie and Ito, 1963, 1965).

Previously, Kikkawa (1950) had reported that indole lactic acid is converted into tryptophan in the silkworm. A more recent study showed, however, that the dietary tryptophan can be only partly replaced by indole lactate (Horie *et al.*, 1970). This compound is considered to have been oxidized to indole pyruvate, which then transaminated to form tryptophan, although the efficiency of these conversions is low. Thus the requirement for tryptophan, as in the case of the requirement for phenylalanine, may be concluded to be at least partly due to the fact that the proper precursors are not readily available in the larval body, or are available only, at a lower level than the quantity normally required by the larva.

DIETARY AMINO ACIDS AND COCOON PRODUCTION

The amino acid composition of silk protein in the silkworm has been analyzed in detail, and the silk protein is known to have a very high proportion of the amino acids, alanine, glycine, serine, and tyrosine. The relationship between silk production, or the cocoon crop, and the food value of mulberry leaves has been studied in some detail, but mostly on the basis of a practical crop increase.

A term, the cocoon-shell ratio, has been important in considering the efficiency of silk production and is calculated from the following formula:

$$\frac{\text{Weight of a cocoon shell}}{\text{Weight of a whole cocoon}} \times 100\ (\%).$$

The ratio of an excellent silkworm race reaches a level of 25% under normal rearing conditions.

The use of an amino-acid diet for the study of the relationship between cocoon production and nutrition was found to be effective. In general, the cocoon-shell ratio as well as the production of silk increases according to the increase in the concentration of dietary amino acids. The relative amount of fibroin versus sericin increases simultaneously. An extra addition of the ten essential amino acids to the standard amino-acid diet, however, results in a drop not only of silk production but of relative fibroin content versus sericin content. On the contrary, an extra addition of the five dispensable amino acids, alanine, cystine, glycine, serine, and tyrosine, or that of the acidic amino acids, clearly accelerates the production of silk (Ito *et al.*, 1967).

The relative dietary amounts of the five dispensable amino acids as opposed to the amounts of the acidic ones influenced silk production. There were also differing ratios for the two groups of amino acids for male and female, respectively (Horie *et al.*, 1970). In the silkworm it is well known that there are many differences between the sexes, such as the ability to digest mulberry leaves or artificial diets, the weight of a cocoon and of a cocoon shell, the cocoon-shell ratio, and so on.

EFFECT OF DIETARY AMINO ACIDS ON THE COMPOSITION OF FREE AMINO ACIDS OF THE HEMOLYMPH

The composition of free amino acids in the hemolymph of the silkworm has been analyzed by the microbiological method, by means of ion-change chromatography, or with an amino-acid analyzer. The composition varies largely during larval and pupal development, and more or less according to rearing seasons or from strain to strain in the silkworm. It is without doubt that the amino acid composition of the hemolymph reflects the physiological status of the silkworm or indicates some aspects of amino acid metabolism. Amino acid composition of the hemolymph of larvae from which the silk glands have been extirpated differs largely from that of the control (Fukuda *et al.*, 1955; see a review by Florkin and Jeuniaux, 1964). This, however, rather gave us a little information about the amino acid nutrition.

Recently it was found that the composition of the free amino acids of the hemolymph is varied specifically by the composition of dietary amino acids. The composition was altered as rapidly as 6 hours after the larvae had been transferred to a diet lacking one of the indispensable amino acids. The essential amino acid omitted from the diet disappeared completely from the hemolymph within 30 hours. The protein level in the hemolymph also decreased, accompanied by the accumulation of ninhydrinpositive substances. When the diet lacked arginine, for instance, free ornithine also disappeared eventually, whereas free lysine, glycine, and valine were accumulated in great quantities. This change in the amino acid composition was specific according to the kind of amino acid omitted. In general, when one of the essential acids was omitted, most of the amino acids, especially glutamic acid, were inclined to increase in quantity, whereas the amino acids considered to be related metabolically to the omitted one showed a tendency to decrease. When the diet lacked either alanine, glycine, or serine, changes in the patterns of amino acids of the hemolymph were smaller than those caused by deletion of the essential ones (Inokuchi, 1970).

An equation has been offered to calculate the coefficient for comparison of the similarity between two patterns of amino acids in various foods (Tamura and Osawa, 1969). The values calculated using the analytical data on the amino acid composition of the hemolymph of larvae on deficient diets are shown in Table 3. When one pattern has a strong resemblance to the

Table 3. Coefficient of pattern similarity of free amino acids of hemolymph of silkworm larvae on the diet lacking single amino acid.

Amino acid omitted	Coefficient of pattern similarity against control	Amino acid omitted	Coefficient of pattern similarity against control
Female larvae used[a]		Male larvae used[b]	
Arginine	0.846	Alanine	0.957
Histidine	0.893	Glycine	0.931
Isoleucine	0.875	Serine	0.940
Leucine	0.869		
Lysine	0.858		
Methionine	0.898		
Phenylalanine	0.928		
Threonine	0.879		
Valine	0.798		

[a,b]Larvae were kept on the deficient diet for 30 hours (a) or for 5 days (b), then analyzed.
Calculated from data by Inokuchi (1970).

other, the value of the coefficient should reach 0.999 or 1.000. In general, a big divergence from the control pattern is found in the absence of an essential amino acid, as indicated by the lower value of the coefficient, or when higher values are obtained in the absence of any dispensable amino acid.

The amino acid pattern as a whole differs significantly between male and female, even during larval stages (Inokuchi, 1972a; Ito and Inokuchi, in preparation). Sexual difference in the pattern is less in the newly ecdysed fifth-instar larvae, and becomes greater with age. It was further shown that both histidine and serine accumulate significantly at higher levels in the female, in addition to the sulfur amino acids.

NUTRITIVE EFFECT OF D-AMINO ACIDS

When each of the ten essential L-amino acids was replaced by the D-isomers singly, D-methionine had a slight sparing effect and D-histidine had the same effect, whereas other D-amino acids used in lieu of the essential L-isomers showed no activity or were sometimes detrimental, such as D-leucine. Furthermore, D-proline did not seem to have any nutritive value, and both D-alanine and D-serine were detrimental at their higher dietary levels.

Nutritional values of D-amino acids were subsequently evaluated by comparing their effects on the composition of free amino acids in the hemolymph, and calculating the coefficient of pattern similarity (Table 4). The value obtained with larvae on the diet containing either D-histidine or

Table 4. Coefficient of pattern similarity of free amino acids of hemolymph of silkworm larvae (female) on the diet either containing D-amino acid or lacking L-isomer.

Diet[a]	Coefficient of pattern similarity against control
D-Histidine addition	0.765
L-Histidine omission	0.723
D-Methionine addition	0.763
L-Methionine omission	0.744
D-Leucine addition	0.627
L-Leucine omission	0.743

[a]Larvae were kept on the experimental diet for 48 hours, then analyzed. From Ito and Inokuchi (in preparation).

D-methionine was higher than that obtained by omitting either of the L-isomers. This indicates that these D-amino acids are utilized slightly by the silkworm. On the other hand, the value obtained with the diet containing D-leucine was lower than that without L-leucine, indicating that D-leucine is toxic, in agreement with the nutritional test of rearing. A comparison of the amino acid pattern of the hemolymph is assumed to be useful in evaluating the efficiency of the diet, either amino-acid or practical artificial for the silkworm.

SUMMARY

In the study on amino acid nutrition of the silkworm, *Bombyx mori*, replacement of a specific nutrient in the diet by other metabolites provided information about what kinds of metabolic pathways are involved. In these cases the use of radioisotopes or other biochemical techniques was very helpful and sometimes decisive in the elucidation of the metabolic pathway involved. Nutritional studies on amino acids seem to have contributed more to the solution of amino acid metabolism than did the biochemistry of amino acids to the solution of amino acid nutrition in the silkworm.

Furthermore, it was demonstrated that the rate of metabolic conversion is influenced by dietary conditions. The quality and concentration of amino acids in the hemolymph were found to depend on the amino acid composition of the diet. Dependence of other kinds of compounds in the larva on the dietary conditions is also reported for this insect (Ito and Nakasone, 1969). Analysis of dietary conditions for the control of metabolism will come up as an important problem.

At present the L-amino acids are considered to be classified in the following groups, judging from their nutritional role and metabolism in the silkworm (Ito, unpublished). 1) essential: the usual ten plus either aspartic or glutamic acid. 2) semi-essential: proline. 3) non-essential: including (a) those readily formed by transamination (alanine, glycine, and serine) and (b) those mainly formed from direct precursors (cystine and tyrosine). D-amino acids are grouped into three categories: those utilized slightly (histidine and methionine); those possessing some toxicity (leucine, alanine, serine, and others possibly); and those not utilized and inert.

REFERENCES

ARAI, N. and ITO, T. (1964). Amino acid requirements of the silkworm, *Bombyx mori* L. *J. Seric. Sci.* Tokyo. 33:107-110.

_____&_____(1967). Nutrition of the silkworm, *Bombyx mori*. XVI. Quantitative requirements for essential amino acids. *Bull. Seric. Exp. Stn. Japan.* 21:373-384.

BRICTEUX-GRÉGOIRE, S., VERLY, W. G. and FLORKIN, M. (1956). Utilization of the carboxyl carbon of L-phenylalanine for the synthesis of the amino-acids of silk by *Bombyx mori. Nature:* (Lond.) 177:1237-1238.

CHEFURKA, W. (1965). Intermediary metabolism of nitrogenous and lipid compounds in insects. In *The Physiology of Insects*, Vol. II (M. Rockstein, Ed.) Academic Press, New York.

FLORKIN, M. and JEUNIAUX, C. (1964). Hemolymph: Composition. In *The Physiology of Insects*, Vol. I (M Rockstein, Ed.). Academic Press, New York.

FUKUDA, T. (1956). Conversion of phenylalanine into tyrosine in the silkworm larva (*Bombyx mori*). *Nature:* (Lond.) **177**:429-430.

_____(1957). Biochemical studies on the formation of the silkprotein. IV. The conversion of pyruvic acid to alanine in the silkworm larva. *J. Biochem.* **44**:505-510.

FUKUDA, T., KIRIMURA, J., MATUDA, M. and SUZUKI, T. (1955). Biochemical studies on the formation of the silkprotein. I. The kinds of free amino acids concerned in the biosynthesis of the silkprotein. *J. Biochem.* **42**:341-346.

GILMOUR, D. (1961). *The Biochemistry of Insects.* Academic Press, New York.

HORIE, Y., INOKUCHI, T. and WATANABE, K. (1970). Effects of dietary amino acids balances on growth and cocoon quality in the silkworm, *Bombyx mori* L. *Bull. Seric. Exp.* Stn. Japan. **24**:345-365.

HORIE, Y. and ITO, T. (1963). Vitamin requirements of the silkworm. *Nature:* (Lond.) **197**:98-99.

_____ &_____(1965). Nutrition of the silkworm, *Bombyx mori.* X. Vitamin B requirements and the effects of several analogues. *J. Insect Physiol.* **11**:1585-1593.

INOKUCHI, T. (1969a). Nutritional studies of amino acids in the silkworm, *Bombyx mori.* I. Effect of dietary cystine on methionine requirement and its metabolism. *Bull. Seric. Exp.* Stn. Japan. **23**:371-387.

_____(1969b). Nutritional studies of amino acids in the silkworm, *Bombyx mori.* II. Effect of several amino acids on proline and arginine requirements. *Bull. Seric. Exp.* Stn. Japan. **23**:389-410.

INOKUCHI, T. (1970). Nutritional studies of amino acids in the silkworm, *Bombyx mori.* III. Effect of dietary amino acid on the concentration of protein and free amino acids in the hemolymph of larvae. *Bull. Seric. Exp.* Sta. **24**:389-408.

_____(1972a). Isolation of lanthionine and its metabolism in the silkworm, *Bombyx mori.* Bull. Seric. Exp. Sta. (in press).

INOKUCHI, T., HORIE, Y. and ITO, T. (1967). Nutrition of the silkworm, *Bombyx mori.* XIX. Effects of omission of essential amino acids in each of the larval instars. *Bull. Seric. Exp.* Stn. Japan. **22**:195-205.

_____,_____ &_____(1969). Urea cycle in the silkworm, *Bombyx mori. Biochem. Biophys. Res. Commun.* **35**:783-787.

INOUE, Y. and KAWAGUCHI, S. (1943). Sulfur-containing amino acid in the egg-shell of the silkworm. Nippon Nogei Kagaku Kaishi. *J. Agric. Soc.* Japan. **21**:385-400.

ITO, T. (1952). Nicotinic acid in the silkworm pupae. *Jap. J. Genet.* **27**:75-78.

_____(1967). Nutritional requirements of the silkworm, *Bombyx mori* L. *Proc. Japan Acad.* **43**:57-61.

_____(1970). Lipid nutrition of the silkworm, *Bombyx mori* L. *Proc. Japan Acad.* **46**:1036-1040.

ITO, T. and ARAI, N. (1965). Nutrition of the silkworm, *Bombyx mori.* VIII. Amino acid requirements and nutritive effects of various proteins. *Bull. Seric. Expt.* Stn. Japan. **19**:345-373.

_____ &_____(1966). Nutrition of the silkworm, *Bombyx mori.* IX. Requirements for aspartic and glutamic acids. *J. Insect Physiol.* **12**:861-869.

_____ &_____(1967). Nutritive effects of alanine, cystine, glycine, serine, and tyrosine on the silkworm, *Bombyx mori. J. Insect Physiol.* **13**:1813-1824.

ITO, T., ARAI, N. and INOKUCHI, T. (1967). Nutrition of the silkworm, *Bombyx mori.* XVII. Effect of dietary levels of amino acids on growth of fifth-instar larvae and on cocoon quality. *Bull. Seric. Exp.* Stn. Japan. **21**:385-400.

ITO, T., HORIE, Y. and ISHIKAWA, S. (1958). Oxidative enzymes of the midgut of the silkworm *Bombyx mori. J. Insect Physiol.* **2**:313-323.

ITO, T. and INOKUCHI. (in preparation). Nutritive effects of D-amino acids in the silkworm, *Bombyx mori.*

ITO, T. and NAKASONE, S. (1969). Effects of dietary fatty acids on the fatty acid composition of larvae of the silkworm, *Bombyx mori* L. *Bull. Seric. Exp.* Stn. Japan. 23:295-311.

KIKKAWA, H. (1941). Mechanism of pigment formation in *Bombyx* and *Drosophila. Genetics.* 26:587-607.

_____(1950). Tryptophane synthesis in insects. *Science.* 111:495-496.

KIKKAWA, H. and KUWANA, H. (1952). Origin of nicotinic acid in *Bombyx mori* (silkworm). *Annotnes Zool. Jap.* 25:30-33.

KOIDE, F., HAGAYAMA, H. and SHIMURA, K. (1955). Transaminases in silkworm tissues. *J. Agric. Soc.* Japan. 29:987-990.

KONDO, Y. (1959). The isolation of L-cystathionine from the body fluid of silkworm larvae, *Bombyx mori. J. Seric. Sci.* Tokyo. 28:1-9.

_____(1960). The isolation of L-cystathionine from silkworm pupae and its distribution in the silkworm, *Bombyx mori. J. Seric. Sci.* Tokyo. 29:149-152.

_____(1962). The cystathionine pathway in the silkworm larva *Bombyx mori. J. Biochem.* 51:188-192.

RAO, D. R., ENNOR, A. H. and THORPE, B. (1967). The isolation and identification of L-lanthionine and L-cystathionine from insect haemolymph. *Biochemistry.* N. Y. 6:1208-1216.

TAMURA, S. and OSAWA, F. (1969). Amino acid pattern similarity between foods in Japan. *J. Jap. Soc. Fd. Nutr.* 22:494-496.

PROTEIN DIGESTION IN HEMATOPHAGOUS INSECTS

Shoshana Akov
Israel Institute for Biological Research
Ness-Ziona, Israel

The literature dealing with various aspects of the digestive process in hematophagous insects has recently been reviewed by Gooding (1972). This paper concerns the factors responsible for stimulating the appearance of midgut protease in blood-sucking insects.

The midguts of unfed (or sugar-fed) mosquitoes contain very little or no proteolytic activity (Fisk, 1950), but there is considerable proteolytic activity in the midguts of teneral tsetse flies (Langley, 1966), in unfed simuliid flies (Yang & Davies, 1968) and in unfed *Rhodnius* (Persaud & Davey, 1971). In all blood-feeding insects, the midgut proteolytic activity increases after engorging on blood and then decreases again.

There is so far no evidence that the proteases in insects are stored in the form of an inactive precursor, like those of the vertebrate pancreas. Attempts to activate mosquito protease by incubating midguts with blood *in vitro* were unsuccessful (Shambaugh, 1954; Chen, 1969). In *Aedes aegypti*, trypsin is synthesized after the ingestion of the blood meal (Gooding, personal communication). Further evidence for *de novo* synthesis of midgut protease comes from studies of the ultrastructure of the mosquito midgut epithelial cells. In unfed *Aedes aegypti* females the rough-surfaced endoplasmic reticulum (rer) is organized in compact finger-print like structures ("whorls"). A few minutes after the intake of blood (but not after feeding on sugar) the "whorls" start to unfold, indicating synthesis of proteins, possibly digestive enzymes. After the blood digestion is completed, the compact structures are restituted (Bertram & Bird, 1961; Staubli *et al.*, 1966).

In newly emerged female *A. aegypti* the fine structure of the midgut epithelium is not yet differentiated (Hecker *et al.*, 1971a). This process takes 2 - 3 days, and is similar to the reconstruction of the "whorls" after a blood meal. Indeed, newly emerged mosquito females do not take up blood. When young *Culex* females were given a blood enema, the blood was excreted completely within a few hours, or remained bright red and undigested until death (Chen, 1969). In the non-hematophagous males of *A. aegypti* the finger-print like structures are never found (Hecker *et al.*, 1971b). *A. aegypti* males fed on plasma or hemoglobin in sugar solution do not produce protease (Table 1).

The mechanism for starting and stopping the production of proteolytic enzymes is not yet known. Attempts to demonstrate hormonal or nervous control were negative or inconclusive (Fisk & Shambaugh, 1952; Langley,

Insect and mite nutrition – North-Holland – Amsterdam, (1972)

1967; Chen, 1969). The removal of neuro-secretory cells does not inhibit the production of proteolytic enzymes in *Rhodnius* (Persaud & Davey, 1971). Ablation of the median neurosecretory cells (mnc) in aedine mosquitoes does not interfere with the digestion and absorption of protein from a blood meal, as measured by net synthesis of triglycerides (Lea, 1967). Ligaturing the abdomens of mosquitoes immediately after blood feeding does not prevent the production of protease in *Aedes aegypti* (personal observation) and in *Culex* (Chen, 1969). Similar results were obtained with decapitated *A. aegypti* (Gooding, 1966a).

Proteolytic activity in various blood-feeding insects was investigated by removing the midgut at various intervals after blood feeding and incubating the homogenates with proteins or synthetic substrates. *A. aegypti* trypsin is the first protease from a blood-sucking insect that has been isolated and purified (Huang, 1971a). A chymotrypsin from *A. aegypti* has also been partially purified. Larvae contain both chymotrypsin and trypsin, but adult females contain very little chymotrypsin activity, as measured by the hydrolysis of specific synthetic esters (Gooding, 1966b; Yang & Davies, 1971).

Protease activity in midgut homogenates of mosquitoes 18 hours after feeding was proportional to the amount of blood ingested (Fisk and Shambaugh, 1952). In order to induce gorging through membranes, washed sheep erythrocytes had to be added. The amount of protease was maximal after feeding on whole blood or when dialyzed plasma proteins were added to the erythrocytes (Shambaugh, 1954).

Langley (1966) found that the feeding of saline and of bovine red blood cells did not increase the amount of protease in *Glossina morsitans* above that of unfed flies. Flies fed on serum alone took only small meals, but produced protease. Langley concluded that the factor necessary for the production of midgut protease is contained in the serum and not in the corpuscles. Protease production in *G. austeni* was also stimulated by bovine serum (Langley, personal communication).

In females of simuliid flies the trypsin activity was increased after feeding on whole blood, blood-sucrose or erythrocyte-sucrose mixture (Yang & Davies, 1968). The erythrocytes stimulated tne enzyme production to the same extent as whole blood. The effect of plasma could not be evaluated, since the flies did not take plasma-sucrose mixtures. Sucrose alone produced no increase in the enzyme activity.

Using ATP (10^{-2}M) as feeding stimulant for mosquitoes and tsetse flies (Galun, 1971) enabled us to dispense with the erythrocytes formerly used in membrane-feeding and feed the insects on various blood components as well as on other solutions.

A. aegypti mated females (virgins refuse to feed on membranes, see Akov, 1966) were fed on various dilutions of defibrinated sheep blood, thrice washed sheep erythrocytes suspended in 12% dextran in physiological saline, sheep plasma, and solutions of sheep hemoglobin, purified bovine serum albumin and other proteins (Akov & Samish, 1971). Mosquitoes fed on 12% dextran in saline served as controls. Only fully engorged females were taken and kept at 28°C until dissection. Proteolytic activity was measured in whole midgut homogenates from mosquitoes dissected at various intervals after feeding. Casein (0.625 mg per ml) served as substrate; incubation was 20 - 30 minutes at 39°C, in 0.05 M Tris buffer pH 8.8. The tyrosine in the trichloroacetic acid supernatant was measured fluorometrically (Waalkes & Udenfriend, 1957).

Diluting the whole blood had little effect on the amount of protease activity 4 and 8 hours after feeding, but at 20 hours the amount of protease is related to the amount of blood in the meal (Fig. 1). Twenty-eight hours

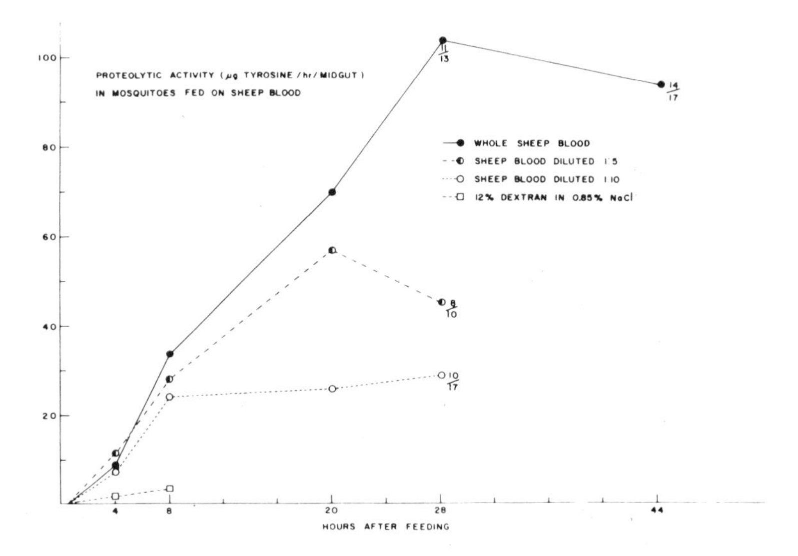

Figure 1. Midgut proteolytic activity in mosquitoes fed on various dilutions of defibrinated sheep blood. The numerator of the fractions indicates the number of mosquitoes with visible amounts of food in their midguts out of the total fed.

after feeding some of the females had already emptied their midguts. (Only midguts with visible amounts of food were taken for protease determination, since empty midguts never contain protease activity). Peak protease activity is attained earlier with more dilute diets, and midguts are also emptied more rapidly. Engorging on 12% dextran in saline induces a small, but measurable amount of protease, but 20 hours after feeding all the dextran-fed females had empty midguts. Feeding on washed sheep erythrocytes induced

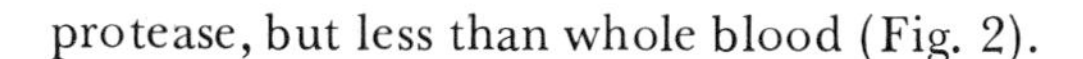

protease, but less than whole blood (Fig. 2).

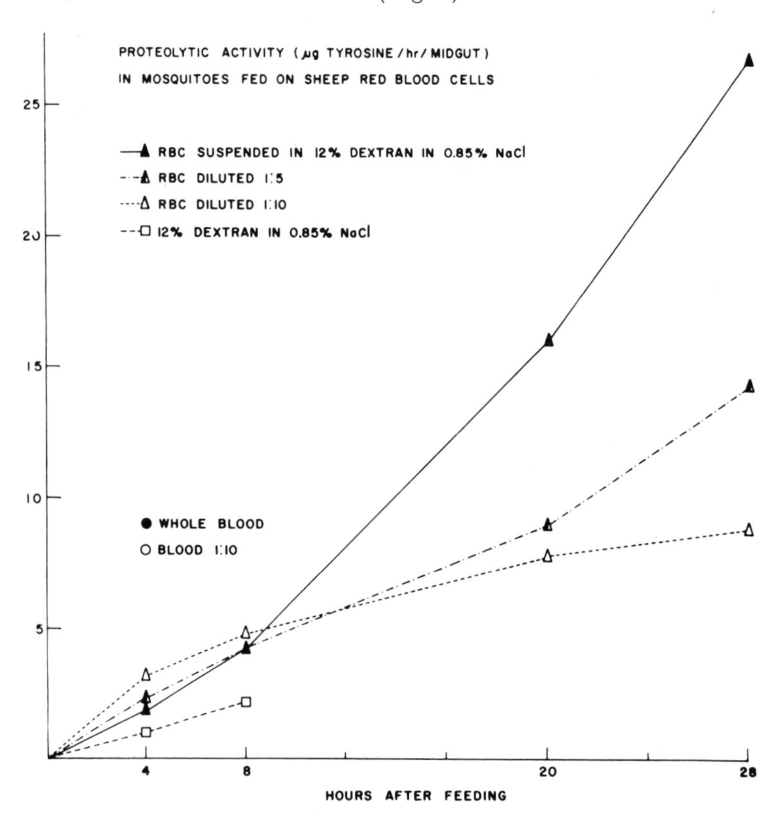

Figure 2. Midgut proteolytic activity in mosquitoes fed on various dilutions of thrice washed sheep erythrocytes.

Four and 8 hours after feeding on sheep plasma the midgut protease activity was lower than in mosquitoes fed on diluted (1:10) plasma (Fig. 3). The inhibitory effect of sheep plasma is not unexpected, since sheep plasma contains relatively more *A. aegypti* protease inhibitors than plasma from other mammals (Huang, 1971a).

Twenty hours after feeding the inhibitory effect of plasma had already been overcome, and the amount of protease was proportional to the amount of protein. Eight hours after feeding on 5% bovine serum albumin (which contains no inhibitor) there is more protease than after feeding on plasma.

Hemoglobin in solution induces less protease than plasma and whole blood, but it is a better inducer than equivalent amounts of washed red blood cells or bovine serum albumin. Casein is also a good inducer of

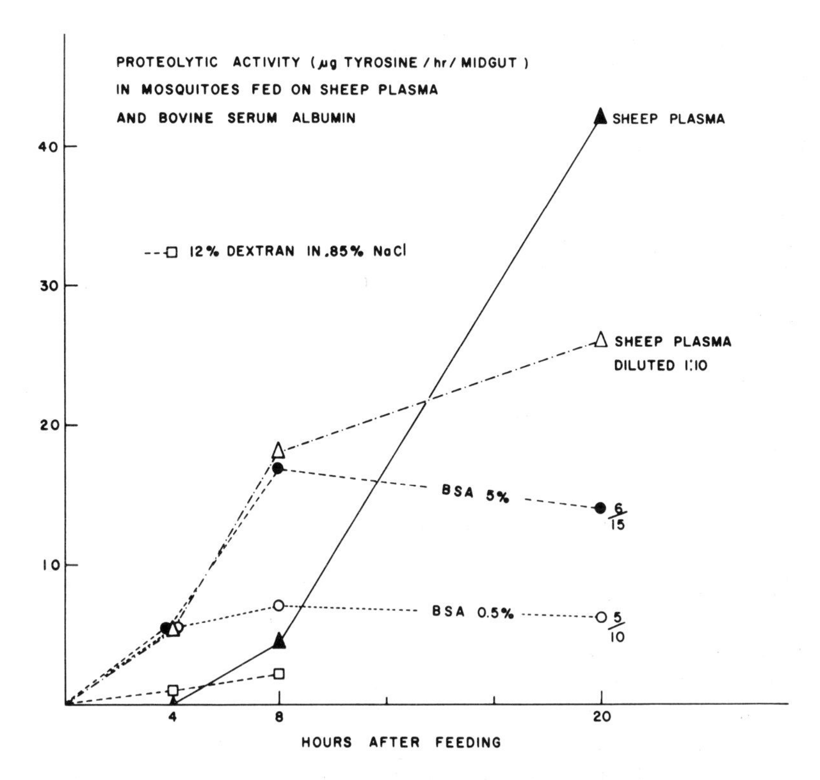

Figure 3. Midgut proteolytic activity in mosquitoes fed on various dilutions of sheep plasma and bovine serum albumin. Fractions as in Fig. 1.

protease, but casein hydrolysate is ineffective.

Females fed on dextran alone never initiated ovary development. No egg development was observed in females fed on erythrocytes or on sheep hemoglobin. All females that had fed on whole sheep blood, whole sheep plasma, and 5% casein initiated egg development, but the diluted blood, diluted plasma, and 1% casein did not initiate ovary development. Clearly, there is no relation between the ability of a certain diet to induce protease activity and its capacity to initiate egg development. It seems that more protein is needed for ovary development than for inducing protease.

The question whether the distention of the abdomen after engorging on a nutritionally inert solution (12% dextran in saline) is the direct cause for the small increase of protease observed cannot yet be answered. However, the distention of the midgut is obviously not necessary for protease production, because mosquito females fed on protein-sugar solutions produce protease. Under these conditions the mosquitoes of both sexes feed continuously, the food first enters the crop and is gradually released into the midgut. The males had full crops, but no protease was found in their midguts (Table 1).

Table 1. Proteolytic activity (μg tyrosin/hr/midgut) in mosquitoes fed on sugar solutions.

Continuous feeding for 18 hr	Females	Males
Water only	0.5	0
9% sucrose	0.25	0
9% sucrose : sheep plasma (1 : 1)	19.4	0
9% sucrose : sheep hemoglobin 12% (1 : 1)	17.6	0

The viviparous tsetse flies have eliminated larval nutrition and for them blood is the sole source of energy and water for both sexes as well as the protein source for reproduction and growth. It has been calculated that 50% of the blood imbibed by fertilized female *Glossina austeni* is for normal energy requirement, the other 50% for production of pupae (Boyle, 1971).

Our work (Akov & Samish, unpublished) was done with teneral *G. austeni* bred at the Langford, Bristol laboratory and shipped to us in the pupal stage. We examined the midgut protease in unfed flies and in flies fed on a rabbit and through membranes. The method for estimating protease activity was the same as with mosquitoes, except for the pH of the buffer (7.9) and the time of incubation (60 min).

Newly emerged flies (dissected within one hour after emergence, and before the wings were fully expanded) contain a small amount of protease. The amount of enzyme increases within the next 24 hours, and remains at that level for the next two days. We found a significant correlation between the weight of the unfed (aged 1 - 2 days) and the amount of midgut protease. There was also a significant correlation between the weight of the flies (of both sexes) and the size of the meal taken on a rabbit or through membranes. However, we did not find a correlation between the size of the first blood meal on a rabbit and the amount of protease 24 hours after feeding. Such a correlation was found by Langley and Abasa (1970) in the same species. The blood meals subsequent to the first one are larger, and the amount of midgut protease 24 hours after feeding is greater. Unmated females, (aged 11 - 13 days) that had each taken 4 blood meals, were dissected 24 hours after their fourth meal. Protease activity in one-week-old males that had taken four meals was also much higher than that after the first feeding.

The protease activity was increased in one day after feeding on diluted sheep blood, sheep hemoglobin and casein (Table 2), but not after feeding on sheep serum. The protease activity in the serum-fed flies was less than that of the unfed controls, indicating that the residual protease was inhibited. Diluted sheep blood (40%) contains only 20% serum, and thus has less inhibitors of protease. Fig. 4 shows that one day after feeding on sheep plasma the protease activity was less than in flies fed on human plasma. Sheep plasma contains more inhibitors than human plasma. Two days after feeding on plasma the inhibitory effect was overcome, and protease activity was increased.

Table 2. Protease activity (μg tyrosine/hr/midgut) in tsetse flies one day after feeding through membranes.

Food	Protease activity	No. of flies
Unfed controls	30.4	10
Sheep serum	16.1	9
40% citrated sheep blood	62.3*	7
12% sheep hemoglobin	73.4*	18
4% casein	52.5*	12

*Significantly more than unfed, P<0.05.

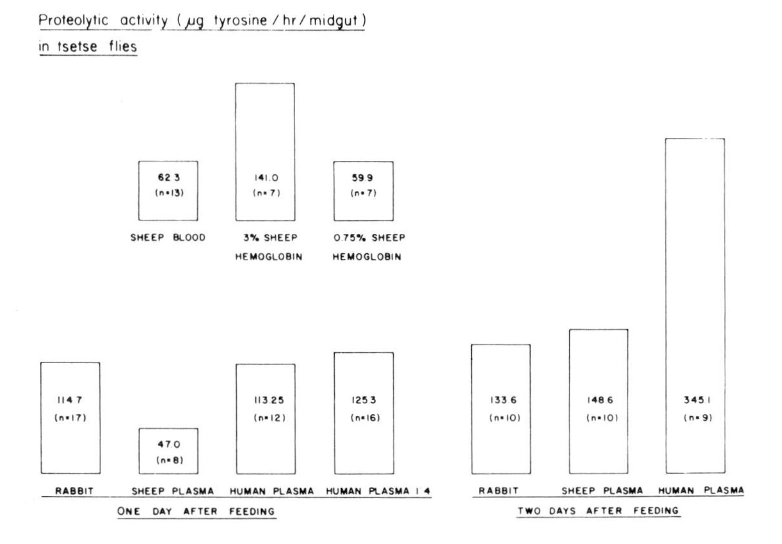

Figure 4. Midgut proteolytic activity in teneral tsetse flies after feeding on various diets.

The inhibitory effect of mammalian serum could also be observed in flies fed on a rabbit. Five hours after feeding the amount of protease was lower than that in unfed flies (Table 3). The increase in protease activity occurred earlier in flies fed on bovine serum albumin, which contains no inhibitor, than in the rabbit-fed flies.

Table 3. Protease activity (μg tyrosine/hr/midgut) in tsetse flies after feeding on bovine serum albumin and on a rabbit.

Food	Time after feeding (hr)		
	5	8	24
5% BSA	59.9(15)	-	114.7(21)
Rabbit	15.6(13)	48.3(6)	115.5(17)

In vitro experiments have shown that tsetse protease activity is strongly inhibited by sheep serum. It would be interesting to find out if the inhibitory effect of bovine serum on *G. austeni* (Langley, personal communication) is due to the same two *A. aegypti* trypsin inhibitors that have recently been isolated from bovine serum (Huang, 1971b).

In summary, our experiments with mosquitoes and tsetse flies show that any protein fed stimulates protease. Plasma protease inhibitors exert their effect *in vivo* for many hours after feeding, but it is not known how this affects the utilization of blood meals from various sources. Chicken serum contains more *A. aegypti* trypsin inhibitor than other vertebrate sera tested (Huang, 1971a), yet chicks are good hosts for rearing these mosquitoes. The trypsin from sheep kids is also inhibited by sheep serum (Gooding, in press).

Tsetse flies reared in the laboratory produced smaller pupae than flies in nature, and flies fed on pregnant guinea pigs produced significantly smaller pupae than flies fed on males (Langley, 1968). It has been suggested that the lower blood hemoglobin content in females was responsible, but it is also possible that the digestion of the blood in flies fed exclusively on pregnant guinea pigs was impaired by the higher amount of proteinase inhibitors that is found in pregnant animals (Vogel *et al.*, 1968). Feeding a natural protease inhibitor to mosquitoes delayed blood digestion and ovarian development (Akov, 1965).

It is hoped that further research on digestion and utilization of food might help to devise a suitable artificial diet for successful maintenance of tsetse flies in laboratory colonies.

REFERENCES

AKOV, S. (1965). Inhibition of blood digestion and oocyte growth in *Aedes aegypti* by 5 fluorouacil. *Biol. Bull. Mar. Biol. Lab. Woods Hole.* **129**:439-453.

_____ (1966). Retention of the blood-meal in *Aedes aegypti* following sterilization by chemicals and irradiation. *Ann. Trop. Med. Parasit.* **60**:482-494.

AKOV, S. and SAMISH, M. (1971). Proteolytic enzymes in mosquitoes. 9th Meeting of the *Assoc. Adv. Sci. Israel*, Haifa July 6-8, p. 123.

_____ & _____ (in preparation). Midgut protease activity in artificially fed tsetse flies, *Glossina austeni.*

BERTRAM, D. S. and BIRD, R. G. (1961). Studies on mosquito-borne viruses in their vectors. I. The normal fine structure of the midgut epithelium of the adult female *Aedes aegypti* (L) and the functional significance of its modification following a blood meal. *Trans. R. Soc. Trop. Med. Hyg.* **55**:404-423.

BOYLE, J. A. (1971). Effect of blood intake of *Glossina austeni* Newst on pupal weights in successive reproductive cycles. *Bull. Ent. Res.* **61**:1-5.

CHEN, S. S. (1969). The neuroendocrine regulation of gut protease activity and ovarian development in mosquitoes of *Culex pipiens* complex and the induction of imaginal diapause. Ph.D. Thesis, Purdue University.

FISK, F. W. (1950). Studies on proteolytic digestion in adult *Aedes aegypti* mosquitoes. *Ann. Ent. Soc. Am.* **43**:555-572.

FISK, F. W. and SHAMBAUGH, G. (1952). Protease activity in adult *Aedes aegypti* mosquitoes as related to feeding. *Ohio J. Sci.* **52**:80-88.

GALUN, R. (1971). Recent developments in the biochemistry and feeding behaviour of haematophagous arthropods as applied to their mass-rearing. In *Sterility Principle for Insect Control or Eradication* International Atomic Energy Agency, Vienna.

GOODING, R. H. (1966a). Physiological aspects of digestion of the blood meal by *Aedes aegypti* (Linnaeus) and *Culex fatigans* Wiedemann. *J. Med. Ent.* 3:53-60.

_____ (1966b). *In vitro* properties of proteinases in the midgut of adult *Aedes aegypti* L. and *Culex fatigans* (Wiedemann). *Comp. Biochem. Physiol.* **17**:115-127.

_____ (1972). Digestive processes of haematophagous insects. I.A. literature review. *Quaest. Ent.* **8**:5-60.

_____ (in press). Digestive processes of haematophagous insects II. Trypsin from Melophagus ovinus (L.) and its inhibition by mammalian sera *Comp. Biochem.*

HECKER, H., FREYVOGEL, T. A., BRIEGEL, H. and STEIGER, R. (1971a). Ultrastructural differentiation of the midgut epithelium in female *Aedes aegypti* (L) (Insecta, Diptera) imagines. *Acta Trop.* **28**:80-104.

HECKER, H., FREYVOGEL, T. A., BRIEGEL, H. and STEIGER, R. (1971b). The ultrastructure of midgut epithelium in *Aedes aegypti* (L) (Insecta, Diptera) males. *Acta Trop.* **28**:275-290.

HUANG, C. T. (1971a). Vertebrate serum inhibitors of *Aedes aegypti* Trypsin. *Insect Biochem.* **1**:27-38.

_____ (1971b). The interactions of *Aedes aegypti* (L) trypsin with its two inhibitors found in bovine serum *Insect Biochem.* **1**:207-227.

LANGLEY, P. A. (1966). The control of digestion in the tsetse fly, *Glossina morsitans.* Enzyme activity in relation to the size and nature of the meal. *J. Insect Physiol.* **12**:439-448.

_____ (1967). Experimental evidence for a hormonal control of digestion in the tsetse fly, *Glossina morsitans* Westwood: A study of the larva, pupa, and teneral adult fly. *J. Insect Physiol.* **13**:1921-1931.

_____ (1968). The effect of host pregnancy on the reproductive capability of the tsetse fly, *Glossina morsitans*, in captivity. *J. Insect Physiol.* **14**:121-133.

LANGLEY, P. A. and ABASA, R. O. (1970). Blood meal utilisation and flight muscle development in the tsetse fly, *Glossina austeni*, following sterilizing doses of gamma irradiation. *Entomologia Exp. & Appl.* **13**:141-152.

LEA, A. O. (1967). The medial neurosecretory cells and egg maturation in mosquitoes. *J. Insect Physiol.* **13**:419-429.

PERSAUD, C. E. and DAVEY, K. G. (1971). The control of protease synthesis in the intestine of adults of *Rhodnius prolixus*. *J. Insect Physiol.* **17**:1429-1440.

SHAMBAUGH, G. F. (1954). Protease stimulation by foods in adult *Aedes aegypti* (L). *Ohio J. Sci.* **54**:151-160.

STAUBLI, W., FREYVOGEL, T. A. and SUTER, J. (1966). Structural modification of the endoplasmic reticulum of midgut epithelial cells of mosquitoes in relation to blood intake. *J. Microscopie.* **5**:189-204.

VOGEL, R., TRAUTSCHOLD, I. and WERLE, E. (1968). *Natural Proteinase Inhibitors.* New York, Academic Press.

WAALKES, T. P. and UDENFRIEND, S. (1957). A fluorometric method for the estimation of tyrosine in plasma and tissues. *J. Lab. Clin. Med.* **50**:733-736.

YANG, Y. J. and DAVIES, D. M. (1968). Digestion, emphasizing trypsin activity, in adult simuliids (Diptera) fed blood, blood-sucrose mixtures, and sucrose. *J. Insect Physiol.* **14**:205-222.

_____ & _____ (1971). Trypsin and chymotrypsin during metamorphosis in *Aedes aegypti* and properties of the chymotrypsin. *J. Insect Physiol.* **17**:117-131.

NEUROENDOCRINE REGULATION OF INSECT METABOLISM AND THE INFLUENCE OF NUTRITION

Larry L. Keeley
Department of Entomology
Texas A&M University
College Station, Texas 77843

Nutrition is the source for the chemical substances required by an organism to produce energy and new, cellular structural components. The molecular interconversions of intermediary metabolism are the means whereby these nutrients are converted into the chemical species for building cell structures or yielding energy. To maintain the dynamic homeostasis of cell metabolism, metabolic pathways are regulated by activating or inhibitory processes. One type of metabolic regulation is by hormonal factors which increase the activity of specific enzyme systems to facilitate a particular metabolic pathway. In insects, the neuroendocrine system regulates aspects of metabolism such as O_2 uptake and the synthesis of trehalose, lipids, and proteins. Furthermore, insect neuroendocrine activity can be correlated to feeding so that nutrient intake may play a role in initiating the synthesis and release of neurohormones affecting metabolic processes. The following will review the studies from our laboratory on neuroendocrine effects on insect oxidative metabolism and will try to correlate our findings with those of other workers to develop a theoretical picture for endocrine regulation of general metabolism in insects. Of particular interest are endocrine effects on O_2 uptake since oxidative metabolism reflects the rate of cellular energy production and hence the capacity for endothermic cellular processes such as protein synthesis.

Endocrine regulation of insect oxidative metabolism has been investigated for some time with few definitive results. Early studies focused on the corpora allata as the source for a metabolism regulating hormone. Thomsen (1949) reported at 24% decrease in whole body respiration for allatectomized, adult, male and female *Calliphora erythrocephala* and demonstrated corpora allata implants increased the O_2 consumption of normal females. It was known at this time that the corpora allata in adult, female insects affected ovarian maturation. However, respiration of ovariectomized, female *Calliphora* did not differ from controls and suggested the decreases in O_2 uptake after allatectomy were effects directly on general metabolism, rather than ovarian development (Thomsen and Hamburger, 1955). Other research also correlated corpora allata activity with O_2 uptake and supported a direct action by these glands on general metabolism (DeWilde and Stegwee, 1958; Roussel, 1963; Sagesser, 1966). However, in a study significant to this subject, Slama (1964) demonstrated that although allatectomy of adult, female *Pyrrhocoris apterus* decreased respiration, cardiacectomy-allatectomy depressed it further. Slama suggested that the corpora allata affect reproduction-related metabolism in adult, female insects; whereas, the brain-corpora cardiaca neuroendocrine complex affects "trophic" metabolism (digestion and food utilization). These findings indicated that the neurosecretory system may play a metabolic role beyond that of the corpora allata alone.

Recent studies have confirmed a role for the neurosecretory system in regulating various general metabolic processes in insects. Steele (1963) reported factors from the corpora cardiaca regulated blood trehalose levels in *Periplaneta americana*, and Mayer and Candy (1969) described an adipokinetic effect by the same glands in *Locusta migratoria*. Hill (1962) and Highnam *et al.* (1967) found blood protein levels decreased after neurosecretory cell cautery in *Schistocerca gregaria*. The blood protein titers increased again after corpora cardiaca implantation (Hill, 1962). Furthermore, Osborne, *et al.* (1968) reported a direct stimulus by corpora cardiaca extracts to both RNA and protein synthesis in *Schistocerca* fat body. Several *in vitro* studies on the fat body of *Leucophaea maderae* have reported increased O_2 uptake after addition of corpora cardiaca extracts (Luscher and Leuthold, 1965; Wiens and Gilbert, 1965; Muller and Engelmann, 1968). In addition, research indicates a direct correlation between intestinal protease levels and neurosecretory activity in several insect species (Thomsen and Moller 1963; Mordue, 1967; Dogra and Gillott, 1971). In conclusion, these studies indicate neurosecretory control of a number of metabolic processes, all of which contribute to the basal metabolism.

Feedback activation of brain neurosecretory cells by the corpora allata, which may explain the early reports indicating corpora allata effects on metabolism, has been reported recently. For example, allatectomy of *Schistocerca* inhibited release of neurosecretory materials from the corpora cardiaca (Highnam *et al.*, 1967). This inhibition was reversed by corpora allata implantation. In *Calliphora* Thomsen and Lea (1969) demonstrated that cyclic increases in the nuclear volume of brain neurosecretory cells were regulated by the corpora allata. In view of these findings, many of the early reports ascribing metabolic effects to the corpora allata may now be explained on the basis of a feedback stimulation to neurosecretory activity.

The results of our studies indicate the neuroendocrine system regulates insect repiratory metabolism in a manner comparable to the thyroid gland in vertebrates. For example, the results in Fig. 1 show cardiacectomy-allatectomy depressed whole-body respiration of adult, male *Blaberus discoidalis* cockroaches. Sham surgery and allatectomy had no effects on O_2 uptake, indicating the effective endocrine system was the brain-corpora cardiaca complex. It is interesting that the amount of respiratory decrease reported for gland deficient *Blaberus* agreed, both with that reported by Thomsen (1949) for allatectomized *Calliphora* and with the 30% decreases in whole

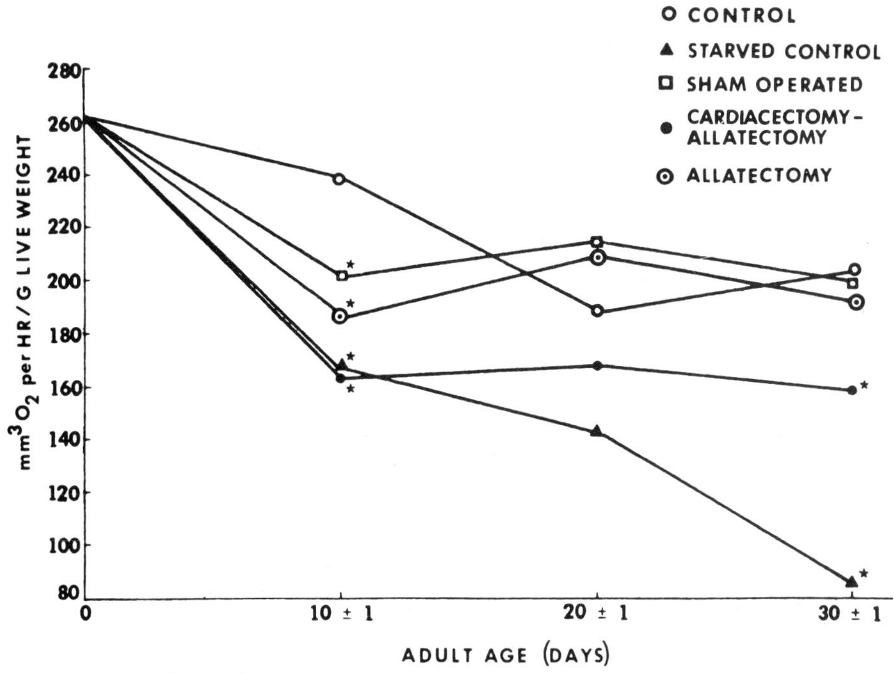

Figure 1. Effect of surgery, starvation and endocrine gland extirpation on whole body oxygen consumption of adult, male *Blaberus discoidalis* at various ages of adult life.

Measurements based on 1 O_2 consumed per hr/g. live wt. Readings were taken by standard Warburg manometry utilizing 140 ml respiration vessels containing enough paraffin to leave an ca. 100 ml volume. Cockroaches were separated by a copper screen from 0.5 ml 20% KOH. Animals were equilibrated for 30 min at 25°C and readings were taken at 15 min intervals if the animals showed no signs of movement. Ref. Keeley and Friedman (1967).

*Indicates significant difference from the normal, control mean value.

body respiration found in thyroidectomized rats (Tata *et al.*, 1963).

Based on the relationship described earlier between neurosecretory activity and intestinal protease, it was considered possible that cardiacectomy-allatectomy of *Blaberus* might result in nutritional imbalance due to altered digestive processes. In this way, respiratory rates might be lowered but not as the result of a direct hormonal influence on basal metabolism *per se*. The data in Fig. 1 shows that the respiration rates of operated and starving control cockroaches were identical at 10 days, but the respiration of the starved animals fell to 50% of that of the operated groups by 30 days. The respiration of cardiacectomized-allatectomized *Blaberus* remained constant and indicated cardiacectomy did not result in extreme starvation. In addition, Fig. 2 shows a continuous weight loss for starving animals over the 30-day study period. All operated animals lost weight comparable to starving animals for the first 10 days, but thereafter, operated groups began to regain weight and were equivalent to normal, unoperated controls by 30 days. In contrast, the starving group steadily declined over the entire study period. The dry weight of 30 day old cardiacectomized-allatectomized animals was unchanged from controls (Keeley and Friedman, 1967) and indicated that the weight gain was not the result of water retention. Therefore, by 30 days after surgery, cardiacectomized animals had recovered from surgical trauma and were apparently feeding and gaining nourishment from their food.

The early allatectomy studies of other workers showed changes in whole-body respiration, but few attempted to determine specific metabolic effects. For our studies *in vitro*, O_2 uptake of *Blaberus* muscle and fat body

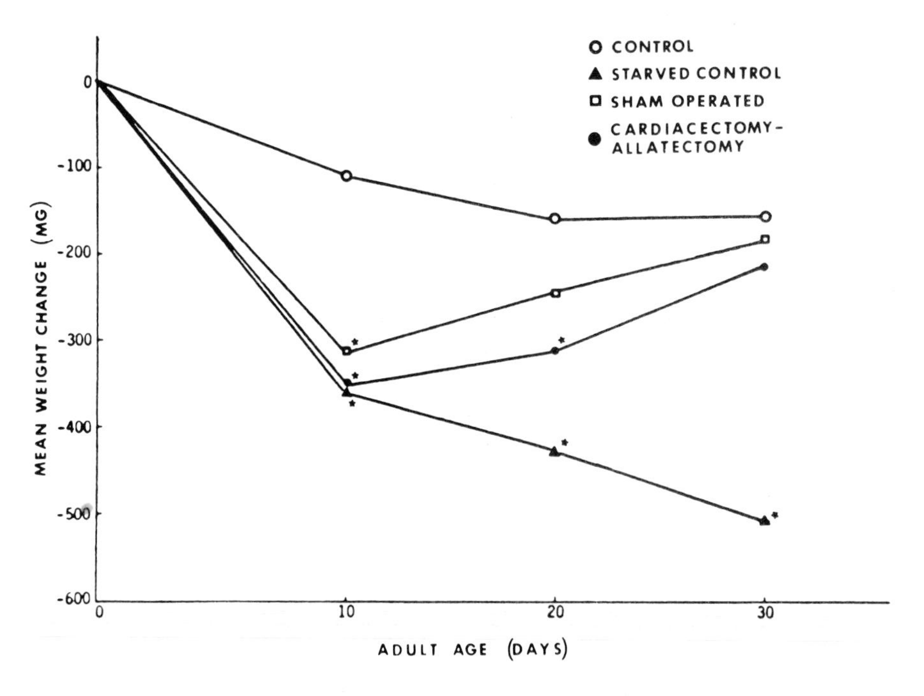

Figure 2. Effects of surgery, starvation, and endocrine gland extirpation on live body weight of adult, male *Blaberus discoidalis* at various ages of adult life.

*Indicates significant difference from the normal, control mean value. Ref. Keeley and Friedman (1967).

was measured to determine if a target organ was involved in the decreased respiratory metabolism that followed corpora cardiaca removal. Manometric studies demonstrated gland deficiency depressed fat body respiration but had no effect on coxal muscles (Keeley and Friedman, 1967). This was true for both intact and homogenized fat bodies. Since α-ketoglutarate, a Krebs cycle donor to electron transport, was added to the homogenized fat body preparations and failed to restore normal O_2 uptake, it was concluded that the effect of corpora cardiaca-deficiency may be at the level of the mitochondria.

Thyroid deficiency is typically reflected in vertebrate tissues by lower mitochondrial respiration, lower electron transport activity and lower tissue cytochrome c content (Tata *et al.*, 1963; Bronk, 1966; Klitgaard, 1966). No comparable relationship linking endocrine functions to mitochondrial activity is identified for active insects, although two conflicting studies have been reported. In *Locusta migratoria*, Clarke and Baldwin (1960) stimulated whole body respiration and that of isolated flight muscle and fat-body mitochondria by the addition of crushed corpora allata. In contrast, Minks (1967) found no decreases in the respiration of either flight muscle or fat-body mitochondria from allatectomized *Locusta*, although additions of corpora allata to isolated mitochondria stimulated both respiratory control and P:O ratios. No evidence was presented in the latter paper to show a correlation between *in vitro* and *in vivo* effects. In adult *Leptinotarsa decemlineata*, structural and enzymatic degeneration of flight muscle mitochondria accompanied allatectomy (Stegwee, 1964; Stegwee *et al.*, 1963); however, this was a case of diapause and was accompanied by degeneration of the tissue as well. In conclusion, the results from these investigations were conflicting and failed to demonstrate a definitive endocrine regulation of mitochondrial functions in active insects.

Our early data indicated corpora cardiaca deficiency in *Blaberus* affected the fat body tissue with possible effects at the mitochondrial level. To examine this, fat-body mitochondria were isolated by differential centrifugation and oxygen consumption measured polarographically (Keeley and Friedman, 1969; Keeley, 1971). The data in Table 1 summarize the results of our studies on the effects of various endocrine deficiencies on fat-body mitochondrial respiration. In general, mitochondrial respiration was decreased 30 to 40 percent by either corpora cardiaca removal or by severance of the nervi corporis cardiaci I (NCCI). Sham surgery and allatectomy, as in the earlier studies on live animal respiration, had no effects on mitochondrial respiration.

Studies were made with gland extracts to determine if a positive correlation existed between the neurosecretory system and fat-body mitochondrial functions. Gland and/or brains were isolated from 5-day old adult, male *Blaberus* and extracts prepared (Keeley and Waddill, 1971). Corpora

Table 1. Effects of endocrine gland extirpation on O_2 consumption by fat body mitochondria from adult, male *Blaberus discoidalis.*

Type of gland extirpation	QO_2 [a] (1 O_2 per hr/mg protein)
None (control)	98 + 6 (11)
Sham surgery	99 + 7 (9)
CC + CA	69* + 5 (11)
CA (only)	90 + 9 (10)
CC (only)	61* + 5 (8)
Severance of NCC I	74* + 5 (12)

[a]Values are means + S.E.; numbers of replicate tests in parenthesis. CA^- = allatectomy; CC^- = cardiacectomy. Incubation medium: 30 mM potassium phosphate, 5 mM $MgCl_2$, 2 mM EDTA, 50 mM tris (hydroxymethyl) aminomethane, 15 mM KCl, 2 mM ADP and 20 mM succinate (pH 7.5). Reaction system consisted of 1.90-1.95 ml incubation medium and mitochondria (200-800 g protein) to a 2 ml total volume. T = 35°C.

*Indicates significantly different from control based on Student's t-test.

cardiaca-allata deficient animals received the total equivalent of 1 gland pair by daily injections between the ages of 20 and 30 days. The results reported in Table 2 demonstrated neither Ringer nor corpora allata extracts had any recovery effects. By comparison, extracts of brain plus corpora cardiaca resulted in O_2 uptake rates comparable to normal animals (Table 1). Brain extracts alone caused only a marginal recovery response; whereas, extracts of corpora cardiaca alone resulted in complete recovery. Since the brain is the same tissue type as the corpora cardiaca, brain extracts were useful as controls to indicate that the respiratory response was gland specific and not the result of administering nonspecific tissue components. Furthermore, boiled corpora cardiaca preparations were inactive and the active agent was nondialysable, suggesting the agent was proteinaceous. Finally, the data in Table 2 also indicate the extract equivalent of a gland pair of corpora cardiaca added directly to a preparation of fat-body mitochondria from corpora cardiaca-allata deficient *Blaberus* had no respiratory stimulus.

The results of the previous studies show the neurosecretory system of *Blaberus* influences the respiratory activity of fat-body mitochondria. Muller and Engelmann (1968) have reported that corpora cardiaca implants in female *Leucophaea* remained active for the secretion of their intrinsic factors, up to 78 days after separation from the brain. Assuming the same was true for *Blaberus*, we concluded the lack of activity after NCCI severance would result from an inability to transfer a brain neurohormone to the

Table 2. Effects of endocrine tissue extracts on the respiratory activity of fat body mitochondria from cardiacectomized-allatectomized *Blaberus discoidalis.*

Extract	QO_2[a] (1 O_2 per hr/mg protein)
In vivo	
Ringer	68 + 5 (6)
Brain + C. cardiaca	94* + 4 (6)
Brain	80 + 3 (4)
C. cardiaca	118* + 8 (6)
C. allata	66 + 2 (4)
C. cardiaca (boiled)	72 + 6 (6)
Brain + C. cardiaca (dialysate)	120* + 11 (5)
In vitro	
Basal rate	72 + 10 (4)
+ C. cardiaca	65 + 10 (4)

[a]Values are mean S. E.; numbers of replicate tests in parenthesis. Assay conditions similar to Table 1.

*Indicates significantly different from Ringer-injected, gland deficient cockroaches.

corpora cardiaca for storage and release, although further studies would be required to confirm or deny this point.

Other workers have also reported effects by the corpora cardiaca on fat-body respiration. Luscher and Leuthold (1965) reported an *in vitro* stimulus to *Leucophaea* fat body respiration by added corpora cardiaca. They also demonstrated a slight brain effect, but no action by the corpora allata or subesophageal ganglion. Weins and Gilbert (1965) confirmed the *in vitro* stimulation of *Leucophaea* fat-body O_2 uptake by the corpora cardiaca extracts. However, in the latter study, evidence was presented that the increased O_2 uptake by the corpora cardiaca-stimulated fat-body was due to a switch from carbohydrate to lipid as the primary metabolic substrate. This change was the result of lipids being utilized as an energy source while carbohydrates were converted to trehalose (Weins and Gilbert, 1965).

Although the studies of others showed the corpora cardiaca affected fat body respiration, their results were not comparable to our studies in 3 ways. First, they were all performed with intact tissues and the respiratory effects observed did not necessarily bear a relationship to mitochondria. Second, the respiratory stimulus by the corpora cardiaca extracts appeared to be direct, in that it was observed immediately after addition of the extracts of the fat body. This was in contrast to our extract, which appeared to have no direct effect on mitochondria but, rather to require a latent period of action in the intact tissue. Thirdly, in the earlier studies, gland extracts were prepared by boiling, which would have destroyed our active principle. Therefore, our data suggest the existence of a hitherto undescribed neurohormone affecting fat-body metabolism at the mitochondrial level. The heat lability of our neurohormone sets it apart from those previously described, including the so-called "brain or activation hormone" which is reported to be heat-stable (Yamazaki and Kobayashi, 1969).

Investigations which show *Blaberus* fat-body mitochondria undergo an increase in respiratory capacity during the first 10 days of adult life have helped to determine the hormone-mitochondria relationship (Keeley, 1970). The results shown in Table 3 indicate that on the day of imaginal ecdysis (day 0) both the QO_2 and succinate-cytochrome c reductase activities of fat-body mitochondria were 1/3 of the 30 day value. At 5 days of adult age, they were equivalent to corpora cardiaca-deficient animals, and by day 10, and after, they compared to 30 day old cockroaches. Cytochrome c oxidase activities showed a different pattern, being low at day 0, remaining constant through day 5, and increasing to developed levels by day 10 where they remained thereafter. Respiratory control capacities, likewise, increased concomitant with respiration, although ADP:O ratios indicated respiration and phosphorylation were coupled throughout the development period. Removal of the corpora cardiaca essentially left the mitochondria at the partially developed 5-day level for all the biochemical parameters except oxidative phosphorylation which appeared normal.

The results of our studies suggest a development in fat-body mitochondrial respiratory capacities in relation to age and neurosecretory activity. The 0-5 day period of development is neurohormone independent; whereas, the 5-10 day period is neurohormone dependent since corpora cardiaca extirpation, at any age, results in respiratory rates comparable to the 5-day rate (Keeley, 1970; 1971). In confirmation of the hormone dependency, injection of corpora cardiaca extracts during the 0-5 day period resulted in precocious 10-30 day levels of respiration by 5 days of age (Keeley, in preparation). As mitochondria and fat body cells appeared competent in their response to the neurohormone between 0-5 days, it appeared that the neuroendocrine system must be inactive for this particular effect during this period.

Feeding has been shown as a frequent stimulus to neurosecretory activity in insects. Wigglesworth (1934) demonstrated feeding as the stimulus initiating the molting cycle of *Rhodnius prolixus*. Van der Kloot (1960) reported feeding by *Rhodnius* caused abdominal distention resulting in the stimulation of abdominal stretch receptors which activated the central nervous system for the release of brain hormone. In *Locusta*, Clarke and Langley (1963) reported that the severance of pharyngeal nerves at the

Table 3. Effects of age and gland deficiencies on respiratory activities of fat body mitochondria of adult, male *Blaberus discoidalis*.

Experimental animal	Age (days)	QO_2[a]	succinate-cytochrome c reductase activity	Cytochrome c oxidase activity	RCR[b]	ADP:O (succinate)
Normal	0	35	68	45	2.12	1.31
Normal	5	57	90	41	2.81	1.25
Normal	10	92	145	82	2.89	1.34
Normal	15	72	160	89	3.13	1.28
Normal	20	88	147	89	-	-
Normal	28-32	87	154	89	3.36	1.65
CC^- + CA^-[c]	29-32	69	95	54	2.41	1.50
CA^-	30-31	90	155	89	-	-

[a]QO_2 = 1 O_2 per hr/mg protein. Reductase and oxidase activities = nanomoles cytochrome c converted per min/mg protein.

[b]RCR = State 3 respiration rate : initial State 4 respiration rate. Respiratory rate assayed as described in Table 1. RCR and ADP:O assayed in an incubation medium (Table 1) containing 0.3% alcohol-extracted bovine albumin and 250 mM sucrose but lacking ADP. ADP (5 1, c 200 moles) added after measurement of State 4.

[c]CA^- = allatectomy; CC^- = cardiacectomy.

Electron transport activities assayed according to Keeley (1970).

frontal ganglion blocked the release of brain neurosecretions. The pharyngeal nerves were stimulated by the mechanical act of swallowing. Likewise, Dogra and Gillott (1971) demonstrated feeding of 3-day starved *Melanoplus sanguinipes* resulted in a release of stored neurosecretory materials within 20-40 min followed by renewed synthesis of neurosecretions 2 hr later. Furthermore, Goldsworthy (1970) reported the corpora cardiaca of adult *Locusta* stimulated a continuous increase in total fat body phosphorylase activity from ecdysis until 10 days. This increase in phosphorylase activity was prevented by starvation.

In Diptera an interaction is recognized between nutrient quality, corpora allata activity and neurosecretory activities. The corpora allata of *Calliphora* increase in volume after feeding on protein but remain small if fed on sugar (Strangeways-Dixon, 1962). Furthermore, Thomsen and Lea (1969) demonstrated that cyclic increases in the nuclear volume of *Calliphora* brain neurosecretory cells were regulated by the corpora allata. Neurosecretory cell nuclear volumes increased only in meat-fed, normal flies that failed to increase either in sugar-fed flies or in meat-fed, allatectomized flies. Implantation of active corpora allata into sugar-fed flies stimulated the nuclear increases.

The results of the feeding experiments suggest that, depending on the species, the mechanical act of feeding or the nutritional status of the insect may function as initiators of neurosecretory activity. The use of either nutritional status or feeding as a stimulus is reasonable, insofar as either would be an indicator that ample nutrition had provided abundant precursors for metabolic processes. Inasmuch as *Blaberus* fat body mitochondria responded to corpora cardiaca extracts between days 0-5, the presence or absence of the hormone appears to be the regulating factor in increasing respiration. Therefore, investigations were undertaken to determine whether feeding and nutrition might function as the initiators for the neurohormone-mediated phase of respiratory development of fat body mitochondria. Adult, male *Blaberus* were presented amaranth dye-stained food or water to determine the onset of feeding, based on the presence of dye in the crop or midgut. The animals were found to begin feeding about day 3, and by day 5 all individuals had fed. Animals receiving dyed food received plain water, but those receiving dyed water were not supplied food. Water consumption followed a similar pattern with no consistent drinking until days 4 and 5, although one individual drank as early as day 1. In general, these results correlated the onset of nutrient consumption with the hormone-dependent increase in mitochondrial respiration. However, other data did not substantiate this concept. Ten days of starvation (water provided) did not prevent the natural development in the respiratory rate of the fat body mitochondria. This latter result suggests food intake was unimportant in initiating the hormone-mediated stage of development of *Blaberus* fat body mitochondria. Presumably, sufficient nutrient reserves are present to utilize for mitochondrial development during the 0-10 day period of adult life and consumed food plays only a negligible role in this aspect of metabolism.

In conclusion, the above research on *Blaberus* fat body indicates a neuroendocrine regulation of basal metabolism exists in insects. The neurohormone functions in a manner analogous to the vertebrate thyroid hormones, in that mitochondrial respiration is altered. The hormone has no direct action on mitochondrial respiration *in vitro* but appears to require a latent period *in vivo* to manifest its action. Studies with the uncoupler 2,4-dinitrophenol have shown electron transport as the rate-limiting process of respiration (Keeley, in preparation) and have suggested the action of the neurohormone may be on a phase in the biosynthesis of a rate-limiting component of electron transport.

Our findings do not identify the primary action of the hormone, but demonstrate the mitochondrial responses are an ultimate aspect of the hormone's actions. It is generally accepted that cellular ADP levels determine the rate of metabolism via respiratory control, and the major metabolic process utilizing ATP is protein biosynthesis. Since the neuroendocrine system of insects does affect fat body protein biosynthesis (Hill 1962; Clark and Gillott, 1967; Osborne *et al.*, 1967), it could be argued that the decreases in fat body respiration following cardiacectomy are reflecting a general decrease in protein biosynthesis. However, this does not account for the reduced capacity for respiratory control shown by mitochondria isolated from gland deficient cockroaches. Rather, a decreased capacity to respond to ADP would appear to result from an intrinsic disruption of mitochondrial integrity, presumably the result of a specific action by the neurohormone on a phase of mitochondrial development and maintenance.

The finding that fat-body mitochondria undergo a development and respond to a neuroendocrine factor indicates 2 ways by which such mitochondria may function in regulating basal metabolism. In the classic scheme, mitochondria respond directly to ADP, by respiratory control, resulting in increased O_2 uptake and ATP generation. This is a direct stimulus to existing mitochondria to meet the immediate metabolic demands of the cell. In addition, our data suggest an endocrine control of fat-body metabolism may exist based on the numbers of functional mitochondria present in the tissue. This control on the numbers of functional mitochondria would result from regulation of the activity level of a rate-limiting component in a key metabolic pathway. For example, if the component were rate-limiting for electron transport, its activity level would determine the rate of energy production and thus establish the limit for the rate of all other endothermic activities.

The following theoretical model might be proposed for integrating environmental conditions, neuroendocrine activity and the regulation of metabolic activity levels in insects. By coupling metabolic regulation to the neuroendocrine system, insects can budget their innate metabolic capacities in response to external conditions, such as climate and seasons which influence food availability and which, in turn, influence insect survival or reproductive potential. By monitoring key environmental variables such as nutrient quantity and quality, humidity, photoperiod, and temperature, the insect could then adjust its endocrine output and metabolic rate for optimal survival. These environmental factors may act independently or synergistically to determine endocrine function, depending on the species. In this way, diapause or the less extreme metabolic decreases found in insects such as *Blaberus* might be considered similar, but graded responses to hormonal inactivity, or deficiency based on the differing physiological and environmental needs of the species.

REFERENCES

BRONK, J. R. (1966). Thyroid hormone: effects on electron transport. *Science* 153:638-639.

CLARKE, K. U. and BALDWIN, R. W. (1960). The effect of insect hormones of 2:4 dinitrophenol on the mitochondrion of *Locusta migratoria* L. *J. Insect Physiol.* 5:37-46.

CLARKE, K. U. and GILLIOTT, C. (1967). Studies on the effect of the removal of the frontal ganglion in *Locusta migratoria* L. I. The effect on protein metabolism. *J. Exp. Biol.* 46:13-25.

CLARKE, K. U. and LANGLEY, P. A. (1963). Studies on the initiation of growth and moulting in *Locusta migratoria migratorioides* R. and F. IV. The relationship between the stomatogastric nervous system and neurosecretion. *J. Insect Physiol.* 9:423-430.

DOGRA, G. S. and GILLOTT, C. (1971). Neurosecretory activity and protease synthesis in relation to feeding in *Melanoplus sanguinipes*. *J. Exp. Zool.* 177:41-50.

GOLDSWORTHY, G. J. (1970). The action of hyperglycaemic factors from the corpus cardiacum of *Locusta migratoria* on glycogen phosphorylase. *Gen. Compar. Endocr.* 14:78-85.

HIGHNAM, K. C., LUSIS, O. and HILL, L. (1967). The role of the corpora allata during oocyte growth in the desert locust, *Schistocerca gregaria* Forsk. *J. Insect Physiol.* 9:587-596.

HILL, L. (1962). Neurosecretory control of haemolymph protein concentration during ovarian development in the desert locust. *J. Insect Physiol.* 8:609-619.

KEELEY, L. L. (1970). Insect fat body mitochondria: endocrine and age effects on respiratory and electron transport activities. *Life Sci.* 9:1003-1011.

_____(1971). Endocrine effects on the biochemical properties of fat body mitochondria from the cockroach, *Blaberus discoidalis*. *J. Insect Physiol.* 17:1501-1515.

KEELEY, L. L. and FRIEDMAN, S. (1967). Corpus cardiacum as a metabolic regulatory in *Blaberus discoidalis* Serville (Blattidae). I. Long-term effects of cardiacectomy on whole body and tissue respiration and trophic metabolism. *Gen. Compar. Endocr.* 8:129-134.

_____ & _____ (1969). Effects of long-term cardiacectomy-allatectomy on mitochondrial respiration in the cockroach, *Blaberus discoidalis*. *J. Insect. Physiol.* **15**:509-518.

KEELEY, L. L. and WADDILL, V. H. (1971). Insect hormones: evidence for a neuroendocrine factor affecting respiratory metabolism. *Life Sci.* **10**:(2), 737-745.

KILTGAARD, H. M. (1966). Effect of thyroidectomy on cytochrome c concentration of selected rat tissues. *Endocr.* **78**:642-644.

LUSCHER, M. and LEUTHOLD, R. (1965). Uber die hormonale Beeinflussung des respiratorischen Stoffwechsels bei der Schabe *Leucophaea maderae* (F.) *Revue suisse Zool.* **72**:618-623.

MAYER, R. J. and CANDY, D. J. (1969). Control of haemolymph lipid concentration during locust flight: an adipokinetic hormone from the corpora cardiaca. *J. Insect Physiol.* **15**:611-620.

MINKS, A. K. (1967). Biochemical aspects of juvenile hormone action in the adult *Locusta migratoria*. *Archs. Neerl. Zool.* **17**:175-258.

MORDUE, W. (1967). The influence of feeding upon the activity of the neuroendocrine system during oocyte growth in *Tenebrio molitor*. *Gen. Compar. Endocr.* **9**:406-415.

MULLER, H. P. and ENGELMANN, F. (1968). Studies on the endocrine control of metabolism in *Leucophaea maderae* (Blattaria) II. The effect of the corpora cardiaca on fat-body respiration. *Gen. Compar. Endocr.* **11**:43-50.

OSBORNE, D. J., CARLISLE, D. B. and ELLIS, P. E. (1968). Protein synthesis in the fat body of the female desert locust, *Schistocerca gregaria* Forsk., in relation to maturation. *Gen. Compar. Endocr.* **11**:347-354.

ROUSSEL, J. (1963). Consommation d'oxygene apres ablation des corpora allata chez des femelles adultes de *Locusta migratoria* L. *J. Insect Physiol.* **9**:721-725.

SAGESSER, H. (1960). Uber die Wirkung der Corpora Allata auf der Sauerstoffverbrauch bei der Schabe *Leucophaea maderae* (F.). *J. Insect Physiol.* **5**:264-285.

SLAMA, K. (1964). Hormonal control of respiratory metabolism during growth, reproduction, and diapause, in female adults of *Pyrrhocoris apterus* L. (Hemiptera). *J. Insect Physiol.* **10**:283-303.

STEELE, J. E. (1963). The site of action of insect hyperglycemic hormone. *Gen. Compar. Endocr.* **3**:46-52.

STEGWEE, D. (1964). Respiratory chain metabolism in the Colorado potato beetle - II. Respiration and oxidative phosphorylation in 'sarcosomes' from diapausing beetles. *J. Insect Physiol.* **10**:97-102.

STEGWEE, D., KIMMEL, E. C., DEBOER, J. A. and HENSTRA, S. (1963). Hormonal control of reversible degeneration of flight muscle in the Colorado potato beetle, *Leptinotarsa decemlineata* Say (Coleoptera). *J. Cell Biol.* **19**:519-528.

STRANGEWAYS-DIXON, J. (1962). The relationship between nutrition, hormones, and reproduction in the blowfly *Calliphora erythrocephala* (Meig.) III. The corpus allatum in relation to nutrition, the ovaries, innervation and the corpus cardiacum. *J. Exp. Biol.* **39**:293-306.

TATA, J. R., ERNSTER, L., LINDBERG, O., ARRHENIUS, E., PEDERSEN, S. and HEDMAN, R. (1963). The action of thyroid hormones at the cell level. *Biochem. J.* **86**:408-428.

THOMSEN, E. (1949). Influence of the corpus allatum on the oxygen consumption of adult *Calliphora erythrocephala* Meig. *J. Exp. Biol.* **26**:137-149.

THOMSEN, E. and HAMBURGER, K. (1955). Oxygen consumption of castrated females of the blowfly, *Calliphora erythrocephala* Meig. *J. Exp. Biol.* **32**:692-699.

THOMSEN, E. and LEA, A. O. (1969). Control of the medial neurosecretory cells by the corpus allatum in *Calliphora erythrocephala. Gen. Compar. Endocr.* **12**:51-57.

THOMSEN, E. and MOLLER, I. (1963). Influence of neurosecretory cells and of corpus allatum on intestinal proteinase activity in the adult *Calliphora erythrocephala. J. Exp. Biol.* **40**:301-322.

VAN DER KLOOT, W. G. (1960). Neurosecretion in insects. *A. Rev. Entomol.* **5**:35-52.

WEINS, A. W. and GILBERT, L. I. (1965). Regulation of cockroach fat body metabolism by the corpus cardiacum *in vitro. Science* **150**:614-616.

WIGGLESWORTH, V. B. (1934). The physiology of ecdysis in *Rhodnius prolixus* (Hemiptera) II. Factors controlling moulting and metamorphosis. *Q. Jl Microsc. Sci.* **77**:191-222.

WILDE, J. DE and STEGWEE, D. (1958). Two major effects of the corpus allatum in the adult Colorado beetle (*Leptinotarsa decemlineata* Say). *Archs. Neerl. Zool.* **13** (suppl.), 277-289.

YAMAZAKI, M. and KOBAYASHI, M. (1969). Purification of the proteinic brain hormone of the silkworm, *Bombyx mori. J. Insect Physiol.* **15**:1981-1990.

NUTRITIONAL ASPECTS OF PEST MANAGEMENT

INTRODUCTION
by

Stanley D. Beck, Section Editor
Department of Entomology
University of Wisconsin
Madison, Wisconsin 53706

Suppression of pest populations through management programs involves manipulation of factors limiting the population-environment equilibrium of the target species. Under old-style pest control methods, the limiting factor was the imposition of a toxic environment by means of chemical pesticides. Because of extensive inimical effects of such an environment on nontarget species, there has been an urgent need for more sophisticated and selective management concepts and practical programs. Modern pest management systems currently include such components as genetic lethals, sterile males, pheromone trapping, sex attractant saturation, predators, parasites, pathogens, cultural and processing practices, resistant crop varieties, and judicious use of chemosterilants and insecticides. Because the greatest proportion of economic damage inflicted by insects is a direct consequence of the insect's efforts to obtain nutrients needed for growth and reproduction, there is interest in the possible application of nutritional knowledge to modern pest management systems.

Knowledge that a given insect species displays a nutritional requirement for a substance or group of compounds has seldom if ever led to a practical method of insect control. Required manipulation of the nutritional quality of the insect's food may be neither possible nor desirable. A crop or manufactured product that would be nutritionally inadequate for the insect might be equally inadequate for man or the animals for whom the product was intended. Animal organisms do not differ so widely in their basic biochemical pathways that such a method could be expected to be feasible in any but the most exceptional instances. Similarly, plant organisms cannot be expected to be capable of basic mutations in their biochemical processes without an accompanying loss of viability; such mutant genes would be uniformly lethal. These obstacles notwithstanding, a limited manipulation of nutritional quality may be possible in the case of manufactured processed food (discussed in the papers by J. J. Pratt, H. L. House, and A. Mansingh and by J. G. Rodriguez).

Manipulation of insect nutrition in the broader sense, including be-

Insect and mite nutrition – North-Holland – Amsterdam, (1972)

havioral and metabolic aspects, holds promise of applicability in pest management systems. Certainly this is the case with host plant resistance, typified by the subsequent papers by F. G. Maxwell and Ernst Horber. Maxwell examines nutritional aspects in relation to mechanisms of resistance and calls attention to the potential that lower levels of resistance may have on pest management programs of the future. A great deal more emphasis needs to be placed on investigations of host resistance (both plant and animal) from the standpoints of behavior, biochemistry, and genetics. Our present understanding of the mechanisms by which resistance works (whether nonpreference, antibiosis, or tolerance) is based almost exclusively on elucidations of existing instances of manifested resistance. There is a great need to advance to a more synthetic approach, in which resistance is designed and incorporated into the host system.

The inability of some stored products insects to cope with certain naturally-occurring molecules in seeds points up an area of great potential in protecting man's food supply. Shalom W. Applebaum and Yehudith Birk suggest the feasibility of increasing total protein production by selecting or breeding legume seed varieties resistant to stored products insects.

Some nongenetic resistance may be induced through manipulation of host physiology, as be means of fertilizers and other soil adjuvants. Insects have not proved to be particularly susceptible to this type of management, but it may yet prove applicable in some special instances (discussed in the papers by P. Harrewijn and by G. G. Shaw and C. H. A. Little).

Research on all aspects of insect nutrition has a very important additional role to play in the development of insect pest management systems. This role is that of informational and methodological inputs; in respect to behavior and its control (see papers by L. M. Schoonhoven and by R. T. Yamamoto and R. Y. Jenkins) and specific biochemical requirements and their application to mass-rearing. Mass-rearing has been essential to the management systems now in use or currently under development, and it seems almost certain that such will continue to be the case. Difficulties encountered in mass-rearing programs--such as gene frequency changes, behavioral aberations, loss of competitive capability--clearly point to the need of continuing vigorous research in this area.

SOME ASPECTS OF HOST SELECTION AND FEEDING IN PHYTOPHAGOUS INSECTS

L. M. Schoonhoven
Department of Entomology, Agricultural University
Wageningen, The Netherlands

INTRODUCTION

Food selection in phytophagous insects may involve three steps: finding the food, trying it and then the actual ingestion. These behaviorally distinct phases are each guided by certain sensory information. Extensive articles, reviewing the literature in this field, are available (Dethier, 1954; Thorstein-son, 1960; Beck, 1965; Schoonhoven, 1968) and the present paper is only an attempt to indicate some current lines of research.

RESEARCH AREAS

Foodplant finding

The finding of food from a distance usually involves orientation on visible and chemical cues. Several aphid species tend to alight on yellow-white surfaces (see e.g. Moericke, 1969), acridids orientate themselves toward vertical lines or objects (Williams, 1954; Mulkern, 1969), caterpillars (Hundertmark, 1937) and beetles (Hierholzer, 1950) are likewise attracted by vertical patterns. Rarely, however, is complete orientation accomplished by sight alone. Odors are also involved and may guide insects to possible food plants. Grasshoppers (Haskell *et al.*, 1962), the Colorado potato beetle (de Wilde *et al.*, 1969) and other insects as well, show a positive anemotaxis when the odor of their host plant is perceived. Most phytophagous insects appear to be able to discriminate many plant odors. Their antennae contain olfactory cells which may be stimulated by volatile factors of plants. The vapors of aldehydes, which typically occur in green leaves, and fatty acids stimulate receptory cells in the migratory locust (Kafka, 1971). Olfactory cells in caterpillar antennae appear to be not narrowly specific nor uniquely tuned to certain odors. Each cell is sensitive to several odors and its response spectrum differs from other cells, but may exhibit overlap. Plant odors are thus coded in activity changes in some cells, whereas others are not affected (Schoonhoven & Dethier, 1966). Odor discrimination, in this case, is based to a very large extent on central decoding of neutral activity patterns rather than on narrow sensory filtering.

Foodplant testing

Once an insect is in actual contact with a possible source of food it is in a position to test it with its contact chemical sense. Insect species with tarsal receptors may detect chemicals on the surface of their substrate. Thus some aphids, occurring on apple, seem to perceive after landing a flavonoid (phloricin) which typically occurs in leaves and other organs of appletrees (Klingauf, 1971). Alkanes, present on the leaf surfaces of *Vicia faba*, exert an "arrestant" effect on the pea aphid (Klingauf *et al.*, 1971). Tarsal receptors of the beetle *Chrysolina brunsvicensis* are stimulated by hypericin, a polynuclear quinone characteristic for *Hypericum* spp. (Rees, 1969). The beetle appears to be monophagous on this plant genus. From these and other observations it must be concluded that several typical secondary plant substances are present in free form on plant surfaces in quantities large enough to be detected by some insects. Locusts, which also have tarsal receptors (Kendall, 1970), may in addition contact such free compounds by a characteristic "drumming" behavior, during which their labial and maxillary palpae appear to test possible food sources (Blaney & Chapman, 1970). The pegs on the tip of these palpae are known to contain chemosensory cells, reacting to various substances (Haskell & Schoonhoven, 1969). In the case of insects with sucking mouth parts, contact chemical stimulation will take place when the probosces setae-like structures (Adams & Fyfe, 1970), which probably represent chemoreceptors, and cotton stainers *(Dysdercus* spp.) bear likewise at the tip of this appendage 24 sensilla basiconica, which function, according to morphological and electrophysiological evidence, as contact chemoreceptors (Schoonhoven & Henstra, 1972). Behavioral observations agree with this assumption (Bongers, 1969).

Once the insect has made "chemical contact" with a possible food source it may be stimulated to perform a test bite or test probe (although such activities may also arise "spontaneously"). Now intimate contact between the plant contents and the insect's chemical sense is secured. In some cases the test bite can be distinguished from regular continuous feeding bites. In locusts the amount of plant material checked in a first bite is smaller than a regular bite, and it is chewed more carefully (Williams, 1954). The pea aphid may execute after arrival on a plant leaf several superficial probings before deciding to penetrate into deeper tissues (Klingauf, 1970). When during this exploratory phase distasteful compounds are met, insects of various orders have been seen not only to stop their investigation but also to spit out the repugnant material and even to regurgitate some of the foregut contents.

Feeding

When the described chain of activities is not interrupted, the insect continues to feed till satiation. During eating the chemical composition of the food is continuously monitored. In some cases different chemicals appear to govern different phases of feeding behavior. Some compounds,

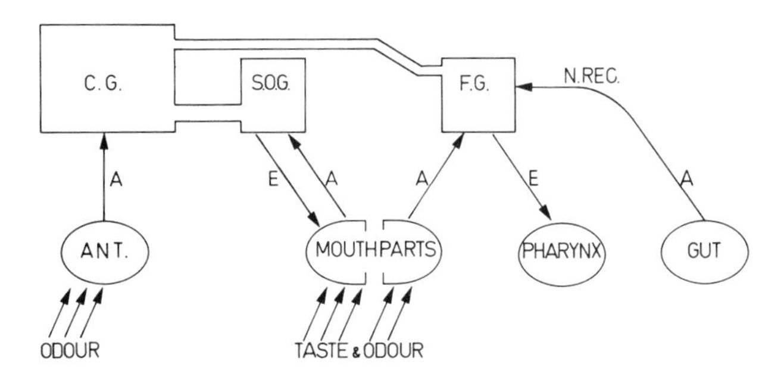

Figure 1. Afferent (A) and efferent (E) neural pathways involved in initiation and termination of food uptake in insects.

often belonging to the so-called secondary plant substances, may function mainly as "incitant" (Beck, 1965), inducing the insect to take a bite, whereas others affect swallowing activity. Thus in the larva of the cabbage white (*Pieris brassicae*) sinigrin serves as an incitant and promotes biting activity, as has been shown in an extensive study by Ma (1972). It is perceived by maxillary receptors, and stimulates feeding at a limited and fairly low concentration range. Higher concentrations (although eliciting higher receptor activity) do not increase food uptake, which is in contrast to the effects of sucrose. In addition to stimulating biting activity, the latter compound also promotes swallowing movements. In addition to sinigrin, the maxillary receptors are stimulated by sucrose. Swallowing is controlled by two sensilla on the epipharynx, which contain sense cells sensitive to sugars, salts and deterrents, frontal ganglion, which regulates the swallowing movements. In locusts two groups of sensilla near the edge of the clypeo-labrum probably are first involved in the exploratory biting procedure, whereas the remaining sensilla, located more dorsad on the clypeo-labrum and on the hypopharynx, act as further monitors and need to be "satisfactorily" stimulated to allow continued feeding (Haskell & Schoonhoven, 1969).

In the larvae of *Pieris brassicae*, as in many other lepidopterous larvae, specialized deterrent receptors detect unpalatable compounds, secondary plant substances, which render the food unacceptable. Inactivation of these cells, as has been done e.g. in the silkworm by Ishikawa *et al.* (1969) leads to an increased feeding upon non-host diets. Not only deterrents acting on taste receptors may prevent feeding on certain plants, but also olfactory organs may be involved. The caterpillars of *Sphinx ligustri* do not accept, or at the most nibble only small pieces from oleander leaves. After amputation of their antennae, however, they will readily feed upon this plant. Fifth instar larvae, one day after antennectomy, eat on the average 170 mm^2 oleander leaf surface during four hours, whereas intact larvae consume on the average only 8 mm^2. The odor of this plant apparently prevents continuous feeding (Schoonhoven, unpublished observations). Interestingly enough this plant, though probably nutritionally inadequate, does not seem to be toxic to *S. ligustri*. Even after 24 hours of feeding from it no deleterious effects could be observed, despite the presence of the notorious cardiac glycosides.

Sensory coding

During the exploratory phase and the subsequent period of feeding, contact chemoreceptors inform the insect about the chemical composition of the substrate. The first question now arising is: which compounds in the food play a role in this respect? Presently this question is often approached by analysing (with electrophysiological methods) the properties of the insect's chemosensors. Once knowing the capacities of this system it has to be concluded from behavioral experiments to what extent the animal makes use of it.

The most extensive analysis in this respect of the gustatory sense in phytophagous insects has been done in lepidopterous larvae, and more or less detailed information is now available for more than 25 species (Ishikawa *et al.*, 1969; Dethier & Kuch, 1971; Schoonhoven, 1972a). Several taste cells appear to be stimulated by groups of chemically related compounds, and sometimes the specificity of these cells seems to be fairly high. Thus cells, each sensitive to one of the following chemical groups have been found: sugars, inositol, glucose, sorbitol, mustard oil glucosides, amino acids, anthocyanins. Also cells reacting to some salts or weak acids occur, though their specificity seems to be less unequivocal. Many chemicals on the other hand (e.g. various amino acids) may affect several cells, either directly or by increasing or diminishing their sensitivity to other substances. As a consequence complex reaction patterns are elicited when combinations of various compounds, imitating a natural situation, are tested.

The receptor systems of different species, although morphologically identical, appear in a physiological sense to be never completely congruent, but on the contrary often show considerable differences. Apparently each species evolved a unique sensory apparatus, which, although based on a common vein (Schoonhoven, 1972a), is tuned in such a way that an optimal discrimination capacity for hosts versus non-host plants is obtained. There seems to be no simple correlation between the type of chemosensors with which a particular species is equipped and the extent of its food plant range (Dethier & Kuch, 1971). As in the olfactory system, the interpretation of the incoming peripheral information by the central nervous system is of paramount importance.

Host selection

Plants are often characterized by particular secondary plant substances.

Insects, with their well-developed chemical senses, appear to use in many cases such characters when selecting their food (Fraenkel, 1959). In the foregoing section, the presence of receptors which are specialized to detect various secondary plant substances, which may be appreciated either as "deterrent" or "phagostimulant", have been mentioned. *Pieris brassicae* larvae, for instance, have both categories of receptors. Jermy (1966) has suggested that most insects select primarily in a negative way, i.e. eat everything not containing particular deterrent secondary plant substances. Thus monophagous insects (Colorado potato beetle, silkworm) differ only from polyphagous species in that the former tolerate fewer secondary plant substances than the latter. Indeed, the polyphagous fall webworm (*Hyphantria cunea*) lacks a "deterrent" receptor, which typically occurs in the silkworm (Ishikawa *et al.*, 1969). However, another polyphagous insect, the gypsy moth, does possess such a receptor (Schoonhoven, 1972a).

The suggestion that in general secondary plant substances act as protecting agents against insect attack, and thus deter and/or poison insects has been tested somewhat more systematically in only a few cases. Some polyphagous insects were examined, such as the two-striped grasshopper *Melanoplus bivittatus* (Harley & Thorsteinson, 1967) and the peach aphid (Schoonhoven & Derksen-Koppers, unpublished observations) as well as oligophagous species, such as the Colorado potato beetle (Buhr *et al.*, 1957) and some cotton stainers (*Dysdercus* spp.) (Schoonhoven & Derksen-Koppers, 1972). The majority of the compounds tested in these cases appear to affect food uptake negatively, thus confirming the views expressed by Fraenkel and Jermy.

However, the problem is more complex, because plant species differ quantitatively in their contents of nutritive elements, and therefore experiments of the type mentioned should be interpreted with care. It is likely that food selection depends on more subtle information than only the presence/absence of these secondary plant substances (Schoonhoven, 1972b). As has been shown above for Lepidoptera larvae, the sensory system allows for an appraisal of the levels of several nutritive compounds, such as sugars, amino acids, sterols, etc. Behavioral evidence suggests the same for other insects (e.g. Hsiao and Fraenkel, 1968). The sweet clover weevil, a monophagous insect, reacts strongly to variations in glucose, fructose and sucrose levels in its hosts (Akeson *et al.*, 1970).

It is easy to demonstrate that insects can discriminate parts of the same plant which differ in their physiological condition. Thus *Pieris brassicae* larvae, in a choice situation, eat 626 mm^2 of young growing leaves of green cabbage per day per larva, and only 242 mm^2 of the full-grown leaves of the same plant. Leaves exposed to light are significantly preferred to leaves which were kept for 24 hours in the dark (566 mm^2/larva/day on previously illuminated leaves and 459 mm^2/larva/day on leaves kept dark). Probably the changes in the leaves concern mainly essential nutrients, which are recognized by the caterpillars. Moreover, from Ma's experiments, cited above, it was concluded that the amount of feeding incitant, provided it exceeds a certain low level, is unimportant. This agrees with the results of our experiment in which larvae were given a choice between normal leaves and leaves with raised sinigrin levels. No significant preference could be found. Larvae of the gypsy moth prefer full-grown elder leaves (572 mm^2/larva/4 hours) to young leaves of the same tree (204 mm^2). Leaves which were previously illuminated are, as in the cabbage worm, preferred to non-exposed leaves (608 mm^2/larva/4 hours compared to 312 mm^2). Thus an epicuristic touch cannot be denied, even in this species, which is notable for its polyphagous habits.

Regulation of food intake

When discussing factors which initiate feeding, some words should be devoted to the mechanism which stops feeding activity. However, at present little information pertaining to phytophagous insects is available.

The duration of a meal in lepidopterous larvae may be of considerable length. Larvae of *Manduca sexta* occasionally eat continuously up to periods of almost one hour, but the average meal length is much shorter (see Table 1). In *Pieris brassicae* there is a correlation between the duration of a resting period and the length of the subsequent meal (Ma, 1972).

Table 1. Feeding time and duration of feeding periods ("meals") of fifth instar larva of *M. sexta* at 25°C.

Fifth instar larva (day)	Fraction of time spent on eating (%)	Average duration of one meal (min)
1	29	6,8
2	49	18,3
3	58	18,9
4	58	18,3

Factors which may, in analogy to the blowfly (Dethier, 1969) regulate meal length are the rate of adaptation of the chemoreceptors on the mouth parts and the degree of filling of the foregut. When the chemoreceptors are continuously stimulated, an adaptation level of a low, but constant activity is reached after 5 - 60 seconds, depending on stimulus concentration. It is questionable, however, whether during a meal the maxillary receptors are continuously stimulated. Detailed observations have shown that very short periods between two bites occur and that the length of these inter-bite periods increases gradually till a maximum is reached every 10 to 15 seconds (Fig. 2). Subsequently a new cycle begins and one meal may include several of such activity cycles. A relation between receptor activity and these cycles is suggested by the observation that maxillectomy abolishes the cyclic activity (Doctors van Leeuwen, pers. comm.). Presumably the periodically slightly reduced biting frequencies during a meal allow the (maxillary) taste receptors to disadapt somewhat. As mentioned before, it is also known that a low activity level in the mustard oil glucoside receptor may still evoke complete feeding reactions in *Pieris brassicae*. In the milkweed bug sensory adaptation appears to limit feeding periods to about 36 minutes per meal (Feir & Beck, 1963).

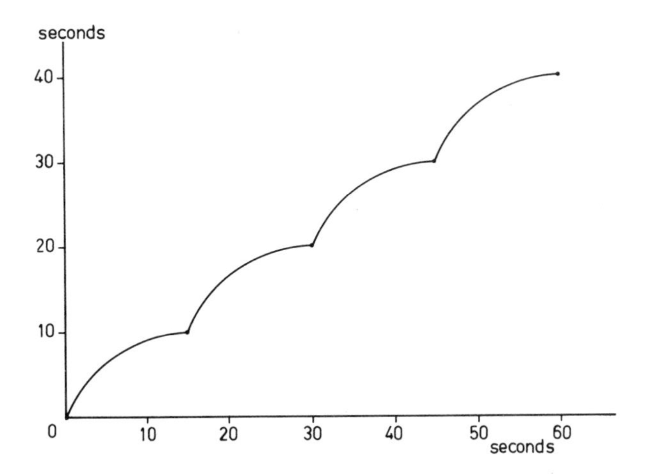

Figure 2. Relation between total time spent between actual bites (abscissa) and time spent on biting activity (ordinate) in *Pieris brassicae* larvae. The period presented forms part of a longer lasting meal (courtesy of Mr. B. Doctors van Leeuwen).

The effect of the distension of the foregut on the regulation of food intake may be ascertained by cutting the recurrent nerve (Dethier, 1969). This operation in locusts results, in a number of cases, in increased food uptake (Fraser Rowell, 1963). Hyperphagia under such conditions, however, is also due to a reduction in the speed of food transfer in the gut.

Whether there exists in some cases, in addition to the two systems mentioned, an endogenous feeding drive, as suggested by Dadd (1960) for *Schistocerca gregaria*, is still an open question.

Besides the regulation of food intake in short periods, there is obviously also a long-term controlling mechanism. Thus *Celerio* larvae (House, 1965) and locusts (Beenakkers *et al.*, 1971) compensate for a nutritionally sub-optimal diet by eating more.

The analysis of the functional pathway which connects this long-term regulation process to the neural system which governs food selection and intake remains one of the many fascinating areas of insect behavior to be elucidated.

SUMMARY

The various chemosensory processes which preceed and attend feeding activity are reviewed. In addition to detecting phagostimulating and deterring secondary plant substances, many insects also obtain sensory information concerning the nutritive quality of their food. This permits a subtle discrimination process, allowing many insects to select particular parts of their host plant.

The chemoreceptory system involved transmits the information of smell and food composition to the CNS, encoded in a fairly complex pattern of nerve activities. Studies on the larvae of more than 20 Lepidoptera species showed that each species has a physiologically characteristic receptory system, which is never completely identical with that of other species.

REFERENCES

ADAMS, J. B. and FYFE, F. W. (1970). Stereoscan views of some aphid mouthparts. *Can. J. Zool.* **48**:1033-1034.

AKESON, W. R., GORZ, H. J. and HASKINS, F. A. (1970). Sweetclover weevil (*Sitona cylindricollis*: Col., Curculionidae) feeding stimulants: variation in levels of glucose, fructose, and sucrose in *Melilotus* leaves. *Crop Sci.*, **10**:477-479.

BECK, S. D. (1965). Resistance of plants to insects. *Ann. Rev. Entomol.* **10**:205-232.

BEENAKKERS, A. M. T., MEISEN, M. A. H. Q. and SCHERES, J. M. J. C. (1971). Influence of temperature and food on growth and digestion in fifth instar larvae and adults of *Locusta*. *J. Insect Physiol.*, **17**:871-880.

BLANEY, W. M. and CHAPMAN, R. F. (1970). The function of the maxillary palmps of Acrididae (Orthoptera). *Ent. Exp. & Appl.*, **13**:363-376.

BONGERS, J. (1969). Zur Frage der Wirtsspezifität bei *Oncopeltus fasciatus* (Heteroptera: Lygacidae) *Ent. Exp. & Appl.* **12**:147-156.

BUHR, H., TOBALL, R. and SCHREIBER, K. (1958). Die Wirkung von einigen pflanzlichen Sonderstoffen, insbesondere von Alkaloiden, auf die Entwicklung der Larven des Kartoffelkäfers (Leptinotarsa decemlineata Say). *Ent. Exp. & Appl.*, **1**:209-224.

DADD, R. H. (1960). Observations on the palatability and utilization of food by locusts, with particular reference to the interpretation of performances in growth trials using synthetic diets. *Ent. Exp. & Appl.* **3**:283-304.

DETHIER, V. G. (1954). Evolution of feeding preferences in phytophagous insects. *Evolution*, **8**:33-54.

DETHIER, V. G. (1969). Feeding behavior of the blowfly. *Adv. Study Behav.*, 2:111-266.

DETHIER, V. G. and KUCH, J. H. (1971). Electrophysiological studies of gustation in lepidopterous larvae. *Z. Vergl. Physiol.* 72:343-363.

FEIR, D. and BECK, S. D. (1963). Feeding behavior of the large milkweed bug *Oncopeltus fasciatus*. *Ann. Ent. Soc. Amer.*, 56:224-229.

FRAENKEL, G. S. (1959). The raison d'être of secondary plant substances. *Science*, 129:1466-1470.

FRASER ROWELL, C. H. (1963). A method for chronically implanting stimulating electrodes into the brains of locusts and some results of stimulation. *J. Exp. Biol.*, 40:271-284.

HARLEY, K. L. S. and THORSTEINSON, A. J. (1967). The influence of plant chemicals on the feeding behavior, development and survival of the two-striped grasshopper, *Melanoplus bivittatus* (Say), Acrididae: Orthoptera. *Can. J. Zool.*, 45:305-319.

HASKELL, P. T., PASKIN, M. W. J. and MOORHOUSE, J. E. (1962). Laboratory observations on factors affecting the movements of hoppers of the desert locust. *J. Insect Physiol.* 8:53-78.

HASKELL, P. T. and SCHOONHOVEN, L. M. (1969). The function of certain mouth part receptors in relation to feeding in *Schistocerca gregaria* and *Locusta migratoria migratorioides*. *Ent. Exp. & Appl.*, 12:423-440.

HIERHOLZER, O. (1950). Ein Beitrag zur Frage der Orientierung von *Ips curvidens* Germ. *Z. Tierpsychologie*, 7:588-620.

HOUSE, H. L. (1965). Effects of low levels of the nutrient content of a food and of nutrient imbalance on the feding and nutrition of a phytophagous larva, *Celerio euphorbiae* (Linnaeus) (Lepidoptera: Sphingidae). *Can. Ent.* 97:62-68.

HSIAO, T. H. and FRAENKEL, G. (1968). The influence of nutrient chemicals on the feeding behavior of the colorado potato beetle, *Leptinotarsa decemlineata* (Coleoptera: Chrysomelidae). *Ann. Ent. Soc. Amer.*, 61:44-54.

HUNDERTMARK, A. (1937). Das Formen unterscheidungsvermögen der Eiraupe der Nonne (*Lymantria monacha* L.). *Z. Vergl. Physiol.*, 24:563-592.

ISHIKAWA, S., HIRAO, T. and ARAI, N. (1969). Chemosensory basis of host plant selection in the silkworm. *Ent. Exp. & Appl.*, 12:544-554.

JERMY, T. (1966). Feeding inhibitors and food preference in chewing phytophagous insects. *Ent. Exp. & Appl.*, 9:1-12.

KAFKA, W. A. (1971). Specificity of odor-molecule interaction in single cells. In: Gustation and Olfaction (Eds: G. Ohloff & A. F. Thomas). Academic Press, London, New York, 61-70.

KENDALL, M. D. (1970). The anatomy of the tarsi of *Schistocerca gregaria* Forskal. *Z. Zellforsch.*, 109:112-137.

KLINGAUF, F. (1970). Zur Wirtswahl der grünen Erbsenlaus, *Acyrthosiphon pisum* (Harris) (Homoptera: Aphididae). *Z. Angew. Ent.*, 65:419-427.

KLINGAUF, F. (1971). Die Wirkung des Glucosids Phlorizin auf das Wirtswahlverhalten von *Rhopalosiphum insertum* (Walk.) und *Aphis pomi*. De Geer (Homoptera: Aphididae). *Z. Angew. Ent.*, 68:41-55.

KLINGAUF, F., NÖCKER-WENZEL, K. and KLEIN, W. (1971). Einfluss einiger Wachskomponenten von *Vicia faba* L. auf das Wirtswahlverhalten von *Acyrthosiphon pisum* (Harris) (Homoptera: Aphididae). *Z. Pflanzenkrankh. U. Pflanzenschutz* 78:641-648.

MA, W. C. (1972). Dynamics of feeding responses in *Pieris brassicae* Linn. as a function of chemosensory input: a behavioural, ultrastructural and electrophysiological study. Mededelingen Landbouwhogeschool, Wageningen 72/11, 1-162.

MOERICKE, V. (1969). Hostplant specific colour behaviour by *Hyalopterus pruni* (Aphididae). *Ent. Exp. & Appl.*, 12:524-534.

MULKERN, G. B. (1969). Behavioral influences on food selection in grasshoppers (Orthoptera: Acrididae). *Ent. Exp. & Appl.*, 12:509-523.

REES, C. J. C. (1969). Chemoreceptor specificity associated with choice of feeding site by the beetle *Chrysolina brunsvicensis* on its foodplant, *Hypericum hirsutum*. *Ent. Exp. & Appl.*, 12:565-583.

SCHOONHOVEN, L. M. (1968). Chemosensory bases of host plant selection. *Ann. Rev. Entomol.* 13:115-136.

SCHOONHOVEN, L. M. (1972a). Plant recognition by lepidopterous larvae. *Proc. Symp. Roy. Ent. Soc. London,* 6:83-93.

SCHOONHOVEN, L. M. (1972b). Secondary plant substances and insects. In: Structural and functional aspects of phytochemistry (Ed: V. C. Runeckless) Academic Press, New York, 197-224.

SCHOONHOVEN, L. M. and DETHIER, V. G. (1966). Sensory aspects of host-plant discrimination by lepidopterous larvae. *Archs. Neerl. Zool.*, **16**:497-530.

SCHOONHOVEN, L. M. and DERKSEN-KOPPERS, I. (1972). Effects of secondary plant on drinking behaviour in some Heteroptera (in preparation).

SCHOONHOVEN, L. M. and HENSTRA, S. (1972). Morphology of some rostrum receptors in *Dysdercus* spp. *Neth. J. Zool.*, 22 (in press).

THORSTEINSON, A. J. (1960). Host selection in phytophagous insects. *Ann. Rev. Entomol.*, 5:193-218.

DE WILDE, J., LAMBERS-SUVERKROPP, K. HILLE RIS and VAN TOL, A. (1969). Responses to air flow and airborne plant odour in the Colorado beetle. *Neth. J. Pl. Path.*, 75:53-57.

WILLIAMS, L. H. (1954). The feeding habits and food preferences of Acrididae and the factors which determine them. *Trans. Roy. Entomol. Soc. London,* **105**:423-454.

HOSTPLANT PREFERENCES OF TOBACCO HORNWORM MOTHS

R. T. Yamamoto and R. Y. Jenkins
North Carolina State University
Raleigh, N. C. 27607

INTRODUCTION

The adults of oligophagous lepidopterous insects like the tobacco hornworm, *Manduca sexta* (L.) and the cabbage butterflies, *Pieris brassicae* L. and *P. rapae* L., and possibly many other species utilize specific contact chemical stimuli to discriminate host- from non-host plants. The stimuli appear to be perceived by chemoreceptors located on the tarsi and the effect of the stimulation is oviposition. In the absence of the appropriate chemical stimuli, eggs are not normally laid. This type of "trigger" mechanism assures that the eggs are laid on the hostplants of the insects concerned. The oviposition stimuli for the cabbage butterflies are sinigrin and related mustard oil glycosides (David and Gardiner, 1962; Hovanitz and Chang, 1964). The oviposition stimulus for the tobacco hornworm has not been identified as yet.

Since mustard oil glycosides are widely distributed among members of the plant family *Cruciferae*, it is in theory possible for these plants to stimulate oviposition if the cabbage butterflies came in contact with them. Indeed it is possible to demonstrate this effect by confining butterflies in cages with various species of cruciferous plants. In nature, however, only a few species of plants are actually chosen for oviposition. These are the "preferred" plants and they usually consist of commercial varieties of cabbage, *Brassica oleracea* L., Radish, *Raphanus sativus* L. and certain species of wild mustards (Genus *Brassica*).

In attempts to explain plant preferences in the cabbage butterfly *P. rapae*, larvae were reared on one of several different test plants. When they emerged as adults, they were given a choice of two or more plants on which to oviposit, one plant being the one on which the adults had been reared on during the larval stages. Results of one experiment of this type was positive, the adults ovipositing more eggs on the plants on which they grew up as larvae (Hovanitz and Chang, 1963). In a very similar type of experiment but using larger cages, the results were negative (Takata, 1961).

Insect and mite nutrition – North-Holland – Amsterdam, (1972)

The above mentioned experiments were not isolated endeavors. The idea that adult insects might choose plants on which they had fed as larvae was first used as a means to explain apparent phytophagic varieties (Walsh, 1864) and later phrased into a hypothesis known as "Hopkins Host-Selection Principle" which states that "an insect species that breeds in two or more hosts will continue to breed in the host to which it has become adapted" (Hopkins 1917, Thompson and Parker, 1928). Evidence in support of this hypothesis was provided by Craighead (1921) who showed that adult Cerambycid beetles which infested various species of trees tended to oviposit on species which they had fed as larvae. However, the findings of other workers have not supported this hypothesis (Larsen, 1927; Thompson and Parker, 1928; Takata and Ishida, 1957; Takata, 1961; Wood, 1963; and Palmiter, 1966).

In a somewhat different context, Thorpe and Jones (1937) showed that a parasitic hymenopteran, *Nemeritis canescens* (Grav.), finds its host, the flour moth *Anagasta (Ephestia) kuhniella* (Zell.), by the host's odor. They also found that *Nemeritis* would preferentially orient to the odor of a non-host insect, the wax moth *Meliphora grisella* F., when the larval stage of the parasite was reared on *Meliphora* or when the newly emerged adult parasites were exposed to the odor of *Meliphora*. Adults from larvae reared on normal hosts showed no preference for the odor of *Meliphora* in a 2-arm olfactometer, which was employed in the experiments. The inductive process involved in the preferential response to the odor of *Meliphora* was termed "pre-imaginal olfactory conditioning." However, it was also shown that when the odor of *Meliphora* was pitted against that of *Anagasta*, conditioned adults preferred the odor of *Anagasta*. Thus a preference for the odor of the natural host existed despite conditioning. This work has stimulated many workers, including those working with insects and plants.

We have undertaken an analysis of the hostplant preference of the tobacco hornworm moths not so much to test theories of learning but rather to seek the kind of information that would assist us in rationally developing resistant varieties of tobacco against this insect. Inevitably, the question of whether the preferences are flexible or fixed (innate) must be answered and thus we looked into the problem of "pre-imaginal olfactory conditioning." The background information we have to date about the tobacco hornworm is that an oviposition stimulus is widely distributed among solanaceous plants and this stimulus thus allows the hornworm moths to discriminate solanaceous from non-solanaceous plants (Yamamoto and Fraenkel, 1960). There is, however, a consistent pattern of preference for tobacco (*Nicotiana tabacum* L.) and tomato (*Lycopersicum esculentum* Mill.) throughout its geographic range in South, Central, and North America. These plants are quite unlike each other in physical appearance and belong to different genera of the family Solanaceae. In North Carolina, only once in 6 years of survey was a larva found on another species, the horsenettle *Solanum carolinense* L. Large stands of potato, *Solanum tuberosum* L., grown in the eastern coastal plains of this state are also ignored. It thus appears on the surface that the world of the hornworm consists of tobacco and tomato against a background of all other plants, including suitable solanaceous plants.

BEHAVIOR OF HORNWORM MOTHS IN CAGES

The behavior of moths in respect to oviposition is analyzed in large cages situated in a greenhouse during May through September when environmental parameters, particularly temperature and photoperiod are optimal (Yamamoto, 1969). The moths fly soon after sunset when the intensity of light changes from light to dark. Since observation is difficult in the dark, the cages are illuminated with a voltage-regulated 15-watt tungsten bulb. The intensity of light is approximately that of a full-moon, however, flight is not inhibited nor are the moths attracted to the bulb. Briefly, the following flight components were observed and blocked out for analysis: dispersal, host-seeking, approach, and landing (Yamamoto *et al.*, 1969). Moths in the host-seeking pattern fly in an undulating fashion, somewhat slow in tempo and take a circular path around the cage or make a to- and fro-movement from wall to wall. The flight is usually about 60 to 90 cm above the floor of the cage although some moths will fly very close to the floor of the cage. Out of this host-seeking flight, the moths will visually approach any clearly delineated object such as a plant, bottle, cardboard leaf, etc. Moths with eyes coated with lampblack cannot locate objects including hostplants placed in the cages.

When a moth approaches an object, it may land or it may veer off about 15 to 20 cm from the object. If the object happens to be a tobacco plant, it lands nearly 100% of the time. If it happens to be an inanimate object, the moth may land from 25 to 50% of the time. Finally, if the object is a non-hostplant such as cabbage, the moth may land from 0 to 10% of the time.

These behavioral observations indicated that short-range attractants emanated from tobacco whereas repellents emanated from cabbage. The absence of odors such as in the artificial leaf did not deter landing albeit the frequency of landing was one-half or less than on tobacco. Short-range attractants can also be demonstrated by applying steam distillates of tobacco on artificial leaves. The frequency of landing and oviposition on these treated leaves are 4 to 5 times greater than on untreated control leaves. Upon landing on tobacco leaf, contact chemoreceptors located in the tarsi are stimulated, resulting in the deposition of an egg on the underside of the leaf. An alcoholic extract of tobacco or other solanaceous plants applied on the leaves of non-hostplants such as bean, *Phaseolus vulgaris* L. or on artificial leaves will also stimulate oviposition if moths land on them. We think that any changes in the behavior of the moths might occur in respect to the attractants rather than the oviposition stimulus.

CHANGES IN FEEDING BEHAVIOR OF TOBACCO HORNWORM LARVAE

Two qualitatively different kinds of effects can be measured in larvae after they are fed on suitable hostplants. The first effect is an inductive change occurring in the 1st instar and causes feeding specificity to solanaceous plants in subsequent instars. If 1st instar larvae are reared on an artificial diet, feeding specificity is not induced, and diet-reared larvae may therefore feed readily on certain non-solanaceous plants, which in turn are rejected by larvae reared on a hostplant (Schoonhoven, 1967). The inducing agent is thought to be a token or feeding stimulant. A more complete exposition on this induction of feeding specificity will be given elsewhere (Yamamoto, 1972). The other change also involves inductive processes but the change is directed towards preference for a single species of plant rather than specificity towards a group of related plants. It appears to be restricted to older larvae or at least it has been measured only in older larvae. When Jermy and co-workers (1968) reared tobacco hornworm larvae through the first 3 instars on diet, then transferred them to tobacco, tomato, and potato during the 4th instar, the larvae reared on tomato and potato showed a significant preference for these plants during the 5th instar. Preference for tobacco was not demonstrable. When either olfactory or gustatory receptors were removed, feeding preferences were estinguished. Larvae therefore discriminated plants by odor and taste qualities (Hanson and Dethier, 1972, private communication). The appropriate question to ask is whether induced larval preferences for particular plants are transferred to the adults. That is to say, do moths utilize the same olfactory and tactile chemical stimuli as the larvae in discriminating plants?

We began our study by measuring preferences of larvae in all 5 instars. The technique employed consisted in recording the initial choice of foodplants by larvae faced with choice situations. Jermy and co-workers (1968) measured the amount of leaves consumed by larvae; however, the results employing the different techniques were very similar. Preference tests with 1st instars showed that they can discriminate various species of suitable hostplants without prior feeding experience. According to the tests, their preference for 4 plants were as follows: jimsonweed 〉 tomato 〉 horsenettle 〉 tobacco. However, induced feeding preference for any of the 4 plants could be demonstrated by rearing larvae on the plants. Induced larval preference for jimsonweed was of particular interest (Fig. 1) because jimsonweed

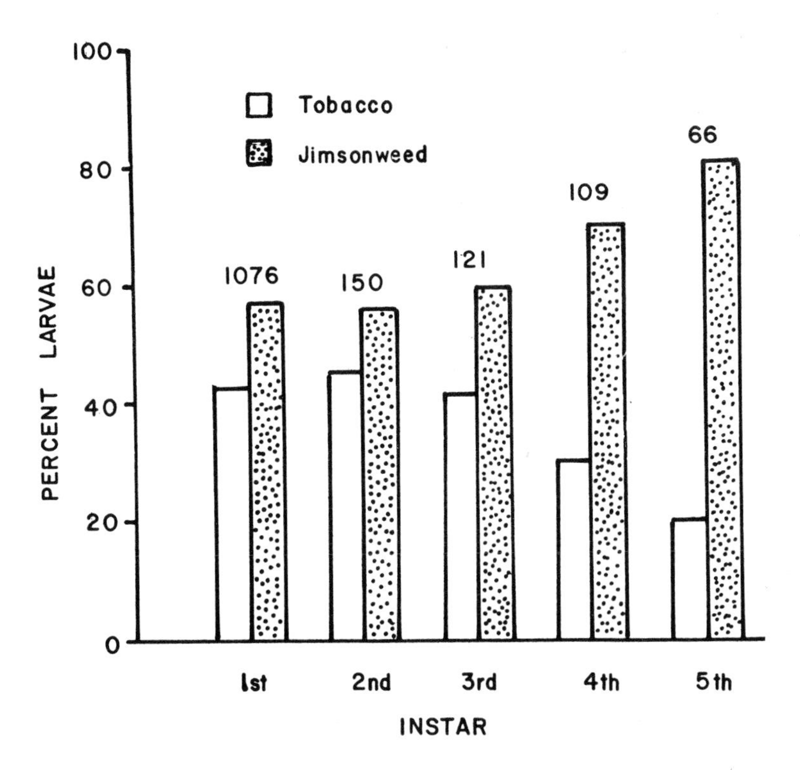

Figure 1. Preferences of jimsonweed-reared larvae for tobacco or jimsonweed during each instar. The number above the bars represent larvae used in the preference tests.

together with tobacco was chosen for more extensive study in respect to adult oviposition behavior and for future analyses of their chemical components.

CHANGES IN OVIPOSITION PREFERENCES OF MOTHS

Oviposition preferences of moths are measured in cages approximately 9 x 9 x 6 ft. The test plants are inserted in 100 ml tubes which are held approximately 2 ft above the floor of the cages by means of ring stands and clamps. The stands, clamps, and tubes are painted black in order to eliminate reflecting surfaces. There are 16 stands altogether, arranged in a block of 4 rows and 4 columns. Only young plants or shoots of plants are tested. The moths are 3-5 days old, and used only once in a test. In certain other tests, they are used several times in order to assess whether their responses vary from night to night. In 2-choice situations, non-hostplants are used to fill in some of the stands, and in 4-choice situations, each hostplant is represented once in each row and column. The number of eggs laid in each stand is recorded, but for statistical analysis, the combined number of eggs laid on each species of plants is used.

Solanaceous plants used in the present study are tobacco, tomato, black nightshade (*Solanum nigrum* L.) and jimsonweed (*Datura stramonium* L.). Tobacco and tomato are natural hostplants whereas jimsonweed and the black nightshade are not. The latter two plants are abundant in North Carolina and they are suitable as food for the larvae. Both species of plants grow in cultivated land as well as on roadsides. Jimsonweed is more abundant in recently plowed fields whereas the black nightshade is prevalent in older, unplowed fields.

Two-choice experiments were conducted with tobacco and jimsonweed. Larvae were reared on diet, jimsonweed, or tobacco and the ensuing moths were tested individually. The distribution of eggs on tobacco and jimsonweed were recorded and analyzed statistically. The results showed that diet-reared moths preferred tobacco, jimsonweed-reared moths preferred jimsonweed, and tobacco-reared moths preferred neither (Table 1). This

Table 1. Preferences of plants for oviposition by tobacco hornworm moths reared on indicated food during the larval stages.

	Number preferring		
Larval food	Tobacco	Jimsonweed	Neither
Artificial diet	18	2	3
Tobacco	10	12	5
Jimsonweed	7	11	4

table indicates the preferences of moth for either tobacco or jimsonweed and is not a table of analysis. Of interest here is that diet-reared moths, considered as unbiased controls, showed a strong preference for tobacco. It is recalled that newly hatched larvae preferred jimsonweed over tobacco and therefore, it would seem that unbiased larvae and moths are responding to different plant factors. At any rate, the fact that tobacco-reared moths did not show a preference for tobacco in the 2-choice tests may not be important at this stage of the study since responses of plant-reared moths, taken together, are significantly different from diet-reared moths.

In 4-choice preference tests, tomato, tobacco, nightshade and jimsonweed are being used and results to date show that diet-reared moths (unbiased controls) oviposited on plants in the following order of preference: tomato=nightshade > tobacco > jimsonweed. As indicated, tobacco is not the preferred plant, however, it is still preferred over jimsonweed, an outcome which was observed in the 2-choice tests. Experiments with insects reared on plants are now being conducted but the entries are insufficient for analysis. It is, however, becoming clear to us that changes in responses of moths can best be demonstrated in choice situations where a test plant, such as tobacco, is pitted against a more preferred plant. For example, tobacco-reared moths might have shown a preference for tobacco if tobacco had been pitted against tomato rather than jimsonweed in the 2-choice tests. The species of plants chosen for analysis is therefore an important ingredient in work of this kind.

SUMMARY AND PROSPECTS FOR MANAGEMENT OF INSECTS

The long-term objective of this study is to obtain information which can be utilized in reducing or eliminating oviposition on tobacco by the tobacco hornworm moths. Chemical factors which regulate preferences are short-range attractants. Our findings to date show that responses to the attractants are variable whereas responses to the oviposition stimulant is fairly rigid. The oviposition stimulant triggers oviposition in diet-reared and plant-reared moths. Its absence from a plant means that oviposition does not occur on the plant. In considering management possibilities, manipulation of the oviposition stimulant would be more likely to confer degrees of resistance against oviposition than manipulation of the attractants. Nevertheless, chemical isolation and identification of these factors are requisites for a plant resistance program which might be initiated as a consequence of this study.

REFERENCES

CRAIGHEAD, F. C. (1921). Hopkins host selection principle as related to certain Cerambycid beetles. *J. Agr. Res.* 22:189-200.

DAVID, W. A. L. and GARDINER, B. O. C. (1962). Oviposition and the hatching of the eggs of *Pieris brassicae* (L.) in a laboratory culture. *Bull. Entomol. Res.* 53:91-109.

HOPKINS, A. D. (1917). A discussion of C. G. Hewitt's paper on "Insect Behavior." *J. Econ. Entomol.* 10:92-93.

HOVANITZ, W. and CHANG, V. C. S. (1963). Ovipositional preference tests with *Pieris*. *J. Res. Lepid.* 2:185-200.

_____(1964). Adult oviposition responses in *Pieris rapae*. *J. Res. Lepid.* 3:159-172.

JERMY, T., HANSON, F. E. and DETHIER, V. G. (1968). *Entomol. Exp. & Appl.* 11:211-230.

LARSEN, A. O. (1927). The host-selection principle as applied to *Bruchus quadrimaculatus* Fab. *Ann. Ent. Soc. Amer.* 20:37-78.

PALMITER, R. D. (1966). Absence of olfactory conditioning in an oligophagous insect,

the corn earworm, *Heliothis zea* (Boddie). *Animal Behav.* 14:236-238.
SCHOONHOVEN, L. M. (1967). Loss of hostplant specificity by *Manduca sexta* after rearing on an artificial diet. *Entomol. Exp. & Appl.* 10:270-272.
TAKATA, N. (1961). Studies on the host plant preference of the common cabbage butterfly, *Pieris rapae crucivora* Boisduval. XII. Successive rearing of the cabbage butterfly larvae with certain host plants and its effect on the oviposition preference of the adult. *Jap. J. Ecol.* 11:147-154.
____ and Ishida, H. (1957). Studies on the host preference of cabbage butterflies (*Pieris rapae* L.). II. Preference between cabbage and radish for oviposition. *Jap. J. Ecol.* 7:56-58.
THOMPSON, W. R. and PARKER, H. L. (1928). Host selection in *Pyrausta nubilalis* Hubn. *Bull. Entomol. Res.* 18:359-364.
THORPE, W. H. and JONES, F. G. W. (1937). Olfactory conditioning in a parasitic insect and its relation to the problem of host selection. *Proc. Roy. Soc. Lond. B.* 124:56-81.
WALSH, B. D. (1864). On phytophagic varieties and phytophagic species. *Proc. Entomol. Soc. Philadelphia.* 3:403-430.
WOOD, D. L. (1963). Studies on host selection by Ips confusus (Leconte) (Coleoptera:- Scolytidae) with special reference to Hopkin's host selection principle. *Univ. Calif. Publ. Entomol.* 27:241-282.
YAMAMOTO, R. L. (1968). Mass rearing of the tobacco hornworm. I. Egg production. *J. Econ. Entomol.* 61:170-174.
____ (1972). Induction of feeding specificity in the tobacco hornworm, *Manduca sexta.* (In preparation).
____ and FRAENKEL, G. (1960). The specificity of the tobacco hornworm, *Protoparce sexta* (Johan.) to solanaceous plants. *Ann. Entomol. Soc. Amer.* 53:503-507.
____, JENKINS, R. Y. and MCCLUSKY, R. K. (1969). Factors determining the selection of plants for oviposition by the tobacco hornworm, *Manduca sexta. Entomol. Exp. & Appl.* 12:504-508.

WING PRODUCTION BY THE APHID *MYZUS PERSICAE* RELATED TO NUTRITIONAL FACTORS IN POTATO PLANTS AND ARTIFICIAL DIETS

P. Harrewijn
Institute of Phytopathological Research (I.P.O.)
Binnenhaven 12, Wageningen, The Netherlands

INTRODUCTION

Several workers have studied nutritional relations between economically important aphids and their host plants. A review of these studies which often lead to contradictory results is given by Pritam Singh (1970). More detailed information on the physiology and ecology of *Myzus persicae* (Sulz.) is summarized by van Emden *et al.* (1969).

In a previous paper (Harrewijn, 1970) we established the relationship between the mineral nutrition of the potato plant and development of *Myzus persicae*. Reproductive capacity of this aphid was found to be correlated to both the composition of the nutrient solutions and to the age of the plants. On leaves of relatively young plants a very low percentage of alatae was formed, even when the number of insects had increased to a density of about $10/cm^2$. The occurrence of a high percentage of alatae on relatively old or senescent leaves has been reported before (Johnson 1966, Lees 1967). Wearing (1972a) found that reproduction as well as wing production of *Brevicoryne brassicae* was positively correlated to the age of leaves of Brussels sprouts. On the other hand it is generally assumed that very few alatae are formed on seedlings (Johnson and Birks 1960, Johnson 1966, Sutherland 1967) although they can be highly suitable for reproduction. This seedling effect may be in concordance with Johnson's (1966) statement that on favourable plants the nutritional status of the host promotes the apterous course of development, the experiments with aging plants and with artificial diets of different nutritional value (Mittler and Kleinjan, 1970) seem to contradict this theory.

Using the technique of Steiner (1961, 1968) it is possible to vary the mineral nutrition of plants considerably without exceeding the physical or physiological limits of solutions or plants. We used this technique to investigate whether wing production and reproduction rate of *M. persicae* were similarly dependent on the nutritional status of the potato plant.

In our previous experiments (Harrewijn, 1970) the phosphate level in the nutrient solutions had to be kept constant at a given pH. Mittler and Kleinjan (1970) found that larval growth drops sharply and the percentages of apterae markedly increase as the KH_2PO_4 content in their artificial diets is reduced to below half the optimum level. In the present study reproductive rate and wing production of *M. persicae* reared on *Solanum tuberosum* is measured in relation to phosphate levels and to the total NPK content of the nutrient solutions used.

MATERIALS AND METHODS

Culture system and plant material

The culture system has been described in an earlier paper (Harrewijn, 1970). For all experiments potato plants of the variety Bintje were used, grown from eyes of tubers. At the start of the experiments the plants had grown for 8 weeks on a gravel culture. For wing production experiments radish seedlings of the variety Cherry Belle were used.

Physical constants

All plant material was reared in a glasshouse with closed air circulation at a temperature of 20°C (± 1°C). Air velocity did not exceed 30 cm/sec. Relative humidity was 55 - 70% and the daily photoperiod 16 hrs. Artificial lighting supplied 1800 mW/m^2. Aphids reared on artificial diet were maintained in a cabinet at 20°C (± 0.2°) and an RH of 80% (± 3%) with a constant air velocity of 15 cm/sec. Artificial light supplied 500 mW/m^2, 16 hrs daily. Refreshing of diet sachets and pooling of larvae was carried out at a laboratory bench, locally air conditioned at 20°C (± 2°) and 60% RH (± 5%). In some experiments diets could be changed automatically without removing the aphids from the membranes using a continuous flow artificial apparatus (Harrewijn, in prep.). Conditions in this apparatus were the same as in the cabinet described above.

Table 1. Characteristics of the nutrient solutions.

nutrient code	solution symbol	anion ratio			cation ratio		
		NO_3^-	$H_2PO_4^-$	$SO_4^=$	K^+	Ca^{++}	Mg^{++}
$N_1P_1K_3$	○	05	03	92	70	15	15
$N_2P_2K_2$	●	60	05	35	35	45	20
$N_2P_3K_2$	☆	60	25	15	35	45	20
$N_2P_3K_1$	▲	50	30	20	15	15	20

Mineral nutrition

Table 1 shows the characteristics of the nutrient solutions. Total ion concentration of all solutions was kept at 30 mg ion/l and the pH at 6.0. Preparation of the solutions was made according to Steiner (1961, 1966).

Demineralized water was used of at least 5 x 10^5 Ω cm/cm^2. The solutions were renewed every 14 days during the first two months, after that every week.

Chemical analyses of plant material

Daily samples of at least 10 potato leaves of the same age as those on which the aphids were reared were taken at 4.00 PM. Leaf material was dried at 60°C for 48 hrs and at 105°C for one hour. Phosphorus content of the samples was estimated with the method of van Belle (1970), adapted to a Carlo Erba automatic analyser at our institute. To determine the free amino acid content fresh samples of 2 or 4 g were ground with 80% ethanol and stored at -20°C. Analyses were carried out with a Beckman multichrome automatic amino acid analyser according to Spackman *et al.* (1958).

Determination of aphid fecundity and wing production

Aphid fecundity was measured as described in a previous paper (Harrewijn, 1970). For experiments on wing production apterous females were reared singly on seedlings of the cotyledon stage. After deposition of at least one larva they were confined to leaves or to artificial diets in groups of 10 to 25 per cage, depending on the required crowding stimulus. Every 24 hrs the larvae deposited by groups of adults, reared at the same density, were pooled in a Petri dish of 10 cm diameter and then allocated to potato leaves, seedlings or artificial diets in groups of 20 to 40. The dimensions of the cages used in experiments with potato leaves and artificial diets were identical. In the case of seedlings the larvae could make free choice between cotelydons and very young first leaves, each cage enveloping one seedling.

RESULTS AND CONCLUSIONS

The first series of experiments was set up with potato plants grown on the combinations of $N_2P_2K_2$ and $N_2P_3K_2$. These solutions differ in $H_2PO_4^-$/ SO_4^{--} rate only; the cation ratio being exactly the same. Fig. 1 shows the mean cumulative offspring of 180 adults and Fig. 2 represents the P-content of leaf samples taken each day during determination of aphid fecundity. On $N_2P_2K_2$ and $N_2P_3K_2$ plants; b - basal artificial diet (Mittler *et al.* 1970, but (Significant with $P<0.01$). There appears to be a marked influence of phosphate in the mineral nutrition of the potato plant on reproduction of *M. persicae*, which seems to be related to the P-content of leaves of the same age.

The arrows in Fig. 2 indicate the start of the wing production experiments. Each 24 hrs the successive batches of larvae from apterous aphids reared in equal densities on potato leaves were caged on: a - potato leaves of

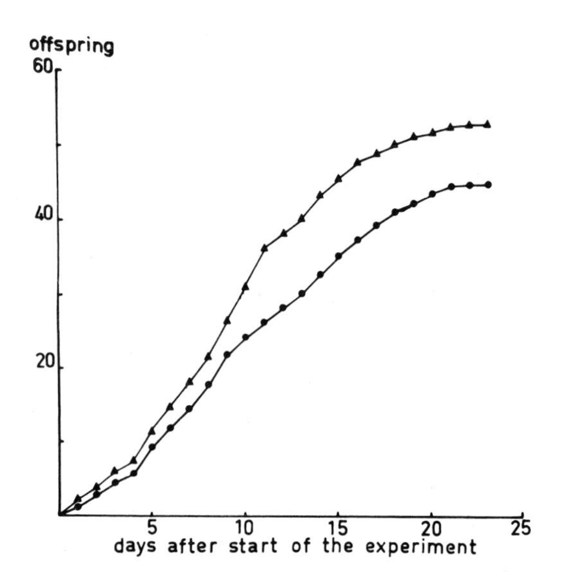

Figure 1. Mean cumulative daily offspring of single apterous aphids on potato plants growing on nutrient solutions with normal (●—●) and high (▲—▲) phosphorus content.

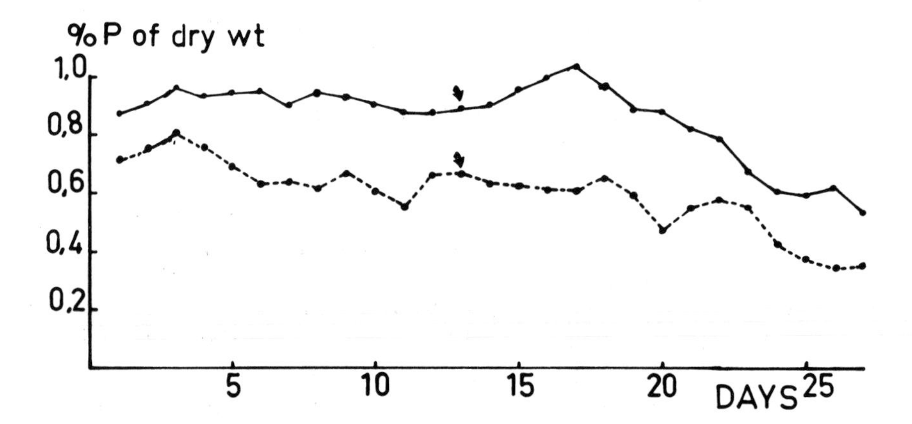

Figure 2. Phosphorus content of leaf samples (➘ start of the wing production experiments).

$N_2P_3K_2$ plants, indicating that reproduction and wing formation are similarly related to the nutritional status of the host (mean weight of 4th batch of larvae after 8 days on $N_2P_2K_2$ plants = 430 μg and on $N_2P_3K_2$ plants = 480
general there was only a low number of alatae present among the first batches of larvae deposited on leaves or diets. Probably the apterizing effect

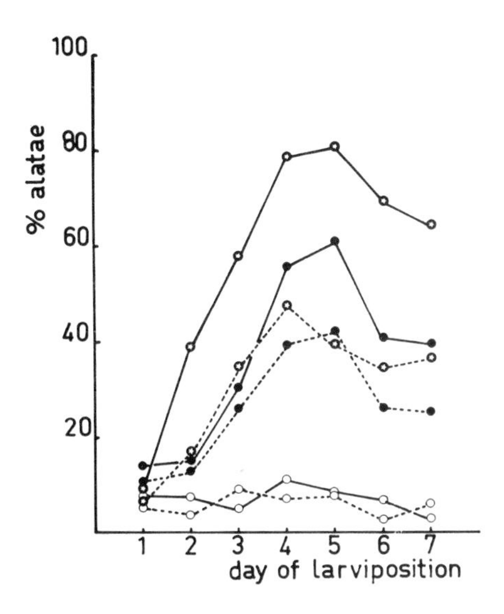

Figure 3. The proportion of alatae among batches of larvae of *Myzus persicae* deposited by grouped adult apterae reared in cages on potato plants growing on nutrient solutions with normal and high phosphorus content: ☆——☆ mothers and larvae on $N_2P_3K_2$ plants; ●——● mothers and larvae on $N_2P_2K_2$ plants; ☆——☆ mothers on $N_2P_3K_2$ plants, larvae on artificial diet; ●——● mothers on $N_2P_2K_2$ plants, larvae on artificial diet; ○——○ mothers on $N_2P_2K_2$ plants, larvae on seedlings; ○——○ mothers on $N_2P_2K_2$ plants, larvae on seedlings.

of the seedlings on which the mothers were reared expressed itself for more than 24 hrs. The proportion of alatae was highest among aphids reared on $N_2P_3K_2$ plants, indicating that reproduction and wing formation are similarly related to the nutritional status of the host (mean weight of 4th batch of larvae after 8 days on $N_2P_2K_2$ plants = 430 μg and on $N_2P_3K_2$ plants = 480 μg). There is an evident larval response to the seedlings, the results clearly show that the apterizing effect of the seedlings overrules the host plant effect on the mothers. A similar, although less pronounced apterizing effect was observed after confining the larvae to an artificial diet. When *M. persicae* were reared continuously on this diet, however, they produced more than 70% alatae. Kunkel and Mittler (1971) indicated that gustatory responses may cause presumptive alatae to become apterae after transferring them to different, but suitable hosts even for a short period. Especially radish cotelydons have a very strong apterizing effect in this respect. Table 2 gives mean values of free amino acid content of different stages of the radish plants.

Table 2. Free amino acid content of 4 developmental stages of radish seedlings (μ mol/g fresh weight).

	cotelydons	first leaves	young leaves	mature leaves
alanine	0.32	0.35	0.42	0.84
asparagine + glutamine	0.67	1.27	1.48	0.91
glutamic a.	2.13	1.83	1.98	2.10
glycine	< 0.10	< 0.10	< 0.10	< 0.10
proline	0.49	0.34	< 0.10	<< 0.10
serine	0.37	0.63	0.59	0.50
tyrosine	0.02	0.01	0.01	0.02
arginine	0.09	0.06	0.08	0.11
histidine	.06	.07	.05	< .01
isoleucine	.04	.04	.04	.03
leucine	.01	.02	.03	< .01
lysine	.12	.09	.06	.04
methionine	.09	.17	.08	.06
phenylalanine	.04	.03	.05	.04
valine	.12	.10	.10	.12

On the "apterizing" parts of the plants a statistically higher ($P<0.05$) amount of proline, histidine and lysine was found; the other amino acids did not show interesting variations. Tryptophan proved to be present but not estimated quantitatively. Though histidine may be an essential amino acid for *M. persicae*, we found it to reduce food uptake in the routine concentration for artificial diets (Harrewijn and Noordink, 1971); proline is not considered to be essential for *M. persicae*. It is difficult to point out a specific morphactive amino acid from the data presented in Table 2.

In a previous paper (Harrewijn, 1970) we reported that on plants growing on nutrient solutions with different nitrogen content up to 60% difference in fecundity can be found. When interaction of ions is ruled out, an additional effect of nitrogen and phosphorus in the mineral nutrition of the potato plant on developmental possibilities of *M. persicae* is to be expected. Balance - studies (Storms *et al.* 1967, Harrewijn in prep.) have shown, however, that at high phosphate levels of nutrient solutions nitrate uptake is partly inhibited. Nevertheless it was regarded of interest to study the effect of different NPK-treatments on the physiology of our plants and aphids.

The second series of experiments was set up with potato plants grown on the combinations of $N_1P_1K_3$, $N_2P_2K_2$ and $N_2P_3K_1$. These solutions enabled us to make a range from low N and P together with low Mg and high K to high N and P with high Mg and low K. Experiments on wing production were carried out following the same procedure as described above, except that mothers and larvae were confined to the same plants. For each treatment the offspring of 120 aphids was counted. On $N_1P_1K_3$ plants a mean total progeny of 41.6 larvae per aphid was counted, on $N_2P_2K_2$ plants 52.0 and on $N_2P_3K_1$ plants 57.2, the greatest difference of 37.5% in reproduction rate was found between $N_1P_1K_3$ and $N_2P_3K_1$ plants, which is about twice as much as obtained in the previous experiment.

As Fig. 4 shows, only a low percentage of alatae was formed on the $N_1P_1K_3$ plants. Small differences are observed between the $N_2P_3K_1$ and $N_2P_2K_2$ treatments; they find a statistical expression (P¡0.05) only among second- and fifth-day larvae. Again on plants less suitable for reproduction equal crowding stimulus seems to be unable to neutralize a predominating apterizing effect of the host. Table 3 gives the amounts of free mainly essential amino acids in leaf samples of different dates and times of the day. The sensitivity of the method needed to analyze minor quantities of amino acids like isoleucine, makes it impossible to include data on contents of amino acids like glutamic acid from the same samples. In previous work (Harrewijn, 1970) we found a positive correlation between nitrogen fertilization and the amount of soluble nitrogen in the leaves of potato plants. Although the total amount of free amino acids may vary considerably due to different treatments, it appears from Table 3 that this is hardly the case for essential amino acids like methionine, histidine and isoleucine. The content of phenylalanine and tyrosine is remarkably constant in all samples. The same is true for different stages of radish seedlings (Table 2). In fact, we found that omission of both phenylalanine and tyrosine from artificial diets results in malformation of wings in *M. persicae* up to 90%, which is not found under field circumstances.

The data of both Table 2 and 3 indicate that unless data on the amino acid content of phloem sap are available, it will be difficult to point out certain morph-active amino acids from leaf material. In general, omission of essential amino acids from artificial diets leads to poor growth, reduced food uptake, and an apterous course of development. Mittler and Kleinjan (1970) repeatedly found a close association between poor growth and the apterous condition. Seedling experiments have demonstrated, however, that excellent

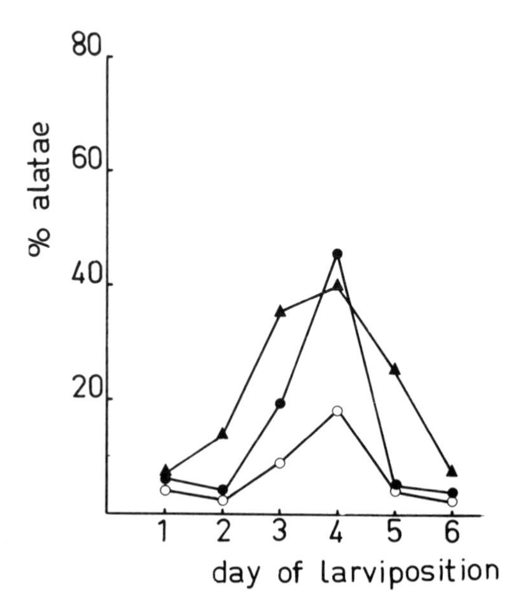

Figure 4. The proportions of alatae among batches of larvae of *Myzus persicae* deposited by grouped adult apterae reared in cages on potato plants growing on nutrient solutions with different NPK-content: ▲—▲ mothers and larvae on $N_2P_2K_1$ plants; ●—● mothers and larvae on $N_2P_2K_2$ plants; ○—○ mothers and larvae on $N_1P_1K_3$ plants.

growth still may be connected to formation of apterae.

On artificial diets as well as on host plants alatiform larvae of *M. persicae* develop a specific reddish colouring, which in the mature aphid is replaced or superposited by the well known dark brown pigmentation. It was noticed, that the reddish colouring is more pronounced in aphids reared on basal artificial diet than on potato plants. Many workers have tried to analyze the red and brown pigments of aphids. They are considered to be synthesized by oxidative coupling of phenols, as prothoaphins (dark pigments) can be derived by coupling of a quinone and a glucoside (Cameron *et al.*, 1964), whereas erythroaphins (red pigments) are regarded to be derivatives of 4-9-dihydroxy-3,10-perylenequinone (Calderbank *et al.*, 1954), derived from perylene. Bowie *et al.* (1966) found colourless glycosides with hydroxynaphthalene spectra in the haemolymph of aphids that may be regarded as precursors of pigments.

Up till now, we found only a few dietary influences on coloring in *M. persicae*. As mentioned above, omission of phenylalanine and tyrosine from the diets, though not affecting the percentage of presumptive alatae, results in abnormal wing formation. Red colouring is still present, but final brown pigmentation is considerably weakened. Probably tyrosine can be spared by phenylalanine, because these effects do not occur when phenylalanine alone is given.

Table 3. Free amino acid content of leaf samples of potato plants, grown on nutrient solutions with different NPK-content. (+: present, but no quantitative analysis possible; -: not present).

	$N_1P_1K_3$			$N_2P_2K_2$			$N_1P_3K_1$		
	7-4 4 PM	10-4 10 PM	10-4 4 PM	7-4 4 PM	10-4 10 AM	10-4 4 PM	7-4 4 PM	10-4 10 AM	10-4 4 PM
alanine	0.38	0.45	0.33	0.21	0.45	0.50	0.15	0.77	0.62
arginine	.01	.01	.03	.02	.02	.01	+	+	.02
cystine	.02	–	.04	+	.04	+	.12	–	+
glycine	.61	.12	.10	.16	.21	.43	.31	.23	.15
histidine	+	+	.01	+	+	+	+	+	+
isoleucine	.08	.03	.03	.03	.03	.03	.05	.03	.03
leucine	.08	.02	.02	.02	.02	.02	.02	.02	.01
lysine	.04	.02	.06	.05	.05	.06	.02	.07	.06
methionine	+	+	+	+	+	+	+	+	–
phenylalanine	.05	.06	.07	.03	.04	.05	.04	.04	.05
tyrosine	.02	.02	.02	.01	.01	.02	.01	.01	.01
tryptophan	.13	.16	.21	.01	.09	.04	.03	.10	.11
valine	.11	.05	.05	.04	.06	.04	.04	.05	.04

A very strong red coloring starting at the second larval stage can be obtained by omitting tryptophan from the diets. In fact, this is the only amino acid of which we found omission to promote the alate course of development, as is shown in Fig. 5. As far as nutritional aspects are concerned, this effect appears to be specific, because food uptake, measured with isotope techniques (Harrewijn and Noordink, 1971) on diets lacking tryptophan differed less than 10% from the basal diet. Though Table 3 shows that a relatively low amount of tryptophan is found on leaves on which a high percentage of alatae is developed, it remains an open question whether this amino acid may be regarded as a morph-active amino acid. From tryptophan a number of important metabolic compounds may derive, including nicotinate (probably not the case in *M. persicae*, as it is an essential vitamin), tryptamine, 3-hydroxykynurenin and 5-hydroxytryptamine (serotonin), which can act as or induce the formation of tissue hormone(s) in animals (Gustafson and Toneby, 1970, Smith *et al.*, 1972). The rate-limiting factor in the conversion of tryptophan to serotonin, the enzyme tryptophan hydrolase, can be inhibited easily with Na-diethyl dithiocarbamate or para-chlorphenylalanine. Fig. 6 shows that the addition of the latter compound to the diets which contain the routine amount of 80 mg tryptophan/100 ml has a

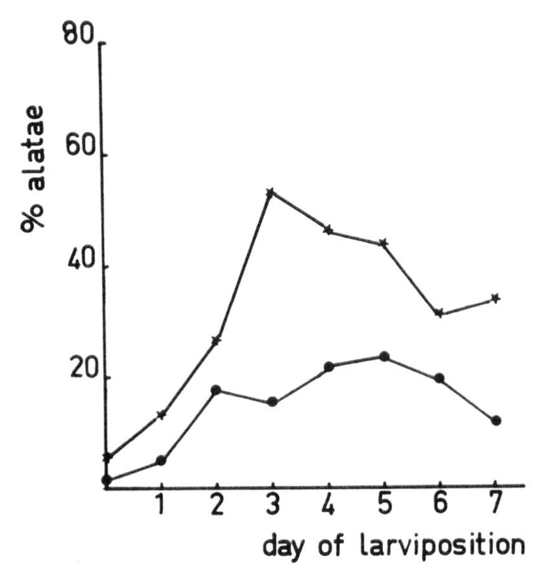

Figure 5. The proportions of alatae of larvae of *Myzus persicae* reared on diets with (●—●) and without (★—★) the amino acid tryptophan.

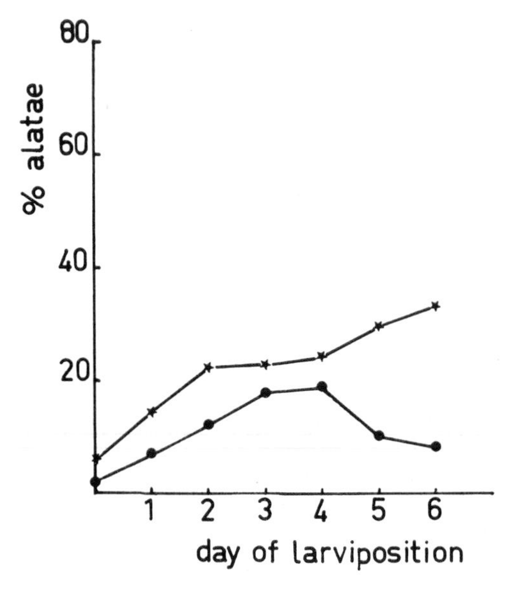

Figure 6. The proportions of alatae of larvae of *Myzus persicae* reared on basal diet (●—●) and on a diet with 5 x 10^{-3} mol para-chlor-phenylalanine (★—★).

similar effect to the omission of tryptophan. Diet uptake showed less than 10% deviation from the diet without serotonin.

The effect of the above mentioned compounds should be distinguished from non-specific reactions of the aphids to alterations of their diets. Addition of 2 x 10^{-3} M kynurenin, (of which the occurrence in *M. persicae* is uncertain) to the diet caused nearly 100% of the presumptive alatae to become apterae. Food uptake (and with that uptake of all essential nutrients), however, is reduced to about 50% of normal food uptake from the basal diet.

Our results indicate that wing production of *M. persicae* is positively correlated with the nutritional status of mature leaves of the potato plant. On very young parts of plants or seedlings wing formation may be regulated by yet unknown morph-active compounds. Tryptophan metabolism is probably involved in wing formation of *M. persicae*.

DISCUSSION

The experiments of Mittler and Kleinjan (1970) suggest that aphids developing on deficient artificial diets tend towards the apterous course of development. Our data on wing production of aphids developing on potato plants are in accord with their results. Because the aphids strain used by Mittler produces mainly alatae on artificial diets it will be difficult to point out dietary factors which increase alatae production. In our experimental set up the aphids produced 30 - 50% alatae on potato plants or diets when a medium crowding stimulus is provided. The amount of alatae decreases sharply on plants or diets with reduced nutritional value.

Our results seem to be in absolute contradiction to those of Raccah *et al.* (1971), who find an increase of alatae on diets of presumably nutritional imbalance. However, they found that larvae reared on diets deficient in phosphate developed mainly into apterae.

We think wing formation in *M. persicae* to be a highly energy-consuming process. Wing muscles as well as wings have to be at the aphid's disposal at the end of the teneral stage. In this respect it is of interest that the phosphate content of the potato plant and of artificial diets is positively correlated with wing production. In fact we found that addition of 2 x 10^{-3} mol of adenosine triphosphate to the diet promotes the alate course of development (Harrewijn, in prep.). Recently we were able to develop a diet with a dramatic improvement on growth compared to the diet used so far (Mittler *et al.* 1970). On this new diet, which differs in amino acid ratio and with a low amount of serine and tryptophan, a high proportion of alatae is obtained, the presumptive alatae showing a very intensive red colouring.

Wearing (1972) states that *M. persicae* produces fewer alatae on parts of plants of Brussels Sprouts that permit higher reproduction. Crowding, however, was not controlled in these experiments. Schaefers and Judge (1971) found that wing production of *Chaetosiphon fragaefolii* was positively correlated with growth on *Fragaria vesca*. We think our results justify the conclusion that well-balanced nutrition, favourable for optimal growth, does not in itself counteract a crowding condition promoting formation of alatae, unless the existence of a specific apterous promoting factor is accepted. Such a factor may be present in seedlings, which generally provide good growth of *M. persicae*. Kunkel and Mittler (1971) suggest that although an apterous promoting principle may be connected with optimal nutrition, an important component operates as a "token stimulus" perceived immediately after confining young larvae to seedlings. This might explain why we did not find important differences in free amino acid content of different stages of radish seedlings. On the other hand, we believe only exact data on phloem sap composition can identify the missing nutritional link between aphids and their hosts.

Kunkel and Mittler (1971) found that short-term food deprivation of crowded larvae of *M. persicae* results in a higher percentage of alatae, probably due to enhanced restlessness of the aphids. To avoid similar effects in which dietary changes had to be made in the early larval life, we used our continuous flow artificial feeding apparatus (Harrewijn, in prep.). We found that when larvae of *M. persicae* are once settled on an artificial diet, it may be exchanged, using a flow system, by almost any non-acid solution and the aphids will continue to feed and to produce honey-dew. This behaviour is also found for a strain of *M. persicae* at Cornell University (Mt. Gomery, pers. comm.).

It might be of practical importance that on mature potato plants reproduction and wing development of *M. persicae* are equally dependent on the nutritional status of the host. Manuring schedules or resistance breeding set up to reduce the amount of phosphates and soluble nitrogen or to shift the ratio of amino acids in the phloem sap may be able to reduce both the reproduction capacity and the proportion of alatae of an aphid population.

SUMMARY

Although it is gradually accepted that growth and reproduction of aphids is dependent on the quality of their host plants, it is generally assumed that the alate course of development is favoured on plants having a suboptimal nutritional value.

The present work describes experiments set up to establish the relationship between the physiological status of the potato plant and the production of alatae by the aphid *Myzus persicae*, reared in constant densities on leaves of potato plants growing on different nutrient solutions.

Wing production of *Myzus persicae* appeared to be positively related to the amount of phosphorus and nitrogen in the mineral nutrition of the potato plant. On leaves less suitable for reproduction aphids caged in a density normally resulting in a high percentage of alatae produced mainly apterous offspring. A similar trend was found on artificial diets with varying nutritional values. A positive effect of a substrate lacking certain amino acids on wing formation was found only in the case of the amino acid tryptophan. A specific reddish colouring normally present in presumptive alatae was also demonstrated in the absence of tryptophan from the diets. The same effect was obtained by administering an inhibitor of the enzyme tryptophan hydrolase, suggesting that tryptophan metabolism is involved in wing formation of *Myzus persicae*.

Provided a crowding stimulus of sufficient intensity was present, "good food" was generally correlated with a high number of alatae, except in the case of seedlings, where larvae from crowded mothers developed mainly into apterae. The existence of a gustatory apterous-inducing factor is discussed.

Acknowledgements.—The author wishes to thank Dr. A. A. Steiner for his advice on the special nutrient solutions, Drs. C. Vonk and W. Mosch for amino acid and phosphate analyses, Mr. C. A. van den Anker for carrying out statistical analyses and Miss. A. Kinderman for her technical assistance.

REFERENCES

BOWIE, J. H., CAMERON, D. W., FINDLAY, J. A. and QUARTEY, J. A. (1966). Haemolymph pigments of aphids. *Nature, Lond.* 210:395-397.

CALDERBANK, A., JOHNSON, A. W., TODD, A. R. (1954). Colouring matters of the aphididae. Part X. Preparation and properties of 4:9-dihydroxyperylene-3:10-quinone. *J. Chem. Soc.* nr. 4925:1285-1289.

CAMERON, D. W., CROMARTIE, R. I. T., KINGSTON, D. G. I. and TODD, A. R. (1964). Colouring matters of the aphididae. Part XVII. The structure and absolute stereochemistry of the protoaphins. *J. Chem. Soc.* 10:51-61.

GUSTAFSON, T. and TONEBY, M. (1970). On the role of serotonin and acetylcholine in sea urchin morphogenesis. *Exp. Cell. Res.* 62:102-117.

HARREWIJN, P. (1970). Reproduction of the aphid *Myzus persicae* related to the mineral nutrition of potato plants. *Ent. Exp. & Appl.* 13:307-319.

HARREWIJN, P. and NOORDINK, J. Ph. W. (1971). Taste perception of *Myzus persicae* in relation to food uptake and developmental processes. *Ent. Exp. & Appl.* 14:413-419.

JOHNSON, B. (1966). Wing polymorphism in aphids III. The influence of the host plant. *Ent. Exp. & Appl.* 9:213-222.

JOHNSON, B. and BIRKS, P. R. (1960). Studies on wing polymorphism in aphids I. The developmental processes involved in the production of the different forms. *Ent. Exp. & Appl.* 3:327-339.

KUNKEL, H. and MITTLER, T. E. (1971). Einflusz der Ernährung bei Junglarven von

Myzus persicae (Sulz.) (Aphididae) auf ihre Entwicklung zu Geflügelton oder Ungeflügelten. *Oecologia* **8**:110-134.

LEES, A. D. (1967). The production of the apterous and alatiforms in the aphid *Megoura viciae* Buckton, with special reference to the role of crowding. *J. Insect Physiol.* **13**:289-318.

MITTLER, T. E. and KLEINJAN, J. E. (1970). Effect of artificial diet composition on wing-production by the ahpid *Myzus persicae. J. Insect Physiol.* **16**:833-850.

MITTLER, T. E., TSITSIPIS, J. A. and KLEINJAN, J. E. (1970). Utilisation of dehydro-ascorbic acid and some related compounds by the aphid *Myzus persicae* feeding on an improved diet. *J. Insect Physiol.* **16**:2315-2326.

RACCAH, B., TAHORI, A, S., APPLEBAUM, S. W. (1971). Effect of nutritional factors in synthetic diet on increase of alate forms in *Myzus persicae. J. Insect Physiol.* **17**:1385-1390.

SCHAEFERS, G. A. and JUDGE, F. D. (1971). Effects of temperature, photoperiod, and host plant on alary molymorphism in the aphid, *Chaetosiphon fragaefolii. J. Insect Physiol.* **17**:365-379.

SINGH, P. (1970). Host-plant nutrition and composition: effects on agricultural pests. *Inf. Bull. Nr. 6, Canada Department of Agriculture:* 102 pp.

SPACKMAN, D., STEIN, W. H. and MOORE, S. (1958). Automatic recording apparatus for use in the chromatografy of amino acids. *Anal. Chem.* **30**:1190-1206.

SMITH, A. R., JONGKIND, J. F. and ARIËNS Kappers, J. (1972). Distribution and quantification of serotonin-containing and autofluorescent cells in the Rabbit pineal organ. *Gen. and Comp. Endocrin.* **18**:364-371.

STEINER, A. A. (1961). A universal method for preparing nutrient solutions of a certain desired composition. *Pl. & Soil* **15**:134-154.

_____(1966). The influence of the chemical composition of a nutrient solution on the production of tomato plants. *Pl. & Soil* **24**:454-465.

_____(1968). Soilless culture. *Proc. 6th Coll. Int. Potash. Inst. Florence:* 324-341.

STORMS, J. H. H., HARREWIJN, P. and NOORDINK, J. Ph. W. (1967). A new approach to the physiological host plant-parasite relationship—a technique in the field of applied entomology. *Neth. J. Pl. Path.* **73**:165-169.

SUTHERLAND, O. R. W. (1967). Role of host plant in production of winged forms by a green strain of pea aphid, *Acyrthosiphon pisum* Harris. *Nature, Lond.* **216**:387-388.

VAN BELLE, H. (1970). New and sensitive reaction for automatic determination of inorganic phosphate and its application to serum. *Anat. Biochem.* **33**:132-142.

VAN EMDEN, H. F., EASTOP, V. F., HUGHES, R. D. and WAY, M. J. (1969). The ecology of *Myzus persicae. A. Rev. Ent.* **14**:197-270.

WEARING, C. H. (1972). Responses of *Myzus persicae* and *Brevicoryne brassicae* to leaf age and water stress in Brussels sprouts grown in pots. *Ent. Exp. & Appl.* **15**:61-80.

EFFECT OF HIGH UREA FERTILIZATION OF BALSAM FIR TREES ON SPRUCE BUDWORM DEVELOPMENT

G. G. Shaw and C. H. A. Little
Maritimes Forest Research Centre
Canadian Forestry Service
Fredericton, New Brunswick, Canada

INTRODUCTION

Choristoneura fumiferana (Clem.) is a major lepidopteran defoliator of balsam fir and spruces in the boreal forest of North America. Although the inset is normally present at low densities, periodic eruptions follow the occurrence of three or four consecutive springs that are warmer and dryer than usual (Greenbank 1956). This association has been elaborated into 'the concept of climatic release' to explain the time and place of budworm outbreaks (Wellington *et al.* 1950).

Apart from the direct effects of high radiant energy on the budworm, *per se*, warm dry seasons may have indirect effects either through biocontrols or nutritional quality of the food supply, or both. Our objective has been to discover the causative role, if any, that nutrition plays in rapid population increase. As a first step, we experimentally altered the nutritional quality of maturing, current-year, fir foliage by differential shading. It was found that shading influenced various budworm and foliage parameters (Table 1), but did not affect ambient temperature or relative humidity. Yet whether the budworm responses were due to shade-induced changes in food quality or microclimate could not be determined conclusively; however, we favored the hypothesis of changed food quality. To test this further, we set up an experiment in which foliar nutritional changes were induced by fertilization. These results are reported here.

METHODS

Two contiguous groups of 6-year-old fir trees, spaced 10 cm apart in three rows 30 cm apart, were used. On 6 May 1971, soon after the nursery soil became bare of snow and had thawed, one group of trees was fertilized with 280, 22, and 50 lb./acre NPK in the form of urea, triple superphosphate, and potassium sulphate, respectively. The aim was to maximize the difference in foliar nitrogen content between this group of trees and a second group that were not fertilized.

On 25 May, second-instar larvae were placed carefully on bursting buds of first-whorl shoots in all trees in both groups. These larvae had been collected the previous day as they emerged from natural hibernacula under controlled conditions.

Table 1. Responses[a] by budworm and foliage to heavy shading and high-urea fertilization in relation to their responses in control treatments.

Response parameter	Treatment	
	Shading	Fertilization
Budworm		
Larval survival	+	+
Larval development time	–	o
♀♀ pupal weight	+	+
Foliage		
Crude fat	+	+
Nitrogen	+	+
Total sugars	–	+
Starch	–	o
Moisture	+	+
Ash	+	+
Heat of combustion	+	+
CHO/N	–	–

[a]+ increase; – decrease; o no change.

One-day-old pupae were collected, weighed, and sexed daily. Development time and survival of larvae, and pupal weight (which correlates with fecundity, Miller 1963), were determined since these parameters together account for much of the variability in generation survival. Female pupae were carried through to moth emergence at 70% R.H. and 22°C; then the adults were weighed. This was done to assess any treatment-induced difference that might have occurred during the pupal stage.

Foliage from first-whorl current shoots was sampled periodically to monitor treatment-induced and seasonal changes in moisture, nitrogen, crude fat, total sugars, starch, ash, and heat of combustion. Immediately after sampling, the shoots were frozen in liquid nitrogen and stored at -15°C until they were freeze dried. Subsequently, the needles were separated from the twigs and ground in a Wiley micro-mill to pass a 20-mesh screen. Moisture, crude fat, starch and total sugar concentrations were determined as outlined by Little (1970a). Nitrogen was measured by the Kjeldahl semi-micro method (McKenzie and Wallace 1954), ash by using a muffle furnace (A.O.A.C. 1965), and heats of combustion were determined with a Parr bomb calorimeter. The 1970 and 1971 leader lengths were measured at the end of the 1971 growing season to assess the effect of the fertilization treatment on tree growth.

RESULTS

Budworm

Larvae reared on fertilized trees had significantly higher survival, and pupal and female moth weights than larvae reared on unfertilized trees (Tables 2 and 3). The weight difference was consistently observed in the daily pupal collections, the difference being occasionally as much as 20 mg for females. The female pupal weight difference remained significant even after correlation for the large weight difference between treatments in pupae dying before moth emergence. Fertilization had no significant effect on pupal mortality (Table 2).

Table 2. Experimental results from spruce budworm reared on unfertilized and high-urea fertilized groups of balsam fir trees.

Parameter	Treatment	
	Unfertilized	Fertilized
No. of trees	78	107
No. of 2nd instars emplaced	567	734
% larval survival[a]	28	37
No. of surviving ♀♀ moths	60	104
% ♀♀ pupal mortality[b]	15	22

[a]Difference between treatments significant ($p = 0.05$).
[b]Difference between treatments not significant.

Table 3. Average weights (mg) of budworms reared on unfertilized and high-urea fertilized balsam fir trees.

Classification	Unfertilized	Fertilized
1-day-old ♀♀ pupae[a]	71.0 (71)[b]	80.4 (133)
♀♀ pupae surviving to adult[a]	71.6 (60)	78.3 (104)
♀♀ pupae dying as pupae[a]	67.6 (11)	87.7 (29)
1-day-old ♂♂ pupae	48.2 (86)	53.9 (137)
♀♀ moths[a]	48.7 (60)	53.6 (104)

[a]Difference between treatments significant ($p = 0.05$) on transformed scale of 0.5 (log x +1).
[b]Values in parentheses are numbers of observations.

There was no difference in length of larval development time between the two treatments, the median date of pupation for females (29 June) and males (28 June) being similar for both treatments.

Foliage

The moisture, ash, crude fat, starch, total sugar, nitrogen, total carbohydrate-nitrogen ratio (CHO/N), and caloric value determinations are presented in Figs. 1 to 3. Values and seasonal trends are similar to those found previously (Little 1970b; unpublished data). To relate the treatment-induced changes in foliar chemical concentrations to those in budworm development, the budworm feeding period was taken from 25 May (date of larval emplacement) to 29 June (median date of pupation for

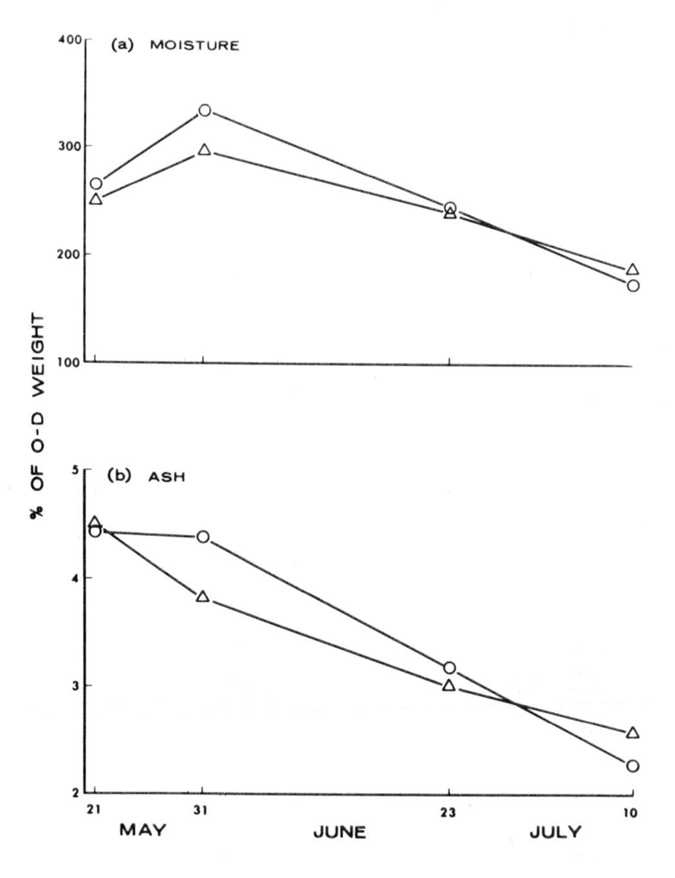

Figure 1. Changes in (a) percentage moisture and (b) percentage ash in maturing current foliage of unfertilized (▲) and high-urea fertilized (0) balsam fir trees.

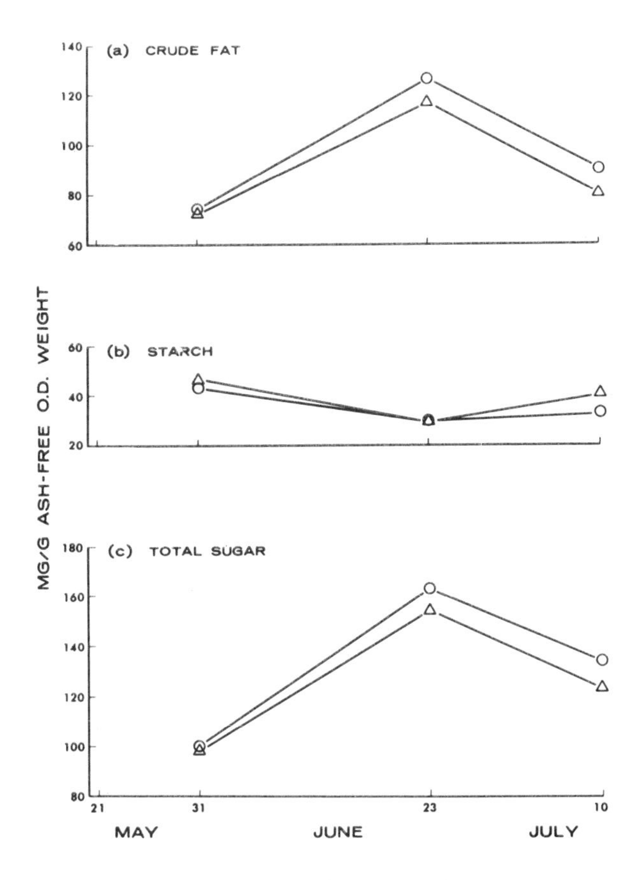

Figure 2. Changes in concentration (mg/g ash-free oven-dry weight) of (a) crude fat, (b) starch, and (c) total sugars in maturing current foliage of unfertilized (▲) and fertilized (0) balsam fir trees.

females). During this period, the current foliage of fertilized trees, compared with that of unfertilized trees, was consistently higher in moisture, ash, crude fat, total sugar, nitrogen, and caloric content, but lower in the CHO/N ratio and, perhaps, starch content. The treatment-induced differences in crude fat, total sugar, nitrogen, and caloric content were statistically significant ($p = 0.05$).

To account for any initial differences in tree size, the 1971 terminal shoot length was calculated as a ratio of the 1970 terminal shoot length, there being a positive correlation between final current shoot length and the length of the preceding internode (Little 1970c). The ratios were 0.65 and 0.60 for fertilized and unfertilized trees, respectively, indicating that the fertilization treatment had little, if any, effect on shoot growth.

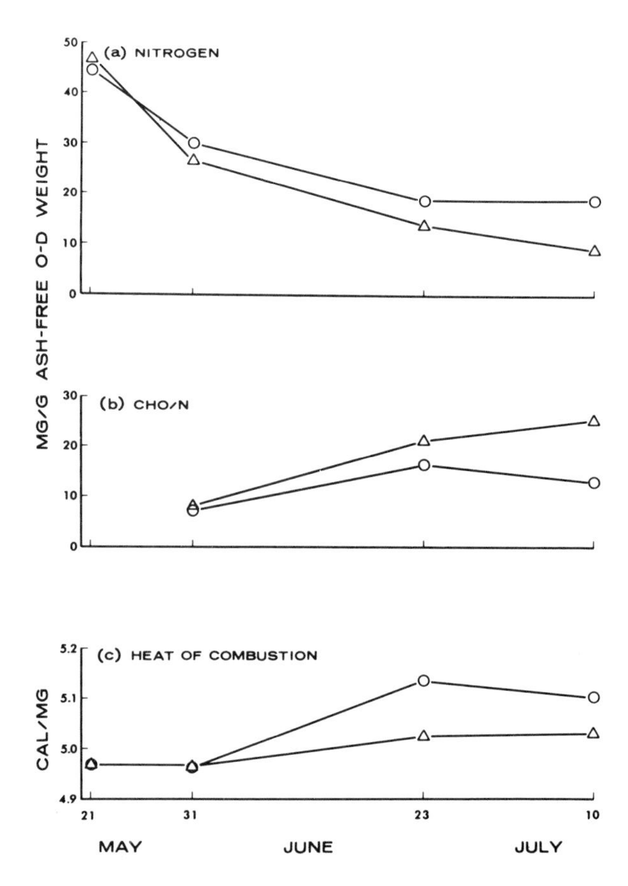

Figure 3. Changes in concentrations (mg/g ash-free oven-dry weight) of (a) nitrogen, (b) ratio of)crude fat + starch + total sugars(to nitrogen, and (c) the heat of combustion (calories/mg) in maturing current foliage of unfertilized (▲) and fertilized (0) balsam fir trees.

DISCUSSION AND CONCLUSIONS

High urea fertilization of fir trees resulted in increased survival and pupal weight of feeding budworm and, at the same time, marked changes in foliar chemical content. Since budworms reared on both fertilized and unfertilized trees experienced the same conditions of weather, predation, parasitism, etc., it follows that the budworm responses to the fertilization treatment must have been caused by the treatment-induced changes in food quality.

The budworm and foliar chemical responses observed in the fertilization experiment, together with those observed in the shading experiment conducted earlier (unpublished report), were summarized in Table 1 to aid discussion of the overall interpretation. In both experiments, treatment similarly influenced larval survival and female pupal weight on the one hand and the contents of crude fat, nitrogen, moisture, and ash, as well as heats of combustion and the CHO/N ratios, on the other hand. However, the effect of treatment on larval development time and on the concentrations of total sugar and starch was dissimilar in the two experiments. Between budworm and foliage parameters, several correlations are apparent, but the relative importance of these is not obvious.

The correlations of budworm response with such general food-quality indices as heat of combustion and CHO/N ratio are as expected. Heat of combustion measures the potential caloric value of a food to the consumer. In both experiments, budworm did better on foliage with the higher potential caloric value. The CHO/N ratio is typically low in young foliage for most plants, becoming higher as the foliage matures. In as much as the budworm is a spring defoliator, it should do better on young foliage than mature foliage (Edel'man 1963). Such was the case in both of our experiments.

The part that individual parameters such as nitrogen or moisture played in the budworm's responses to treatment is speculative at best. Differences in moisture would affect foliage digestibility while the quality and quantity of amino and fatty acids probably would affect the efficiency with which digested foliage was converted to body matter (House 1965). It is entirely possible that the role of any particular parameter was appreciably different in the two types of experiments.

For many insects, carbohydrates are not essential dietary substances (Gilmour 1961; House 1965). It may be that total sugars are not very important to budworm, since survival and pupal weight increased while total sugars decreased when foliage was heavily shaded. In artificial diets, however, budworms use the principal sugars found in fir foliage (Dr. G. Harvey, Can. Forest. Serv., Sault Ste. Marie, Ont. Personal Commun.) and it is possible that the foliar concentrations of these sugars may have been affected differently in the two experiments. The starch responses in the two experiments, coupled with the finding of Harvey that budworm larvae reared on artificial diet apparently do not use starch extracted from fir foliage, suggests that starch as a single substrate is of low importance to budworms.

The fact remains that we do not know the actual significance of any of the individual nutrient parameters, nor the importance of interactions between single parameters, nor have we monitored all possible chemical (or physical) parameters; hence, further speculation is unwarranted at present.

Altering the nutrient quality of fir foliage via high-urea fertilization increased budworm larval survival and pupal weight. Since it had been found previously that artificial shading also affected foliar chemical content and budworm survival and pupal weight, it is now clear that budworm development can be influenced significantly by changes in the quality of its natural food. However, further field research is needed before assessing the importance of individual nutrient parameters, alone and in combination, to budworm development.

SUMMARY

Spruce budworm larvae were reared on developing current foliage of balsam fir trees fertilized with 280, 22 and 50 lb/acre of N, P and K, respectively. Current foliage of fertilized trees, compared with that of unfertilized trees, was higher in water, ash, crude fat, total sugar, nitrogen and heat of combustion, but lower in the total carbohydrate to nitrogen ratio. Larvae reared on fertilized trees had higher survival and pupal and female moth weights than larvae reared on unfertilized trees. Similar results were found previously when growing trees were heavily shaded. From these two experiments it was concluded that changes in the quality of natural food could affect the development of spruce budworm.

Acknowledgements.—We are grateful to Dr. M. K. Mahendrappa, Maritimes Forest Research Centre, for recommending the fertilization regime used in this investigation.

REFERENCES

A.O.A.C. (1965). Official methods of analysis of the Association of Official Agricultural Chemists, 10th ed. (Edited by W. Horowitz). Assoc. Offic. Agr. Chem., Washington, D.C.

EDEL'MAN, N. M. (1963). Age changes in the physiological condition of certain arbivorous larvae in relation to feeding conditions. *Entomol. Rev.* **42**:4-9.

GILMOUR, D. (1961). The biochemistry of insects. Academic Press, N. Y.

GREENBANK, D. O. (1956). The role of climate and dispersal in the initiation of outbreaks of the spruce budworm in New Brunswick. I. The role of climate. *Can. J. Zool.* **34**:453-476.

HOUSE, H. L. (1965). Insect nutrition. In: The Physiology of Insecta. Vol. II, pp. 769-813, (Edited by M. Rockstein). Academic Press, N. Y.

LITTLE, C. H. A. (1970a). Derivation of the springtime starch increase in balsam fir *(Abies balsamea). Can. J. Bot.* **48**:1995-1999.

LITTLE, C. H. A. (1970b). Seasonal changes in carbohydrate and moisture content in needles of balsam fir *(Abies balsamea). Can. J. Bot.* **48**:2021-2028.

LITTLE, C. H. A. (1970c). Apical dominance in long shoots of white pine *(Pinus strobus). Can. J. Bot.* **48**:239-253.

MCKENZIE, H. A. and WALLACE, H. D. (1954). The Kjeldahl determination of nitrogen: A critical study of digestion conditions - temperature, catalyst, and oxidizing agent. *Australian J. Chem.* **7**:50-70.

MILLER, C. H. (1963). The analysis of female proportion in the unsprayed area. In: The dynamics of epidemic spruce budworm populations (Edited by R. F. Morris). *Mem. Entomol. Soc. Can.* 31:75-87.

WELLINGTON, W. G., FETTES, J. J., TURNER, K. B. and BELYEA, R. M. (1950). Physical and biological indicators of the development of outbreaks of the spruce budworm. *Can. J. Res. D.* 28:308-331.

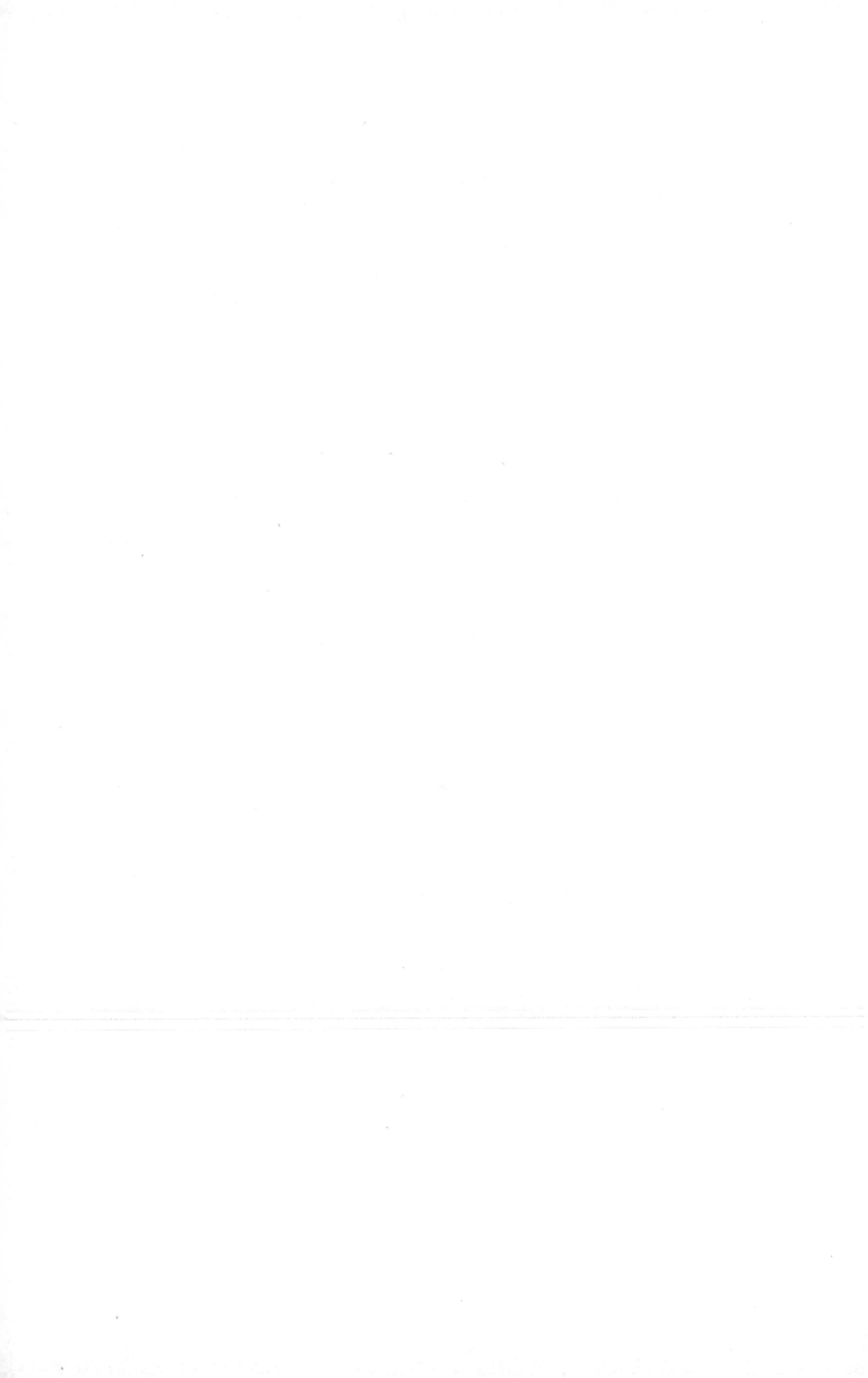

HOST PLANT RESISTANCE TO INSECTS–NUTRITIONAL AND PEST MANAGEMENT RELATIONSHIPS

Fowden G. Maxwell
Department of Entomology
Mississippi State University
State College, Mississippi 39762

INTRODUCTION

Plants that are inherently less severely damaged or less infested by a phytophagous pest under comparable environments in the field are termed resistant (Painter 1951, 1958). Resistance as expressed in the field is usually complicated, involving in most instances more than one of the three components of resistance as defined by Painter (1951). In addition there are many interactions occurring between the pest, the plant and the environment. Scientists have only recently begun to understand how complicated some of these responses are. Because of the complexity of most cases of resistance, very little is understood concerning the basis of resistance, particularly the biochemistry aspect. Fortunately, entomologists and plant breeders have been able to make good progress in developing resistant varieties without knowing the biochemical basis of resistance.

A plant may be deficient in certain nutritional requirements for the insect and thereby be resistant. However, there are few or no documented cases where the basic cause of resistance can be attributed to the lack of a nutritional substance alone. The insect is dependent on the plant for much more than nutrients; chemostimulation, physical factors, and a satisfactory micro-environment are all interacting factors which play a role in determining whether a plant may be susceptible or resistant (Beck 1965).

The objective of this paper is to examine nutritional aspects in relation to the mechanisms of resistance and to generally discuss the role of host plant resistance to insects in pest management programs.

NUTRITIONAL RELATIONSHIPS

Antibiosis

Antibiosis is the adverse effect of a plant on some aspect of the insect's biology (Painter 1951, 1958, 1969). More often this type of resistance stems from toxins or other such antibiotic agents. From a nutritional standpoint, although there are few documented examples, antibiosis may occur from one or more of the following reasons (Painter 1969): (1) the absence of some nutritional materials such as vitamins or essential amino acids in the plant;

(2) the deficiency of certain nutritional materials, especially amino acids or specific sterols; (3) the imbalance in available nutrients, especially sugar-protein or sugar-fat ratios.

Sugar content of plants may be very important to insect pests, usually because of feeding stimulation but may also be limiting in proper growth and survival other than regulating quantitative intake. For example low levels of soluble sugar content in the host plant of *Brevicoryne brassicae* (Linnaeus) limits reproduction and development of winged forms in the aphid (Evans 1938). Certain pentoses in the bean, *Phaseolus vulgaris*, have been shown to inhibit the growth of the weevil, *Callosobruchus chinensis* L. (Friend 1958).

Some insects requirements for sugars may vary with age. Reduction in sugar content of the plant at the more critical stages may cause resistance. For example larvae of European corn borer *Ostrinia nubilalis* (Hubner) require glucose until the fourth instar and are capable of minutely differentiating between different concentrations (Beck 1957). Knapp *et al.* (1966) also showed that the corn earworm, *Heliothis zea* Boddie could discriminate between concentrations of sugars and that the sugar balance was important in the resistance of certain lines of corn to this insect.

Most phytophagous insects studied thus far have shown the same requirements for amino acids as the rat. In pea aphid (*Acyrthosiphon pisum* (Harris) resistant plants of *Pisum sativum* L., lower concentrations of amino acids occur in resistant than in susceptible lines (Auclair 1957). The silks of certain corn lines resistant to the corn earworm also have lower concentrations of amino acids than those of susceptible lines (Knapp 1966). It has not been shown, however, that this is the principal factor in resistance in either of the previous cases.

Parrott *et al.* (1969) conducted an extensive test for qualitative and quantitative differences in amino acids of the various species of *Gossypium* in the hopes of finding a species that would be deficient or void of an essential amino acid required for development of the boll weevil, *Anthonomus grandis* Boheman. Unfortunately, no great differences quantitatively were found and no qualitative differences.

Most insects utilize dietary fat or fatty acids for energy, for a source of metabolic water, and for building reserves of depot fat and glycogen. The body fat of insects is affected quantitatively and qualitatively by the host on which they feed. Plants that do not provide sufficient amounts of fats in the insects' diet may have a detrimental effect on the insects' biology. Mature larvae and pupae of bollworm (*Heliothis zea* (Boddie), contained more fat when the larvae were reared on a high-fat, dough-stage corn than on lower-fat, milk-stage corn; although the iodine and saponification values of the corn lipids at both stages remained almost constant (Friend 1958). Cholesterol has been used in most chemical diets developed for plant feeding insects (Vanderzant *et al.* 1956), although it is not the characteristic sterol of the higher plants. When the plant sterol ergosterol, sitosterol, or stigmasterol was substituted for cholesterol in the artificial diet for the pink bollworm, *Pectinophora gossypiella* (Saunders), the larvae grew faster and were larger (Vanderzant *et al.* 1956). Lecithin, a phosphoric acid ester of a diglyceride and choline, increases the fecundity of *Leptinotarsa decemlineata* (Say). Females laid an average of one egg per day when fed young potato leaves painted with sucrose solution and three per day when fed leaves painted with a lecithin-sucrose emulsion (Grison 1948). Carnitine, lysine, linoleic acid, and inositol are other examples of substances reported affecting the biology of particular insects, when deficient in amount (Painter 1969). Dadd (1957) speculated that a derivative common to both carotene and vitamin A constituted a growth factor for *Schistocerca* and *Locusta* spp. Hagan (1958) reported that of nine B vitamins tested, lack of thiamin, nicotinic acid, folic acid, or choline caused a significant decrease in fecundity of *Dacus dorsalis* Hendel adults. Waites and Gothilf (1959) stated that when thiamin, riboflavin, pyridoxine, Ca-pantothenate, folic acid, biotin, and nicotinic acid were individually omitted from the diet, high mortality of the almond moth larvae, *Cadra cautella* (Walker) resulted. There are many other similar studies that have been done which show the importance of vitamins and related compounds in the diets of insects. Unfortunately, few studies have been done which involve the analysis of the insect host plant. One of the few studies conducted in recent years is that of Hudspeth *et al.* (1969) who surveyed a cross section of cotton lines for vitamin C content. Vitamin C is necessary for boll weevil development. Tremendous quantitative differences were evident between the investigated cotton lines, but lines containing very small quantities of Vitamin C were found to support excellent boll weevil growth and development.

There is much more evidence in the literature on the importance of minerals in plants as they relate to resistance. Plants deficient in essential minerals needed by insects may in addition contain atypical concentrations of organic compounds that can affect the growth or reproductive capacity of the insects feeding on them. Barker and Tauber (1964) reported that the pea aphid exhibits lower reproductive capacity on plants deficient in Ca, Mg, N, P, and K. Allen and Selman (1955) observed the oviposition rate of the mustard beetle, *Phadeon viridus* (Melsheimer) on watercress leaves decreased when N, P, K, or Fe concentrations were lowered. Cannon and Terriere (1966) found that the number of eggs laid by females of the two spotted spider mite, *Tetranychus urticae* (Koch), was not affected for at least 7 days on leaf discs from plants that received excess amounts of Fe, Mn, Zn, or Co. Also a wide range of Fe in the leaves appeared to be tolerated by the females without affecting the quantity of eggs laid. The lack of Fe or Zn in diets for the green peach aphid, *Myzus persicae* (Sulzer) increased the percentage of alates formed, whereas larvae deprived of P will develop mainly to apterae

(Raccah *et al.* 1971). The imported cabbageworm (*Pieris rapae* (L.), has been observed to be affected in different ways on leaves deficient in N, P, K, and Fe (Allen and Selman 1956). High water content of rice plants has been generally associated with fast rates of development and reproduction of several species of rice weevils. *Aphis fabae* Scopoli, the bean aphid, decreased feeding activities and larve-position rates and produced mostly alate forms on water deficient leaves.

The effects of various chemical minerals applied to plants may affect the nutrition of the pest insect. For example, copper supplied to growing wheat and soybeans improved the nutritional value of their products for *Tribolium confusum* Jacquelin deVal (Chirigos 1958).

Maxwell and Harwood (1958) reported that soil insecticides may change the nutritional value of the host plants. They also reported that some herbicides applied to plants increased reproduction in feeding aphids, whereas others increased mortality and reduced fecundity. Fertility in *Leptinotarsa decemlineata* (Say) has been observed to vary with both species and variety of host plant. This difference was affected on potatoes more by the lecithin content of the leaves than by the carbon-nitrogen ratio (Grison 1958). Varieties of peas observed to be most susceptible to the aphid, *Acyrthosiphon pisum*, contained more N and less sugar than did resistant varieties. Branson and Simpson (1966) allowed corn leaf aphid, *Rhopalosiphum maidis* (Fitch) to feed on sorghum plants grown under high and low levels of nitrogen. Results showed that more than twice as many aphids were to be found on plants receiving nitrogen than on nitrogen-deficient plants. Kindler and Staples (1970) studied the effect of excess, medium, and deficient amounts of calcium, magnesium, nitrogen, potassium, phosphorus, and sulfur on spotted alfalfa aphid, *Therioaphis maculata* (Buckton) feeding on alfalfa clones resistant and susceptible to the aphid. None of the treatments made the susceptible clone more resistant. Resistance was significantly decreased but not eliminated when the resistant clone was treated with deficient levels of calcium or potassium or excess levels of magnesium or nitrogen, but resistance was significantly increased in plants receiving deficient levels of phosphorus. Sulfur did not affect resistance.

Similar type studies conducted by Rodriguez *et al.* (1970) with susceptible and resistant strawberry clones at different nitrogen levels demonstrated significant correlation between foliage nitrogen and mite injury. Studies of the survival of European corn borer larvae on resistant and susceptible corn plants supplied different levels of nitrogen and phosphorus showed that increased nitrogen did not affect degree of resistance, but amount of nitrogen in the soil influenced survival in the field.

In general, parallel tests of resistant and susceptible plant varieties under different fertility conditions show that resistant and susceptible plants tend to be affected in the same way. Under no conditions of soil fertility has a resistant plant been susceptible or a susceptible one become highly resistant. Because each species of insect, each host plant species, and often each type of soil constitutes a separate problem, no general conclusion regarding the importance of particular nutrient elements can be advanced. However, it is well known that certain plant species and varieties have the ability to extract and utilize various chemicals from the soil with greater efficiency than other species and varieties. If the increased absorption and use of the element contributes to resistance, the ability indicated may be a basis for resistance.

Preference

Preference or nonpreference is defined as those chemical or morphological host plant characteristics and insect responses that lead away from the selection and use of a particular plant for food, oviposition, shelter, or a combination of the three (Painter 1969).

The morphology of the host plant may affect the nutrition of the insect in the following ways: (a) it may limit the amount of feeding because of texture, shape, or color which would reduce amount of nutritive material being ingested and (b) may limit the digestibility and utilization of the food by the insect (Kasting and McGinnis 1958, 1959, 1961).

If the morphological and physical characteristics of the plant are suitable for the insect, chemicals having nutritive characteristics may also play an important role in initial attraction, feeding and oviposition. In many instances, the nutritive relationship is not with the chemicals themselves as with the feeding behavior that it may in catalyze the insect. For example, synthetic diets of some Cruciferae feeders gain in their acceptability after adding mustard oil glucosides (Nayar and Thorsteinson 1963). Feeding by the spotted cucumber beetle (*Diabrotica undecimpunctata howardi* Barber) is stimulated by cucurbitacins (Chambliss and Jones 1966). Boll weevil feeding on artificial diet is increased by the addition of gossypol (Maxwell *et al.* 1963). The point to be made with the above examples and many others that could be cited is simply that these secondary plant materials added to the diet probably have little or no nutritive value in themselves, but do act as feeding stimulants to increase intake of food materials by the insect which in most cases improves the general nutrition of the insect. The reverse may be said of materials in plants which are classed as feeding deterrents or repellents which may drastically limit the potential food intake by the insect. There are many examples in the literature of insects actually starving on a substrate that would adequately sustain them. For a recent comprehensive treatment of the whole area of biologically active materials as they relate to insect behavior see Dethier (1970), Beck (1965), Beck and Maxwell (1972), Maxwell (1972), and Hedin *et al.* (1972).

Some plant substances may play a dual role in that they stimulate as well as being essential nutritionally. Proteins, sugars, phospholipids, inorganic salts, minerals, vitamins, etc. are examples of materials that many times function in dual roles of acting as feeding stimulants and also being essential for proper development. Essential oils, glycosides and other secondary substances as previously stated are usually involved more in a singular role--that of attraction and feeding stimulation, and less so from a strictly nutritive standpoint.

NEED FOR BASIC RESEARCH ON RESISTANCE AND NUTRITIONAL RELATIONSHIPS OF HOST PLANTS

Plant chemicals affect the nutrition of the insect, directly, through nutritive properties for completion of life cycle and indirectly through modification of behavior associated with normal attraction, feeding and oviposition on the host plant. Concentrations of these substances in plants can be generally affected by genetic manipulation and by cultural management. Proper manipulation through selection and plant breeding techniques can produce plants that carry measurable resistance. This has been demonstrated over the years with the release of many varieties of plants resistant to insect attack. Because of the lack of research on the basic mechanisms of resistance in these resistant varieties, it is impossible to say to what degree nutrition is directly involved. With the more sophisticated techniques and instrumentation being developed by biochemical analysis, it is likely that with increased research much more can be learned about the bases of resistance and interrelationship directly and indirectly with insect nutrition. This type of research involves expensive basic biochemical techniques, but it is only through this type of approach that we can systematically work toward better artificial diets for mass rearing insects and adequately define the important chemicals relating to nutrition and behavior. Spin-off from this type of research is often very helpful in identifying and isolating biologically active materials from plants that can be utilized in developing unique methods for manipulating, managing, trapping, or killing the pest insect. Certainly a basic, as well as an applied approach is desirable from the standpoint of achieving maximum utilization of resistant mechanisms in plants, especially those mechanisms that may affect the nutritive value of the plant. Basic approaches should supplement but not replace our traditional field and greenhouse methods of screening and selecting for resistance.

USE OF PLANT RESISTANCE TO INSECTS IN PEST MANAGEMENT

Principal control measure

Most resistant varieties that have been released to date have been developed primarily as a principal control. Outstanding examples are wheat stem sawfly, *Cephus cinctus* Norton (seven resistant varieties of wheat); Hessian fly, *Mayetiola destructor* (Say) (24 resistant varieties of wheat); European corn borer, *Ostrinia nubilalis* (Hubner) (many resistant hybrids); grape phylloxera, *Phylloxera vitifoliae* (Fitch); spotted alfalfa aphid, *Therioaphis maculata* (Buckton) (17 resistant varieties); pea aphid, *Acyrthosiphon pisum* (Harris) (five resistant varieties); and corn earworm, *Heliothis zea* (Boddie) (several resistant hybrids). For many years corn, especially sweet corn could not be grown economically in the South until moderate resistance was introduced. There are many other examples that could be cited where resistance has been utilized as a primary control; however, these examples are sufficient to illustrate the point.

Component in large suppressant or eradication schemes

Resistant varieties carrying moderate to high levels of resistance can be very effectively used alone or in conjunction with other methods to reduce the initial population of a pest to a level that would enable the effective and economical use of the sterile male technique or some other feasible method for eliminating the pest from a geographical area. An example of this type of approach is the use of frego cotton (77% reduction in oviposition) in the pilot boll weevil eradication experiment in South Mississippi. The variety in this experiment promises to have a powerful suppressant effect on the total seasonal population of the boll weevil.

A resistant variety can also play a vital role after eradication or general suppression in an area by helping to prevent reestablishment. If reinfestation occurs, a resistant variety may in itself prevent the population from ever building to economic proportions.

As a supplement to biological and chemical control
(Biological Control).

Many times resistance, even at low levels may be very beneficial in increasing the effectiveness of biological and chemical agents in the environment. This may happen through (1) decreasing vigor or general physiological condition of the pest insect as a result of toxins or inadequate nutrition which would improve the efficiency of the predator or parasite in finding and attacking the host insect. The effectiveness of pathogens to infect and kill the insect is very often associated with the physiological condition of the host. A host that is weakened or has been affected adversely by plant substances, inadequate nutrition or toxins may be more susceptible, (2) altering configuration or morphological characteristics of the plant through genetic manipulations may improve the ability of the predator or parasite to seek and find its host. Leaf shapes, type of fruiting body, color and stem characteristics are usually easily altered. This type of change may also improve penetration and coverage of pathogens on the plant surface. One should keep in mind that certain types of morphological changes may be advantageous to the pest insect rather than being beneficial. Plants that are modified should be checked thoroughly to determine fully the impact of such change on beneficial insects and their efficiency.

Tolerance has been the least studied mechanism of resistance, and for this reason has not been utilized to any great extent. Plant varieties developed with relatively high levels of tolerance may prove to have an additional advantage in pest management in that these plants could maintain high levels of pest insects which would in turn provide high levels of predators and parasites which would help control other harmful species that the crop is not tolerant or resistant to.

Chemical control

The most obvious advantage that resistance can impart to insecticide control is a reduction in numbers of the insect to control. Secondly, the resistance may affect the insect, making it more susceptible to the insecticide, thus providing better control. In some instances, satisfactory insect control cannot be obtained by either method alone, but where combined, excellent results have been obtained. North Carolina has reported numerous vegetable crops where insecticide control has been improved vastly on crops containing low levels of resistance. In many cases insecticide rates have been dropped to a fraction of that normally recommended and there is still excellent control where applied on crops containing some resistance. Some specific examples are *Heliothis zea* on sweet corn, *Myzus persicae* on Irish potatoes, and squash and cucumber varieties to cucumber beetles, *Diabrotica* spp. (Brett 1970). Little research has been conducted in this area and much needs to be done. The value of low levels of resistance can be greatly amplified if interactions detrimental to the insect can be demonstrated between the insecticide and the host plant. For years it has been known that LD_{50}'s of various insecticides can vary up to 100%, depending on the food source or nutritional diet of the insect. We need to explore this more fully as we look to methods for reducing amounts of pesticides used on our crops.

As mentioned earlier, with biological agents, the efficiency of insecticides may likewise be increased through morphological change of the plant. Excellent examples of changes in morphology that have improved insecticide coverage and efficiency are the super okra leaf and frego bract condition in cotton. Improved efficiency in killing boll weevils and bollworms has been demonstrated on cottons containing these characters. These characters are also beneficial in reducing boll rot and providing good natural resistance to boll weevil.

A point that should be made is that we might well look for morphological types of plants that might be developed especially for improving efficiency of biological and chemical control agents and reducing rates of pesticides. In some instances this might be as useful an approach as breeding for disease or insect resistance.

Host plant resistance as a supplement to cultural control methods

The use of resistant crops in providing physical barriers to migrating insects has been effectively used in the past and may have limited utilization in pest management systems. For example, Atlas sorgo, highly resistant to chinch bug, *Blissus leucopterus leucopterus* (Say), has been used successfully to protect susceptible yellow dwarf milo by planting strips several rows wide around the milo field. Chinch bugs migrate on the ground and are decimated by the resistant sorgo before reaching the milo. In the mid-west, some years ago corn was particularly protected from grasshoppers by planting resistant sorghum adjacent to grassland from which the grasshoppers migrated into the corn fields.

Strip cropping of a more susceptible plant species to manage or attract the pest insect out of a crop is a technique that may have wide application in pest management. It has been used successfully in California and is being explored in Mississippi in cotton production. The use of alfalfa and sesame in cotton to help manage plant bugs, *Lygus* spp., and *Heliothis virescens* is but one example. Strip cropping also provides a reservoir for predators and parasites to increase and spread into adjacent crops.

More attention in the future should be devoted to planting resistant crops to limit population spill-over into more susceptible crops that follow in the season, i.e., resistant corn to limit bollworm in cotton, resistant cotton to limit bollworm in soybeans, etc.

In summary, host plant resistance can and will have a vital role in pest control in the coming years. It should provide the foundation on which to build pest management programs. Implementation of pest management programs will by necessity, accentuate the importance of host plant resistance, particularly the utilization of lower levels of resistance that have not been properly emphasized in the past. Wiser use of fertilizers, herbicides, soil insecticides, irrigation, and other cultural practices as they affect nutrition of the insect and expression of insect resistance in the plant will need to be emphasized strongly as we work toward sound pest management for the future.

REFERENCES

ALLEN, M. D. and SELMAN, I. W. (1957). The response of larvae of the large white butterfly (*Pieris brassicae* (L.) to diets of mineral-deficient leaves. *Bull. Entomol. Res.* **48**:229-242.

AUCLAIR, J. L. (1957). Developments in resistance of plants to insects. *Ann. Rept. Entomol. Soc. Ontario* **88**:7-17.

BARKER, J. S. and TAUBER, O. E. (1954). Fecundity of the pea aphid on garden pea under various combinations of light, moisture, and nutrients. *J. Econ. Entomol.* **47**:113-116.

BECK, S. D. (1957). The European corn borer, *Pyrausta nubilalis* (Hbn.) and its principal host plant. IV. Larval saccharotrophism and host plant resistance. *Ann. Entomol. Soc. Am.* **50**:247-250.

BECK, S. D. (1965). Resistance of plants to insects. *Ann. Rev. Entomol.* **10**:207-232.

BECK, S. D. and MAXWELL, F. G. (1972). Use of plant resistance. In Theories and Practices of Biological Control, ed. by Carl Huffaker. Academic Press, New York. (in press).

BRANSON, T. F. and SIMPSON, R. G. (1966). Effects of a nitrogen-deficient host and crowding on the corn leaf aphid. *J. Econ. Entomol.* **59**(2):290-293.

BRETT, C. H. and SULLIVAN, M. J. (1970). Vegetable insect control. North Carolina Exp. Sta. Bull. 440.

CANNON, W. N., JR. and TERRIERE, L. C. (1966). Egg production of the twospotted spider mite on bean plants supplied nutrient solutions containing various concentrations of iron, manganese, zinc and cobalt. *J. Econ. Entomol.* **59**:89-93.

CHAMBLISS, O. L. and JONES, C. M. (1966). Cucurbitacins: specific insect attractants in Cucurbitaceae. *Sci.* **153**:1392-1393.

CHIRIGOS, M. A. (1958). Nutritional studies with the insect, *Tribolium confusum* (Duvel). *Diss. Abs.* **19**:216-217.

DADD, R. H. (1957). Ascorbic acid and carotene in the nutrition of the desert locust, *Schistocerca gregaria* Forsk. *Nature* **179**:427-428.

DETHIER, V. G. (1970). Some general considerations of insects' responses to the chemicals in food plants. In Control of Insect Behavior by Natural Products, ed. by D. L. Wood, R. M. Silverstein and M. Nakajima. Academic Press, New York, pp. 21-28.

EVANS, A. C. (1938). Physiological relationships between insects and their host plants. I. The effect of the chemical composition of the plant on reproduction and production of winged forms in *Brevicoryne brassicae* L. (Aphididae). *Ann. Appl. Biol.* **25**:558-572.

FRIEND, W. G. (1958). Nutritional requirements of phytophagous insects. *Ann. Rev. Entomol.* 3:57-74.

GRISON, P. (1958). Action des lecithines sur la fecondite du Doryphore. *Academie des Sciences Compt. Rend.* **227**:1172-1174.

HEDIN, P. A., MAXWELL, F. G., and JENKINS, J. N. (1972). Insect plant attractants, repellents, deterrents, feeding and other related factors. Proc. Inst. Bio. Contr. Pl. Insects and Diseases.

HUDSPETH, W. N., JENKINS, J. N., and MAXWELL, F. G. (1969). Ascorbic acid impractical on a character for resistance in cotton to the boll weevil. *J. Econ. Entomol.* **62**(3):583-584.

KASTING, R., MCGINNIS, A. J., and LONGAIR, E. L. (1958). Moisture and nitrogen contents of stem tissues from solid- and the hollow-stemmed varieties of spring wheat and their relation to sawfly resistance. *Can. J. Plant Sci.* **38**:287-291.

KASTING, R., and MCGINNIS, A. J. (1959). Nutrition of the pale western cutworm, *Agrotis orthogonia* Morr. (Lepidoptera: Noctuidae). II. Dry matter and nitrogen economy of larvae fed on sprouts of a hard red spring and a durum wheat. *Can. J. Zool.* **37**:713-720.

KASTING, R., and MCGINNIS, A. J. (1961). Comparison of tissues from solid- and hollow-stemmed spring wheats during growth. II. Food values determined with the pale western cutworm, *Agrotis orthogonia* Morr. (Lepidoptera: Noctuidae). *Can. J.*

Zool. 39:273-280.

KINDLER, S. D. and STAPLES, R. (1970). Nutrients and the reaction of two alfalfa clones to the spotted alfalfa aphid. *J. Econ. Entomol.* 63(3):938-940.

KNAPP, J. L. (1966). A comparison of resistant and susceptible dent corn single crosses to damage by the corn earworm, *Heliothis zea* Boddie. *Diss. Abs.* 26(9):4918.

KNAPP, J. L., HEDIN, P. A., and DOUGLAS, W. A. (1966). A chemical analysis of corn silk from silk from single crosses of dent corn rated as resistant, intermediate, and susceptible to the corn earworm. *J. Econ. Entomol.* 59(5):1062-1064.

MAXWELL, F. G. (1972). Morphological and chemical changes that evolve in the development of host plant resistance to insects. *J. Environ. Qual.* In press.

MAXWELL, F. G., JENKINS, J. N., KELLER, J. C., and PARROTT, W. L. (1963). An arrestant and feeding stimulant for the boll weevil in water extracts of cottonplant parts. *J. Econ. Entomol.* 56:449-454.

MAXWELL, R. C., and HARWOOD, R. F. (1958). Increased reproduction of aphids on plants affected by the herbicide 2,4-dichlorophenoxyacetic acid. *Bull. Entomol. Soc. Amer.* 4:100.

NAYAR, J. K. and THORSTEINSON, A. J. (1963). Further investigations into the chemical basis of insect-host relationships in an oligophagous insect, *Plutella maculipennis* (Curtis) (Lepidoptera: Plutellidae). *Can. J. Zool.* 41:923-929.

PAINTER, R. H. (1951). Insect resistance in crop plants. Macmillan, New York. 520 pp.

PAINTER, R. H. (1958). Resistance of plants to insects. *Ann. Rev. Entomol.* 3:267-290.

PAINTER, R. H. (1969). Plant and animal resistance to insects. In Principles of Plant and Animal Pest Control, Vol. 3:64-99. *Natl. Acad. Sci. Publ.* 1965.

PARROTT, W. L., MAXWELL, F. G., JENKINS, J. N., and HARDEE, D. D. (1969). Preference studies with hosts and non-hosts of the boll weevil, *Anthonomus grandis. Ann. Entomol. Soc. Amer.* 62:261-264.

RACCAH, B., TAHORI, A. S., and APPLEBAUM, S. W. (1971). Effect of nutritional factors in synthetic diet on increase of alate forms in *Myzus persicae. J. Insect Physiol.* 17:1385-1390.

ROGRIGUEZ, J. G., CHAPLIN, E. E., STOLTZ, L. P., and LASHEEN, A. M. (1970). Studies on resistance of strawberries to mites. I. Effects of plant nitrogen. *J. Econ. Entomol.* 63(6):1855-1858.

VANDERZANT, E. S., and REISER, R. (1956). Studies of the nutrition of the pink bollworm using purified casein media. *J. Econ. Entomol.* 49:4-10, 454-458.

WAITES, R. E. and GOTHILF, S. (1969). Nutrition of the almond moth. I. Analysis and improvements of the experimental diet. *J. Econ. Entomol.* 62:301-305.

ALFALFA SAPONINS SIGNIFICANT IN RESISTANCE TO INSECTS

Ernst Horber
Department of Entomology
Kansas State University
Manhattan, Kansas, U.S.A.

INTRODUCTION

Saponin content in alfalfa

Alfalfa (*Medicago sativa* L.) varieties differ in content and composition of saponins by varying amounts of the same or several different saponins. Content and composition appear to be genetically controlled and modified by selection and breeding (Hanson *et al.*, 1963; Pedersen *et al.*, 1966). Pedersen and Wang (1971) reported progress in selection for both low and high saponin content in alfalfa. Selection for saponin content significantly affected protein and forage yield, but not fiber, fat, ash, N-free extract, or seed yield.

Saponins are found in leaves, stems, roots, and flowers of the alfalfa plant (Cole *et al.*, 1945, Pedersen and Taylor, 1962; Morris *et al.*, 1961; Morris and Hussey, 1965). Content varied with location, cutting, variety, and other variables, when Buffalo, Ranger, Lahontan, Vernal, and DuPuits varieties were compared at 8 locations in the United States (Hanson and Kohler, 1961; Hanson *et al.*, 1963). Lahontan was lowest; DePuits highest in saponin content, which ranged from 2 to more than 3%. Contents of first cuttings averaged lower than those of second and third cuttings, saponin content being significantly and positively correlated with protein, oil, fat, and N-free extract, and negatively with crude fiber and high yield. Leaves contained twice as much as stems, older plants contained decreasingly less.

Chemistry

Since alfalfa saponins are complex compounds and difficult to separate, progress on their chemistry has been slow (Coulson, 1958; Van Atta and Guggolz, 1958; Hanson *et al.*, 1963). Four constituent saponins were separated by paper electrophoresis and at least ten by one-dimensional ascending paper chromatography with multiple development (Coulson, 1958; Coulson and Davis, 1962). Pedersen *et al.* (1966) separated 5 fractions each from different alfalfa varieties and found those of Lahontan and DuPuits differing in Rf values, color reactions, and in biological assays.

Insect and mite nutrition – North-Holland – Amsterdam, (1972)

Applying thin-layer chromatography, Birk (1969) showed that some saponins isolated from both tops and roots could be separated into 8 distinct fractions. Separation of alfalfa saponins is complicated because of similarity of their components. Upon hydrolysis of the saponins, a mixture of triterpenoid sapogenins, sugars, and uronic acids is obtained. By column- and paper-chromatography four components have been isolated and identified or characterized as sapogenols A, B, C, and medicagenic acid (Potter and Kummerow, 1954; Djerassi *et al.*, 1951; Van Atta and Guggolz, 1958; Walter *et al.*, 1955). A fifth, lucernic acid, has been isolated and partially characterized but its complete structure is not yet known (Livingston, 1959). Soyasapogenol E has been demonstrated in acid hydrolyzates of alfalfa saponins by Birk (1969), who identified soyasapogenols A, B, C, D, and E as well as glucose, galactose, arabinose, xylose, rhamnose, and glucuronic acid in saponins from alfalfa top and roots.

Besides variation in chemical composition of saponins, marked differences occur also in physical and biological properties, e.g., complexing with cholesterol, inhibiting proteases, being toxic to fish, and by having hemolytic activity and an antibiotic effect on insects. Root saponins demonstrated a higher surface activity and were more hemolytic than those from tops, the hemolytic index (H. I.) for roots ranged from 13,000 to 20,000 and for tops around 2,000 (Birk, 1969).

Besides differences among species, the properties of saponins vary also among alfalfa varieties: DuPuits-derived saponins inhibited chick growth more than those from Lahontan. Water extracts from DuPuits, Lahontan, Ranger, and Uinta gave strikingly different growth curves when tested with *Trichoderma* (Pedersen *et al.*, 1966).

Saponins have been related to bloat in ruminants. Chemical, physiological, pharmacological, physical, and toxicological properties of alfalfa saponin in relation to ruminant bloat were studied at the Western Utilization Research Laboratory of the USDA (Lindahl *et al.*, 1957). The foam strength of alfalfa saponins is pH dependent with an optimum between 4.5 and 5.0, according to Mangan (1958, 1959), who suggested a possible significance of pH in the problem of bloat. Since alfalfa saponins have a triterpenoid structure exhibiting free carboxyl groups (Walter *et al.*, 1954), the state of ionization of these groups may be related to their surface activity.

Apparently saponins do not have the same importance in bloat promotion as proteins, pectins and polyuronides, which constitute the favorable media for rapid and intensive CO_2 production, foam promotion, and suppression of eructation (Maymone, 1963).

Effects on growth of animals

Alfalfa saponins appear to be partly responsible for the depressing effect of alfalfa on the growth, diet consumption, and diet utilization of chicks (Draper, 1948; Cooney *et al.*, 1948; Lepkovsky *et al.*, 1950). Both dehydrated and sun-cured alfalfa meals contain, as Heywang (1950) showed, a factor that retards growth of young chicks as well as egg production of layers, when alfalfa meal is included in the diet at levels of 10% and above. Peterson (1950a, b) attributed the effect of saponins because fractions from a water extract of alfalfa, which depressed growth of chicks, foamed and caused hemolysis. A 0.5% level of saponin fed to chicks by Heywang and Bird (1954) was the least saponin that unmistakably inhibited growth, while 0.4% inhibited growth more. When Pedersen *et al.* (1966) compared saponins from four different alfalfa varieties, DuPuits, Uinta, Ranger, and Lahontan, added at 0.3% to a basic chick ration, all rations containing saponins produced significantly less gain than the control, with rations containing DuPuits saponins producing the least gain.

Alfalfa saponins' preventing normal weight gains by chicks can be partly reversed by concurrently feeding cholesterol or phytosterols, principally β-sitosterol, and can be fully reversed by including a mixture of cholesterol and cottonseed oil in the diet (Peterson, 1950b). By adding 1% cholesterol to the diet containing 0.3% saponin, Anderson (1957) overcame growth depression. When as much saponin (0.3%) was fed to laying hens, egg production decreased immediately after first feeding, but returned to normal or higher within 10 days. Adding saponins to hens' ration did not affect growth of their chicks.

Extending Anderson's (1957) studies on laying hens, Heywang *et al.* (1959) fed diets containing 0 and 0.4% saponin extracted from alfalfa or 0.26% saponin supplied from sun-cured alfalfa meal, which was 20% of the diet. Both saponins depressed egg production all the time they were fed but production returned to normal soon after saponin was discontinued. Both saponins depressed feed consumed but neither affected egg weight. Those data do not reflect on practical situations because alfalfa meal normally does not exceed 5% of a ration.

The nutritional significance of saponins in alfalfa needs further study. The reversal of the saponin effect by adding cholesterol to the diet may be explained by cholesterol forming an insoluble, unabsorbed complex of saponin-cholesterol similar to that observed *in vitro*. As Potter and Kummerow (1954) pointed out, sapogenins derived from alfalfa saponins did not affect growth of chicks.

Wilson *et al.* (1957), studying pharmacological properties and toxicity of alfalfa saponins, found that toxicity by mouth was much less than by parenteral administration to rats. No subacute oral toxicity was noticed, only that rats disliked the treated food but rabbits and guinea pigs did not object to food containing saponins.

Alfalfa saponins, saponin samples derived from Ladino clover, and commercial saponin varied widely in toxicity and in hemolytic activity.

According to these authors no correlation was found between acute toxicity to small laboratory animals, hemolytic action, and bloat promotion in ruminants.

Effects on growth of plants

Mishustin and Naumova (1955) reported that alfalfa root saponin was toxic to cottonseed and reduced germination. Their data supported the view that alfalfa root saponins accumulated in the soil were harmful to cotton germination and reduced the population of certain microorganisms. Pedersen (1965), measuring effects of alfalfa meal and alfalfa saponins on cottonseed germination, found that germination percentage, weight, and length of the radicle all were reduced in proportion to alfalfa meal added to the experimental medium. Alfalfa saponin also reduced radicle length in proportion to saponin used but effects on germination were not conclusive with the concentrations examined.

Pedersen *et al.* (1966) found that saponins also reduced growth of fungi (*Trichoderma* sp.). Fractions extracted from DuPuits variety were much more active than those from Lahontan (Zimmer *et al.*, 1967).

Alfalfa saponins could be isolated from dry and sieved soil where alfalfa had been grown (Birk, 1969). Germination of cotton seeds on the soil was 75% compared with 92.5% on the same soil after sorghum had grown. When 20% of alfalfa meal or milled alfalfa roots were added to the soil, germination decreased by 50 and 65%. Radicle length was also reduced. Guenzi *et al.* (1964) found water extracts of alfalfa forage inhibitory to corn germination, but they did not identify the toxic compound. Bioassaying lettuce seeds, Pedersen *et al.* (1966) found differences among saponins from alfalfa varieties DuPuits, Ranger, Lahontan, and Uinta significant for radicle length and seedling weight, but not for germination percentages.

Effects of feeding alfalfa saponins to insects

Working with whitegrubs (*Melolontha vulgaris* F.) on alfalfa, Horber (1964) found antibiosis and nonpreference related to high saponin content in roots of resistant strains, which suggested a possible relationship between saponins and resistance to phytophagous insects.

Alfalfa root saponins, but not alfalfa top saponins, are strikingly toxic to *Tribolium castaneum* larvae; they inhibit growth about 75% when the diet is supplemented with 0.1% alfalfa root saponin, which is almost fully overcome by 0.5% cholesterol added to the diet probably because a saponin-cholesterol complex is formed (Birk, 1969).

Alfalfa saponins, like soybean saponins, inhibit the enzymes α-chymotrypsin, proteases from midguts of *Tribolium* larvae, and cholinesterase (Ishaaya and Birk, 1965). Incubating alfalfa saponins with casein fully or partly counteracts inhibitions of those enzymes. The same nonspecific type of inhibition has been proved for soybean saponins. Similar trials with cholesterol, a well known complexing agent for alfalfa saponins but not for soybean saponins, did not prevent alfalfa saponins from inhibiting enzymes when incubated with the saponin before the enzymatic reaction began. Birk (1969) concluded that the sites differ for interactions between saponin-protein and for saponin-cholesterol.

Roof *et al.* (1972) described a bioassay technique for testing saponin fractions or other secondary plant substances extracted from different alfalfa varieties on *Empoasca fabae* (Harris). Nymphs and adults fed readily on a diet of 1% sucrose + 0.5% agar in water contained in a "Parafilm" sachet. Mortality of leafhoppers feeding on the 5% and 1%, respectively, saponin diets was rapid, within 2 and 3 days, respectively. Total nymphal days were significantly fewer on 0.1% than on 0.01 or 0% saponin diets. The threshold of sensitivity of *E. fabae* to the saponin extracted from *Yucca* spp. (J. T. Baker, lot 90248) lies between 0.01 and 0.1%. Survival differences of *E. fabae* on the 0.01% saponin diet and of those on the control diet were nonsignificant.

The same technique was applied to bioassaying saponin (J. T. Baker, lot #90248) with the first and second instar nymphs of the pea aphid (*Acyrthosiphon pisum* Harris). A 20% sucrose + 0.5% agar diet was used to test a series of concentrations: 5, 1, 0.1, 0.01 to 0% saponin. The diets containing 5, 1, and 0.1% saponin eliminated more than 50% of the nymphs between the second and third day, and 100% by the fourth day; mean nymphal days differed significantly from nymphal days by those on the control diet. The sensitivity threshold to Baker's saponin was lowered in pea aphid nymphs than in leafhopper nymphs, demonstrating that pea aphids are more appropriate test organisms than potato leafhoppers for saponin bioassay.

Currently saponin fractions extracted from DuPuits and Lahontan alfalfa cultivars and purified by Dr. K. H. Davis, Jr., Research Triangle Institute, North Carolina, are being bioassayed; the results will be published later.

FEEDING EXPERIMENTS USING ALFALFA ROOTS AND EXTRACTS ON WHITEGRUBS (*MELOLONTHA VULGARIS* F.)

Observations and results from four feeding experiments from 1960 to 1964 at the former Swiss Federal Agricultural Experiment Station, Zurich-Oerlikon, follow.

Screening for resistant alfalfa

Individual alfalfa plants apparently little damaged by heavy, natural infestation were collected from grassland in several regions in Switzerland. To eliminate evaded, pseudoresistant plants, we replanted each plant singly in concrete pipes and introduced enough whitegrubs (40-60/m^2) to stimulate heavy infestation. Seed of surviving plants was collected, planted in clay pots or trays made of asbestos cement, and seedlings were retested. A severe infestation was maintained with whitegrubs in various developmental stages. Commercial cultivars were used as standards (Horber, 1961).

Feeding experiments to determine type of resistance

Progeny of the most resistant plants were used to study preference during feeding and effects of feeding on resistant alfalfa roots and on the whitegrubs. During the feeding experiments the whitegrubs were either exposed to preference or no-choice situations using the rearing technique described by Horber (1959): In preference tests whitegrubs were singly exposed in 10 or 50 ml uncoated aluminum cartridges to groups of dandelion root slices (*Taraxacum officinale* L.) as the preferred standard and alfalfa roots as the tested entries. In no-choice tests only one kind of root slice at a time was offered to the same larvae.

In the first experiment we observed feeding behavior of young, first-stage larvae immediately after hatching, before they had access to other food. The experimental diet consisted of ca. 2-cm-long root sections of two alfalfa strains, selected for resistance in the field. Each group was provided roots of one of the resistant selections, either #91 or #92. A third group had dandelion; a fourth, a choice between root sections of dandelion and resistant alfalfa selection #91. Young larvae were first kept separately in 10 ml cartridges. Those that survived the second moulting were transferred to 50 ml cartridges. The feeding period extended 24 months. Records were taken on feeding preferences between alfalfa and dandelion, on development of whitegrubs, and on number of moults from the first instar to pupal and adult stages and on mortality.

The results of the first feeding experiment follows: A clear preference for dandelion and nonpreference for alfalfa #91 and #92 were evident in both preference and no-choice situations. However, it appeared that larvae with access to both alfalfa and dandelion in the same cartridge consumed more dandelion than those on only dandelion. Alfalfa selection #91 was less preferred than #92. Many more moults occurred in containers where grubs had access to both alfalfa and dandelion. No larvae moulted when kept on alfalfa alone. Mortality was highest on alfalfa selections #91 and #92 alone and lowest among grubs on dandelion and alfalfa together.

However, it was not possible to distinguish between nonpreference and antibiosis, as refusal of nonpreferred food may have reduced weight gain and moulting and increased mortality, thus simulating antibiosis.

Attempts were made to isolate the factors responsible for antibiosis or nonpreference. The roots of the most resistant plants were cut, freeze-dried, and extracted with water, alcohol, acetone, and benzene. The extracts or residues were added to ovendried vermiculite. After moistening to 100% RH, first and second stage grubs were introduced. Dandelion roots served as standard diet. Every 10 days the cartridges were inspected and, when necessary, food and vermiculite were replaced. Weight increase, number of moults, and mortality were recorded. Owing to long life span (3 years in the field), feeding experiments were extended several weeks, months, or years. For each treatment, groups of 33 to 66 white grubs of the same stage or approximate weight were used.

During the *second feeding experiment* the water extract and residue of the resistant selection #93 were added to dried and ground roots of dandelion. For each entry, 120 g fresh roots were ground and freeze-dried, and those of alfalfa were extracted with water. Five groups of 33 white grubs each of the second larval stage were fed according to the following plan:

Group	Diet
1	fresh dandelion root sections.
2	48 g dried ground dandelion roots.
3	48 g dried ground roots of alfalfa selection #93.
4	diet 2 + water extract of 48 g alfalfa selection #93.
5	diet 2 + the residue of 40 g alfalfa selection #93 extracted with water. The ground plant material, extracts, or residues for groups 2 to 5 were mixed with 100 g dry vermiculite.

The feeding was 6 weeks; observations continued 10 weeks.

The results of the second feeding experiment follow: Feeding, weight increase, moulting, and mortality weie recorded. The first group fed fresh dandelion roots ate the most and gained the most. The third group fed dried alfalfa roots ate the least and gained the least. Groups 4 and 5 exposed to water extract of alfalfa selection #93 or its residue gained little (see Table 1). Most moults occurred on dried dandelion roots; fewest moults, on dried roots of alfalfa. Fewer moults occurred on water extract and residue of the alfalfa selection #93 than on dried dandelion but more than on dried alfalfa roots. Highest mortality after 8 weeks was in the group on dried alfalfa; lowest mortality, among the group on dried dandelion with alfalfa residue added. Mortality on water extract of alfalfa #93 was slightly lower than on dried alfalfa roots.

In the *third experiment*, root extracts obtained from two commercial cultivars, the Hungarian MV 129, highly resistant to whitegrubs, and the moderately resistant German variety, Frankenwarte, were compared with dandelion root extracts. Fresh dandelion roots were provided as supplementary food in all treatments, to incite feeding response.

The roots were freeze-dried, extracted with water, the first residue extracted with ethanol; the second, with acetone; the third, with benzene. The extracts were added to vermiculite in the ratio of 16 g fresh roots to 100 g vermiculite. Each group consisted of 40 whitegrubs in the second larval stage. Dandelion root sections were added to the vermiculite containing the different extracts. The control group got dandelion instead of extracts. Feeding period was 3 months; observations, 24 months.

Table 1. Second feeding experiment with whitegrubs.

Diet	Composition	No. of whitegrubs surviving	No. of whitegrubs reaching 3rd stage	Mortality %	Weight gain mg	Weight gain %
1	Dandelion, fresh roots	27[a]	73[b]	12[c]	714[d]	101.7[e]
2	Dandelion, dried roots	30	91	6	211	27.0
3	Alfalfa, dried roots	24	52	27	116	2.1
4	Diet 2 + alfalfa root extract	27	73	18	372	4.4
5	Diet 2 + alfalfa root residue	32	85	3	751	9.5

[a]Final size of groups surviving at end of 43-day feeding period after initial exposure of 33 whitegrubs.
[b]Percent of whitegrubs that moulted during feeding experiment and reached 3rd stage after being exposed in 2nd stage.
[c]Percent of whitegrubs that died during feeding experiment including those obviously infested with diseases.
[d]Weight gain in milligrams equals average final weight minus average initial weight of groups that survived feeding period.
[e]Weight gain in % of initial weight.

The results of the third feeding experiment. After 12 weeks approximately 80% of the control whitegrubs had moulted and attained the 3rd stage. That percentage was surpassed only by the group fed water extract of dandelion and the group exposed to alcohol extract of Hungarian MV 129. Among the groups fed water extracts, fewest moulted on Hungarian MV 129 as expected from a resistant cultivar when water-soluble growth inhibitors are involved. Among the groups fed alcohol extracts, MV 129 gave the most moults; Frankenwarte, the fewest. Among groups exposed to acetone extracts, those on dandelion extract moulted least. Averages from all four extracts showed dandelion and MV 129 with the most moults; Frankenwarte with the fewest. Mortality after 24 weeks was highest on resistant MV 129, water extracts only. In the groups exposed to alcohol extracts, Frankenwarte had the highest mortality, while among groups fed acetone and benzene extracts, those on dandelion extracts had the highest mortality.

In the fourth feeding experiment, extracts prepared from roots of the resistant alfalfa selection #93, of the moderately resistant cultivar Frankenwarte, and of dandelion were compared. From each entry 600 g fresh material was ground, freeze-dried, and extracted with a mixture of ethanol-water. The first residue was extracted with acetone; the second, with benzene. The extracts were added to vermiculite in the ratio of 20 g fresh roots to 100 g vermiculite whereas the residues were added in the ratio of 7.5 g fresh root to 100 g vermiculite. After the solvents were evaporated, the vermiculite was moistened to 100% RH. Each group consisted of 33 white-grubs in the second larval stage and each cartridge was provided with a section of dandelion root as supplementary food and feeding incitant. The control group obtained untreated vermiculite and dandelion roots only. The feeding period was 4 months; the observation period, 20 months. Food consumption, weight, gain, development, moults, and mortality were recorded.

The results of the fourth feeding experiment (Table 2). Weight increased 100% after six weeks of feeding fresh roots of dandelion, whereas those on alcohol-water extract (1) gained only 60%, the increase was less on Frankenwarte than on selection #93. Gains by groups on the first residue did not differ from gains by those on fresh dandelion. The acetone extracts (II) showed some antibiotic effect–about 70% lower weight gains by both alfalfa entries. The second residues gave more gain than fresh dandelion roots alone. The groups on benzene extracts (III) gained little, whereas gains by groups on the third residue were comparable to gains by those on fresh dandelion roots. Growth was inhibited about 40% on water-ethanol extract, about 30% on acetone extract, and slightly over 20% on benzene extract without dandelion. Groups on the first and on the second residues of Frankenwarte essentially failed to gain. The repellent or antibiotic factors in alfalfa roots apparently were only partially removed by extraction with the ethanol-water mixture. Extraction procedures may have depleted nutrients or affected digestibility of residues.

The most moults were recorded from groups fed, respectively, dandelion, fresh roots, or residues after acetone or benzene extractions. The residues after ethanol-water extraction still showed some residual growth inhibition. More moults occurred in groups fed with ethanol-water or acetone extracts of the moderately resistant Frankenwarte than in groups fed the highly resistant selection #93, but the reverse was true with the benzene extracts. No moults occurred in groups fed with residues not supplemented with dandelion roots.

Mortality was zero after six weeks in the dandelion control group. The first alfalfa extract with water-alcohol produced 21% mortality compared with 30% from selection #93. The first residue of both alfalfas gave somewhat higher mortalities than did the benzene extract of selection #93. Residues of all extractions produced higher mortality when not supplemented with fresh sections of dandelion. The first residue of Frankenwarte produced a 50% mortality but with selection #93 it was the second residue that produced mortality exceeding 60%.

Results of the fourth feeding experiment indicated that 2 alfalfas, the

Table 2. Fourth feeding experiment with whitegrubs.

Diet	Extract or residue	Composition	Survivors	Moultings %	Mortality %	Weight gain mg	Weight gain %
1		Dandelion, fresh roots	33[a]	45[b]	0[c]	388.0[d]	100.9[e]
2	I	Extract Frankenwarte	26	20	21	211.4	55.6
3	I	Extract 93	23	9	30	243.0	66.6
4	I	Residue Frankenwarte	30	27	9	357.7	94.6
5	I	Residue 93	26	35	21	386.3	104.0
6	II	Extract Frankenwarte	30	28	9	303.5	72.0
7	II	Extract 93	32	19	3	268.8	68.0
8	II	Residue Frankenwarte	28	54	15	497.2	122.7
9	II	Residue 93	28	54	15	457.5	113.0
10	III	Extract Frankenwarte	29	28	12	301.8	78.1
11	III	Extract 93	27	43	18	329.6	76.2
12	III	Residue Frankenwarte	28	43	15	396.2	100.5
13	III	Residue 93	29	25	12	371.5	93.9
14	I	Residue Frankenwarte[f]	13	0	61	80.4	20.3
15	I	Residue 93[f]	25	0	24	7.3	1.6
16	II	Residue Frankenwarte[f]	24	0	27	17.9	4.6
17	II	Residue 93[f]	12	0	63	-7.3	-2.2
18	III	Residue Frankenwarte[f]	21	5	36	-1.3	-0.33
19	III	Residue 93[g]	13	0	32	-9.5	-2.00

[a]Final size of group = number of surviving whitegrubs at end of 6-week feeding period.
[b]Number of whitegrubs that developed and moulted during feeding period.
[c]Percent whitegrubs that died during experiment.
[d]Weight gain in mg equals final weight minus initial weight of whitegrubs that survived 45-day feeding period.
[e]Weight gain expressed in % (average initial weight = 100%).
[f]Groups fed with residues after extraction without dandelion as supplementary food.
[g]19 whitegrubs instead of 33 as in other groups.

moderately resistant Frankenwarte and highly resistant selection #93, differ in chemical compositions of the antibiotic factor.

DISCUSSION OF RESISTANT FACTORS IN ALFALFA ROOTS

Several properties distinguish alfalfa roots from those of other plants, e.g., dandelion roots: alfalfa roots have a strongly bitter taste and they resist cutting, breaking, tearing, or grinding. Ground alfalfa roots produce a fibrous, light, creamy flour. Considerable differences were also noted among roots of the various alfalfa varieties used. When roots of highly resistant selection #93 were extracted with water-ethanol and the extract was shaken 15 seconds and left to settle 15 minutes, a copious, stiff foam remained on top. The foam of the highly resistant selection #93 persisted several days, whereas the extract of the moderately resistant Frankenwarte produced little weak foam that lasted only a few hours. Cholesterol added to the water-ethanol extract produced a pronounced precipitation in selection #93 but not in Frankenwarte.

When ground roots of alfalfa selection #93 were added to the diet of *Drosophila funebris* L., no larvae were produced. In a preference test with cockroaches (*Blatta orientalis* L.) the water-ethanol extract of selection #93 showed pronounced repellent action.

Resistance of alfalfa roots to the polyphagous whitegrub *(Melolontha vulgaris* F.) may be explained by several modalities: antibiosis, nonpreference, and tolerance. Though tolerance may be of great practical importance in the field, the reported investigations are intended to explain observed varietal differences in terms of antibiosis and nonpreference. In highly resistant, nearly immune varieties (as MV 129 and selections #91) and #93) both types of resistance were expressed by less feeding, inhibited growth, reduced moulting, and increased mortality. Nonpreference was demonstrated in preference tests and may be explained by repellency of the roots. Reduced feeding and weight gain and increased mortality also were apparent in no-choice situations when whitegrubs were exposed to root sections or water or water-ethanol extracts of the highly resistant MV 129, selections #91 and #93, but somewhat less than exposed to the moderately resistant Frankenwarte. Feeding of residues of those same selections after water or water-ethanol extraction was less detrimental to whitegrubs. Additional evidence on different chemical compositions of the roots of alfalfa varieties, demonstrating various degrees of resistance to whitegrubs, was obtained by extracting resistant selection #93 and moderately resistant Frankenwarte with dilute ethanol. The foam produced on the extract of the resistant selection persisted for hours or days, while that on the extract of moderately resistant Frankenwarte was weak and did not persist. When cholesterol was added to those extracts, only the extract of the resistant selection produced substantial precipitation. From results of feeding tests with extracts from alfalfa roots and additional evidence, I concluded that antibiosis and non-preference in roots of resistant alfalfa were caused by saponin content. Because extracts and residues after treatment with other solvents showed some residual inhibitory effect, the possibility that factors other than saponins are detrimental to whitegrub feeding on roots of resistant alfalfa could not be excluded.

Saponins might also cause resistance to whitegrubs in such other legumes as red clover *(Trifolium pratense* L.), white clover *(T. repens* L.), birdsfoot trifoil *(Lotus corniculatus* L.), etc. and in root crops like sugar beets and potatoes or various forbs and herbs like *Silene vulgaris* L., which like *Saponaria officinalis* L. is a Cariophyllaceae. *Silene vulgaris* L. sometimes took over as a weed after pastures were destroyed by whitegrubs.

The physiological effect of saponins on whitegrubs was not elucidated. Saponins may block sterol uptake essential for the development of phytophagous insects. Growth inhibition in *Tribolium castaneum* larvae induced by alfalfa root saponins but not by alfalfa top saponins, was almost fully overcome when 0.5% cholesterol was added to the diet (Birk *et al* 1968), which indicates that young larvae should be more susceptible to saponins than older larvae because the older ones should have already obtained supplement of cholesterol from other food sources.

Weight gains recorded in the second feeding experiment were analyzed using initial weights of whitegrubs when feeding started (Table 3). Those with initial weight below average lost weight and were more inhibited by dried alfalfa roots than were those with above average initial weight.

Another possibility is protease inhibition as Ishaaya and Birk (1965) demonstrated in midguts of *Tribolium* larvae where alfalfa saponins inhibited the enzyme α-chymotrypsin.

Another implied effect of saponin is on symbionts or on the metabolism of insects that depend on symbionts to break down indigestible substances (such as cellulose) or to synthesize essential nutrients.

Alfalfa saponins may be significant in hostplant resistance to such polyphagous insects as whitegrubs, pea aphid, leafhoppers, etc., whereas oligophagous insects, narrowly specialized on alfalfa, sometime during coadaptive evolution may have found a way to overcome saponins as a barrier or even to use saponins as a guide to lock into their host. They may also use saponins as energy source. Sapogenins' rich supply of sugar molecules unquestionably could supply energy. Growth of the highly specialized alfalfa weevil (*Hypera postica* Gyll.) was not inhibited by any of 16 saponin fractions isolated from alfalfa leaves incorporated into artificial diets (Hsiao, 1969). Instead, some of the fractions improved larval feeding and growth.

Attention should be focused on the possible role of saponins on other insects such as the spotted alfalfa aphid *(Therioaphis maculata* Buck.) and

Table 3. Weight gains in relation to initial weight of whitegrubs.

Treatment	Composition of diet	Average weight gain mg	r_1	Weight gains of whitegrubs: Below average initial weight	r_2	Above average initial weight	r_3
1	Dandelion, fresh roots	500.8[b]	c	307.5[d]	e	782.0[f]	g
2	Dandelion, dried roots	225.5	2.2	216.3	1.4	237.6	3.3
3	Alfalfa[a], dried roots	37.3	13.4	-21.9	15.0	120.2	6.5
4	Alfalfa[a], water extract of roots	180.9	2.8	96.6	3.1	271.8	2.8
5	Alfalfa[a], residue after roots were extracted with water	109.4	4.6	61.0	5.0	152.1	5.1

[a] Highly resistant alfalfa selection 93.
[b] Average weight gain of all individuals of the same treatment.
[c] r_1 = ratio of weight gain between treatment and standard (= 1st diet).
[d] Average weight gain of all whitegrubs with initial weight below average.
[e] r_2 = ratio of weight gain between treatment and standard for whitegrubs with initial weight below average.
[f] Average weight gain of all whitegrubs with initial weight above average.
[g] r_3 = ratio of weight gain between treatment and standard for whitegrubs with initial weight above average.

the seed chalcids (*Bruchophagus roddi* Guss.). Applebaum *et al.* (1965, 1969) have drawn attention to saponins as possible factors of resistance by legume seeds to attack of insects. The role of saponins in nectar and pollen and their effects on attractiveness and wholesomeness to pollinators of alfalfa also remains unknown.

Once the role of saponins in insect-hostplant relationship and resistance to insects has been elucidated, and how it is inherited is better understood, resistant cultivars may be bred more rapidly.

In the case of whitegrubs, the combination of factors associated with nonpreference, antibiosis, and tolerance evident in established alfalfa stands, contributed to resistance not only to feeding damage but also to drought, which sometimes accompanied and aggravated yield losses. Well established, resistant alfalfa fields would afford best insurance to farmers of uninterrupted forage production in areas exposed to recurrent whitegrub infestations.

Alfalfa is important in crop rotation because it contributes nitrogen and organic matter to the soil, increases water infiltration, and improves soil structure.

The alfalfa ecosystem is unique among other field crops because it represents, if it is well established, a relatively long lasting perennial across a wide variety of climatic, geographic, and edaphic conditions. But that enables insect populations, pests as well as beneficial predators and parasites, to build up during several seasons.

Control of soil insects with insecticides is not only expensive but also creates difficult and often persistent residue problems that affect quality of harvested products, edaphon, wildlife, and water. Biological control of soil insects is complicated because they are almost inaccessible to biological agents and, therefore, has been limited to only one widely used such control, milky disease, for only one species of whitegrubs (*Popillia japonica* Neum.). Breeding highly resistant alfalfa varieties would therefore be an ideal solution. Because alfalfa root saponins are particularly active in insect repellency and antibiosis, growing resistant alfalfa could be effective against various soil insects, e.g., whitegrubs, wireworms, and rootworms. Resistant alfalfa's potential usefulness as a biological control agent could extend to other difficult-to-control soil insect pests to the benefit of following crops and should, therefore, challenge more detailed studies.

SUMMARY

Individual alfalfa plants apparently resistant to heavy, natural infestations by the polyphagous whitegrub (*Melolontha vulgaris* F.) were selected in several regions in Switzerland. To eliminate pseudoresistant plants, each was replanted singly in concrete pipes, where heavy whitegrub infestation was maintained. Progeny of survivors were retested in the seedling stage and compared with commercial cultivars. Feeding experiments with root sections of alfalfa to determine type of resistance were performed in preference or no-choice tests. Records were taken on feeding preference, development, number of moults, and mortality. Nonpreference and antibiosis were observed in various degrees among alfalfa selections and cultivars. However, it was difficult to distinguish between nonpreference and antibiosis, as refusal of food uptake resulted in reduced weight gain, fewer moults, and higher mortality. Attempts were made to isolate factors responsible for antibiosis or nonpreference by extracting lyophilized roots of more or less resistant alfalfa with water, water-ethanol, acetone, and benzene.

Growth inhibition and mortality were higher on water or water-ethanol extracts of the highly resistant selections than on those of the moderately resistant, on the subsequent extracts, or on residues after extraction. Since foaming properties and precipitation with cholesterol were also more intensive in the highly resistant selections and varieties, saponin was considered responsible for the high degree of nonpreference and antibiosis in alfalfa roots. The current state of knowledge on alfalfa saponins, their chemical, physical, and biological properties is reviewed and implications for resistance and insect control discussed.

Acknowledgements.—This work, contribution No. 1088, Kansas State Agricultural Experiment Station, was financed by a grant from the Division of Agriculture of the Department of Public Economy of the Swiss Federal Government. Dr. F. Bachmann prepared chemical extracts and residues, Mrs. A. Kundig and Mr. A. Fekti maintained the whitegrub cultures in the laboratory. Dr. K. Kemp, Department of Statistics, analyzed the data found in Tables 1 to 3.

REFERENCES

ANDERSON, J. O. (1957). Effect of alfalfa saponin on the performance of chicks and laying hens. *Poultry Sci.* **36**:873-6.

APPLEBAUM, S. W., GESTETNER, B., and BIRK, Y. (1965). Physiological aspects of host specificity in the Bruchidae. IV. Developmental incompatibility of soybeans for *Callosobruchus. J. Insect Physiol.* **11**:611-16.

APPLEBAUM, S. W., MARCO, S., and BIRK, Y. (1969). Saponins as possible factors of resistance of legume seeds to the attack of insects. *J. Agr. Food Chem.* 17:618-22.

BARTLEY, E. E. (1965). Bloat in cattle. VI. Prevention of legume bloat with a nonionic surfactant. *Jour. Dairy Sci.* **48**:102-4.

BIRK, Y. Saponins, Chapter 7, 169-210 *in* Liener, I. E., Toxic Constituents of Plant Foodstuffs. 1969. Academic Press, New York, and London. 500 p.

COLE, H. H., HUFFMAN, C. F., KLEIBER, M., OLSON, T. M., and SHALK, A. F.

(1945). A review of bloat in ruminants. *J. Animal Sci.* **4**:183-236.

COONEY, W. T., BUTTS, J. S., and BACON, L. E. (1948). Alfalfa meal in chick rations. *Poultry Sci.* **27**:828-30.

COULSON, C. B. (1958). Saponins. I. Triterpenoid saponins from lucerne and other species. *J. Sci. Food Agr.* **9**:281-88.

COULSON, C. B. and DAVIES, T. (1962). Saponins. II. Fractionation and pharmacological properties of lucerne saponins. The possible relation of these to bloat. *J. Sci. Food Agr.* **13**:53-57.

DJERASSI, C., THOMAS, D. B., LIVINGSTON, A. L., and THOMPSON, C. R. (1957). Terpenoids. XXXI. The structure and stereochemistry of medicagenic acid. *J. Am. Chem. Soc.* **79**:5292-7.

DRAPER, C. I. (1948). A comparison of sun-cured and dehydrated alfalfa meal in the diet of the chick. *Poultry Sci.* **27**:659.

GUENZI, W. D., KEHR, W. R. and MCCALLA, T. M. (1964). Water soluble phytotoxic substances in alfalfa forage. Variation with variety, cutting, year, and stage of growth. *Agron. J.* **56**:499-500.

HANSON, C. H. and KOHLER, G. O. (1961). Progress report on a study of cultural factors related to estrogen and saponin content of alfalfa. Proc. 7th Tech. Alfalfa Conf., Albany, California, July 1961, p. 46. U. S. Dept. Agr.

HANSON, C. H., KOHLER, G. O., DUDLEY, J. W., SORENSEN, E. L., VAN ATTA, G. R., TAYLOR, K. W., PEDERSEN, M. W., CARNAHAN, H. L., WILSIE, C. P., KEHR, W. R., LOWE, C. C., STANDFORD, E. H., and YUNGEN, J. A. (1963). Saponin content of alfalfa as related to location, cutting, variety, and other variables. *Agr. Res. Serv. U. S. Dept. Agr. Bul.* 33-44.

HEYWANG, B. W. (1950). High levels of alfalfa meal in diets for chickens. *Poultry Sci.* **29**:804-11.

HEYWANG, B. W. and BIRD, H. R. (1954). The effect of alfalfa saponin on the growth, diet consumption, and efficiency of diet utilization of chicks. *Poultry Sci.* **33**:239-41.

HEYWANG, B. W., THOMPSON, C. R., and KEMMERER, A. R. (1959). Effect of alfalfa saponin on laying chickens. *Poultry Sci.* **38**:268-71.

HORBER, E. (1959). Verbesserte Methode zur Aufzucht und Haltung von Engerlingen des Feldmaikäfers (*Melolontha vulgaris* F.) im Laboratorium. *Landw. Jahrb. Schweiz. N. F.* **8**:361-70.

HORBER, E. (1961). Versuche zur Verhinderung der vom Maikäferengerling (*Melolontha vulgaris* F.), von der Fritfliege (*Oscinella frit* L.) and vom Maiszünsler (*Pyrausta nubilalis* Hbn.) verursachten Schäden mittels resistenter Sorten. *Landw. Jahrb. Schweiz. N. F.* **9**:635-69.

HORBER, E. (1964). Isolation of components from the roots of alfalfa (*Medicago sativa* L.) toxic to white grubs (*Melolontha vulgaris* F.). *Proc. XIIth Intern. Cong. Entomol. London*, 540-1.

HSIAO, T. H. (1969). Chemical basis of host selection and plant resistance in oligophagous insects. *Proc. 2nd Intern. Symp. "Insect and Hostplant," Wageningen*, 777-88.

ISHAAYA, I. and BIRK, Y. (1965). Soybean saponins. IV. The effect of proteins on the inhibitory activity of soybean saponins on certain enzymes. *J. Food Sci.* **30**:118-20.

LEPKOVSKY, S., SHAELEFF, W., PETERSON, D., and PERRY, R. (1950). Alfalfa inhibitor in chick rations. *Poultry Sci.* **29**:208-13.

LINDAHL, I. L., SHALKOP, W. T., DOUGHERTY, R. W., THOMPSON, C. R., VAN ATTA, G. R., BICKOFF, E. M., WALTER, E. D., LIVINGSTON, A. G., GUGGOLZ, J., WILSON, R. H., SIDEMAN, M. B., and DE EDS, F. (1957). Alfalfa saponins. Studies on their chemical, pharmacological, and physiological properties in relation to ruminant bloat. *U. S. Dept. Agr. Tech. Bull.* 1161.

LIVINGSTON, A. L. (1959). Lucernic acid, a new triterpene from alfalfa. *J. Org. Chem.* 24:1567-8.

MANGAN, J. L. (1958). Bloat in cattle. VII. The measurement of foaming properties of surface-active compounds. *New Zealand J. Agr. Res.* 1:140-7.

MANGAN, J. L. (1959). Bloat in cattle. XI. The foaming properties of proteins, saponins, and rumen liquor. *New Zealand J. Agr. Res.* 2:47-61.

MAYMONE, B. (1963). Saponine delle leguminose e loro effetto nell'alimentazione animale. *Annale Della Sperimentazione Agrarie N. S.* 17:1-2, 1-20.

MISHUSTIN, E. N. and NAUMOVA, A. N. (1955). Secretion of toxic substances by alfalfa and their effect on cotton and soil microflora. *Izv. Akad. Nauk. SSSR Ser. Biol.* 6:3-9.

MORRIS, R. J., DYE, W. B., and GISLER, P. S. (1961). Isolation, purification, and structural identity of an alfalfa root saponin. *J. Org. Chem.* 26:1241-3.

MORRIS, R. J. and HUSSEY, E. W. (1965). A natural glycoside of medicagenic acid. An alfalfa blossom saponin. *J. Org. Chem.* 30:166-8.

PEDERSEN, M. W. and TAYLOR, G. A. (1962). Varietal differences in the saponin content of alfalfa. *Proc. 7th Conf. Rumen Function.*

PEDERSEN, M. W. (1965). Effect of alfalfa saponin on cottonseed germination. *Agron. J.* 57:516-7.

PEDERSEN, M. W., ZIMMER, D. E., ANDERSON, J. O., and MCGUIRE, C. F. (1966). A comparison of saponins from DuPuits, Lahontan, Ranger and Uinta alfalfas. *Proc. 10th Intern. Grassland Cong. Helsinki*, pp. 266-9.

PEDERSEN, M. W. and WANG, LI-CHUNG. (1971). Modification of saponin content of alfalfa through selection. *Crop Sci.*, 11:833-5.

PETERSON, D. W. (1950a). Some properties of a factor in alfalfa meal causing depression of growth in chicks. *J. Biol. Chem.* 183:647-53.

PETERSON, D. W. (1950b). Effect of sterols on the growth of chicks fed high alfalfa diets or a diet containing Quillaja saponin. *J. Nutr.* 42:597-608.

POTTER, G. C. and KUMMEROW, F. A. (1954). Chemical similarity and biological activity of the saponins isolated from alfalfa and soybeans. *Science* 120:224-5.

ROOF, M., HORBER, E. and SORENSEN, E. L. (1972). Bioassay technique for the potato leafhopper *Empoasca fabae* (Harris). *Proc. North Central Branch ESA Meeting,* Kansas City (in press).

VAN ATTA, G. R. and GUGGOLZ, J. (1958). Forage constituents: detection of saponins and sapogenins on paper chromatograms by Liebermann-Burchard reagent. *J. Agr. Food Chem.* 6:849-50.

WALTER, E. D., VAN ATTA, G. R., THOMPSON, C. R., and MACLAY, W. D. (1954). Alfalfa saponin. *J. Am. Chem. Soc.* 76:2271-3.

WALTER, E. D., BICKOFF, E. M., THOMPSON, C. R., ROBINSON, C. H., and DJERASSI, C. (1955). Saponin from ladino clover (*Trifolium repens). J. Am. Chem. Soc.* 77:4936-7.

WILSON, R. H., SIDEMAN, M. B., and DE EDS, F. (1957). Some pharmacological effects of alfalfa saponin on nonruminants and on isolated muscle strips. *U. S. Dept. Agr. Tech. Bull.* 1161:70-81.

ZIMMER, D. E., PEDERSEN, M. W., and MCGUIRE, D. F. (1967). A bioassay for alfalfa saponins using the fungus *Trichoderma viride* pers. ex. fr. *Crop Sci.* 7:223-224.

NATURAL MECHANISMS OF RESISTANCE TO INSECTS IN LEGUME SEEDS

Shalom W. Applebaum and Yehudith Birk
Departments of Entomology and Biochemistry
Faculty of Agriculture, The Hebrew University, Rehovot, Israel

The interest in control of stored products insects which cause extensive damage to legume seeds stems from the importance of these seeds as a potential source of plant proteins for foods and feeds. It is our contention that insect control in the less developed countries cannot successfully advance beyond the general level of sophistication in agricultural practice which the farming community of those countries has attained. The more elaborate methods of seed storage in controlled atmospheres or by fumigation are barely acceptable culturally and difficult to implement institutionally in exactly those countries which suffer most from protein malnutrition. This is without taking into account in addition the long-term detrimental effect of chemical control on the total environment, which is today a basic issue in the more advanced countries. We feel that judicious choice of improved legume seed varieties which—among other attributes—exhibit multiple mechanisms of resistance to insects, could improve the protein situation in countries which are the main producers of legume seeds, and that this manner of insect control would be much simpler to institute. This approach is the rationale behind our concerted efforts to identify and evaluate the relative merit of the multiple mechanisms of resistance which are present in legume seeds and which do not adversely affect the nutritional value of these seeds for humans and farm animals.

The development of insects on viable legume seeds is dependent on the chemical and physical composition of these seeds, which has evolved as a consequence of the selective inter-relations between legume plants and insect populations. Among general chemical characteristics of legume seeds are the presence of proteinaceous factors which affect the digestibility of the bulk of legume seed protein by inhibiting proteolytic enzymes (Laskowski & Laskowski, 1954; Pusztai, 1967; Birk, 1968) and are involved in pancreatic hypertrophy and enzyme synthesis (Gertler *et al.*, 1967; Konijn *et al.*, 1970). Extensive research has been carried out on soybeans, resulting in the isolation and characterization of several distinct factors: A trypsin-inhibitor SBTI (Kunitz, 1947), a trypsin - and chymotrypsin - inhibitor AA (Birk, 1961; Birk *et al.*, 1963a; Birk & Gertler, 1968), and a *Tribolium* larval protease inhibitor (Birk *et al.*, 1963b). *Tribolium* larvae are insensitive, and their proteolytic enzymes unaffected, by SBTI and AA. We have recently turned to methods of affinity chromatography in order to facilitate separation of trypsin and/or chymotrypsin inhibitors from *Tribolium* protease inhibitors of various legume seeds. Partially purified preparations exhibiting mixed inhibitory activities are passed through a sepharose-trypsin or chymotrypsin column prepared as described by Feinstein (1971). The effluent does not contain trypsin and/or chymotrypsin inhibitors which are selectively bound to the respective column; and it exhibits inhibitory capacity against *Tribolium* proteases. By this alternate method we can easily separate the *Tribolium* protease inhibitor of soybeans from SBTI and AA, and we have isolated a distinct *Tribolium* protease inhibitor of groundnuts from a trypsin - and chymotrypsin - inhibitor. We have also found *Tribolium* protease inhibitor activity in chickpeas.

In order to elucidate the specificity of *Tribolium* protease inhibitors, we are presently attempting the characterization of *Tribolium* larval proteases and the comparison of their mode of action to what is known of mammalian proteolytic enzymes. The present work, of a preliminary nature, is on an enzyme preparation after centrifugation at 12,000 rpm and precipitation with ammonium sulfate to 90% saturation. *Tribolium* protease shows no esterolytic activity on benzoyl-arginine-ethyl ester or acetyl-tyrosine-ethyl ester - the specific substrates for trypsin and chymotrypsin, respectively - thus indicating the absence of trypsin and chymotrypsin - like enzymes in this insect.

However, it should be kept in mind that *Tribolium* protease does exhibit lysine polypeptidase activity (Applebaum & Konijn, 1966) with an action pattern similar to that of mammalian trypsin (Katchalski *et al.*, 1961) and *Tenebrio* trypsin (Applebaum *et al.*, 1964). Proteolytic activity was assayed on casein (Kunitz, 1947). *Tribolium* protease is activated twofold by 10^{-3}M 2-mercaptoethanol. A similar activation is achieved with cysteine or dithiothreitol. This finding indicates the presence of essential -SH group(s) in or close to the active site. No activation was encountered in the presence of different concentrations of Na^{+}, K^{+}, Ca^{++}, Zn^{++}, Mg^{++}, Cl^{-} and EDTA. Since activity was found to be -SH dependent, the effect of mercurials was studied. A concentration of 5 x 10^{-4}M p-hydroxymercuribenzoate completely inhibited activity. Di-isopropyl - fluoro-phosphate, a serine-protease reagent, completely inhibited activity at 10^{-2}M.

We were particularly attracted by two general observations: Firstly, that *Tribolium* protease inhibitors are evident in legume species which are normally regarded as susceptible to damage by *Tribolium*. Secondly, that the separate inhibitors are invariably exclusive in regard to their effect on *Tribolium in vivo* and *in vitro*. The first observation encourages us to suggest that selection or breeding of seed varieties (e.g. groundnuts or chickpeas) containing higher concentration of the specific *Tribolium* protease inhibitors would afford at least partial resistance to damage in storage, while the second observation indicates that increasing the level of these specific factors in seeds would in no way be detrimental to human or farm animal nutrition.

Similar observations are relevant to the resistance to haricot beans to the bruchid beetle *Callosobruchus chinensis.* We have found that a specific heteropolysaccharide, containing arabinose, xylose, glucose, galactose and rhamnose residues, and normally comprising about 1% of the dry weight of the bean, is responsible for this resistance (Applebaum *et al.*, 1970). Higher concentrations of this factor impart resistance to a second bruchid beetle - *Acanthoscelides obtectus* - which is regarded as a major pest of stored haricot beans (Applebaum & Guez, 1972). The difference between the response of the two beetle species appears to relate to the importance of some complex structure of the heteropolysaccharide for its biological activity, and to the ability of *Acanthoscelides* to hydrolyse the heteropolysaccharide to a smaller molecular core than *Callosobruchus* is able to. This more thorough hydrolysis partly, albeit not totally, neutralizes the toxicity of the heteropolysaccharide for *Acanthoscelides.* Here again it would seem logical to breed for higher concentrations of this character in haricot beans.

Among other natural compounds occurring in legume seeds in small concentrations, are the saponins. Soybean saponins which have been more extensively investigated than other legume seed saponins, are triterpenoid glycosides of which five aglycones have been isolated and their chemical structure determined (Birk, 1969). A comparative survey of similar, although not necessarily identical compounds, has been carried out on chickpeas, garden peas, broad beans, haricot beans, lentils and groundnuts (Applebaum *et al.*, 1969). There appears to be a correlation between the toxicity of certain legume saponin fractions to *Callosobruchus* larvae, and between the relative resistance of these different legume seeds to damage by this beetle. We are guiding our recent investigations in several directions with regard to the possible role that saponins play in host compatibility:

(1) To what extent the integral saponin structure may, or may not be, mandatory for biological activity. We had previously demonstrated that this integrity was necessary for toxicity towards *Callosobruchus.* Larvae of this beetle do not appreciably hydrolyse soybean saponins to their complement of free sapogenins and individual sugars, although they do possess α-galactosidase and α-glucosidase activities (Applebaum & Tadmor, 1971). We have recently demonstrated α-mannosidase, and to a lesser extent β-mannosidase, β-glucosidase and β-xylosidase activities. Incorporation into diets of either soybean sapogenins alone, sugars alone, or sapogenins plus sugars in proportions and quantities in accordance with their content in soybean saponins had no adverse effect on development of *Callosobruchus* (Applebaum *et al.*, 1965).

In contrast, the negative effects of alfalfa saponins on *Tribolium* larvae appear to be due mainly to the action of their aglycone moiety. Partly degraded saponins are more toxic than the intact extracts, and isolated alfalfa sapogenins—in particular medicagenic acid—are most toxic (Shany *et al.*, 1970).

We are presently working on the isolation of two saponin preparations from soybeans and groundnuts, each containing a single sapogenin residue, and we intend to determine at which stage sequential removal of sugar residues from the end of the constituent oligosaccharides results in loss of toxicity towards *Callosobruchus*. We anticipate that *Callosobruchus* will be unable to sufficiently hydrolyse these saponin molecules and we feel that lack of appropriate hydrolytic enzyme activity—as a basis for host incompatibility—is not the sort of resistance easily overcome. On the other hand, we intend to study the toxic effect of released sapogenins on susceptible insect species, which do hydrolyse saponins.

(2) That legume saponins are sapogenins might conceivably cause hormonal imbalance. This line of thought arose as a consequence of the observation that soybean saponin toxicity towards *Callosobruchus* is manifested in the failure of fully-grown larvae to pupate (Applebaum *et al.*, 1965). In order to assay one aspect of this possibility, we determined the effect of soybean seed saponins on diapause pupae of the cynthia silkmoth. Diapause pupae of saturniids are in a state of arrested development due to hormonal deficiency. Pharate adult development can be initiated by injection of suitable amounts of synthetic molting hormones (α- or β - ecdysone). Three groups of 10 diapausing *Phylosamia cynthia* pupae were injected with (a) 10μl soybean saponin (mg/ml 50% ethanol) + 10 μl 50% ethanol (b) 10 μl soybean saponin + 10 μl β-ecdysone (mg/ml 50% ethanol). (c) 10 μl β-ecdysone + 10 μl 50% ethanol). None of the pupae injected with saponin alone terminated diapause. Injection of saponin in addition to β-ecdysone did not potentiate or inhibit induction of development, compared to the control injection of β-ecdysone alone.

(3) That legume saponins may in some cases act by altering the membrane permeability of the insect gut. This line of thought does not offer an explanation for the intimation of hormonal imbalance implicit in the previous paragraph. It does take into account the general characteristics of saponins as surfactants exhibiting both hydrophobic and hydrophilic properties. We are attempting to determine whether saponins affect active transport of water in the insect gut, or other membrane characteristics. Alfalfa saponins, for example, impair oxygen diffusion through

the cotton seed-coat and membrane and thus, inhibit germination (Marchaim *et al.*, In Press).

(4) That legume saponins may in some cases act as deterrents. We have recently found that this is so in preference assays performed on synthetic diets with the polyphagous aphid *Myzus persicae,* which in nature does not choose to develop on leguminous plants.

High concentrations of common and uncommon amino acids and peptides are often found in dormant legume seeds (Bell, 1971) and this source of potentially available nitrogen has been implicated in the host selection of bruchid beetles (Applebaum, 1964). Several non-protein amino acids present in legume seeds are selectively lathyrogenic or neurotoxic to various vertebrates (Ressler, 1964; Murti *et al.*, 1964). It was anticipated that some of these, or others not recognized as neurotoxic to higher animals, might be selectively toxic to insects, and thus to some extent determine the resistance of legume seeds to various stored products insects.

We have been recently investigating the biological activities of β-cyano-alanine (BCNA) and 2,4 - diaminobutyric acid (DABA), which accumulate in some legume seeds. As a working hypothesis we would like to present the following scheme for detoxification of these two compounds and their possible metabolic conversions. The conversion of BCNA - an intermediate of cyanide metabolism in higher plants - to asparagine by the action of nitrilase (Fowden & Bell, 1965; Ressler *et al.*, 1969) or directly to aspartic acid by asparaginase (Lauinger & Ressler, 1970) has previously been shown to occur in plants and micro-organisms.

$$\begin{array}{ccccc}
CONH_2 & & C\equiv N & & CH_2\text{-}NH_2 \\
| & & | & & | \\
CH_2 & & CH_2 & & CH_2 \\
| & & | & \text{reductase} & | \\
CHNH_2 & \xleftarrow{\text{nitrilase}} & CHNH_2 & \rightleftharpoons & CHNH_2 \\
| & & | & \text{dehydrogenase} & | \\
COOH & & COOH & & COOH \\
 & & & & \\
\text{asparagine} & & \text{BCNA} & & \text{DABA}
\end{array}$$

Preliminary enzymic experiments with fully grown *Tribolium* larvae indicate the presence of nitrilase activity: asparagine accumulates when tissue homogenates are incubated with BCNA as substrate. Similar activities were found in larvae of *Lasioderma serricorne, Tenebrio molitor, Acanthoscelides obtectus* and *Callosobruchus chinensis.* It is not clear at this stage whether susceptibility of specific insect larvae to BCNA is correlated with their inability to convert this compound to asparagine. We have looked for, but have not yet demonstrated, interconversion of BCNA and DABA.

BCNA appears to be more potent when incorporated in artificial diets. A level of 1% BCNA drastically reduced the number of developing *Callosobruchus* larvae, while at this concentration neonate larvae of *Tribolium* did not survive at all. In comparison, mortality of *Tribolium* on 1% DABA resulted in 50% mortality (determined after 12 days on the diet). When 8-day old larvae were transferred from control diets to 1% BCNA diets, 50% died within 4 days, whereas no significant mortality was evident when larvae as young as 4 days old were transferred to 1% BCNA diets. Their rate of development was initially depressed, and although surviving larvae did finally attain maximal weight - about 2 mg per larva after 26 days (which is twice the developmental period on control diets) - none pupated.

In conclusion, we see that multiple mechanisms of potential resistance to insects are present in legume seeds and that in many cases each factor alone could presumably cope with specific insects. This multiplicity presumably stems from the logical attribute of "not putting all the eggs in one basket." We are obviously interested in those mechanisms of resistance which have no detrimental effect on human and farm animal nutrition, for otherwise we would be defeating our purpose in this research. We suggest that it is feasible to increase total protein production by selecting or breeding legume seed varieties resistant to stored products insects.

Acknowledgements.—The studies performed in our laboratories were carried out with the assistance and collaboration of D. Feinerman, S. Khalef, P. Leckstein, Na. Levin, Ni. Levin, H. Z. Levinsky, H. Podoler, B. Raccah, C. M. Schlesinger, U. Tadmor, J. Torten and D. Zach, and are supported in part by U.S. Department of Agriculture grant No. FG-Is-295. The assays on diapausing *cynthia* pupae were performed in the laboratory of L. I. Gilbert, Department of Biological Sciences, Northwestern University, Evanston, Illinois, U.S.A.

REFERENCES

APPLEBAUM, S. W. (1964). Physiological aspects of host specificity in the Bruchidae - I. General considerations of developmental compatibility. *J. Insect Physiol.* **10**:783-788.

APPLEBAUM, S. W. and GUEZ, M. (1972). Comparative resistance of *Phaseolus vulgaris* beans to *Callobruchus chinensis* and *Acanthoscelides obtectus* (Coleoptera:Bruchidae) : The differential digestion of soluble heteropolysaccharide. *Ent. Exp. Et Appl.* **15**:203-207.

APPLEBAUM, S. W. and KONIJN, A. M. (1966). The presence of a *Tribolium*-protease inhibitor in wheat. *J. Insect Physiol.* **12**:665-669.

APPLEBAUM, S. W. and TADMOR, U. (1971). The α-galactosidase activity of *Callosobruchus chinensis. Israel J. Entomol.* **6**:71-80.

APPLEBAUM, S. W., GESTETNER, B. and BIRK, Y. (1965). Physiological aspects of host specificity in the *Bruchidae* - IV. Developmental incompatibility of soybeans for *Callosobruchus. J. Ins. Physiol.* **11**:611-616.

APPLEBAUM, S. W., MARCO, S. and BIRK, Y. (1969). Saponins as possible factors of resistance of legume seeds to the attack of insects. *J. Agric. Fd. Chem.* **17**:618-622.

APPLEBAUM, S. W., TADMOR, U. and PODOLER, H. (1970). The effect of starch and of a heteropolysaccharide fraction from *Phaseolus vulgaris* on development and fecundity of *Callosobruchus chinensis* (Coleoptera:Bruchidae). *Ent. Exp. Et Appl.* **13**:61-70.

APPLEBAUM, S. W., BIRK, Y., HARPAZ, I. and BONDI, A. (1964). Comparative studies on proteolytic enzymes of *Tenebrio molitor* L. *Comp. Biochem. Physiol.* **11**:85-103.

BELL, E. A. (1971). "Comparative biochemistry of non-protein amino acids", In: Chemotaxonomy of the Leguminosae (J. B. Harbourne, D. Boulter and B. L. Turner, eds.), Academic Press, pp. 179-206.

BIRK, Y. (1961). Purification and some properties of a highly active inhibitor of trypsin and α-chymotrypsin from soybeans. *Biochim. Biophys. Acta* **54**:378-381.

BIRK, Y. (1968). "Chemistry and nutritional significance of proteinase inhibitors from plant sources." In: Chemistry, Pharmacology and Clinical Applications of Proteinase Inhibitors. *Ann. N.Y. Acad. Sci.* **146**, (N. Back and E. M. Weyer, eds.) pp. 388-399.

BIRK, Y. (1969). "Saponins". In: Toxic Constituents of Plant Foodstuffs. (I. E. Liener, ed.) Academic Press, pp. 169-210.

BIRK, Y. and GERTLER, A. (1968). An inhibitor of trypsin and α-chymotrypsin, *Biochemical Preparations,* 12 (W.E.M. Lands, ed.) John Wiley & Sons, Inc., pp. 25-29.

BIRK, Y., GERTLER, A. and KHALEF, S. (1963a). A pure trypsin inhibitor from soya beans. *Biochem. J.* **87**:281-284.

BIRK, Y., GERTLER, A. and KHALEF, S. (1963b). Separation of a *Tribolium*-protease inhibitor from soybeans on a calcium phosphate column. *Biochim. Biophys. Acta,* **67**:326-328.

FEINSTEIN, G. (1971). Isolation of chick ovoinhibitor by affinity chromatography on chymotrypsin-sepharose. *Biochim. Biophys. Acta* **236**:74-77.

FOWDEN, L. and BELL, E. A. (1965). Cyanide metabolism by seedlings. *Nature* (Lond.) **206**:110-112.

GERTLER, A., BIRK, Y. and BONDI, A. (1967). A comparative study of the nutritional and physiological significance of pure soybean trypsin inhibitors and of ethanol-extracted soybean meals in chicks and rats. *J. Nutrition* **91**:358-370.

KATCHALSKI, E., LEVIN, Y., NEUMANN, H., RIESEL, E. and SHARON, N. (1961). Studies on the enzymatic hydrolysis of poly- α-amino acids. *Bull. Res. Counc. Israel* **10A**:159-171.

KONIJN, A. M., BIRK, Y. and GUGGENHEIM, K. (1970). *In vitro* synetheis of pancreatic enzymes: Effect of soybean trypsin inhibitor. *Amer. J. Physiol.* **218**:1113-1117.

KUNITZ, M. (1947). Crystalline soybean trypsin inhibitor - II. General properties. *J. Gen. Physiol.* **30**:291-310.

LASKOWSKI, M. and LASKOWSKI, Jr., M. (1954). Naturally occurring trypsin inhibitors. *Adv. Protein Chem.* **9**:203-242.

LAUINGER, C. and RESSLER, C. (1970). β-cyanoalanine as a substrate for asparaginase. Stoichiometry, kinetics and inhibition. *Biochim. Biophys. Acta* 198:316-323.

MARCHAIM, U., BIRK, Y., DOVRAT, A. and BERMAN, T. (In Press). Lucerne saponins as inhibitors of cotton seed germination: Their effect on diffusion of oxygen through the seed coats. *J. of Exp. Bot.*

MURTI, V. V. S., SESHADRI, J. R. and VENKITASUBRAMANIAN, T. A. (1964). Neurotoxic compounds of the seeds of *Lathyrus sativus*. *Phytochem.* 3:73-78.

PUSZTAI, A. (1967). Trypsin inhibitors of plant origin, their chemistry and potential role in animal nutrition. *Nutr. Abstr. and Rev.* 37:1-9.

RESSLER, C. (1964). Neurotoxic amino acids of certain species of *Lathyrus* and vetch (1964). *Fed. Proc.* 23:1350-1353.

RESSLER, C., NAGARAJAN, G. R. and LAUINGER, C. (1969). Biosynthesis of asparagine from -L-[$^{14}C^{15}N$] Cyanoalanine in *Lathyrus sylvestris* W. seedlings. Origin of the amide nitrogen. *Biochim. Biophy. Acta* 184:578-582.

SHANY, S., GESTETNER, B., BIRK, Y. and BONDI, A. (1970). Lucerne saponins. III. Effect of lucerne saponins on larval growth and their detoxification by various sterols. *J. Sci. Fd. Agric.* 21:508-510.

INHIBITION OF ACARID MITE DEVELOPMENT BY FATTY ACIDS

J. G. Rodriguez
Department of Entomology
University of Kentucky
Lexington, Kentucky 40506

INTRODUCTION

The acarid mite, *Tyrophagus putrescentiae* (Schrank) is a cosmopolitan species capable of infesting many types of foods and stored products (Chmielewski, 1969; Hughes, 1961; Rivard, 1961; Sasa, 1964; Sinha, 1963; Sinha, 1964). It thrives especially on food products containing a high protein and fat content such as cheeses and certain pet foods. The moisture content of the food and/or the relative humidity of its habitat become very important in the population dynamics of this species. This acarid mite has been cultured on many natural or processed foods and it is cultured in our laboratory on dog food conditioned to contain approximately 25% water in an environment of 80% R.H.

Another mite that belongs to the cheese mite group that may be encountered in certain processed foods is *Caloglyphus berlesei* (Michael), a much larger mite than *T. putrescentiae*, and one that requires very high moisture conditions to thrive. It is also encountered in places where food residues may accumulate and remain wet as in out-of-the-way places in food processing plants. In the laboratory, it can be cultured in wet dog food, wet brewers yeast, or on a meridic diet we have developed.

In the protection of processed foods against these mites, or other members of the "cheese mite" group, nothing is as effective a preventive measure as an air tight package, but there are many processed foods that are marketed in packages not hermatically sealed. Although the control of mites infesting food products can be successfully achieved with certain fumigants (Barker, 1969; Jalil, *et al.*, 1970), this measure of control clearly has many limitations and serious drawbacks. Even more serious drawbacks would probably be encountered with residual chemical sprays even though they would not necessarily be applied directly on the package (Dicke *et al.*, 1953; Marzke and Dicke, 1959).

The possibilities of using fatty acids as insecticides have been pointed up by a small number of investigators. The toxic effects of fatty acids on muscid larvae were brought to light by Levinson and Ascher (1954) and more recently by House (1967), House and Graham (1967a, b) and Maw and House (1971).

Our research in the nutritional physiology of acarines has led us into the possible approaches of using a naturally occurring fatty acid in the food as a supplement.

The objectives in the studies presented here were to investigate the potential of fatty acids as growth inhibiting nutrients of *T. putrescentiae* and *C. berlesei* when added as supplements to their food. It was also our especial desire to compare these species. In order to do this without the possible influence of different microorganisms, comparison studies would utilize axenic culture and a chemically defined diet.

MATERIALS AND METHODS

Experiments on processed food

A commercial dog food was oven dried and ground to a fine powder (40 mesh) in a Wiley mill. This pulverized food was treated with varying levels of a number of fatty acids that had been qualitatively and quantitatively predetermined by gas chromatography analyses as constituents of the product. These fatty acids comprised approximately 25% of the total fatty acids of the dog food. Palmitic, stearic and oleic acids were included although these 3 fatty acids constituted about 75% of the fatty acid total. Fatty acids occurring only in trace amounts in the product, were arbitrarily included. The fatty acids, all included at 2% concentration, were dissolved in ethyl ether and were added to 5 g of the pulverized food and mixed thoroughly after which the ether was driven off under gentle heat in a chemical hood. The food was then moistened by addition of 2 ml water which was again thoroughly mixed and placed in 10 microcups, each containing 0.7 g per cup. The food was packed into each microcup so that the top surface was quite smooth, each cup was then inoculated with 20 *T. putrescentiae* eggs or 10 adult female mites and then capped with parafilm M. These were then cultured at 27°C$\pm$ 1°C and 80% $\pm$2.5 R.H., and counts of mites and eggs were made after an interval of 21 days.

The sequel to the first series of experiments was a study to define further the inhibitory characteristics of a selected group of the most effective fatty acids. The short chain fatty acids, propionic, butyric, caproic, caprylic and capric were tested at 0.5, 1.0 and 2.0 percent concentration in the same food as before. They were incorporated into the food, however, by emulsification with Tween 80 (0.2 mg) in the water (2 ml) used in moistening the food (5 g). Other aspects of procedure were the same as in the first series of experiments.

Experiments with artificial diets

The first diet studied utilized the meridic diet that was developed for *T. putrescentiae* (Rodriguez and Lasheen, 1971). To use this diet for *T. putrescentiae* the free moisture that condensed inside the vials was dissipated overnight; this meant a loss of about 1.8% of water, most of it in the form of condensate. It was found that this diet could also be utilized for *C. berlesei* if this free water was allowed to remain in the diet. Axenic cultures of both of these species were started by surface sterilizing eggs with 2% formalin for 5 min. The eggs were rinsed in sterile distilled water and inoculated into 3 dram vials containing the test diet. The larvae (20) were transferred to test diets (5 replicates) after they emerged. Two such experiments were conducted. Propionic, butyric, caproic, caprylic, and capric acids were added to this diet and their effect on *T. putrescentiae* and *C. berlesei* were compared.

This type of experiment was repeated using only *T. putrescentiae* as the test mite and omitting butyric acid from the treatments. The rationale being that butyric was objectionable because of its bad odor.

It was deemed important to test other formulations of selected fatty acids hence the cupric salts of propionic, caproic, caprylic and capric acids as well as various methyl esters were tested. For this series of tests, a second diet was developed in which casein (vitamin-free) was substituted for the amino acids. For the sake of having a defined lipid fraction, the lipids listed in the chemically defined diet for *T. putrescentiae* culture (Rodriguez and Lasheen, 1971) were used in these diets.

RESULTS AND DISCUSSION

Some 25 fatty acids were evaluated for their relative effectiveness in inhibiting *T. putrescentiae* when cultured in a dog food product (Table 1). Also included in this list were 3 fatty acids that comprise approximately 75% of the total fatty acid entity of the food; namely, palmitic, stearic and oleic. All treatments were 2% supplements.

The data show that the short chain fatty acids were relatively more effective in inhibiting than the longer chain fatty acids.

In the follow-up experiments the emulsification method of incorporating the fatty acid into the food was an improvement over incorporation with ether in that the substrate appeared physically superior. This was perhaps reflected in the larger *T. putrescentiae* population that developed using the latter technique (Table 2). Propionic, butyric, caproic, caprylic and capric acids were again quite effective at 2% concentration. At 0.5% concentration, however, only capric acid proved to be relatively free of *T. putrescentiae* development after 3 weeks. All treatments, however, were significantly different than the control.

It was deemed important to work also with an artificial diet in order to better standardize variables. Hence, a chemically defined diet that had been previously developed for *T. putrescentiae*, was utilized in axenic culture (Table 3).

Table 1. Relative inhibition of *T. putrescentiae* by fatty acids incorporated in food.

Fatty acid (2%)	No. carbons	Inhibition
Propionic	3	+++
Butyric	4	+++
Caproic	6	+++
Caprylic	8	+++
Pelargonic	9	+++
Capric	10	+++
Undecanoic	11	++
Lauric	12	++
Tridecanoic	13	+
Isomyristic	14	+
Myristic	14	+
Myristoleic	14 1=	-
Tetradecadienoic	14 2=	-
Pentadecadienoic	15	-
Isopalmitic	16	-
Palmitic	16	-
Palmitoleic	16 1=	-
Hexadecadienoic	16 2=	-
Heptadecanoic	17	-
Heptadecenoic	17 1=	-
Stearic	18	-
Oleic	18 1	-
Linoleic	18 1=	-
Linolenic	18 3=	-
Eicosanoic	20 1=	-
Control	-	-

Table 4A summarizes data obtained when propionic, caproic, caprylic, and capric acids were added at concentrations ranging from 0.01 to 0.50 percent. It is apparent that all of these saturated fatty acids were quite inhibiting to *T. putrescentiae* growth and development. Dosages as low as 0.05% of caproic, caprylic and capric were significantly different than the control. The weight data show that growth is definitely affected; females weighed significantly less (30%) when fed 0.5% supplementary caproic or capric in their chemically defined diet (Table 4B).

Table 2. Inhibition of *T. putrescentiae* by selected fatty acids incorporated in food.[a]

	Concentration (%)			
	0	0.5	1.0	2.0
Propionic				
Mites	410	150	120	14
Eggs	120	80	50	0
Butyric				
Mites	428	142	78	16
Eggs	132	76	28	0
Caproic				
Mites	406	110	48	0
Eggs	124	64	18	0
Caprylic				
Mites	454	90	36	0
Eggs	120	50	16	0
Capric				
Mites	420	25	10	5
Eggs	118	4	0	0

[a]Mite and egg counts made after 3 weeks development. Initial inoculation was 10 larvae per vial, 10 vials per treatment; data represent mean of 2 experiments.

It had been previously determined that it was possible to culture *C. berlesei* on the *T. putrescentiae* diet by simply allowing the condensate water that formed in the vials to remain there. These species were compared side by side for their reaction to 5 fatty acids and the results are shown in Figs. 1-3. It was apparent that *C. berlesei* was more tolerant to propionic acid and less tolerant to caprylic and butyric than *T. putrescentiae.* The differences were not significant on other comparisons between species or of concentrations of treatment when evaluated at 28 days.

Because acarid mites react more favorably to casein based diets than to amino acid formulations, and because this type of diet is relatively simple to prepare, the aforementioned chemically defined diet was converted to a "casein diet" by substituting 8 g vitamin-free casein for 5.22 g amino acids and wheat germ for the lipid fraction (Table 5). For these particular experiments, however, the lipid fraction from the former diet was retained.

Table 3. Composition of the axenic diet used in feeding *T. putrescentiae*.[a]

Compound	g/100 ml	Compound	g/100 ml
L-Alanine	0.26	Alphacel	1.0
L-Arginine	0.34		
L-Aspartic	0.60	Wesson's Salts (Mix W)	2.5
L-Cysteine	0.09		
L-Glutamic acid	0.94	Myristic acid	0.056
Glycine	0.28	Palmitic acid	0.494
L-Histidine	0.18	Stearic acid	0.268
L-Isoleucine	0.23	Oleic acid	0.494
L-Leucine	0.40	Linoleic acid	0.070
L-Lysine	0.31	Linolenic	0.028
L-Methionine	0.07	Cholesterol	0.500
L-Phenylalanine	0.25		
L-Proline	0.33	Tween 80	0.01
L-Serine	0.26		
L-Threonine	0.20	Vitamin Diet Fortification	
L-Tryptophan	0.08	Mixture[b]	0.40
L-Tyrosine	0.15		
L-Valine	0.25	Streptomycin	0.05
		Methyl-p-hydroxybenzoate	
RNA	0.10	(15% in 95% ethanol)	1.0 ml
		Formalin (10%)	0.1 ml
Agar	3.0	KOH (22.5% in H_2O)	0.5 ml
Sucrose	8.0		

[a]From Rodriguez, J. G. and Lasheen, A. M., 1971.
[b]Nutritional Biochemicals Corporation. Mixture total includes 0.339 g dextrose.

Selected fatty acid salts of copper were studied as supplements to the casein-wheat germ diet (Table 6). Cupric salts of propionic, caproic, caprylic and capric were all extremely effective as growth inhibitors, only a relatively small number of adults developed after 28 days in one fatty acid, caproic. Three short chain methyl esters were also tested as supplements to the same casein diet (Table 7A). The mites demonstrated great sensitivity to these compounds at 0.5 and 1.0%, but methyl caprate was the most inhibitory. Laurate, myristate and oleate were the longest chain methyl esters tested (Table 7B). These acids were effective in inhibiting mite development at the lowest concentration tested, 0.5 percent.

Table 4. Summary of 3 experiments with various levels of fatty acids. Average number of *T. putrescentiae* adults, immatures and eggs after 3 weeks of development in an axenic chemically defined diet (see Table 3).

Fatty acid	Concentration of fatty acid					
	.00	.01	.05	.10	.25	.50
A.			Average counts of all stages			
Propionic	40.3	27.3	24.4	21.9	14.5*	13.6
Caproic	31.2	26.6	16.1*	11.3**	7.8**	2.0**
Caprylic	38.5	30.4	32.2*	22.4*	15.6**	6.8**
Capric	31.3	22.5	21.3*	18.9*	13.1*	2.9**
B.			Weight µg/female			
Propionic	10.1	9.5	9.9	9.2	7.9*	7.9*
Caproic	9.4	8.0	4.4*	7.3*	6.6*	5.8*
Capric	9.0	7.1*	7.0*	7.3	5.8*	5.6*

*Significant at P (from Blank) = 0.05 level; ** = 0.01 level.

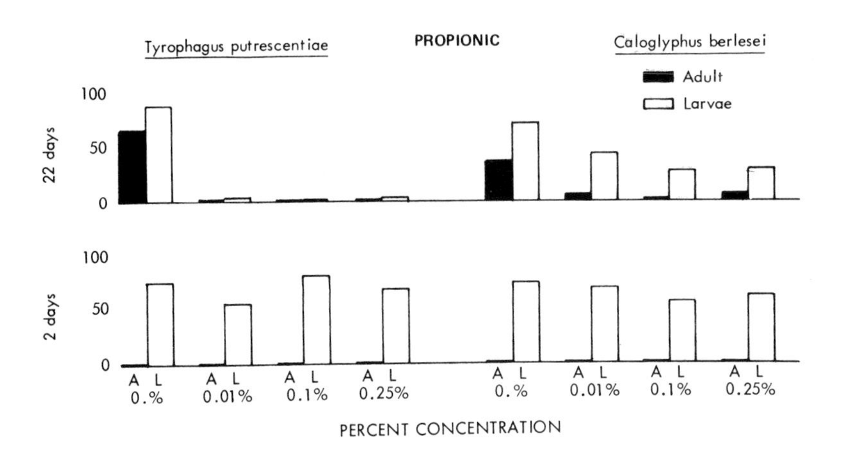

Figure 1. Comparison of *T. putrescentiae* and *C. berlesei* survival and development when propionic acid was added to chemically defined diet (listed in Table 3) in concentrations of 0.01 to 0.25 percent. Data represent mean of 2 experiments, each having treatments composed of 5 tubes, each inoculated with 20 larvae/tube.

Table 5. Casein and wheat germ diet for culture of *T. putrescentiae* and *C. berlesei.*[a]

Component	g/100 ml
Wesson's Salt Mixture	2.5 g
Sucrose	8.0 g
Wheat germ[b]	7.0 g
Alphacel	1.0 g
Agar	3.0 g
KOH (22.5% in H_2O)	0.5 ml
Casein (vitamin free)	8.0 g
Vitamin mixture	0.4 g
Ascorbic acid	1.0 g
Streptomycin	0.050 g
Formalin (10%)	0.1 ml
Methyl-p-hydroxybenzoate (15% in 95% ethanol)	1.0 ml

[a]Diet must be adjusted to contain about 1.8% water for *T. putrescentiae* culture by allowing free condensate in vials to dissipate overnight.
[b]Lipid fraction from Table 3 (Rodriguez, J. G. and Lasheen, A. M., 1971) may be substituted for wheat germ.

Fatty acids of 10 carbon atoms or less are rarely present in animal lipids, and when they occur, they occur in minute quantities (White *et al.*, 1968). When oleic acid in the free form did not cause inhibition of mites at 2% concentration in the dog food substrate in the preliminary testing, we attributed this reaction to an innocuous level of oleic acid. In a casein diet, however, the methyl esters of laurate, myristate, and oleate were more inhibitory than methyl esters of caproate, caprylate and caprate. To be sure, these would be evaluated as supplements to a processed food.

McFarlane and Henneberry (1965) reported that the growth of the cricket *Gryllodes sigillatus* was inhibited by fatty acids and their methyl esters and the mode of entry appeared to be through the external body wall. Lauric, myristic, stearic and behenic acids were effective inhibitors, as well as the methyl esters of palmitic acid, especially, but also of myristic, stearic, and oleic acids. Lauric was the only free acid that inhibited growth when fed in a diet at 1 percent. When straight chain saturated fatty acids containing 8-11 carbon atoms were applied to house fly, *Musca domestica* L. larvae, they were all found toxic. Lauric acid was toxic only when the larvae were wetted with water (Quraishi and Thorsteinson, 1965).

Table 6. Effect of cupric salts of selected fatty acids on *T. putrescentiae* cultured in meridic diet. Survival and development of larvae (L), nymphs (N), and development of adults (A) from initial inoculation of 200 newly emerged larvae, 20 larvae per tube, 10 tubes per treatment. Mean of 2 experiments.

Cupric salt of fatty acid & conc. (%)	Survival/development–days					
	7		14		28	
	A	L	A	L-N	A	L-N
Propionic						
0.5	0	99	0	16	0	0
1.0	0	68	0	0	0	0
1.5	0	57	0	0	0	0
Caproic						
0.5	0	126	0	60	16	2
1.0	0	82	2	84	0	0
1.5	0	77	0	58	0	0
Caprylic						
0.5	0	91	0	44	1	0
1.0	0	121	0	38	0	0
1.5	0	104	0	32	0	0
Capric						
0.5	0	85	0	24	0	0
1.0	0	61	0	5	0	0
1.5	0	57	0	1	0	0
Control	0	158	92	86	284	370

Levinson and Ascher (1954) placed larvae of the muscid *Musca vicina* on filter paper impregnated with fatty acids having 6 to 18 carbon atoms. Caproic, caprylic and capric acids had the most rapid lethal effect on the larvae. Linoleic and oleic acid gave 100% mortality after a contact of 3-12 hr. The higher saturated fatty acids, from lauric to stearic acid, were practically non-toxic. When fed to fly larvae at 5% concentration the short chain, $\langle$ 10 carbon atoms, fatty acids produced mortalities ranging from 87.5 to 94 percent in 48 hours. The toxicity of lauric was increased from 42 to 100% by feeding a medium free of lipids except cholesterol.

House (1967) found that certain free fatty acids in the diet of *Pseudosarcophaga affinis* were toxic. On diets that contained 2.5% of the fatty acid, and in order of ascending toxicity, caproic, caprylic, or capric acid, the larvae died within a few minutes of exposure. Lauric was slower acting but

Table 7. Effect of methyl esters of selected fatty acids on *T. putrescentiae* cultured in meridic diet survival and development of larvae (L)[a], nymphs (N)[1] and development of adults (A) from initial inoculation of 100 newly emerged larvae, 20 larvae per tube, 5 tubes per treatment. Mean of 2 experiments.

Methyl Esters (%)	Mite survival and development (days)			
	5		21	
	A	L-N	A	L-N
A. Caproate				
0.5	-	56	42 ± 10.1	78
1.0	-	75	41 ± 10.4	72
Caprylate				
0.5	-	63	40 ± 10.0	49
1.0	-	42	6 ± 0.8*	28
Caprate				
0.5	-	47	5 ± 0.6*	16
1.0	-	15	0 - *	4
Control	-	79	43 ± 9.7	176
	7		30	
	A	L-N	A	L-N
B. Laurate				
0.5	-	32	0 -	0
1.0	-	26	0 -	0
Myristate				
0.5	-	58	10 ± 0.9*	24
1.0	-	43	3 ± 0.1*	17
Oleate				
0.5	-	50	5 ± 0.8*	17
1.0	-	48	5 ± 0.7*	7
Control	-	81	55 ± 10.2	186

[a]Immature forms, all inclusive.
*Significant difference between control at P=0.01 and the mean.

detrimental nonetheless; myristic acid had no apparent effect.

In a study involving the confused flour beetle, *Tribolium confusum* and the possibility of its control with capric acid, House and Graham (1967) showed that *T. confusum* could indeed be controlled with capric acid, 2.5%,

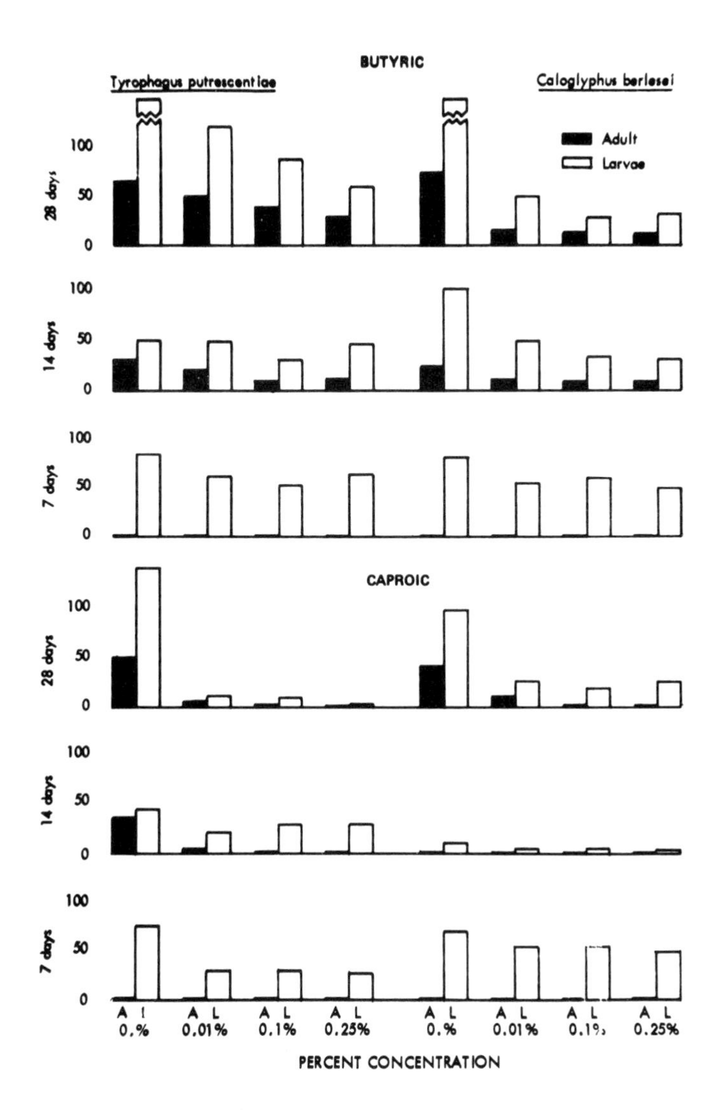

Figure 2. Comparison of *T. putrescentiae* and *C. berlesei* survival and development when butyric and caproic acids were added to chemically defined diet (listed in Table 3) in concentrations of 0.01 to 0.25 percent. Data represent mean of 2 experiments, each having treatments composed of 5 tubes, each inoculated with 20 larvae/tube.

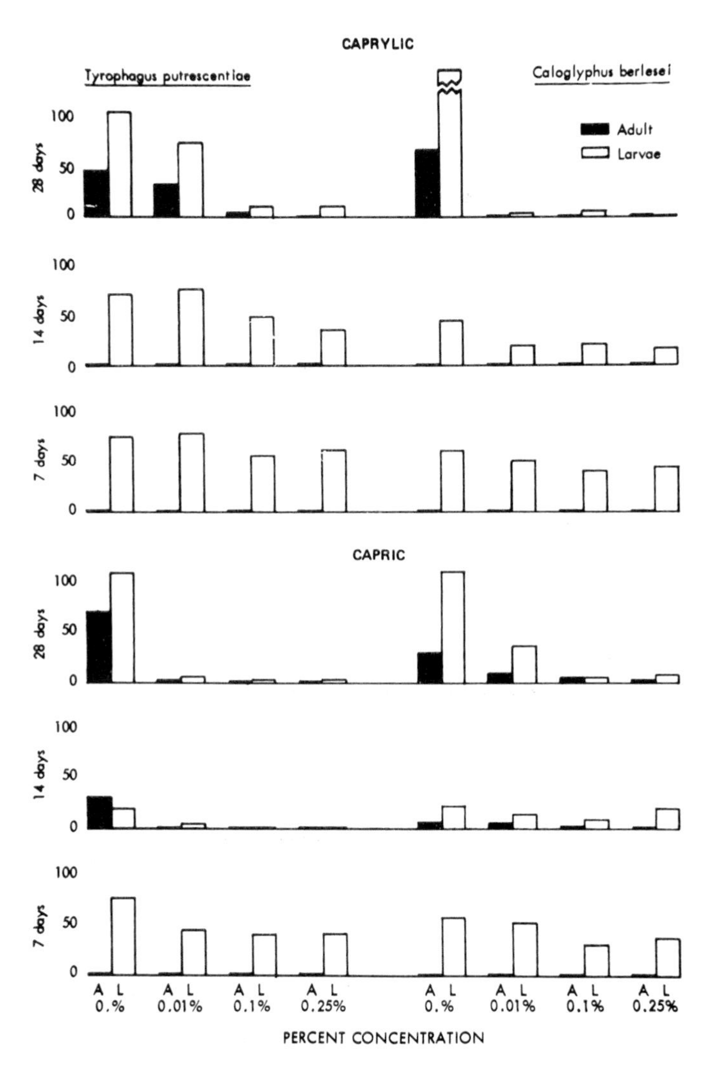

Figure 3. Comparison of *T. putrescentiae* and *C. berlesei* survival and development when caprylic and capric acids were added to chemically defined diet (listed in Table 3) in concentrations of 0.01 to 0.25 percent. Data represent mean of 2 experiments, each having treatments composed of 5 tubes, each inoculated with 20 larvae/tube.

blended into a foodstuff such as whole wheat flour.

It would appear that the routes of entry of fatty acids into organisms such as larvae of various flies, beetles, crickets, acarid mites, etc., are by ingestion, absorption through the integument, and in some cases vapor. In the case of *T. putrescentiae* and *C. berlesei* entry probably occurred by all of the aforementioned routes. The concentrations of fatty acids utilized in the various treatments were deliberately aimed as low as possible - not to kill the organism quickly but rather slowly and to block metabolic processes and prevent oviposition. An indication that this was achieved was reflected by the weight data shown in Table 4B; in this instance propionic, caproic and capric acids retarded growth to the extent of significant loss of weight.

Cupric salts of propionic, caproic, caprylic and capric acids were all effective at 0.5 percent. Methyl esters of caproate and caprylate were not effective at 0.5% although caprylate was effective at 1 percent. Caprate, however, was significantly inhibitory at 0.5 percent. This was also true of the longer chain methyl esters tested, laurate, myristate and oleate.

To conclude, there are marked differences in the degree of inhibition of growth and development in these acarid mites that are produced by free fatty acids, cupric salts of some selected fatty acids and methyl esters. At present supplementation of a processed food with a free fatty acid or a methyl ester, at growth inhibitating levels, would appear to be a practical approach for "nutritional control" of acarid mites that attack food products.

SUMMARY

The short chain fatty acids, propionic, butyric, caproic, caprylic and capric acids, were especially effective in inhibiting *T. putrescentiae* when incorporated into a commercial dog food. Very few eggs were produced by mites cultured in a dog food preparation containing 0.5% capric acid, and none were oviposited at 1.0% concentration. All other fatty acids required 2% to produce this level of inhibition. When the fatty acids were incorporated into an agar based chemically defined diet, they were effective at 0.05%. Cupric salts of propionic, caproic, caprylic and capric acids were all extremely effective as growth inhibitors. Methyl esters of caproate, caprylate, caprate, laurate, myristate and oleate were also studied and found to be quite inhibitory; all were effective at 0.5% except caproate and caprylate, the latter being effective at 1%.

REFERENCES

BARKER, P. S. (1969). Susceptibility of the mushroom mite to phosphine and ethylene dibromide. *J. Econ. Ent.* **62**:145-146.

DICKE, R. J., IHDE, K. D., and PRICE, W. V. (1953). Chemical control of cheese mites. *J. Econ. Ent.* **46**:844-849.

CHIMIELEWSKI, WIT. (1969). Fauna of mites in stored seeds of sugar beets. *Bul. Ent. de Pologne* 39:619-628 (English summary).

HOUSE, H. L. (1967). The nutritional status and larvicidae activities of C6- to C14-saturated fatty acids in *Pseudosarcophaga* affinis. *Can. Ent.* **99**:384-392.

HOUSE, H. L., and GRAHAM, A. R. (1967). Capric acid blended into foodstuff for control of an insect pest, *Tribolium confusum. Can. Ent.* **99**:994-999.

HUGHES, A. M. (1961). The mites of stored food Tech. Bul. No. 9 London. Her Majesty's Stationery Office, 287 p.

JALIL, M., ROSS, I. J., and RODRIGUEZ, J. G. (1970). Methyl bromide and phosphine as fumigants for some acarid mites. *J. Stored Prod. Res.* **6**:33-37.

LEVINSON, Z. H. and ASCHER, K. R. S. (1954). Chemicals affecting the preimaginal stages of the housefly. IV. The fatty acids. *Riv. Parassit.* **15**:111-119.

MARZKE, F. O., and DICKE, R. J. (1959). Laboratory evaluations of various residual sprays for the control of cheese mites. *J. Econ. Ent.* **52**:237-240.

MAW, M. G. and HOUSE, H. L. (1971). On capric acid and potassium capricate as mosquito larvicides in laboratory and field. *Can. Ent.* **103**:1435-1440.

MCFARLANE, J. E. and HENNEBERRY, G. O. (1965). Inhibition of the growth of an insect by fatty acids. *J. Insect Physiol.* **11**:1247-1252.

QUARISHI, M. S. and THORSTEINSON, A. J. (1965). Toxicity of some straight chain saturated fatty acids to housefly larvae. *J. Econ. Ent.* **58**:400-402.

RIVARD, I. (1961). Influence of temperature and humidity on mortality and rate of development of immature stages of the mite *Tyrophagus putrescentiae* (Schrank) (Acarina:Acaridae) reared on mold cultures. *Can. J. Zool.* **39**:419-426.

RODRIGUEZ, J. G. and LASHEEN, L. M. (1971). Axenic culture of *Tyrophagus putrescentiae* in a chemically defined diet and determination of essential amino acids. *J. Insect Physiol.* **17**:979-985.

SASA, MANABU. (1964). Special problems of mites in stored food and drugs in Japan *Acarologia 1st Internat'l. Congr. Acarology Proc.*, 438 p.

SINHA, R. N. (1963). Stored product acarology in Canada. *Adv. in Acarology* Cornell Univ. Press, 480 p.

SINHA, R. N. (1964). Ecological relationships of stored-products mites and seed-borne fungi *Acarologia, 1st Internat'l. Congr. Acarology Proc.*, 439 p.

WHITE, A., HANDLER, P. and SMITH, E. L. (1968). *Principles of Biochemistry* McGraw Hill Publ. Co., 1187 p.

INSECT CONTROL STRATEGIES BASED ON NUTRITIONAL PRINCIPLES: A PROSPECTUS

J. J. Pratt, Jr.
Pioneering Research Laboratory, U.S. Army Natick Laboratories
Natick, Massachusetts 01760

H. L. House
Smithfield Experimental Farm, Canada Department of Agriculture
Trenton, Ontario

A. Mansingh
Department of Biology, Queen's University
Kingston, Ontario

INTRODUCTION

We propose new insect control strategies for certain constituted foods based upon the manipulation of food components and upon nutritional principles. We admit, of course, to the speculativeness of our proposals, though there is a rationale for them. In general the application of nutritional principles may be a useful adjunct in control practices designed to decrease undesirable toxic pollutants.

RATIONALE

As Geier (1966) stated, one way to manage pests is to modify intrinsically favorable habitats in such a way that they no longer provide adequate environments for the population of the pest involved, for example, by providing unsuitable sources of food. That nutritional factors can act detrimentally on the nutrition of pest insects with respect to plant resistance, as shown by various workers, was pointed out by Painter (1958) and Beck (1965). There are many examples in which agricultural chemicals – fertilizers, insecticides, etc. – changed the proportions of nutrient constituents in plant tissues, and consequently their nutritional value for pests, sometimes selectively, (Rodriguez, 1960; Beck, 1965; Singh, 1970). For instance, by varying the amount of 4-8-12 fertilizer applied to tobacco plants, Wooldridge & Harrison (1968) found varying effects on the nitrogen, phosphorus, and potassium content of the leaf and on the rate of population increase of *Myzus persicae* (Sulzer). Van Emden (1966), upon applying different rates of nitrogen and potassium to Brussels sprouts plants showed that soluble nitrogen in leaves could be increased, which resulted in greater fecundity in *M. persicae* than in *Brevicoryne brassicae* (Linnaeus). Differences in the relative concentrations of amino acids and glucose found between varieties of pea plants correlated with resistance of the plants to the pea aphid, *Acyrthosiphon pisum* (Harris) (Auclair *et al.*, 1957; Maltais & Auclair, 1957); and it was clearly shown on synthetic diets that this aphid chooses a diet having a certain level of amino acids and sucrose (Cartier, 1968). Friend *et al.* (1957), recognizing the importance of amino acid relationships, suggested that varieties of plants might be developed that would be nutritionally less suitable than present varieties are as food for their pests. Nevertheless, no serious attempts made on a practical scale to alter plants nutritionally for pest management are known to us.

No doubt nutritional principles can be used most readily to control insects in synthetic food-environments, such as processed, or manufactured food products. Smallman (1945) showed that the sodium bicarbonate in biscuits destroyed sufficient thiamine during baking to decrease markedly normal growth and reproduction in a pest, *Tribolium confusum* Jacquelin du Val. Moreover, increased fat levels and decreased water content in biscuits were detrimental to the insect (Smallman & Aitkin, 1944). Inclusion of certain inorganic salts and of sugars, qualitatively and quanitatively, in food rendered it unsuitable for normal growth of insects (Majumder & Bano, 1964; Pant & Dang, 1965; Pant & Gabrani, 1963). Some food components normally possess toxic substances; for example, growth inhibiting factors in soybean and corn germ (Lipke *et al.*, 1954; Lipke & Fraenkel, 1954). Some fatty acids, carbohydrates, food additives, etc. may also be detrimental. For as Lipke & Fraenkel (1956) pointed out, a given carbohydrate may be inert nutritionally, or satisfactory as a carbon source but unacceptable in a gustatory sense, or toxic *per se*. Certain vegetable gums, agar, etc., are used as food additives in food manufacturing as thickeners etc. (Furia, 1968). Among these some include substances or their components, such as pentosans, which can be detrimental to insects (Lipke & Fraenkel, 1956); presumably owing to their indigestibility, interference with nutrient absorption, or perhaps toxicity *per se*. Some fatty acids may be dietary requirements – for instance, linoleic, linolenic acids (Fast, 1964), but certain ones, such as capric, caprylic, are quite toxic (House, 1967a). The principle of 'nutritional pest control' is in evidence in some human foods where in processing the food is inadvertently rendered inadequate for growth and development of *T. confusum* (House & Graham, 1967b).

Especially for the protection of processed foods, the application of nutritional principles offers prospects of being a useful tactic; inasmuch as improper ratios of different food constituents result in impairment, arrestment or death of insects, or discrimination in choice of food (Gordon, 1959; House, 1959, 1965a, 1966, 1967b, 1971). However, the qualitative requirements of vertebrates and insects are much the same, and the principles of nutrition that apply to one usually apply equally to the other. For example, both require and utilize much the same proteins, carbohydrates, fats, vitamins and minerals; both require these nutrients in some suitable amounts and proper proportional relationships for normal nutrition (Maynard, 1937;

House, 1965b, 1969; Dadd, 1970). This being so, to alter the nutrient composition of a foodstuff to the extent that it is nutritionally detrimental for insects may likewise be so for man or other vertebrates. But in fact, most insect pests limit themselves to one food and one food niche, whereas man and his domestic animals are polyphagous. This suggests that if an 'altered food' is involved – let us say for our purpose a dehydrated cake-mix – the insect, which limits itself to this foodstuff will suffer, but the polyphagous habits of man will enable him to overcome, or 'patch-up', readily the nutritionally significant faults that were built into the altered food. The nutritional value of commercial food products that House & Graham (1967b) showed were quite nutritionally inadequate for *T. confusum*, would be enhanced for humans when finally milk is added or otherwise prepared for the table.

NUTRITION: NORMAL AND ABNORMAL

Nutrition involves the nutrients digested out of foodstuffs and the physiological and biochemical processes by which these nutrients are transformed into body tissues and energy for all the activities attributed to life. In other words: For nutrition two major variables are (1) the digestibility (and consequently availability of nutrients) and the nutritive value of the foodstuff, and (2) the qualitative and quantitative nutritional requirements of the organism. Digestibility varies between foodstuff and with the organism, nutritional requirements (especially quantitative) may vary with the organism. Normal nutrition as referred to here is that which permits efficient metabolism resulting in unimpaired life processes. Abnormal nutrition therefore is that which, owing to dietary faults, does not permit efficient metabolism and is manifest by impairment of life processes. For normal nutrition, food (perhaps aided by symbiotic microorganisms) has to supply all of the nutrients needed in adequate amounts and in suitable relative proportions to each other to provide for the normal functioning of the organism. This is so because the efficiency of utilization of metabolizable nutrients depends on how closely their amounts and proportions correspond to those that are metabolically optimum for the body function in question. In other words: Intake should closely approximately demand. When these two variables are not coordinated suitably metabolic injury, or malnutrition, results. Of course, for pest management we need concern ourselves only with nutritional faults, or the factors that give rise to abnormal nutrition.

Nutritional faults may be owing to unavailability (redigestion) or lack of an essential nutrient (qualitative factors) or to unsuitable amounts and proportions of nutrients ingested (quantitative factors). Should a food lack an essential nutrient, some metabolic function is deprived of essential components for its activities. A deficiency of a nutrient may be regarded really as an abnormal proportion of nutrients, whereby there is provided too little of one in relation to some other. With both a lack of an essential nutrient or a deficiency, or imbalance, of one or more nutrients, metabolism can be impaired and the organism affected accordingly. Thus abnormal nutrition in insects can result in disease and mortality as shown by various workers reviewed by House (1963). An imbalance of nutrients, for example, can result in discrimination in the choice of diets by insects (House, 1967b, 1971), slow growth and development (House, 1966), curtailed feeding (House, 1965a), and low population density on a given amount of food (Gordon 1959).

From the standpoint of nutritional control of insects in processed foods, one might readily alter qualitatively and quantitatively the composition of the food. Introduction of qualitative nutritional faults possibly holds some promise. But the quantitative approach, particularly concerning nutrient imbalances, is likely to be the most fruitful area for research on nutritional principles to control insects in processed foods. One can find many examples in work on insect nutrition of qualitatively and quantitatively induced nutritional faults. Although perhaps needing further elaboration, many of these would serve to discover useful techniques and strategies for applying nutritional means of pest control. Several such heuristic examples follow.

HEURISTIC EXAMPLES

Qualitatively, amino acids, carbohydrates, lipids, vitamins, minerals, and other nutrients were tabulated and noted as being individually required, utilized, nor not (Altman & Dittmer, 1968). Reviews (Lipke & Fraenkel, 1956; House, 1965b, 1972; Dadd, 1970) provide discussions of both qualitative and quantitative aspects of insect nutrition that are pertinent.

Qualitative

Generally speaking: The qualitative requirements are much the same among insects. In some cases a given requirement can be met by several substances, for instance, a carbohydrate requirement by several sugars. In other cases a requirement can be met by some but not others of a class of nutrients, for instance, among sterols. This suggests that perhaps a substance that is poorly utilized by the insect concerned can be substituted for an utilizable one in manufactured food thereby disadvantaging the insect. Such substances may be very detrimental to insects but less so to humans owing to perhaps different rates of consumption, body mass etc. Certain fatty acids (including capric, caprylic, lauric, myristic, stearic, palmitic) commonly found in foodstuffs, certain sterols. the D-isomers of certain amino acids, decreased the rate of growth, development, and survival in various insects

(McFarlane & Henneberry, 1965; House & Graham, 1967b; Pausch & Fraenkel, 1967; Hinton *et al.* 1951; McGinnis *et al.* 1956; Davis, 1959). Because the nutritional value of a protein depends on its amino acid composition, lack of any essential amino acid decreases very much the biological value of the protein. Such proteins of low biological value include, for example, gelatin and zein. Lack of an essential acid, of course, results in cessation of growth and ultimately in death of the insect. Generally aspartic and glutamic acids are not considered essential. Nevertheless, Cheldelin & Newburgh (1959) showed that, though omission of either of these alone had no effect on *Phormia regina* (Meigen), omission of them together resulted in death of the larva. The nutritional value of a number of proteins was tested in different insects and found to vary (Chirigos *et al.*, 1960; Geer, 1966; Naylor, 1963). Similarly, the nutritional value of a number of carbohydrates was tested and found to vary, as shown by a number of workers (Bernard & Lemonde, 1949; Lemonde & Bernard, 1953; Pausch & Fraenkel, 1967).

Of course, insects require a number of vitamins (Altman & Dittmer, 1968. House, 1972). And some insects require for normal development certain polyunsaturated fatty acids (Fraenkel & Blewett, 1946; Dadd, 1960; Rock *et al.*, 1965; Vanderzant *et al.*, 1957; Gordon, 1959). All insects require a sterol (House, 1972). Lack of any of these essential substances results in serious disabilities or death. However, it is doubtful that food products made of a blend of natural foodstuffs – milk, flour, eggs, etc. – could be composed to lack an essential amino acid, vitamin, fatty acid, mineral element, etc.

Quantitative

The chances are greatest that natural foodstuffs can be blended to effect quantitatively the nutrition of an insect pest. The dominant quantitative factor in a diet is the balance of nutrients, as Gordon (1959) explained. Provided a suitable balance of nutrients is maintained, dilution of the entire nutrient content (by water or undigestibles) has little effect, but when the total nutrient content level is maintained and the proportions of nutrients unsuitably disarranged, or imbalanced, an insect decreases its rate of feeding and growth (House, 1965a). Low levels of salts of various minerals deleteriously affected *T. confusum* (Medici & Taylor, 1966). On the other hand, some of these minerals can be toxic at levels several orders of magnitude above the nutritionally required level (Medici & Taylor, 1966). Sastry & Sarma (1958) suggested from their work that metabolic interactions among minerals might cause imbalances, for copper influences iron and zinc metabolism in *Coryra cephalonica* (Stainton). It is difficult to distinguish between nutritional and toxic effects of minerals. Effects of minerals are shown in Table 1.

Table 1. Deleterious effects of certain mineral salts on the growth of some insect species.

Salts	Concentration (%)	Species	Effects	Reference[a]
$Ca_3(PO_4)_2$	1 – 2	*Sitophilus oryza* (L.)	100% mortality	1
"	1 – 3	*Callosobruchus chinensis* (L.)	"	1
"	3	*Trogoderma granarium* Everts.	growth retardation	1
"	1 – 2	*Laemophloeus minutus* (Oliv.)	100% mortality	1
"	1 – 2	*Ephestia cautella* (Wlk.)	"	1
"	1 – 2	*Oryzaephilus surinamensis* (L.)	"	1
"	3	*Stegobium panicium* (L.)	growth retardation	1
"	1 – 3	*Tribolium castaneum* (Hbst.)	100% mortality	1
$Ca(H_2PO_4)_2H_2O$	>1.5	*Tribolium confusum* Duv.	growth retardation	2
$CuSO_4,5H_2O$	0.01	*Corcyra cephalonica* St.	"	3
"	>0.15	*Tribolium confusum*	"	2
$FeSO_4,7H_2O$	>0.5	"	"	2
KH_2PO_4	>5.0	"	"	2
$MgSO_4$	>2.0	"	"	2
$MnSO_4,H_2O$	>0.9	"	"	2
NaCl	>0.6	"	"	2
$ZnCl_2$	>0.2	"	"	2
$ZnSO_4,7H_2O$	>0.8	*Corcyra cephalonica*	high mortality	4

[a]References:
1. Majumder & Bano (1964); 2. Medici & Taylor (1966); 3. Sastry & Sarma (1958); 4. Sastry *et al.* (1958).

Table 2. Deleterious effects of certain amino acids on some insect species.

Kind	Configuration	Concentration[a]	% Amino acids or casein in dry diet	Insect species	Growth response	Mortality	References[d]
Alanine	DL	1.3	amino acids, 20	*Oryzaephilus surinamensis* (L.)	—	high	3
		3.5[b] – 7	casein, 97	*Xenopsilla cheopis* (Rothsch.)	retarded	″	8
		—	—	*Musca domestica* L.	″	—	2
	L	—	—	*Drosophila melanogaster* (Meig.)	″	—	10
Arginine	L	9.5[c]	casein, 97	*X. cheopis*	slightly retarded	—	8
Aspartic acid	L	3.6 – 7.2[b]	″	″	—	high	8
Cysteine	L	—	—	*D. melanogaster*	retarded	—	11
		8.9	amino acids, 20	*O. surinamensis*	—	high	4
Cystine	L	—	—	*M. domestica*	retarded	—	2
Glutamic acid	L	6.0 – 24.0[b]	casein, 97	*X. cheopis*	″	high	8
Isoleucine	DL	—	—	*D. melanogaster*	″	″	6
		5.0	amino acids, 45.5	*Calliphora erythrocephala* (Meig.)	″	—	9
	L	7.5[b]	casein, 97	*X. cheopis*	slightly retarded	—	8
			″	″			
Lysine	L	4.2 – 8.5[b]	″	″	retarded	high	8
		8.5 – 11.0	amino acids, 20	*O. surinamensis*	″	″	4
		—	—	*M. domestica*	″	—	1
Methionine	DL	—	—		″	—	1
	L	—	—	*D. melanogaster*	″	—	11
	L	1.0 – 4.0[b]	casein, 97	*X. cheopis*	—	very high	8

Table 2. (cont'd.)

Kind	Configuration	Concen- tration[a]	% Amino acids or casein in dry diet	Insect species	Growth response	Mor- tality	References[d]
Ornithine	L	1.6 – 2.2	amino acids, 20	*O. surinamensis*	—	″	5
Phenylalanine	DL	—	—	*M. domestica*	retarded	—	1
		6.5 – 13.0	casein, 97	*X. cheopis*	—	high	8
Proline	L	″	″	″	—	″	8
Serine	DL	—	—	*D. melanogaster*	—	″	6
		2.3 – 4.5	amino acids 45.5	*C. erythrocephala*	—	″	9
		—	″	*Phormia regina* (Meig.)	—	″	7
		—	—	*M. domestica*	retarded	—	1
		3.5 – 7.0[b]	casein, 97	*X. cheopis*	″	moderate	8
	L	—	—	*D. melanogaster*	—	″	6
Threonine	DL	—	—	*M. domestica*	retarded	—	1
		4.5[b]	casein, 97	*X. cheopis*	″	high	8
	L	″	″	″	″	″	8
Tryptophan	DL	—	—	*D. melanogaster*	″	—	6
	L	—	—	″	″	—	6
Tyrosine	L	6.6[b]	casein, 97	*X. cheopis*	—	high	8
Valine	DL	10.5[c]	″	″	retarded	″	8
	L	″	″	″	″	—	8

[a]% of the total amino acid mixture or casein in the diet.
[b]Equivalent to the normal level of the L–amino acid in casein.
[c]Twice the normal level of the L–amino acid in casein.
[d]References:
1. Brookes (1956); 2. Brookes & Fraenkel (1958); 3. 4. 5. Davis (1959, 1961, 1962); 6. Hinton *et al.* (1951); 7. McGinnis *et al.* (1956); 8. Pausch & Fraenkel (1967); 9. Sedee (1956); 10. 11. Wilson (1945, 1946).

Because the nutritional value of a protein depends on its amino acid composition, we can explain the low biological value of some proteins as being due to a deficiency, i.e., imbalance, of some amino acids making up the protein: see Chirigos *et al.* (1960), Geer (1966), Naylor (1963), House (1959). That one protein may be less suitable than another for a given insect was clearly shown by Vanderzant (1958) using amino acid mixtures based on casein and on a cotton protein, respectively. Both contained the same amino acids, but in one (casein) less suitably balanced than in the other (cotton protein) for a cotton insect. Michelbacher *et al.*, (1932) showed that, for *Lucilia sericata* (Meigen), casein was deficient in cystine. Ingestion of excessive amounts of amino acids can be detrimental to mammals (Harper, 1959; Russell *et al.*, 1952) and insects (Hammen, 1956; House & Barlow, 1956; House, 1966; Melvin & Bushland, 1940; Sang, 1956; Sedee, 1956): see Table 2. In fact, the effects of dietary levels of protein, carbohydrate, vitamins, etc. depend on the levels of other nutrients in the diet; in other words, suitable ratios, or proportions, define what is deficient or excessive levels; for examples see Sang (1956), House (1966, 1967b), Fraenkel (1955). An imbalance resulting predominantly from high levels of certain amino acids accompanied by low levels of vitamins impaired the nutrition of *Celerio euphorbia* (Linnaeus) (House, 1965a). The effects of amino acid imbalance on *Drosophila melanogaster* Meigen, as affected by amino acid additives to supplement casein, and the effects of balance between casein and vitamins, were demonstrated by Sang (1959, 1962): see Tables 3 and 4. As inferred above, excessive amounts of certain amino acids, and the D-isomers of some which are not utilized, may be toxic: amino acid toxicities may be severe when superimposed upon some amino acid deficiency (Harper, 1959).

Apparently insects are not affected by levels of vitamins considerably in excess of amounts needed for optimum nutrition as shown by Brust & Fraenkel (1955). However, the amount of each vitamin required for optimum nutrition must be relative to other dietary components (Sang, 1962; Fraenkel & Stern, 1951). Thus a given vitamin may become deficient if the level of an amino acid or of protein is increased. In addition to familiar effects on growth and metamorphosis, various gross physiochemical and histopathological conditions of avitaminosis occur (House, 1963).

Of course, water plays an important role in feeding and nutrition. Some insects drink water (Mellanby & French, 1958). The flea *Orchopeas howardi* (Baker), requires food with a water content of 15 to 28% (Sikes, 1931). Fraenkel (1955) pointed out that some sugars may be nutritionally useful, however, their hygroscopicity renders the food medium too wet for the larva, *Tenebrio molitor* Linnaeus, to exist. Grain beetles cannot breed in grain with a moisture content below 9% and propagate poorly in grain having up to 11% moisture (Cotton & Ashby, 1952). In general, 6% moisture in food is the lowest limit that will support growth of most food-infesting insects (Blake & Russell, 1943).

Table 3. Effects of dietary imbalance created by the addition of certain amino acids on the vitamin requirements and growth of *Drosophila melanogaster* (Data calculated from Sang, 1962).

% Protein	L-Amino acids added (% dry diet)	Vitamin concentration (μg/100 ml of diet)		Growth response
Casein, 4.0	none	nicotinic acid,	90.0	normal
	tryptophan, 0.1		130.0	"
			176.0	"
	glutamic acid, 0.7		160.0	retarded
	glutamic acid, 0.7 + arginine, 0.1 + proline, 0.25		250.0	"
Casein, 5.5	none		110.0	normal
	glutamic acid, 0.7		"	retarded
	α-ketoglutarate, 0.7		"	"
Casein, 4.0	none	ca-pantothenate,	90.0	normal
	leucine, 0.35 + phenylalanine, 0.15 + tyrosine, 0.29		$>$ 300.0	retarded
	none	biotin,	1.0	normal
	aspartic acid, 0.2		$<$ 16.0	retarded
	aspartic acid, 0.2 + glutamic acid, 0.7		$<$ 16.0	normal
	none	folic acid,	3.6	normal
	glycine, 0.025		18.0	accelerated
	glycine, 0.2		$<$ 64.0	retarded

Table 4. Minimal vitamin requirement for normal growth of *Drosophila melanogaster* in diets containing different amounts of casein (Data from Sang, 1962).

Vitamins	Vitamin requirement (μg/100 ml diet) at 4% Casein	7% Casein	Increased requirement (%)
Thiamine-HCI	32.0	32.0	0
Riboflavin	85.0	85.0	0
Biotin	2.0	2.4	20
Nicotinic acid	86.0	260.0	202
Ca-pantothenate	100.0	190.0	90
Pyridoxine-HCI	18.0	90.0	344
Folic acid	3.6	20.0	456

NUTRITIONAL CONTROL: SUGGESTED STRATEGIES

The basic strategy of nutritional control that we propose is to alter the normal composition of a food to render it less suitable as a food for insects and mites. Suggested strategies include omission, introduction or substitution of a specific nutrient; varying the proportions of some nutrients; adding non-nutrients, or inert substances; and altering physical characteristics. By doing this the food admittedly may be likewise reduced in nutritional value for humans. But whereas humans eat an almost endless variety of foods prepared according to almost endless recipes, insects usually are limited solely to rather species-specific foods of limited variety. Therefore if a food were altered – say to be deficient in an essential nutrient – an insect feeding exclusively on it would suffer nutritional impairment. But man feeding on it and on a variety of other foods as well would overcome the deficiency in his diet. This need not seem too infeasible. Indeed many familiar and often used foods are inherently faulty for human nutrition. For example, gelatin lacks tryptophan, which is essential to both insects and vertebrates, and is deficient in several other amino acids. So use of gelatin by humans as a sole source of protein would be disastrous. However, with some color and flavoring added it is a common dessert. And, of course, it is not the only protein eaten by man, and so its deficiencies are not serious. Rice, which is commonly polished, is in fact rendered deficient in the essential vitamin thiamine and thus is decreased in nutritional value. And many processed breakfast cereals are not nutritionally adequate by themselves, but their nutritional value becomes somewhat enhanced when milk, sugar, fruits, etc. are added at the table. It is probably owing to such faults in a nutritional sense that many processed foods are "self-protected", as shown by House & Graham (1967a). Maintaining a moisture content below 6% protects some dehydrated foods from insect infestation (Blake & Russell, 1943).

That a number of such foods used by humans do not support insect growth and development suggests that slight modifications in the composition and physical characteristics of many other processed foods might achieve in these the same end. It seems feasible to employ food components and nutritional principles to impinge disruptively on the metabolism and behavior of the insect concerned. Although an effective factor may be manipulated alone, probably several factors could be brought to bear in unison to greater advantage. Fig. 1 shows schematically many of the factors that might be employed.

Generally speaking, the seemingly easiest way to create a dietary fault would be the omission or deletion of an essential nutrient. But in such foods as prepared cake mixes and dehydrated foods, for example, which contain naturally constituted ingredients, such as flour, eggs, milk, etc., all of the

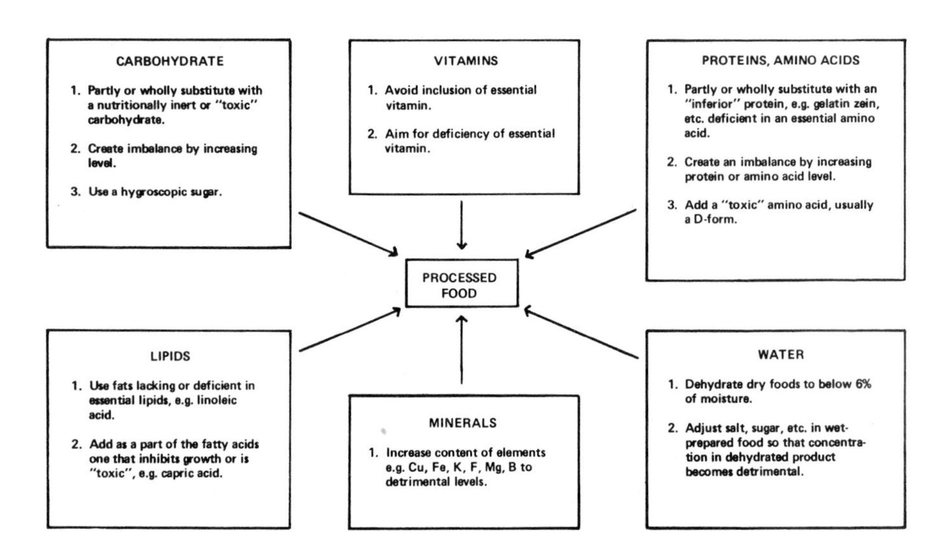

Figure 1. Generalization of some suggested strategies of nutritional insect control in manufactured food.

nutrients qualitatively required by insects likely to infest the product are present; and probably in sufficient amounts for insect nutrition. Very likely, therefore, it would be impractical, even impossible, to remove an essential nutrient from such food mixtures. It would be feasible, however, to modify a processed food by the addition of a nutrient so as to derange, or imbalance, the ratio of nutrients to one less suitable for insect nutrition. Or, inasmuch as many edible substances such as emulsifiers, stabilizers, glazes etc. are used in the food industry (Furia, 1968), it may be feasible to make use of those that disadvantage insects. For example, whereas dextrin and sugars are used for glazes, one might wholly or partly substitute poly- or oligosaccharides that are nutritionally inert or toxic to insects. Possibly a toxic fatty acid, such as capric which can control insects in food (House & Graham, 1967a), might be added to a pet food and possibly human food. Any resulting off-taste or odor might be overcome by added characteristic flavors and aromas associated with the normal food.

Moreover, many processed foods are marketed in dehydrated form to facilitate handling, storage, and preservation. And, as mentioned above, dehydration, of course, can be a means of controlling food-infesting insects. To manage insects, however, dehydration of food to extreme dryness might not be necessary. For by taking into account nutritional factors, perhaps foods need be dehydrated only to the point that certain nutrients reach detrimental concentrations for the insect concerned, but when reconstituted and prepared for the table the nutrients fall within normal concentrations for human needs and taste. For example, a beef stew may contain 1.27% of sodium chloride when prepared (Anon, 1964), but if it were dehydrated to a water content of 12% the proportion of sodium chloride would become 4.2% of the dehydrated mass – a level above that (0.6%) which Medici & Taylor (1966) found could greatly impair development of *T. confusum.*

It may be theoretically possible to manipulate and blend components in a constituted or processed food so as to achieve ratios of specific nutrients in metabolically inefficient unharmonious imbalances for normal insect nutrition. But admittedly it may not be possible to involve imbalances very widely as a means of nutritional control at present. Lacking is a sufficiently precise understanding of the quantitative nutritional requirements of insects (except in a very few examples), especially in terms of the range of proportionality that governs the utilization of various nutrients with respect to permitting or preventing normal nutrition. Also lacking are sufficiently detailed analyses of foods in terms (not of chemical analysis) of their composition as sources of available amino acids, vitamins, etc. after the food is digested by the insect. Chemical analyses of uneaten food make little or no distinction between whether the data represent strictly nutritive substances, or include substances that are not made available by the digestive processes of the insect, or in any case would not be a nutritionally useful substance. It is necessary to evaluate food of insects in much the same way as feeds for livestock are evaluated – for example, by their nutritive ratio values concerning digestible protein, fats, and carbohydrates (see: Morrison, 1941). We can expect nutritive ratios to differ between foods and among insects perhaps as much as they do between specific mammals and insects. For instance, irrespective of the chemical analyses of a cabbage, the food value of a cabbage, measured as nutritive ratio, in the cow is quite different from that in the cabbageworm (House, 1969). Therefore, available data with respect to man and domestic animals is not likely to be of much use in dealing with insects. In such matters differences in the rate of digestion play a very important part, though one to which we have made scarcely much direct reference. The fact is that the nutritional value of common foodstuffs for insects is scarcely known. This despite the fact that the nutritional value of a synthetic diet and the nutritional requirements of an insect pest, say *T. confusum,* may be known. And the nutritional value of a constituted food, say a cake-mix for humans, may be well-determined. What is needed is determination of the nutritional value of the cake-mix and of other processed foods likely to be attacked by *T. confusum*, for example.

In prospect: The qualitative requirements of insects are well known. Abnormal proportions of nutrients probably provide the most opportune means of nutritional control in processed food. It is immaterial whether the nutritional factors involved result in metabolic complications or toxic activities. It seems feasible in preparing foods consisting of flour, eggs, butter, sugar, milk, etc. to create a protein/vitamin imbalance, for example, by the addition of more protein, such as casein or gelatin or by decreasing vitamin sources. One can infer from House's (1971) work that detrimental imbalances can be rather easily effected by additions of amino acids (or protein) to a diet already containing ample levels; additions over a given range, however, may have little effect (see Fig. 1 in House, 1971). This suggests that effective imbalances for nutritional control in processed food can be best created where the level of a nutrient in the food is high in relation to others but at least marginally within the range for normal nutrition. Of course, relations between nutrient proportions in foods with respect to normal/abnormal nutrition of the insect are relative to the quantitative nutritional requirements of the species of insect involved. Research, therefore, is needed to determine to what extent one must vary the proportions of nutrients in order to cause serious disabilities in economically important insects pests versus the common practice of determining the nutritional factors that permit normal nutrition.

We explored the possibilities of manipulating the protein, carbohydrate, fat, and vitamin content in a tapioca pudding, which, incidently, proved to be already deficient in niacin. The results, to be published elsewhere, indeed showed in *T. confusum* and *Oryzaephilus surinamensis* (Linnaeus) that it was possible to accentuate or nullify the niacin imbalance, as well as a general vitamin imbalance, by varying the protein content of the tapioca pudding with casein. We also observed retarded growth and increased mortality in *T. confusum* reared on whole wheat flour mixed with various mineral salts.

We may conclude that these proposed strategies leave many questions unanswered, and that only extensive research will provide insights into what is required to translate the strategies into practical applications. Not the least of these applications would be finding means of preventing physical, organoleptic and, possibly, nutritional changes in the food that could be detrimental to its use by humans. It appears that if representative processed foods and manufactured food products were examined for their ability to support insect growth and development, many would fail partly or wholly to do so. Detailed analyses of the nutritive components, digestibility, etc. of the foods could well reveal the reasons for their inadequacies when examined in the light of our present knowledge and understanding of insect nutrition. It is probable that in a few food types an examination of their nutrient components, quantitatively, would suffice to determine whether such foods would support growth and development of certain insects, and perhaps suggest what might be done simply to disadvantage the insect.

SUMMARY

It should be possible to control insects in various constituted foods by use of procedures based upon nutritional principles of manipulation of food components. Changes in nutrients in plant tissues caused by fertilizers or other means are known to affect growth and development of insects. Certain human processed foods are inadvertently made inadequate for insect growth and development. When current knowledge of normal and abnormal insect nutrition is studied, it becomes apparent that nutritional faults can be readily created by qualitative and quantitative alterations in the composition of an insect's food. There are many available examples which demonstrate various adverse effects on insects of changes in specific nutrients. The possibility thus exists of developing strategies for controlling insects in certain human foods by developing nutritional faults detrimental to insects but harmless to humans.

REFERENCES

ALTMAN, P. L. and DITTMER, D. S., Eds. (1968). *Metabolism.* Bethesda. *Fed. Am. Socs. Exp. Biol.,* 737 pp.

ANONYMOUS. (1964). *Heinz Nutritional Data.* 5th Ed. Pittsburg, H. G. Heinz Co., 143 pp.

AUCLAIR, J. L., MALTAIS, J. B. and CARTIER, J. J. (1957). Factors in resistance of peas to the pea aphid, *Acyrthosiphon pisum* (Harr.) (Homoptera: Aphididae) – II. Amino Acids. *Can. Ent.* **89**:457-464.

BECK, S. D. (1965). Resistance of plants to insects. *Ann. Rev. Ent.* **10**:207-232.

BERNARD, R. and LEMONDE, A. (1949). Aspect qualitatifs des besoins en glucides de *Tribolium confusum* Duval. *Revue Cannadienne Biologie* **8**:498-503.

BLAKE, C. H. and RUSSELL, H. D. (1943). Insects and other animals of interest to the Quartermaster Corps. Report No. 2091. Washington Office of Sci. Res. & Dev. U.S. Army. 342 pp.

BROOKES, V. J. (1956). The nutrition of the larva of the house fly (*Musca domestica*) L. Muscidae, Diptera). *Ph.D. Thesis,* University of Illinois: 111 pp.

BROOKES, V. J. and FRAENKEL, G. (1958). The nutrition of the larva of the house fly *Musca domestica* L. *Physiol. Zool.* **31**:208-223.

BRUST, M. and FRAENKEL, G. (1955). The nutritional requirements of the larvae of a blow fly *Phormia regina* (Meig). *Physiol. Zool.* **28**:186-204.

CARTIER, J. J. (1968). Factors of host-plant specificity and artificial diets. *Bull. Ent. Soc. Am.* **14**:18-21.

CHELDELIN, V. H. and NEWBURGH, R. W. (1959). Nutritional studies on the blow fly. *Ann. N.Y. Acad. Sci.* **77**:373-383.

CHIRIGOS, M. A., MEISS, A. N., PISSANO, J. J. and TAYLOR, W. M. (1960). Growth response of the confused flour beetle *Tribolium confusum* (Duval) to six selected protein sources. *J. Nutr.* **72**:121-130.

COTTON, R. T. and ASHBY, A. D. (1952). Insect pests of stored grains and seeds. *In: Insects: The Year Book of Agriculture.* Washington, D.C., USDA, pp. 629-639.

DADD, R. H. (1960). The nutritional requirements of locusts. I. Development of synthetic diets and lipid requirements. *J. Insect Physiol.* 4:319-347.

_____(1970). Arthropod Nutrition. *In: Chemical Zoology* Vol. 5:35-95.

DAVIS, G. R. F. (1959). Alanine and proline in the diet of larvae of *Oryzaephilus surinamensis* (L.). (Coleoptera: Silvanidae). *Ann. Ent. Soc. Am.* 52:164-167.

_____(1961). Sulfur-containing amino acids in the nutrition of the saw-toothed grain beetle, *Oryzaephilus surinamensis. J. Nutr.* 75:275-279.

_____(1962). Quantitative L-arginine requirements of larvae of the saw-toothed grain beetle, *Oryzaephilus surinamensis* (L.). (Coleoptera: Silvanidae). *J. Insect Physiol.* 8:377-382.

FAST, P. C. (1964). Insect Lipids: A Review. *Mem. Ent. Soc. Can.* No. 37, 50 pp.

FRAENKEL, G. (1955). Inhibitory effects of sugars on the growth of the mealworm *Tenebrio molitor* L. *Jour. Cell Comp. Physiol.* 45:393-408.

FRAENKEL, G. and BLEWETT, M. (1946). Linoleic acid, vitamin E and other fat-soluble substances in the nutrition of certain insects, *(Ephestia kuehniella, E. elutella, E. cautella* and *Plodia interpunctella* (Lep.) *J. Exptl. Biol.* 22:172-190.

FRAENKEL, G. and STERN, H. R. (1951). The nicotinic acid requirements of two insect species in relation to the protein content of their diets. *Arch Biochem.* 30:438-444.

FRIEND, W. G., BOCKS, R. H. and CASS, L. M. (1957). Studies on the amino acid requirements of larvae of the onion maggot, *Hylemya antiqua* (MG), under aseptic condition. *Can. J. Zool.* 35:535-543.

FURIA, T. E. (1968). *Handbook of food additives.* Cleveland, Ohio. Chemical Rubber Co. 771 pp.

GEER, B. W. (1966). Comparison of some amino acid mixtures and proteins for the diet of *Drosophila melanogaster. Trans. Illinois State Acad. Sci.* 59:3-10.

GEIER, P. W. (1966). Management of Insect Pests. *Ann. Rev. Ent.* 11:471-490.

GORDON, H. T. (1959). Minimal nutritional requirements of the German roach *Blattella germanica* L. *Ann. N. Y. Acad. Sci.* 77:290-351.

HAMMEN, C. S. (1956). Nutrition of *Musca domestica* in single-pair culture. *Ann. Ent. Soc. Am.* 49:365-368.

HARPER, A. E. (1959). Balance and imbalance of amino acids. *Ann. N. Y. Acad. Sci.* 69:1025-1041.

HINTON, T., NOYES, D. T. and ELLIS, J. (1951). Amino acids and growth factors in a chemically defined medium for *Drosophila. Physiol. Zool.* 24:335-353.

HOUSE, H. L. (1959). Nutrition of the parasitoid *Pseudosarcophaga affinis* (Fall.) and other insects. *Ann. N. Y. Acad. Sci.* 77:394-405.

_____(1963). Nutritional Diseases. *In: Insect Pathology* (Steinhaus, E.A., Ed.) Vol. 1:133-160. New York, Academic Press.

_____(1965a). Effects of low levels of nutrient content of a food and of nutrient imbalance on the feeding and the nutrition of a phytophagous larva *Celerio euphorbiae* (Linnaeus) (Lepidoptera: Sphingidae). *Can. Ent.* 97:62-68.

_____(1965b). Insect Nutrition. *In: The Physiology of Insecta* (Rockstein, M., Ed.) Vol. 2:815-858. New York, Academic Press.

_____(1966). Effects of varying the ratio between the amino acids and the other nutrients in conjunction with a salt mixture on the fly *Agria affinis* (Fall.). *J. Insect Physiol.* 12:299-310.

_____(1967a). Nutritional status and larvicidal activities of C_6 – to C_{14} – saturated fatty acids in *Pseudosarcophaga affinis* (Diptera: Sarcophagidae). *Can. Ent.* 99:384-392.

_____(1967b). The role of the nutritional factors in food selection and preference, as related to larval nutrition of an insect *Pseudosarcophaga affinis* (Diptera: Sarcophagidae), on synthetic diets. *Can. Ent.* 99:1310-1321.

_____(1969). Effect of different proportions of nutrients on insects. *Ent. Exp. Appl.* 12:651-669.

_____(1971). Relations between dietary proportions of nutrients, growth rate, and choice of food in the fly larva *Agria affinis* (Fall.). *J. Insect Physiol.* 17:1225-1238.

_____(1972). Insect Nutrition. *In: Biology of Nutrition* (Fiennes, R.N., Ed.), Vol. 18:513-573. International Encyclopedia of Food and Nutrition. Oxford, Pergamon Press.

HOUSE, H. L. and BARLOW, J. S. (1956). Nutritional studies with *Pseudosarcophaga affinis* (Fall.), a dipterous parasite of the spruce Budworm *Choristoneura fumiferana* (Clem.) V. Effects of various concentrations of the amino acids mixture, dextrose, potassium ion, the salt mixture, and lard on growth and development; and a substitute for lard. *Can. J. Zool.* 34:182-189.

HOUSE, H. L. and GRAHAM, A. R. (1967a). Capric acid blended into foodstuff for control of an insect pest, *Tribolium confusum* (Coleoptera: Tenebrionidae). *Can. Ent.* 99:994-999.

_____&_____(1967b). Nutritional pest control: "The self-protection" of foodstuffs against *Tribolium confusum* (Coleoptera: Tenebrionidae) often presumably through nutritional factors. *Can. Ent.* 99:1082-1087.

LEMONDE, A. and BERNARD, R. (1953). Aspects nutritifs des larvae de *Stegobium panicium* L. Anobiidae et. D. *Oryzaephilus surinamensis* L. *Naturaliste Canadien* 80:125-142.

LIPKE, H. and FRAENKEL, G. (1954). The toxicity of corn germ to the meal worm *Tenebrio molitor. J. Nutr.* 55:165-178.

_____&_____(1956). Insect nutrition. *Ann. Rev. Ent.* 1:17-44.

LIPKE, H. G., FRAENKEL, G. and LIENER, I. E. (1954). Effect of soybean inhibitors on growth of *Tribolium confusum. J. Agr. Food Sci.* 99:410-414.

MAJUMDER, S. K. and BANO, A. (1964). Toxicity of calcium phosphate to some pests of stored grain. *Nature Lond.* 202:1359-1360.

MALTAIS, J. B. and AUCLAIR, J. L. (1957). Factors in resistance of peas to the pea aphid, *Acyrthosiphon pisum* (Harr.) (Homoptera: Aphidae) I. The sugar-nitrogen ratio. *Can. Ent.* 89:365-370.

MAYNARD, L. A. (1937). *Animal Nutrition.* New York, McGraw-Hill, 300 pp.

MEDICI, J. C. and TAYLOR, M. W. (1966). Mineral requirements of the confused flour beetle, *Tribolium confusum* Duval. *J. Nutr.* 88:181-186.

MCFARLANE, J. E. and HENNEBERRY, G. O. (1965). Inhibition of the growth of an insect by fatty acids. *J. Insect. Physiol.* 11:1247-1252.

MCGINNIS, A. J., NEWBURGH, R. W. and CHELDELIN, V. H. (1956). Nutritional studies on the blow fly *Phormia regina* (Meig.). *J. Nutr.* 58:309-324.

MELLANBY, K. and FRENCH, R. A. (1958). The importance of drinking water to larval insects. *Ent. Exp. Appl.* 1:116-124.

MELVIN, R. and BUSHLAND, R. C. (1940). The nutritional requirements of screwworm larvae. *J. Econ. Ent.* 33:850-852.

MICHELBACHER, A. E., HOSKINS, W. M. and HERMS, W. B. (1932). The nutrition of flesh fly larvae, *Lucilia sericata* (Meig.). I. The adequacy of sterile synthetic diets. *J. Exp. Zool.* 64:109-128.

MORRISON, F. B. (1941). *Feeds and Feeding, Abridged.* Ithaca, Morrison, 503 pp.

NAYLOR, A. F. (1963). Possible value of casein, gluten, egg ablumin, or fibrin as whole proteins in the diet of two strains of the flour beetle, *Tribolium confusum*

(Tenebrionidae). *Can. J. Zool.* **42**:1-9.
PAINTER, R. H. (1958). Resistance of plants to insects. *Ann. Rev. Ent.* 3:267-290.
PANT, N. C. and DANG, K. (1965). Effect of different carbohydrates on the growth and development of *Tribolium castaneum* Herbst. *Indian J. Ent.* **27**:432-441.
PANT, N. C. and GABRANI, K. (1963). Nutritive role of several carbohydrates in the larval growth and development of *Tribolium castaneum* Hbst. *Indian J. Ent.* **25**:172-174.
PAUSCH, R. D. and FRAENKEL, G. (1967). The nutrition of the larva of the oriental rat flea *Xenopsilla cheopis* (Rothschild). *Physiol. Zool.* **39**:202-222.
ROCK, G. C., PATTON, R. L. and GLASS, E. H. (1965). Studies of the fatty acid requirements of *Argyrotaenia velutinana* (Walker). *J. Insect Physiol.* **11**:91-101.
RODRIGUEZ, J. G. (1960). Nutrition of the host and reaction to pests. *In: Biological and Chemical Control of Plants and Animal Pests. Publ. Am. Ass. Advmt. Sci.* **61**:149-167.
RUSSELL, W. C., TAYLOR, M. W. and HOGAN, J. M. (1952). Effect of excess essential amino acids on growth of the white rat. *Arch. Biochem. Biophys.* **39**:249-255.
SANG, J. H. (1956). The quantitative nutritional requirements of *Drosophila melanogaster. J. Expt. Biol.* **33**:45-72.
______(1959). Circumstances affecting the nutritional requirements of *Drosophila melanogaster. Ann. N.Y. Acad. Sci.* **77**:352-365.
______(1962). Relationship between protein supplies and B-vitamin requirements, in axenically cultured *Drosophila. J. Nutr.* **77**:355-368.
SASTRY, K. V. and SARMA, P. S. (1958). Effect of copper on growth and catalase levels of *Corcyra cephalonica* St. *In:* Zinc Toxicity. *Nature Lond.* **182**:533.
SASTRY, K. V., MURTY, R. R. and SARMA, P. S. (1958). Studies on zinc toxicity in the larvae of rice moth *Corcyra cephalonica* St. *Biochem. Jour.* **69**:425-428.
SEDEE, J. W. (1956). *Dietetic Requirements and Intermediary Protein Metabolism of the Larva of Calliphora erythrocephala* (Meig.). Assen. Netherlands, Van Gorcum & Co. Ltd., 130 pp.
SIKES, E. L. (1931). Notes on breeding fleas with reference to humidity and feeding. *Parasitology* **23**:243-249.
SINGH, P. (1970). Host-plant nutrition and composition: Effects on agricultural pests. *Information Bull.* No. 6, Belleville, Canada Dept. of Agric. 102 pp.
SMALLMAN, B. N. (1945). Relation of insect damage to thiamine content of biscuits. *J. Econ. Ent.* **38**:106-110.
SMALLMAN, B. N. and AITKEN, T. R. (1944). Susceptibility of biscuits to insect damage. *Cereal Chemistry* **26**:499-510.
VANDERZANT, E. S. (1958). The amino acid requirements of the pink bollworm. *J. Econ. Ent.* **51**:309-314.
VANDERZANT, E. S., KERUR, D. and REISER, R. (1957). The role of dietary fatty acids in the development of the pink bollworm. *J. Econ. Ent.* **50**:606-608.
VAN EMDEN, H. F. (1966). Studies on the relations of insects and host-plant. II. A comparison of the reproduction of *Brevicoryne brassicae* and *Myzus persicae* (Hemiptera: Aphididae) on brussels sprout plants supplied with different rates of nitrogen and potassium. *Ent. Exp. Appl.* **9**:444-460.
WILSON, L. P. (1945). Tolerance of larvae of *Drosophila* for amino acids: tyrosine, phenylalanine and alanine. *Growth* **9**:341-352.
(1946). Tolerance of larvae of *Drosophila* for amino acids: methionine, cystine and cysteine. *Growth* **10**:361-373.
WOOLDRIDGE, A. W. and HARRISON, F. P. (1968). Effects of soil fertility on abundance of green peach aphids on Maryland tobacco. *J. Econ. Ent.* **61**:387-391.

Author Index

Numbers in parentheses indicate page on which complete reference is listed.

H

I

J

K

T

U

V

Y

Z

Subject Index

A

D

E

Y